Building Construction Illustrated

Building Construction Illustrated

Fifth Edition

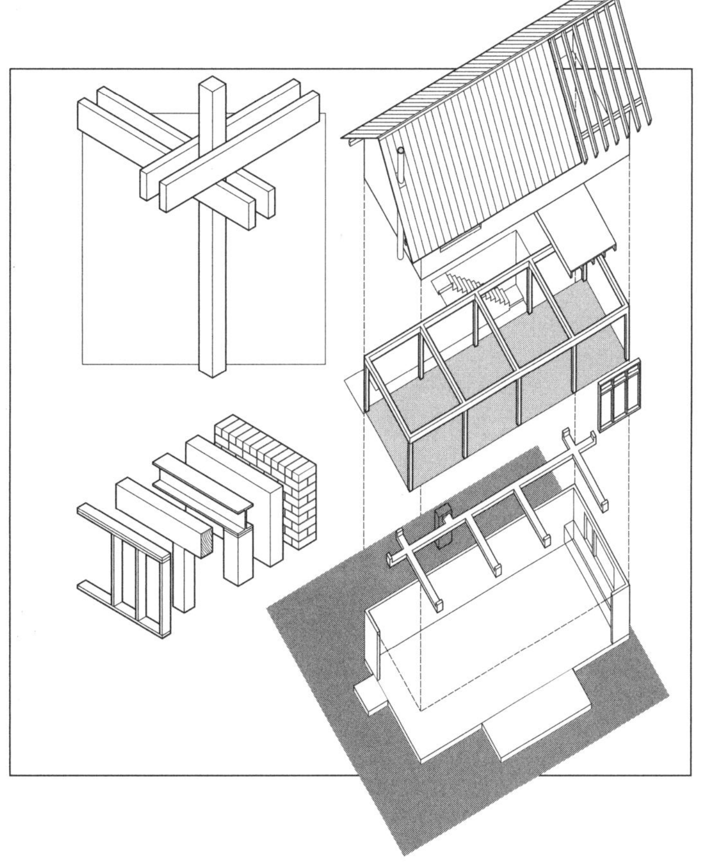

Francis D.K. Ching

WILEY

Cover Design: Wiley
Cover Image: courtesy of Francis D.K. Ching

This book is printed on acid-free paper. ∞

Published by John Wiley & Sons, Inc., Hoboken, New Jersey
Published simultaneously in Canada

For general information about our other products and services, please contact our Customer Care Department within the United States at (800) 762-2974, outside the United States at (317) 572-3993 or fax (317) 572-4002.

Wiley publishes in a variety of print and electronic formats and by print-on-demand. Some material included with standard print versions of this book may not be included in e-books or in print-on-demand. If this book refers to media such as a CD or DVD that is not included in the version you purchased, you may download this material at http://booksupport.wiley.com. For more information about Wiley products, visit www.wiley.com.

Library of Congress Cataloging-in-Publication Data:

Ching, Frank, 1943-
Building construction illustrated / Francis D. K. Ching. -- Fifth edition.
pages cm
Includes index.
ISBN 978-1-118-45834-1 (pbk.)
1. Building. 2. House construction. I. Title.
TH146.C52 2014
690--dc23

2013016213

Printed in the United States of America

10 9 8 7 6 5 4 3

The first edition of this illustrated guide to building construction appeared in 1975, introducing students and builders of architecture to the fundamental principles that govern how buildings are erected. It marked the emergence of a visual approach to understanding the relationship between design and construction.

In 1991, the second edition provided a more expansive survey of building construction by adding coverage of structural steel, reinforced concrete, and curtain wall systems. The third edition in 2001 remained a comprehensive introduction to the principles underlying building construction while refining the graphic format and organization of the first two editions, incorporating an expanded discussion of structural principles, elements, and systems and referencing the Americans with Disabilities Act Accessibility Guidelines and the MasterFormat™ system established by the Constructions Specifications Institute (CSI) for organizing construction information.

The fourth edition in 2008 introduced the LEED® Green Building Rating System™ in Chapter One and referenced specific LEED criteria wherever appropriate; updated section numbers to correspond to the 2004 edition of the CSI MasterFormat™ system; and complied with the requirements of the 2006 edition of the International Building Code®.

A common thread that wove itself through the first four editions and continues in this fifth edition is the attitude that buildings and sites should be planned and developed in an environmentally sensitive manner, responding to context and climate to reduce their reliance on active environmental control systems and the energy they consume. This edition therefore continues to reference the latest edition of the LEED® Green Building Rating System™ criteria and the section numbers of the 2012 CSI MasterFormat™ system wherever appropriate. Many of the changes and additions in this fifth edition, such as updating information in lighting technologies and ways in which to reduce energy usage in buildings, are incremental and often subtle, but together they comprise a continuing commitment to build wisely and sustainably.

It would be nearly impossible to cover all building materials and construction techniques, but the information presented herein should be applicable to most residential and commercial construction situations encountered today. Construction techniques continue to adjust to the development of new building materials, products, and standards. What does not change are the fundamental principles that underlie building elements and the ways in which systems are constructed. This illustrated guide focuses on these principles, which can serve as guideposts when evaluating and applying new information encountered in the planning, design, and construction of a building.

Each building element, component, or system is described in terms of its end use. The specific form, quality, capability, and availability of an element or component will vary with manufacturer and locale. It is therefore important to always follow the manufacturer's recommendation in the use of a material or product and to pay careful attention to the building code requirements in effect for the use and location of a planned building. It is the user's responsibility to ascertain the appropriateness of the information contained in this handbook and to judge its fitness for any particular purpose. Seek the expert advice of a professional when needed.

Metric Equivalents

The International System of Units is an internationally accepted system of coherent physical units, using the meter, kilogram, second, ampere, kelvin, and candela as the base units of length, mass, time, electric current, temperature, and luminous intensity. To acquaint the reader with the International System of Units, metric equivalents are provided throughout this book according to the following conventions:

- All whole numbers in parentheses indicate millimeters unless otherwise noted.
- Dimensions 3 inches and greater are rounded to the nearest multiple of 5 millimeters.
- Nominal dimensions are directly converted; for example, a nominal 2 x 4 is converted to 51 x 100 even though its actual 1-1/2" x 3-1/2" dimensions would be converted to 38 x 90.
- Note that 3487 mm = 3.487 m.
- In all other cases, the metric unit of measurement is specified.
- Refer to the Appendix for metric conversion factors.

1

THE BUILDING SITE

Buildings do not exist in isolation. They are conceived to house, support, and inspire a range of human activities in response to sociocultural, economic, and political needs, and are erected in natural and built environments that constrain as well as offer opportunities for development. We should therefore carefully consider the contextual forces that a site presents in planning the design and construction of buildings.

The microclimate, topography, and natural habitat of a site all influence design decisions at a very early stage in the design process. To enhance human comfort as well as conserve energy and material resources, responsive and sustainable design respects the indigenous qualities of a place, adapts the form and layout of a building to the landscape, and takes into account the path of the sun, the rush of the wind, and the flow of water on a site.

In addition to environmental forces, there exist the regulatory forces of zoning ordinances. These regulations take into account existing land-use patterns and prescribe the acceptable uses and activities for a site as well as limit the size and shape of the building mass and where it may be located on the site.

Just as environmental and regulatory factors influence where and how development occurs, the construction, use, and maintenance of buildings inevitably place a demand on transportation systems, utilities, and other services. A fundamental question we face is how much development a site can sustain without exceeding the capacity of these service systems, consuming too much energy, or causing environmental damage.

Consideration of these contextual forces on site and building design cannot proceed without a brief discussion of sustainability.

In 1987, the United Nations World Commission on Environment and Development, chaired by Gro Harlem Brundtland, former Prime Minister of Norway, issued a report, *Our Common Future*. Among its findings, the report defined sustainable development as "a form of development that meets the needs of the present without compromising the ability of future generations to meet their own needs."

Increasing awareness of the environmental challenges presented by climate change and resource depletion has pushed sustainability into becoming a significant issue shaping how the building design industry operates. Sustainability is necessarily broad in scope, affecting how we manage resources as well as build communities, and the issue calls for a holistic approach that considers the social, economic, and environmental impacts of development and requires the full participation of planners, architects, developers, building owners, contractors, manufacturers, as well as governmental and non-governmental agencies.

In seeking to minimize the negative environmental impact of development, sustainability emphasizes efficiency and moderation in the use of materials, energy, and spatial resources. Building in a sustainable manner requires paying attention to the predictable and comprehensive outcomes of decisions, actions, and events throughout the life cycle of a building, from conception to the siting, design, construction, use, and maintenance of new buildings as well as the renovation process for existing buildings and the reshaping of communities and cities.

Principles
- Reduce resource consumption
- Reuse resources
- Recycle resources for reuse
- Protect nature
- Eliminate toxics
- Apply life-cycle costing
- Focus on quality

Framework for Sustainable Development
In 1994 Task Group 16 of the International Council for Research and Innovation in Building and Construction proposed a three-dimensional framework for sustainable development.

Resources
- Land
- Materials
- Water
- Energy
- Ecosystems

Phase
- Planning
- Development
- Design
- Construction
- Use & Operation
- Maintenance
- Modification
- Deconstruction

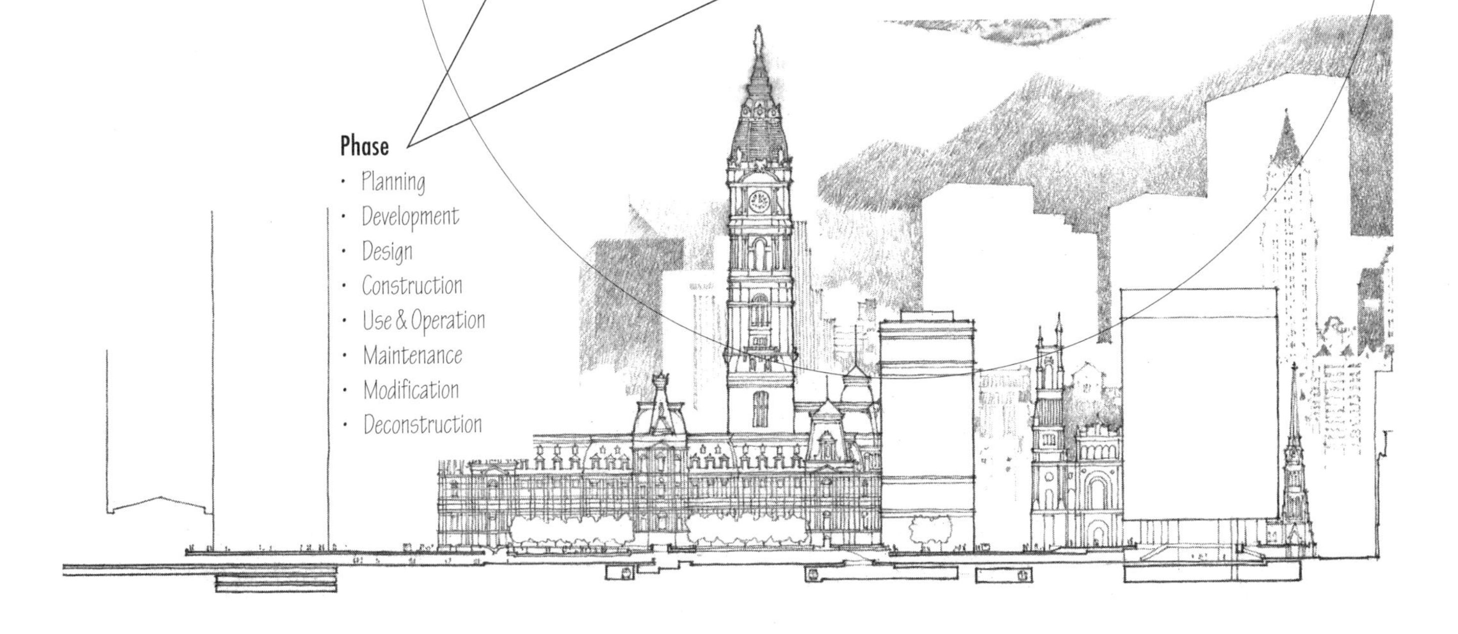

The terms "green building" and "sustainable design" are often used interchangeably to describe any building designed in an environmentally sensitive manner. However, sustainability calls for a whole-systems approach to development that encompasses the notion of green building but also addresses broader social, ethical, and economic issues, as well as the community context of buildings. As an essential component of sustainability, green building seeks to provide healthy environments in a resource-efficient manner using ecologically based principles.

Green building is increasingly governed by standards, such as the Leadership in Energy and Environmental Design (LEED®) Green Building Rating System™, which provides a set of measurable criteria that promote environmentally sustainable construction. The rating system was developed by the U.S. Green Building Council (USGBC) as a consensus among its members—federal/state/local agencies, suppliers, architects, engineers, contractors, and building owners—and is continually being evaluated and refined in response to new information and feedback. In July 2003 Canada obtained a license from the USGBC to adapt the LEED rating system to Canadian circumstances.

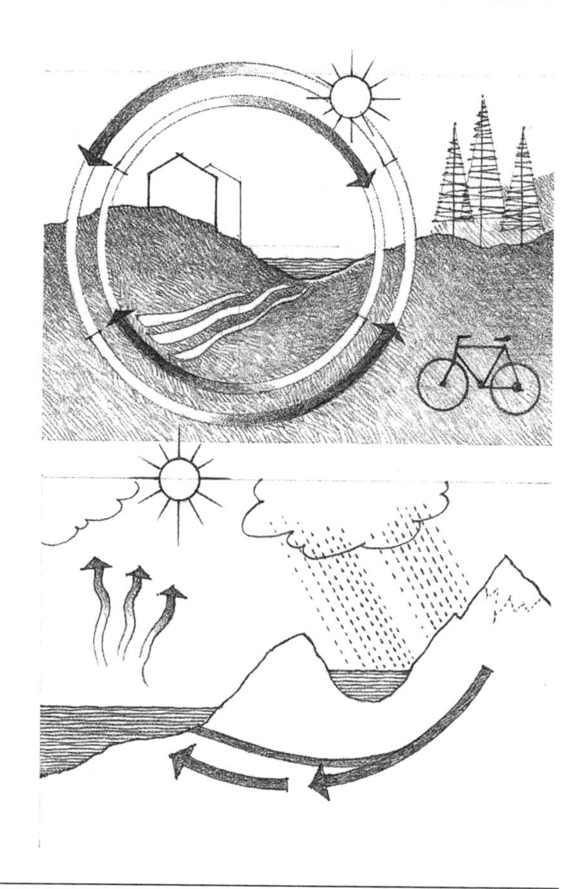

LEED®

To aid designers, builders, and owners achieve LEED certification for specific building types and phase of a building life cycle, the USGBC has developed a number of versions of the LEED rating system:

- LEED for New Construction and Major Renovations
- LEED for Existing Buildings: Operations & Maintenance
- LEED for Commercial Interiors
- LEED for Core & Shell
- LEED for Schools
- LEED for Retail
- LEED for Healthcare
- LEED for Homes
- LEED for Neighborhood Development

The LEED rating system for new construction addresses seven major areas of development.

1. **Sustainable Sites**
   deals with reducing the pollution associated with construction activity, selecting sites appropriate for development, protecting environmentally sensitive areas and restoring damaged habitats, encouraging alternative modes of transportation to reduce the impact of automobile use, respecting the natural water hydrology of a site, and reducing the effects of heat islands.

2. **Water Efficiency**
   promotes reducing the demand for potable water and the generation of wastewater by using water-conserving fixtures, capturing rainwater or recycled graywater for conveying sewage, and treating wastewater with on-site systems.

3. Energy and Atmosphere
encourages increasing the efficiency with which buildings and their sites acquire and use energy, increasing renewable, nonpolluting energy sources to reduce the environmental and economic impacts associated with fossil fuel energy use, and minimizing the emissions that contribute to ozone depletion and global warming.

4. Materials and Resources
seeks to maximize the use of locally available, rapidly renewable and recycled materials, reduce waste and the demand for virgin materials, retain cultural resources, and minimize the environmental impacts of new buildings.

5. Indoor Environmental Quality
promotes the enhanced comfort, productivity, and well-being of building occupants by improving indoor air quality, maximizing daylighting of interior spaces, enabling user control of lighting and thermal comfort systems to suit task needs and preferences, and minimizing the exposure of building occupants to potentially hazardous particulates and chemical pollutants, such as the volatile organic compounds (VOC) contained in adhesives and coatings and the urea-formaldehyde resins in composite wood products.

6. Innovation in Design
rewards exceeding the requirements set by the LEED Green Building Rating System and/or demonstrating innovative performance in Green Building categories not specifically addressed by the LEED Green Building Rating System.

7. Regional Priority
provides incentives for practices that address geographically-specific environmental priorities.

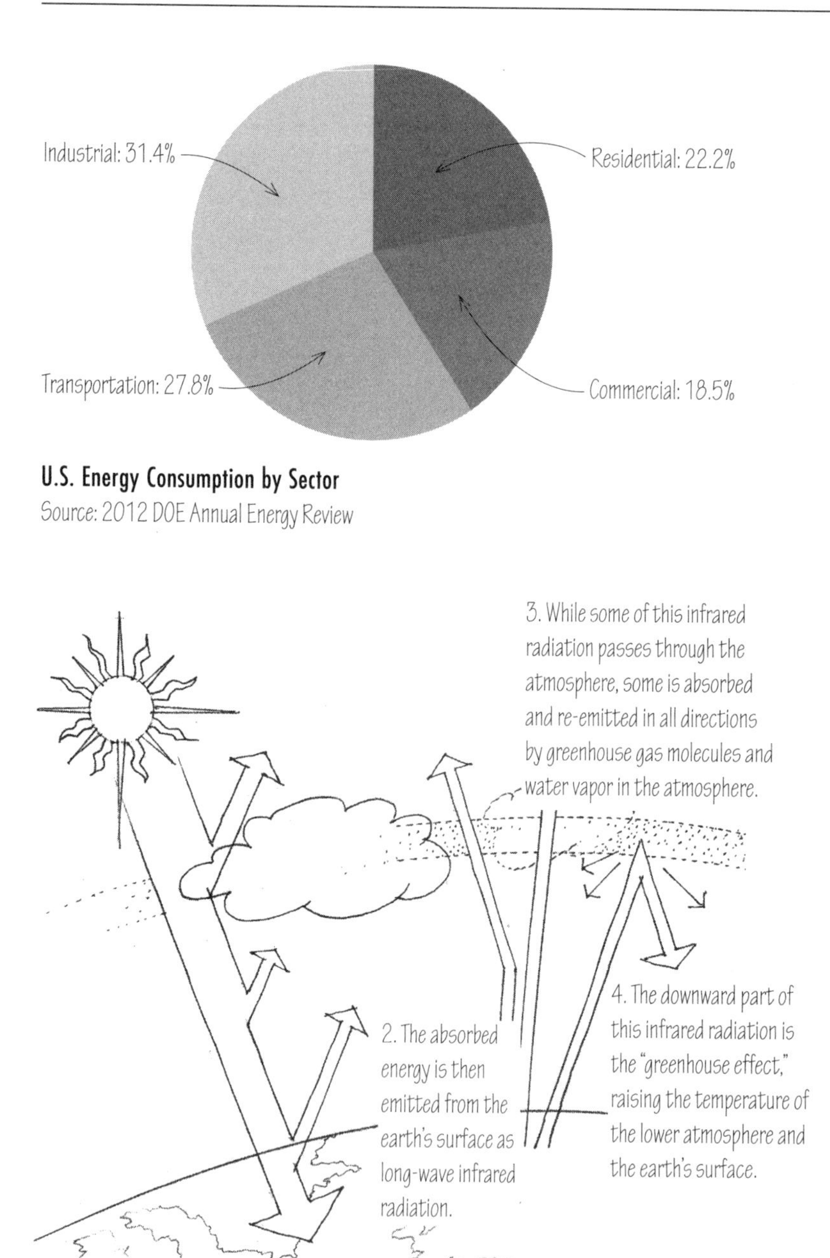

U.S. Energy Consumption by Sector
Source: 2012 DOE Annual Energy Review

Climate Change & Global Warming
Greenhouse gases, such as carbon dioxide, methane, and nitrous oxide, are emissions that rise into the atmosphere. CO_2 accounts for the largest share of U.S. greenhouse gas emissions. Fossil fuel combustion is the main source of CO_2 emissions.

Architecture 2030 is an environmental advocacy group whose mission is "to provide information and innovative solutions in the fields of architecture and planning, in an effort to address global climate change." Its founder, New Mexico architect Edward Mazria, points to data from the U.S. Energy Information Administration that indicates buildings are responsible for almost half the total U.S. energy consumption and greenhouse gas (GHG) emissions annually; globally, Mazria believes the percentage is even greater.

What is relevant to any discussion of sustainable design is that most of the building sector's energy consumption is not attributable to the production of materials or the process of construction, but rather to operational processes—the heating, cooling, and lighting of buildings. This means that to reduce the energy consumption and GHG emissions generated by the use and maintenance of buildings over their life span, it is necessary to properly design, site, and shape buildings and incorporate natural heating, cooling, ventilation, and daylighting strategies.

The 2030 Challenge issued by Architecture 2030 calls for all new buildings and developments to be designed to use half the fossil fuel energy they would typically consume, and that an equal amount of existing building area be renovated annually to meet a similar standard. Architecture 2030 is further advocating that the fossil fuel reduction standard be increased from 70% in 2015 to 80% in 2020 and 90% in 2025, and that by 2030, all new buildings be carbon-neutral (using no fossil-fuel GHG-emitting energy to build and operate).

There are two approaches to reducing a building's consumption of GHG-emitting fossil fuels. The passive approach is to work with the climate in designing, siting, and orienting a building and employ passive cooling and heating techniques to reduce its overall energy requirements. The active approach is to increase the ability of a building to capture or generate its own energy from renewable sources (solar, wind, geothermal, low-impact hydro, biomass, and bio-gas) that are available locally and in abundance. While striking an appropriate, cost-effective balance between energy conservation and generating renewable energy is the goal, minimizing energy use is a necessary first step, irrespective of the fact that the energy may come from renewable resources.

Site analysis is the process of studying the contextual forces that influence how we might situate a building, lay out and orient its spaces, shape and articulate its enclosure, and establish its relationship to the landscape. Any site survey begins with the gathering of physical site data.

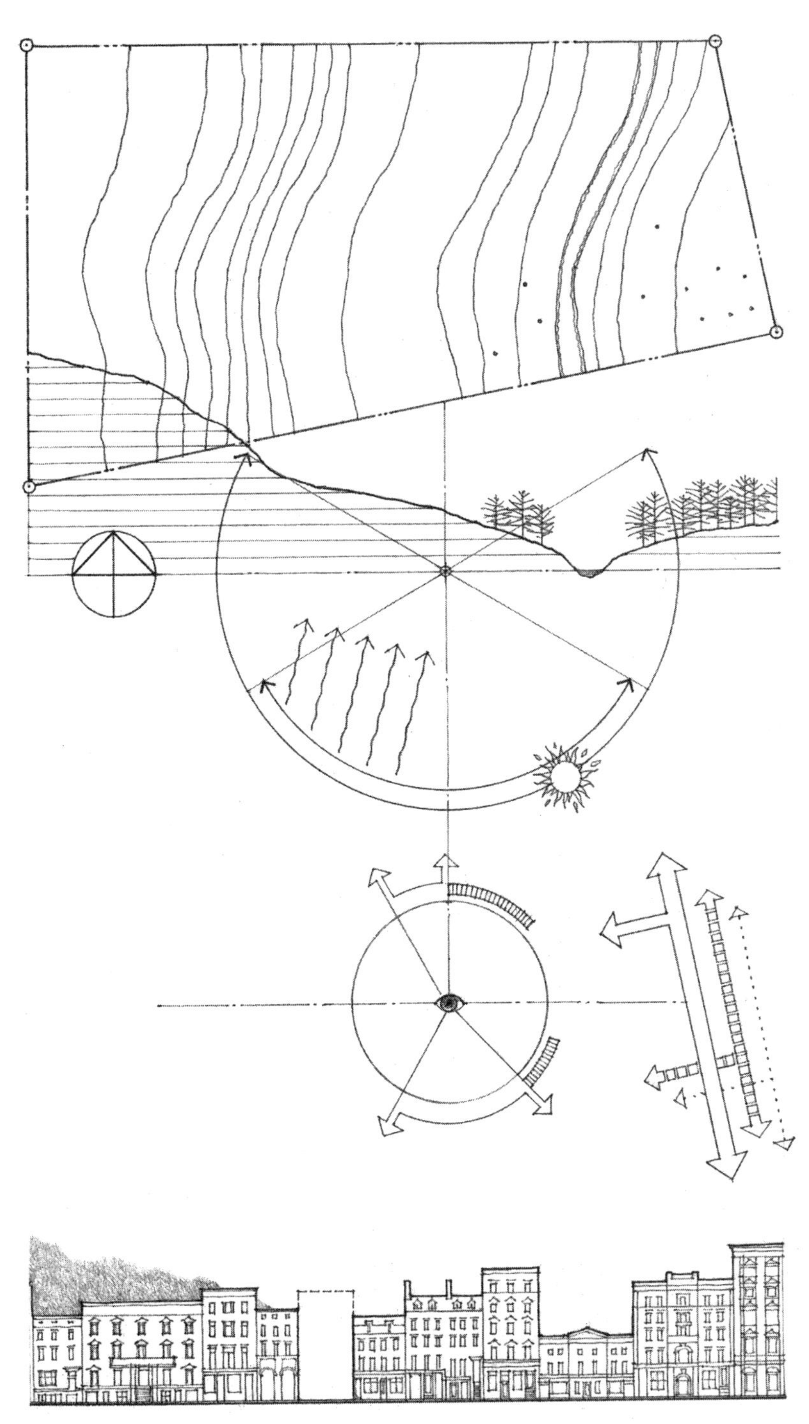

- Draw the area and shape of the site as defined by its legal boundaries.
- Indicate required setbacks, existing easements, and rights-of-way.
- Estimate the area and volume required for the building program, site amenities, and future expansion, if desired.

- Analyze the ground slopes and subsoil conditions to locate the areas suitable for construction and outdoor activities.
- Identify steep and moderate slopes that may be unsuitable for development.
- Locate soil areas suitable for use as a drainage field, if applicable.
- Map existing drainage patterns. (LEED SS Credits 6.1, 6.2: Stormwater Design)
- Determine the elevation of the water table.
- Identify areas subject to excessive runoff of surface water, flooding, or erosion.

- Locate existing trees and native plant materials that should be preserved.
- Chart existing water features, such as wetlands,streams, watersheds, flood plains, or shorelines that should be protected. (LEED SS Credit 5.1: Site Development—Protect or Restore Habitat)

- Map climatic conditions: the path of the sun, the direction of prevailing winds, and the expected amount of rainfall.
- Consider the impact of landforms and adjacent structures on solar access, prevailing winds, and the potential for glare.
- Evaluate solar radiation as a potential energy source.

- Determine possible points of access from public roadways and public transit stops. (LEED SS Credit 4.1: Alternative Transportation—Public Transportation Access)
- Study possible circulation paths for pedestrians and vehicles from these access points to building entrances.
- Ascertain the availability of utilities: water mains, sanitary and storm sewers, gas lines, electrical power lines, telephone and cable lines, and fire hydrants.
- Determine access to other municipal services, such as police and fire protection.

- Identify the scope of desirable views as well as objectionable views.
- Cite potential sources of congestion and noise.
- Evaluate the compatibility of adjacent and proposed land uses.
- Map cultural and historical resources that should be preserved.
- Consider how the existing scale and character of the neighborhood or area might affect the building design.
- Map the proximity to public, commercial, medical, and recreational facilities. (LEED SS Credit 2: Development Density and Community Connectivity)

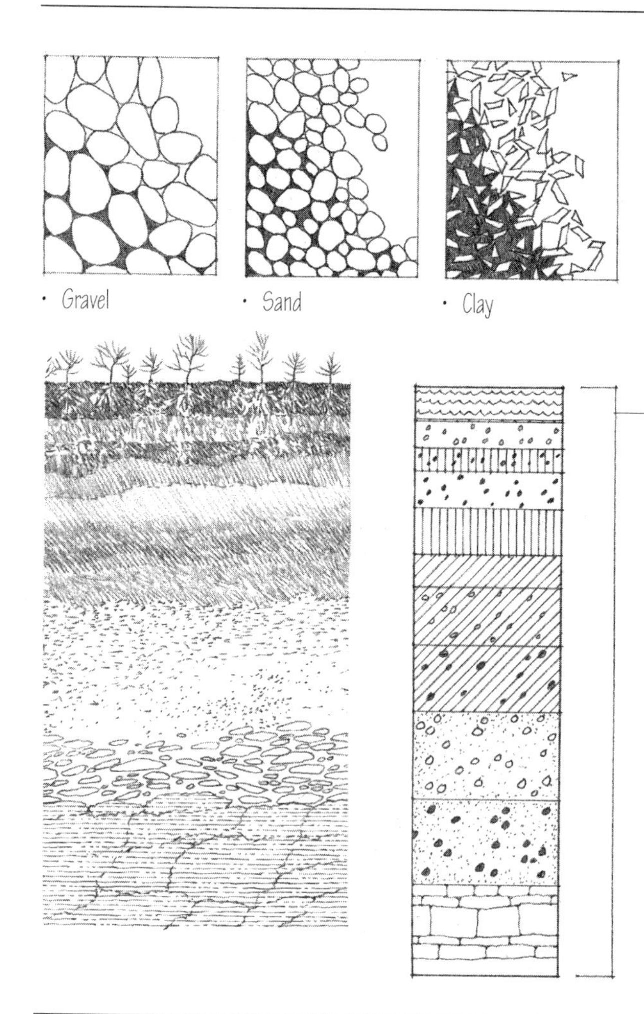

There are two broad classes of soils—coarse-grained soils and fine-grained soils. Coarse-grained soils include gravel and sand, which consist of relatively large particles visible to the naked eye; fine-grained soils, such as silt and clay, consist of much smaller particles. The American Society for Testing and Materials (ASTM) Unified Soil Classification System further divides gravels, sands, silts, and clays into soil types based on physical composition and characteristics. See table that follows.

The soil underlying a building site may actually consist of superimposed layers, each of which contains a mix of soil types, developed by weathering or deposition. To depict this succession of layers or strata called horizons, geotechnical engineers draw a soil profile, a diagram of a vertical section of soil from the ground surface to the underlying material, using information collected from a test pit or boring.

The integrity of a building structure depends ultimately on the stability and strength under loading of the soil or rock underlying the foundation. The stratification, composition, and density of the soil bed, variations in particle size, and the presence or absence of groundwater are all critical factors in determining the suitability of a soil as a foundation material. When designing anything other than a single-family dwelling, it is advisable to have a geotechnical engineer undertake a subsurface investigation.

A subsurface investigation (CSI MasterFormat™ 02 32 00) involves the analysis and testing of soil disclosed by excavation of a test pit up to 10' (3 m) deep or by deeper test borings in order to understand the structure of the soil, its shear resistance and compressive strength, its water content and permeability, and the expected extent and rate of consolidation under loading. From this information, the geotechnical engineer is able to gauge the anticipated total and differential settlement under loading by a proposed foundation system.

Soil Classification*		Symbol	Description	Presumptive Bearing Capacity†		Susceptibility to Frost Action	Permeability & Drainage
				psf‡	kPa		
Gravels	Clean gravels	GW	Well-graded gravel	10000	479	None	Excellent
6.4–76.2 mm		GP	Poorly graded gravel	10000	479	None	Excellent
	Gravels w/ fines	GM	Silty gravel	5000	239	Slight	Poor
		GC	Clayey gravel	4000	192	Slight	Poor
Sands	Clean sands	SW	Well-graded sand	7500	359	None	Excellent
0.05–6.4 mm		SP	Poorly graded sand	6000	287	None	Excellent
	Sands w/ fines	SM	Silty sand	4000	192	Slight	Fair
		SC	Clayey sand	4000	192	Medium	Poor
Silts	LL>50§	ML	Inorganic silt	2000	96	Very high	Poor
0.002–0.05 mm		CL	Inorganic clay	2000	96	Medium	Impervious
& Clays	LL<50§	OL	Organic silt-clay		Very poor	High	Impervious
<0.002 mm		MH	Elastic inorganic silt	2000	96	Very high	Poor
		CH	Plastic inorganic clay	2000	96	Medium	Impervious
		OH	Organic clay and silt		Very poor	Medium	Impervious
Highly organic soils		Pt	Peat		Unsuitable	Slight	Poor

* Based on the ASTM Unified Soil Classification System

† Consult a geotechnical engineer and the building code for allowable bearing capacities.

‡ 1 psf = 0.0479 kPa

§ LL = liquid limit: the water content, expressed as a percentage of dry weight, at which a soil passes from a plastic to a liquid state.

The allowable bearing capacity of a soil is the maximum unit pressure a foundation is permitted to impose vertically or laterally on the soil mass. In the absence of geotechnical investigation and testing, building codes may permit the use of conservative load-bearing values for various soil classifications. While high-bearing-capacity soils present few problems, low-bearing-capacity soils may dictate the use of a certain type of foundation and load distribution pattern, and ultimately, the form and layout of a building.

Density is a critical factor in determining the bearing capacity of granular soils. The Standard Penetration Test measures the density of granular soils and the consistency of some clays at the bottom of a borehole, recording the number of blows required by a hammer to advance a standard soil sampler. In some cases, compaction, by means of rolling, tamping, or soaking to achieve optimum moisture content, can increase the density of a soil bed.

Coarse-grained soils have a relatively low percentage of void spaces and are more stable as a foundation material than silt or clay. Clay soils, in particular, tend to be unstable because they shrink and swell considerably with changes in moisture content. Unstable soils may render a site unbuildable unless an elaborately engineered and expensive foundation system is put in place.

The shearing strength of a soil is a measure of its ability to resist displacement when an external force is applied, due largely to the combined effects of cohesion and internal friction. On sloping sites, as well as during the excavation of a flat site, unconfined soil has the potential to displace laterally. Cohesive soils, such as clay, retain their strength when unconfined; granular soils, such as gravel, sand, or some silts, require a confining force for their shear resistance and have a relatively shallow angle of repose.

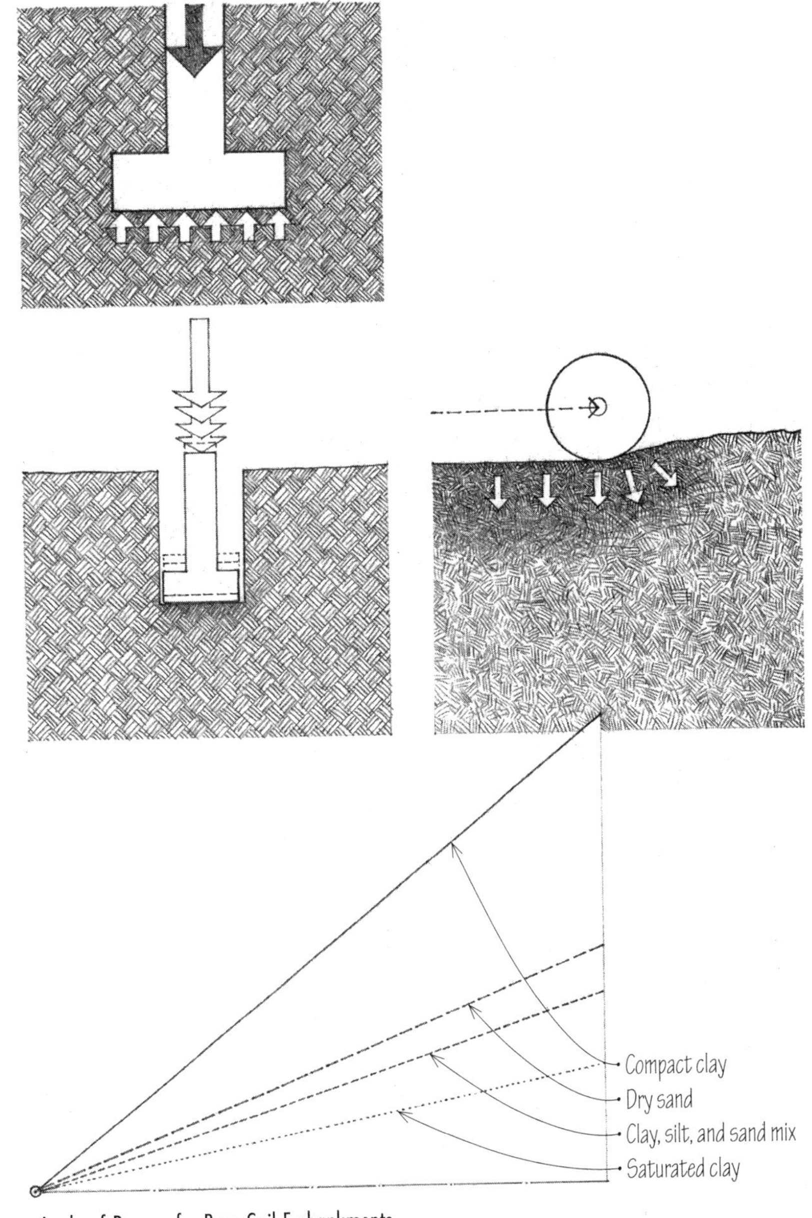

Angle of Repose for Bare Soil Embankments

The water table is the level beneath which the soil is saturated with groundwater. Some building sites are subject to seasonal fluctuations in the level of groundwater. Any groundwater present must be drained away from a foundation system to avoid reducing the bearing capacity of the soil and to minimize the possibility of water leaking into a basement. Coarse-grained soils are more permeable and drain better than fine-grained soils, and are less susceptible to frost action.

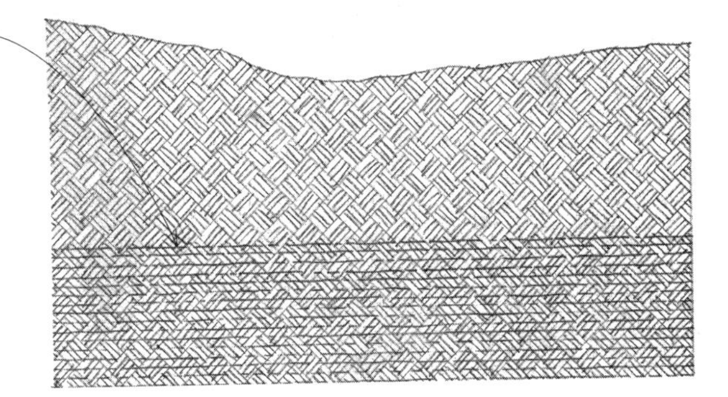

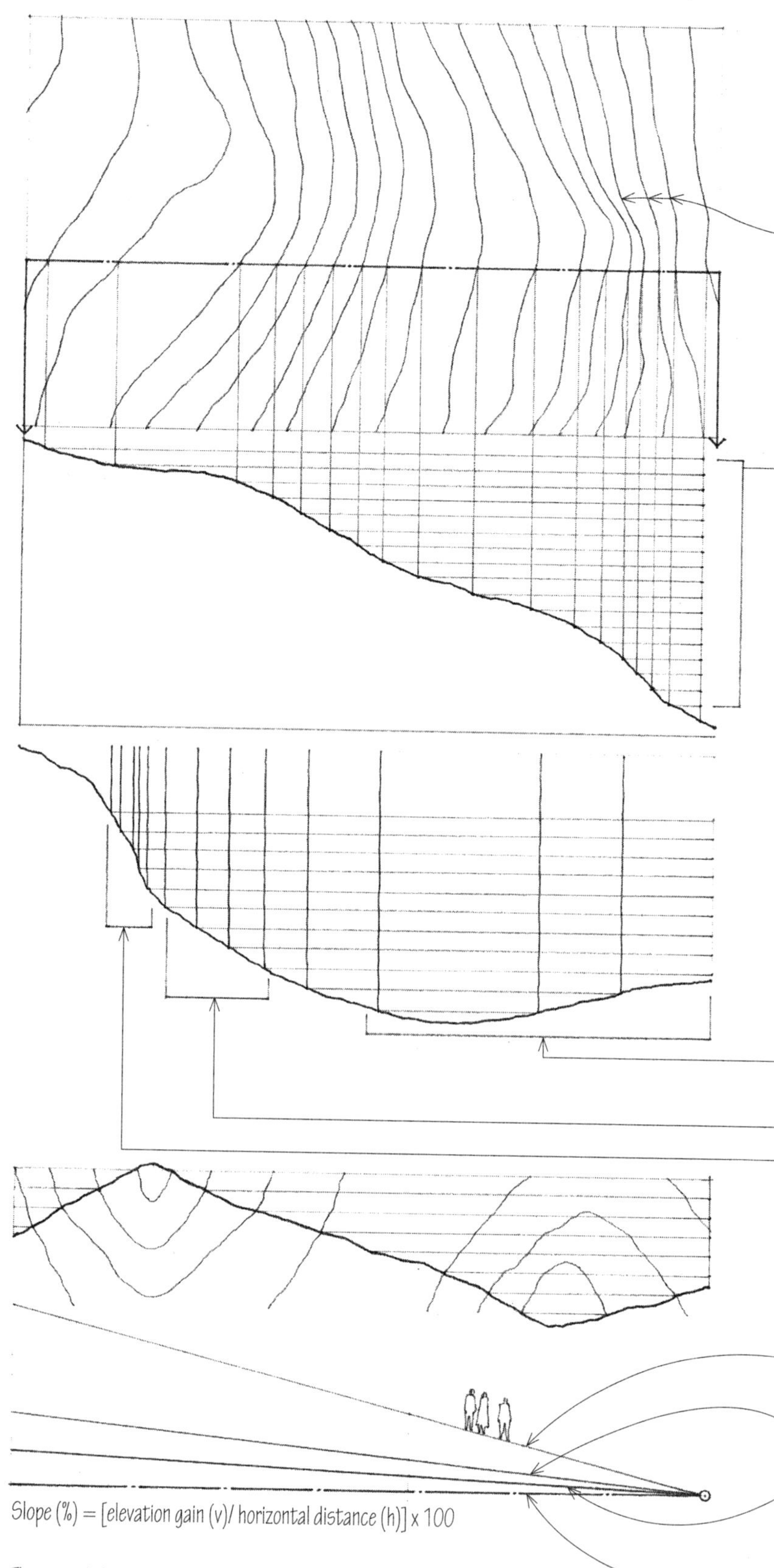

Topography refers to the configuration of surface features of a plot of land, which influences where and how to build and develop a site. To study the response of a building design to the topography of a site, we can use a series of site sections or a site plan with contour lines.

Contour lines are imaginary lines joining points of equal elevation above a datum or bench mark. The trajectory of each contour line indicates the shape of the land formation at that elevation. Note that contour lines are always continuous and never cross one another; they coincide in a plan view only when they cut across a vertical surface.

Contour interval refers to the difference in elevation represented by any two adjacent contour lines on a topographic map or site plan. The interval used is determined by the scale of a drawing, the size of the site, and the nature of the topography. The larger the area and the steeper the slopes, the greater the interval between contours. For large or steeply sloping sites, 20' or 40' (5 or 10 m) contour intervals may be used. For small sites having relatively gradual slopes, 1', 2', or 5' (0.5 or 1.0 m) contours may be necessary.

We can discern the topographical nature of a site by reading the horizontal spacing and shape of contour lines.

- Contours spaced far apart indicate a relatively flat or gently sloping surface.
- Equally spaced contours denote a constant slope.
- Closely spaced contours disclose a relatively steep rise in elevation.
- Contour lines represent a ridge when pointing toward lower elevations; they represent a valley when pointing toward higher elevations.

- Ground slopes over 25% are subject to erosion and are difficult to build on.
- Ground slopes over 10% are challenging to use for outdoor activities and are more expensive to build on.
- Ground slopes from 5% to 10% are suitable for informal outdoor activities and can be built on without too much difficulty.
- Ground slopes up to 5% are usable for most outdoor activities and are relatively easy to build on.

- Slope (%) = [elevation gain (v)/ horizontal distance (h)] x 100

The ground slope between any two contour lines is a function of the total change in elevation and the horizontal distance between the two contours.

For aesthetic and economic, as well as ecological reasons, the general intent in developing a site should be to minimize the disturbance of existing landforms and features while taking advantage of natural ground slopes and the microclimate of the site.

- Site development and construction should minimize disrupting the natural drainage patterns of the site and adjacent properties.
- When modifying landforms, include provisions for the drainage of surface water and groundwater.
- Attempt to equalize the amount of cut and fill required for construction of a foundation and site development.
- Avoid building on steep slopes subject to erosion or slides.
- Wetlands and other wildlife habitats may require protection and limit the buildable area of a site.
- Pay particular attention to building restrictions on sites located in or near a flood plain.

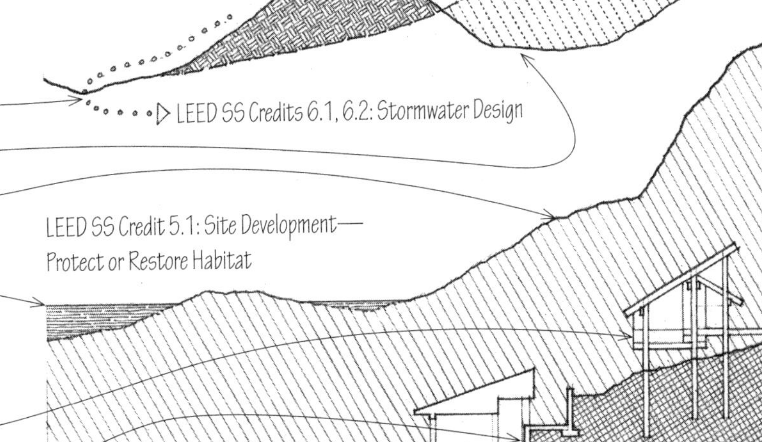

- Elevating a structure on poles or piers minimizes disturbance of the natural terrain and existing vegetation.
- Terracing or stepping a structure along a slope requires excavation and the use of retaining walls or bench terracing.
- Cutting a structure into a slope or locating it partially underground moderates temperature extremes and minimizes exposure to wind, and heat loss in cold climates.

The microclimate of a site is influenced by the ground elevation, the nature and orientation of landforms, and the presence of bodies of water.

- The temperature in the atmosphere decreases with altitude—approximately 1°F (0.56°C) for every 400' (122 m) in elevation.
- Warm air rises.
- Heavier cool air settles into low-lying areas.
- Solar radiation warms southern slopes, creating a temperate zone.
- Daytime breezes, which replace updrafts of warm air over land, can have a cooling effect of up to 10°F (5.6°C).
- Grass and other ground covers tend to lower ground temperatures by absorbing solar radiation and encouraging cooling by evaporation.
- Hard surfaces tend to elevate ground temperatures.
- Light-colored surfaces reflect solar radiation; dark surfaces absorb and retain the radiation.

LEED SS Credits 7.1, 7.2: Heat Island Effect

Large bodies of water:

- Act as heat reservoirs and moderate variations in local temperature;
- Are generally cooler than land during the day and warmer at night, generating offshore breezes;
- Are generally warmer than land in winter and cooler in summer.
- In hot-dry climates, even small bodies of water are desirable, both psychologically and physically, for their evaporative cooling effect.

CSI MasterFormat 31 10 00: Site Clearing
CSI MasterFormat 31 20 00: Earth Moving
CSI MasterFormat 32 70 00: Wetlands

Plant materials provide aesthetic as well as functional benefits in conserving energy, framing or screening views, moderating noise, retarding erosion, and visually connecting a building to its site. Factors to consider in the selection and use of plant materials in landscaping include the:

- Tree structure and shape
- Seasonal density, texture, and color of foliage
- Speed or rate of growth
- Mature height and spread of foliage
- Requirements for soil, water, sunlight, and temperature range
- Depth and extent of the root structure

- Trees and other plant life adapt their forms to the climate.

LEED SS Credits 6.1, 6.2: Stormwater Design
LEED SS Credit 7.1: Heat Island Effect—Nonroof
LEED WE Credit 1: Water Efficient Landscaping

- Existing healthy trees and native plant materials should be preserved whenever possible. During construction and when regrading a site, existing trees should be protected for an area equal to the diameter of their crowns. The root systems of trees planted too close to a building may disturb the foundation system. Root structures can also interfere with underground utility lines.
- To support plant life, a soil must be able to absorb moisture, supply the appropriate nutrients, be capable of aeration, and be free of concentrated salts.

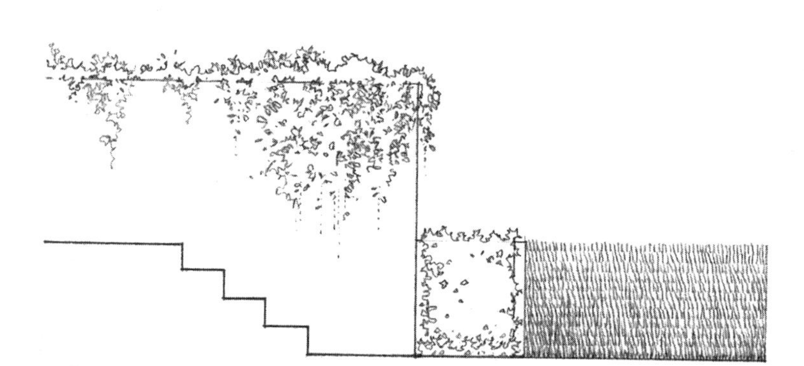

Grass and other ground covers:
- Can reduce air temperature by absorbing solar radiation and encouraging cooling by evaporation;
- Aid in stabilizing soil embankments and preventing erosion;
- Increase the permeability of soil to air and water.

- Vines can reduce the heat transmission through a sunlit wall by providing shade and cooling the immediate environment by evaporation.

CSI MasterFormat 32 90 00: Planting

Trees affect the immediate environment of a building in the following ways:

Providing Shade

The amount of solar radiation obstructed or filtered by a tree depends on its:

- Orientation to the sun
- Proximity to a building or outdoor space
- Shape, spread, and height
- Density of foliage and branch structure

- Trees shade a building or outdoor space most effectively from the southeast during the morning and the southwest during the late afternoon when the sun has a low altitude and casts long shadows.
- South-facing overhangs provide more efficient shading during the midday period when the sun is high and casts short shadows.
- Deciduous trees provide shade and glare protection during the summer and allow solar radiation to penetrate through their branch structures during the winter.
- Evergreens provide shade throughout the year and help reduce snow glare during the winter.

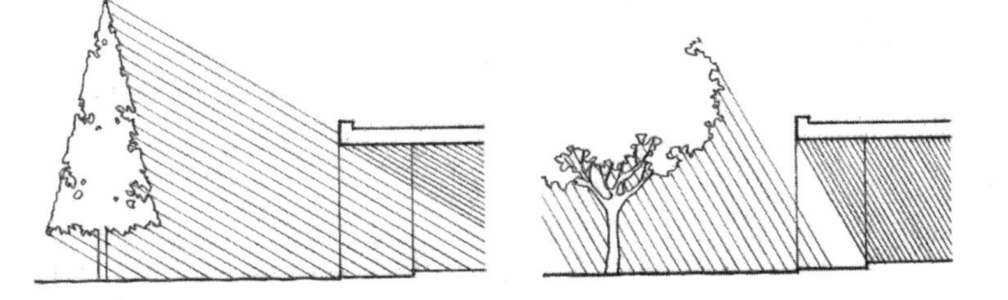

Serving as Windbreak

- Evergreens can form effective windbreaks and reduce heat loss from a building during the winter.
- The foliage of plant materials reduces wind-blown dust.

Defining Space

- Trees can shape outdoor spaces for activity and movement.

Directing or Screening Views

- Trees can frame desirable views.
- Trees can screen undesirable views and provide privacy for outdoor spaces.

Attenuating Sound

- A combination of deciduous and evergreen trees is most effective in intercepting and attenuating airborne sound, especially when combined with earth mounds.

Improving Air Quality

- Trees trap particulate matter on their leaves, which is then washed to the ground during rainfall.
- Leaves can also assimilate gaseous and other pollutants.
- Photosynthetic process can metabolize fumes and other odors.

Stabilizing Soil

- The root structures of trees aid in stabilizing soil, increasing the permeability of the soil to water and air, and preventing erosion.

The location, form, and orientation of a building and its spaces should take advantage of the thermal, hygienic, and psychological benefits of sunlight. Solar radiation, however, may not always be beneficial, depending on the latitude and climate of the site. In planning the design of a building, the objective should be to maintain a balance between underheated periods when solar radiation is beneficial and overheated periods when radiation should be avoided.

The path of the sun through the sky varies with the seasons and the latitude of a building site. The range of solar angles for a specific site should be obtained from a weather almanac or service bureau before calculating the potential solar heat gain and shading requirements for a building design.

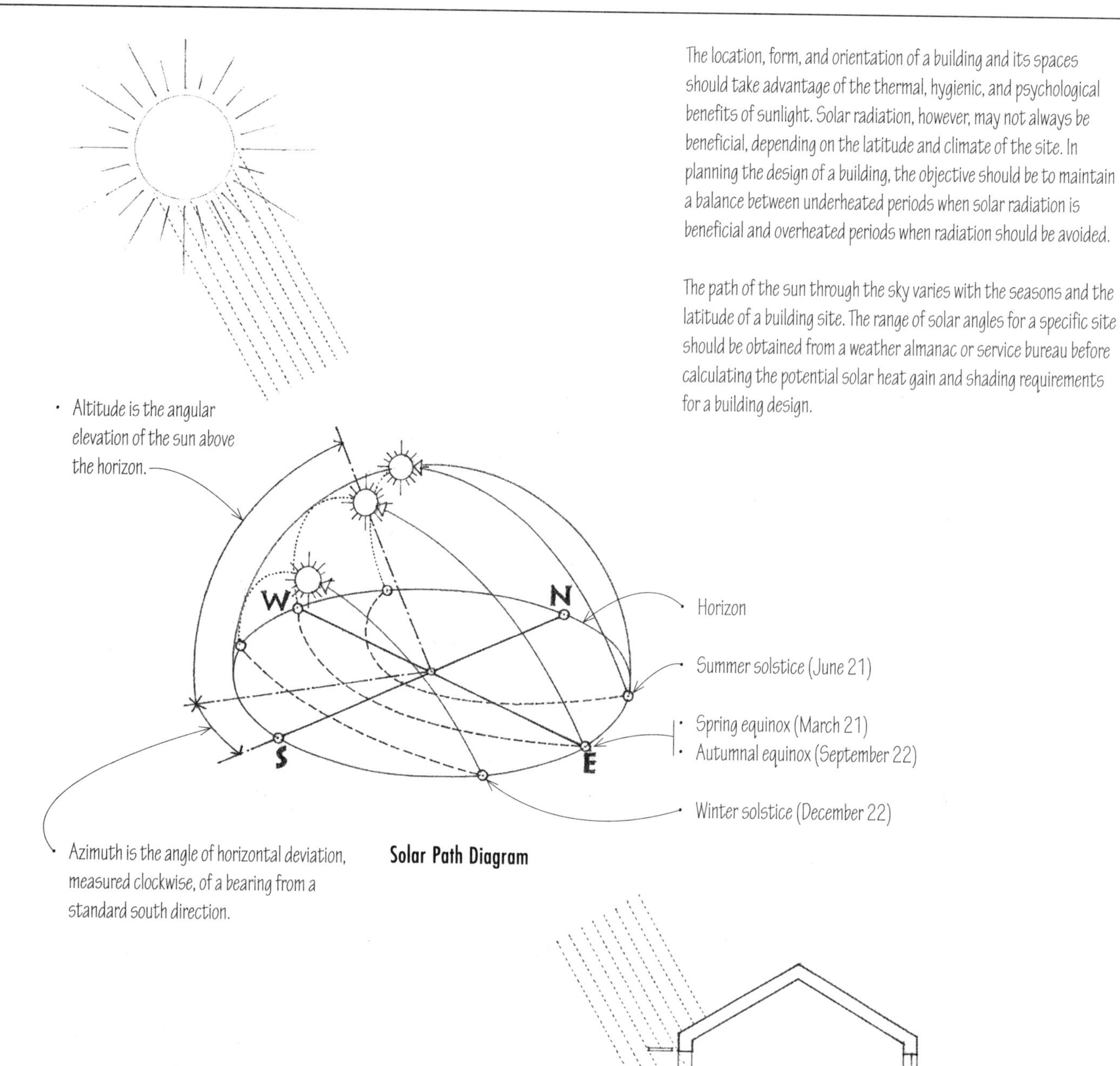

Solar Path Diagram

Representative Solar Angles

North Latitude	Representative City	Altitude at Noon		Azimuth at Sunrise & Sunset*	
		Dec. 22	Mar. 21/Sept. 22	Dec. 22	June 21
48°	Seattle	18°	42°	54°	124°
44°	Toronto	22°	46°	56°	122°
40°	Denver	26°	50°	58°	120°
36°	Tulsa	30°	54°	60°	118°
32°	Phoenix	34°	58°	62°	116°

* Azimuth is east of south for sunrise, and west of south for sunset.

The following are recommended forms and orientations for isolated buildings in different climatic regions. The information presented should be considered along with other contextual and programmatic requirements.

Cool Regions

Minimizing the surface area of a building reduces exposure to low temperatures.

- Maximize absorption of solar radiation.
- Reduce radiant, conductive, and evaporative heat loss.
- Provide wind protection.

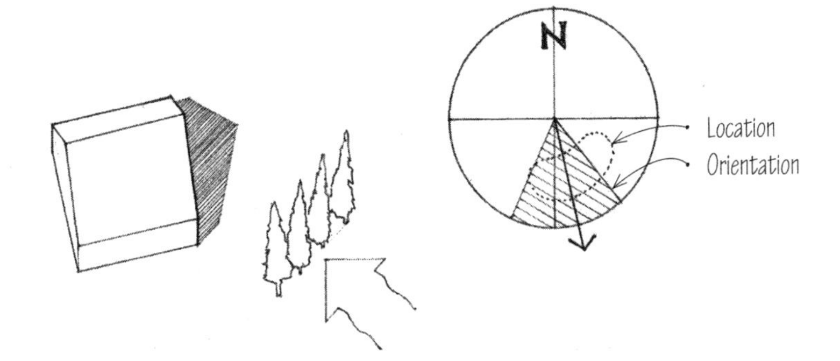

Temperate Regions

Elongating the form of a building along the east-west axis maximizes south-facing walls.

- Minimize east and west exposures, which are generally warmer in summer and cooler in winter than southern exposures.
- Balance solar heat gain with shade protection on a seasonal basis.
- Encourage air movement in hot weather; protect against wind in cold weather.

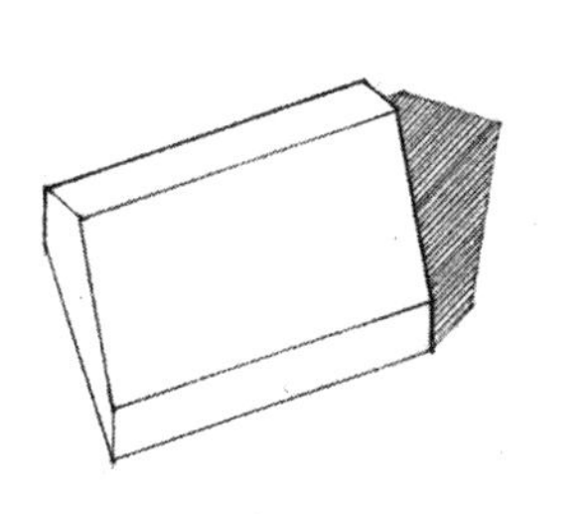

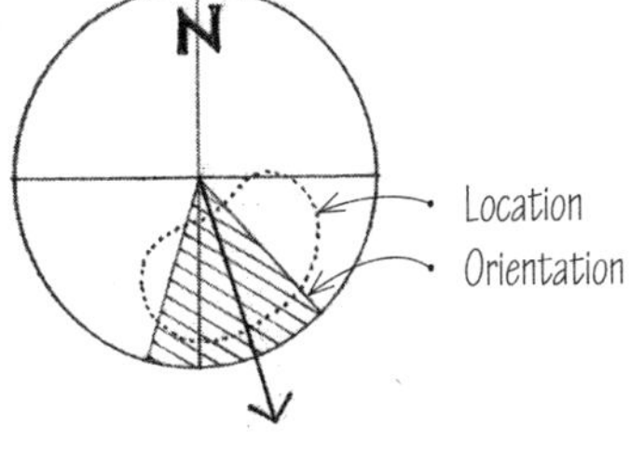

Hot-Arid Regions

Building forms should enclose courtyard spaces.

- Reduce solar and conductive heat gain.
- Promote cooling by evaporation using water features and plantings.
- Provide solar shading for windows and outdoor spaces.

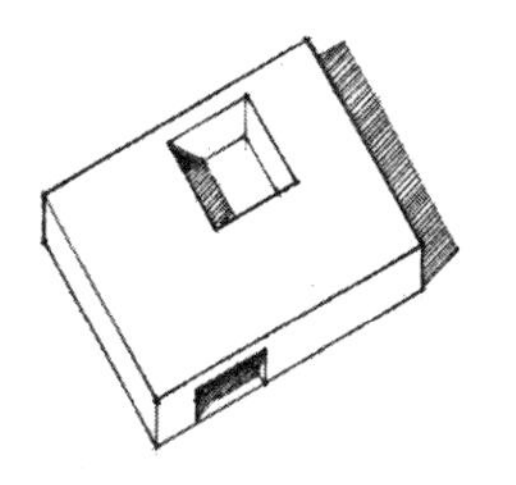

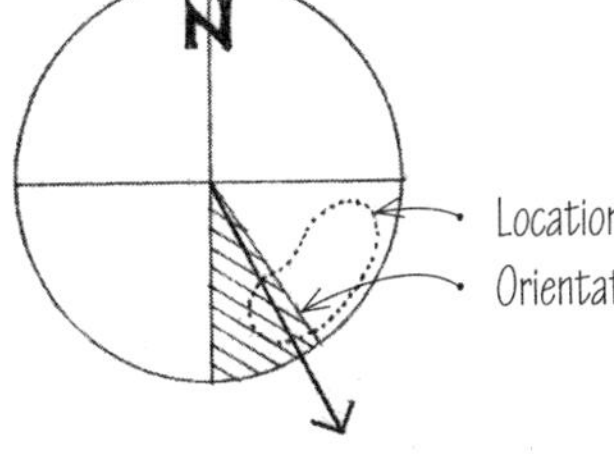

Hot-Humid Regions

Building form elongated along the east-west axis minimizes east and west exposures.

- Reduce solar heat gain.
- Utilize wind to promote cooling by evaporation.
- Provide solar shading for windows and outdoor spaces.

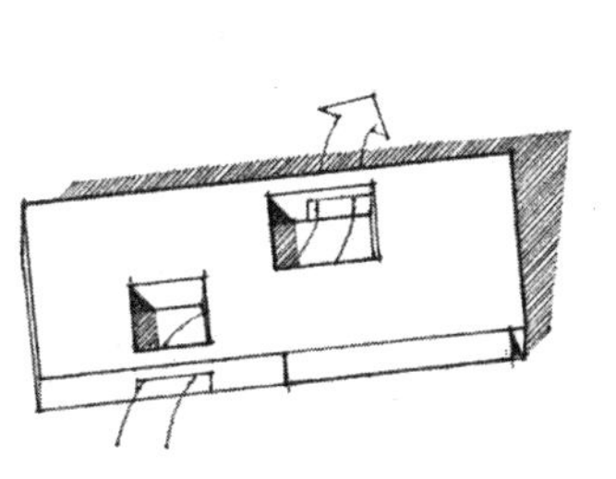

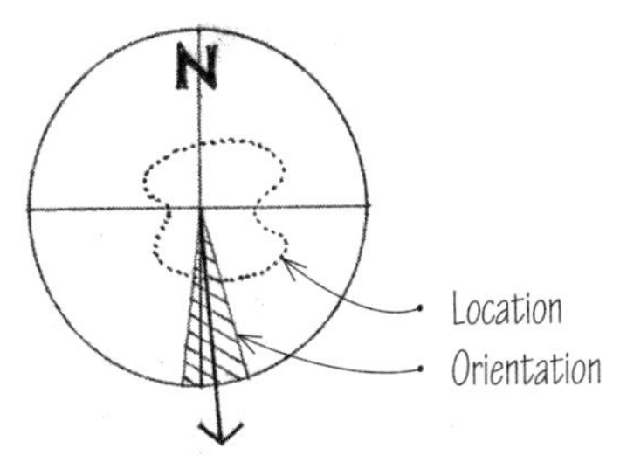

LEED EA Credit 1: Optimize Energy Performance

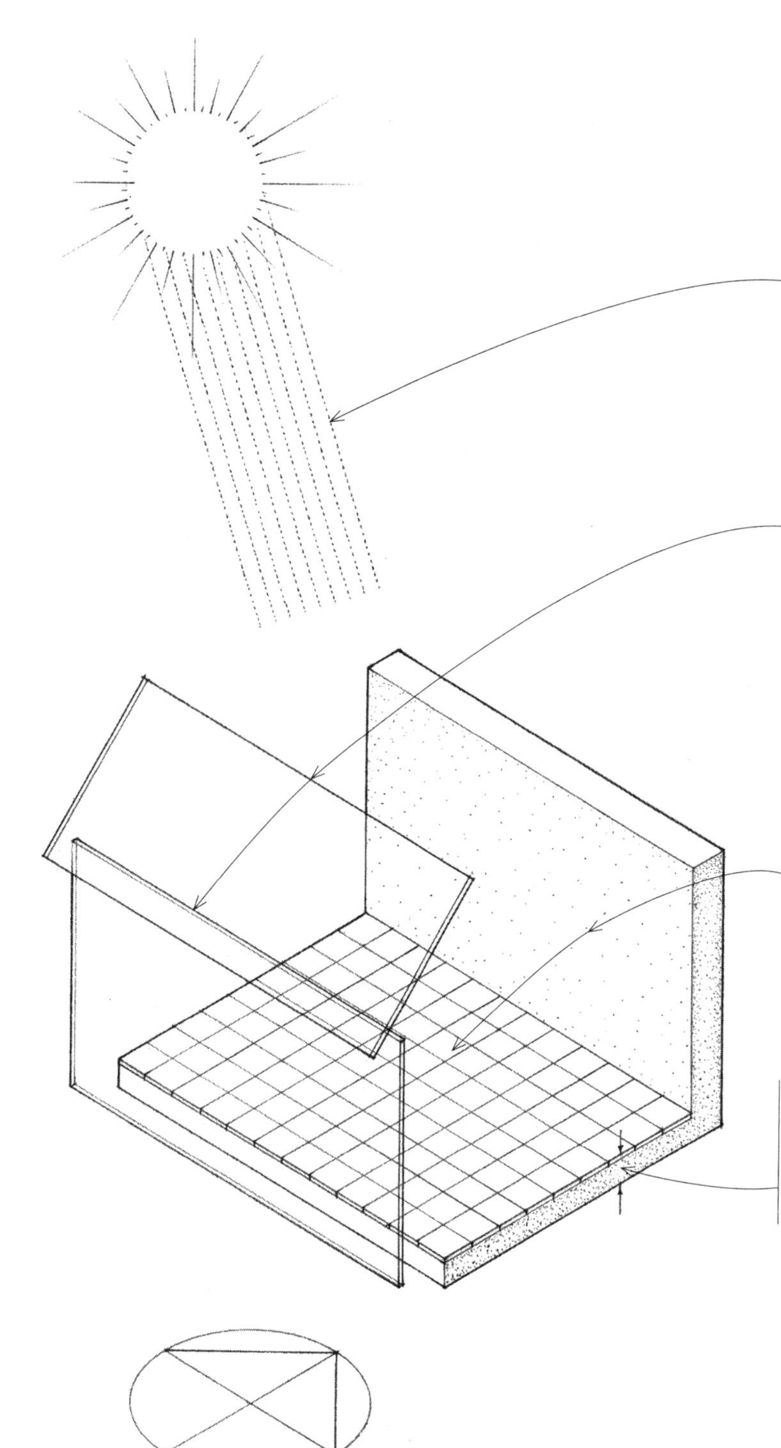

Passive solar heating refers to using solar energy to heat the interior spaces of a building without relying on mechanical devices that require additional energy. Passive solar systems rely instead on the natural heat transfer processes of conduction, convection, and radiation for the collection, storage, distribution, and control of solar energy.

- The solar constant is the average rate at which radiant energy from the sun is received by the earth, equal to 430 Btu per square foot per hour (1353 W/m²/hr), used in calculating the effects of solar radiation on buildings.

There are two essential elements in every passive solar system:

1. South-facing glass or transparent plastic for solar collection
- Area of glazing should be 30% to 50% of floor area in cold climates and 15% to 25% of floor area in temperate climates, depending on average outdoor winter temperature and projected heat loss.
- Glazing material should be resistant to the degradation caused by the ultraviolet rays of the sun.
- Double-glazing and insulation are required to minimize nighttime heat loss.

2. A thermal mass for heat collection, storage, and distribution, oriented to receive maximum solar exposure
- Thermal storage materials include concrete, brick, stone, tile, rammed earth, sand, and water or other liquid. Phase-change materials, such as eutetic salts and paraffins, are also feasible.
- Concrete: 12" to 18" (305 to 455)
- Brick: 10" to 14" (255 to 355)
- Adobe: 8" to 12" (200 to 305)
- Water: 6" (150) or more
- Dark-colored surfaces absorb more solar radiation than light-colored surfaces.

- Vents, dampers, movable insulation panels, and shading devices can assist in balancing heat distribution.

Based on the relationship between the sun, the interior space, and the heat collection system, there are three ways in which passive solar heating can be accomplished: direct gain, indirect gain, and isolated gain.

LEED EA Credit 2: On-Site Renewable Energy
LEED EA Credit 6: Green Power

CSI MasterFormat 23 56 00: Solar Energy Heating Equipment

Direct Gain

Direct gain systems collect heat directly within an interior space. The surface area of the storage mass, which is incorporated into the space, should be 50% to 66% of the total surface area of the space. During the cooling season, operable windows and walls are used for natural or induced ventilation.

Indirect Gain

Indirect gain systems control heat gain at the exterior skin of a building. The solar radiation first strikes the thermal mass, either a concrete or masonry Trombe wall, or a drumwall of water-filled barrels or tubes, which is located between the sun and the living space. The absorbed solar energy moves through the wall by conduction and then to the space by radiation and convection.

Sunspace

A sunroom or solarium is another medium for indirect heat gain. The sunspace, having a floor of high thermal mass, is separated from the main living space by a thermal storage wall from which heat is drawn as needed. For cooling, the sunspace can be vented to the exterior.

Roof Pond

Another form of indirect gain is a roof pond that serves as a liquid mass for absorbing and storing solar energy. An insulating panel is moved over the roof pond at night, allowing the stored heat to radiate downward into the space. In summer, the process is reversed to allow internal heat absorbed during the day to radiate to the sky at night.

Isolated Gain

Isolated gain systems collect and store solar radiation away from the space to be heated. As air or water in a collector is warmed by the sun, it rises to the served space or is stored in the thermal mass until needed. Simultaneously, cooler air or water is pulled from the bottom of the thermal storage, creating a natural convection loop.

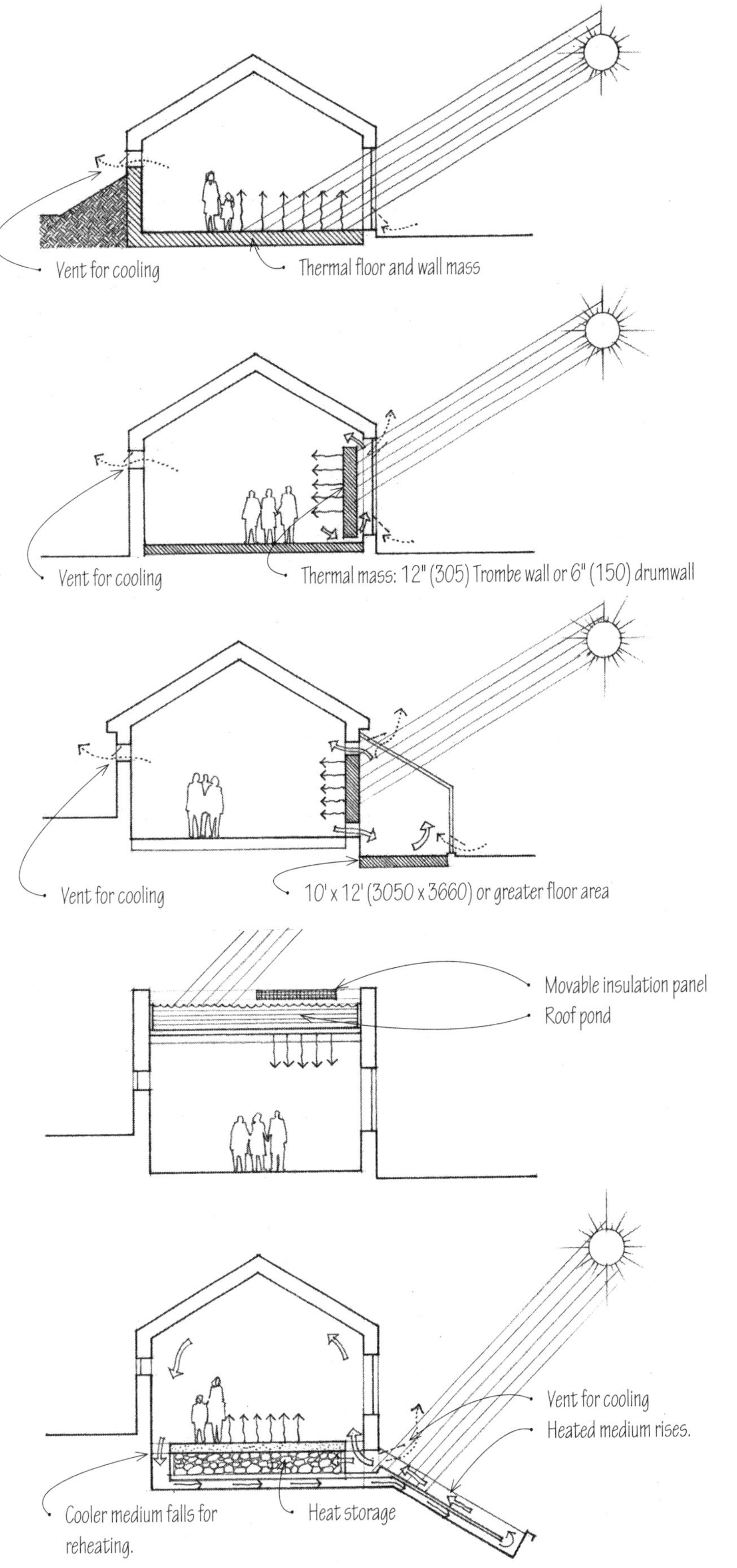

Shading devices shield windows and other glazed areas from direct sunlight in order to reduce glare and excessive solar heat gain in warm weather. Their effectiveness depends on their form and orientation relative to the solar altitude and azimuth for the time of day and season of the year. Exterior devices are more efficient than those located within interior spaces because they intercept solar rays before they can reach an exterior wall or window.

Illustrated are basic types of solar shading devices. Their form, orientation, materials, and construction may vary to suit specific situations. Their visual qualities of pattern, texture, and rhythm, and the shadows they cast, should be considered when designing the facades of a building.

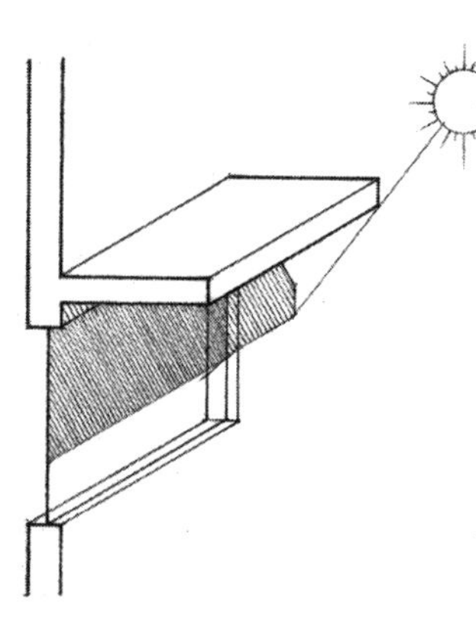

- Horizontal overhangs are most effective when they have southern orientations.

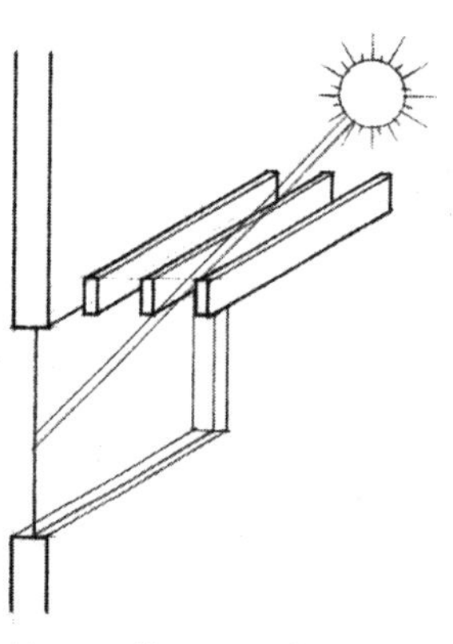

- Horizontal louvers parallel to a wall permit air circulation near the wall and reduce conductive heat gain.
- Louvers may be operated manually or controlled automatically with time or photoelectric controls to adapt to the solar angle.

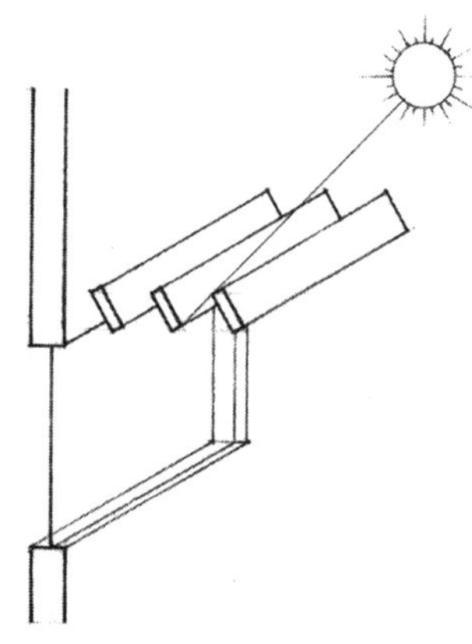

- Slanted louvers provide more protection than those parallel to a wall.
- Angle varies according to the range of solar angles.

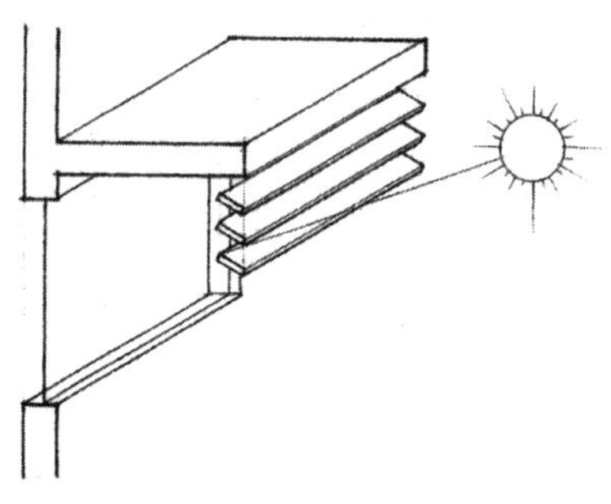

- Louvers hung from a solid overhang protect against low sun angles.
- Louvers may interfere with view.

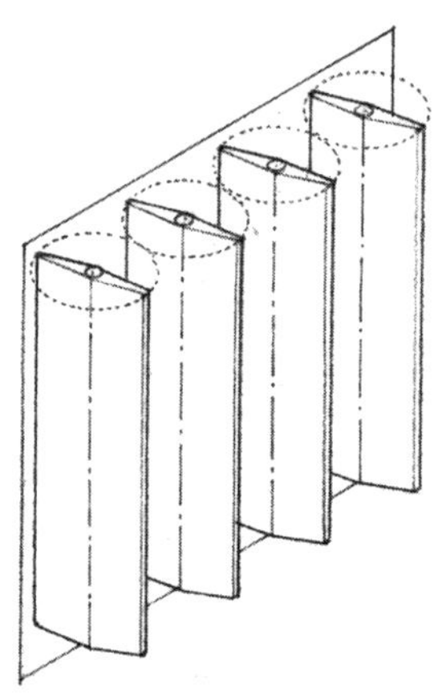

- Vertical louvers are most effective for eastern or western exposures.
- Louvers may be operated manually or controlled automatically with time or photoelectric controls to adapt to solar angle.
- Separation from wall reduces conductive heat gain.

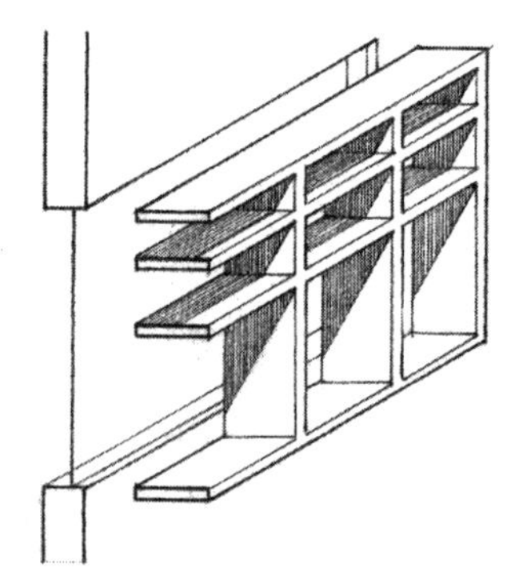

- Eggcrates combine the shading characteristics of horizontal and vertical louvers and have a high shading ratio.
- Eggcrates, sometimes referred to as brise-soleil, are very efficient in hot climates.

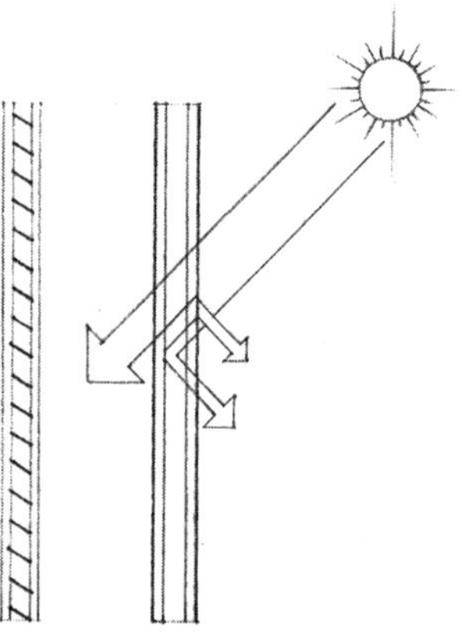

- Solar blinds and screens can provide up to a 50% reduction in solar radiation, depending on their reflectivity.
- Heat-absorbing glass can absorb up to 40% of the radiation reaching its surface.

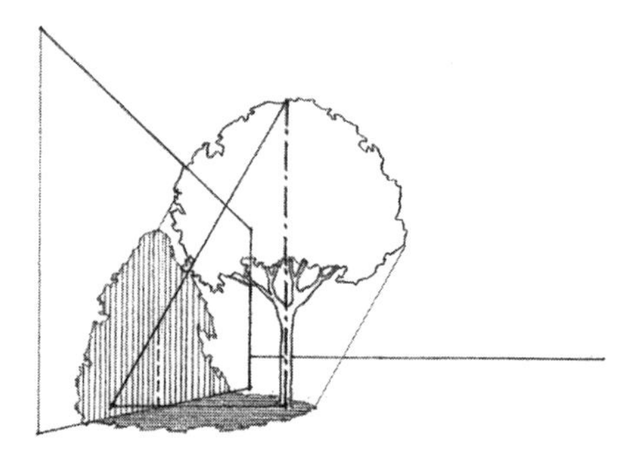

- Trees and adjacent structures may provide shade depending on their proximity, height, and orientation.

CSI MasterFormat 10 71 13: Exterior Sun Control Devices

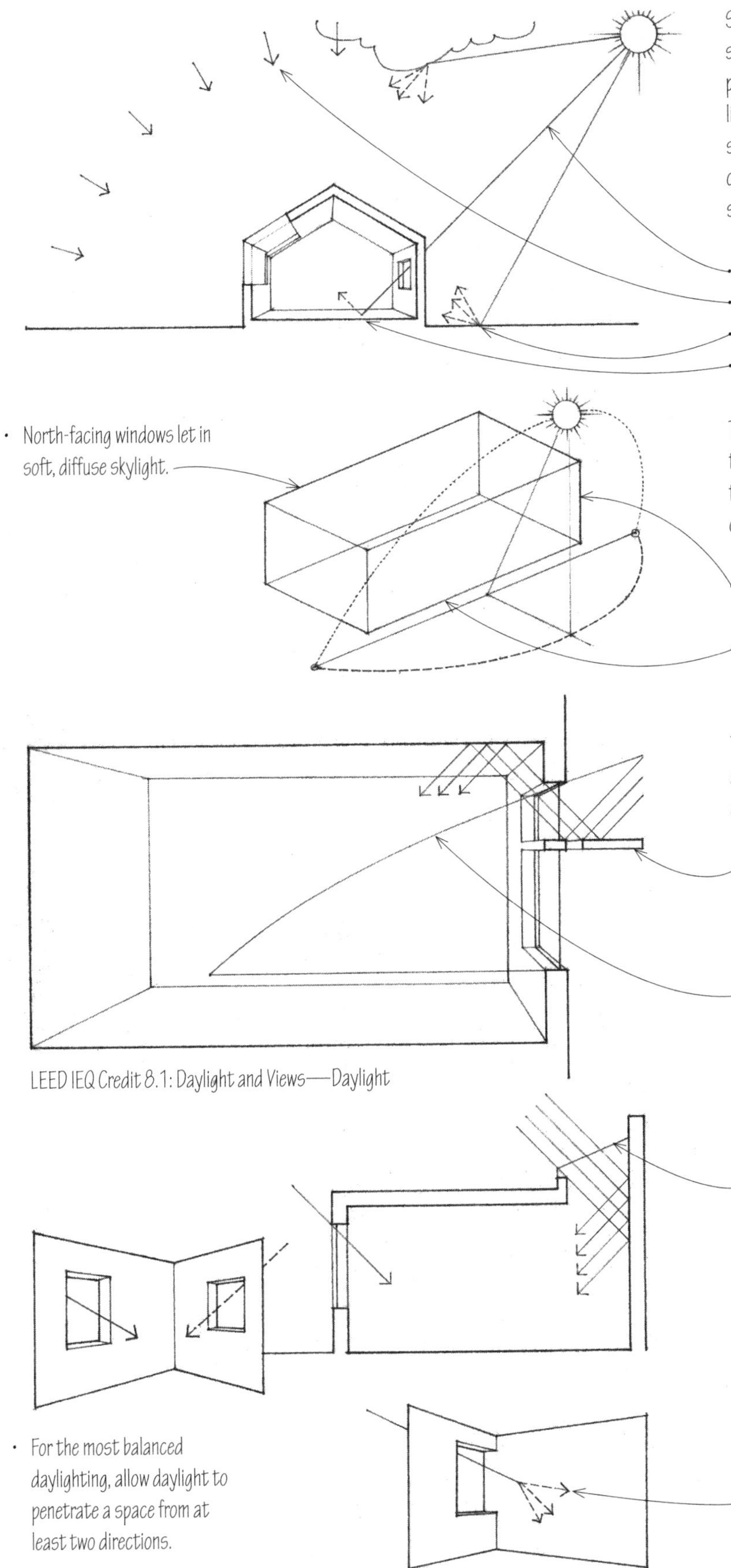

Solar radiation provides not only heat but also light for the interior spaces of a building. This daylight has psychological benefits as well as practical utility in reducing the amount of energy required for artificial lighting. While intense, direct sunlight varies with the time of day, from season to season, and from place to place, it can be diffused by cloud cover, haze, and precipitation, and reflected from the ground and other surrounding surfaces.

- Direct sunlight
- Skylight reflected and diffused by air molecules
- External reflectance from the ground and adjacent structures
- Internal reflectance from room surfaces

The quantity and quality of daylighting in a space are determined by the size and orientation of its window openings, transmittance of the glazing, reflectance of room surfaces and outdoor surfaces, and obstructions of overhangs and nearby trees.

- East- and west-facing windows require shading devices to avoid the bright early-morning and late-afternoon sun.
- South-facing windows are ideal sources for daylight if horizontal shading devices can control excessive solar radiation and glare.

The level of illumination provided by daylight diminishes as it penetrates an interior space. Generally, the larger and higher a window is, the more daylight will enter a room.

- Light shelves shade glazing from direct sunlight while reflecting daylight onto the ceiling of a room. A series of parallel, opaque white louvers can also provide solar shading and reflect diffused daylight into the interior.
- A useful rule of thumb is that daylighting can be effective for task illumination up to a depth of twice the height of a window.
- The ceiling and back wall of a space are more effective than the side walls or the floor in the reflection and distribution of daylight; light-colored surfaces reflect and distribute light more efficiently, but large areas of shiny surfaces can cause glare.
- Skylights with translucent glazing can effectively daylight a space from above without excessive heat gain.
- Roof monitors are another means of reflecting daylight into a space.

Excessive brightness ratios can lead to glare and impairment of visual performance. Glare can be controlled by the use of shading devices, the proper orientation of task surfaces, and allowing daylight to enter a space from at least two directions.

- Place windows adjacent to side walls for additional reflectance and illumination.

The amount of annual and seasonal precipitation expected for a building site should influence the design and construction of the roof structure of a building, the choice of building materials, and the detailing of its exterior wall assemblies. Furthermore, the runoff of rain and melting snow from constructed roof areas and paved surfaces increases the amount of storm water that must be drained from the site.

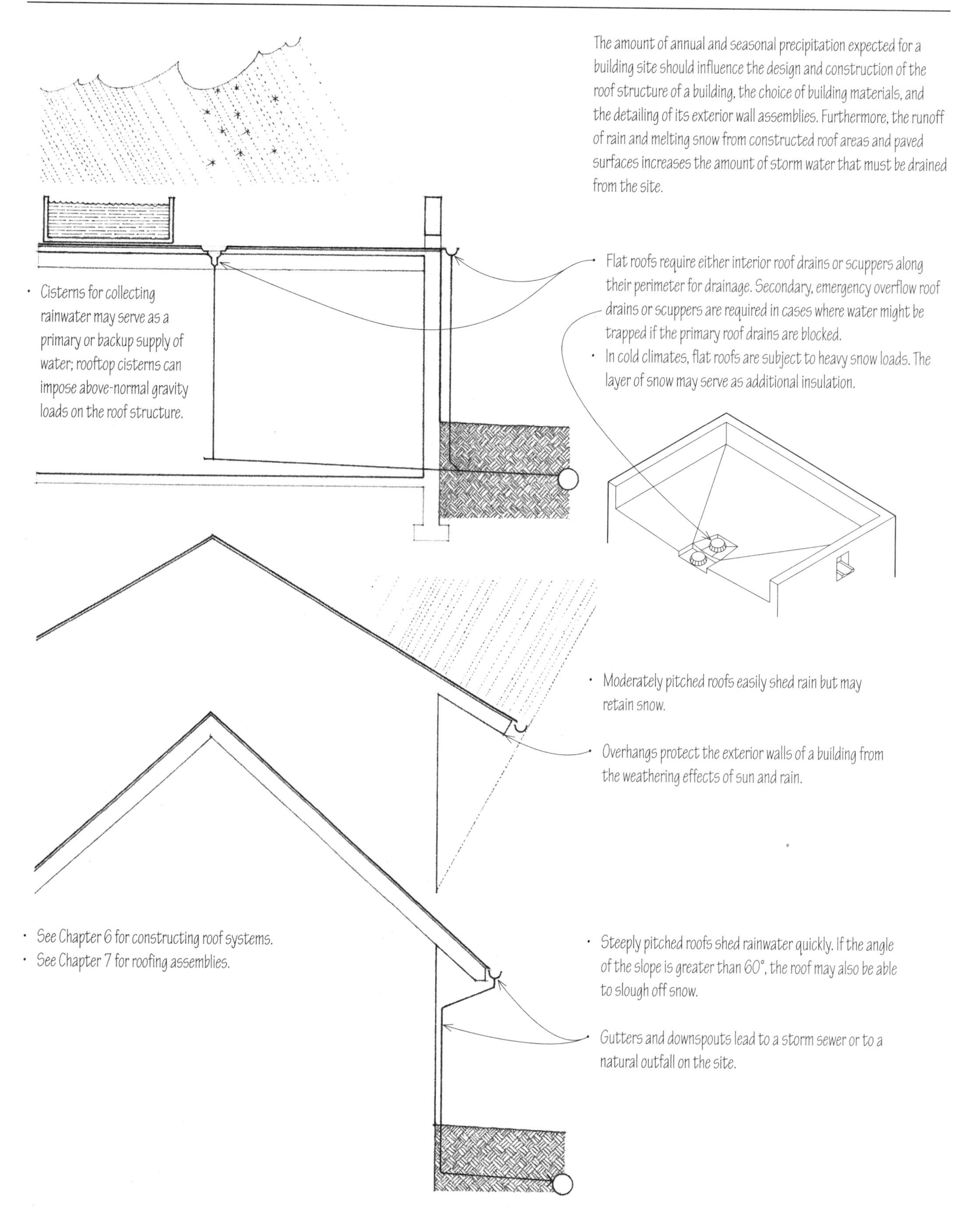

Development of a site can disrupt the existing drainage pattern and create additional water flow from constructed roof areas and paved surfaces. Limiting disruption of a site's natural water hydrology and promoting infiltration by such means as pervious paving and vegetated roofs is advisable. Site drainage is necessary to prevent erosion and the collection of excess surface water or groundwater resulting from new construction.

There are two basic types of site drainage: subsurface and surface drainage systems. Subsurface drainage consists of an underground network of piping for conveying groundwater to a point of disposal, as a storm sewer system or a natural outfall at a lower elevation on the site. Excess groundwater can reduce the load-carrying capacity of a foundation soil and increase the hydrostatic pressure on a building foundation. Waterproofing is required for basement structures situated close to or below the water table of a site.

Surface drainage refers to the grading and surfacing of a site in order to divert rain and other surface water into natural drainage patterns or a municipal storm sewer system. A holding pond may be necessary when the amount of surface runoff exceeds the capacity of the storm sewer system.

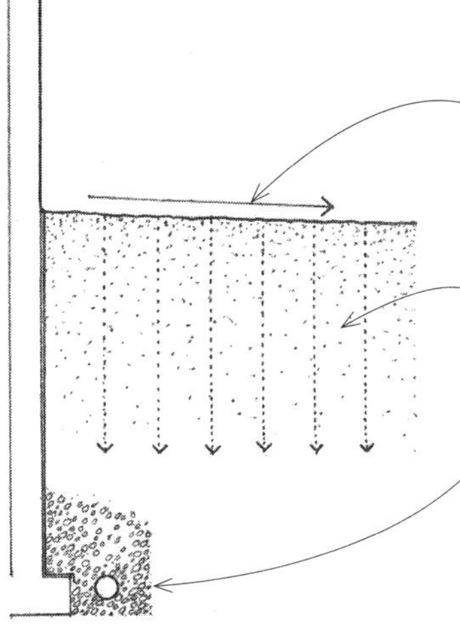

- Finish grades should be sloped to drain surface water away from a building: 5% minimum; 2% minimum for impervious surfaces.
- Groundwater consists largely of surface water that has seeped down through porous soil.
- Foundation drain system; see 3.14.

Surface Drainage Slopes

- Grass lawns and fields: 1.5% to 10% recommended
- Paved parking areas: 2% to 3% recommended

- Swales are shallow depressions formed by the intersection of two ground slopes, designed to direct or divert the runoff of surface water. Vegetated swales can increase infiltration.
- Grass swales: 1.5% to 2% recommended
- Paved swales: 4% to 6% recommended
- Area drains collect surface water from a basement floor or paved area.
- Dry wells are drainage pits lined with gravel or rubble to receive surface water and allow it to percolate away to absorbent earth underground.

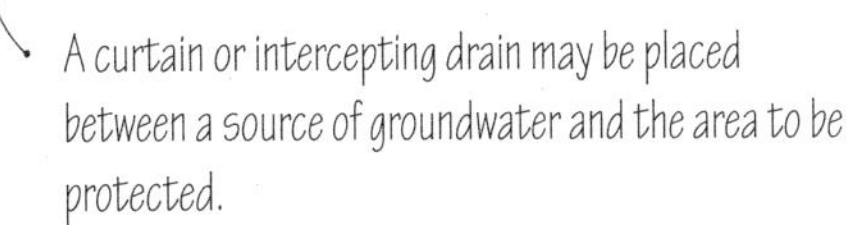

- A curtain or intercepting drain may be placed between a source of groundwater and the area to be protected.
- One type of curtain drain is a French drain, which consists of a trench filled to ground level with loose stones or rock fragments.

- Catch basins are receptacles for the runoff of surface water. They have a basin or sump that retains heavy sediment before it can pass into an underground drainpipe.
- Culverts are drains or channels passing under a road or walkway.

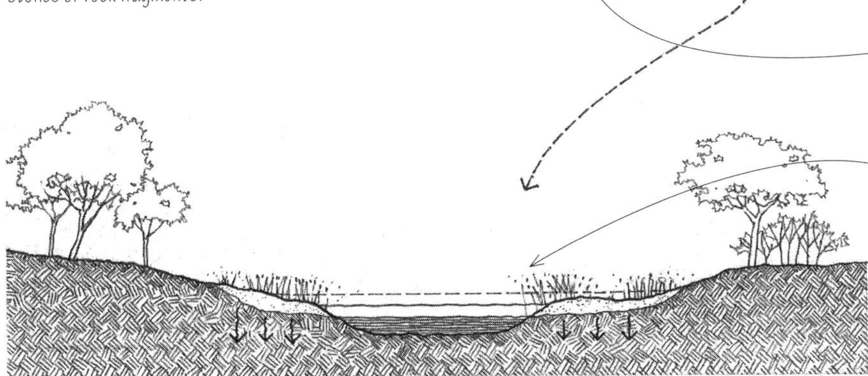

- Catchment areas can be designed to look like and function as ponds and marshes.
- Constructed wetlands are engineered, designed, and constructed to utilize natural processes in treating wastewater and improving water quality.

LEED SS Credits 6.1, 6.2: Stormwater Design
LEED WE Credit 2: Innovative Wastewater Technologies

CSI MasterFormat 32 70 00: Wetlands
CSI MasterFormat 33 40 00: Storm Drainage Utilities

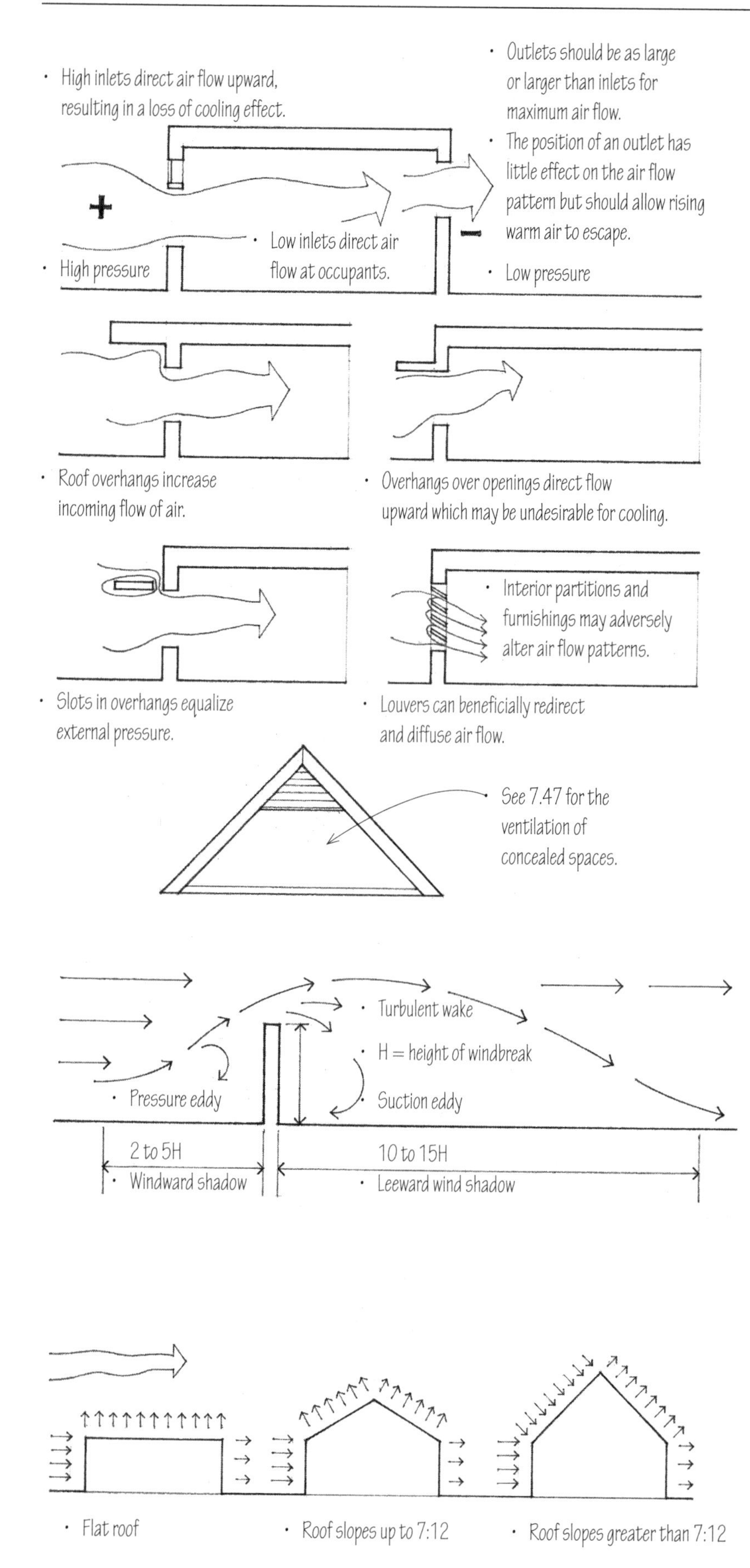

The direction and velocity of prevailing winds are important site considerations in all climatic regions. The seasonal and daily variations in wind should be carefully considered in evaluating its potential for ventilating interior spaces and outdoor courtyards in warm weather, causing heat loss in cold weather, and imposing lateral loads on a building structure.

Wind-induced ventilation of interior spaces aids in the air exchange necessary for health and odor removal. In hot weather, and especially in humid climates, ventilation is beneficial for convective or evaporative cooling. Natural ventilation also reduces the energy required by mechanical fans and equipment. (LEED IEQ Credit 2: Increased Ventilation)

The movement of air through a building is generated by differences in air pressure as well as temperature. The resulting patterns of air flow are affected more by building geometry and orientation than by air speed.

The ventilation of concealed roof and crawl spaces is required to remove moisture and control condensation. In hot weather, attic ventilation can also reduce overhead radiant heat gain.

In cold climates, a building should be buffered against chilling winds to reduce infiltration into interior spaces and lower heat loss. A windbreak may be in the form of an earth berm, a garden wall, or a dense stand of trees. Windbreaks reduce wind velocity and produce an area of relative calm on their leeward side. The extent of this wind shadow depends on the height, depth, and density of the windbreak, its orientation to the wind, and the wind velocity.

- A partially penetrable windscreen creates less pressure differential, resulting in a large wind shadow on the leeward side of the screen.

The structure, components, and cladding of a building must be anchored to resist wind-induced overturning, uplift, and sliding. Wind exerts positive pressure on the windward surfaces of a building and on windward roof surfaces having a slope greater than 30°. Wind exerts negative pressure or suction on the sides and leeward surfaces and normal to windward roof surfaces having a slope less than 30°. See 2.09 for more information on wind forces.

Sound requires a source and a path. Undesirable exterior sounds or noise may be caused by vehicular traffic, aircraft, and other machinery. The sound energy they generate travels through the air outward from the source in all directions in a continuously expanding wave. This sound energy, however, lessens in intensity as it disperses over a wide area. To reduce the impact of exterior noise, therefore, the first consideration should be distance—locating a building as far from the noise source as possible. When the location or dimensions of a site do not make this possible, then the interior spaces of a building may be screened from the noise source in the following ways.

- Use building zones where noise can be tolerated, for example, mechanical, service, and utility areas, as a buffer.
- Employ building materials and construction assemblies designed to reduce the transmission of airborne and structure-borne sound.
- Orient door and window openings away from the sources of undesirable noise.
- Place physical mass, such as earth berms, between the noise source and the building.
- Utilize dense plantings of trees and shrubs, which can be effective in diffusing or scattering sound.
- Plant grass or other ground cover, which is more absorptive than the hard, reflective surfaces of pavements.

An important aspect of site planning is orienting the interior spaces of a building to the amenities and features of a site. Given the appropriate orientation, window openings in these spaces should be positioned not only to satisfy the requirements for natural light and ventilation, but also to reveal and frame desirable views. Depending on the location of the site, these views may be close or distant in nature. Even when desirable views are nonexistent, a pleasant outlook can often be created within a building site through landscaping.

A window may be created within a wall in a number of ways, depending on the nature of the view and the way it is framed in the wall construction. It is important to note that the size and location of windows also affect the spatial quality and daylighting of a room, and the potential for heat loss or gain.

- South-facing windows can be effectively shaded while admitting daylight.
- North-facing windows are exposed to winter winds in cool climates.
- East- and west-facing windows are sources of overheating and are difficult to shade effectively.

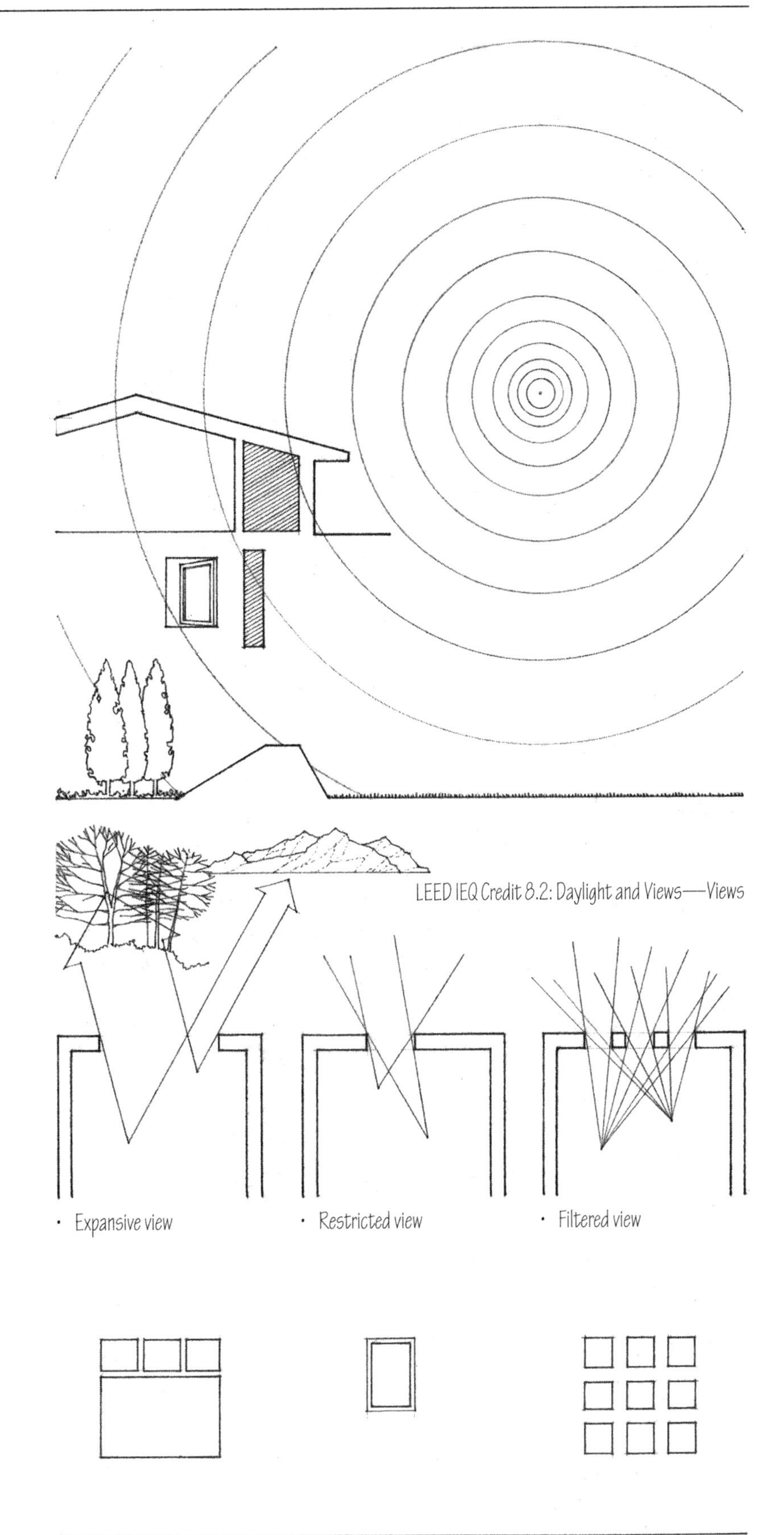

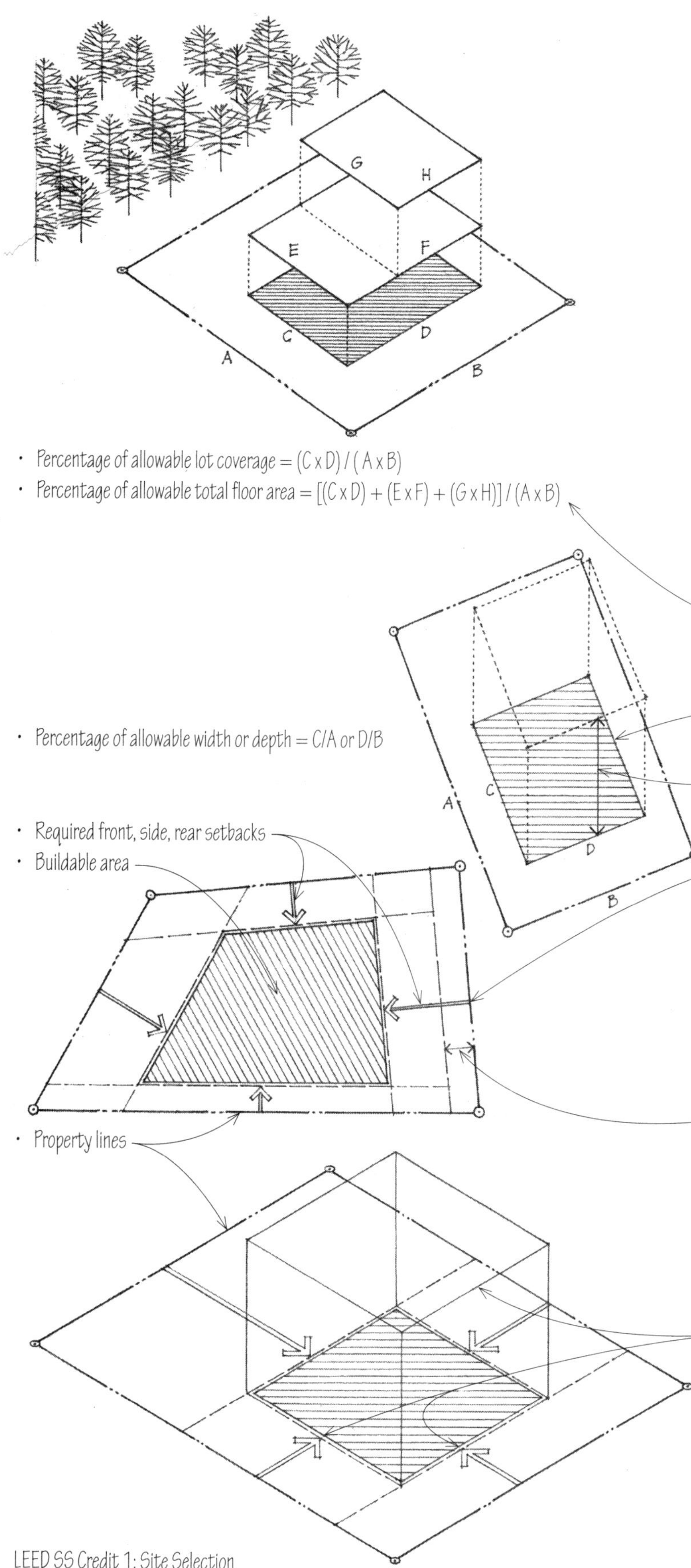

Zoning ordinances are enacted within a municipality or land-use district to manage growth, regulate land-use patterns, control building density, direct development to areas with adequate services and amenities, protect environmentally sensitive areas, and conserve open space.

For any single building site, a zoning ordinance will regulate both the types of activity that may occur on it and the location and bulk of the building or buildings constructed to house such activities. A special type of zoning ordinance is the Planned Unit Development, which allows a fairly large tract of land to be developed as a single entity for added flexibility in the placement, grouping, size, and use of structures.

It is important to understand how a zoning ordinance might constrain the allowable size and shape of a building. The bulk of a building is regulated directly by specifying various aspects of its size.

- How much of the land can be covered by a building structure and the total floor area that may be constructed are expressed as percentages of the lot area.
- The maximum width and depth a building may have are expressed as percentages of the dimensions of the site.
- Zoning ordinances also specify how tall the building structure can be.

The size and shape of a building are also controlled indirectly by specifying the minimum required distances from the structure to the property lines of the site in order to provide for air, light, solar access, and privacy.

Existing easements and rights-of-way may further limit the buildable area of a site.

- An easement is a legal right held by one party to make limited use of the land of another, as for a right-of-way or for access to light and air.
- A right-of-way is a legal right granted to a single party or the public to traverse another's land, as for access to or the construction and maintenance of utility lines.

All of the above requirements, together with any restriction on type and density of use, define a three-dimensional envelope beyond which the volume of a building may not extend. Refer to the applicable zoning ordinance for specific requirements.

LEED SS Credit 1: Site Selection
LEED SS Credit 2: Development Density and Community Connectivity

Exclusions to the general requirements of a zoning ordinance may exist in the form of exceptions or allowances. Exceptions to the normal setback requirements may be made for:

- Projections of architectural features such as roof overhangs, cornices, bay windows, and balconies
- Accessory structures such as low-level decks, fences, and detached carports or garages
- Precedents set by existing, neighboring structures

Exceptions are often made for sloping sites, or for sites adjacent to public open spaces.

- Sloping roofs, chimneys, and other roof projections may be allowed to extend beyond the normal height limitation.
- The height limit may be directly related to the slope of a site.
- A reduction in the setback requirements may be made for sloping sites or for sites fronting on open space.

In order to provide for adequate light, air, and space, and to enhance the streetscape and pedestrian environment, requirements may exist for:

- Open spaces accessible to the public (LEED SS Credit 5.2: Site Development—Maximize Open Space)
- Additional setbacks if a structure rises above a certain height
- Modulation of the facade of a building fronting a public space
- Vehicular access and off-street parking

Zoning ordinances may also contain requirements that apply only to specific use categories as well as procedures for requesting a variance from the regulations.

- Restrictive covenants are provisions in a deed that restrict the action of any party to it, as an agreement among property owners specifying the use to which a property can be put. Racial and religious restrictions are legally unenforceable.

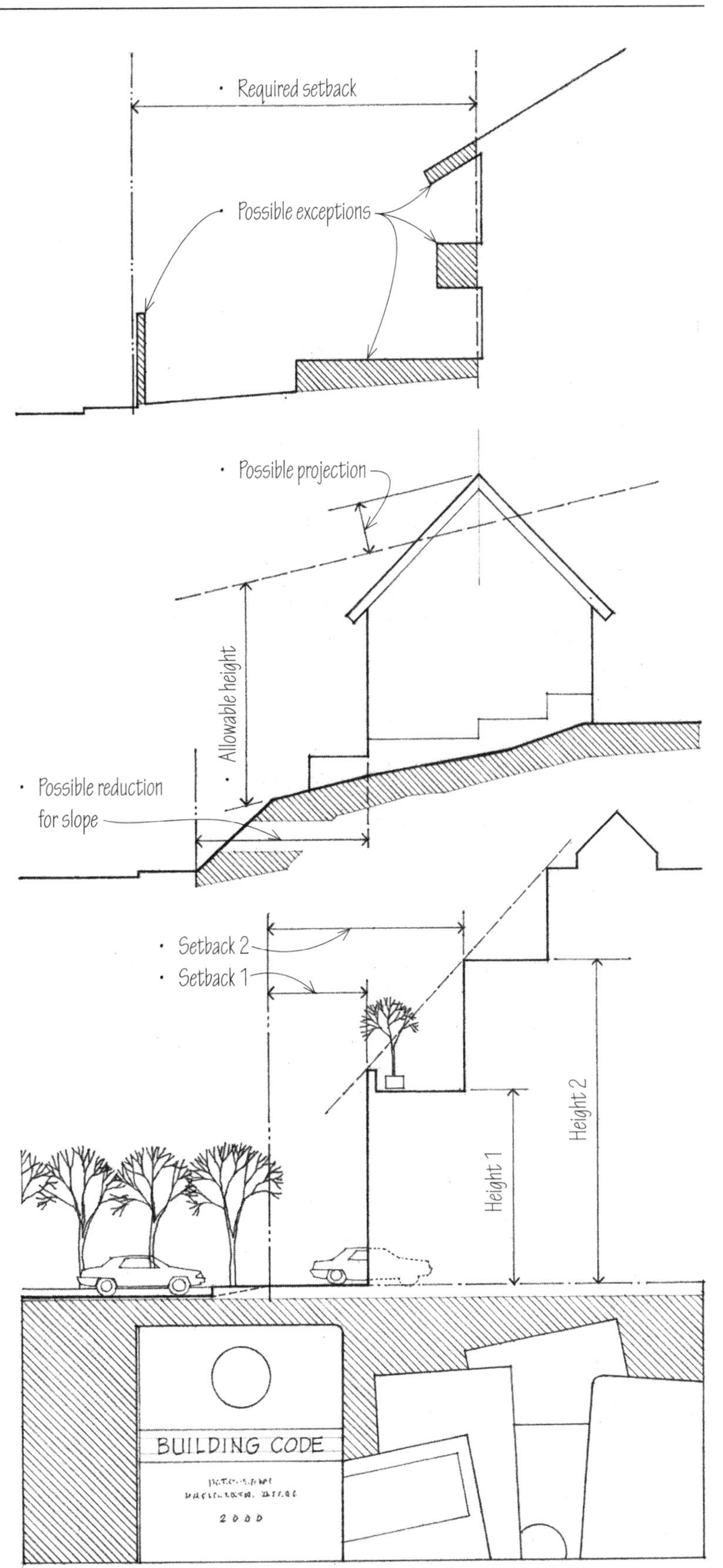

Other regulatory instruments exist that affect the way buildings are sited and constructed. These statutes—commonly referred to as the building code—establish the relationship between:

- The type of occupancy a building houses
- The fire-resistance rating of its structure and construction
- The allowable height and floor areas of the building, and its separation from neighboring structures

- See 2.05 for more information on building codes.

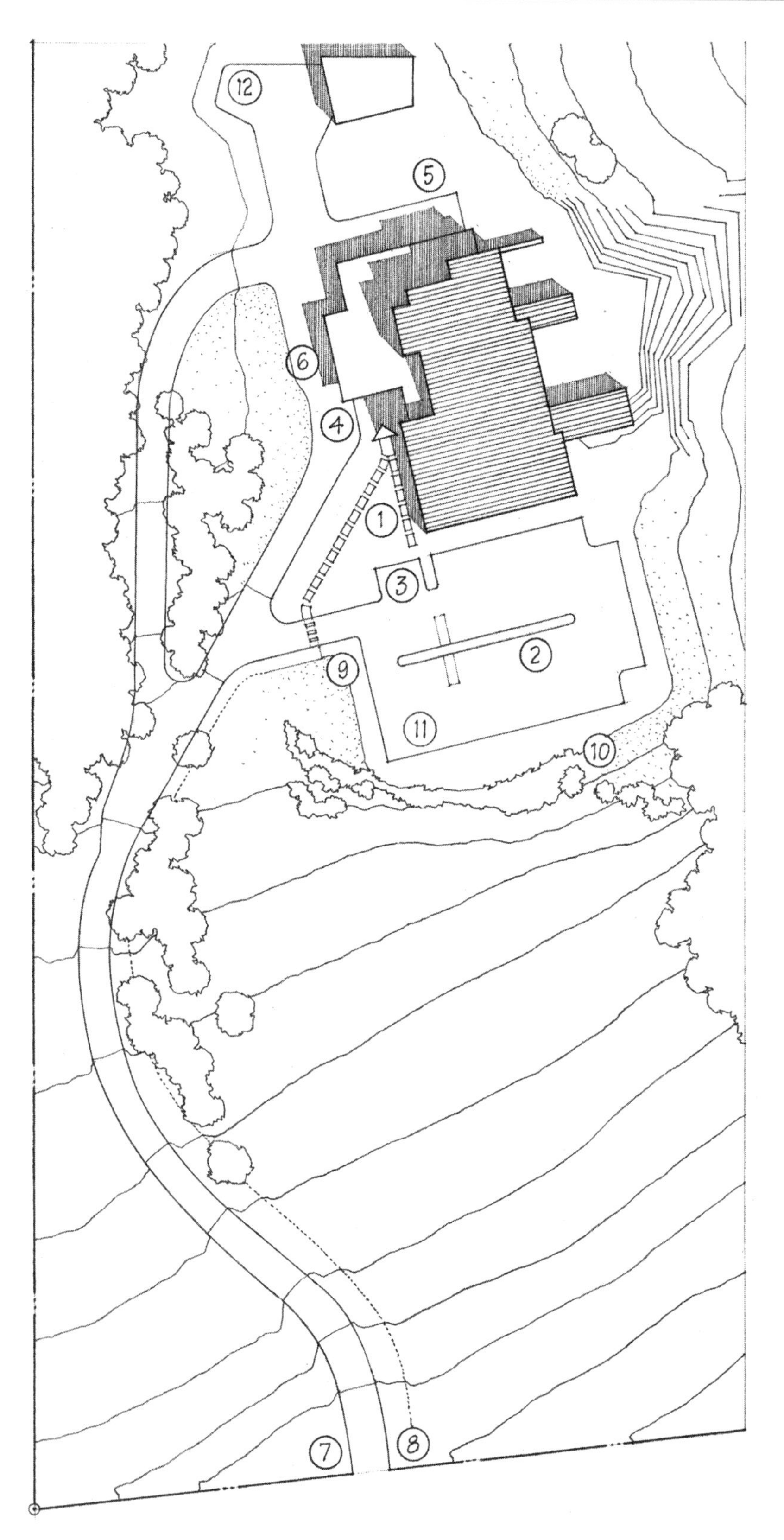

Providing for access and circulation for pedestrians, automobiles, and service vehicles is an important aspect of site planning, which influences both the location of a building on its site and the orientation of its entrances. Outlined here and on the following pages are fundamental criteria for estimating and laying out the space required for walkways, roadways, and surface parking.

1. Provide for safe and convenient pedestrian access and movement to building entrances from parking areas or public transit stops with minimal crossing of roadways.
2. Determine the number of parking spaces required by the zoning ordinance for the type of occupancy and total number of units or floor area of the building.
3. Determine the number of accessible parking spaces as well as curb cuts, ramps, and paths to accessible building entrances required by local, state, or federal law.
4. Provide loading zones for buses and other public transportation vehicles where applicable.
5. Separate service and truck loading areas from pedestrian and automobile traffic.
6. Furnish access for emergency vehicles such as fire trucks and ambulances.
7. Establish the required width and location of curb cuts and their proper distance from public street intersections.
8. Ensure clear sight lines for vehicles entering public roadways.
9. Plan for control of access to parking areas where required.
10. Provide space for landscaping; screening of parking areas may be required by zoning ordinance.
11. Slope paved walkways and parking areas for drainage.
12. Provide space for snow removal equipment in cold climates.

- Illustration adapted from the site plan for the Carré House, designed by Alvar Aalto.

CSI MasterFormat 32 10 00: Bases, Ballasts, and Paving
CSI MasterFormat 32 30 00: Site Improvements

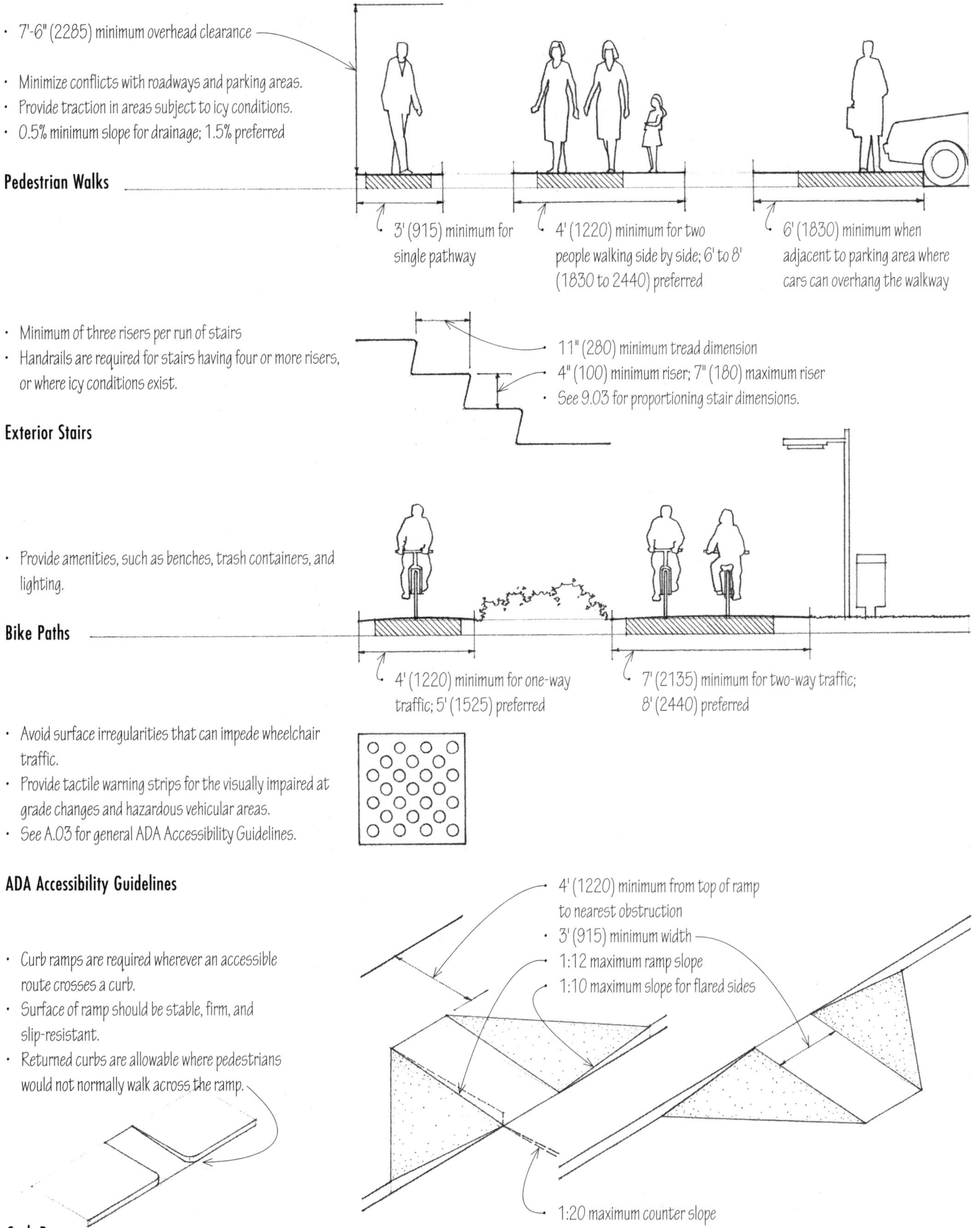

• 7'-6" (2285) minimum overhead clearance
• Minimize conflicts with roadways and parking areas.
• Provide traction in areas subject to icy conditions.
• 0.5% minimum slope for drainage; 1.5% preferred
Pedestrian Walks
3' (915) minimum for single pathway
4' (1220) minimum for two people walking side by side; 6' to 8' (1830 to 2440) preferred
6' (1830) minimum when adjacent to parking area where cars can overhang the walkway
• Minimum of three risers per run of stairs
• Handrails are required for stairs having four or more risers, or where icy conditions exist.
Exterior Stairs
11" (280) minimum tread dimension
4" (100) minimum riser; 7" (180) maximum riser
• See 9.03 for proportioning stair dimensions.
• Provide amenities, such as benches, trash containers, and lighting.
Bike Paths
4' (1220) minimum for one-way traffic; 5' (1525) preferred
7' (2135) minimum for two-way traffic; 8' (2440) preferred
• Avoid surface irregularities that can impede wheelchair traffic.
• Provide tactile warning strips for the visually impaired at grade changes and hazardous vehicular areas.
• See A.03 for general ADA Accessibility Guidelines.
ADA Accessibility Guidelines
4' (1220) minimum from top of ramp to nearest obstruction
• 3' (915) minimum width
• 1:12 maximum ramp slope
• 1:10 maximum slope for flared sides
• Curb ramps are required wherever an accessible route crosses a curb.
• Surface of ramp should be stable, firm, and slip-resistant.
• Returned curbs are allowable where pedestrians would not normally walk across the ramp.
1:20 maximum counter slope
Curb Ramps

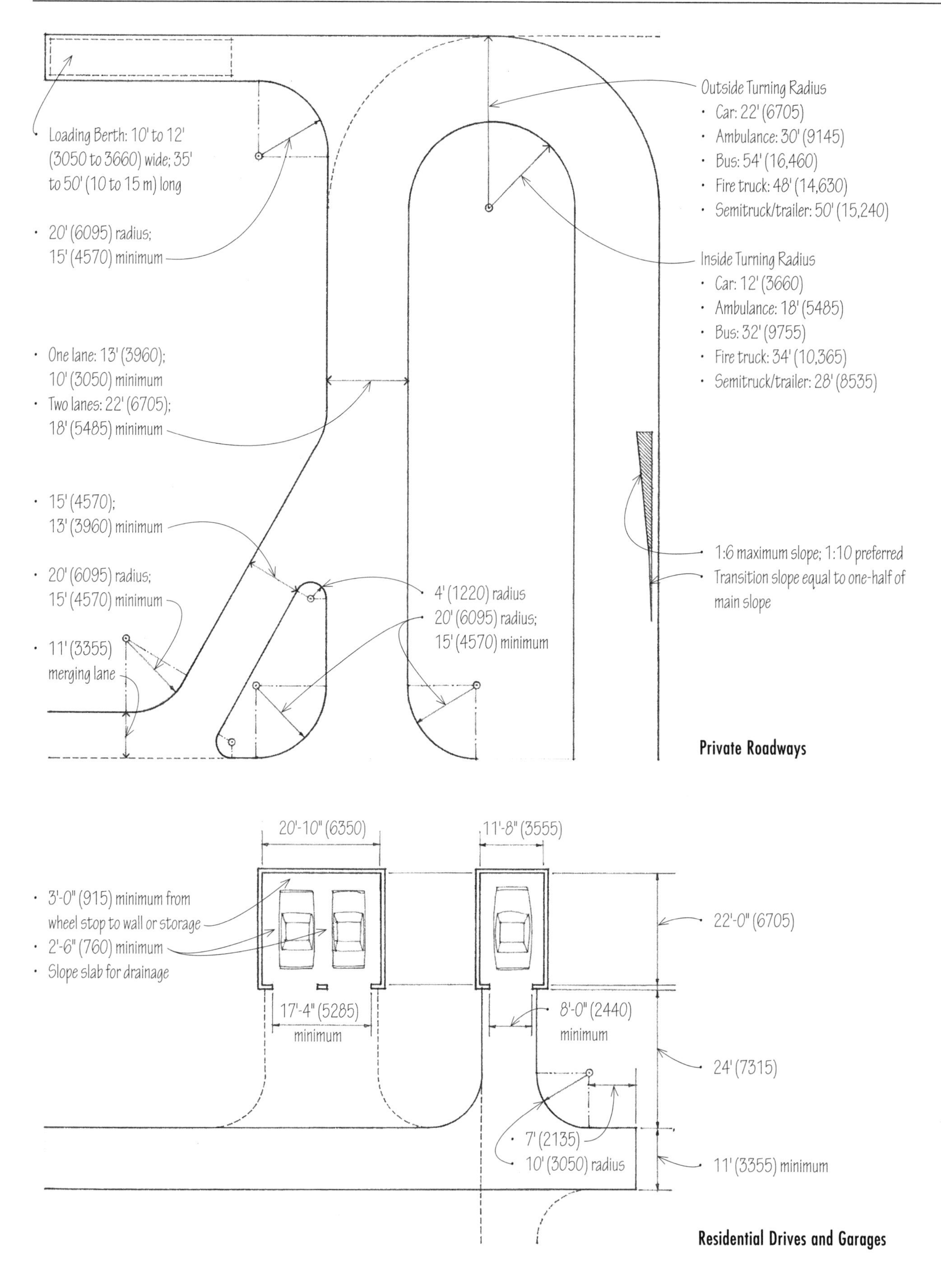

Private Roadways

Residential Drives and Garages

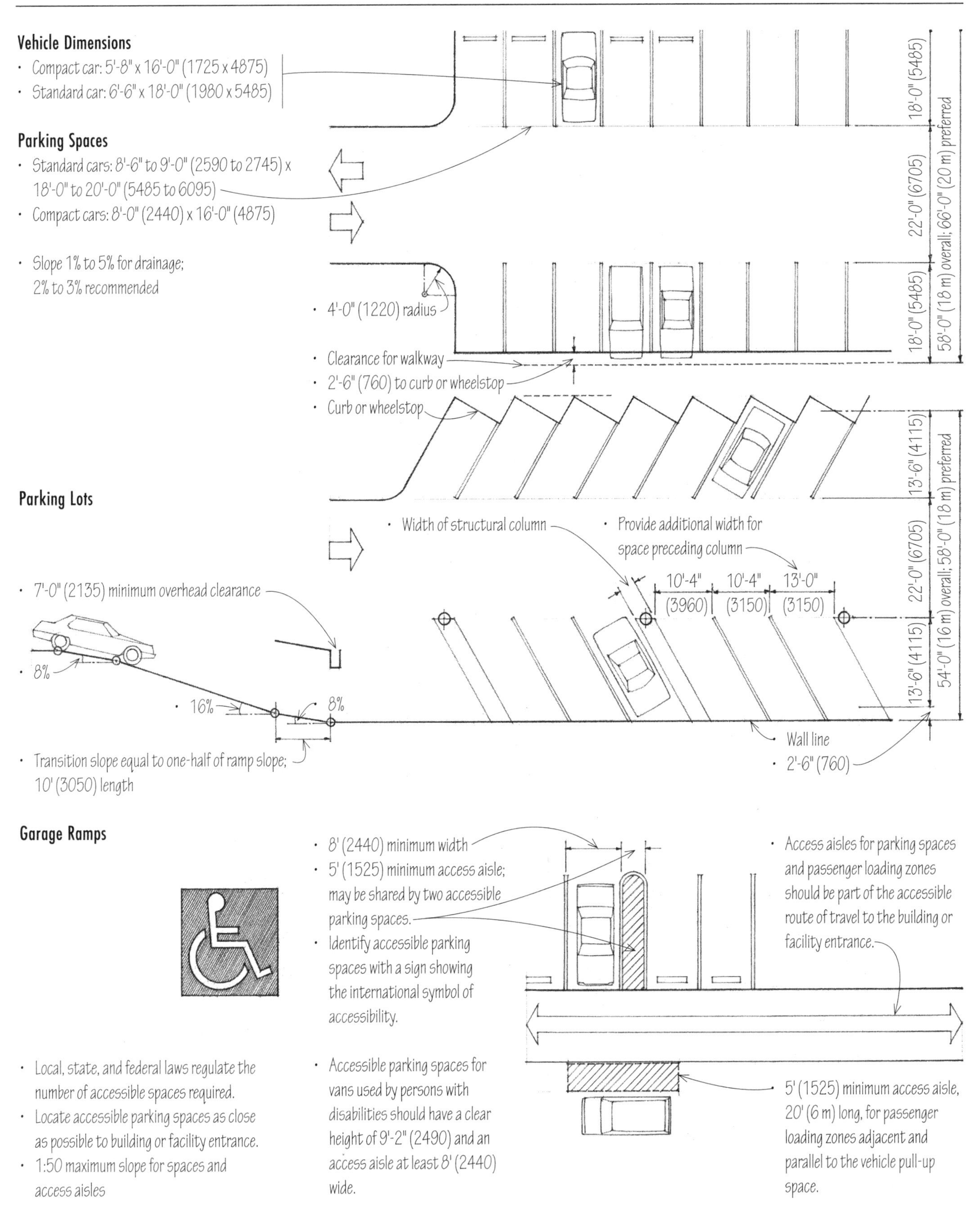
Vehicle Dimensions
• Compact car: 5'-8" x 16'-0" (1725 x 4875)
• Standard car: 6'-6" x 18'-0" (1980 x 5485)
Parking Spaces
• Standard cars: 8'-6" to 9'-0" (2590 to 2745) x 18'-0" to 20'-0" (5485 to 6095)
• Compact cars: 8'-0" (2440) x 16'-0" (4875)
• Slope 1% to 5% for drainage; 2% to 3% recommended
• 4'-0" (1220) radius
• Clearance for walkway
• 2'-6" (760) to curb or wheelstop
• Curb or wheelstop
18'-0" (5485)
22'-0" (6705)
18'-0" (5485)
58'-0" (18 m) overall; 66'-0" (20 m) preferred
Parking Lots
13'-6" (4115)
22'-0" (6705)
13'-6" (4115)
54'-0" (16 m) overall; 58'-0" (18 m) preferred
• Width of structural column
• Provide additional width for space preceding column
10'-4" (3960)
10'-4" (3150)
13'-0" (3150)
• Wall line
• 2'-6" (760)
• 7'-0" (2135) minimum overhead clearance
• 8%
• 16%
8%
• Transition slope equal to one-half of ramp slope; 10' (3050) length
Garage Ramps
• 8' (2440) minimum width
• 5' (1525) minimum access aisle; may be shared by two accessible parking spaces.
• Identify accessible parking spaces with a sign showing the international symbol of accessibility.
• Access aisles for parking spaces and passenger loading zones should be part of the accessible route of travel to the building or facility entrance.
• Local, state, and federal laws regulate the number of accessible spaces required.
• Locate accessible parking spaces as close as possible to building or facility entrance.
• 1:50 maximum slope for spaces and access aisles
• Accessible parking spaces for vans used by persons with disabilities should have a clear height of 9'-2" (2490) and an access aisle at least 8' (2440) wide.
• 5' (1525) minimum access aisle, 20' (6 m) long, for passenger loading zones adjacent and parallel to the vehicle pull-up space.
ADA Accessibility Guidelines

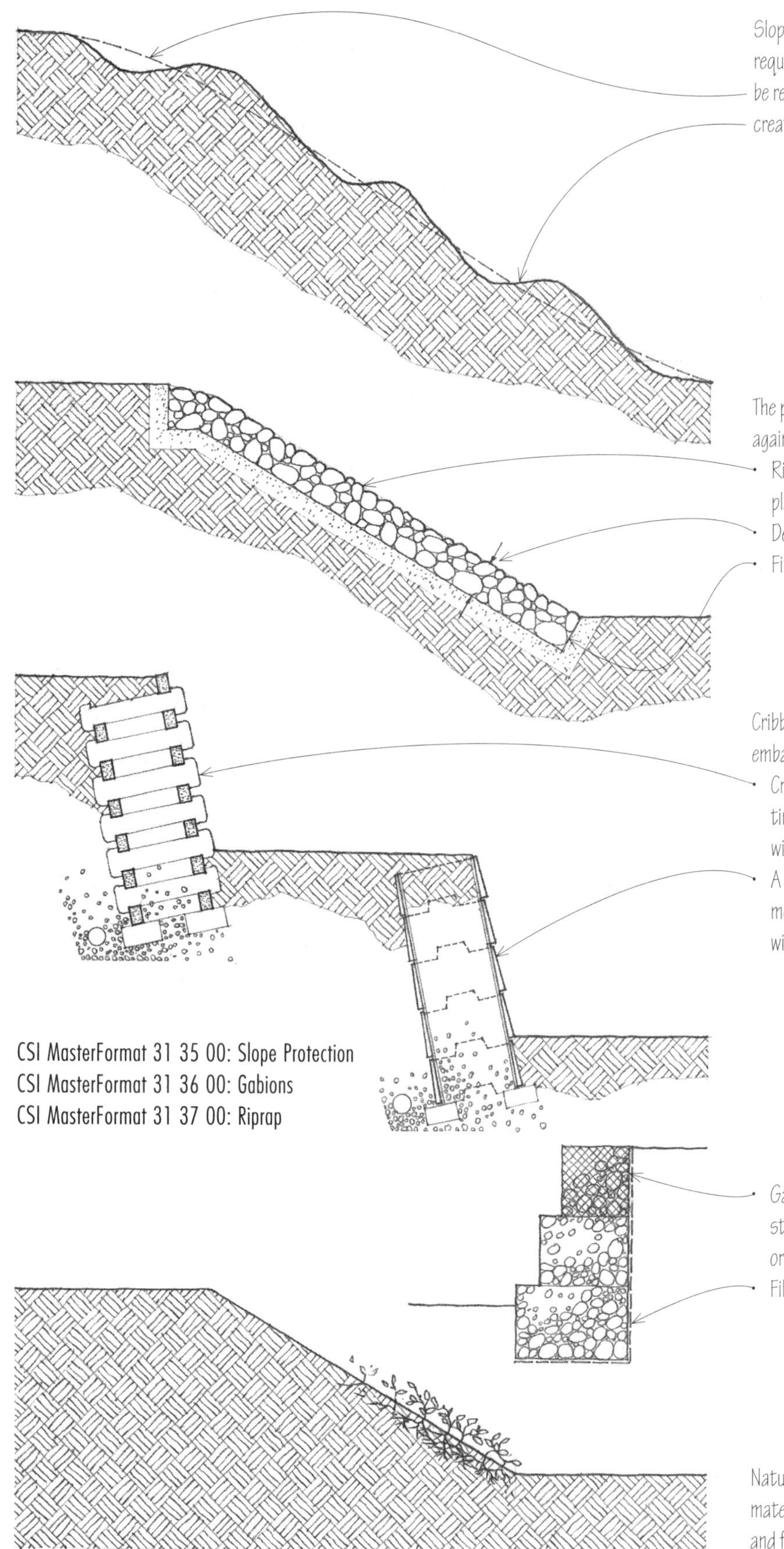

Slopes that are subject to erosion from the runoff of surface water require some means of stabilization. The need for stabilization can be reduced by diverting the runoff at the top of the slope, or by creating a series of terraces to reduce the velocity of the runoff.

The principal mechanical means of protecting an embankment against erosion is a revetment of riprap or gabions.

- Riprap is a layer of irregularly broken and random-sized stones placed on the slope of an embankment to prevent erosion.
- Depth of layer should be greater than the maximum size of stone.
- Filter fabric or graded sand and gravel for drainage

Cribbing or bin walls may also be used to hold back and protect steep embankments.

- Cribbing is a cellular framework of squared steel, concrete, or timber members, assembled in layers at right angles, and filled with earth or stones.
- A bin wall is a type of gravity retaining wall formed by stacking modular, interlocking precast concrete units and filling the voids with crushed stone or gravel.

CSI MasterFormat 31 35 00: Slope Protection
CSI MasterFormat 31 36 00: Gabions
CSI MasterFormat 31 37 00: Riprap

- Gabions are galvanized or PVC-coated wire baskets filled with stones and stacked to form an abutment or retaining structure, or as riprap to stabilize an embankment.
- Filter fabric or graded sand and gravel for drainage

Natural means of stabilization include soil binders—plant materials that inhibit or prevent erosion by providing a ground cover and forming a dense network of roots that bind the soil.

When a desired change in ground elevation exceeds the angle of repose of the soil, a retaining wall becomes necessary to hold back the mass of earth on the uphill side of the grade change.

A retaining wall must be designed and constructed to resist the lateral pressure of the soil being retained. This active pressure increases proportionally from zero at the upper grade level to a maximum value at the lowest depth of the wall. The total pressure or thrust may be assumed to be acting through the centroid of the triangular distribution pattern, one-third above the base of the wall.

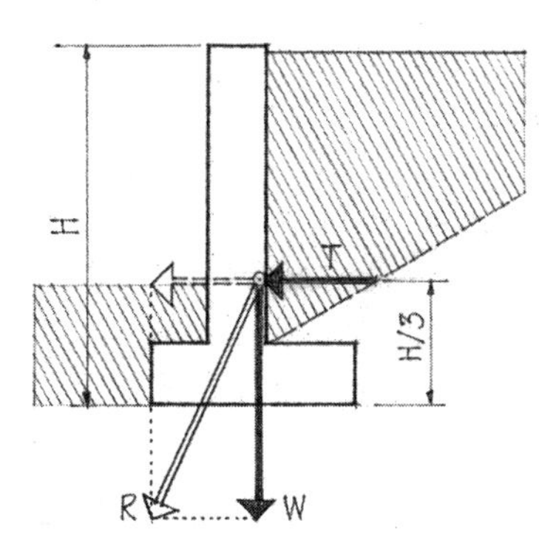

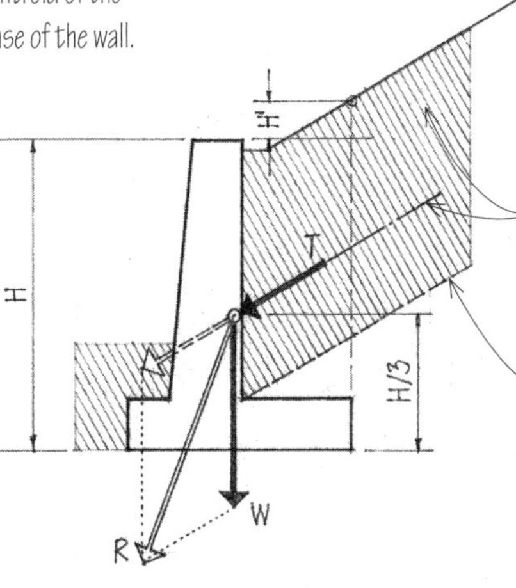

- Surcharge is an additional load, as that of the earth above a retaining wall. The line of thrust parallels the slope of the surcharge.
- Assume 33° for the angle of repose of most soils. See 1.09 for the angle of repose for bare soil embankments.

- $T = 0.286 \times SH^2 / 2$
- $T = 0.833 \times S(H + H')^2 / 2$ (for a retaining wall with surcharge)
- T = total pressure or thrust
- S = weight of retained soil; 100 pcf (1600 kg/m^3) typical
- W = composite weight of wall acting through centroid of the section
- R = resultant of T and W

A retaining wall may fail by overturning, horizontal sliding, or excessive settling.

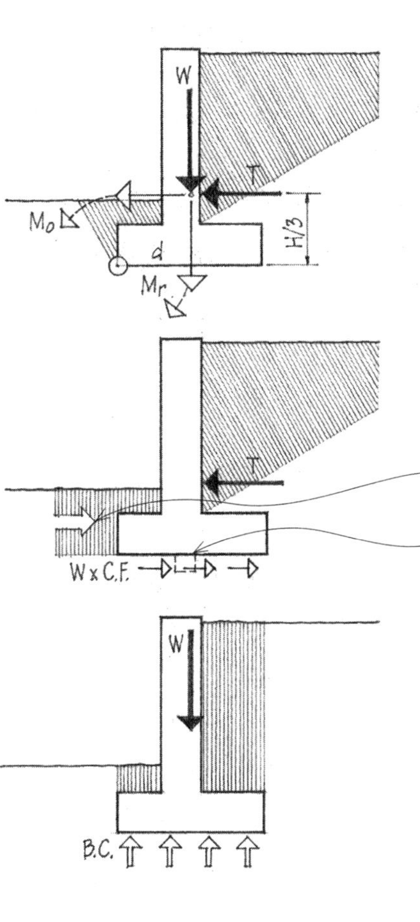

- Thrust tends to overturn wall about toe of base.
- To prevent a retaining wall from overturning, the resisting moment (M_r) of the composite weight of the wall and any soil bearing on the heel of the base (W x d) must counter the overturning moment (M_o) created by the soil pressure (T x H/3). Using a safety factor of 2, $M_r \geq 2M_o$.

- To prevent a retaining wall from sliding, the composite weight of the wall times the coefficient of friction for the soil supporting the wall (W x C.F.) must counter the lateral thrust on the wall (T). Using a safety factor of 1.5, $W \times C.F. \geq 1.5T$.
- The passive pressure of the soil abutting the lower level of the wall aids in resisting the lateral thrust (T).
- A key also increases the resistance of the wall to sliding.
- Average coefficients of friction: gravel, 0.6; silt/dry clay, 0.5; sand, 0.4; wet clay, 0.3

- To prevent a retaining wall from settling, the vertical force (W) must not exceed the bearing capacity of the soil (B.C.), where W = weight of the wall and any soil bearing on the base plus the vertical component of the soil thrust for a wall with surcharge. Using a safety factor of 1.5, $B.C. \geq 1.5\ W/A$.

CSI MasterFormat 32 32 00: Retaining Walls

Reinforced Concrete Retaining Walls

The proportioning guidelines below are for preliminary design only. Consult a structural engineer for final design, especially when a retaining wall is to be built on poor soil or is subject to surcharge or live loads.

Gravity Wall

A gravity retaining wall resists overturning and sliding by the sheer weight and volume of its mass. Gravity walls may be used for retaining structures less than 10' (3048) high.

T-Type Cantilevered Wall

Cantilevered walls of reinforced concrete are used for retaining walls up to 20' (6095) high. Above this height, counterfort walls are employed.

Counterfort Wall

A counterfort wall utilizes triangular-shaped cross walls to stiffen the vertical slab and add weight to the base. The counterforts are spaced at regular intervals equal to one-half the wall height.

L-Type Cantilevered Wall

This type of retaining wall is used when the wall abuts a property line or other obstruction.

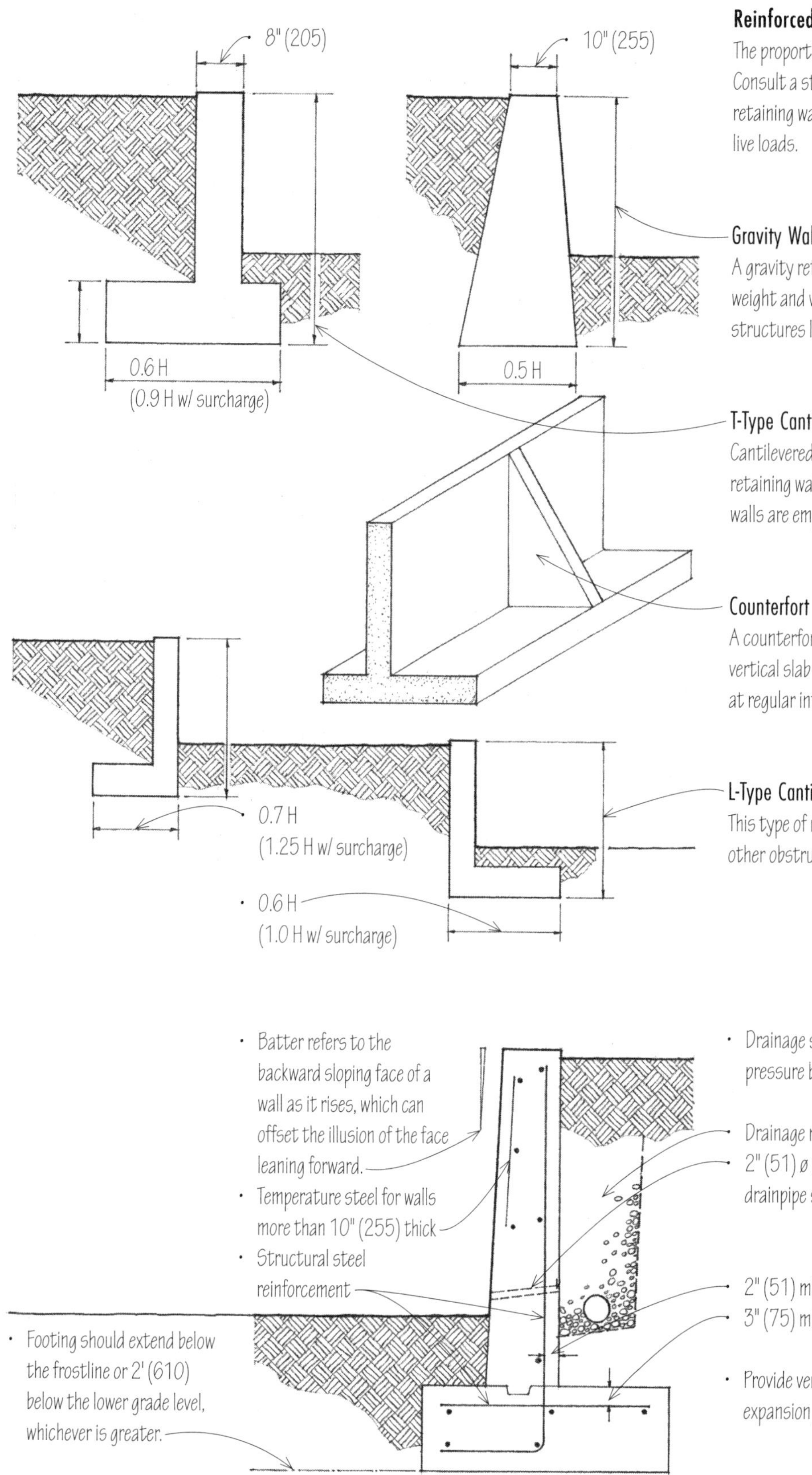

- Drainage system may be required to relieve the buildup of water pressure behind the wall.
- Drainage mat w/ filter fabric or porous gravel backfill
- 2" (51) ø weepholes @ 4'–6' (1220–1830) o.c., or perforated drainpipe sloped to outlet away from wall
- 2" (51) minimum
- 3" (75) minimum
- Provide vertical control joints @ 25' (7620) o.c., and vertical expansion joints every fourth control joint.

CSI MasterFormat 32 32 13: Cast-in-Place Concrete Retaining Walls

Timber and concrete, brick, or stone masonry may be used for relatively low retaining walls.

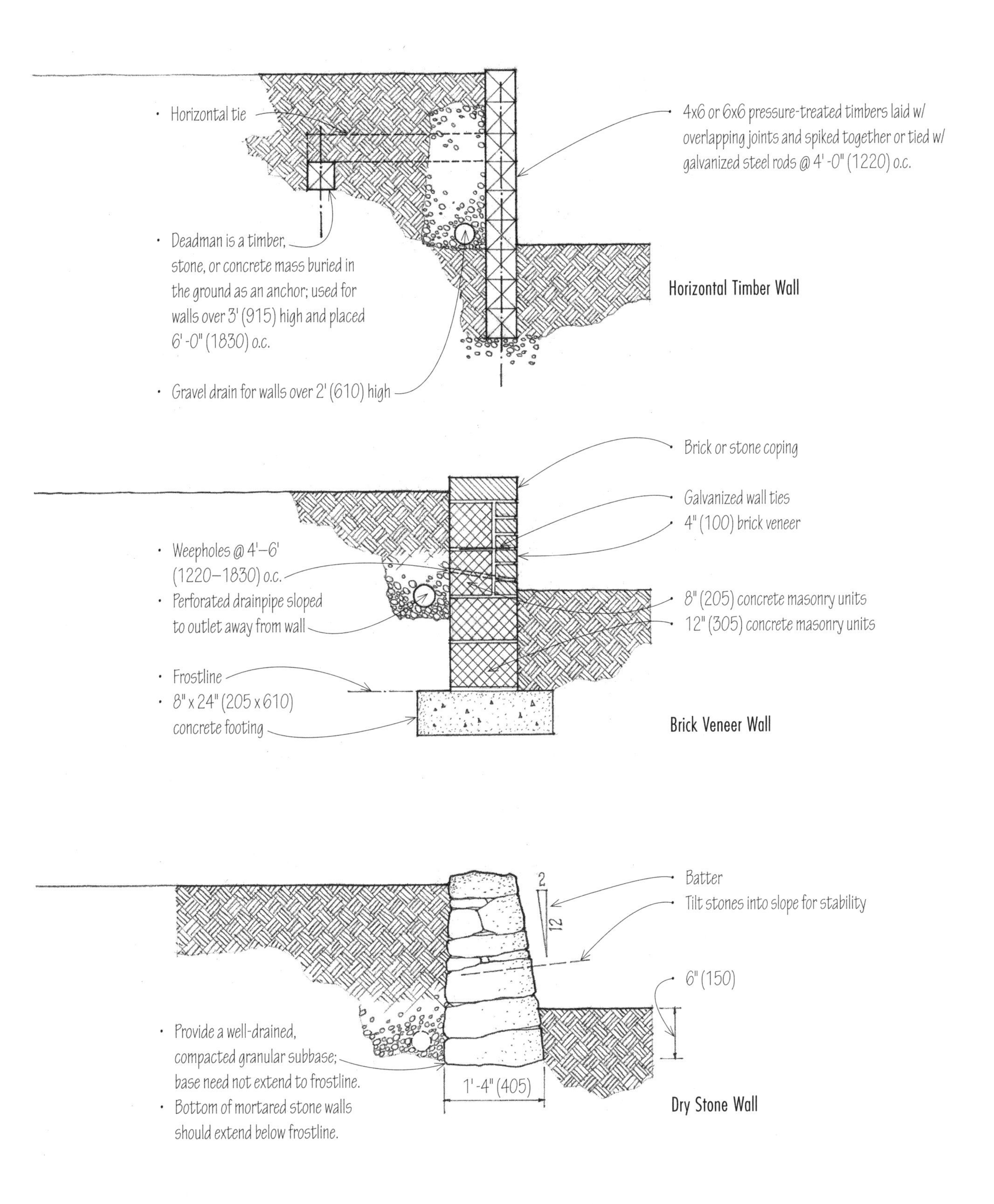

CSI MasterFormat 32 32 19: Unit Masonry Retaining Walls
CSI MasterFormat 32 32 29: Timber Retaining Walls

Paving provides a wearing surface for pedestrian or vehicular traffic on a site. It is a composite structure whose thickness and construction are directly related to the type and intensity of traffic and loads to be carried, and the bearing capacity and permeability of the subgrade.

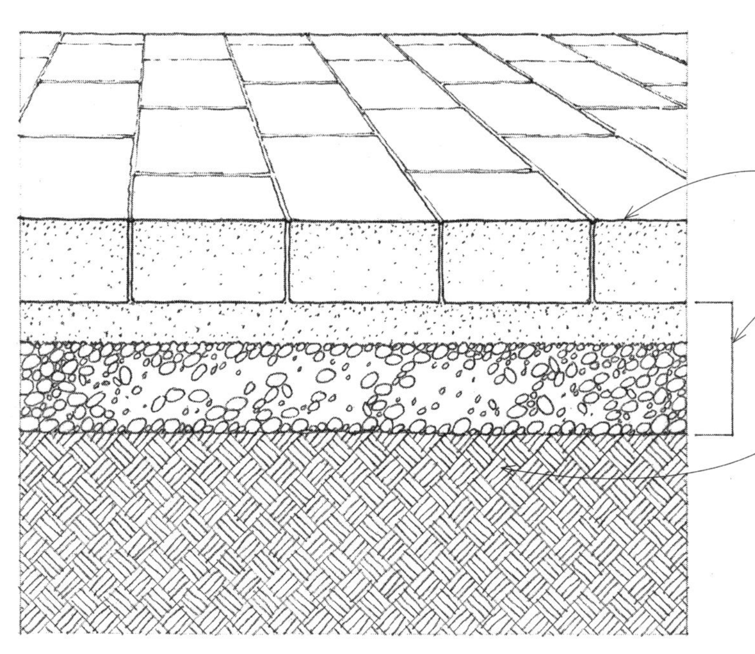

- The pavement receives the traffic wear, protects the base, and transfers its load to the base structure. There are two types of pavement: flexible and rigid.
- The base is a foundation of well-graded aggregate that transfers the pavement load to the subgrade. It also prevents the upward migration of capillary water. Heavy-duty loads may require an additional layer—a subbase of coarser aggregate such as crushed stone.
- The subgrade, which must ultimately carry the pavement load, should be undisturbed soil or compacted fill. Because it may receive moisture from infiltration, it should be sloped to drain.

CSI MasterFormat 32 10 00: Bases, Ballasts, and Paving

Flexible pavements, consisting of concrete, brick, or stone unit pavers laid on a sand setting bed, are somewhat resilient and distribute loads to the subgrade in a radiating manner. They require wood, steel, stone, masonry, or concrete edging to restrain the horizontal movement of the paving material. Specially designed unit pavers may qualify as permeable or pervious paving that allows rainfall and stormwater to percolate to an underlying reservoir base where the runoff is either infiltrated to underlying soils or removed by a subsurface drain.

(LEED SS Credits 6.1, 6.2: Stormwater Design)

Rigid pavements, such as reinforced concrete slabs or paving units mortared over a concrete slab, distribute their loads internally and transfer them to the subgrade over a broad area. They require reinforcement and an extension of the base material along their edges.

- 1% minimum slope for drainage; highly textured paving may require a steeper slope.

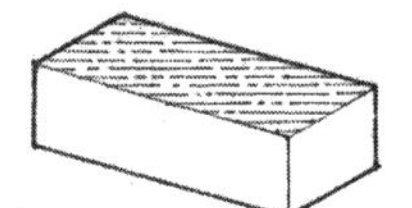

- Brick paver: 4" x 4", 8", 12"; 1"–2" thick (100 x 100, 205, 305; 25–57 thick)

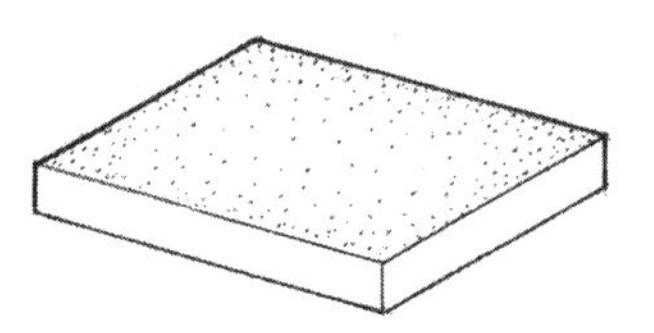

- Concrete unit paver: 12", 18", 24" square; 1 1/2"–3" thick (305, 455, 610 square; 38–75 thick)

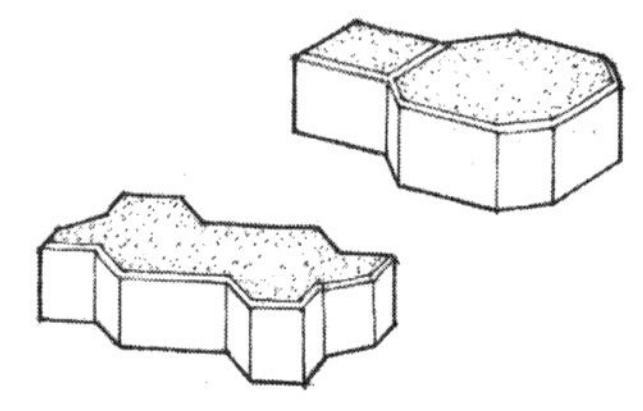

- Interlocking pavers: 2 1/2"–3 1/2" thick (64–90 thick)

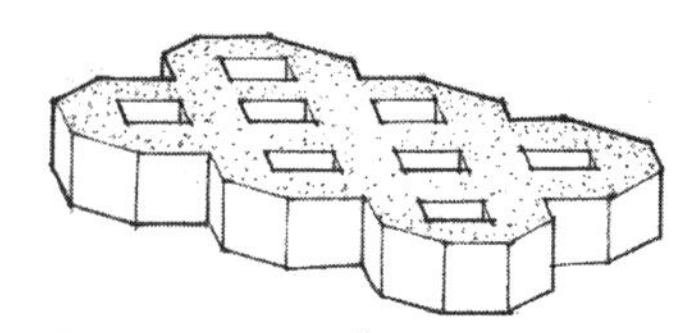

- Grid or turf block: 3 1/2" thick (90 thick)

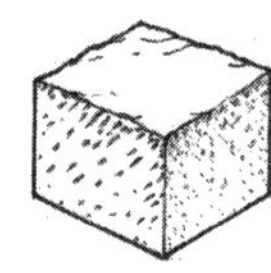

- Granite cobble: 4" or 6" square; 6" thick (100 or 150 square; 150 thick)

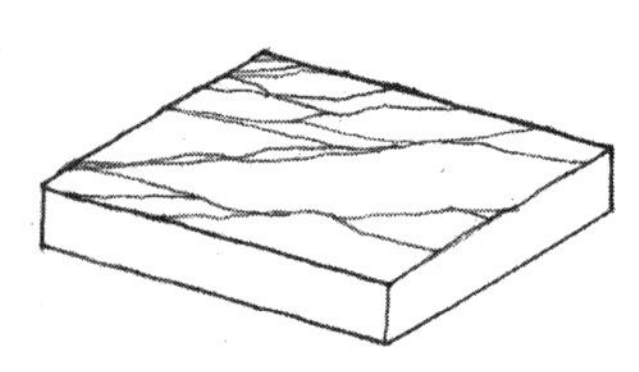

- Cut stone: width and length varies; 1"–2" thick (25–51 thick)

Paving Materials

- Consult local supplier for availability of shapes, sizes, colors, textures, absorption properties, compressive strength, and installation recommendations.

· Unit running bond

· Stack bond

· Unit basketweave

· Interlocking basketweave

· Octagon and dot

· Roman cobble

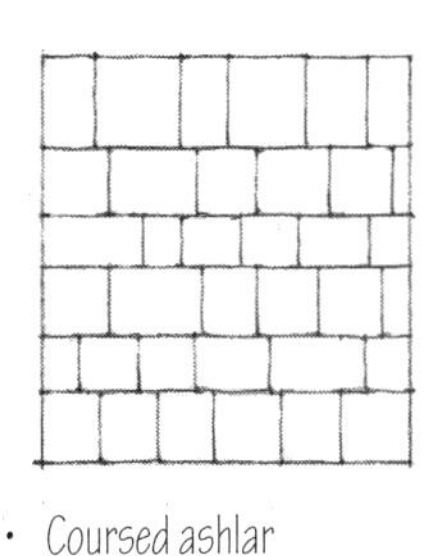

· Coursed ashlar

· Unit herringbone

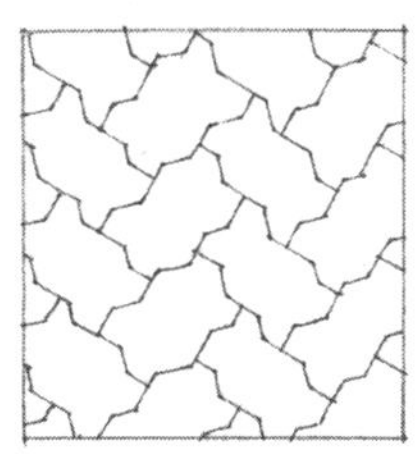

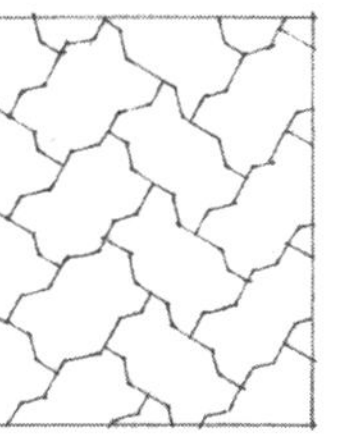

· Interlocking herringbone

· Interlocking basketweave

· Turf block

· Random stone

Paving Patterns

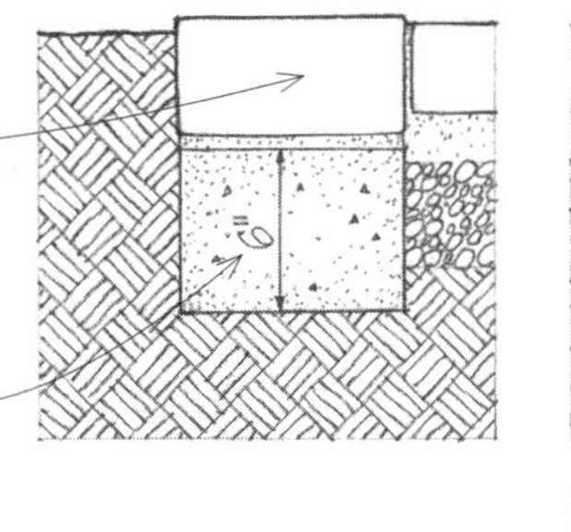

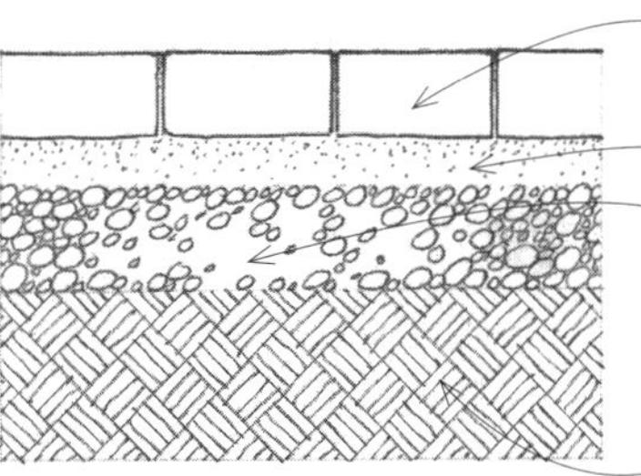

Flexible Base

· Paving unit set vertically on mortar bed; unit may extend up to 1/2 of paver height to form curb.

· Concrete footing

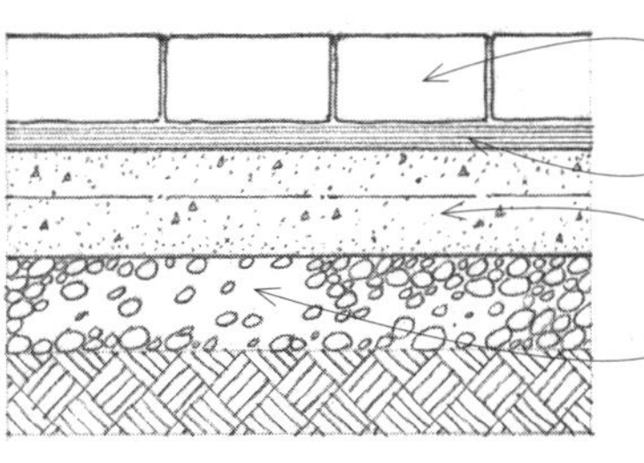

Rigid Base

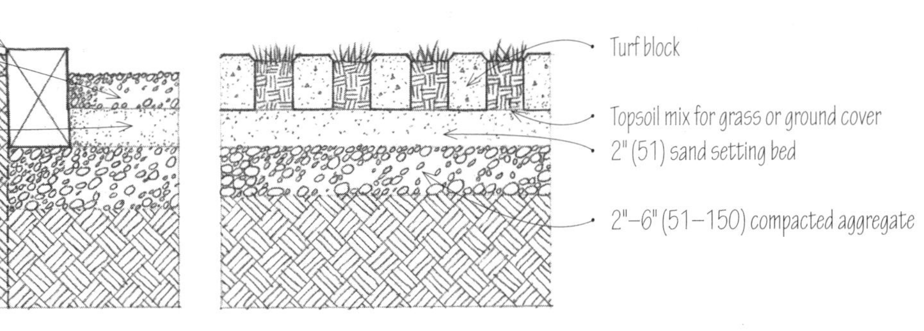

Edge Conditions

Paving Details

The site plan illustrates the existing natural and built features of a site and describes proposed construction in relation to these existing features. Usually based on an engineer's land survey, the site plan is an essential piece of a set of construction documents. A completed site plan should include the following items:

1. Name and address of property owner
2. Address of property, if different from owner's address

3. Legal description of property
4. Source and date of land survey
5. Description of the site boundaries: dimensions of property lines, their bearing relative to north, angles of corners, and radii of curves
6. Contract or project limits, if different from site boundaries
7. North arrow and scale of drawing
8. Location and description of bench marks that establish the reference points for the location and elevations of new construction

9. Identification and dimensions of adjacent streets, alleys, and other public rights-of-way
10. Location and dimensions of any easements or rights-of-way that cross the site
11. Dimensions of setbacks required by the zoning ordinance

12. Location and size of existing structures and a description of any demolition required by the new construction
13. Location, shape, and size of structures proposed for construction, including roof overhangs and other projections
14. Location and dimensions of existing and proposed paved walkways, drives, and parking areas
15. Location of existing utilities: water mains, sanitary and storm sewers, gas lines, electrical power lines, telephone and cable lines, fire hydrants, as well as proposed points of connections
16. Existing contour lines, new contour lines and the finish grades of drives, walks, lawns, or other improved surfaces after completion of construction or grading operations
17. Existing plant materials to remain and those to be removed
18. Existing water features, such as drainage swales, creeks, flood plains, watersheds, or shorelines
19. Proposed landscaping features, such as fencing, retaining walls, and plantings; if extensive, landscaping and other site improvements may be shown on a separate site plan.

20. References to related drawings and details

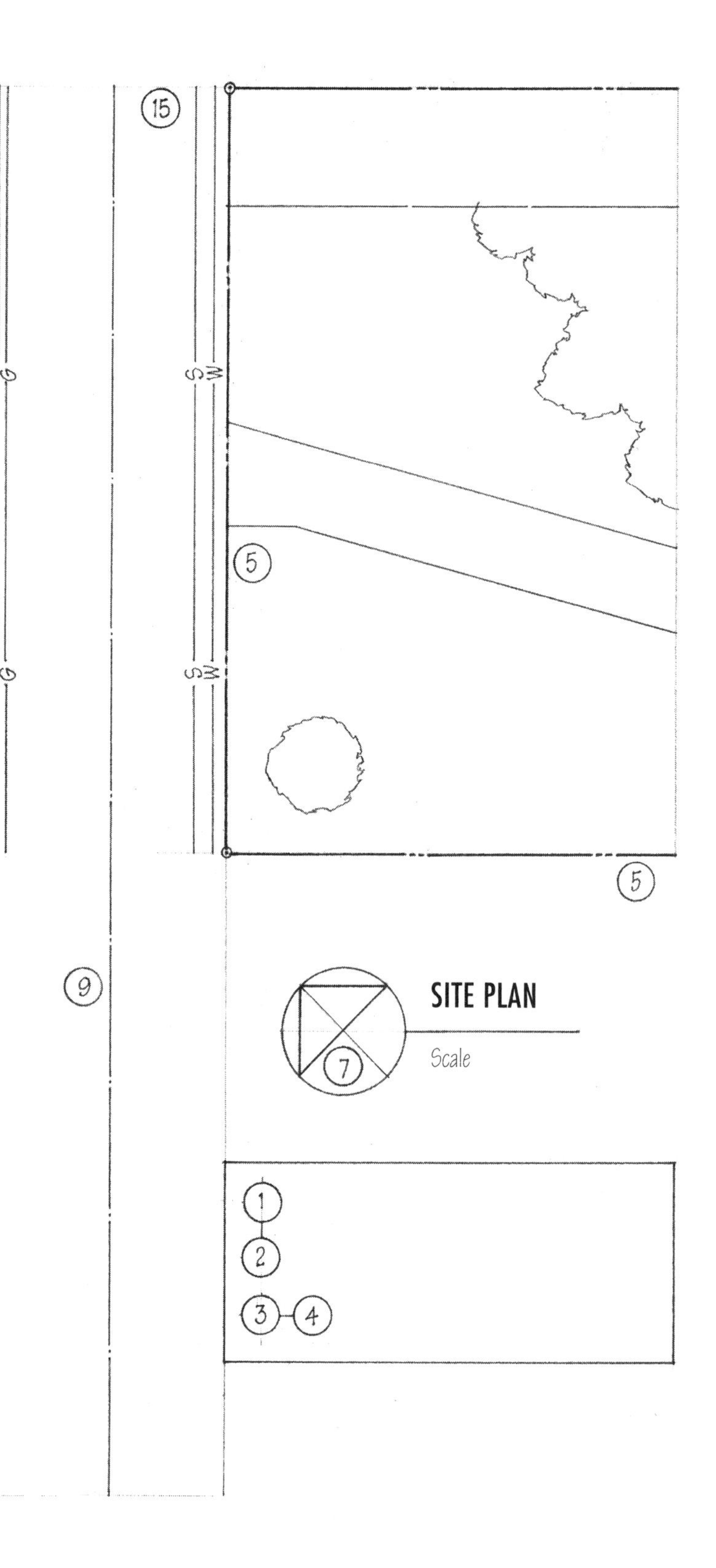

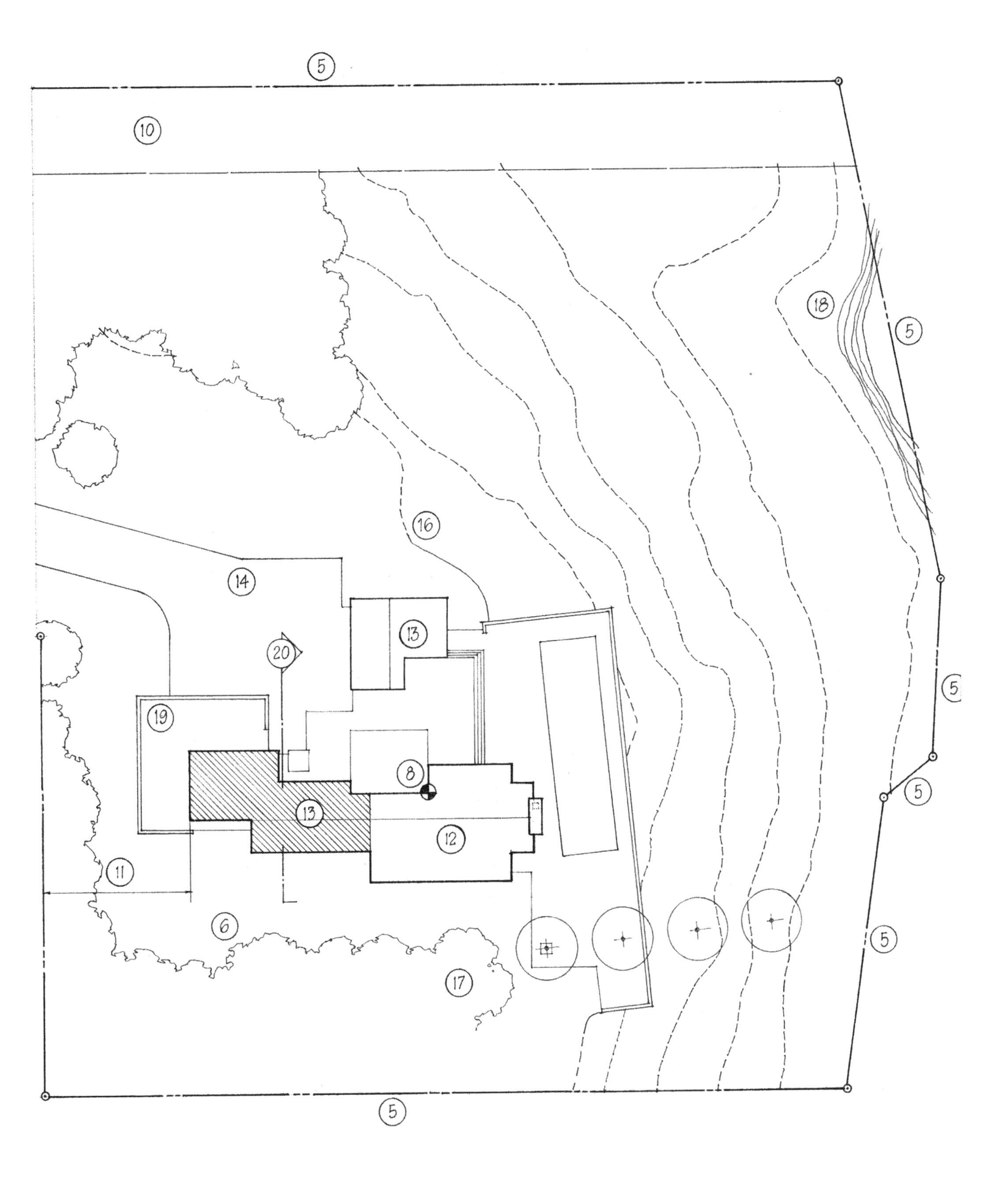

5
10
18
5
16
14
13
20
5
19
8
5
13
12
11
6
5
17
5

A legal description of a site consists of the location and boundaries of a specific parcel of land, based on a metes-and-bounds survey or a rectangular system of survey, or made with reference to a recorded plat.

- A metes-and-bounds survey calls out the course and length of each boundary line of a parcel of land, starting at a known reference point and working around the periphery of the plat until returning to the place of beginning.
- A survey plat is a legal document describing the location, boundaries, and dimensions of a tract or parcel of land, including zoning and planning commission approvals, easements and restrictions, and, for a subdivision, the dividing lines of street, blocks, and lots, and the numbering and dimensions of each lot.

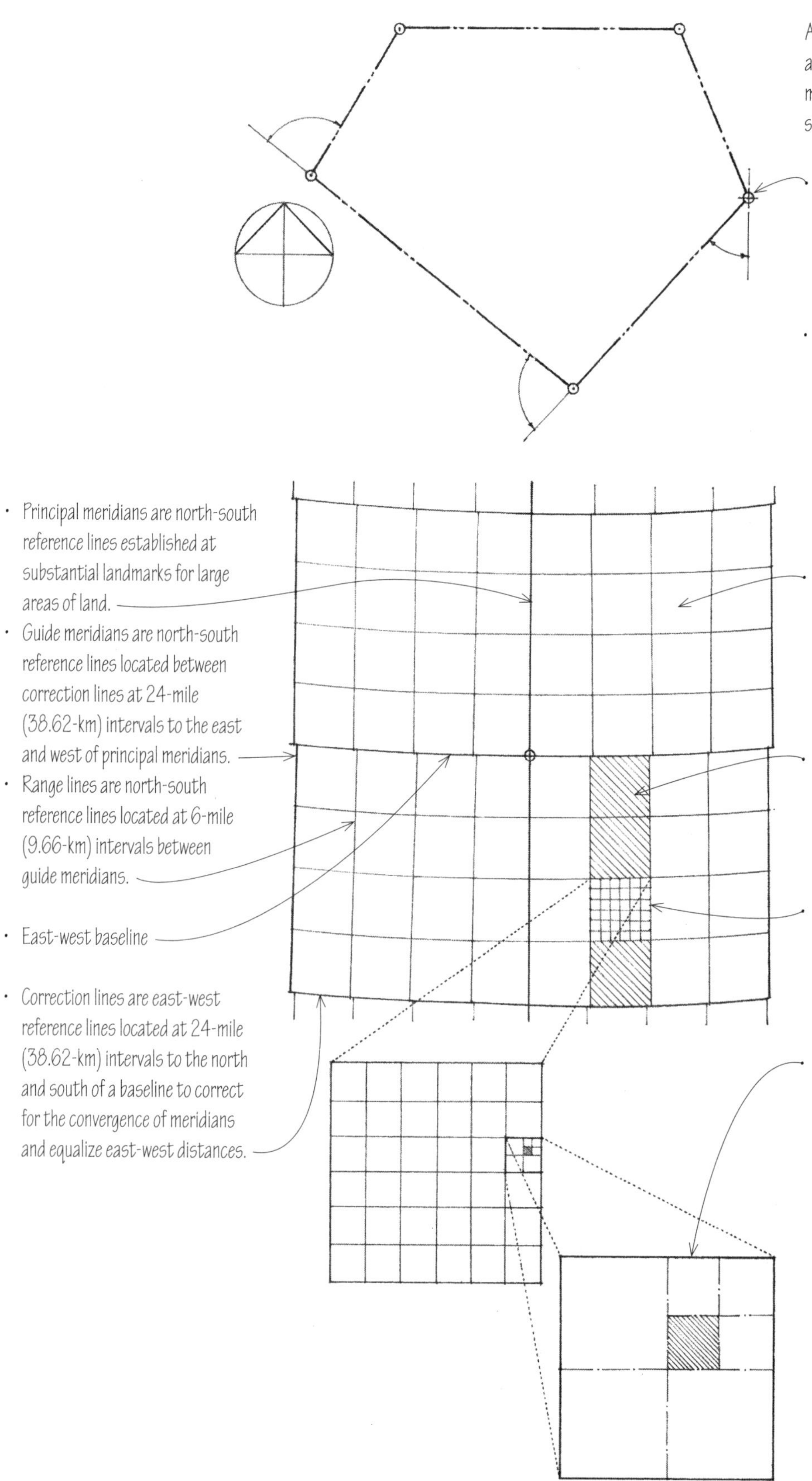

- Principal meridians are north-south reference lines established at substantial landmarks for large areas of land.
- Guide meridians are north-south reference lines located between correction lines at 24-mile (38.62-km) intervals to the east and west of principal meridians.
- Range lines are north-south reference lines located at 6-mile (9.66-km) intervals between guide meridians.
- East-west baseline
- Correction lines are east-west reference lines located at 24-mile (38.62-km) intervals to the north and south of a baseline to correct for the convergence of meridians and equalize east-west distances.

- The rectangular system of survey is based on a modified grid of north-south principal and guide meridians and east-west baselines.
- Range is one of a series of divisions numbered east or west from a guide meridian and consisting of a row of townships that are numbered north or south from a baseline.
- Township is a unit of land area, approximately six miles square (93.2 km^2) containing 36 sections.
- Section is one of the 36 numbered subdivisions of a township, each approximately one square mile (2.59 km^2 or 640 acres) and further subdivided into halves, quarters, and quarter-quarters.

CSI MasterFormat 02 21 13: Site Surveys

2

THE BUILDING

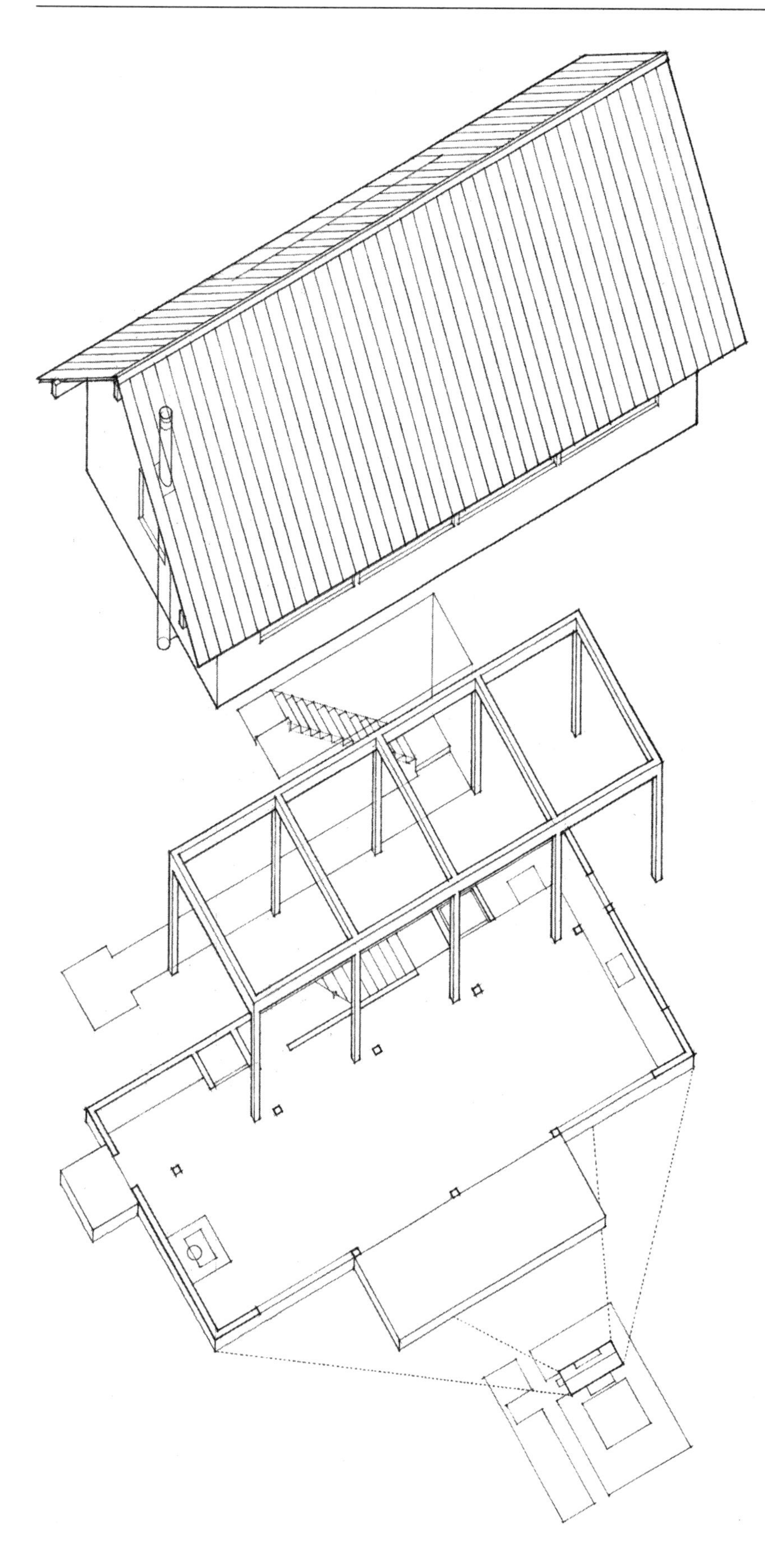

Architecture and building construction are not necessarily one and the same thing. An understanding of the methods for assembling various materials, elements, and components is necessary during both the design and the construction of a building. This understanding, however, while it enables one to build architecture, does not guarantee it. A working knowledge of building construction is only one of several critical factors in the execution of architecture. When we speak of architecture as the art of building, we should consider the following conceptual systems of order in addition to the physical ones of construction:

- The definition, scale, proportion, and organization of the interior spaces of a building
- The ordering of human activities by their scale and dimension
- The functional zoning of the spaces of a building according to purpose and use
- Access to the horizontal and vertical paths of movement through the interior of a building
- The sensible qualities of a building: form, space, light, color, texture, and pattern
- The building as an integrated component within the natural and built environment

Of primary interest to us in this book are the physical systems that define, organize, and reinforce the perceptual and conceptual ordering of a building.

A system can be defined as an assembly of interrelated or interdependent parts forming a more complex and unified whole and serving a common purpose. A building can be understood to be the physical embodiment of a number of systems and subsystems that must necessarily be related, coordinated, and integrated with each other as well as with the three-dimensional form and spatial organization of the building as a whole.

Structural System

The structural system of a building is designed and constructed to support and transmit applied gravity and lateral loads safely to the ground without exceeding the allowable stresses in its members.

- The superstructure is the vertical extension of a building above the foundation.
- Columns, beams, and loadbearing walls support floor and roof structures.
- The substructure is the underlying structure forming the foundation of a building.

Enclosure System

The enclosure system is the shell or envelope of a building, consisting of the roof, exterior walls, windows, and doors.

- The roof and exterior walls shelter interior spaces from inclement weather and control moisture, heat, and air flow through the layering of construction assemblies.
- Exterior walls and roofs also dampen noise and provide security and privacy for the occupants of a building.
- Doors provide physical access.
- Windows provide access to light, air, and views.
- Interior walls and partitions subdivide the interior of a building into spatial units.

Mechanical Systems

The mechanical systems of a building provide essential services to a building.

- The water supply system provides potable water for human consumption and sanitation.
- The sewage disposal system removes fluid waste and organic matter from a building.
- Heating, ventilating, and air-conditioning systems condition the interior spaces of a building for the environmental comfort of the occupants.
- The electrical system controls, meters, and protects the electric power supply to a building, and distributes it in a safe manner for power, lighting, security, and communication systems.
- Vertical transportation systems carry people and goods from one level to another in medium- and high-rise buildings.
- Fire-fighting systems detect and extinguish fires.
- Structures may also require waste disposal and recycling systems.

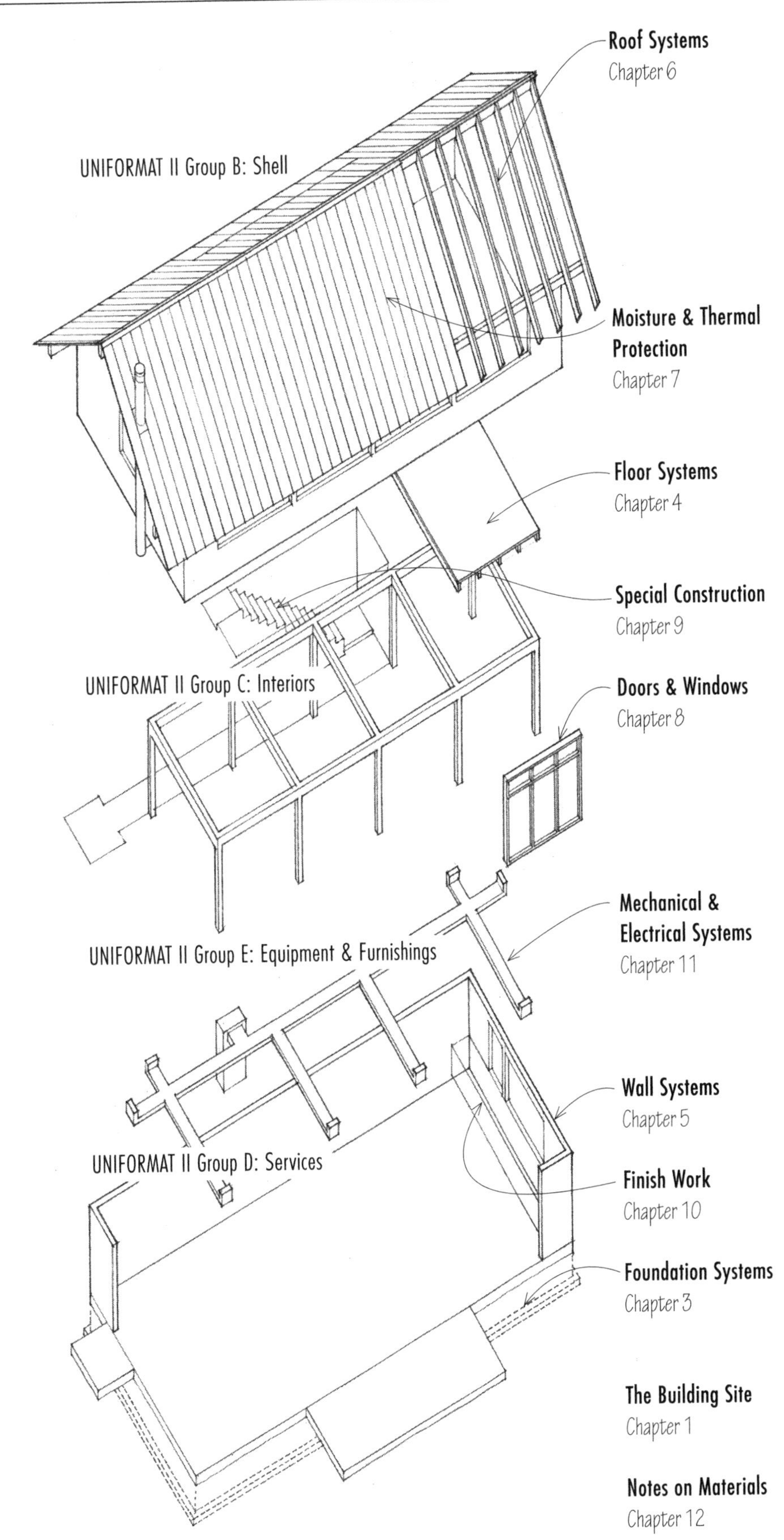

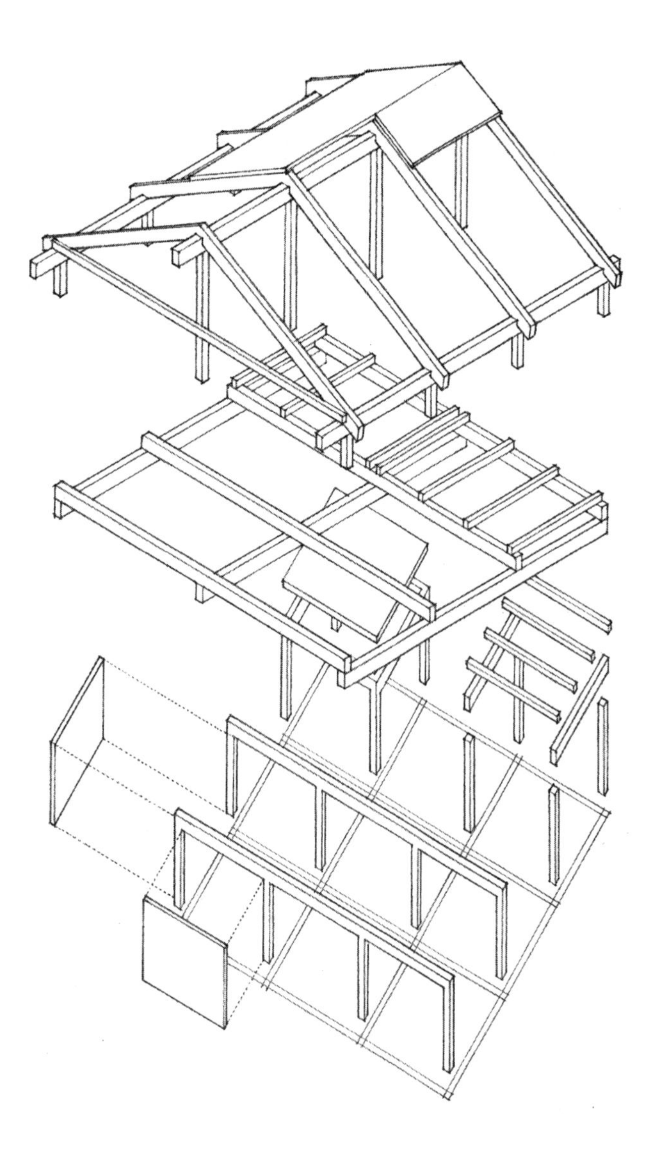

The manner in which we select, assemble, and integrate the various building systems in construction should take into account the following factors:

Performance Requirements
- Structural compatibility, integration, and safety
- Fire resistance, prevention, and safety
- Allowable or desirable thickness of construction assemblies
- Control of heat and air flow through building assemblies
- Control of migration and condensation of water vapor
- Accommodation of building movement due to settlement, structural deflection, and expansion or contraction with changes in temperature and humidity
- Noise reduction, sound isolation, and acoustical privacy
- Resistance to wear, corrosion, and weathering
- Finish, cleanliness, and maintenance requirements
- Safety in use

Aesthetic Qualities
- Desired relationship of building to its site, adjacent properties, and neighborhood
- Preferred qualities of form, massing, color, pattern, texture, and detail

Regulatory Constraints
- Compliance with zoning ordinances and building codes

Economic Considerations
- Initial cost comprising material, transportation, equipment, and labor costs
- Life-cycle costs, which include not only initial cost, but also maintenance and operating costs, energy consumption, useful lifetime, demolition and replacement costs, and interest on invested money

Environmental Impact
- Conservation of energy and resources through siting and building design
- Energy efficiency of mechanical systems
- Use of resource-efficient and nontoxic materials
- See 1.03–1.06

Construction Practices
- Safety requirements
- Allowable tolerances and appropriate fit
- Conformance to industry standards and assurance
- Division of work between the shop and the field
- Division of labor and coordination of building trades
- Budget constraints
- Construction equipment required
- Erection time required
- Provisions for inclement weather

- The U.S. Occupational Health and Safety Act (OSHA) regulates the design of workplaces and sets safety standards under which a building must be constructed.

Building codes are adopted and enforced by local government agencies to regulate the design, construction, alteration, and repair of buildings in order to protect the public safety, health, and welfare. The codes generally establish requirements based on the type of occupancy and construction of a building, minimum standards for materials and methods of construction, and specifications for structural and fire safety. While codes are primarily prescriptive in nature, they also contain performance criteria, stipulating how a particular component or system must function without necessarily giving the means to be employed to achieve the results. The codes often reference standards established by the American Society for Testing and Materials (ASTM), the American National Standards Institute (ANSI), and other technical societies and trade associations, to indicate the properties desired in a material or component and the methods of testing required to substantiate the performance of products.

Model Codes

Model codes are building codes developed by national organizations of code officials, materials and life safety experts, professional societies, and trade associations for adoption by local communities as legally enforceable regulations. If certain provisions need to be modified or added to address local requirements or concerns, a model code may be enacted by a municipality with amendments.

International Building Code®

Since the early part of the last century, three major model codes have been developed for use in various parts of the U.S. by the Building Officials and Code Administrators International, Inc. (BOCA), the International Conference of Building Officials (ICBO), and the Southern Building Code Conference (SBCC). In 1994, these model-code groups merged to form the International Code Council (ICC) with the goal of developing a comprehensive and coordinated set of national model codes. In 2000, the ICC published the first edition of the *International Building Code®* (IBC).

As with the model codes that preceded it, the IBC begins by defining categories of use or occupancy and setting height and area limitations in relation to the occupancy of a building and the type of construction employed, and then prescribes five types of construction according to degree of fire resistance and combustibility. The code also contains provisions for interior finishes, fire protection systems, emergency egress, accessibility, interior environment, energy efficiency, exterior walls and roofs, structural design, building materials, elevators and conveying systems, and existing structures.

Companion Codes

The *International Residential Code®* (IRC) regulates the construction of detached one- and two-family dwellings and townhouses not more than three stories in height and having a separate means of egress. The *International Existing Building Code®*, which regulates the alteration, repair, and renovation of existing facilities, emerged with the increasing importance of historic preservation and sustainable reuse of existing buildings. Other companion codes include the *International Energy Conservation Code®*, *International Fire Code®*, *International Mechanical Code®*, and *International Plumbing Code®*.

Other Important Codes

The National Fire Protection Association (NFPA) has developed a new model building code, NFPA 5000, as an alternative to the *International Building Code*. The NFPA also publishes other code documents.

- NFPA-70, the *National Electric Code*, ensures the safety of persons and the safeguarding of buildings and their contents from hazards arising from the use of electricity for light, heat, and power.
- NFPA-101, the *Life Safety Code*, establishes minimum requirements for fire safety, the prevention of danger from fire, smoke and gases, fire detection and alarm systems, fire extinguishing systems, and emergency egress.
- NFPA-13 governs the installation of fire sprinklers.

Federal Requirements

In addition to the locally adopted version of the model codes, there are specific federal requirements that must be considered in the design and construction of facilities.

- The *Americans with Disabilities Act* (ADA) of 1990 is federal civil-rights legislation requiring that buildings be made accessible to persons with physical disabilities and certain defined mental disabilities. The *ADA Accessibility Guidelines* (*2010 ADA Standards for Accessible Design*) are maintained by the U.S. Access Board, an independent governmental agency, and the regulations are administered by the U.S. Department of Justice. Federal facilities must comply with standards issued under the *Architectural Barriers Act* (ABA). In its last update, the Access Board harmonized the ADA guidelines with its guidelines for facilities covered by the ABA and published them jointly as the *ADA-ABA Accessibility Guidelines*. In addition, the Board and the International Code Council (ICC) worked cooperatively to coordinate the ADA and ABA guidelines and access provisions in the *International Building Code*.
- The Federal Fair Housing Act (FFHA) of 1988 includes Department of Housing and Urban Development (HUD) regulations requiring all residential complexes of four or more dwelling units constructed after March 13, 1991 to be adaptable for use by persons with disabilities.

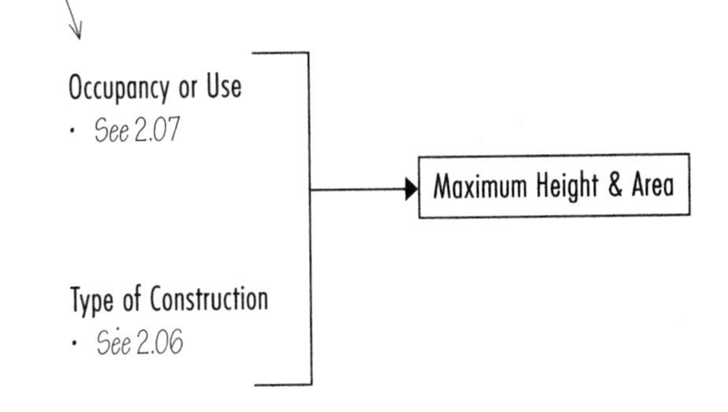

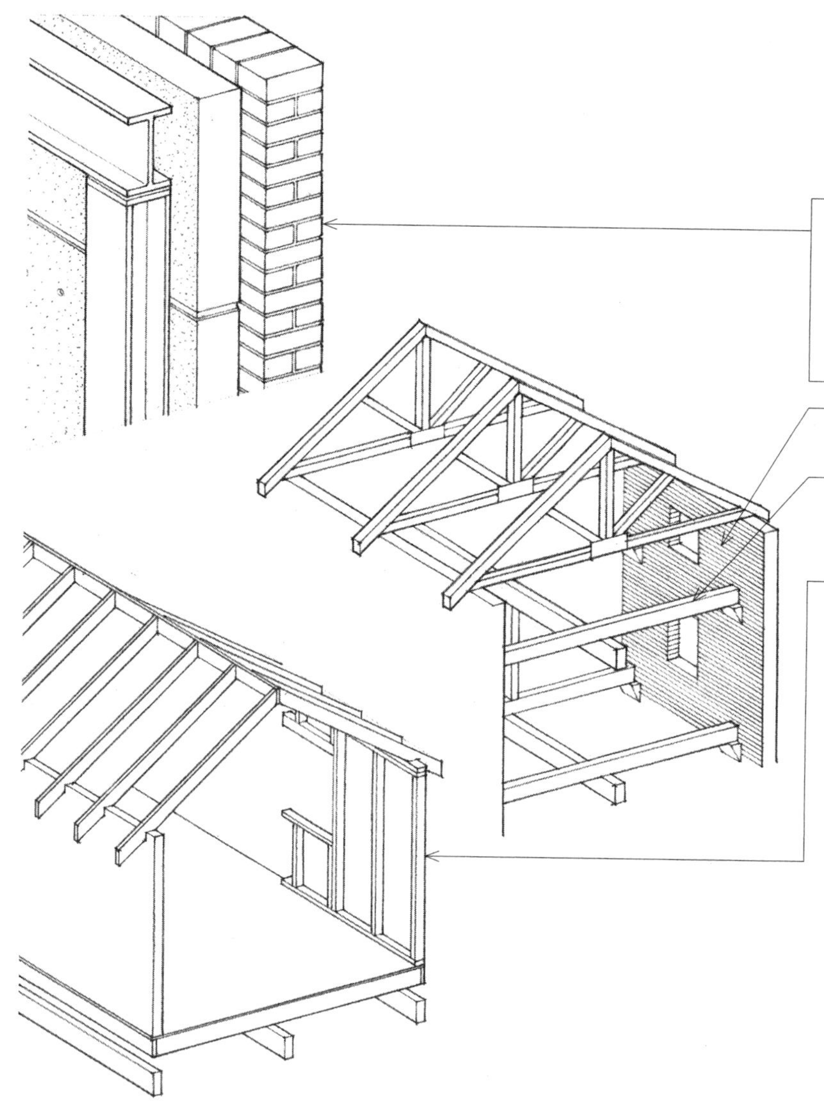

The IBC classifies the construction of a building according to the fire resistance of its major elements: structural frame, exterior and interior bearing walls, nonbearing walls and partitions, and floor and roof assemblies.

- Type I buildings have their major building elements constructed of noncombustible materials, such as concrete, masonry, or steel. Some combustible materials may be allowed if they are ancillary to the primary structure of the building. Type II buildings are similar to Type I buildings except for a reduction in the required fire-resistance ratings of the major building elements.
- Type III buildings have noncombustible exterior walls and major interior elements of any material permitted by the code.
- Type IV buildings (Heavy Timber, HT) have noncombustible exterior walls and major interior elements of solid or laminated wood of specified minimum sizes and without concealed spaces.
- Type V buildings have structural elements and exterior and interior walls of any material permitted by the code.
- Type V-Protected construction requires all major building elements, except for nonbearing interior walls and partitions, to be of one-hour fire-resistive construction.
- Type V-Unprotected construction has no requirements for fire-resistance except for when the code requires protection of exterior walls due their proximity to a property line or to adjacent buildings on the same site.

- The table below outlines the required fire-resistive ratings of major building elements for the various types of construction. Consult Table 601 of the *International Building Code* for more specific requirements.
- See Appendix for the fire-resistance ratings of representative construction assemblies.

Fire-Resistance Rating Requirements (hours)

Construction Type	Type I		Type II		Type III		Type IV	Type V	
A	B	A	B	A	B	HT	A	B	
Building Element									
Structural frame	3	2	1	0	1	0	HT	1	0
Bearing walls									
Exterior	3	2	1	0	2	2	2	1	0
Interior	3	2	1	0	1	0	1/HT	1	0
Nonbearing walls	Varies with occupancy, type of construction, location on property line, and distance to adjacent structures								
Floor construction	2	2	1	0	1	0	HT	1	0
Roof construction	$1\frac{1}{2}$	1	1	0	1	0	HT	1	0

The IBC limits the maximum height and area per floor of a building according to construction type and occupancy group, expressing the intrinsic relationship between degree of fire resistance, size of a building, and nature of an occupancy. The larger a building, the greater the number of occupants, and the more hazardous the occupancy, the more fire-resistant the facility should be. The intent is to protect a building from fire and to contain a fire long enough for the safe evacuation of occupants and for a firefighting response to occur. The limitation on size may be exceeded if the building is equipped with an automatic fire sprinkler system, or if it is divided by fire walls into areas not exceeding the size limitation.

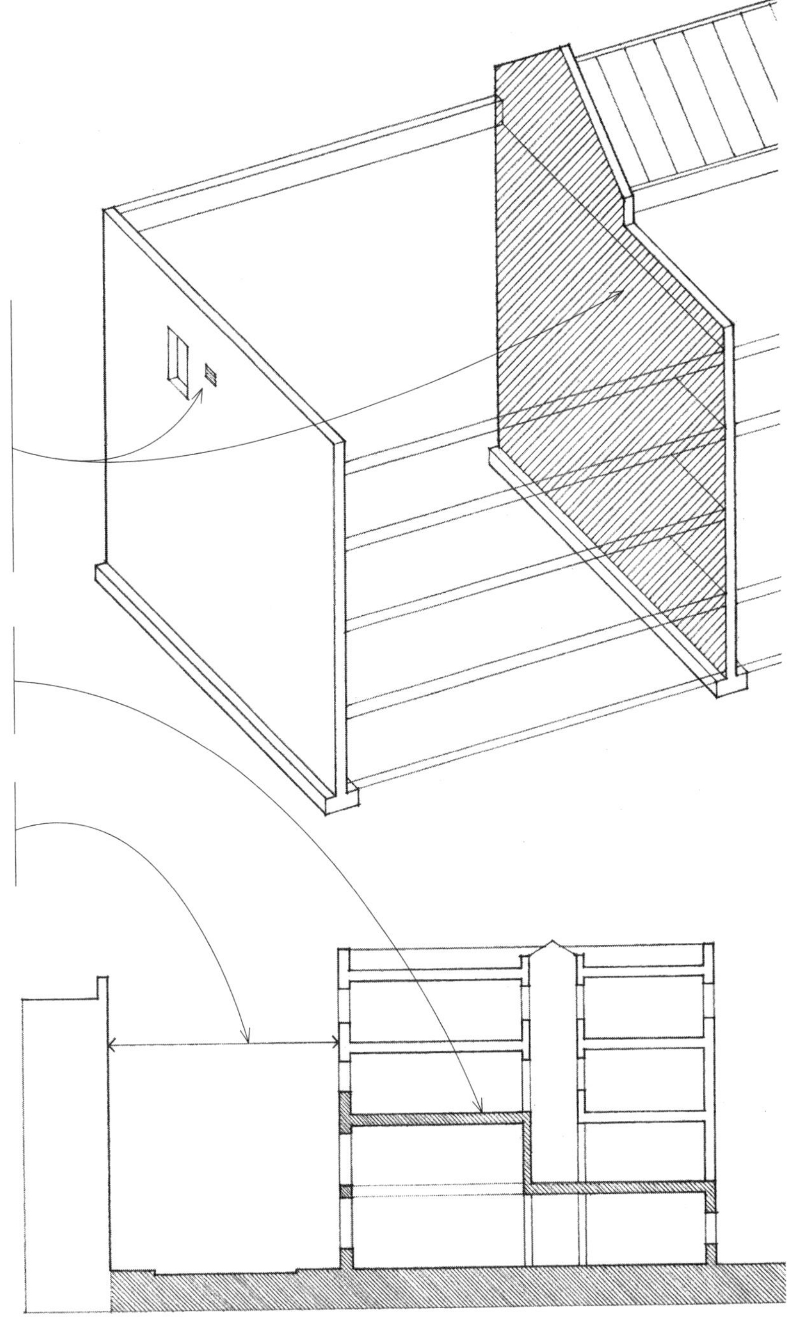

- Fire walls are required to have a fire-resistance rating sufficient to prevent the spread of fire from one part of a building to another. They must extend in a continuous manner from the foundation to a parapet above the roof of a building, or to the underside of a noncombustible roof. All openings in fire walls are restricted to a certain percentage of the wall length and must be protected by self-closing fire doors, fire-rated window assemblies, and, in the case of air ducts, by fire and smoke dampers.
- Occupancy separations refer to fire-resistive vertical or horizontal constructions required to prevent the spread of fire from one occupancy to another in a mixed-occupancy building.
- Fire separation distance refers to the space required between a property line or adjacent building and an exterior wall having a specified fire-resistance rating.

Examples of Occupancy Classifications

A Assembly
Auditoriums, theaters, and stadiums

B Business
Offices, laboratories, and higher education facilities

E Educational
Child-care facilities and schools through the 12th grade

F Factories
Fabricating, assembling, or manufacturing facilities

H Hazardous uses
Facilities handling a certain nature and quantity of hazardous materials

I Institutional
Facilities for supervised occupants, such as hospitals, nursing homes, and reformatories

M Mercantile
Stores for the display and sale of merchandise

R Residential
Homes, apartment buildings, and hotels

S Storage
Warehousing facilities

In enclosing space for habitation, the structural system of a building must be able to support two types of loads—static and dynamic.

Static Loads

Static loads are assumed to be applied slowly to a structure until it reaches its peak value without fluctuating rapidly in magnitude or position. Under a static load, a structure responds slowly and its deformation reaches a peak when the static force is maximum.

- Dead loads are static loads acting vertically downward on a structure, comprising the self-weight of the structure and the weight of building elements, fixtures, and equipment permanently attached to it.

- Live loads comprise any moving or movable loads on a structure resulting from occupancy, collected snow and water, or moving equipment. A live load typically acts vertically downward but may act horizontally as well to reflect the dynamic nature of a moving load.

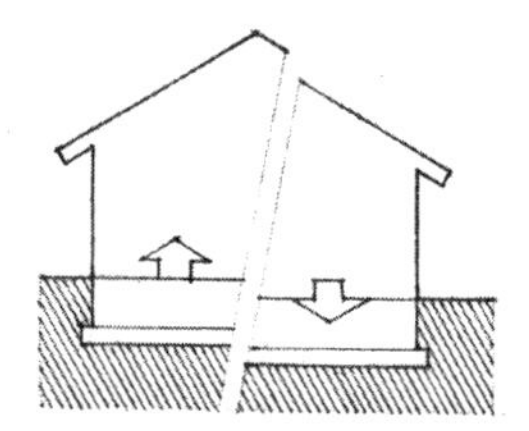

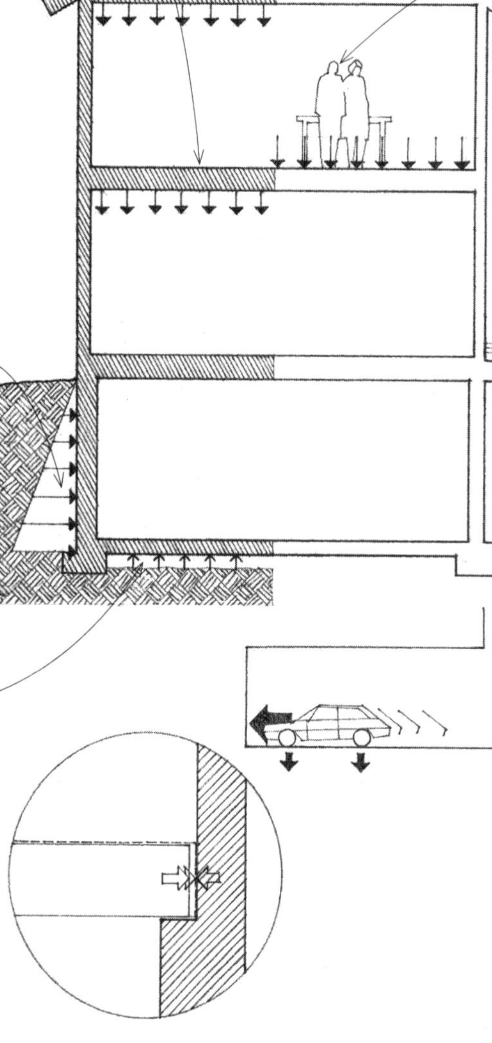

- Occupancy loads result from the weight of people, furniture, stored material, and other similar items in a building. Building codes specify minimum uniformly distributed unit loads for various uses and occupancies.
- Snow loads are created by the weight of snow accumulating on a roof. Snow loads vary with geographic location, site exposure, wind conditions, and roof geometry.
- Rain loads result from the accumulation of water on a roof because of its form, deflection, or the clogging of its drainage system.

- Settlement loads are imposed on a structure by subsidence of a portion of the supporting soil and the resulting differential settlement of its foundation.

- Ground pressure is the horizontal force a soil mass exerts on a vertical retaining structure.

- Water pressure is the hydraulic force groundwater exerts on a foundation system.

- Thermal stresses are the compressive or tensile stresses developed in a material constrained against thermal expansion or contraction.

- Impact loads are kinetic loads of short duration due to moving vehicles, equipment, and machinery. Building codes treat this load as a static load, compensating for its dynamic nature by amplifying the static load.

Dynamic Loads

Dynamic loads are applied suddenly to a structure, often with rapid changes in magnitude and point of application. Under a dynamic load, a structure develops inertial forces in relation to its mass and its maximum deformation does not necessarily correspond to the maximum magnitude of the applied force. The two major types of dynamic loads are wind loads and earthquake loads.

Wind loads are the forces exerted by the kinetic energy of a moving mass of air, assumed to come from any horizontal direction.

- The structure, components, and cladding of a building must be designed to resist wind-induced sliding, uplift, or overturning.

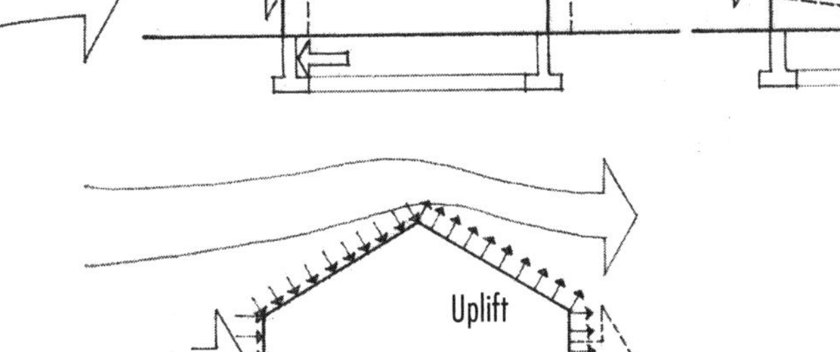

- Total wind loads are determined by taking the product of the wind load per square foot multiplied by the area of building or structure projected on a vertical plane normal to the wind direction.
- Wind is to be assumed to come from any horizontal direction and wind pressures are to be assumed to act normal to the surface considered.
- Because wind can create positive pressure as well as suction or negative pressure on a building, the force is to be resisted in either direction normal to the surface.

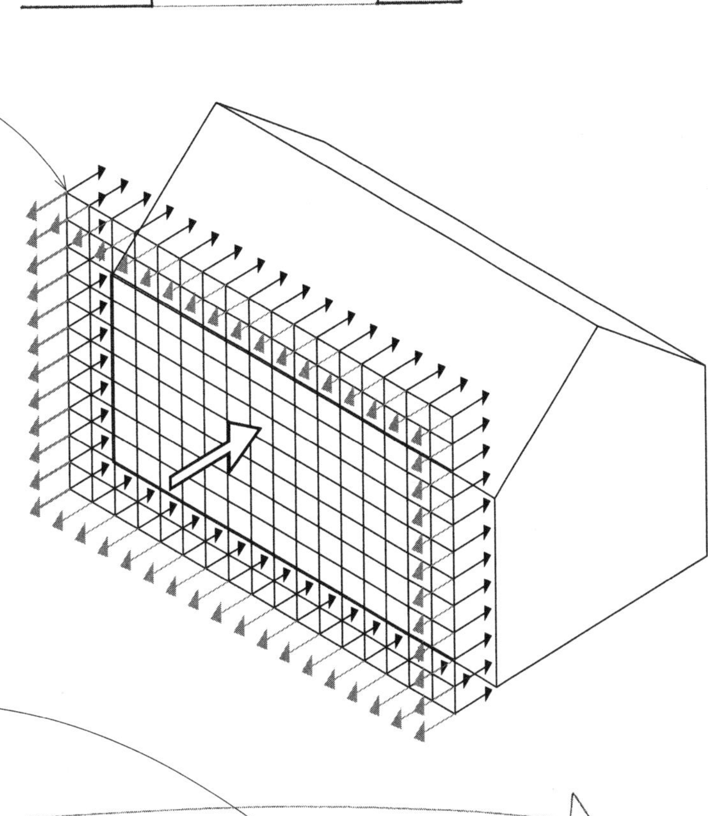

- Design wind pressure is a minimum design value for the equivalent static pressure on the exterior surfaces of a structure resulting from a critical wind velocity, equal to a reference wind pressure measured at a height of 33' (10 m), modified by a number of coefficients to account for the effects of exposure condition, building height, wind gusts, and the geometry and orientation of the structure to the impinging air flow.
- An importance factor may increase the design values for wind or seismic forces on a building because of its large occupancy, its potentially hazardous contents, or its essential nature in the wake of a hurricane or earthquake.

- Flutter refers to the rapid oscillations of a flexible cable or membrane structure caused by the aerodynamic effects of wind.
- Tall, slender buildings, structures with unusual or complex shapes, and lightweight, flexible structures subject to flutter require wind tunnel testing or computer modeling to investigate how they respond to the distribution of wind pressure.

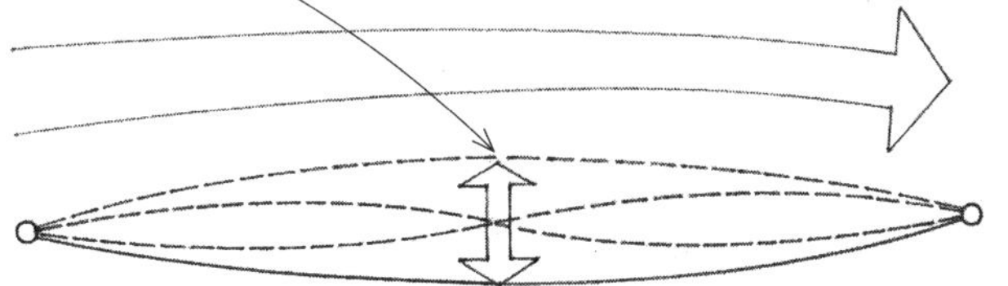

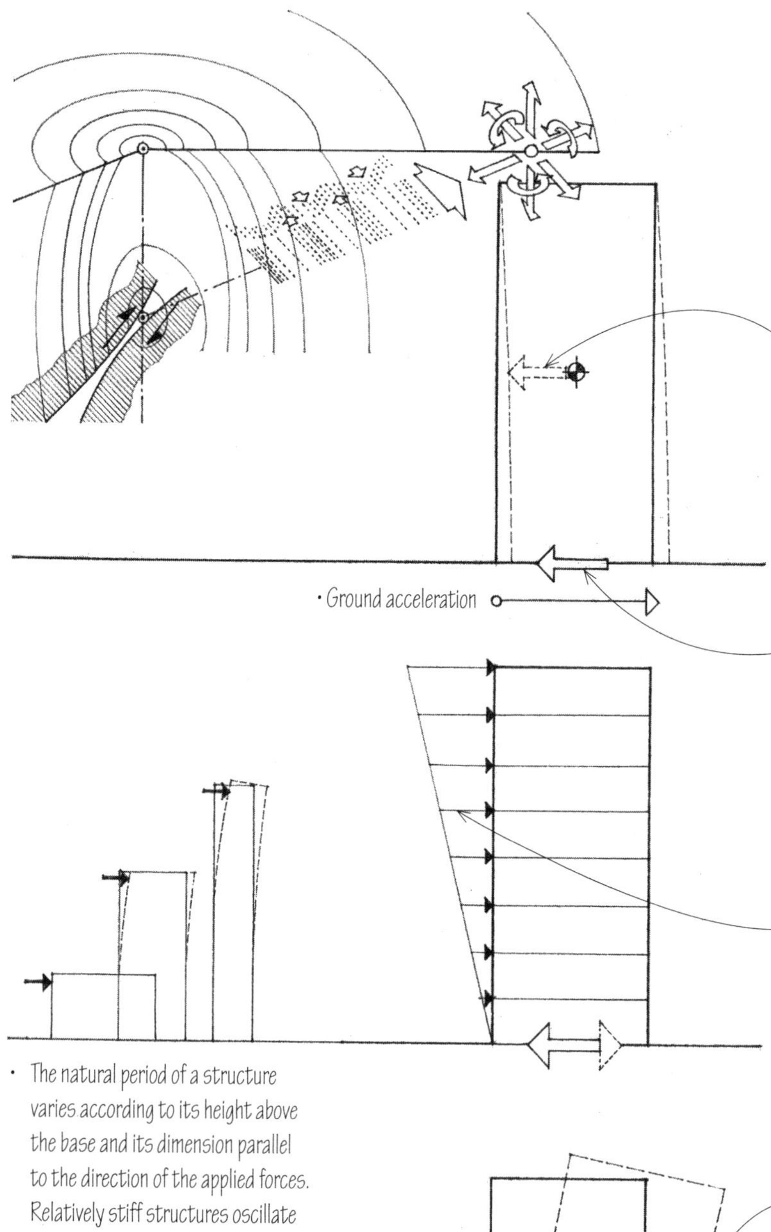

An earthquake consists of a series of longitudinal and transverse vibrations induced in the earth's crust by the abrupt movement of plates along fault lines. The shocks of an earthquake propagate along the earth's surface in the form of waves and attenuate logarithmically with distance from its source. While these ground motions are three-dimensional in nature, their horizontal components are considered to be the more critical in structural design; the vertical load-carrying elements of a structure usually have considerable reserve for resisting additional vertical loads.

- The upper mass of a structure develops an inertial force as it tends to remain at rest while the base is translated by the ground motions of an earthquake. From Newton's second law, this force is equal to the product of mass and acceleration.
- A statically equivalent lateral force, base shear, may be computed for regular structures less than 240' (73 m) in height, irregular structures not more than five stories high, and structures at low seismic risk.
- Base shear is the minimum design value for the total lateral seismic force on a structure assumed to act in any horizontal direction. It is computed by multiplying the total dead load of the structure by a number of coefficients to reflect the character and intensity of the ground motions in the seismic zone, the soil profile type underlying the foundation, the type of occupancy, the distribution of the mass and stiffness of the structure, and the natural period—the time required for one complete oscillation—of the structure.
- Base shear is distributed to each horizontal diaphragm above the base of regular structures in proportion to the floor weight at each level and its distance from the base.
- A more complex dynamic analysis is required for high-rise structures, structures with irregular shapes or framing systems, or for structures built on soft or plastic soils susceptible to failure or collapse under seismic loading.

- The natural period of a structure varies according to its height above the base and its dimension parallel to the direction of the applied forces. Relatively stiff structures oscillate rapidly and have short periods while more flexible structures oscillate more slowly and have longer periods.

- Any lateral load applied at a distance above grade generates an overturning moment at the base of a structure. For equilibrium, the overturning moment must be counterbalanced by an external restoring moment and an internal resisting moment provided by forces developed in column members and shear walls.
- A restoring moment is provided by the dead load of a structure acting about the same point of rotation as the overturning movement. Building codes usually require that the restoring moment be at least 50% greater than the overturning moment.

The following is a brief introduction to the way a structural system must resolve all of the forces acting on a building and channel them to the ground. For more complete information on the structural design and analysis of buildings, see Bibliography.

A force is any influence that produces a change in the shape or movement of a body. It is considered to be a vector quantity possessing both magnitude and direction, represented by an arrow whose length is proportional to the magnitude and whose orientation in space represents the direction. A single force acting on a rigid body may be regarded as acting anywhere along its line of action without altering the external effect of the force. Two or more forces may be related in the following ways:

- Collinear forces occur along a straight line, the vector sum of which is the algebraic sum of the magnitudes of the forces, acting along the same line of action.

- Concurrent forces have lines of action intersecting at a common point, the vector sum of which is equivalent to and produces the same effect on a rigid body as the application of the vectors of the several forces.
- The parallelogram law states that the vector sum or resultant of two concurrent forces can be described by the diagonal of a parallelogram having adjacent sides that represent the two force vectors being added.
- In a similar manner, any single force can be resolved into two or more concurrent forces having a net effect on a rigid body equivalent to that of the initial force. For convenience in structural analysis, these are usually the rectangular or Cartesian components of the initial force.
- The polygon method is a graphic technique for finding the vector sum of a coplanar system of several concurrent forces by drawing to scale each force vector in succession, with the tail of each at the head of the one preceding it, and completing the polygon with a vector that represents the resultant force, extending from the tail of the first to the head of the last vector.

- Nonconcurrent forces have lines of action that do not intersect at a common point, the vector sum of which is a single force that would cause the same translation and rotation of a body as the set of original forces.
- A moment is the tendency of a force to produce rotation of a body about a point or line, equal in magnitude to the product of the force and the moment arm and acting in a clockwise or counterclockwise direction.
- A couple is a force system of two equal, parallel forces acting in opposite directions and tending to produce rotation but not translation. The moment of a couple is equal in magnitude to the product of one of the forces and the perpendicular distance between the two forces.

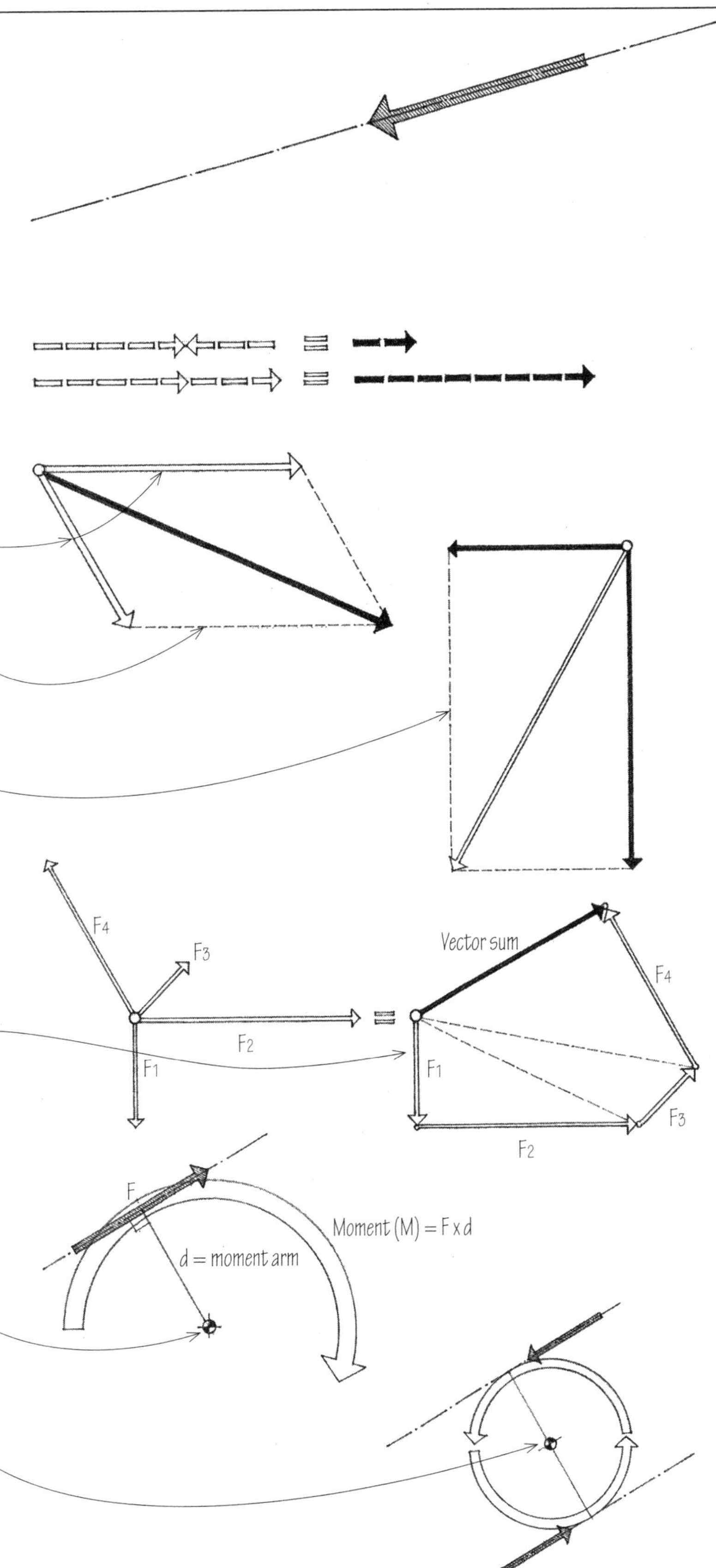

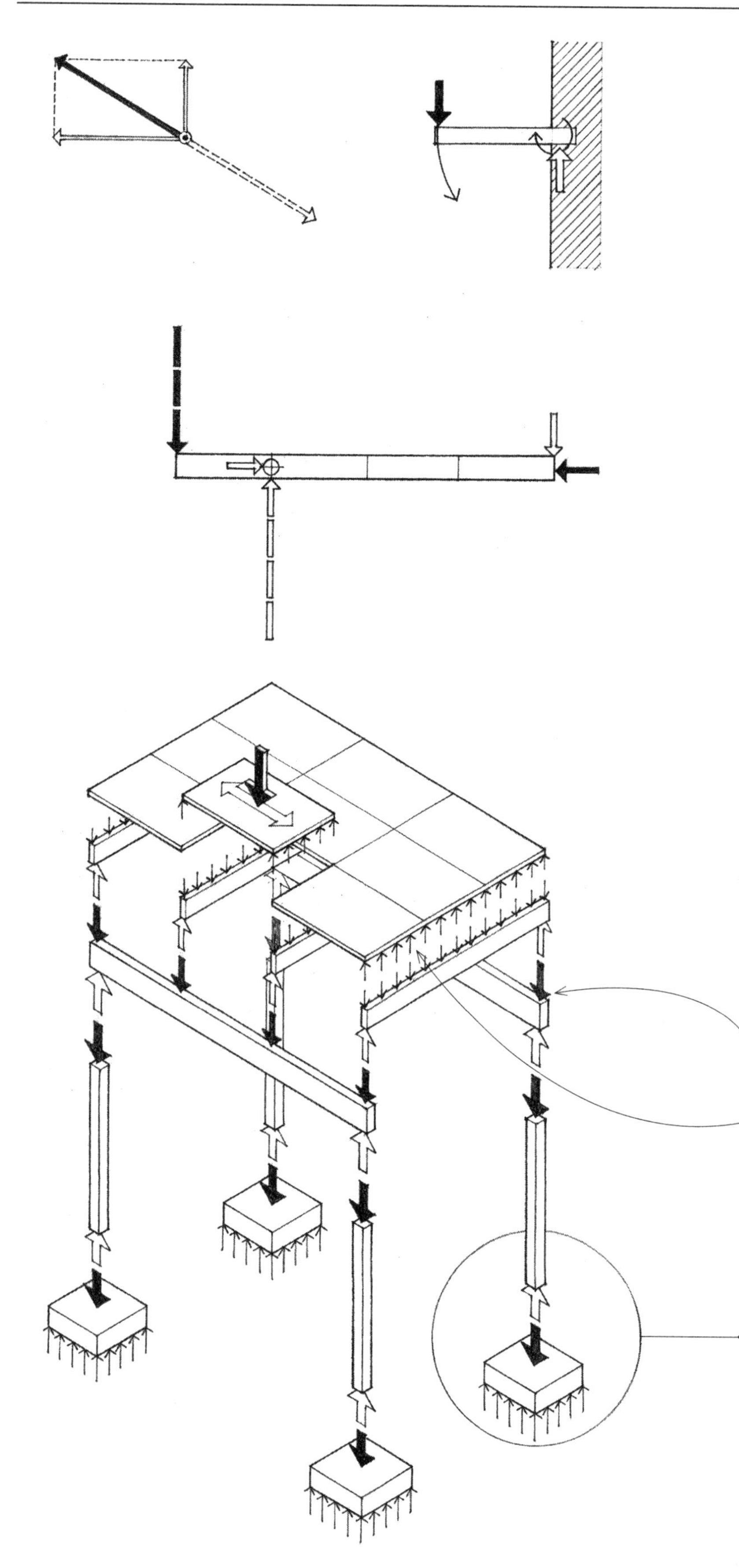

In both structural design and analysis, we are concerned first with the magnitude, direction, and point of application of forces, and their resolution to produce a state of equilibrium. Equilibrium is a state of balance or rest resulting from the equal action of opposing forces. In other words, as each structural element is loaded, its supporting elements must react with equal but opposite forces. For a rigid body to be in equilibrium, two conditions are necessary.

- First, the vector sum of all forces acting on it must equal zero, ensuring translational equilibrium: $\sum F_x = 0; \sum F_y = 0; \sum F_z = 0$.
- Second, the algebraic sum of all moments of the forces about any point or line must equal zero, ensuring rotational equilibrium: $\sum M = 0$.

- Newton's third law of motion, the law of action and reaction, states that for every force acting on a body, the body exerts a force having equal magnitude in the opposite direction along the same line of action as the original force.

- A concentrated load acts on a very small area or particular point of a supporting structural element, as when a beam bears on a post or a column bears on its footing.
- A uniformly distributed load is a load of uniform magnitude extending over the length or area of the supporting structural element, as in the case of the live load on a floor deck or joist, or a wind load on a wall.

- A free-body diagram is a graphic representation of the complete system of applied and reactive forces acting on a body or an isolated part of a structure. Every elementary part of a structural system has reactions that are necessary for the equilibrium of the part, just as the larger system has reactions at its supports that serve to maintain the equilibrium of the whole.

Columns are rigid, relatively slender structural members designed primarily to support axial compressive loads applied to the ends of the members. Relatively short, thick columns are subject to failure by crushing rather than by buckling. Failure occurs when the direct stress from an axial load exceeds the compressive strength of the material available in the cross section. An eccentric load, however, can produce bending and result in an uneven stress distribution in the section.

- Kern area is the central area of any horizontal section of a column or wall within which the resultant of all compressive loads must pass if only compressive stresses are to be present in the section. A compressive load applied beyond this area will cause tensile stresses to develop in the section.

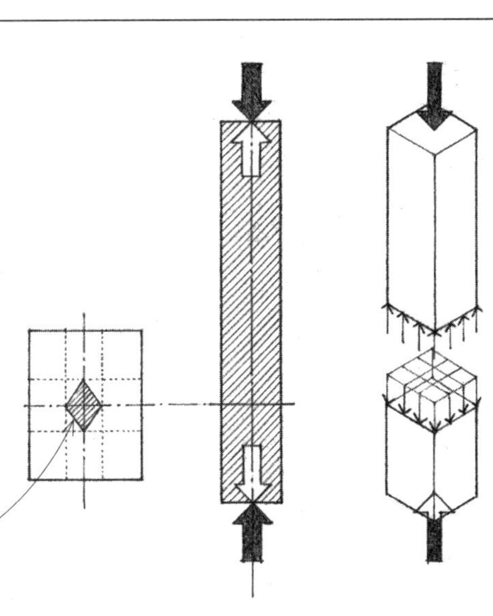

- External forces create internal stresses within structural elements.

Long, slender columns are subject to failure by buckling rather than by crushing. Buckling is the sudden lateral or torsional instability of a slender structural member induced by the action of an axial load before the yield stress of the material is reached. Under a buckling load, a column begins to deflect laterally and cannot generate the internal forces necessary to restore its original linear condition. Any additional loading would cause the column to deflect further until collapse occurs in bending. The higher the slenderness ratio of a column, the lower is the critical stress that will cause it to buckle. A primary objective in the design of a column is to reduce its slenderness ratio by shortening its effective length or maximizing the radius of gyration of its cross section.

- The slenderness ratio of a column is the ratio of its effective length (L) to its least radius of gyration (r). For asymmetrical column sections, therefore, buckling will tend to occur about the weaker axis or in the direction of the least dimension.

- Effective length is the distance between inflection points in a column subject to buckling. When this portion of a column buckles, the entire column fails.
- The effective length factor (k) is a coefficient for modifying the actual length of a column according to its end conditions in order to determine its effective length. For example, fixing both ends of a long column reduces its effective length by half and increases its load-carrying capacity by a factor of 4.

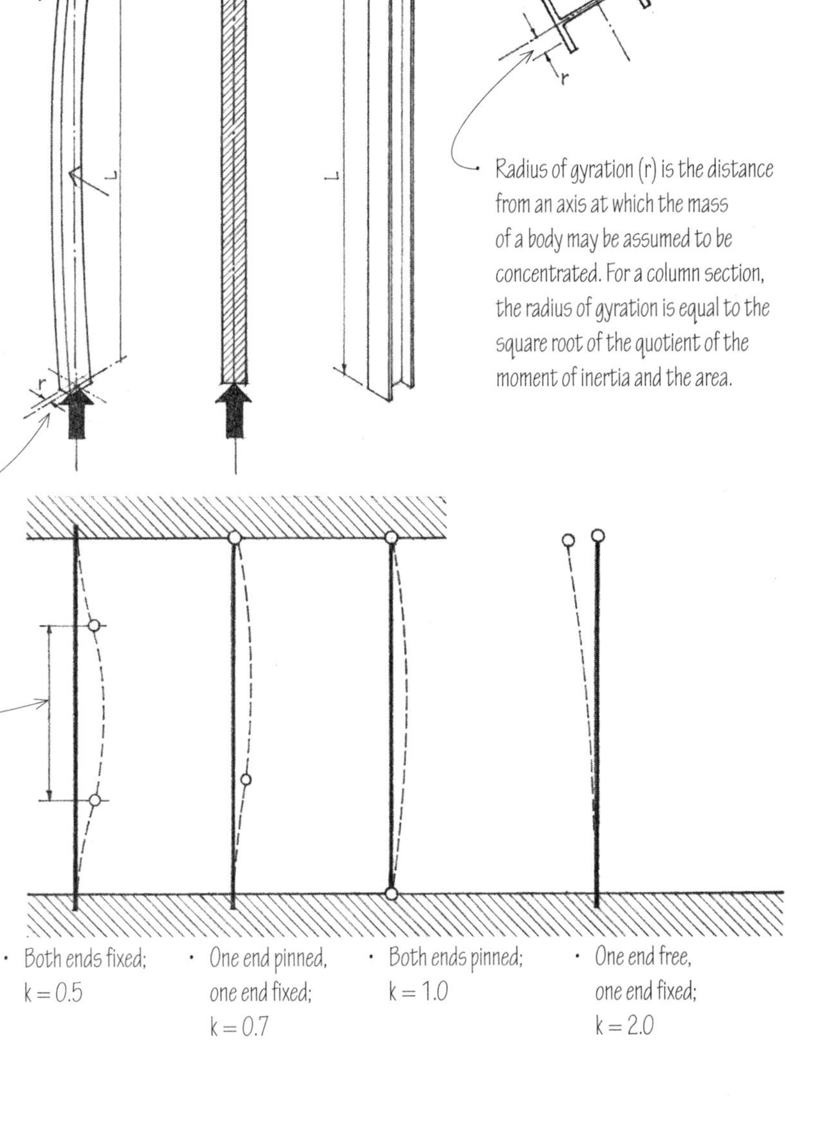

- Radius of gyration (r) is the distance from an axis at which the mass of a body may be assumed to be concentrated. For a column section, the radius of gyration is equal to the square root of the quotient of the moment of inertia and the area.

- Both ends fixed; k = 0.5
- One end pinned, one end fixed; k = 0.7
- Both ends pinned; k = 1.0
- One end free, one end fixed; k = 2.0

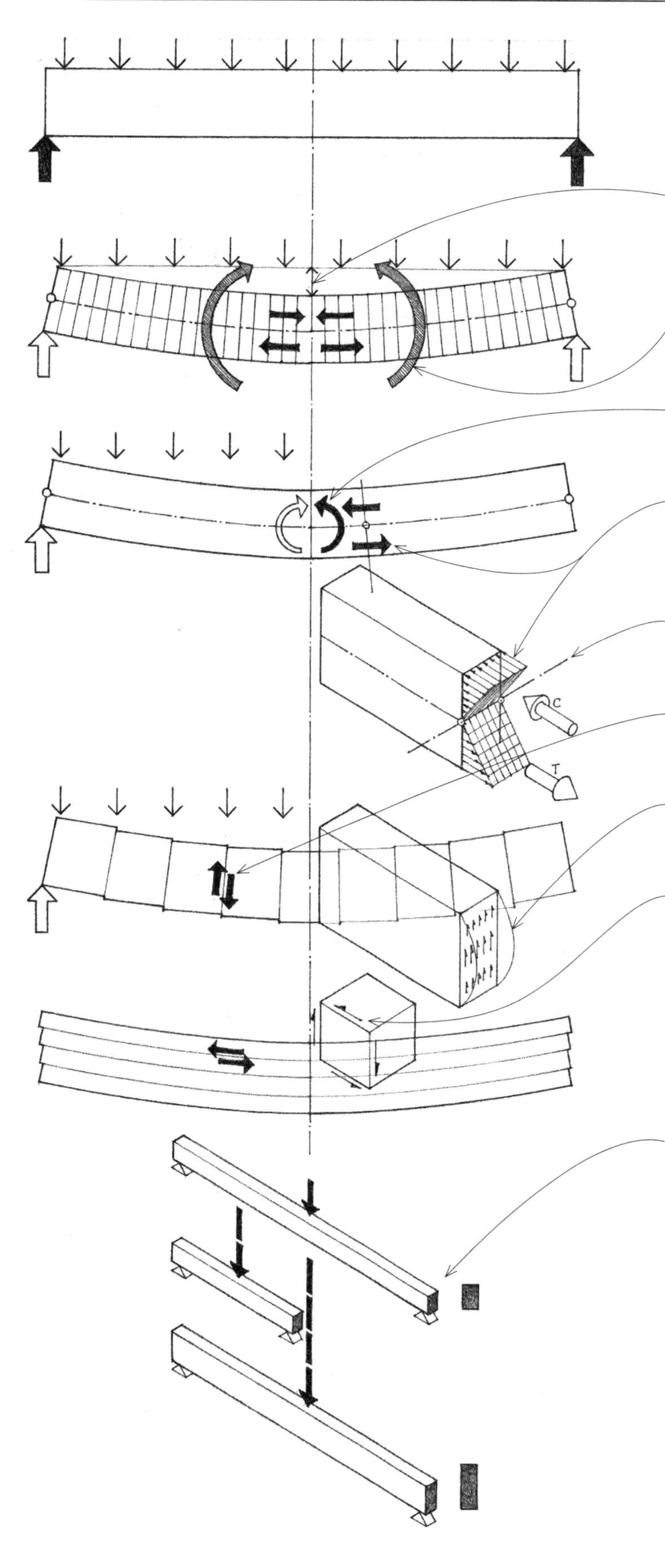

Beams are rigid structural members designed to carry and transfer transverse loads across space to supporting elements. The nonconcurrent pattern of forces subjects a beam to bending and deflection, which must be resisted by the internal strength of the material.

- Deflection is the perpendicular distance a spanning member deviates from a true course under transverse loading, increasing with load and span, and decreasing with an increase in the moment of inertia of the section or the modulus of elasticity of the material.
- Bending moment is an external moment tending to cause part of a structure to rotate or bend, equal to the algebraic sum of the moments about the neutral axis of the section under consideration.
- Resisting moment is an internal moment equal and opposite to a bending moment, generated by a force couple to maintain equilibrium of the section being considered.
- Bending stress is a combination of compressive and tension stresses developed at a cross section of a structural member to resist a transverse force, having a maximum value at the surface farthest from the neutral axis.
- The neutral axis is an imaginary line passing through the centroid of the cross section of a beam or other member subject to bending, along which no bending stresses occur.
- Transverse shear occurs at a cross section of a beam or other member subject to bending, equal to the algebraic sum of transverse forces on one side of the section.
- Vertical shearing stress develops to resist transverse shear, having a maximum value at the neutral axis and decreasing nonlinearly toward the outer faces.
- Horizontal or longitudinal shearing stress develops to prevent slippage along horizontal planes of a beam under transverse loading, equal at any point to the vertical shearing stress at that point.

The efficiency of a beam is increased by configuring the cross section to provide the required moment of inertia or section modulus with the smallest possible area, usually by making the section deep with most of the material at the extremities where the maximum bending stresses occur. For example, while halving a beam span or doubling its width reduces the bending stresses by a factor of 2, doubling the depth reduces the bending stresses by a factor of 4.

- Moment of inertia is the sum of the products of each element of an area and the square of its distance from a coplanar axis of rotation. It is a geometric property that indicates how the cross-sectional area of a structural member is distributed and does not reflect the intrinsic physical properties of a material.
- Section modulus is a geometric property of a cross section, defined as the moment of inertia of the section divided by the distance from the neutral axis to the most remote surface.

- A simple beam rests on supports at both ends, with the ends free to rotate and having no moment resistance. As with any statically determinate structure, the values of all reactions, shears, and moments for a simple beam are independent of its cross-sectional shape and material.

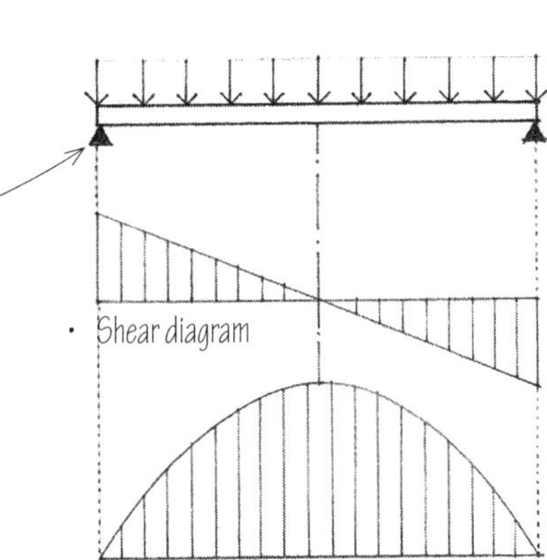

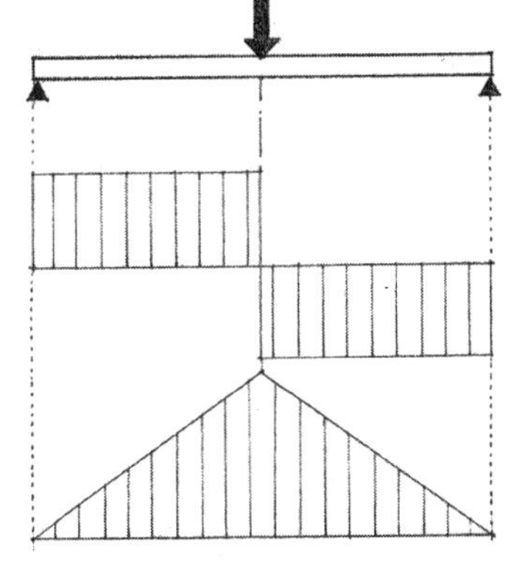

- Moment diagram

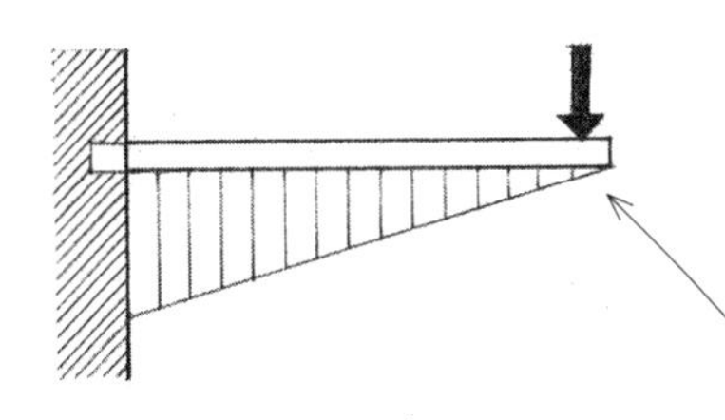

- A cantilever is a projecting beam or other rigid structural member supported at only one fixed end.
- An overhanging beam is a simple beam extending beyond one of its supports. The overhang reduces the positive moment at midspan while developing a negative moment at the base of the cantilever over the support. Assuming a uniformly distributed load, the projection for which the moment over the support is equal and opposite to the moment at midspan is approximately 3/8 of the span.

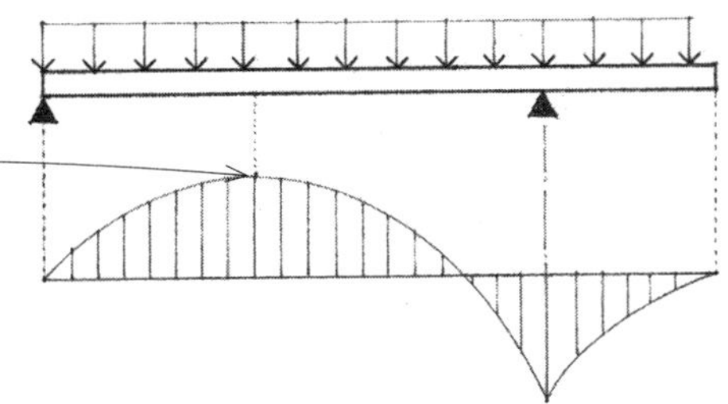

- A double overhanging beam is a simple beam extending beyond both of its supports. Assuming a uniformly distributed load, the projections for which the moments over the supports are equal and opposite to the moment at midspan are approximately 1/3 of the span.

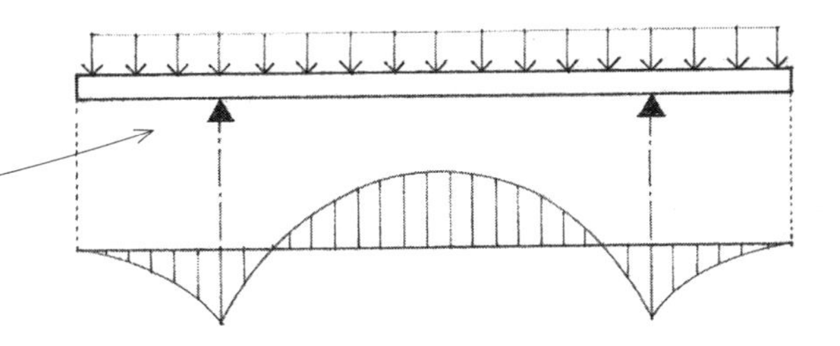

- A fixed-end beam has both ends restrained against translation and rotation. The fixed ends transfer bending stresses, increase the rigidity of the beam, and reduce its maximum deflection.

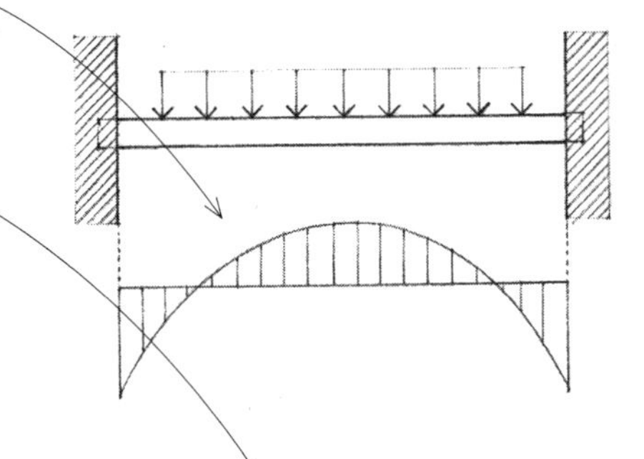

- A suspended span is a simple beam supported by the overhangs of two adjoining spans with pinned construction joints at points of zero moment.
- A continuous beam extends over more than two supports in order to develop greater rigidity and smaller moments than a series of simple beams having similar spans and loading. Both fixed-end and continuous beams are indeterminate structures for which the values of all reactions, shears, and moments are dependent not only on span and loading but also on the cross-sectional shape and material of the beam.

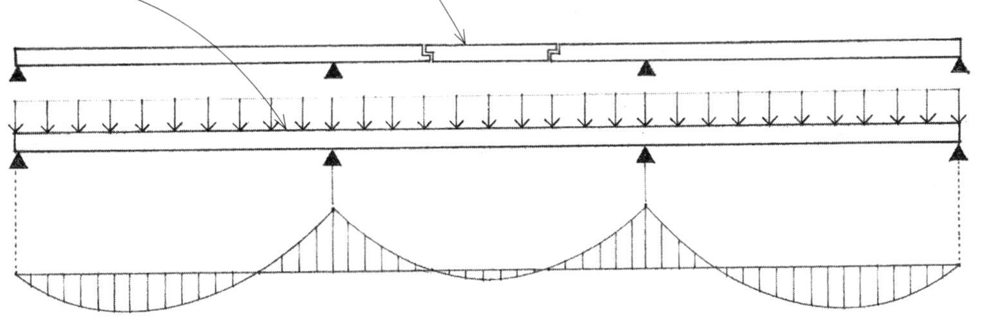

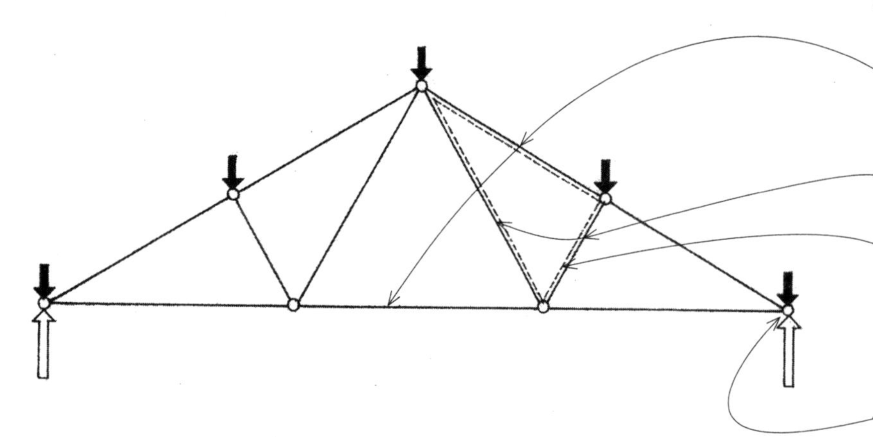

A truss is a structural frame based on the geometric rigidity of the triangle and composed of linear members subject only to axial tension or compression.

- Top and bottom chords are the principal members of a truss extending from end to end and connected by web members.
- Web is the integral system of members connecting the upper and lower chords of a truss.
- Panel refers to any of the spaces within the web of a truss between any two panel points on a chord and a corresponding joint or pair of joints on an opposite chord.
- Heel is the lower, supported end of a truss.

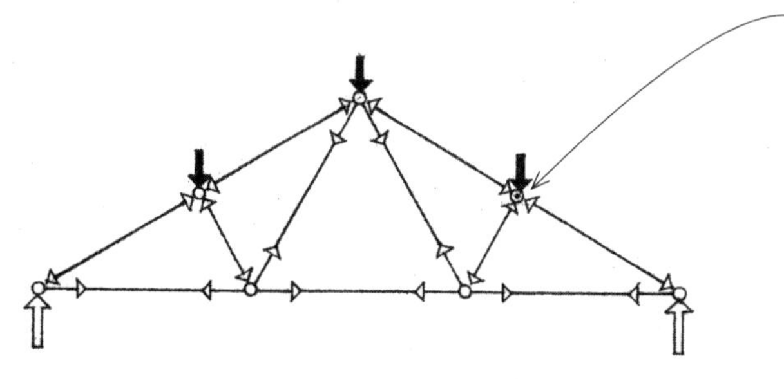

- Panel point is any of the joints between a principal web member and a chord. A truss must be loaded only at its panel points if its members are to be subject only to axial tension or compression. To prevent secondary stresses from developing, the centroidal axes of truss members and the load at a joint should pass through a common point.

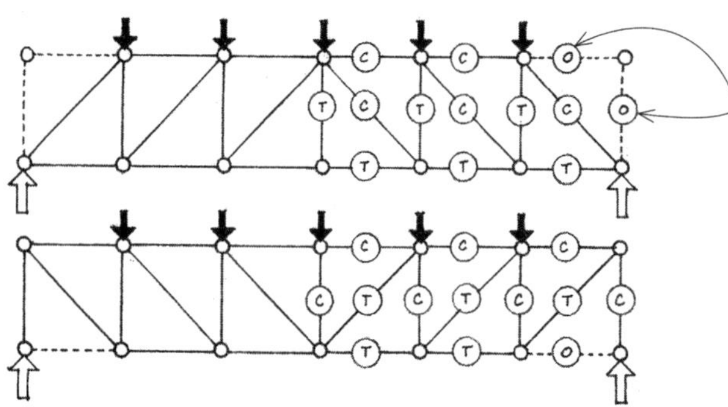

- Zero-force members theoretically carry no direct load; their omission would not alter the stability of the truss configuration.

- See 6.09 for types of trusses and truss configurations.

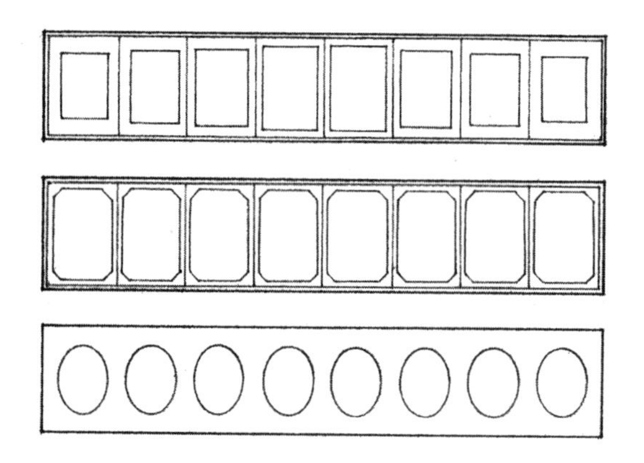

- Vierendeel trusses are framed beam structures having vertical web members rigidly connected to parallel top and bottom chords. Vierendeel trusses are not true trusses because their members are subject to nonaxial bending forces.

A beam simply supported by two columns is not capable of resisting lateral forces unless it is braced. If the joints connecting the columns and beam are capable of resisting both forces and moments, then the assembly becomes a rigid frame. Applied loads produce axial, bending, and shear forces in all members of the frame because the rigid joints restrain the ends of the members from rotating freely. In addition, vertical loads cause a rigid frame to develop horizontal thrusts at its base. A rigid frame is statically indeterminate and rigid only in its plane.

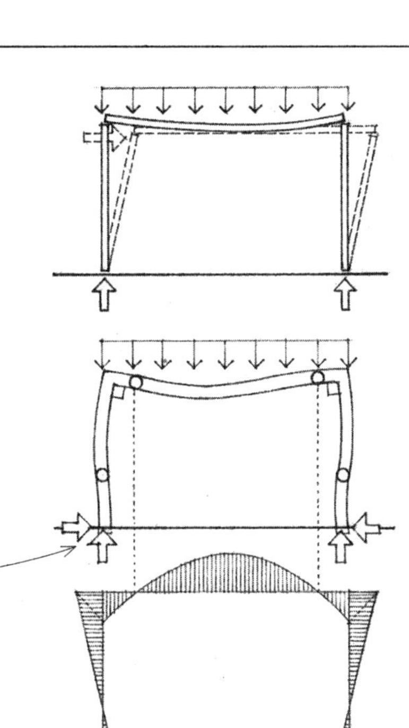

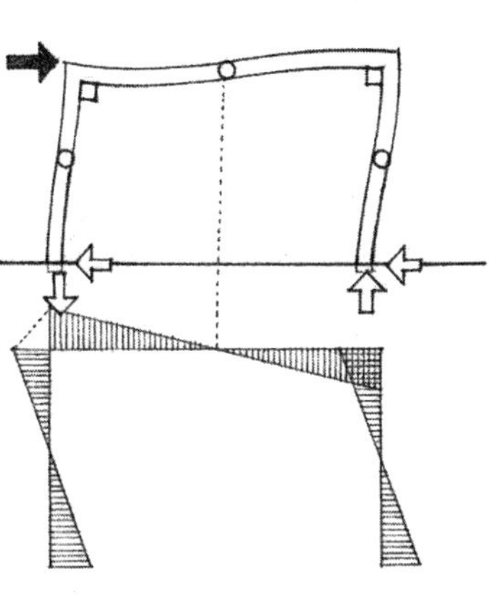

- A fixed frame is a rigid frame connected to its supports with fixed joints. A fixed frame is more resistant to deflection than a hinged frame but also more sensitive to support settlements and thermal expansion and contraction.
- A hinged frame is a rigid frame connected to its supports with pin joints. The pin joints prevent high bending stresses from developing by allowing the frame to rotate as a unit when strained by support settlements, and to flex slightly when stressed by changes in temperature.
- A three-hinged frame is a structural assembly of two rigid sections connected to each other and to its supports with pin joints. While more sensitive to deflection than either the fixed or hinged frame, the three-hinged frame is least affected by support settlements and thermal stresses. The three pin joints also permit the frame to be analyzed as a statically determinate structure.

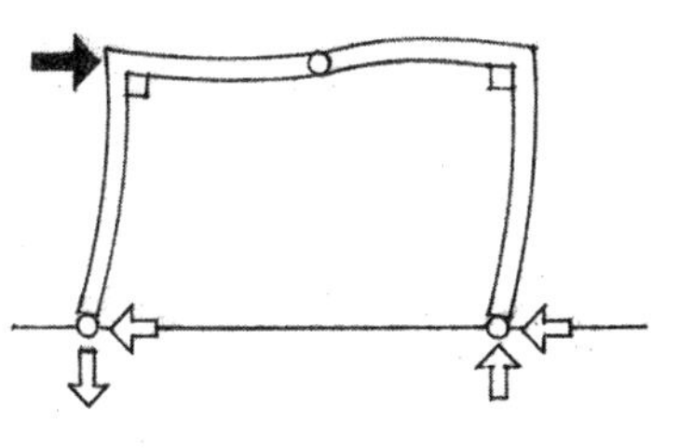

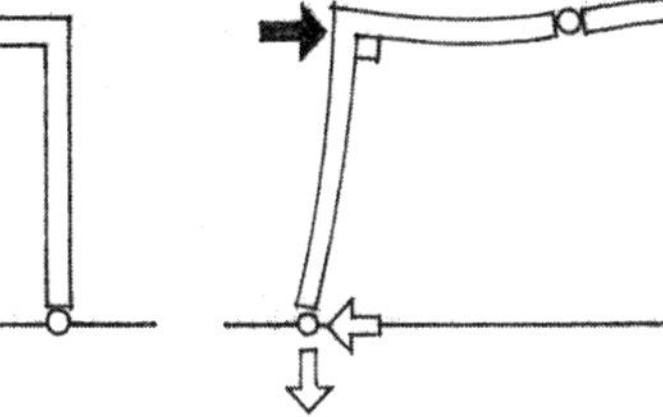

If we fill in the plane defined by two columns and a beam, it becomes a loadbearing wall that acts as a long, thin column in transmitting compressive forces to the ground. Loadbearing walls are most effective when carrying coplanar, uniformly distributed loads and most vulnerable to forces perpendicular to their planes. For lateral stability, loadbearing walls must rely on buttressing with pilasters, cross walls, transverse rigid frames, or horizontal slabs.

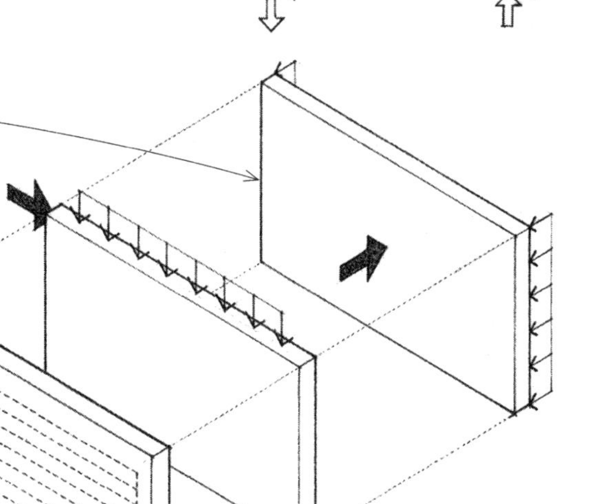

Any opening in a loadbearing wall weakens its structural integrity. A lintel or arch must support the load above a door or window opening and allow the compressive stresses to flow around the opening to adjacent sections of the wall.

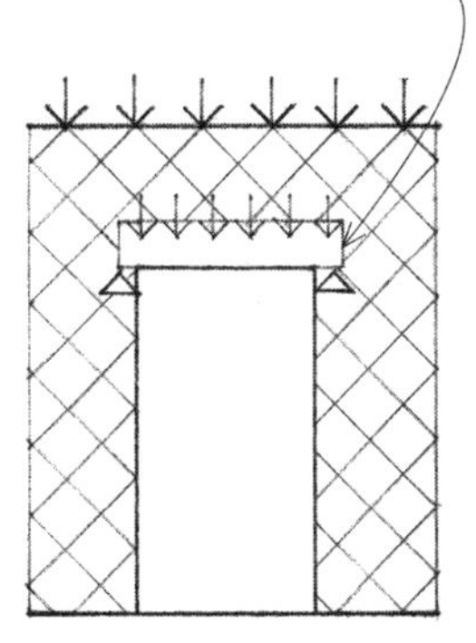

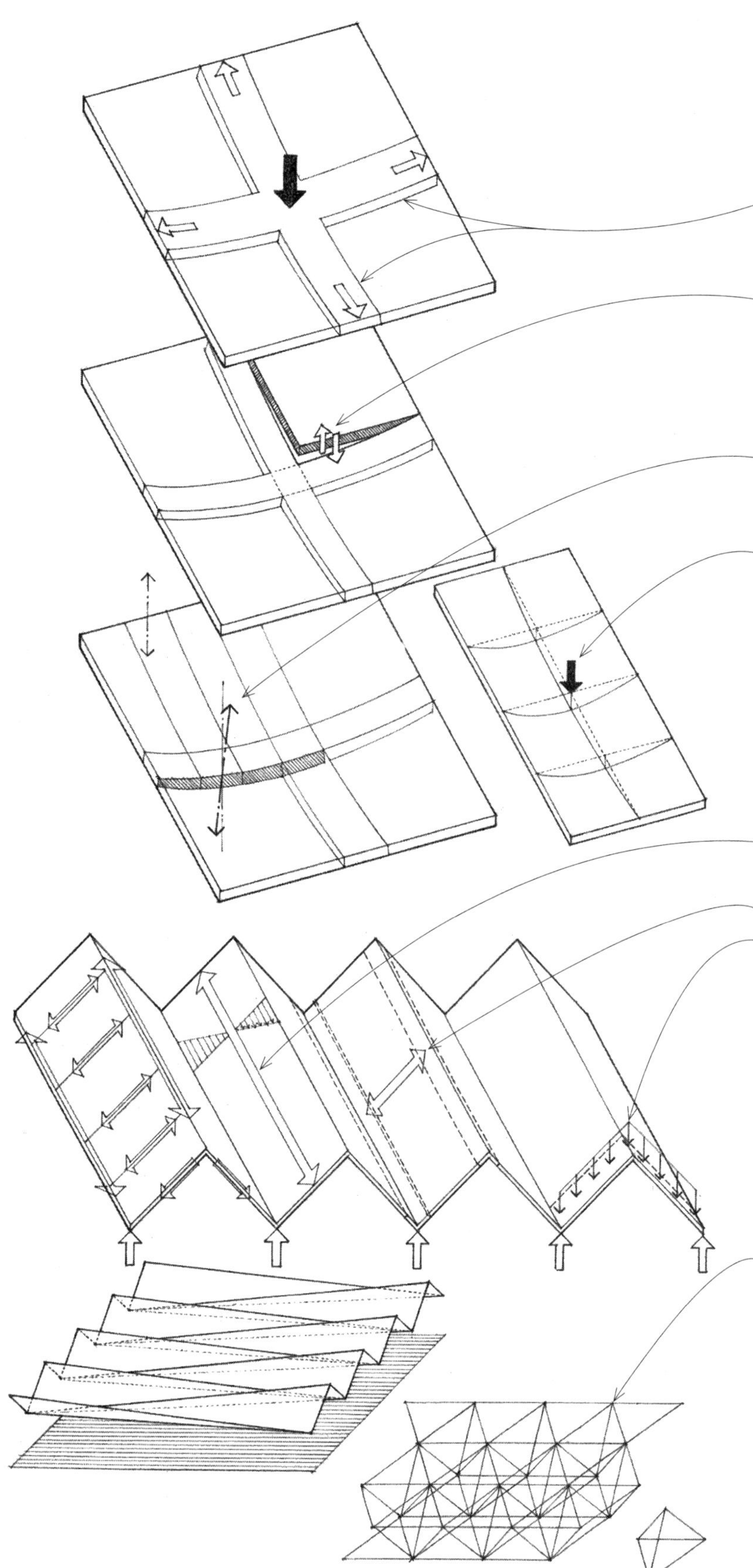

Plate structures are rigid, planar, usually monolithic structures that disperse applied loads in a multidirectional pattern, with the loads generally following the shortest and stiffest routes to the supports. A common example of a plate structure is a reinforced concrete slab.

A plate can be envisioned as a series of adjacent beam strips interconnected continuously along their lengths. As an applied load is transmitted to the supports through bending of one beam strip, the load is distributed over the entire plate by vertical shear transmitted from the deflected strip to adjacent strips. The bending of one beam strip also causes twisting of transverse strips, whose torsional resistance increases the overall stiffness of the plate. Therefore, while bending and shear transfer an applied load in the direction of the loaded beam strip, shear and twisting transfer the load at right angles to the loaded strip.

A plate should be square or nearly square to ensure that it behaves as a two-way structure. As a plate becomes more rectangular than square, the two-way action decreases and a one-way system spanning the shorter direction develops because the shorter plate strips are stiffer and carry a greater portion of the load.

Folded plate structures are composed of thin, deep elements joined rigidly along their boundaries and forming sharp angles to brace each other against lateral buckling. Each plane behaves as a beam in the longitudinal direction. In the short direction, the span is reduced by each fold acting as a rigid support. Transverse strips behave as a continuous beam supported at fold points. Vertical diaphragms or rigid frames stiffen a folded plate against deformation of the fold profile. The resulting stiffness of the cross section enables a folded plate to span relatively long distances.

A space frame is composed of short rigid linear elements triangulated in three dimensions and subject only to axial tension or compression. The simplest spatial unit of a space frame is a tetrahedron having four joints and six structural members. Because the structural behavior of a space frame is analogous to that of a plate structure, its supporting bay should be square or nearly square to ensure that it acts as a two-way structure. Enlarging the bearing area of the supports increases the number of members into which shear is transferred and reduces the forces in the members. See 6.10 for more information on space frames.

CSI MasterFormat™ 13 32 00: Space Frames

With the principal structural elements of column, beam, slab, and loadbearing wall, it is possible to form an elementary structural unit capable of defining and enclosing a volume of space for habitation. This structural unit is the basic building block for the structural system and spatial organization of a building.

- Horizontal spans may be traversed by reinforced concrete slabs or by a layered, hierarchical arrangement of girders, beams, and joists supporting planks or decking.
- The vertical support for a structural unit may be provided by loadbearing walls or by a framework of columns and beams.

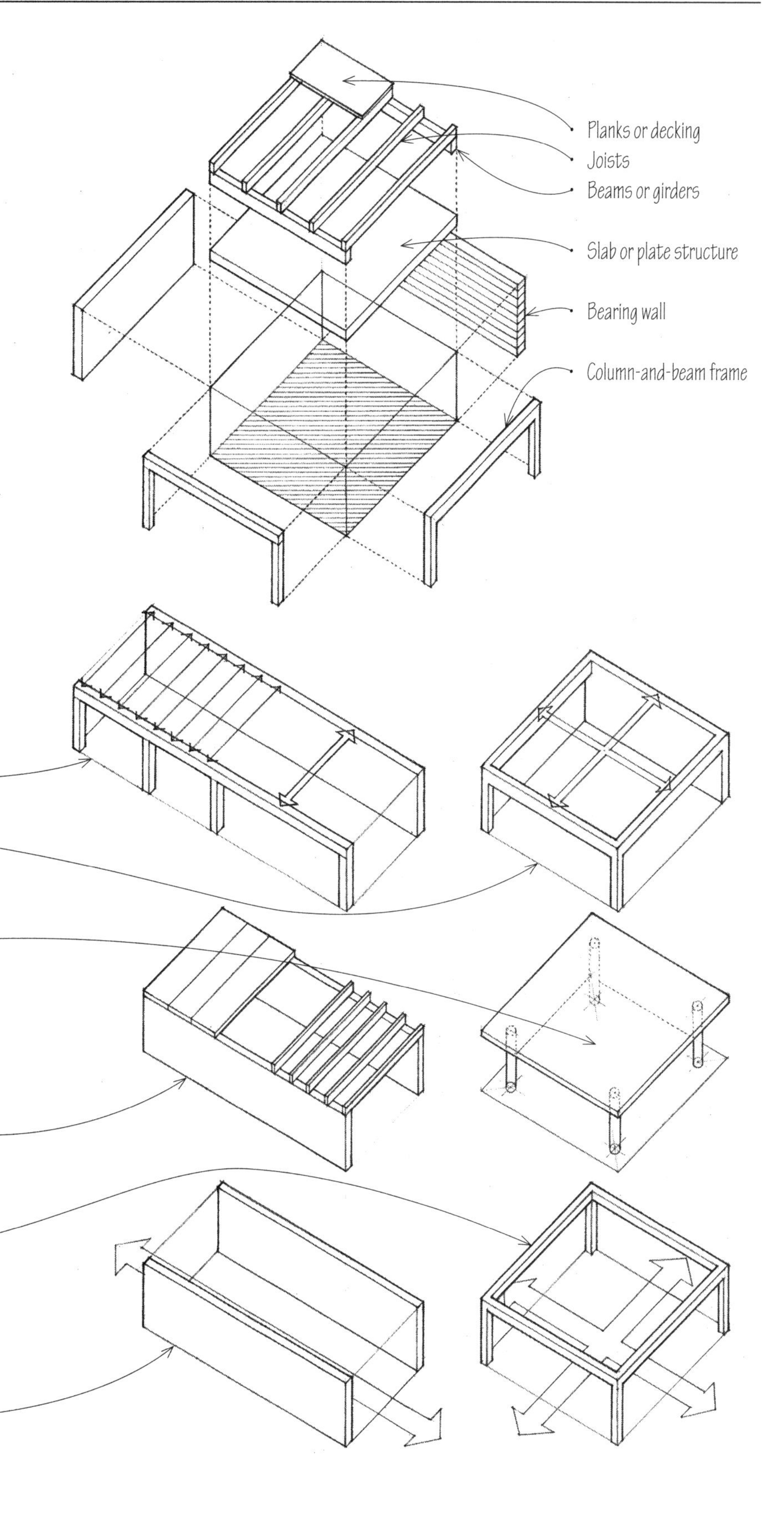

The dimensions and proportions of a structural unit or bay influence the selection of an appropriate spanning system.

- One-way systems of joists, planks, or slabs are more efficient when structural bays are rectangular—that is, when the ratio of the long to the short dimensions is greater than 1.5:1—or when the structural grid generates a linear pattern of spaces.
- Two-way systems of beams and slabs are more effective for square or nearly square bays.

- A two-way slab supported by four columns defines a horizontal layer of space.

- The parallel nature of loadbearing walls leads naturally to the use of one-way spanning systems.
- Because loadbearing walls are most effective when supporting a uniformly distributed load, they typically support a series of joists, planks, or a one-way slab.

- A linear framework of columns and beams defines a three-dimensional module of space capable of being expanded both horizontally and vertically.

- Two loadbearing walls naturally define an axial, bidirectional space. Secondary axes can be developed perpendicular to the primary axis with openings within the loadbearing walls.

The spanning capability of horizontal elements determines the spacing of their vertical supports. This fundamental relationship between the span and spacing of structural elements influences the dimensions and scale of the spaces defined by the structural system of a building. The dimensions and proportions of structural bays, in turn, should be related to the programmatic requirements of the spaces.

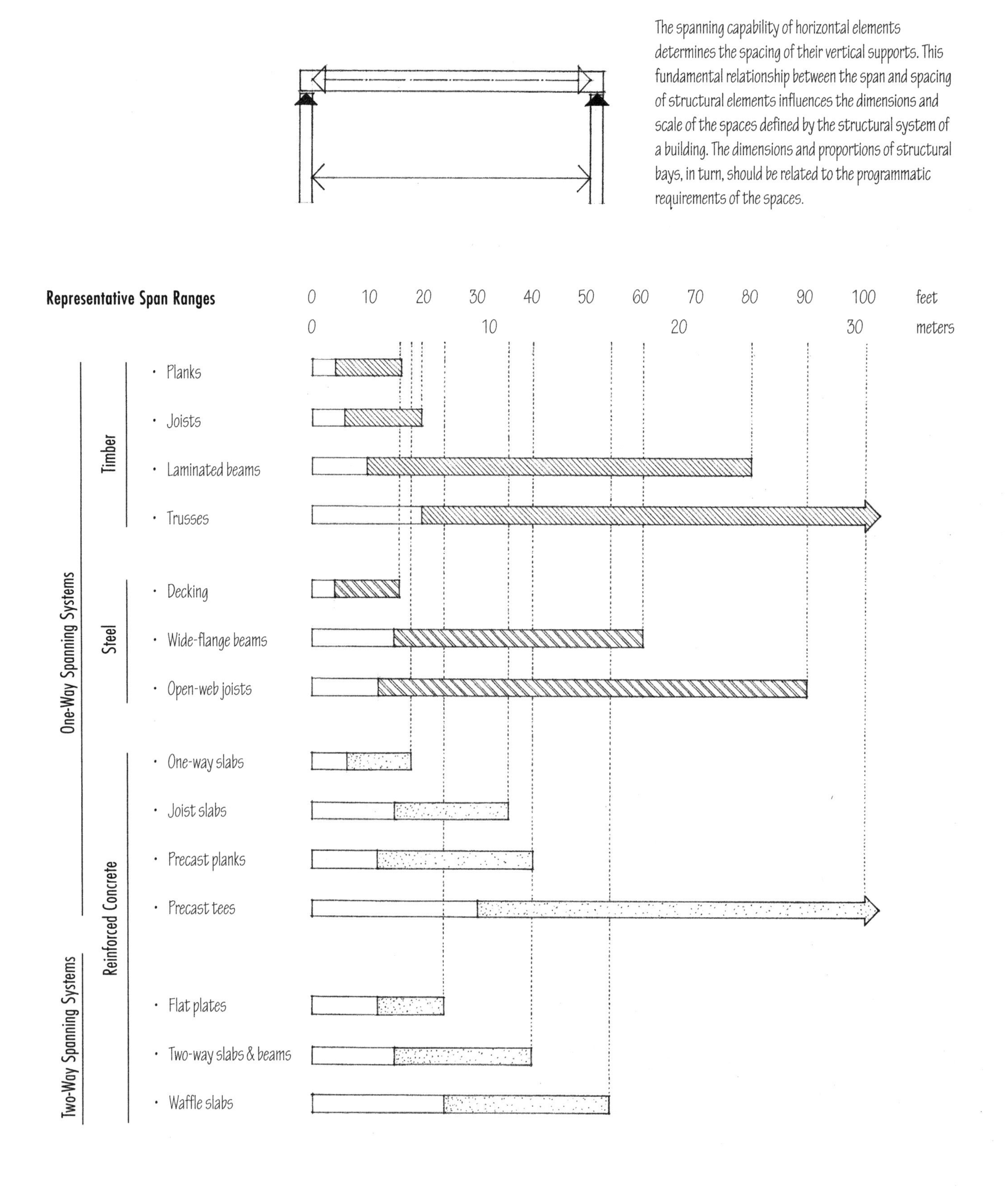

The arrangement of principal vertical supports not only regulates the selection of a spanning system, it also establishes the possibilities for the ordering of spaces and functions in a building.

The principal points and lines of support for a structural system typically define a grid. The critical points of the grid are those at which columns and loadbearing walls collect loads from beams and other horizontal spanning elements and channel these loads vertically to the ground foundation.

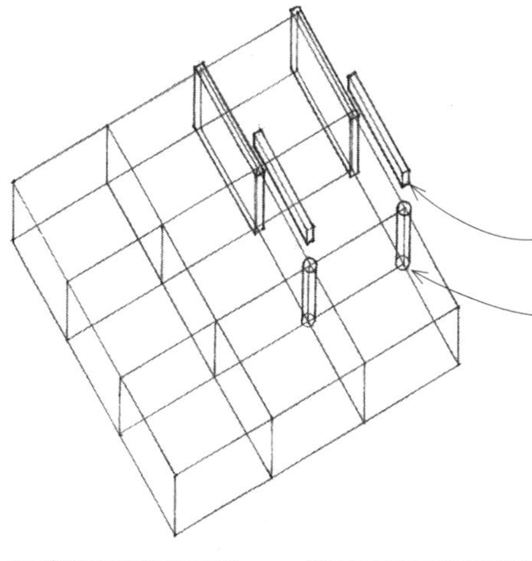

- Grid lines represent horizontal beams and loadbearing walls.
- Intersections of grid lines represent the locations of columns or concentrated gravity loads.
- A basic structural unit or bay can be logically extended vertically along the axes of columns and horizontally along the spans of beams and loadbearing walls.

The inherent geometric order of a grid can be used in the design process to initiate and reinforce the functional and spatial organization of a building design.

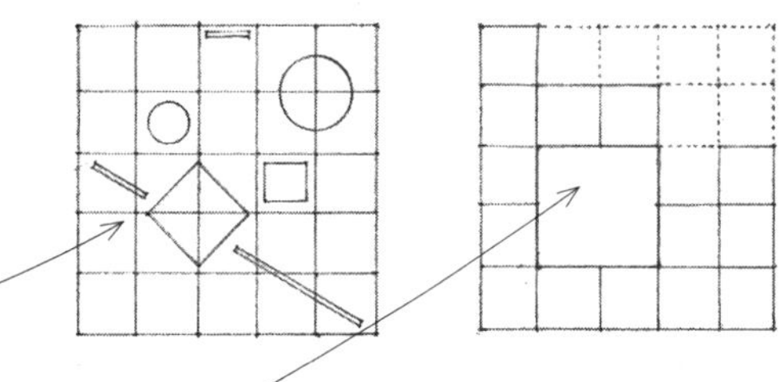

- Nonloadbearing walls may be placed to define a variety of spatial configurations and allow a building to be more flexible in responding to the programmatic requirements of its spaces.
- A structural grid can be modified by addition or subtraction to accommodate special needs, such as large spaces or unusual site conditions.
- A grid may be irregular in one or two directions to accommodate the dimensional requirements of program spaces.
- A portion of the grid can be dislocated and rotated about a point in the basic pattern.
- Two parallel grids can be offset from each other to develop intervening or interstitial spaces that define patterns of movement, mediate between a series of larger spaces, or house mechanical services.
- When two structural patterns cannot be conveniently aligned, a third element, such as a loadbearing wall, a mediating space, or a finer-grained spanning system, can be used.
- Nonuniform or irregular grids can be employed to reflect the hierarchical or functional ordering of spaces within a building.

The structural elements of a building must be sized, configured, and joined to form a stable structure under any possible load conditions. Therefore, a structural system must be designed to not only carry vertical gravity loads, but also withstand lateral wind and seismic forces from any direction. The following are the basic mechanisms for ensuring lateral stability.

Horizontal Diaphragm

- A rigid floor structure, acting as a flat, deep beam, transfers lateral loads to vertical shear walls, braced frames, or rigid frames.

Rigid Frame

- A steel or reinforced concrete frame with rigid joints capable of resisting changes in angular relationships

Shear Wall

- A wood, concrete, or masonry wall capable of resisting changes in shape and transferring lateral loads to the ground foundation

Braced Frame

- A timber or steel frame braced with diagonal members

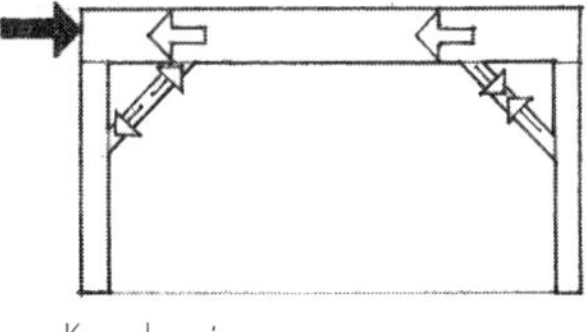

- Knee bracing

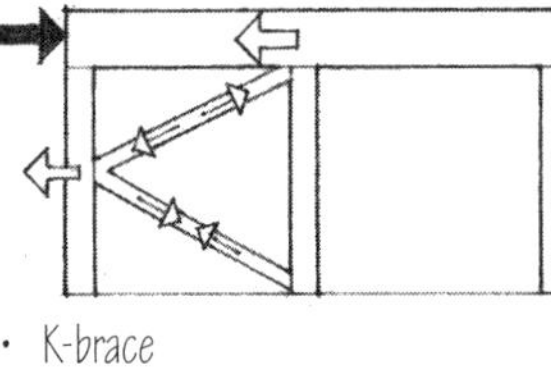

- K-brace

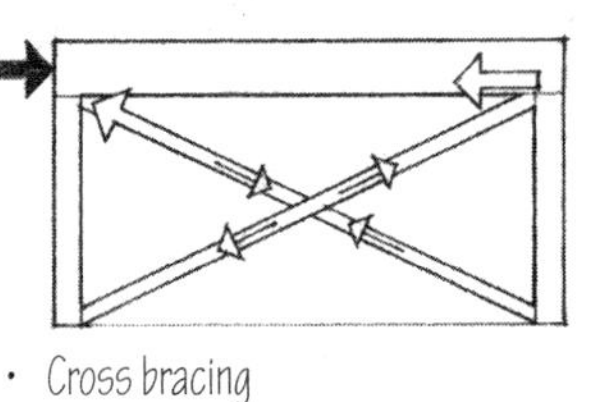

- Cross bracing

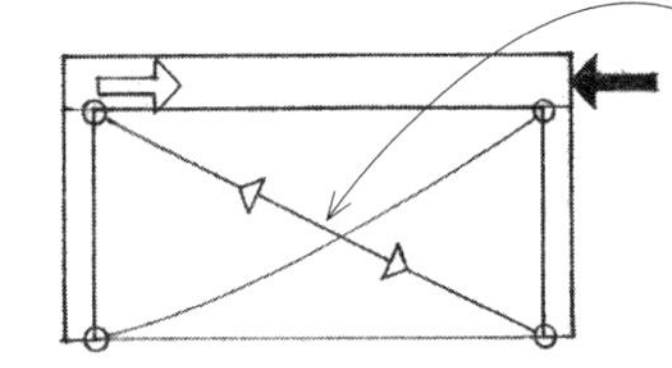

- When using cable bracing, two are necessary to stabilize the structure against lateral forces from either direction. For each direction, one cable will operate effectively in tension while the other would simply buckle. If rigid bracing is used, a certain degree of redundancy is involved because a single member is capable of stabilizing the structure.

Any of these systems may be used singly or in combination to stabilize a structure. Of the three vertical systems, a rigid frame tends to be the least efficient. However, rigid frames can be useful when employing braced frames or shear walls would form undesired barriers between adjacent spaces.

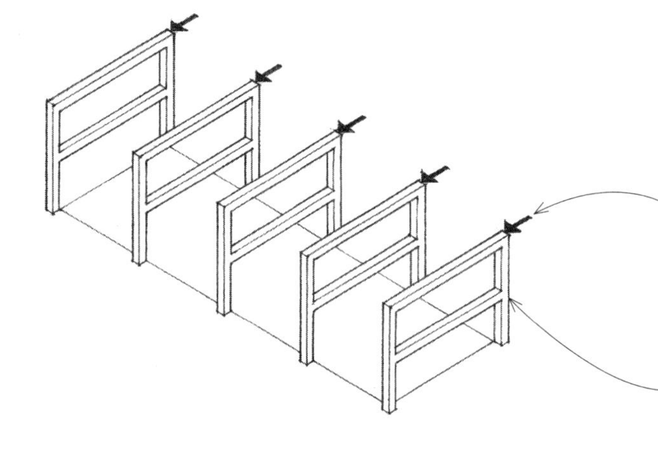

Lateral forces tend to be more critical in the short direction of rectangular buildings, and more efficient shear walls or braced frames are typically used in this direction. In the long direction, any of the lateral force–resisting elements may be used.

- Bents are braced or rigid frames designed to carry vertical and lateral loads transverse to the length of a framed structure.

To avoid destructive torsional effects, structures subject to lateral forces should be arranged and braced symmetrically with centers of mass and resistance as coincident as possible. The asymmetrical layout of irregular structures generally requires dynamic analysis in order to determine the torsional effects of lateral forces.

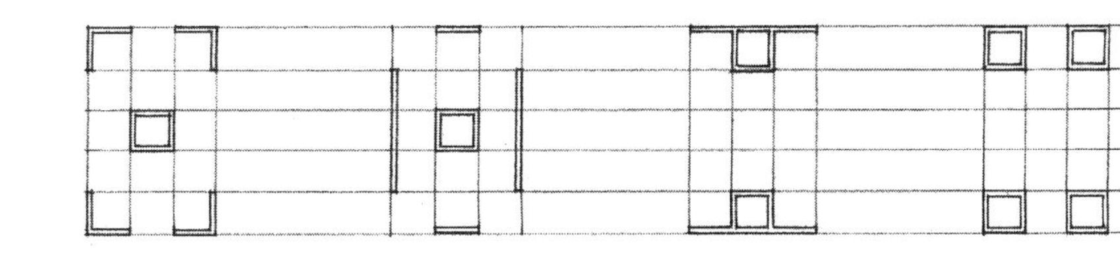

Irregular structures are characterized by any of various plan or vertical irregularities, such as the asymmetrical layout of mass or lateral force–resisting elements, a soft or weak story, or a discontinuous shear wall or diaphragm.

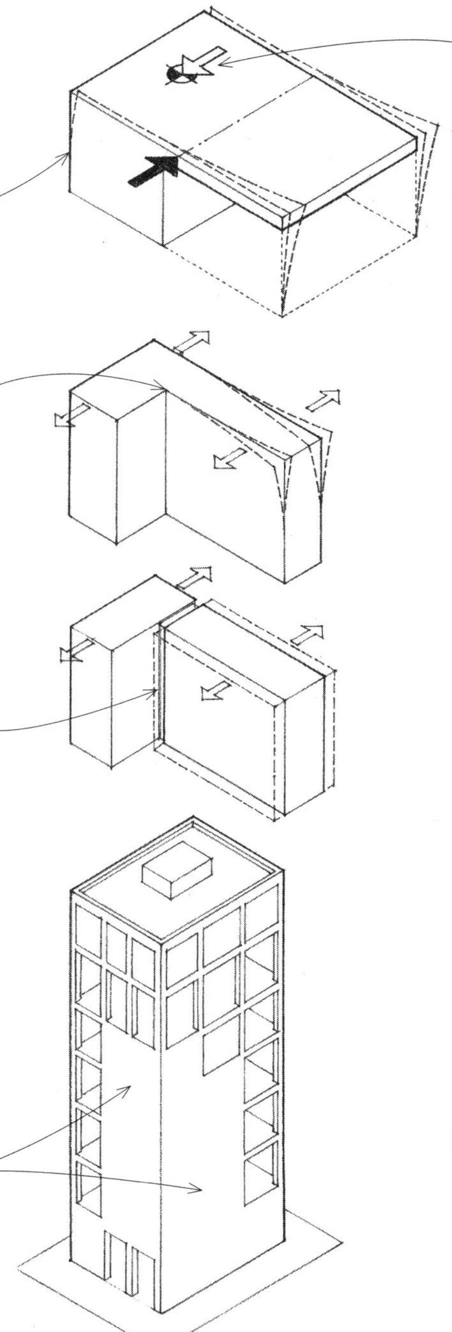

- Torsional irregularity refers to the asymmetrical layout of mass or lateral force–resisting elements, resulting in noncoincident centers of mass and resistance.
- The center of resistance is the centroid of the vertical elements of a lateral force–resisting system, through which the shear reaction to lateral forces acts.
- A reentrant corner is a plan configuration of a structure having projections beyond a corner significantly greater than the plan dimension in the given direction. A re-entrant corner tends to produce differential motions between different portions of the structure, resulting in local stress concentrations at the corner. Solutions include providing a seismic joint to separate the building into simpler shapes, tying the building together more strongly at the corner, or splaying the corner.
- Seismic joints physically separate adjacent building masses so that free vibratory movement in each can occur independently of the other.

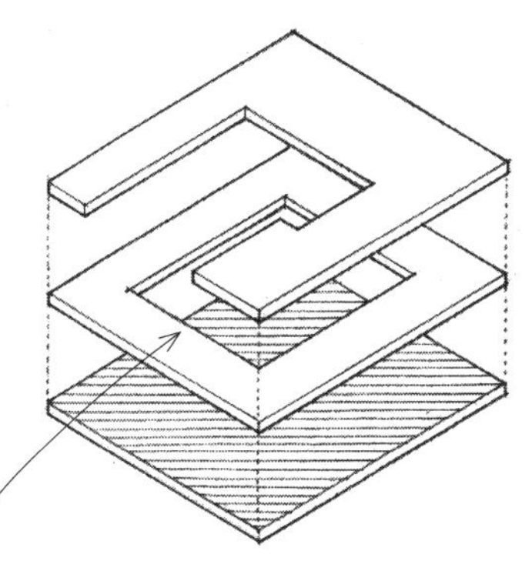

- A discontinuous diaphragm is a horizontal diaphragm having a large cutout or open area, or a stiffness significantly less than that of the story above or below.
- A discontinuous shear wall has a large offset or a significant change in horizontal dimension.

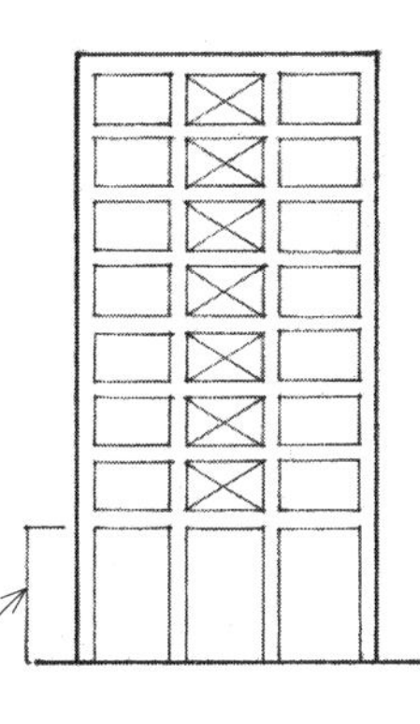

- A soft or weak story has lateral stiffness or strength significantly less than that of the stories above.

CSI MasterFormat 13 48 00: Sound, Vibration, and Seismic Control

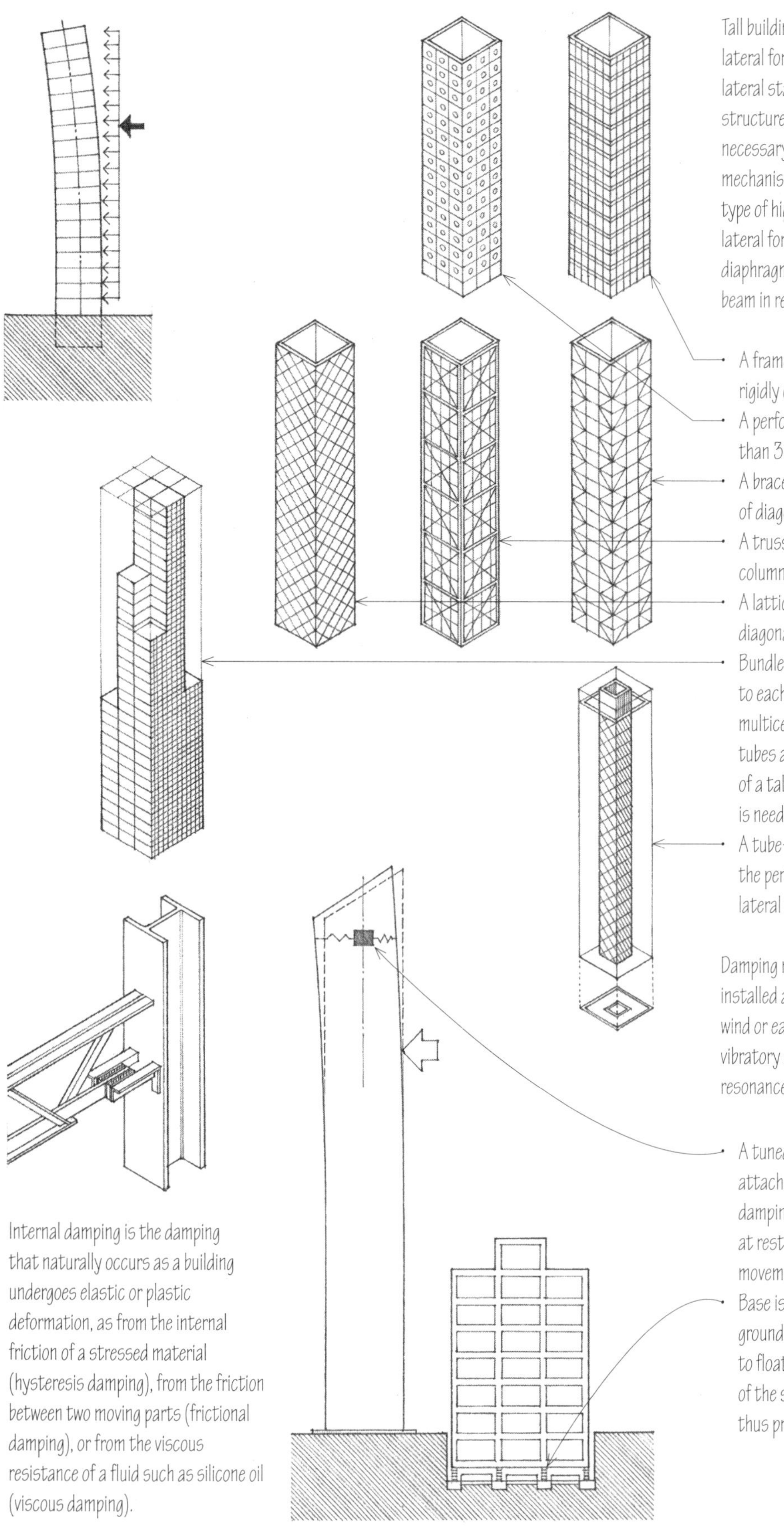

Tall buildings are particularly susceptible to the effects of lateral forces. A rigid frame is the least efficient way to achieve lateral stability and is appropriate only for low- to medium-rise structures. As the height of a building increases, it becomes necessary to supplement a rigid frame with additional bracing mechanisms, such as diagonal bracing or a rigid core. An efficient type of high-rise structure is a tube structure that has perimeter lateral force–resisting systems internally braced by rigid floor diaphragms. The structure acts essentially as a cantilevered box beam in resisting lateral forces.

- A framed tube has closely spaced perimeter columns rigidly connected by deep spandrel beams.
- A perforated shell tube has perimeter shear walls with less than 30% of the surface area perforated by openings.
- A braced tube is a framed structure tied together by a system of diagonal braces.
- A trussed tube has trussed wall frames of widely spaced columns tied together by diagonal or cross bracing.
- A latticed truss tube has perimeter frames of closely spaced diagonals with no vertical columns.
- Bundled tubes is an assembly of narrow tubes tied directly to each other to form a modular structure that behaves like a multicellular box girder cantilevering out of the ground. More tubes are sometimes provided in the lower portion of a tall structure where greater lateral force resistance is needed.
- A tube-in-tube structure has an inner braced core added to the perimeter tube to improve its shear stiffness in resisting lateral forces.

Damping mechanisms are viscoelastic devices that are typically installed at structural joints to absorb the energy generated by wind or earthquake forces, progressively diminish or eliminate vibratory or oscillatory motions, and prevent destructive resonances from occurring.

- Internal damping is the damping that naturally occurs as a building undergoes elastic or plastic deformation, as from the internal friction of a stressed material (hysteresis damping), from the friction between two moving parts (frictional damping), or from the viscous resistance of a fluid such as silicone oil (viscous damping).
- A tuned mass damper is a heavy mass mounted on rollers and attached to the upper portion of a tall building with spring damping mechanisms, having an inertial tendency to remain at rest and thus counteracting and dissipating any building movements.
- Base isolation refers to isolating the base of a building from the ground with damping mechanisms to allow the superstructure to float as a rigid body and alter the natural period of vibration of the structure so that it is different from that of the ground, thus preventing destructive resonances from occurring.

Columns, beams, slabs, and bearing walls are the most common structural elements because of the rectilinear building geometry they are capable of generating. There are, however, other means of spanning and enclosing space. These are generally form-active elements that, through their shape and geometry, make efficient use of their material for the distances spanned. While beyond the scope of this book, they are briefly described in the following section.

Arches are curved structures for spanning an opening, designed to support a vertical load primarily by axial compression. They transform the vertical forces of a supported load into inclined components and transmit them to abutments on either side of the archway.

- Masonry arches are constructed of individual wedge-shaped stone or brick voussoirs; for more information on masonry arches, see 5.20.
- Rigid arches consist of curved, rigid structures of timber, steel, or reinforced concrete capable of carrying some bending stresses.

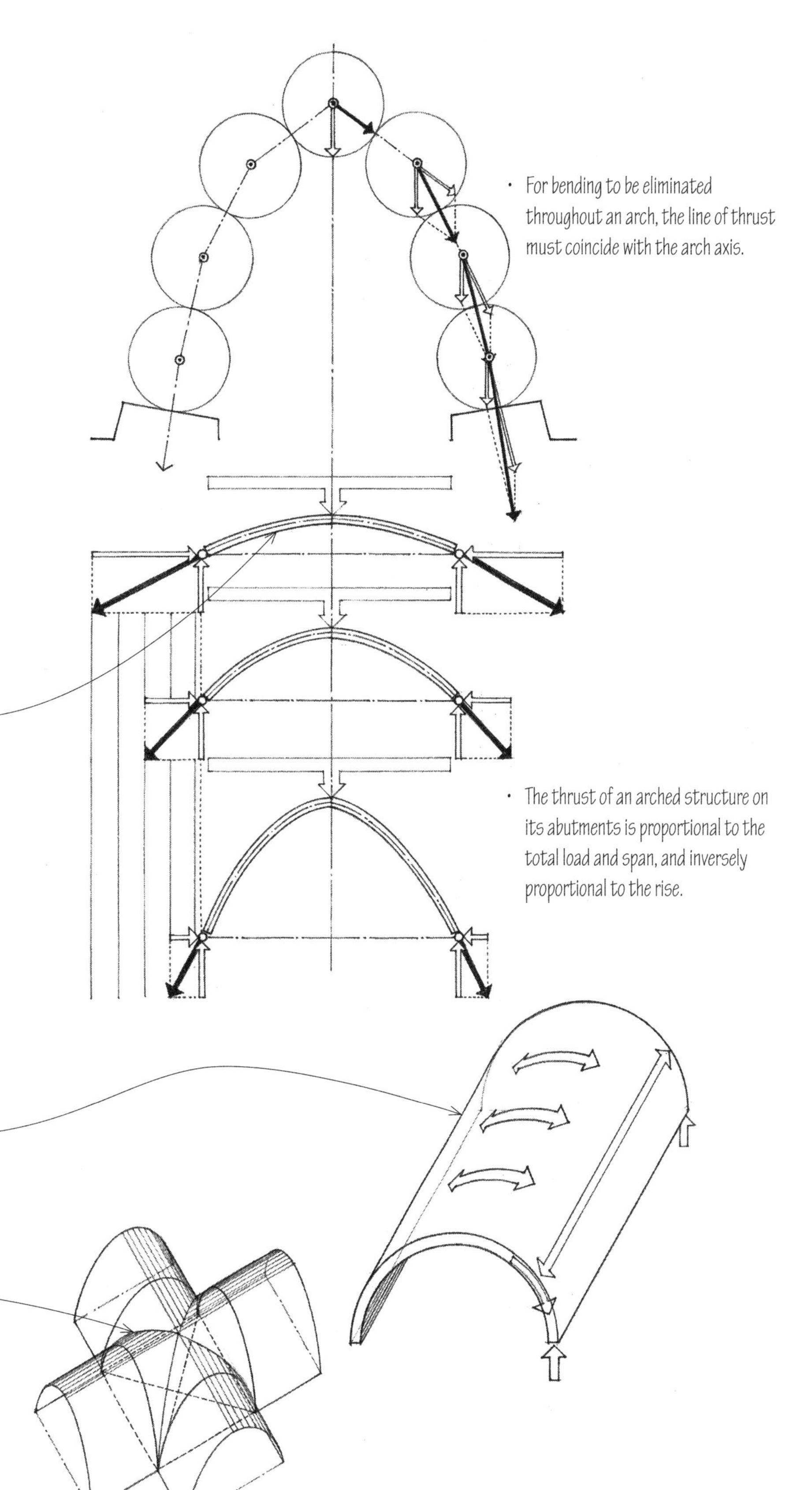

- For bending to be eliminated throughout an arch, the line of thrust must coincide with the arch axis.
- The thrust of an arched structure on its abutments is proportional to the total load and span, and inversely proportional to the rise.

Vaults are arched structures of stone, brick, or reinforced concrete, forming a ceiling or roof over a hall, room, or other wholly or partially enclosed space. Because a vault behaves as an arch extended in a third dimension, the longitudinal supporting walls must be buttressed to counteract the outward thrusts of the arching action.

- Barrel vaults have semicircular cross sections.

- Groin or cross vaults are compound vaults formed by the perpendicular intersection of two vaults, forming arched diagonal arrises called groins.

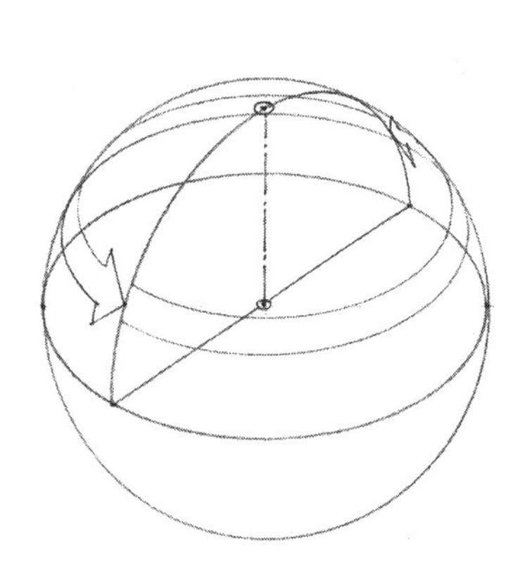

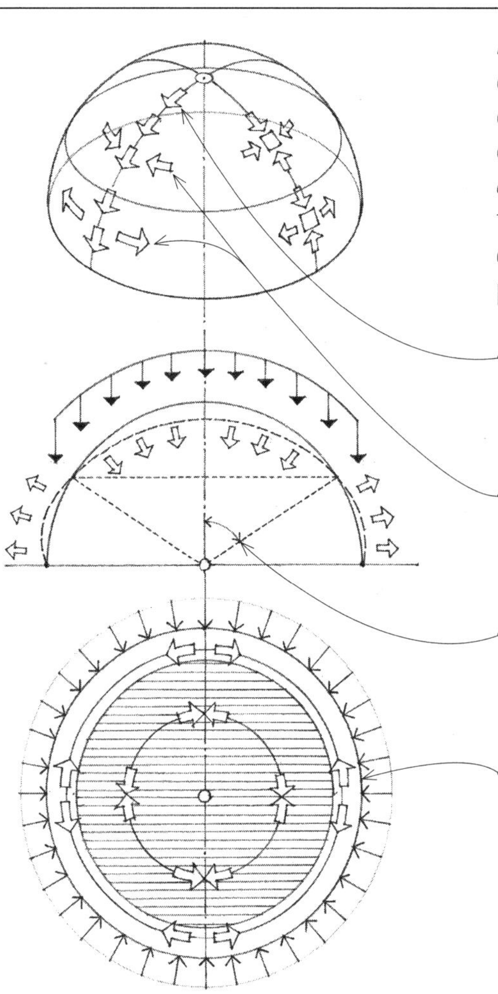

A dome is a spherical surface structure having a circular plan and constructed of stacked blocks, a continuous rigid material like reinforced concrete, or of short, linear elements, as in the case of a geodesic dome. A dome is similar to a rotated arch except that circumferential forces are developed that are compressive near the crown and tensile in the lower portion.

- Meridional forces acting along a vertical section cut through the surface of the dome are always compressive under full vertical loading.
- Hoop forces, restraining the out-of-plane movement of the meridional strips in the shell of a dome, are compressive in the upper zone and tensile in the lower zone.
- The transition from compressive hoop forces to tensile hoop forces occurs at an angle of from 45° to 60° from the vertical axis.
- A tension ring encircles the base of a dome to contain the outward components of the meridional forces. In a concrete dome, this ring is thickened and reinforced to handle the bending stresses caused by the differing elastic deformations of the ring and shell.

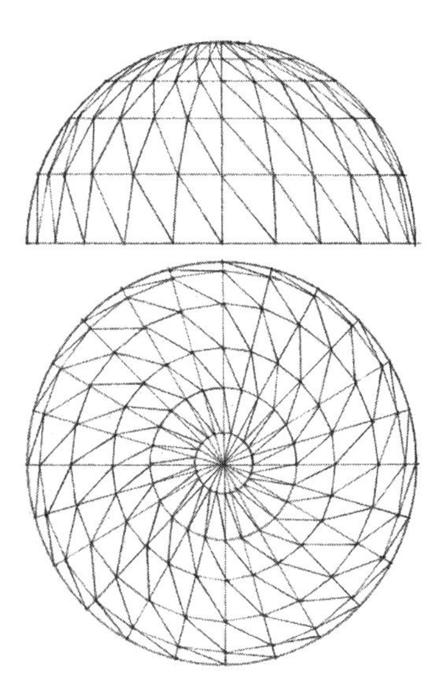

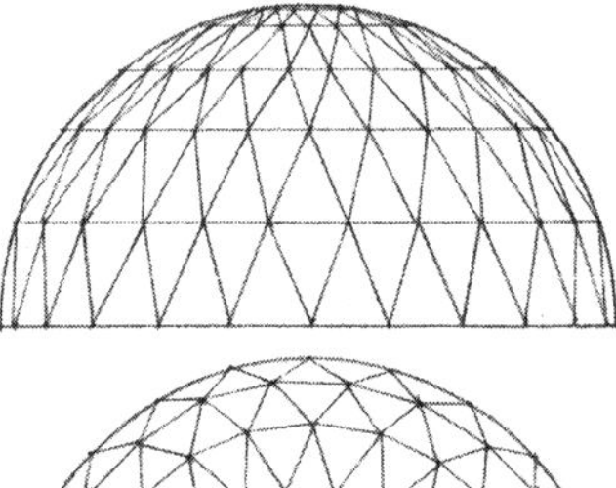

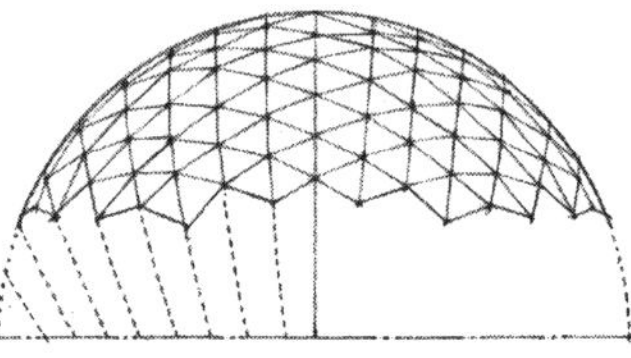

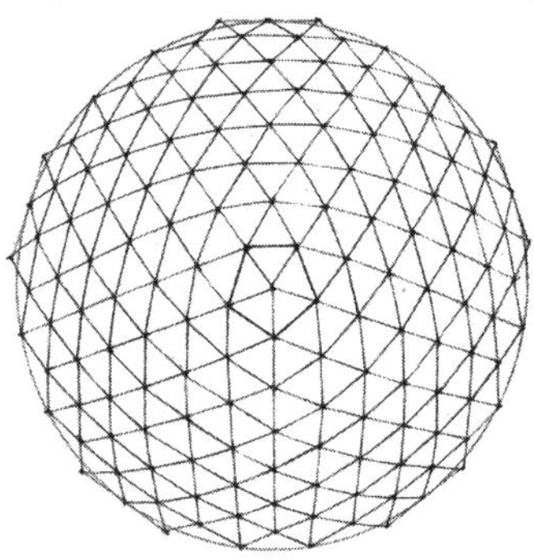

- Schwedler domes are steel dome structures having members that follow the lines of latitude and longitude, and a third set of diagonals completing the triangulation.
- Lattice domes are steel dome structures having members that follow the circles of latitude, and two sets of diagonals forming a series of isosceles triangles.
- Geodesic domes are steel dome structures having members that follow three principal sets of great circles intersecting at 60°, subdividing the dome surface into a series of equilateral spherical triangles.

CSI MasterFormat 13 33 00: Geodesic Structures

Shells are thin, curved plate structures typically constructed of reinforced concrete. They are shaped to transmit applied forces by membrane stresses—the compressive, tensile, and shear stresses acting in the plane of their surfaces. A shell can sustain relatively large forces if uniformly applied. Because of its thinness, however, a shell has little bending resistance and is unsuitable for concentrated loads.

- Translational surfaces are generated by sliding a plane curve along a straight line or over another plane curve.

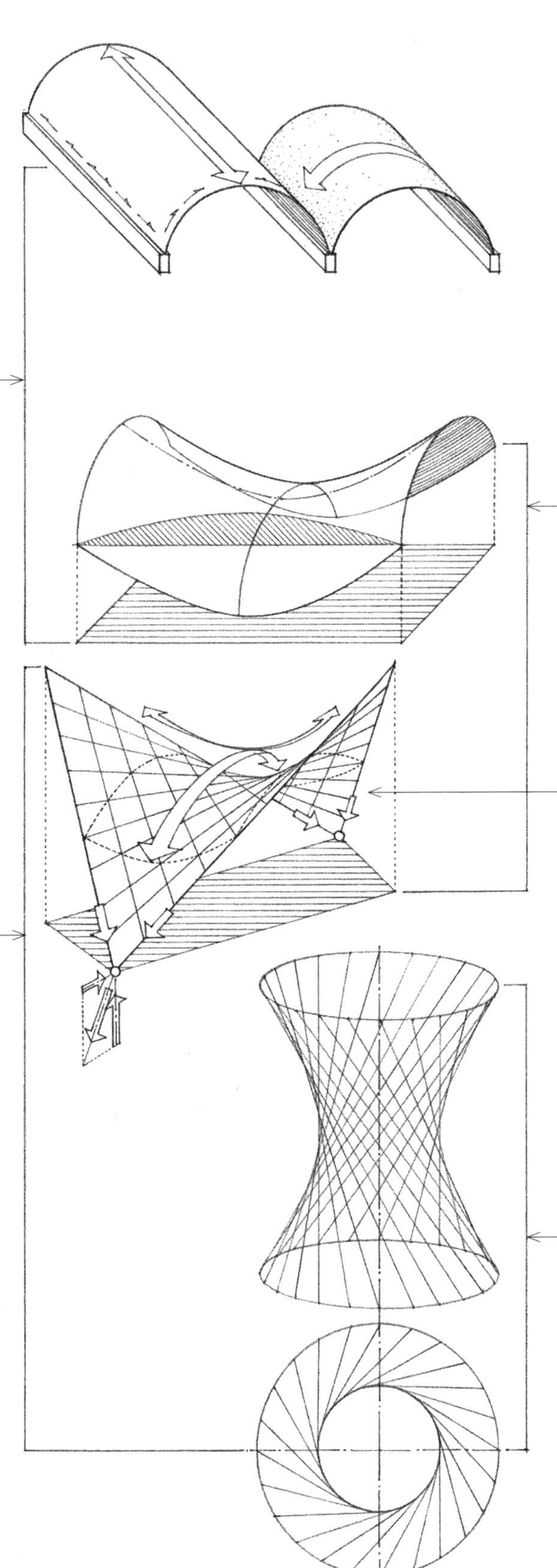

- Barrel shells are cylindrical shell structures. If the length of a barrel shell is three or more times its transverse span, it behaves as a deep beam with a curved section spanning in the longitudinal direction. If it is relatively short, it exhibits archlike action. Tie rods or transverse rigid frames are required to counteract the outward thrusts of the arching action.

A hyperbolic paraboloid is a surface generated by sliding a parabola with downward curvature along a parabola with upward curvature, or by sliding a straight line segment with its ends on two skew lines. It can be considered to be both a translational and a ruled surface.

- Ruled surfaces are generated by the motion of a straight line. Because of its straight-line geometry, a ruled surface is generally easier to form and construct than a rotational or translational surface.

Saddle surfaces have an upward curvature in one direction and a downward curvature in the perpendicular direction. In a saddle-surfaced shell structure, regions of downward curvature exhibit archlike action, while regions of upward curvature behave as a cable structure. If the edges of the surface are not supported, beam behavior may also be present.

A one-sheet hyperboloid is a ruled surface generated by sliding an inclined line segment on two horizontal circles. Its vertical sections are hyperbolas.

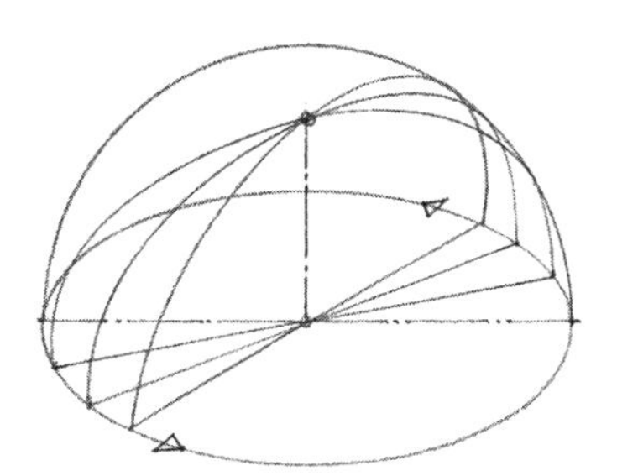

- Rotational surfaces are generated by rotating a plane curve about an axis. Spherical, elliptical, and parabolic dome surfaces are examples of rotational surfaces.

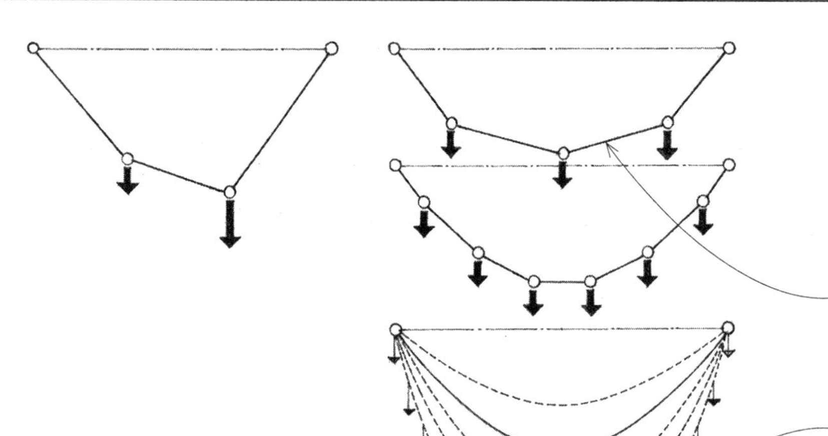

Cable structures utilize the cable as the principal means of support. Because cables have high tensile strength but offer no resistance to compression or bending, they must be used purely in tension. When subject to concentrated loads, the shape of a cable consists of straight-line segments. Under a uniformly distributed load, it will take on the shape of an inverted arch.

- A funicular shape is the shape assumed by a freely deforming cable in direct response to the magnitude and location of external forces. A cable always adapts its shape so that it is in pure tension under the action of an applied load.
- A catenary is the curve assumed by a perfectly flexible, uniform cable suspended freely from two points not in the same vertical line. For a load that is uniformly distributed in a horizontal projection, the curve approaches that of a parabola.

Suspension structures utilize a network of cables suspended and prestressed between compression members to directly support applied loads.

- Guy cables absorb the horizontal component of thrust in a suspension or cable-stayed structure and transfer the force to a ground foundation.
- The mast is a vertical or inclined compression member in a suspension or cable-stayed structure, supporting the sum of the vertical force components in the primary and guy cables. Inclining the mast enables it to pick up some of the horizontal cable thrust and reduces the force in the guy cables.

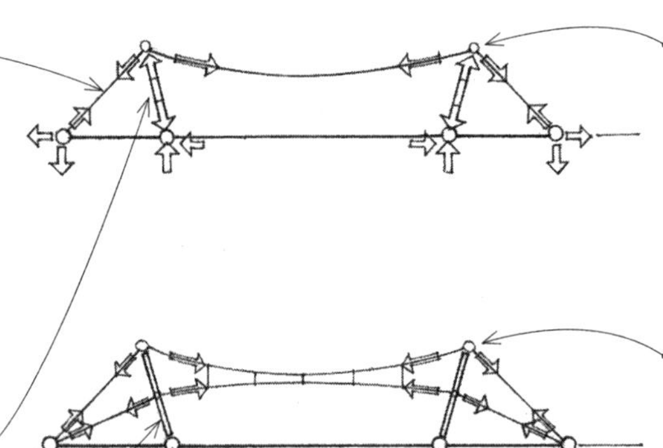

- Single-curvature structures utilize a parallel series of cables to support surface-forming beams or plates. They are susceptible to flutter induced by the aerodynamic effects of wind. This liability can be reduced by increasing the dead load on the structure or by anchoring the primary cables to the ground with transverse guy cables.
- Double-cable structures have upper and lower sets of cables of different curvatures, pretensioned by ties or compression struts to make the system more rigid and resistant to flutter.
- Double-curvature structures consist of a field of crossed cables of different and often reverse curvatures. Each set of cables has a different natural period of vibration, thus forming a self-dampening system that is more resistant to flutter.

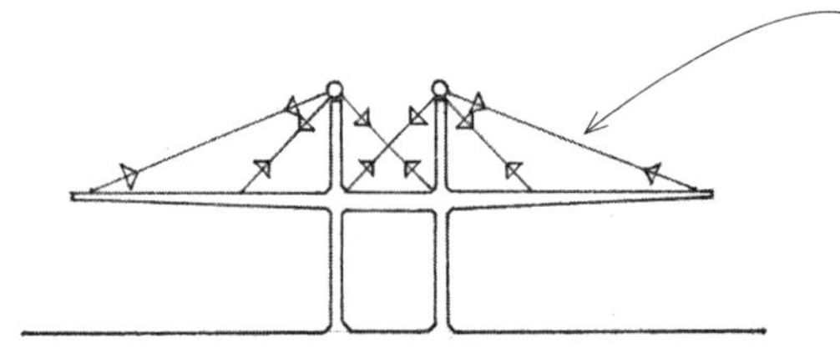

- Cable-stayed structures have vertical or inclined masts from which cables extend to support horizontally spanning members arranged in a parallel or radial pattern.

CSI MasterFormat 13 31 23: Tensioned Fabric Structures

Membranes are thin, flexible surfaces that carry loads primarily through the development of tensile stresses. They may be suspended or stretched between posts, or be supported by air pressure.

Tent structures are membrane structures that are prestressed by externally applied forces and held completely taut under all anticipated load conditions. To avoid extremely high tensile forces, membrane structures should have relatively sharp curvatures in opposite directions.

- Membrane and steel cables transmit external loads to masts and ground anchors by means of tensile forces.
- Reinforcing edge cables stiffen the free edges of a tent structure.
- The membrane may be tied to the mast supports by a reinforcing cable loop or be stretched over a distribution cap.
- The masts are designed to resist buckling under compressive loading.

Pneumatic structures are membrane structures that are placed in tension and stabilized against wind and snow loads by the pressure of compressed air. The membrane is usually a woven textile or glass-fiber fabric coated with a synthetic material such as silicone. Translucent membranes provide natural illumination, gather solar radiation in the winter, and cool the interior space at night. Reflective membranes reduce solar heat gain. A fabric liner can capture air space to improve the thermal resistance of the structure.

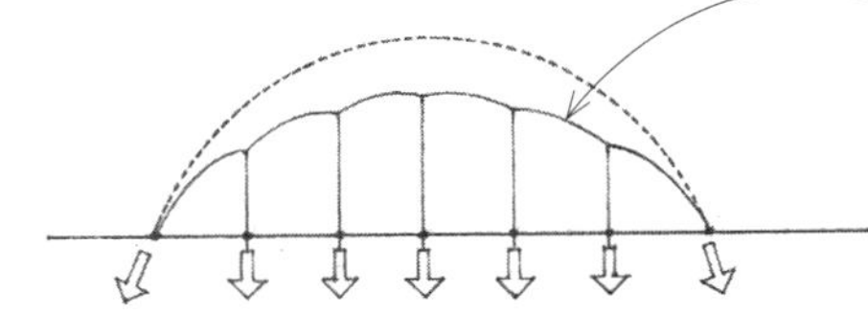

- Some air-supported structures use a net of cables placed in tension by the inflating force to restrain the membrane from developing its natural inflated profile.

There are two kinds of pneumatic structures: air-supported structures and air-inflated structures.

- Air-supported structures consist of a single membrane supported by an internal air pressure slightly higher than normal atmospheric pressure, and securely anchored and sealed along the perimeter to prevent leaking. Air locks are required at entrances to maintain the internal air pressure.
- Air-inflated structures are supported by pressurized air within inflated building elements. These elements are shaped to carry loads in a traditional manner, while the enclosed volume of building air remains at normal atmospheric pressure. The tendency for a double-membrane structure to bulge in the middle is restrained by a compression ring or by internal ties or diaphragms.

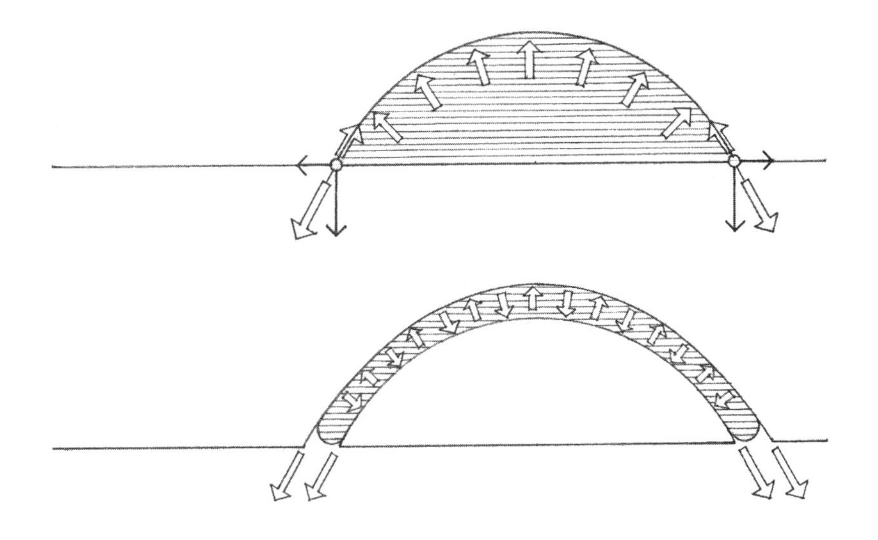

CSI MasterFormat 13 31 13: Air-Supported Fabric Structures

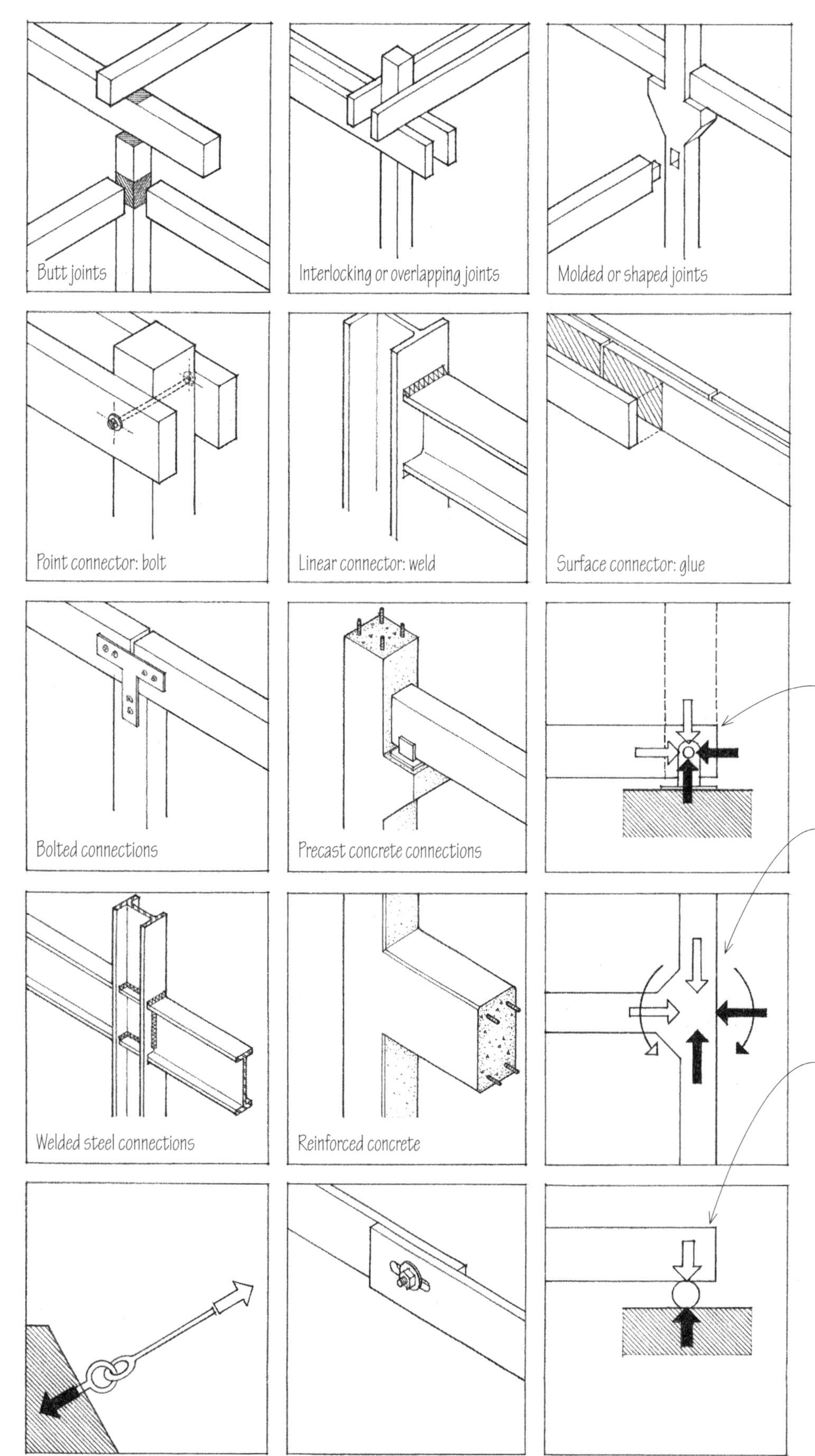

The manner in which forces are transferred from one structural element to the next and how a structural system performs as a whole depend to a great extent on the types of joints and connections used. Structural elements can be joined to each other in three ways. Butt joints allow one of the elements to be continuous and usually require a third mediating element to make the connection. Overlapping joints allow all of the connected elements to bypass each other and be continuous across the joint. The joining elements can also be molded or shaped to form a structural connection.

The connectors used to join the structural elements may be in the form of a point, a line, or a surface. While linear and surface types of connectors resist rotation, point connectors do not unless a series of them is distributed across a large surface area.

- Pinned joints theoretically allow rotation but resist translation in any direction.
- Rigid or fixed joints maintain the angular relationship between the joined elements, restrain rotation and translation in any direction, and provide both force and moment resistance.
- Roller joints allow rotation but resist translation in a direction perpendicular into or away from their faces. They are not employed in building construction as often as pinned or fixed connections but they are useful when a joint must allow expansion and contraction of a structural element to occur.
- A cable anchorage allows rotation but resists translation only in the direction of the cable.

3

FOUNDATION SYSTEMS

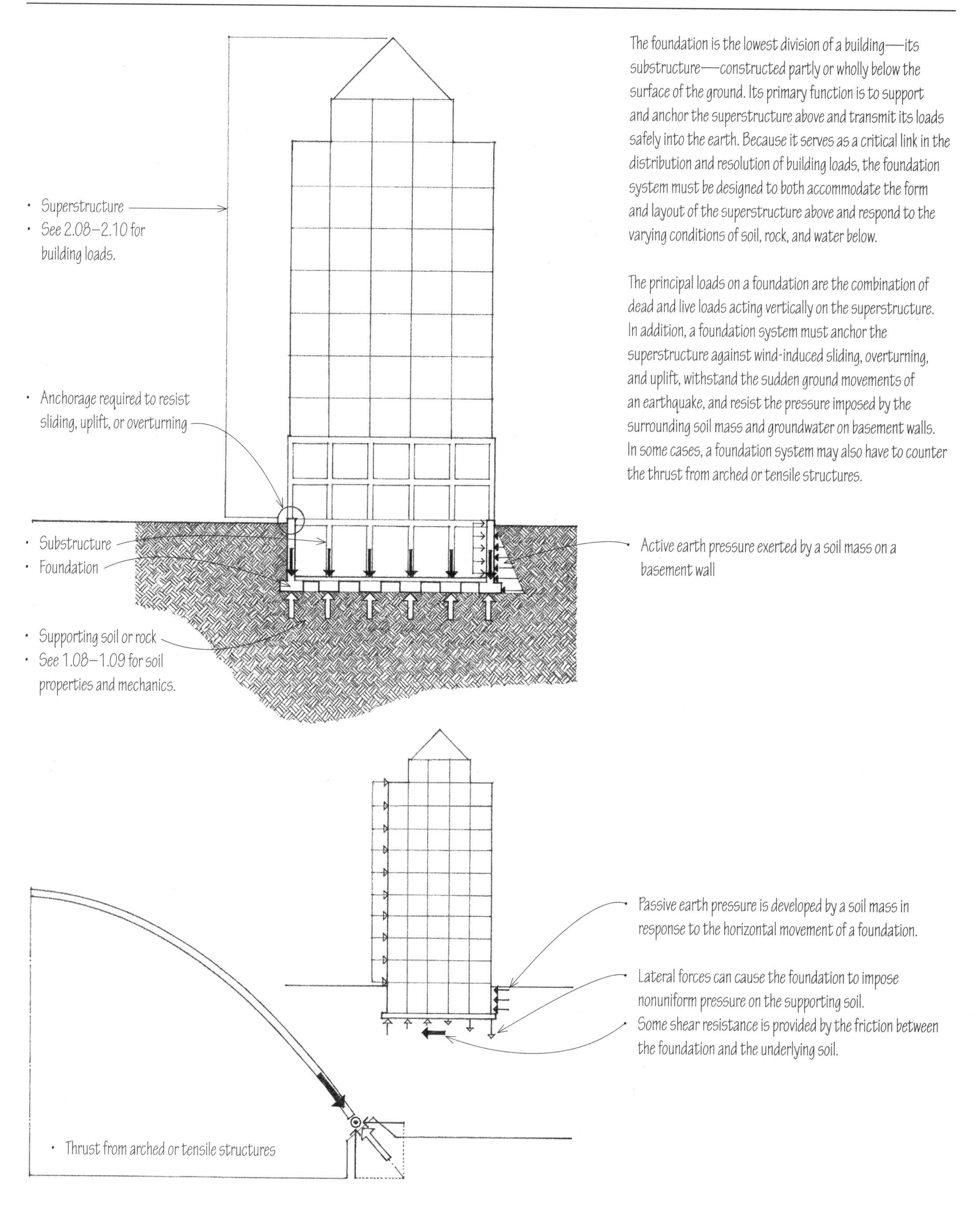

The foundation is the lowest division of a building—its substructure—constructed partly or wholly below the surface of the ground. Its primary function is to support and anchor the superstructure above and transmit its loads safely into the earth. Because it serves as a critical link in the distribution and resolution of building loads, the foundation system must be designed to both accommodate the form and layout of the superstructure above and respond to the varying conditions of soil, rock, and water below.

The principal loads on a foundation are the combination of dead and live loads acting vertically on the superstructure. In addition, a foundation system must anchor the superstructure against wind-induced sliding, overturning, and uplift, withstand the sudden ground movements of an earthquake, and resist the pressure imposed by the surrounding soil mass and groundwater on basement walls. In some cases, a foundation system may also have to counter the thrust from arched or tensile structures.

Settlement is the gradual subsiding of a structure as the soil beneath its foundation consolidates under loading. As a building is constructed, some settlement is to be expected as the load on the foundation increases and causes a reduction in the volume of soil voids containing air or water. This consolidation is usually slight and occurs rather quickly as loads are applied on dense, granular soils, such as coarse sand and gravel. When the foundation soil is a moist, cohesive clay, which has a scale-like structure and a relatively large percentage of voids, consolidation can be quite large and occur slowly over a longer period of time.

A properly designed and constructed foundation system should distribute its loads so that whatever settlement occurs is minimal or is uniformly distributed under all portions of the structure. This is accomplished by laying out and proportioning the foundation supports so that they transmit an equal load per unit area to the supporting soil or rock without exceeding its bearing capacity.

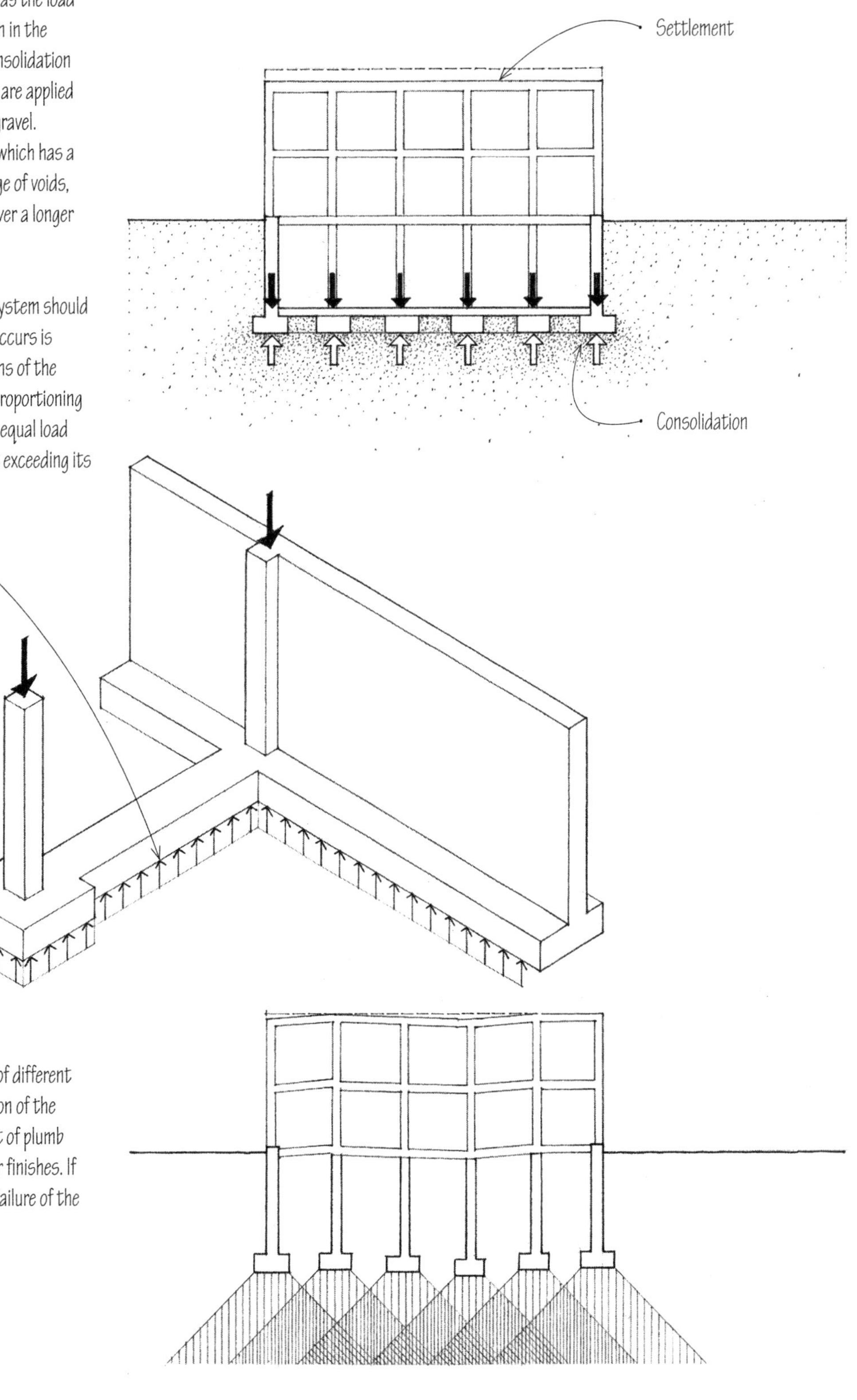

Differential settlement—the relative movement of different parts of a structure caused by uneven consolidation of the foundation soil—can cause a building to shift out of plumb and cracks to occur in its foundation, structure, or finishes. If extreme, differential settlement can result in the failure of the structural integrity of a building.

Foundations utilize a combination of bearing walls, columns, and piers to transmit building loads directly to the earth. These structural elements can form various types of substructures:

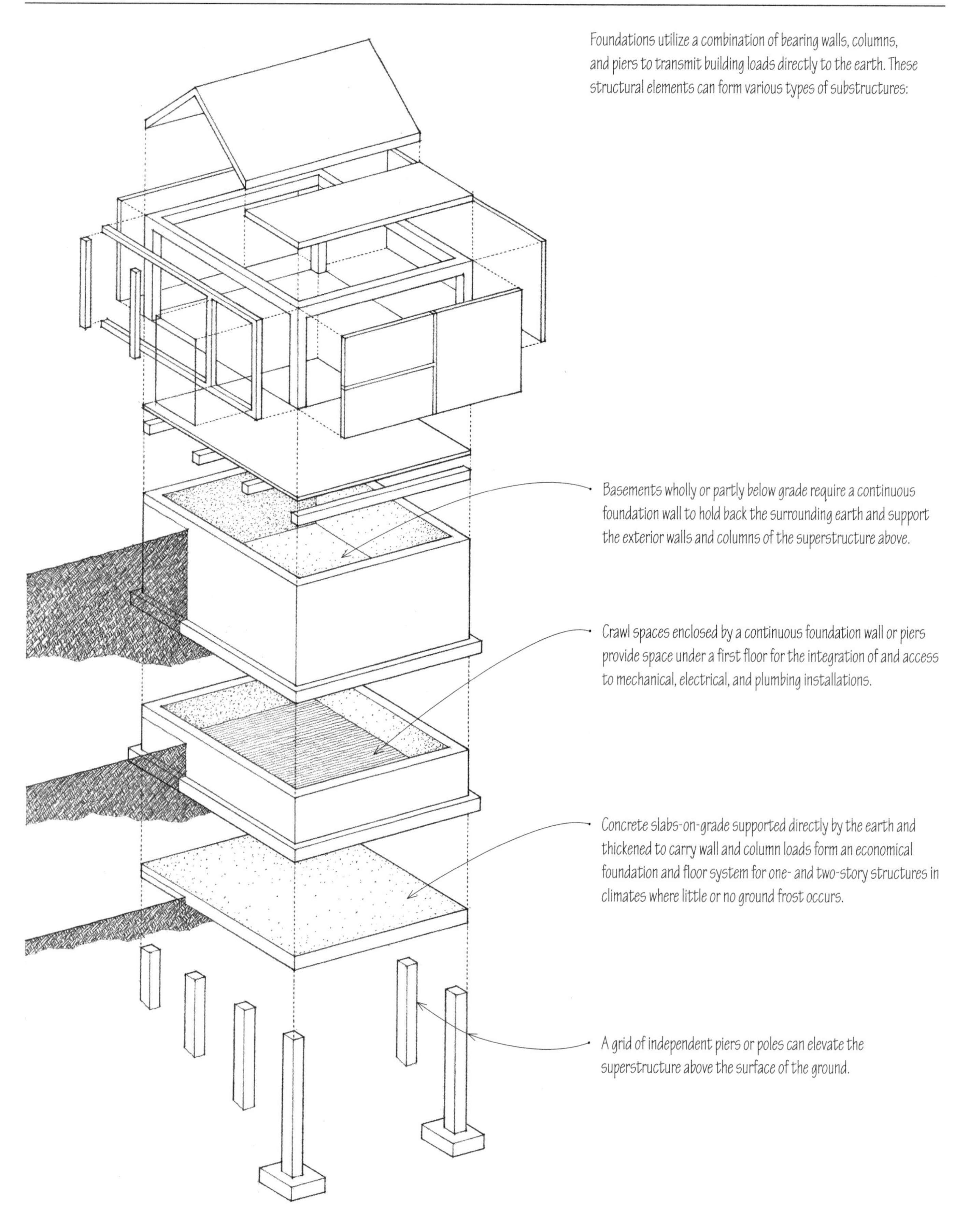

We can classify foundation systems into two broad categories—shallow foundations and deep foundations.

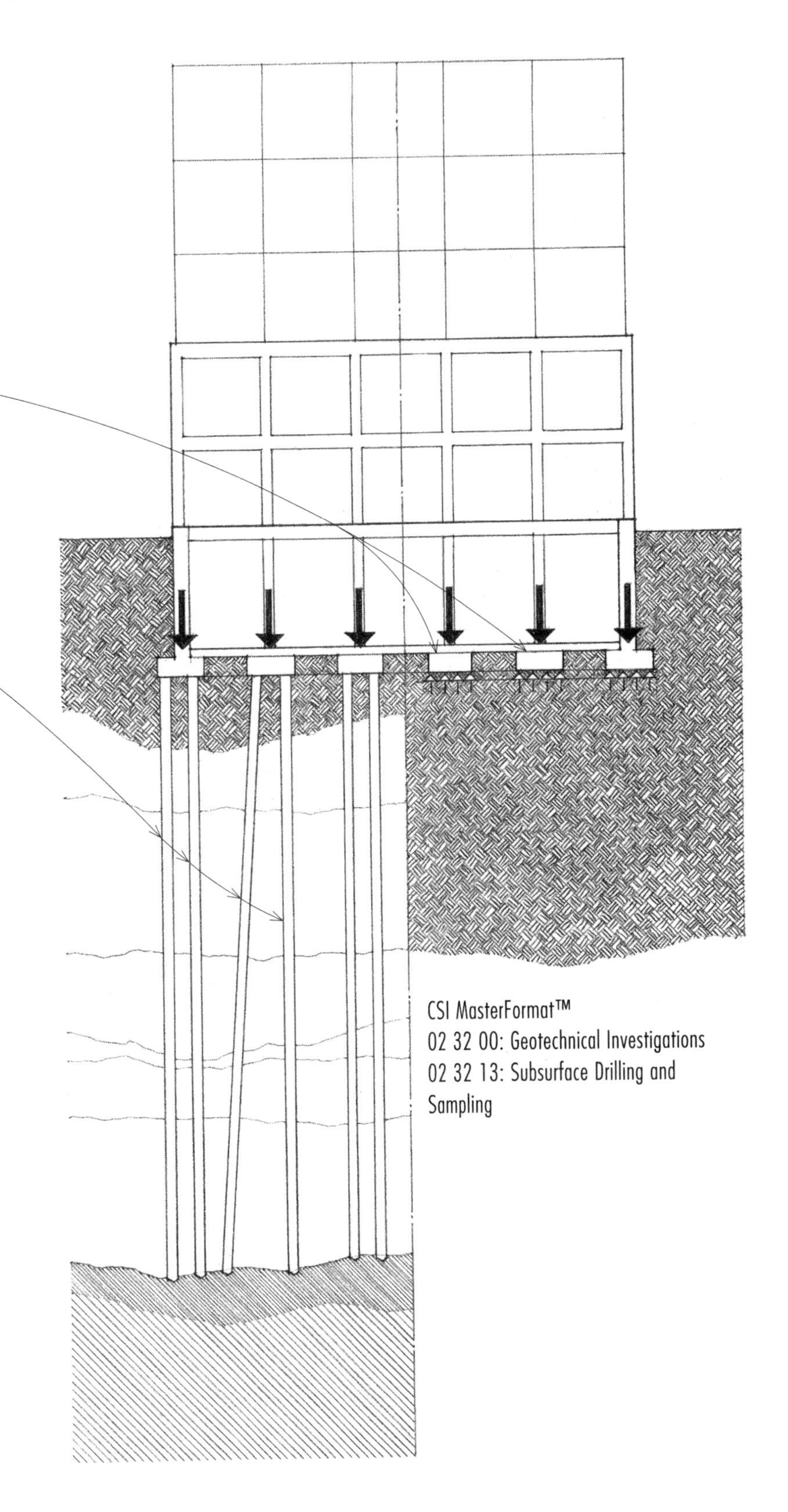

Shallow Foundations

Shallow or spread foundations are employed when stable soil of adequate bearing capacity occurs relatively near to the ground surface. They are placed directly below the lowest part of a substructure and transfer building loads directly to the supporting soil by vertical pressure.

Deep Foundations

Deep foundations are employed when the soil underlying a foundation is unstable or of inadequate bearing capacity. They extend down through unsuitable soil to transfer building loads to a more appropriate bearing stratum of rock or dense sands and gravels well below the superstructure.

Factors to consider in selecting and designing the type of foundation system for a building include:

- Pattern and magnitude of building loads
- Subsurface and groundwater conditions
- Topography of the site
- Impact on adjacent properties
- Building code requirements
- Construction method and risk

The design of a foundation system requires professional analysis and design by a qualified structural engineer. When designing anything other than a single-family dwelling on stable soil, it is also advisable to have a geotechnical engineer undertake a subsurface investigation in order to determine the type and size of foundation system required for the building design.

CSI MasterFormat™
02 32 00: Geotechnical Investigations
02 32 13: Subsurface Drilling and Sampling

Underpinning refers to the process of rebuilding or strengthening the foundation of an existing building, or extending it when a new excavation in adjoining property is deeper than the existing foundation.

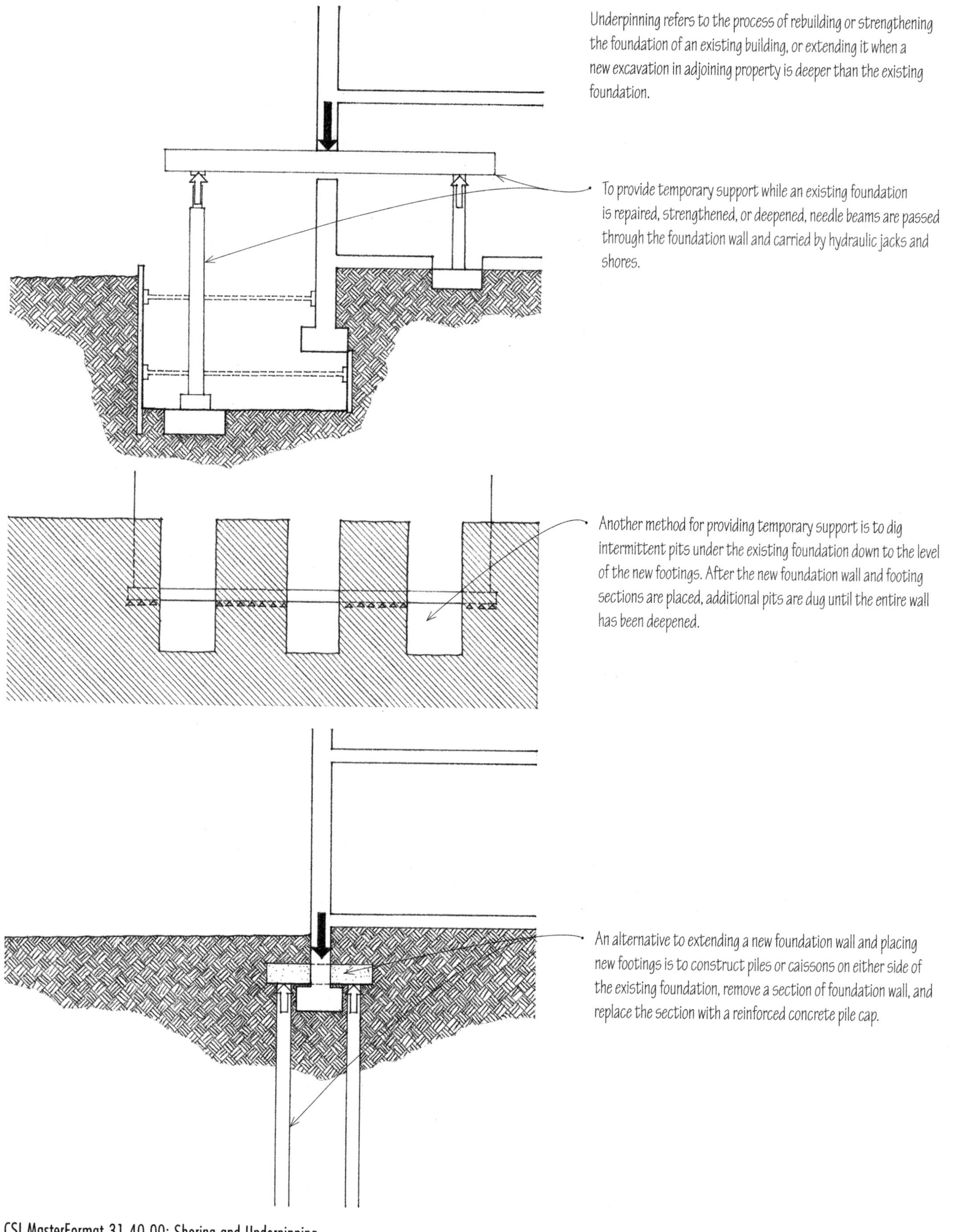

CSI MasterFormat 31 40 00: Shoring and Underpinning

When the building site is sufficiently large that the sides of an excavation can be bench terraced or sloped at an angle less than the angle of repose for the soil, no supporting structure is necessary. When the sides of a deep excavation exceed the angle of repose for the soil, however, the earth must be temporarily braced or shored until the permanent construction is in place.

- Sheet piling consists of timber, steel, or precast concrete planks driven vertically side by side to retain earth and prevent water from seeping into an excavation. Steel and precast concrete sheet piling may be left in place as part of the substructure of a building.

- Soldier piles or beams are steel H-sections driven vertically into the ground to support horizontal lagging.
- Lagging refers to the heavy timber planks joined together side by side to retain the face of an excavation.

- Sheet piling and soldier beams with lagging are supported with continuous horizontal wales braced by horizontal steel crossbracing or by diagonal steel rakers bearing on heel blocks or footings.

- Tiebacks secured to rock or soil anchors (CSI: 31 51 00) may be used if crossbracing or rakers would interfere with the excavation or construction operation. The tiebacks consist of steel cables or tendons that are inserted into holes predrilled through the sheet piling and into rock or a suitable stratum of soil, grouted under pressure to anchor them to the rock or soil, and post-tensioned with a hydraulic jack. The tiebacks are then secured to continuous, horizontal steel wales to maintain the tension.

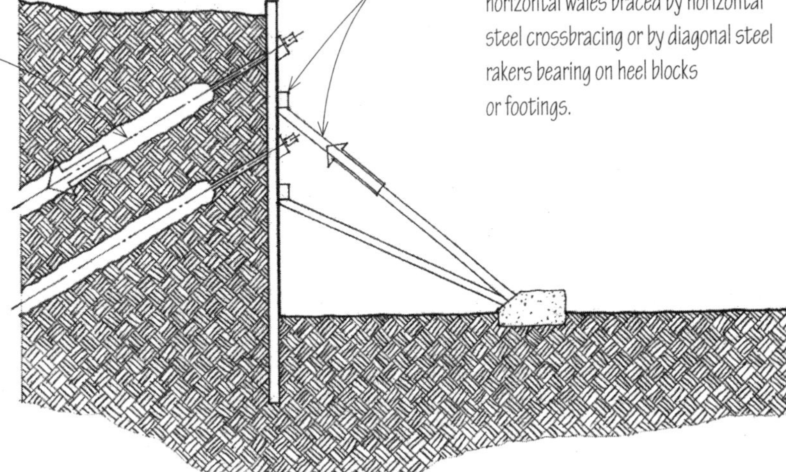

- A slurry wall (CSI: 31 56 00) is a concrete wall cast in a trench to serve as sheeting and often as a permanent foundation wall. It is constructed by excavating a trench in short lengths, filling it with a slurry of bentonite and water to prevent the sidewalls from collapsing, setting reinforcement, and placing concrete in the trench with a tremie to displace the slurry.

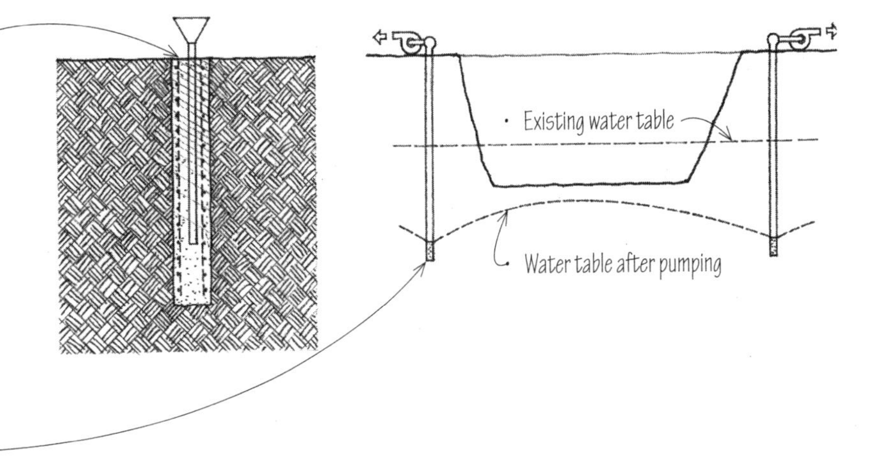

- Dewatering (CSI: 31 23 19) refers to the process of lowering a water table or preventing an excavation from filling with groundwater. It is accomplished by driving perforated tubes called wellpoints into the ground to collect water from the surrounding area so it can be pumped away.

CSI MasterFormat 31 50 00: Excavation Support and Protection

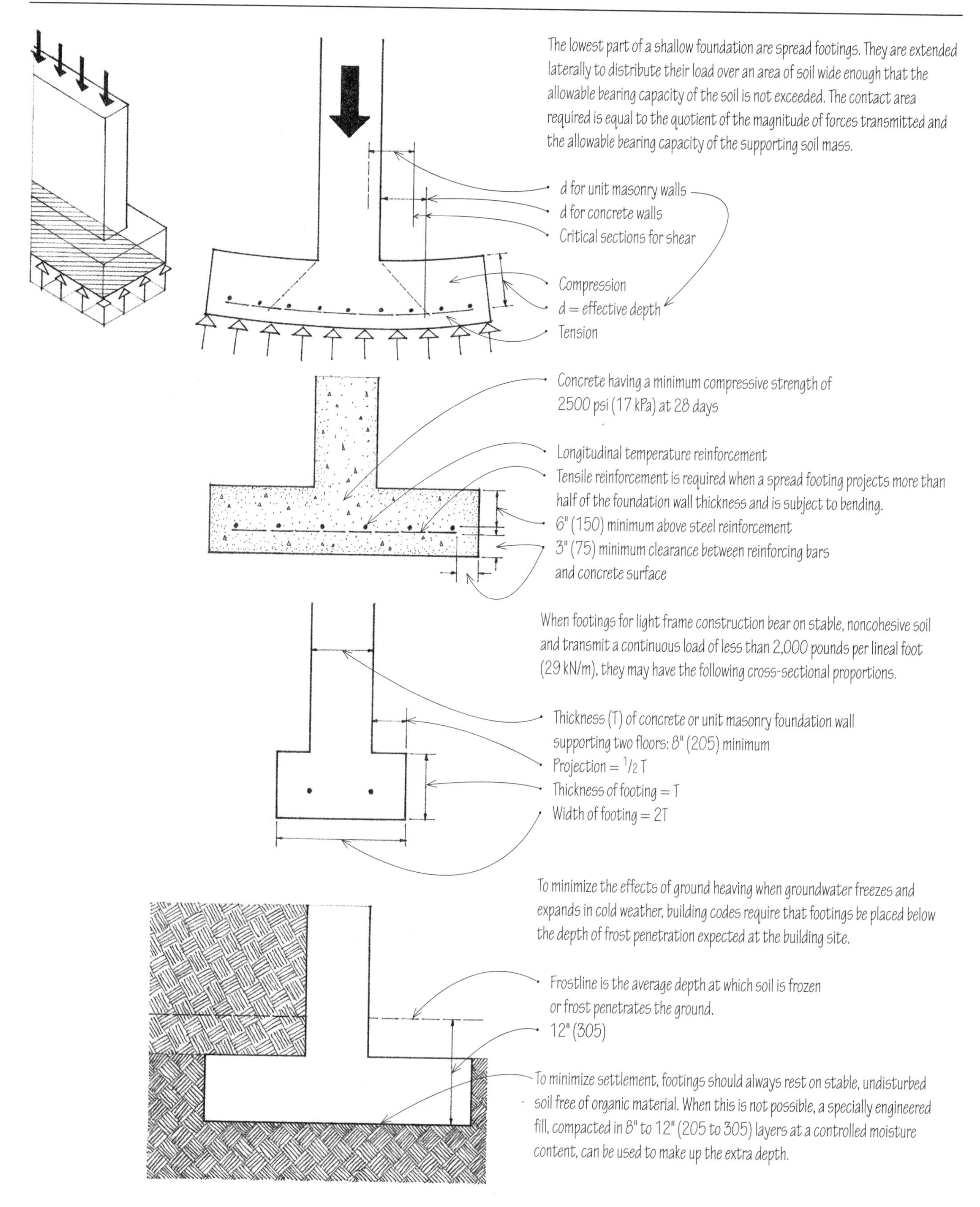
The lowest part of a shallow foundation are spread footings. They are extended laterally to distribute their load over an area of soil wide enough that the allowable bearing capacity of the soil is not exceeded. The contact area required is equal to the quotient of the magnitude of forces transmitted and the allowable bearing capacity of the supporting soil mass.
d for unit masonry walls
d for concrete walls
Critical sections for shear
Compression
d = effective depth
Tension
Concrete having a minimum compressive strength of 2500 psi (17 kPa) at 28 days
Longitudinal temperature reinforcement
Tensile reinforcement is required when a spread footing projects more than half of the foundation wall thickness and is subject to bending.
6" (150) minimum above steel reinforcement
3" (75) minimum clearance between reinforcing bars and concrete surface
When footings for light frame construction bear on stable, noncohesive soil and transmit a continuous load of less than 2,000 pounds per lineal foot (29 kN/m), they may have the following cross-sectional proportions.
Thickness (T) of concrete or unit masonry foundation wall supporting two floors: 8" (205) minimum
Projection = 1/2 T
Thickness of footing = T
Width of footing = 2T
To minimize the effects of ground heaving when groundwater freezes and expands in cold weather, building codes require that footings be placed below the depth of frost penetration expected at the building site.
Frostline is the average depth at which soil is frozen or frost penetrates the ground.
12" (305)
To minimize settlement, footings should always rest on stable, undisturbed soil free of organic material. When this is not possible, a specially engineered fill, compacted in 8" to 12" (205 to 305) layers at a controlled moisture content, can be used to make up the extra depth.

The most common forms of spread footings are strip footings and isolated footings.

- Strip footings are the continuous spread footings of foundation walls.

Other types of spread footings include the following:

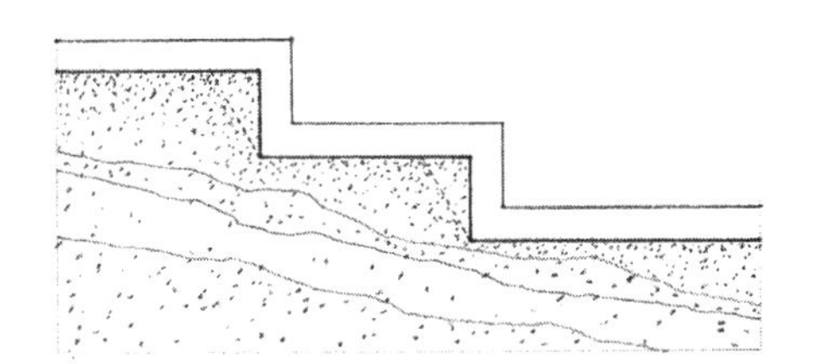

- Stepped footings are strip footings that change levels in stages to accommodate a sloping grade and maintain the required depth at all points around a building.

- A cantilever or strap footing consists of a column footing connected by a tie beam to another footing in order to balance an asymmetrically imposed load.
- A combined footing is a reinforced concrete footing for a perimeter foundation wall or column extended to support an interior column load.
- Cantilever and combined footings are often used when a foundation abuts a property line and it is not possible to construct a symmetrically loaded footing. To prevent the rotation or differential settlement that an asymmetrical loading condition can produce, continuous and cantilever footings are proportioned to generate uniform soil pressure.

- A mat or raft foundation is a thick, heavily reinforced concrete slab that serves as a single monolithic footing for a number of columns or an entire building. Mat foundations are used when the allowable bearing capacity of a foundation soil is low relative to building loads and interior column footings become so large that it becomes more economical to merge them into a single slab. Mat foundations may be stiffened by a grid of ribs, beams, or walls.
- A floating foundation, used in yielding soil, has for its footing a mat placed deep enough that the weight of the excavated soil is equal to or greater than the weight of the construction being supported.

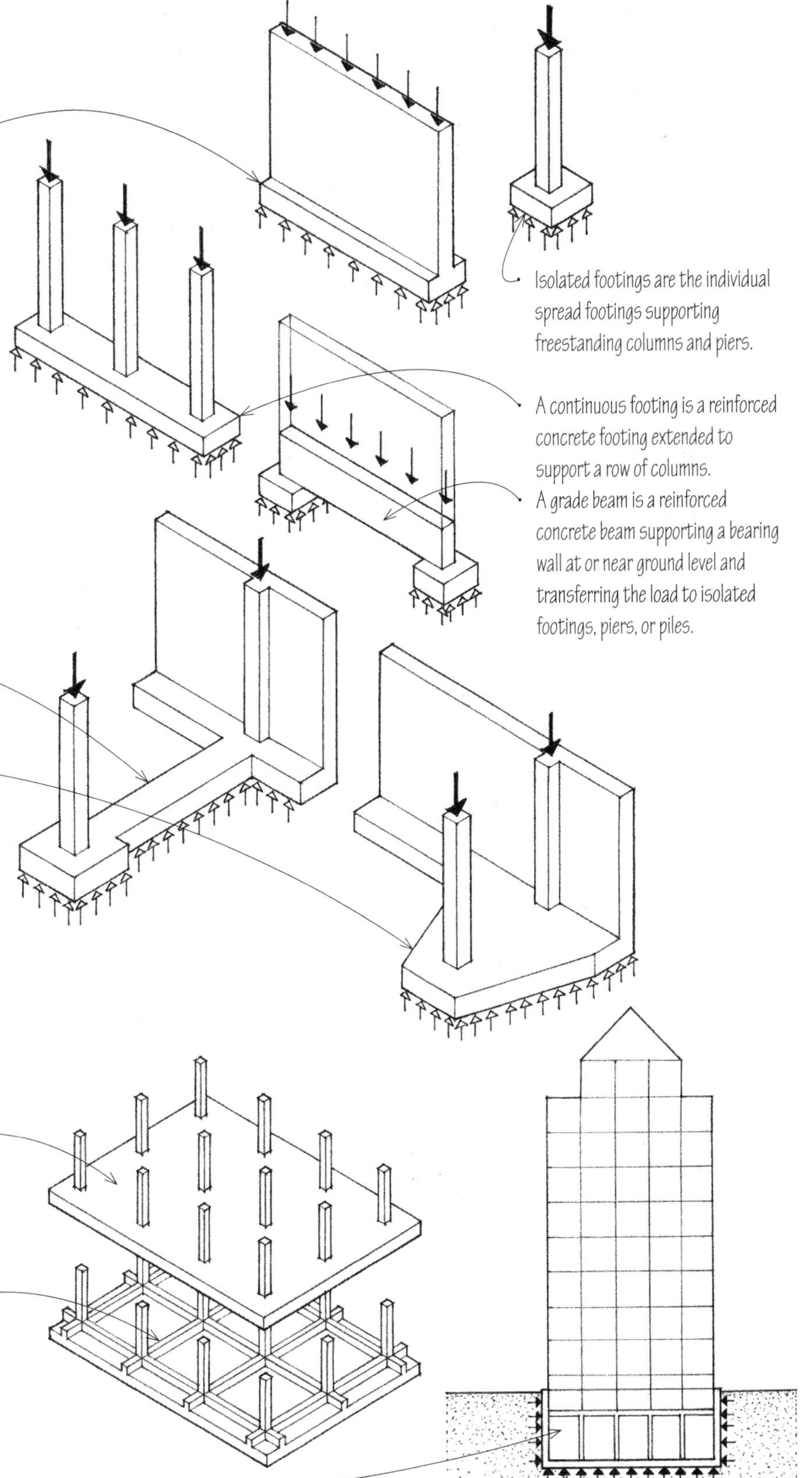

- Isolated footings are the individual spread footings supporting freestanding columns and piers.

- A continuous footing is a reinforced concrete footing extended to support a row of columns.
- A grade beam is a reinforced concrete beam supporting a bearing wall at or near ground level and transferring the load to isolated footings, piers, or piles.

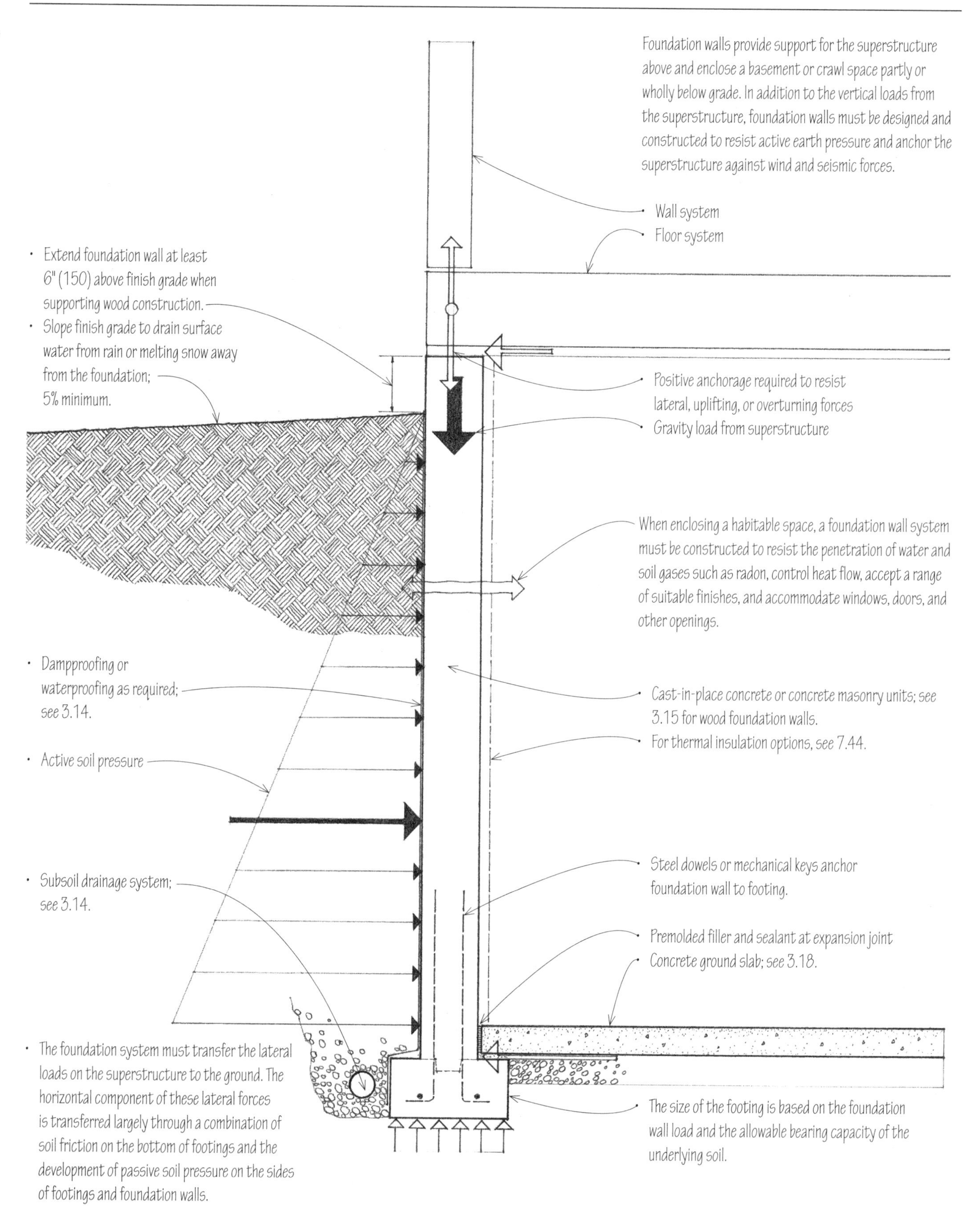
Foundation walls provide support for the superstructure above and enclose a basement or crawl space partly or wholly below grade. In addition to the vertical loads from the superstructure, foundation walls must be designed and constructed to resist active earth pressure and anchor the superstructure against wind and seismic forces.
Wall system
Floor system
Extend foundation wall at least 6" (150) above finish grade when supporting wood construction.
Slope finish grade to drain surface water from rain or melting snow away from the foundation; 5% minimum.
Positive anchorage required to resist lateral, uplifting, or overturning forces
Gravity load from superstructure
When enclosing a habitable space, a foundation wall system must be constructed to resist the penetration of water and soil gases such as radon, control heat flow, accept a range of suitable finishes, and accommodate windows, doors, and other openings.
Dampproofing or waterproofing as required; see 3.14.
Cast-in-place concrete or concrete masonry units; see 3.15 for wood foundation walls.
For thermal insulation options, see 7.44.
Active soil pressure
Steel dowels or mechanical keys anchor foundation wall to footing.
Subsoil drainage system; see 3.14.
Premolded filler and sealant at expansion joint
Concrete ground slab; see 3.18.
The foundation system must transfer the lateral loads on the superstructure to the ground. The horizontal component of these lateral forces is transferred largely through a combination of soil friction on the bottom of footings and the development of passive soil pressure on the sides of footings and foundation walls.
The size of the footing is based on the foundation wall load and the allowable bearing capacity of the underlying soil.

Crawl spaces enclosed by a continuous foundation wall or piers provide space under a first floor for the integration of and access to mechanical, electrical, and plumbing installations.

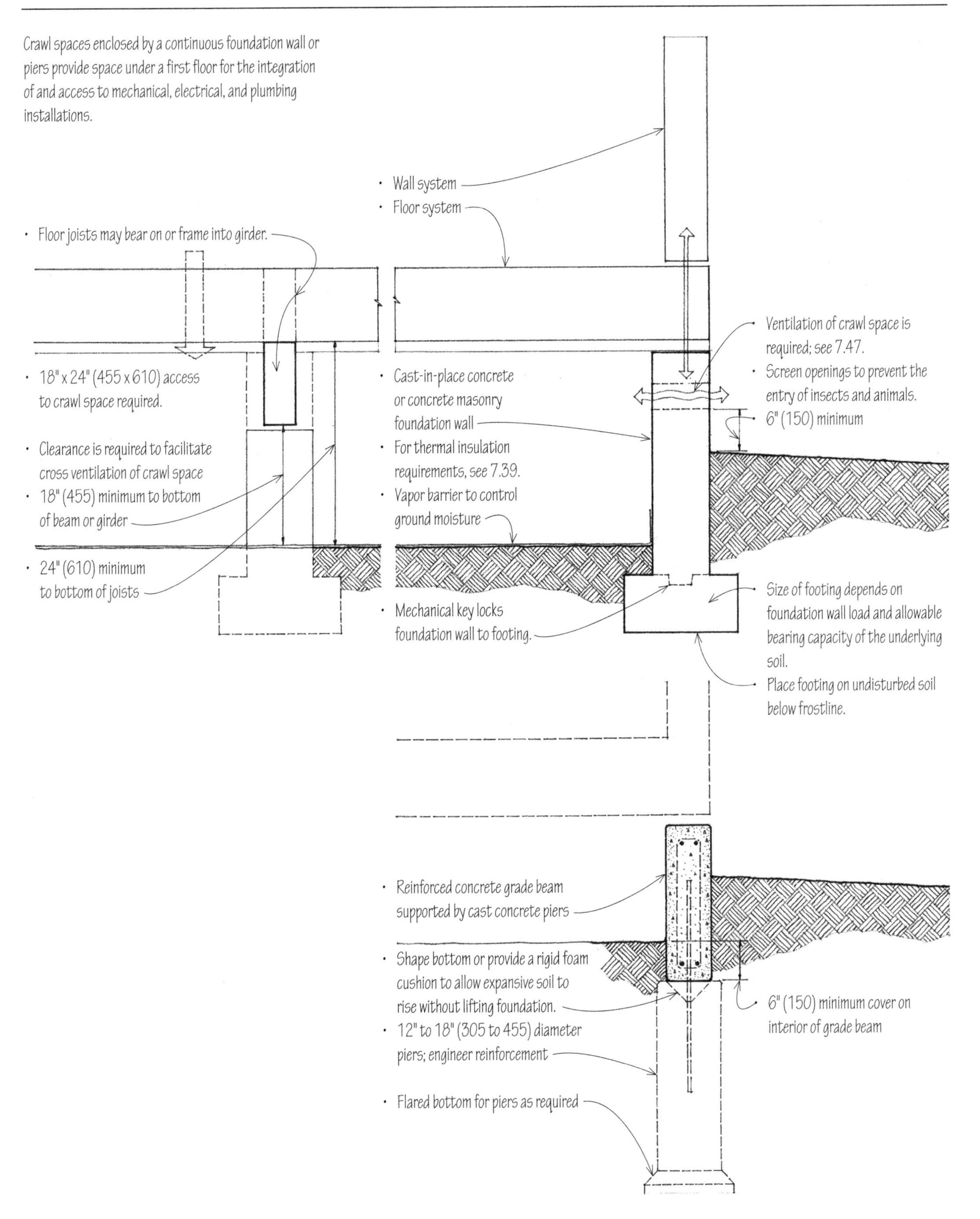

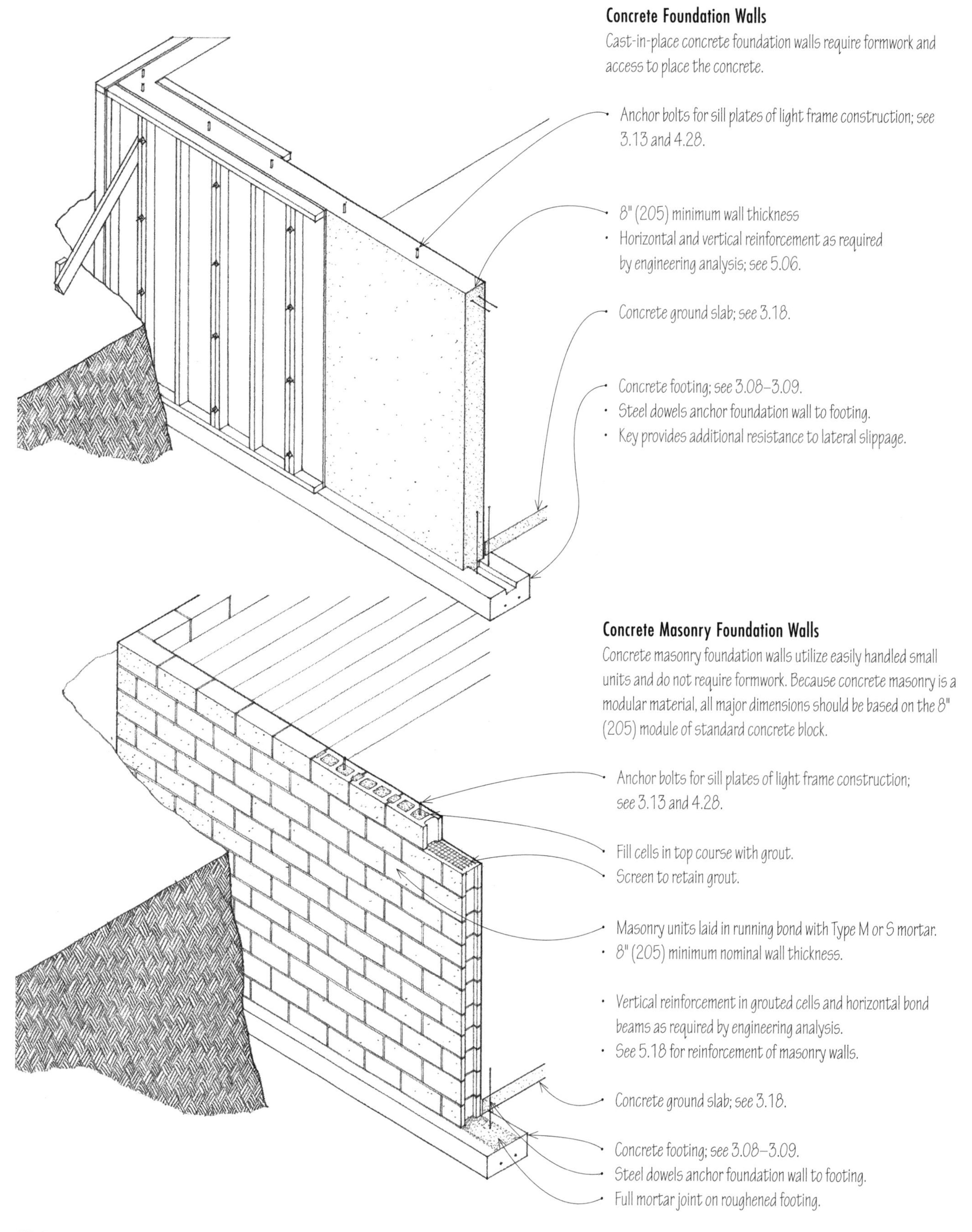

Concrete Foundation Walls

Cast-in-place concrete foundation walls require formwork and access to place the concrete.

Concrete Masonry Foundation Walls

Concrete masonry foundation walls utilize easily handled small units and do not require formwork. Because concrete masonry is a modular material, all major dimensions should be based on the 8" (205) module of standard concrete block.

CSI MasterFormat 03 30 00: Cast-in-Place Concrete
CSI MasterFormat 04 20 00: Unit Masonry

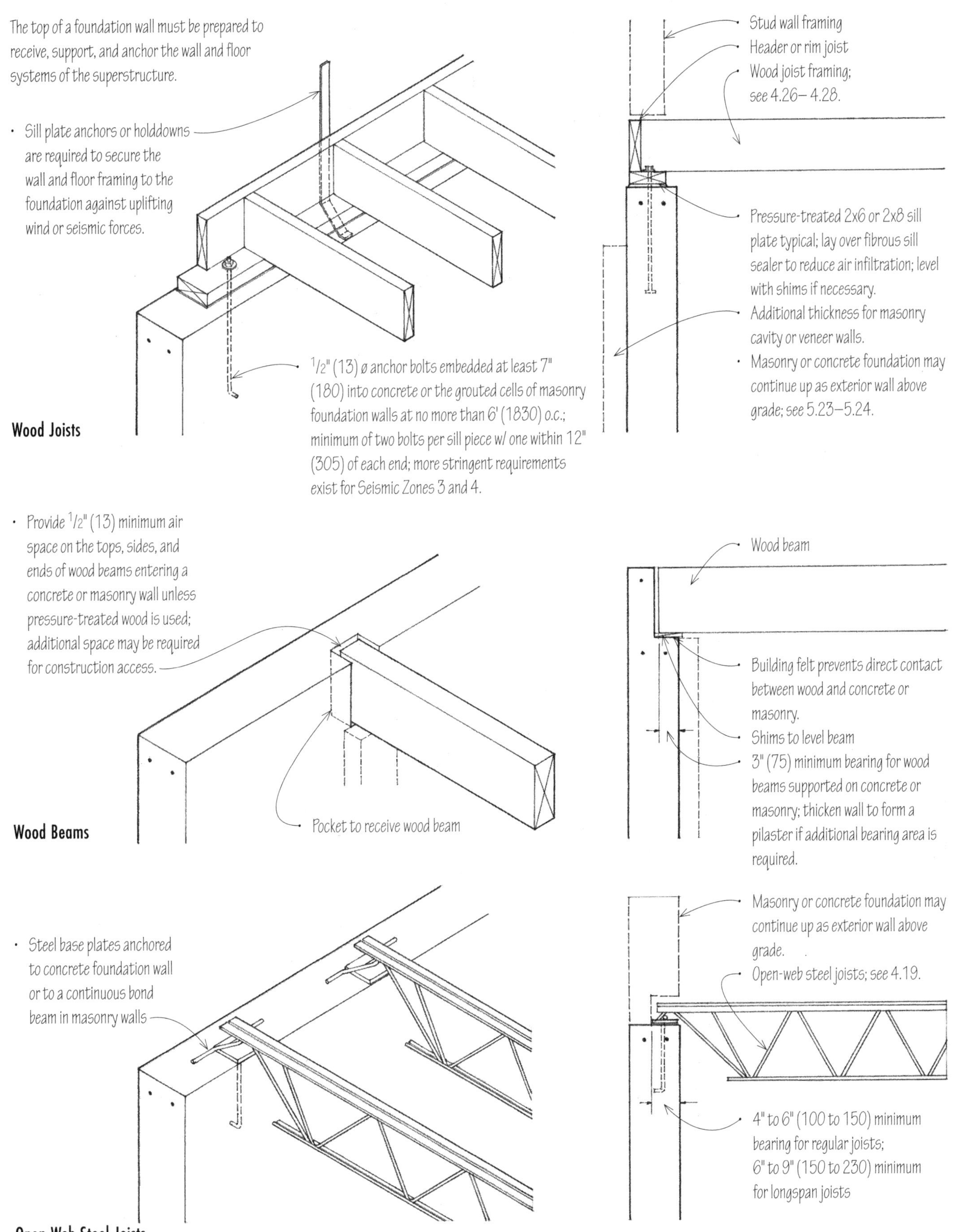

The top of a foundation wall must be prepared to receive, support, and anchor the wall and floor systems of the superstructure.
• Sill plate anchors or holddowns are required to secure the wall and floor framing to the foundation against uplifting wind or seismic forces.
1/2" (13) ø anchor bolts embedded at least 7" (180) into concrete or the grouted cells of masonry foundation walls at no more than 6' (1830) o.c.; minimum of two bolts per sill piece w/ one within 12" (305) of each end; more stringent requirements exist for Seismic Zones 3 and 4.
Wood Joists
• Stud wall framing
• Header or rim joist
• Wood joist framing; see 4.26–4.28.
• Pressure-treated 2x6 or 2x8 sill plate typical; lay over fibrous sill sealer to reduce air infiltration; level with shims if necessary.
• Additional thickness for masonry cavity or veneer walls.
• Masonry or concrete foundation may continue up as exterior wall above grade; see 5.23–5.24.
• Provide 1/2" (13) minimum air space on the tops, sides, and ends of wood beams entering a concrete or masonry wall unless pressure-treated wood is used; additional space may be required for construction access.
• Pocket to receive wood beam
Wood Beams
• Wood beam
• Building felt prevents direct contact between wood and concrete or masonry.
• Shims to level beam
• 3" (75) minimum bearing for wood beams supported on concrete or masonry; thicken wall to form a pilaster if additional bearing area is required.
• Steel base plates anchored to concrete foundation wall or to a continuous bond beam in masonry walls
Open-Web Steel Joists
• Masonry or concrete foundation may continue up as exterior wall above grade.
• Open-web steel joists; see 4.19.
• 4" to 6" (100 to 150) minimum bearing for regular joists; 6" to 9" (150 to 230) minimum for longspan joists

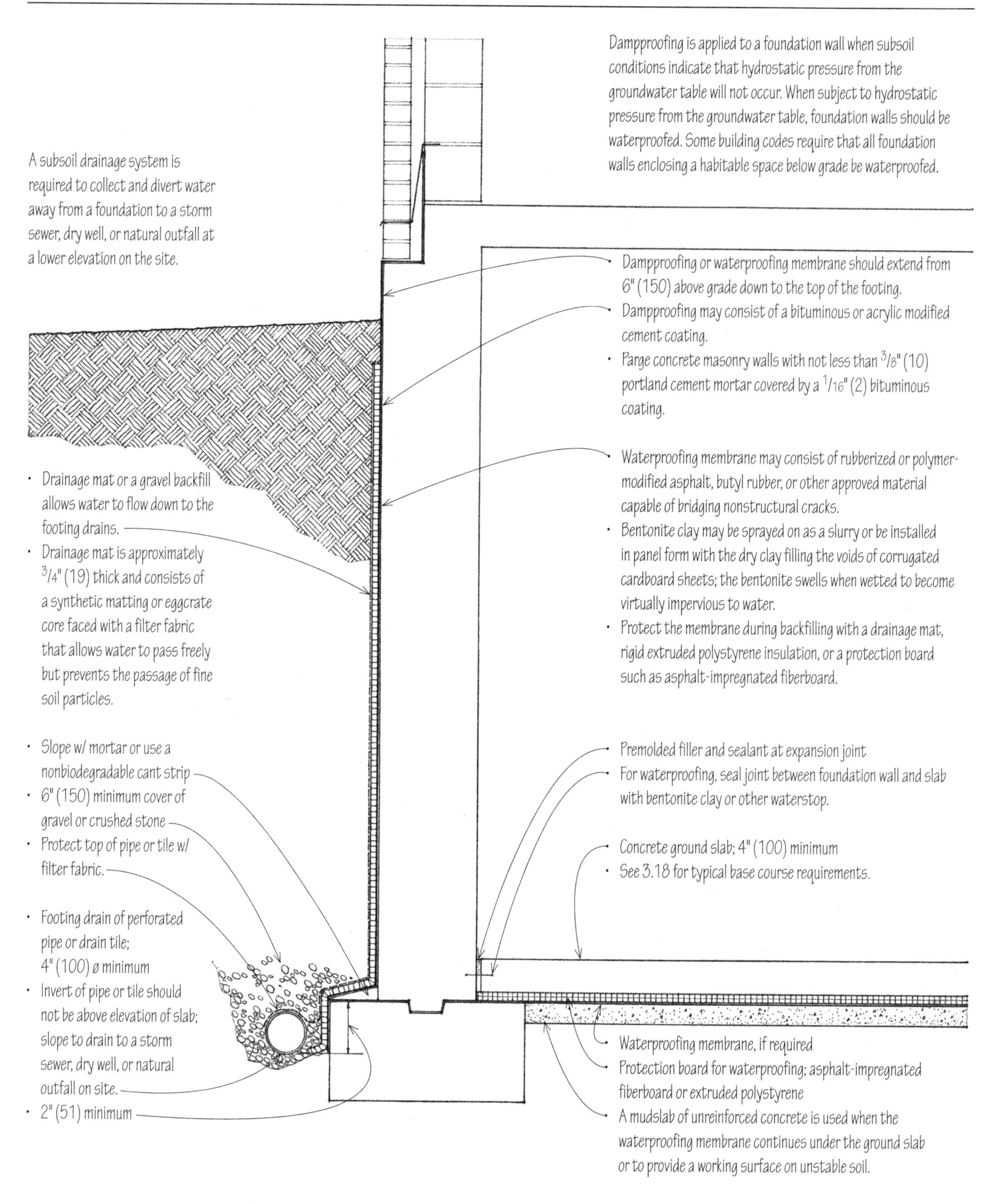

CSI MasterFormat 07 10 00: Dampproofing and Waterproofing
CSI MasterFormat 33 46 13: Foundation Drainage

Treated wood foundation systems can be used for both basement and crawl space construction. The wall sections may be built on-site or be factory-fabricated to reduce erection time. All wood and plywood used to fabricate a foundation system must be pressure-treated with a preservative approved for ground contact use; all field cuts should be treated with the same preservative. All metal fasteners should be of stainless steel or hot-dipped zinc-coated steel.

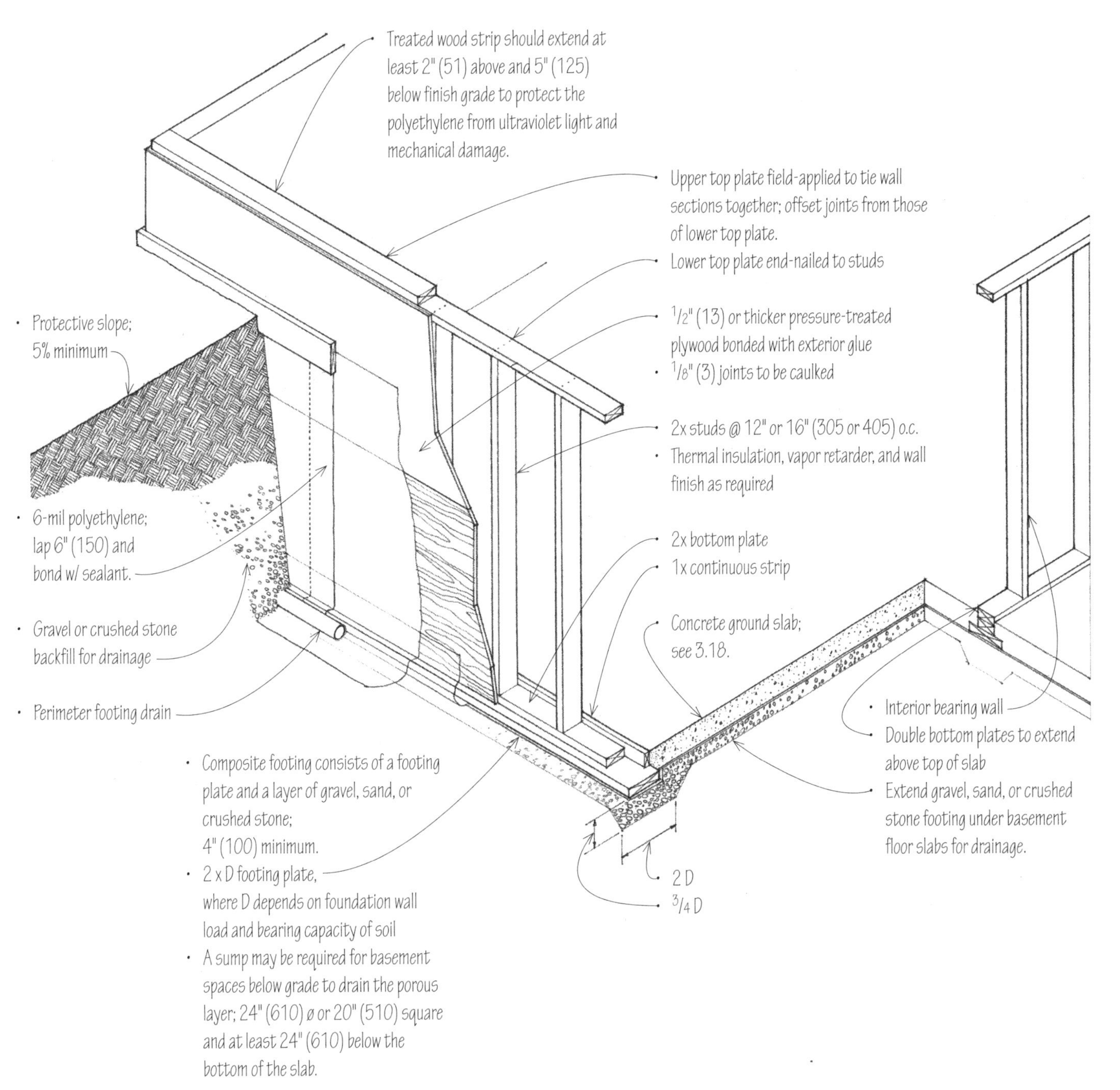

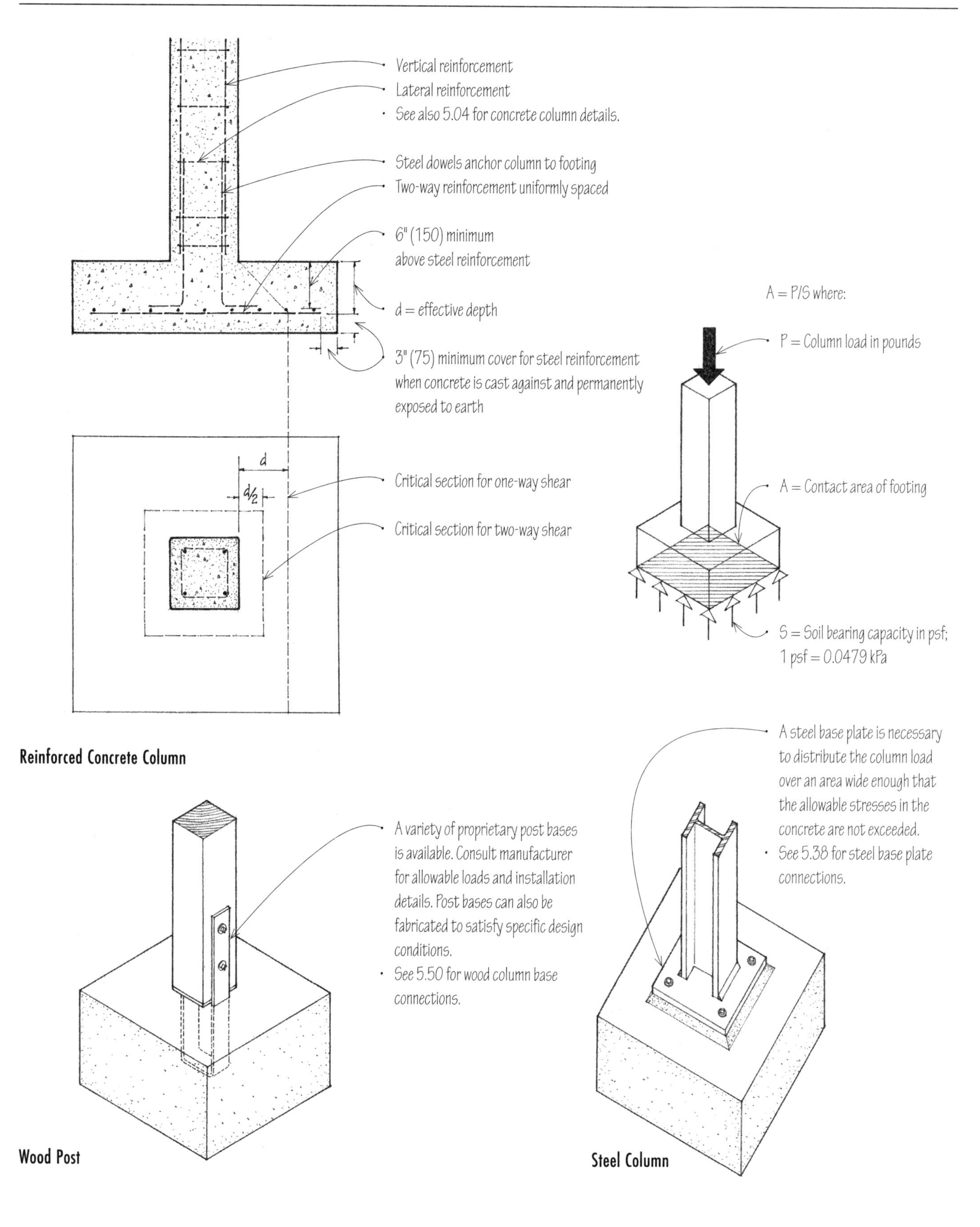

CSI MasterFormat 03 30 00: Cast-in-Place Concrete
CSI MasterFormat 03 31 00: Structural Concrete

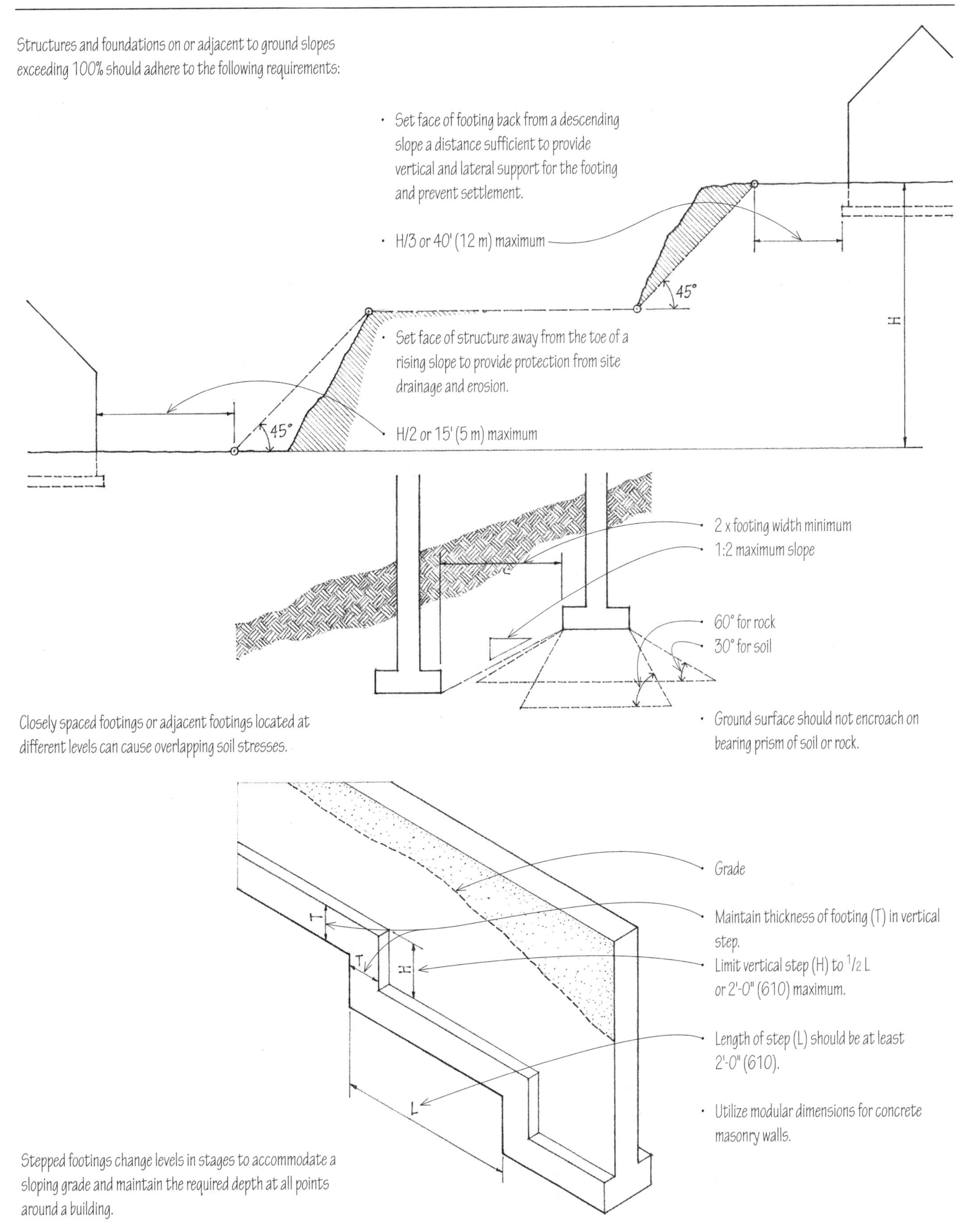
Structures and foundations on or adjacent to ground slopes exceeding 100% should adhere to the following requirements:
• Set face of footing back from a descending slope a distance sufficient to provide vertical and lateral support for the footing and prevent settlement.
• H/3 or 40' (12 m) maximum
45°
H
• Set face of structure away from the toe of a rising slope to provide protection from site drainage and erosion.
45°
• H/2 or 15' (5 m) maximum
• 2 x footing width minimum
• 1:2 maximum slope
• 60° for rock
• 30° for soil
Closely spaced footings or adjacent footings located at different levels can cause overlapping soil stresses.
• Ground surface should not encroach on bearing prism of soil or rock.
• Grade
T
T
H
• Maintain thickness of footing (T) in vertical step.
• Limit vertical step (H) to 1/2 L or 2'-0" (610) maximum.
• Length of step (L) should be at least 2'-0" (610).
L
• Utilize modular dimensions for concrete masonry walls.
Stepped footings change levels in stages to accommodate a sloping grade and maintain the required depth at all points around a building.

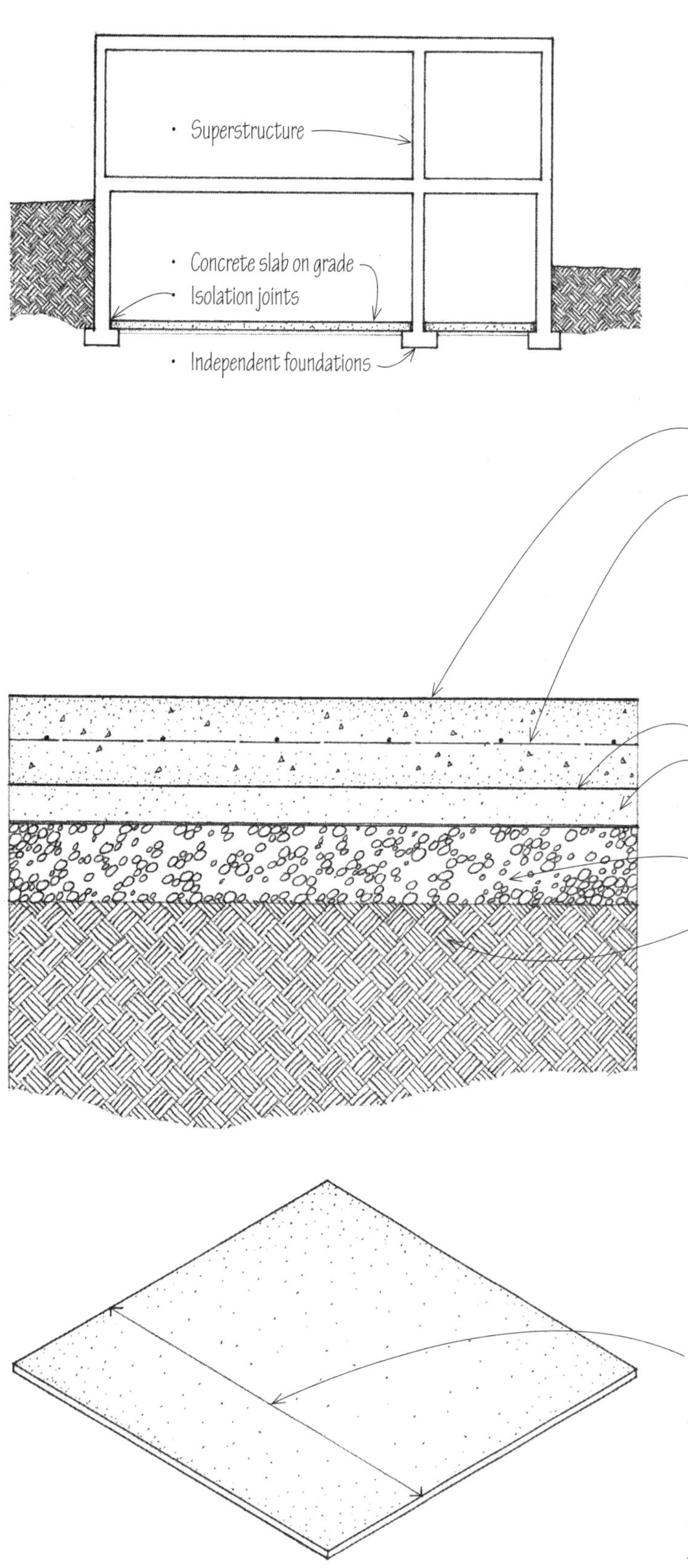

A concrete slab may be placed at or near grade level to serve as a combined floor and foundation system. The suitability of a concrete slab for such use depends on the geographic location, topography, and soil characteristics of the site, and the design of the superstructure.

Concrete slabs on grade require the support of a level, stable, uniformly dense or properly compacted soil base containing no organic matter. When placed over soil of low bearing capacity or over highly compressible or expansive soils, a concrete ground slab must be designed as a mat or raft foundation, which requires professional analysis and design by a qualified structural engineer.

- 4" (100) minimum slab thickness; thickness required depends on expected use and load conditions.
- Welded wire fabric reinforcement set at or slightly above the mid-depth of the slab controls thermal stresses, shrinkage cracking, and slight differential movement in the soil bed; a grid of reinforcing bars may be required for slabs carrying heavier-than-normal floor loads.
- Admixture of glass, steel, or polypropylene fibers may be added to concrete mix to reduce shrinkage cracking.
- Concrete additives can increase surface hardness and abrasion resistance.
- 6-mil (0.15 mm) polyethylene moisture barrier
- The American Concrete Institute recommends a 2" (51) layer of sand be placed over the moisture barrier to absorb excess water from the concrete during curing.
- Base course of gravel or crushed stone to prevent the capillary rise of groundwater; 4" (100) minimum
- Stable, uniformly dense soil base; compaction may be required to increase soil stability, loadbearing capacity, and resistance to water penetration.

Maximum Slab Dimensions feet (m)	Wire Spacing inches (mm)	Wire Size (number)
Up to 45 (14)	6 x 6 (150 x 150)	W1.4 x W1.4
45–60 (14–18)	6 x 6	W2.0 x W2.0
60–75 (18–22)	6 x 6	W2.9 x W2.9

CSI MasterFormat 03 30 00: Cast-in-Place Concrete
CSI MasterFormat 03 31 00: Structural Concrete

Three types of joints may be created or constructed in order to accommodate movement in the plane of a concrete slab on grade—isolation joints, construction joints, and control joints.

Isolation Joints

Isolation joints, often called expansion joints, allow movement to occur between a concrete slab and adjoining columns and walls of a building.

Construction Joints

Construction joints provide a place for construction to stop and then continue at a later time. These joints, which also serve as isolation or control joints, can be keyed or doweled to prevent vertical differential movement of adjoining slab sections.

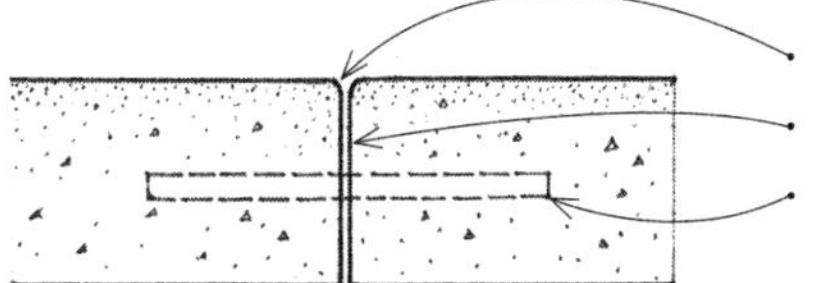

Control Joints

Control joints create lines of weakness so that the cracking that may result from tensile stresses occurs along predetermined lines. Space control joints in exposed concrete 15' to 20' (4570 to 6100) o.c., or wherever required to break an irregular slab shape into square or rectangular sections.

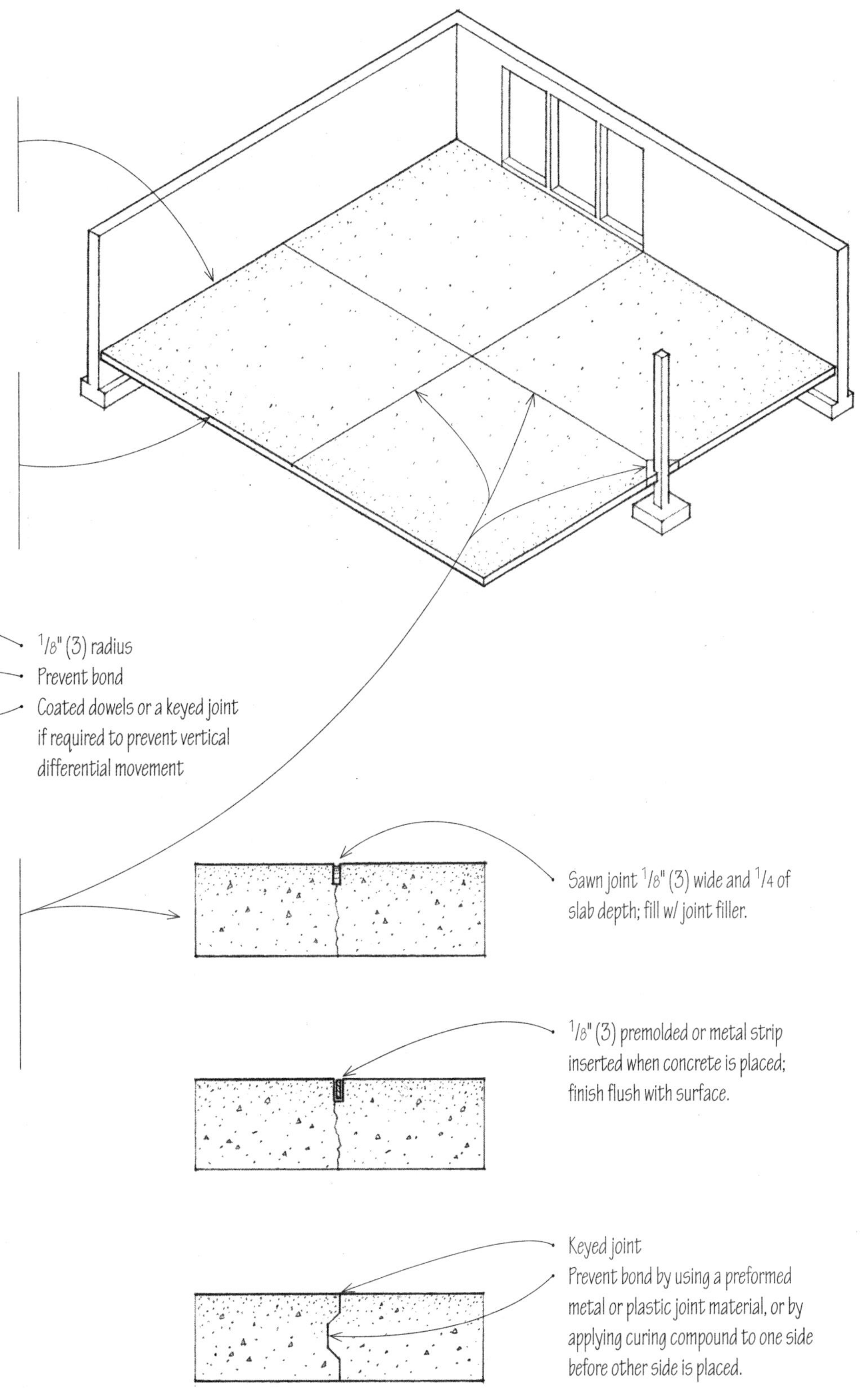

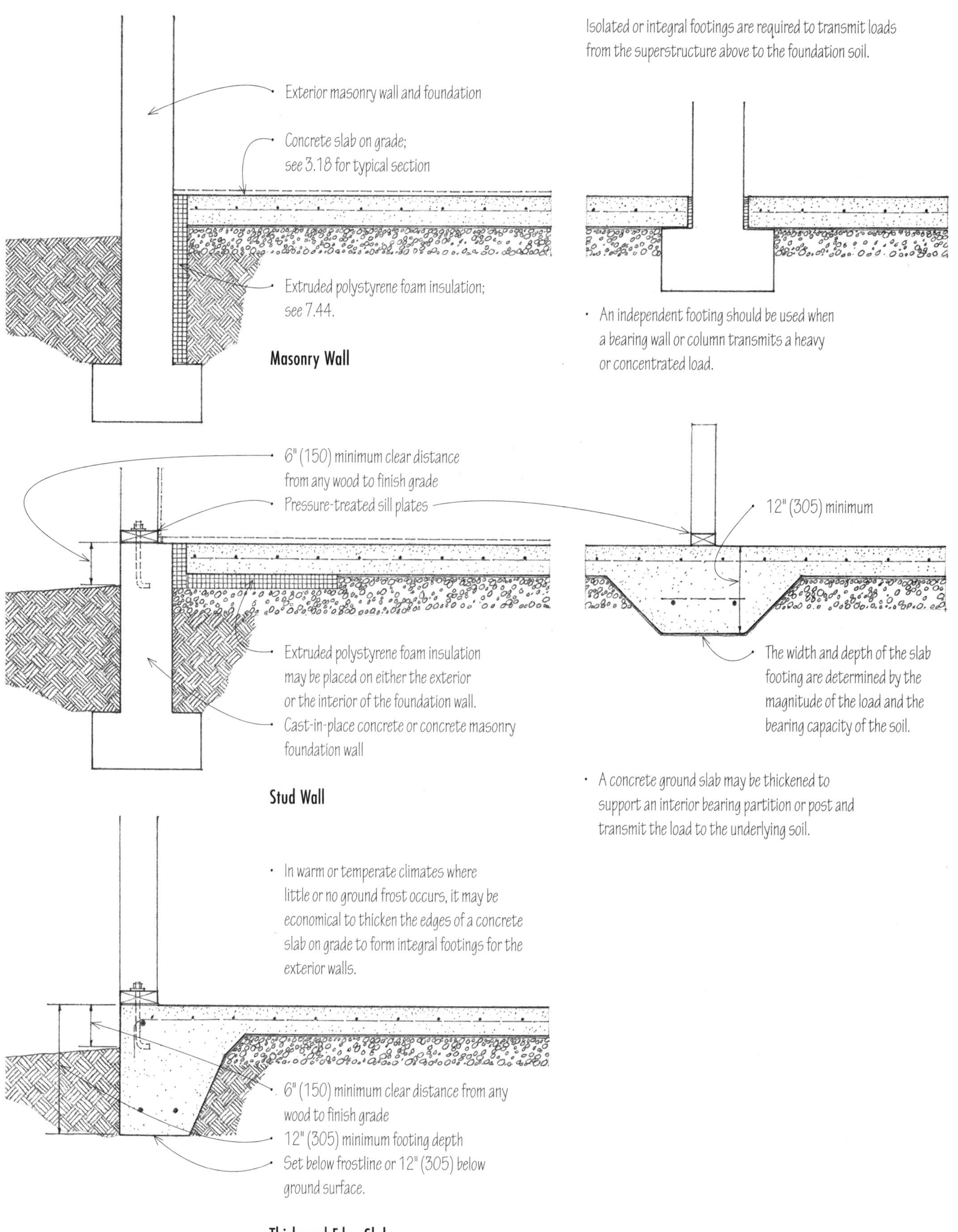

Exterior masonry wall and foundation
Concrete slab on grade; see 3.18 for typical section
Extruded polystyrene foam insulation; see 7.44.
Masonry Wall
Isolated or integral footings are required to transmit loads from the superstructure above to the foundation soil.
• An independent footing should be used when a bearing wall or column transmits a heavy or concentrated load.
6" (150) minimum clear distance from any wood to finish grade
Pressure-treated sill plates
12" (305) minimum
• Extruded polystyrene foam insulation may be placed on either the exterior or the interior of the foundation wall.
• Cast-in-place concrete or concrete masonry foundation wall
Stud Wall
The width and depth of the slab footing are determined by the magnitude of the load and the bearing capacity of the soil.
• A concrete ground slab may be thickened to support an interior bearing partition or post and transmit the load to the underlying soil.
• In warm or temperate climates where little or no ground frost occurs, it may be economical to thicken the edges of a concrete slab on grade to form integral footings for the exterior walls.
• 6" (150) minimum clear distance from any wood to finish grade
• 12" (305) minimum footing depth
• Set below frostline or 12" (305) below ground surface.
Thickened Edge Slab

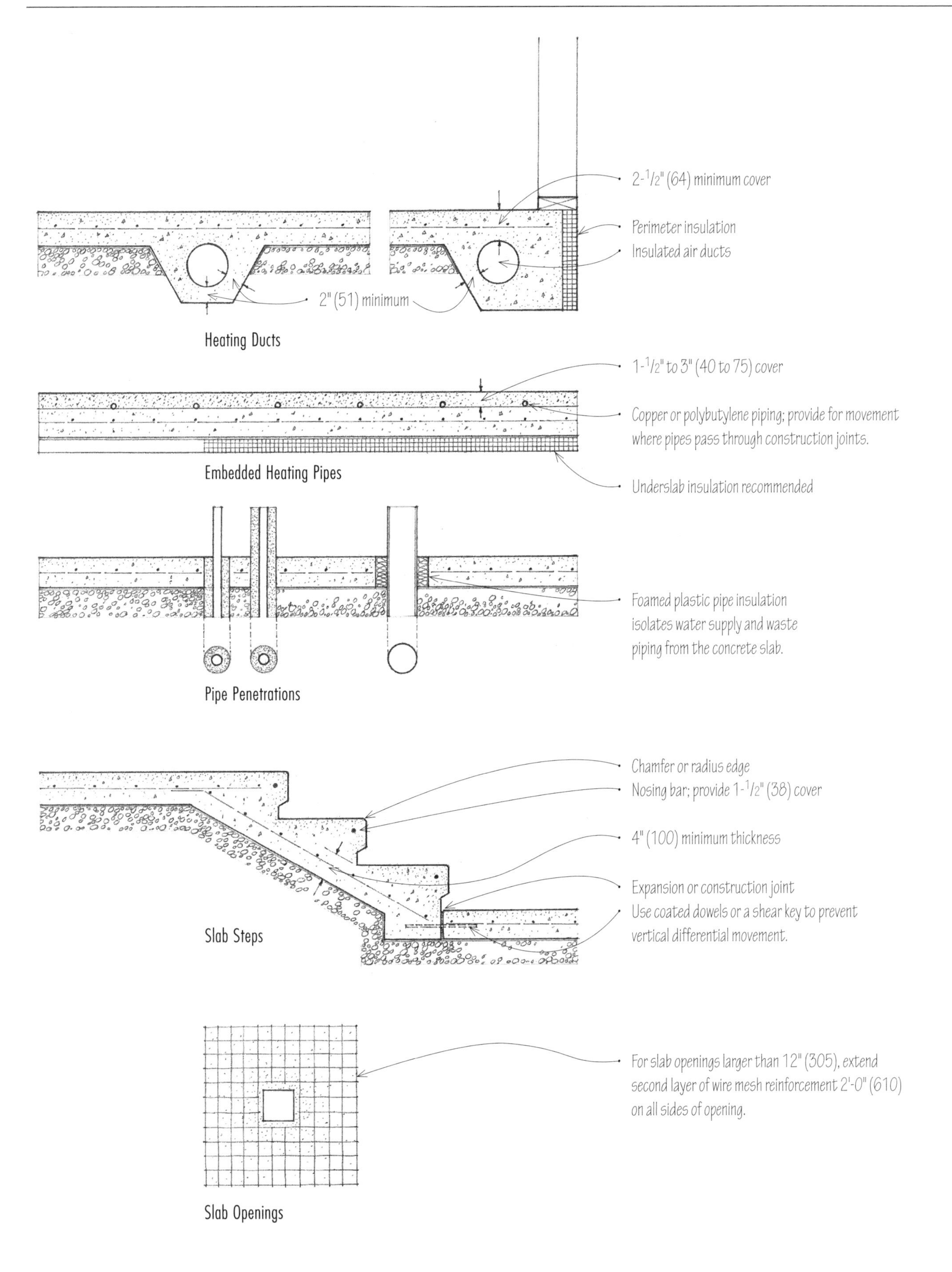
2-1/2" (64) minimum cover
Perimeter insulation
Insulated air ducts
2" (51) minimum
Heating Ducts
1-1/2" to 3" (40 to 75) cover
Copper or polybutylene piping; provide for movement where pipes pass through construction joints.
Embedded Heating Pipes
Underslab insulation recommended
Foamed plastic pipe insulation isolates water supply and waste piping from the concrete slab.
Pipe Penetrations
Chamfer or radius edge
Nosing bar; provide 1-1/2" (38) cover
4" (100) minimum thickness
Expansion or construction joint
Use coated dowels or a shear key to prevent vertical differential movement.
Slab Steps
For slab openings larger than 12" (305), extend second layer of wire mesh reinforcement 2'-0" (610) on all sides of opening.
Slab Openings

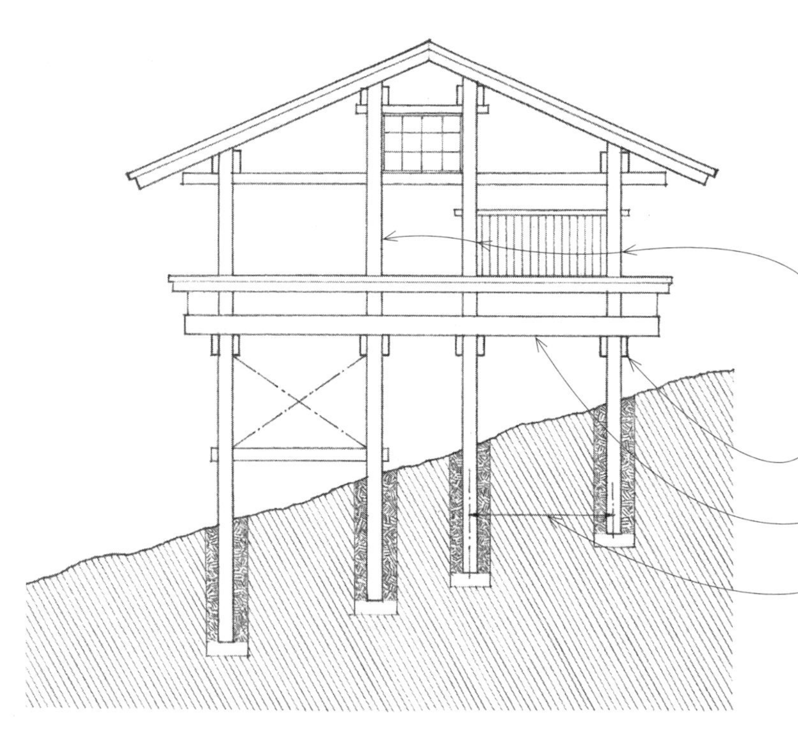

Pole foundations elevate timber structures above the ground plane, require minimal excavation, and preserve the natural features and existing drainage patterns of a site. They are particularly useful when building on steep slopes and in areas subject to periodic flooding.

The treated poles are usually laid out along a grid defined by the beam-and-joist framing pattern. Their spacing determines both the beam-and-joist spans and the vertical loads they must support.

- Poles 6" to 12" (150 to 305) in diameter; treat with a preservative to protect against decay and insect infestation. The treated poles may extend vertically to form the loadbearing frame of the superstructure or terminate at the first-floor level to support a conventional platform frame.
- Solid, built-up, or spaced wood beams; limit overhangs to 1/4 of the backspan.
- Insulate floors, walls, and roof according to local climatic conditions.
- Poles are spaced 6' to 12' (1830 to 3660) apart to support floor and roof areas up to 144 sf (13.4 m^2)

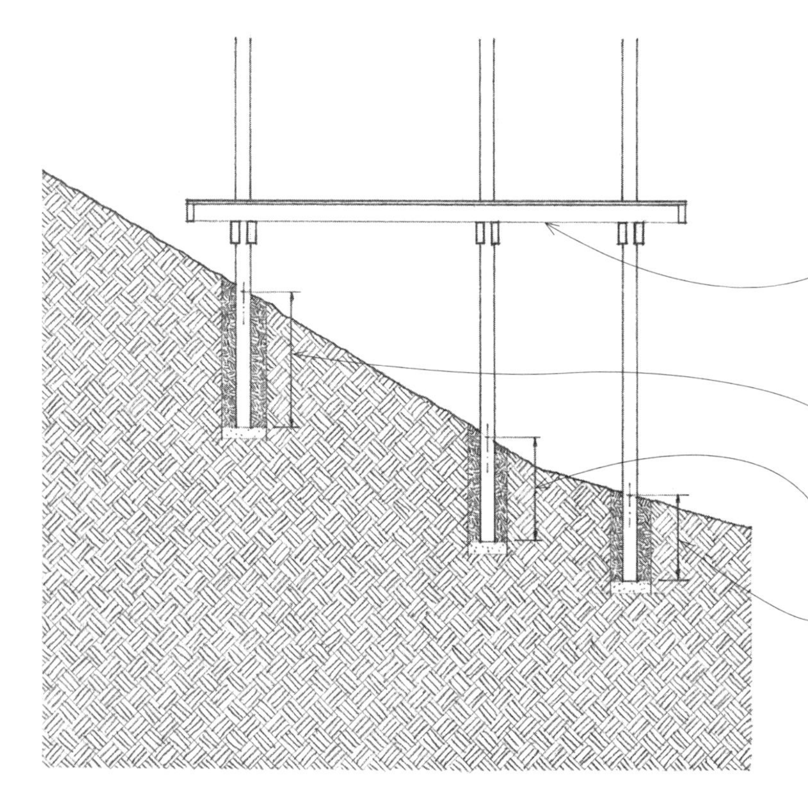

Poles are set in holes dug by hand or by a power auger. Adequate embedment length, suitable backfilling, and proper connections are required for a pole structure to develop the necessary rigidity and resistance to lateral wind and seismic forces. The required embedment length varies according to:

- Slope of the site
- Subsurface soil conditions
- Pole spacing
- Unsupported height of the poles
- Seismic zone

- Floors should be designed and constructed as a diaphragm to transfer the rigidity of uphill poles to the rest of the structure.

Embedment Length for Steep Slopes

- 5' to 8' (1525 to 2440) for uphill poles; uphill poles have shorter unsupported heights but require deeper embedment in order to provide the necessary rigidity for the structure.
- 4' to 7' (1220 to 2135) for downhill poles

Embedment Length for Flat Slopes

- 4' to 5' (1220 to 1525)

When the necessary embedment is not possible, such as on a rocky slope, steel rod crossbracing with turnbuckles or shear walls of concrete or masonry can be used to provide lateral stability.

- Consult a qualified structural engineer when designing and constructing a pole structure, especially when building on a steeply sloping site subject to high winds or flooding.

CSI MasterFormat 06 13 16: Pole Construction

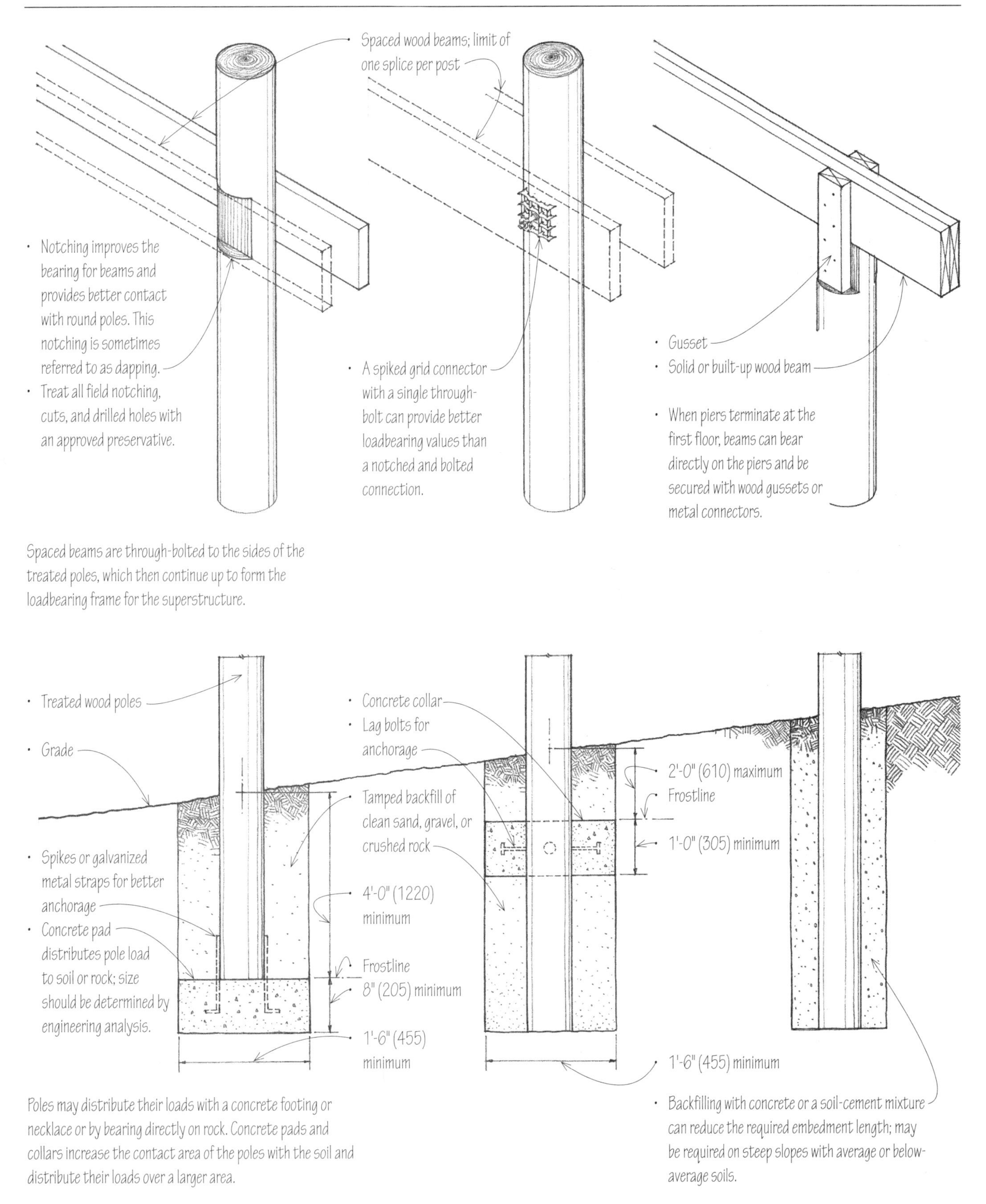

Poles may distribute their loads with a concrete footing or necklace or by bearing directly on rock. Concrete pads and collars increase the contact area of the poles with the soil and distribute their loads over a larger area.

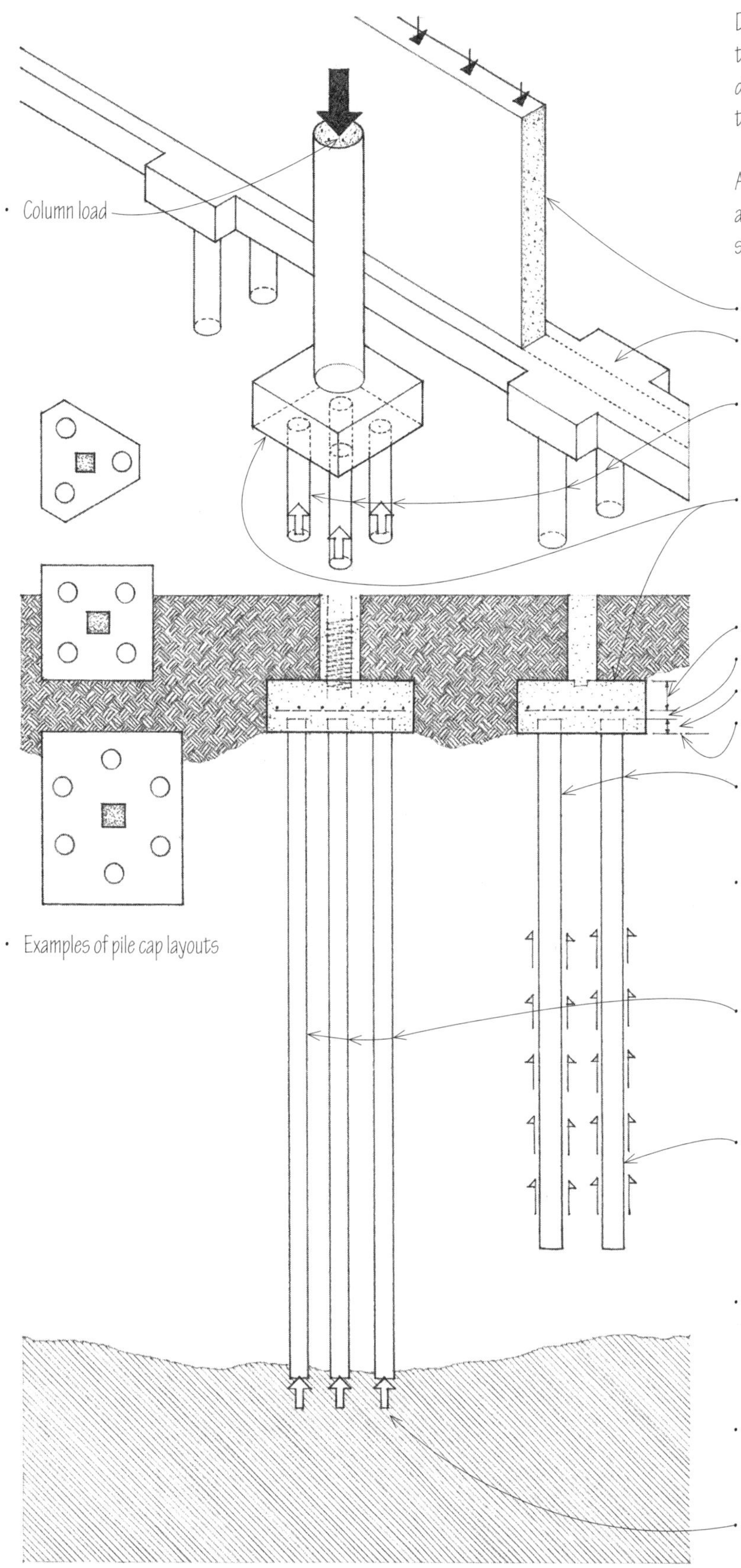

Deep foundations extend down through unsuitable or unstable soil to transfer building loads to a more appropriate bearing stratum of rock or dense sands and gravels well below the superstructure. The two principal types of deep foundations are pile foundations and caisson foundations.

A pile foundation is a system of end-bearing or friction piles, pile caps, and tie beams for transferring building loads down to a suitable bearing stratum.

- Loadbearing wall
- Reinforced concrete grade or tie beam with integral pile caps
- Piles are usually driven in clusters of two or more, spaced 2'-6" to 4'-0" (760 to 1220) o.c.
- A reinforced concrete pile cap joins the heads of a cluster of piles in order to distribute the load from a column or grade beam equally among the piles.
- Varies with column load; 12" (305) minimum
- 3" (75)
- 6" (150)
- Place below frostline
- Piles may be of treated timber poles, but for large buildings, steel H-sections, concrete-filled pipes, or precast reinforced or prestressed concrete are more common.
- Piles are driven into the earth by a pile driver, composed of a tall framework supporting machinery for lifting the pile in position before driving, a driving hammer, and vertical rails or leads for guiding the hammer.
- End-bearing piles depend principally on the bearing resistance of soil or rock beneath their feet for support. The surrounding soil mass provides a degree of lateral stability for the long compression members.
- Friction piles depend principally on the frictional resistance of a surrounding earth mass for support. The skin friction developed between the sides of a pile and the soil into which the pile is driven is limited by the adhesion of soil to the pile sides and the shear strength of the surrounding soil mass.
- The allowable pile load is the maximum axial and lateral loads permitted on a pile, as determined by a dynamic pile formula, a static load test, or a geotechnical investigation of the foundation soil.
- Pile eccentricity, the deviation of a pile from its plan location or from the vertical, can result in a reduction of its allowable load.

- Bearing stratum of soil or rock

- Timber piles are logs driven usually as a friction pile. They are often fitted with a steel shoe and a drive band to prevent their shafts from splitting or shattering.
- Composite piles are constructed of two materials, such as a timber pile having a concrete upper section to prevent the portion of the pile above the water table from deteriorating.
- H-piles are steel H-sections, sometimes encased in concrete to a point below the water table to prevent corrosion. H-sections can be welded together in the driving process to form any length of pile.
- Pipe piles are heavy steel pipes driven with the lower end either open or closed by a heavy steel plate or point and filled with concrete. An open-ended pipe pile requires inspection and excavation before being filled with concrete.
- Precast concrete piles have round, square, or polygonal cross sections and sometimes an open core. Precast piles are often prestressed.

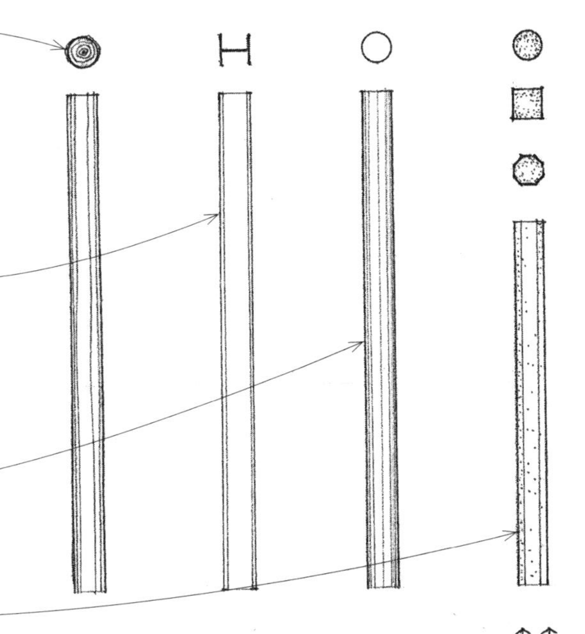

- Cast-in-place concrete piles are constructed by placing concrete into a shaft in the ground. The concrete piles may be cased or uncased.

- Cased piles are constructed by driving a steel pipe or casing into the ground until it meets the required resistance and then filling it with concrete. The casing is usually a cylindrical steel section, sometimes corrugated or tapered for increased stiffness. A mandrel consisting of a heavy steel tube or core may be inserted into a thin-walled casing to prevent it from collapsing in the driving process, and then withdrawn before concrete is placed in the casing.

- Uncased piles are constructed by driving a concrete plug into the ground along with a steel casing until it meets the required resistance, and then ramming concrete into place as the casing is withdrawn.
- A pedestal pile is an uncased pile that has an enlarged foot to increase the bearing area of the pile and strengthen the bearing stratum by compression. The foot is formed by forcing concrete out at the bottom of the casing into the surrounding soil.

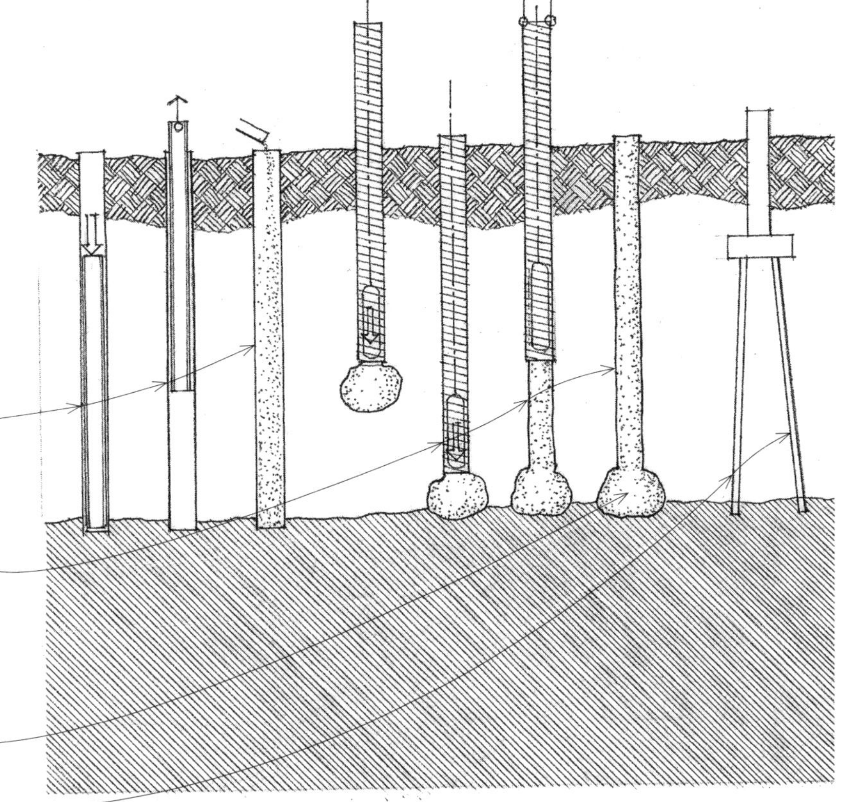

- Micropiles are high capacity, small diameter [5" to 12" (125 to 305)], drilled and grouted in-place piles that are typically reinforced. They are often used for foundations in urbanized areas or in locations with restricted access, and for underpinning or emergency repairs because they can be installed in virtually any ground condition with minimal vibration and disturbance to existing structures.

CSI MasterFormat 31 62 00: Driven Piles
CSI MasterFormat 31 63 00: Bored Piles

Caissons are cast-in-place, plain or reinforced concrete piers formed by boring with a large auger or excavating by hand a shaft in the earth to a suitable bearing stratum and filling the shaft with concrete. For this reason, they are also referred to as drilled piles or piers.

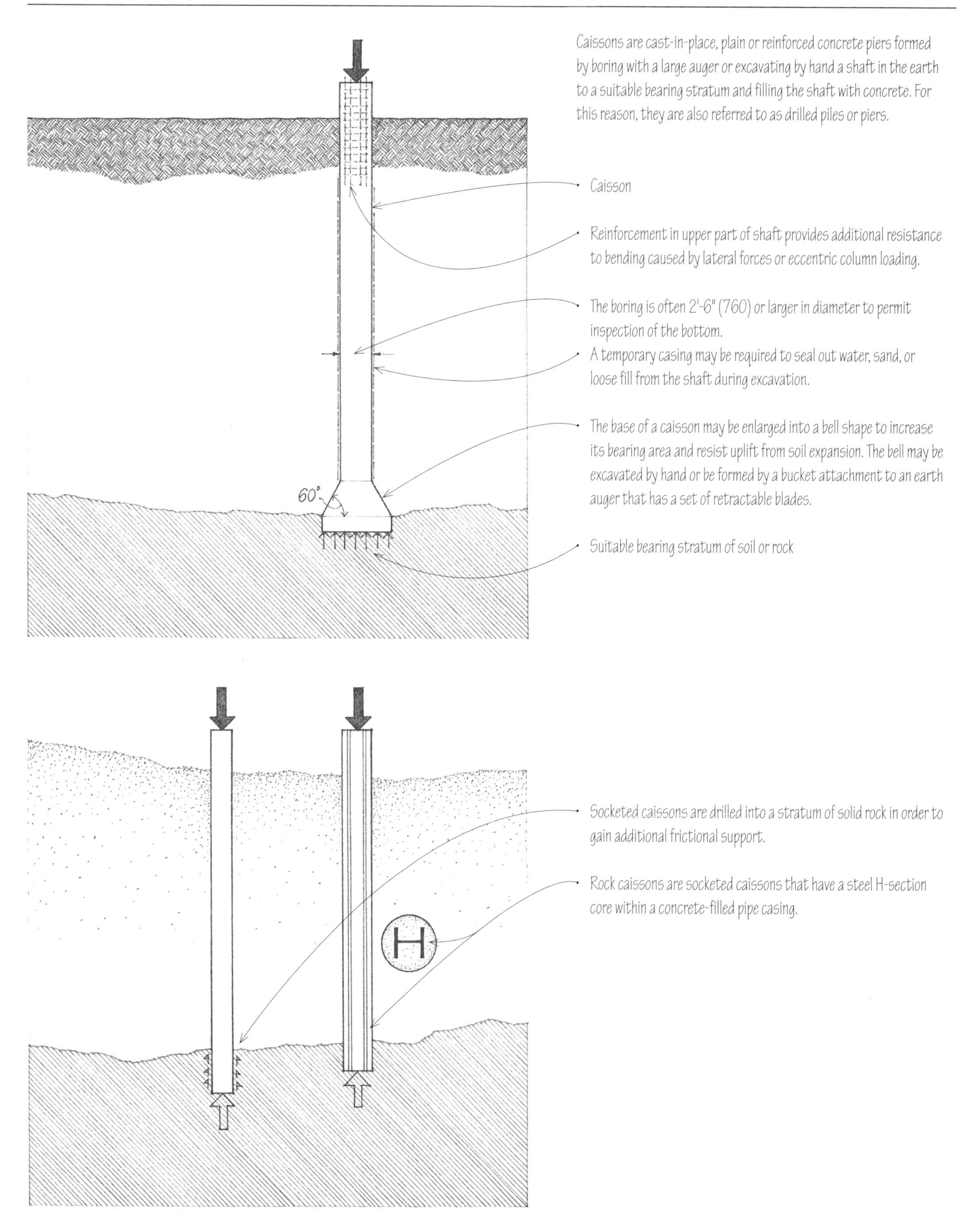

CSI MasterFormat 31 64 00: Caissons

4

FLOOR SYSTEMS

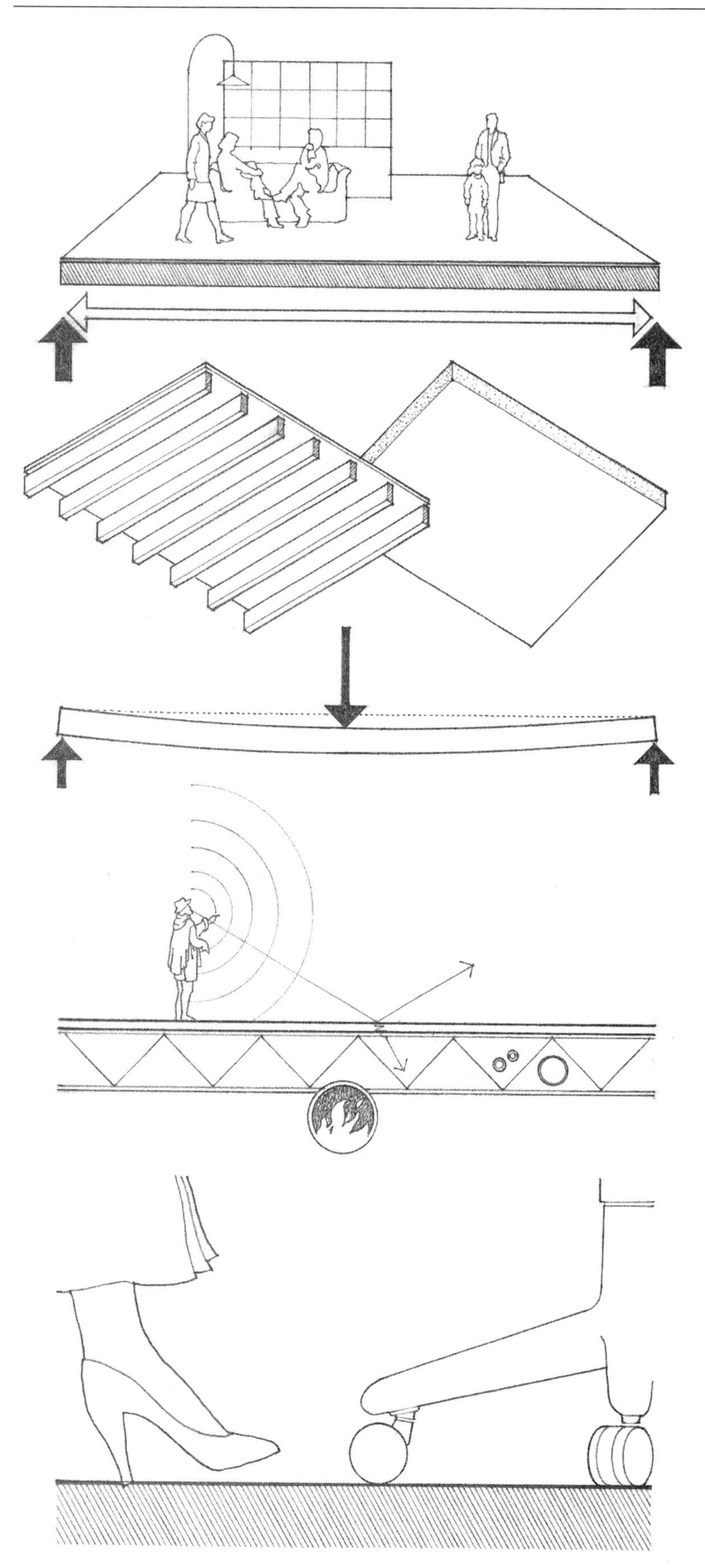

Floor systems are the horizontal planes that must support both live loads—people, furnishings, and movable equipment—and dead loads—the weight of the floor construction itself. Floor systems must transfer their loads horizontally across space to either beams and columns or to loadbearing walls. Rigid floor planes can also be designed to serve as horizontal diaphragms that act as thin, wide beams in transferring lateral forces to shear walls.

A floor system may be composed of a series of linear beams and joists overlaid with a plane of sheathing or decking, or consist of a nearly homogeneous slab of reinforced concrete. The depth of a floor system is directly related to the size and proportion of the structural bays it must span and the strength of the materials used. The size and placement of any cantilevers and openings within the floor plane should also be considered in the layout of the structural supports for the floor. The edge conditions of the floor structure and its connection to supporting foundation and wall systems affect both the structural integrity of a building and its physical appearance.

Because it must safely support moving loads, a floor system should be relatively stiff while maintaining its elasticity. Due to the detrimental effects that excessive deflection and vibration would have on finish flooring and ceiling materials, as well as concern for human comfort, deflection rather than bending becomes the critical controlling factor.

The depth of the floor construction and the cavities within it should be considered if it is necessary to accommodate runs of mechanical or electrical lines within the floor system. For floor systems between habitable spaces stacked one above another, additional factors to consider are the blockage of both airborne and structure-borne sound and the fire-resistance rating of the assembly.

Except for exterior decks, floor systems are not normally exposed to weather. Because they all must support traffic, however, durability, resistance to wear, and maintenance requirements are factors to consider in the selection of a floor finish and the system required to support it.

Concrete

- Cast-in-place concrete floor slabs are classified according to their span and cast form; see 4.05–4.07.
- Precast concrete planks may be supported by beams or loadbearing walls.

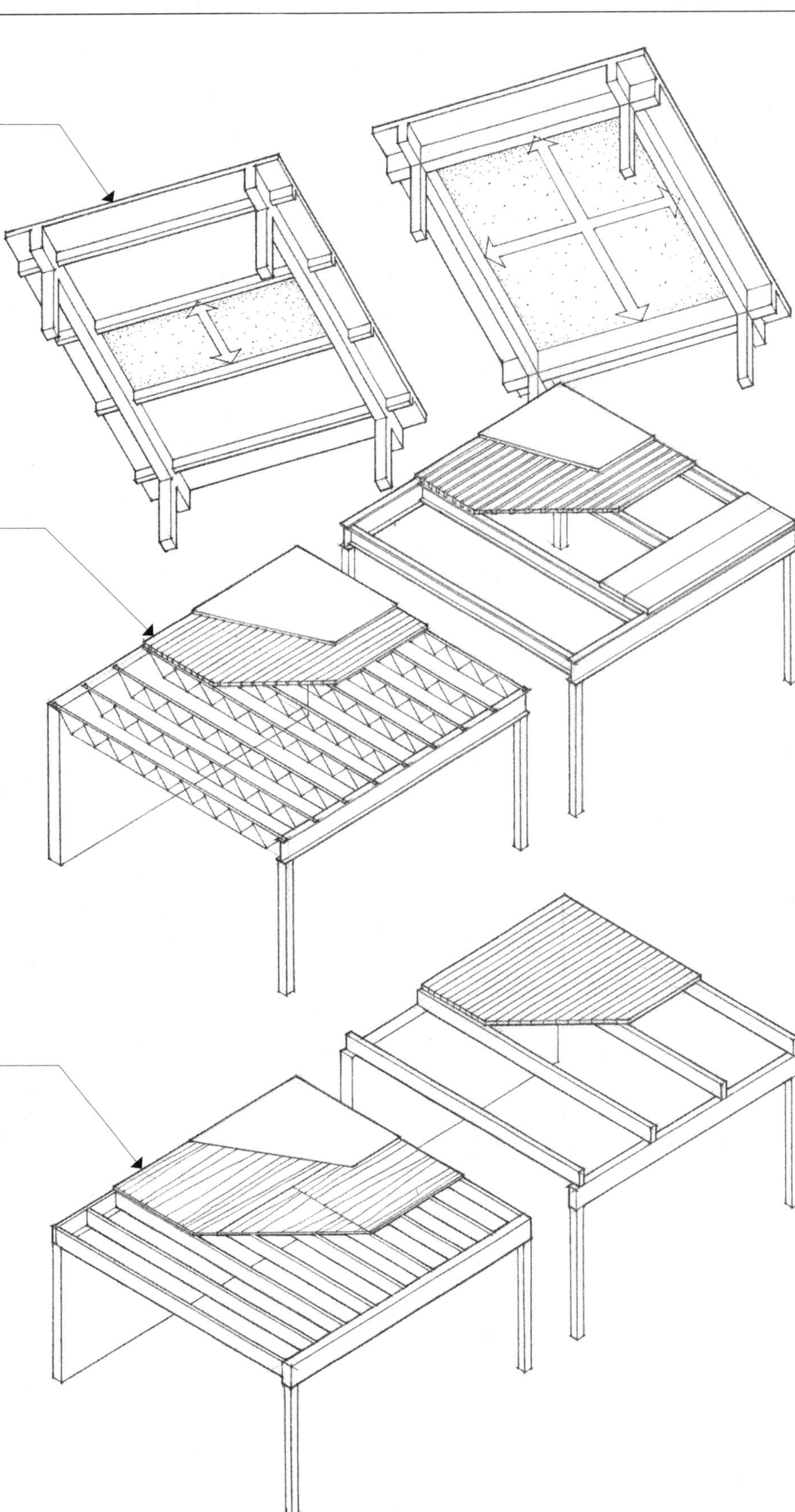

Steel

- Steel beams support steel decking or precast concrete planks.
- Beams may be supported by girders, columns, or loadbearing walls.
- Beam framing is typically an integral part of a steel skeleton frame system.

- Closely spaced light-gauge or open-web joists may be supported by beams or loadbearing walls.
- Steel decking or wood planks have relatively short spans.
- Joists have limited overhang potential.

Wood

- Wood beams support structural planking or decking.
- Beams may be supported by girders, posts, or loadbearing walls.
- Concentrated loads and floor openings may require additional framing.
- Underside of floor structure may be left exposed; an applied ceiling is optional.

- Relatively small, closely spaced joists may be supported by beams or loadbearing walls.
- Subflooring, underlayment, and applied ceiling finishes have relatively short spans.
- Joist framing is flexible in shape and form.

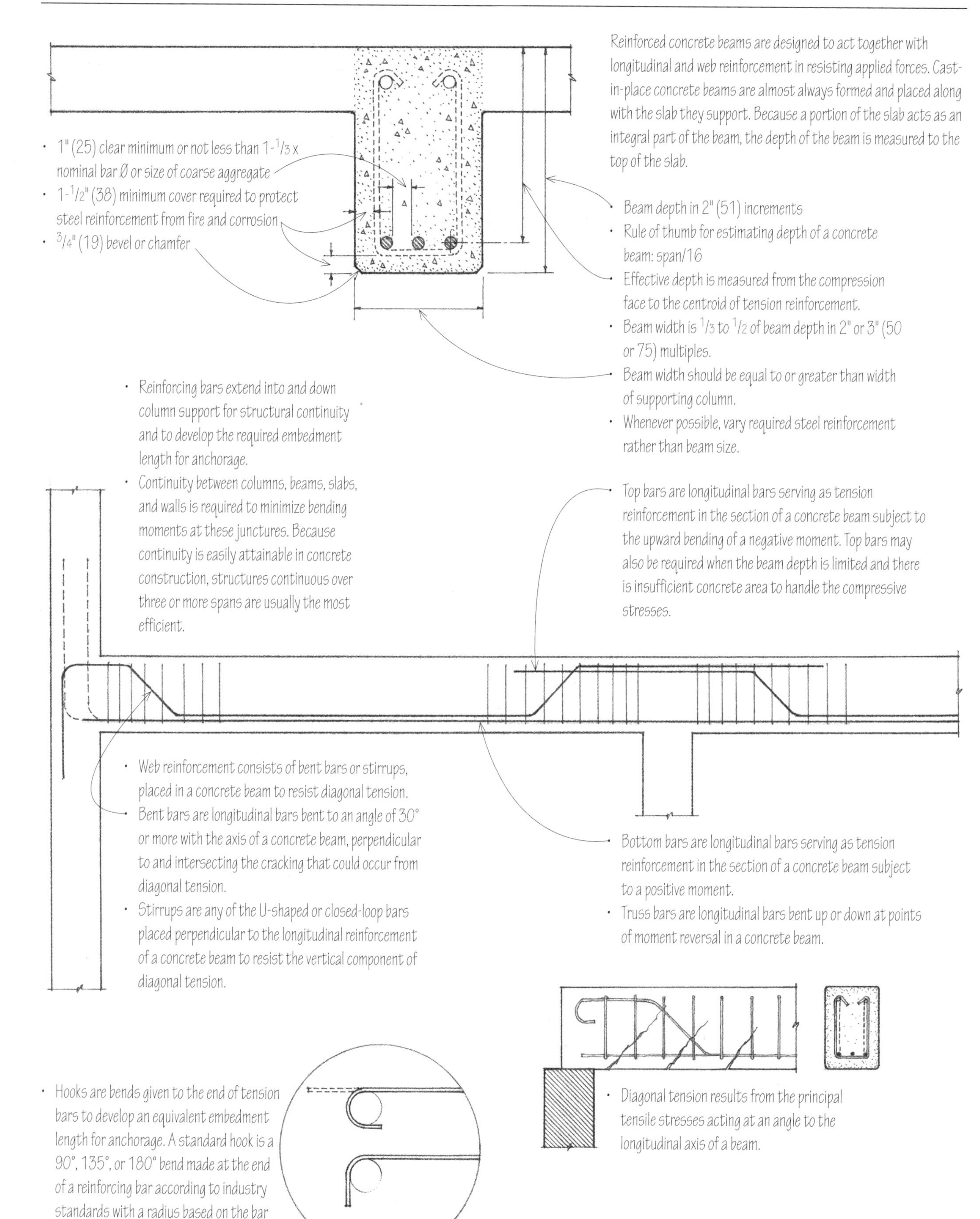
Reinforced concrete beams are designed to act together with longitudinal and web reinforcement in resisting applied forces. Cast-in-place concrete beams are almost always formed and placed along with the slab they support. Because a portion of the slab acts as an integral part of the beam, the depth of the beam is measured to the top of the slab.
• 1" (25) clear minimum or not less than 1-1/3 x nominal bar Ø or size of coarse aggregate
• 1-1/2" (38) minimum cover required to protect steel reinforcement from fire and corrosion
• 3/4" (19) bevel or chamfer
• Beam depth in 2" (51) increments
• Rule of thumb for estimating depth of a concrete beam: span/16
• Effective depth is measured from the compression face to the centroid of tension reinforcement.
• Beam width is 1/3 to 1/2 of beam depth in 2" or 3" (50 or 75) multiples.
• Beam width should be equal to or greater than width of supporting column.
• Whenever possible, vary required steel reinforcement rather than beam size.
• Reinforcing bars extend into and down column support for structural continuity and to develop the required embedment length for anchorage.
• Continuity between columns, beams, slabs, and walls is required to minimize bending moments at these junctures. Because continuity is easily attainable in concrete construction, structures continuous over three or more spans are usually the most efficient.
• Top bars are longitudinal bars serving as tension reinforcement in the section of a concrete beam subject to the upward bending of a negative moment. Top bars may also be required when the beam depth is limited and there is insufficient concrete area to handle the compressive stresses.
• Web reinforcement consists of bent bars or stirrups, placed in a concrete beam to resist diagonal tension.
• Bent bars are longitudinal bars bent to an angle of 30° or more with the axis of a concrete beam, perpendicular to and intersecting the cracking that could occur from diagonal tension.
• Stirrups are any of the U-shaped or closed-loop bars placed perpendicular to the longitudinal reinforcement of a concrete beam to resist the vertical component of diagonal tension.
• Bottom bars are longitudinal bars serving as tension reinforcement in the section of a concrete beam subject to a positive moment.
• Truss bars are longitudinal bars bent up or down at points of moment reversal in a concrete beam.
• Hooks are bends given to the end of tension bars to develop an equivalent embedment length for anchorage. A standard hook is a 90°, 135°, or 180° bend made at the end of a reinforcing bar according to industry standards with a radius based on the bar diameter.
• Diagonal tension results from the principal tensile stresses acting at an angle to the longitudinal axis of a beam.

Concrete slabs are plate structures that are reinforced to span either one or both directions of a structural bay. Consult a structural engineer and the building code for the required size, spacing, and placement of all reinforcement.

CSI MasterFormat™ 03 20 00: Concrete Reinforcing
CSI MasterFormat 03 30 00: Cast-in-Place Concrete
CSI MasterFormat 03 31 00: Structural Concrete

One-Way Slab

A one-way slab is uniformly thick, reinforced in one direction, and cast integrally with parallel supporting beams.

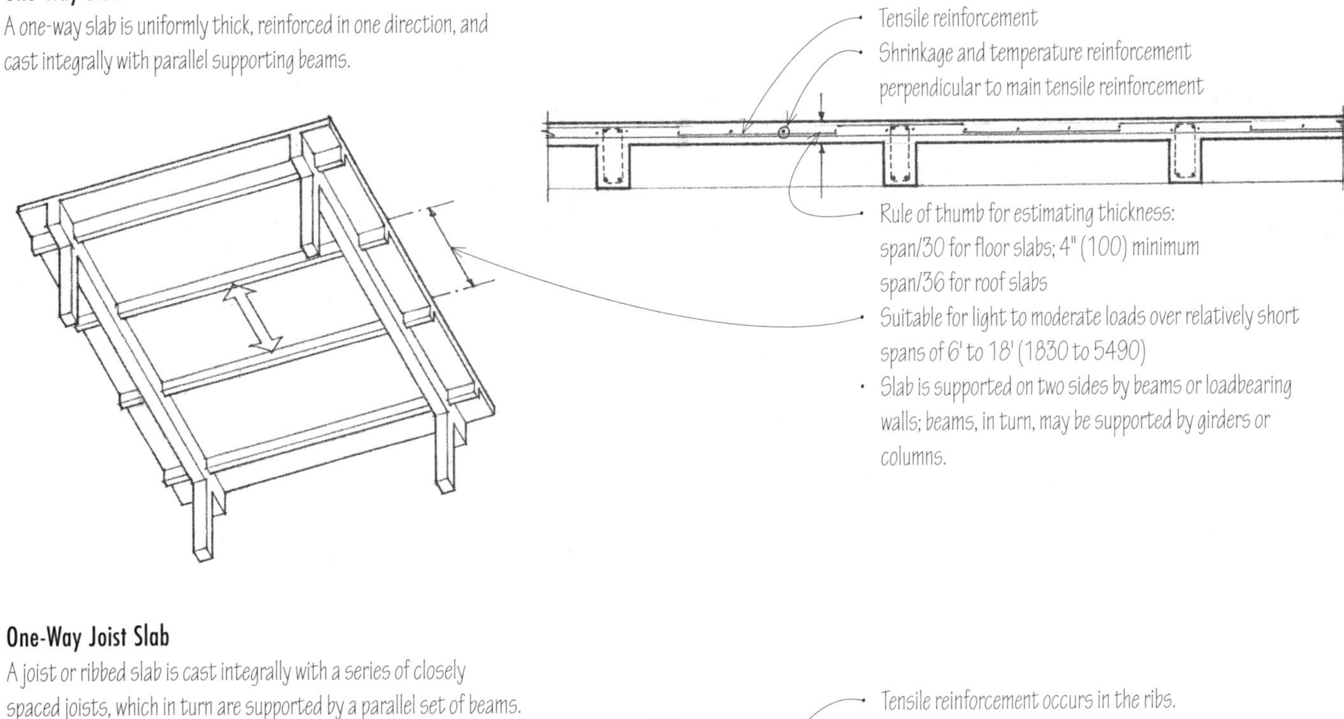

One-Way Joist Slab

A joist or ribbed slab is cast integrally with a series of closely spaced joists, which in turn are supported by a parallel set of beams. Designed as a series of T-beams, joist slabs are more suitable for longer spans and heavier loads than one-way slabs.

- Tensile reinforcement occurs in the ribs.
- Shrinkage and temperature reinforcement is placed in the slab.
- 3" to 4 -1/2" (75 to 115) slab depth: rule of thumb for total depth: span/24
- 5" to 9" (125 to 230) joist width
- Pans are reusable metal or fiberglass molds, available in 20" and 30" (510 and 760) widths and from 6" to 20" (150 to 510) depths in 2" (51) increments. Tapered sides allow for easier removal.
- Tapered endforms are used to thicken joist ends for greater shear resistance.
- Distribution rib is formed perpendicular to the joists in order to distribute possible load concentrations over a larger area: one required for spans between 20' and 30' (6 and 9 m), and not more than 15' (4.5 m) o.c. for spans over 30' (9 m).
- Joist band is a broad, shallow supporting beam that is economical to form because its depth is the same as that of the joists.
- Suitable for light to medium live loads over spans of 15' to 36' (4 to 10 m); longer spans may be possible with posttensioning.

- See 12.04–12.05 for a discussion of concrete as a construction material.

Two-Way Slab and Beam

A two-way slab of uniform thickness may be reinforced in two directions and cast integrally with supporting beams and columns on all four sides of square or nearly square bays. Two-way slab and beam construction is effective for medium spans and heavy loads, or when a high resistance to lateral forces is required. For economy, however, two-way slabs are usually constructed as flat slabs and plates without beams.

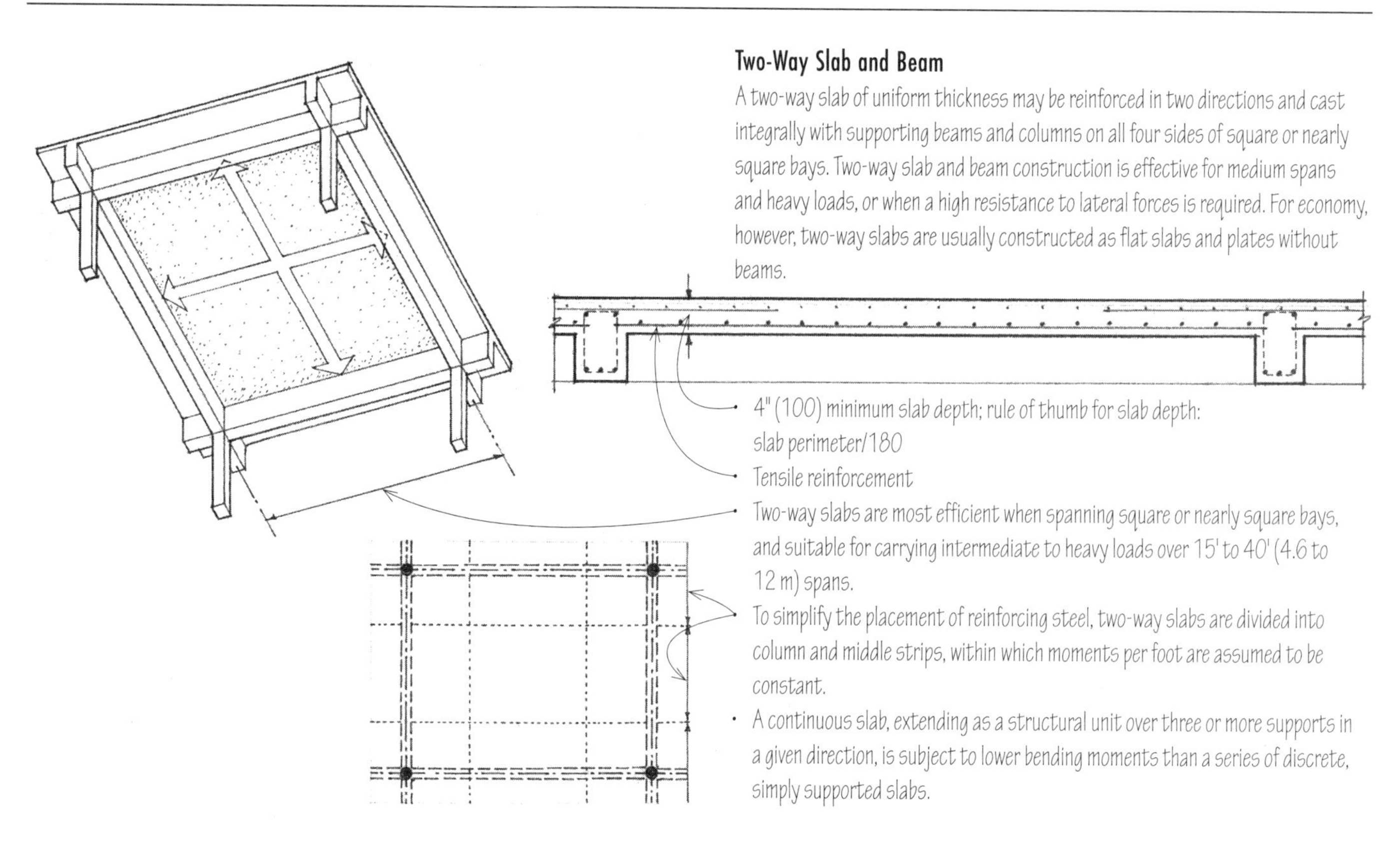

- 4" (100) minimum slab depth; rule of thumb for slab depth: slab perimeter/180
- Tensile reinforcement
- Two-way slabs are most efficient when spanning square or nearly square bays, and suitable for carrying intermediate to heavy loads over 15' to 40' (4.6 to 12 m) spans.
- To simplify the placement of reinforcing steel, two-way slabs are divided into column and middle strips, within which moments per foot are assumed to be constant.
- A continuous slab, extending as a structural unit over three or more supports in a given direction, is subject to lower bending moments than a series of discrete, simply supported slabs.

Two-Way Waffle Slab

A waffle slab is a two-way concrete slab reinforced by ribs in two directions. Waffle slabs are able to carry heavier loads and span longer distances than flat slabs.

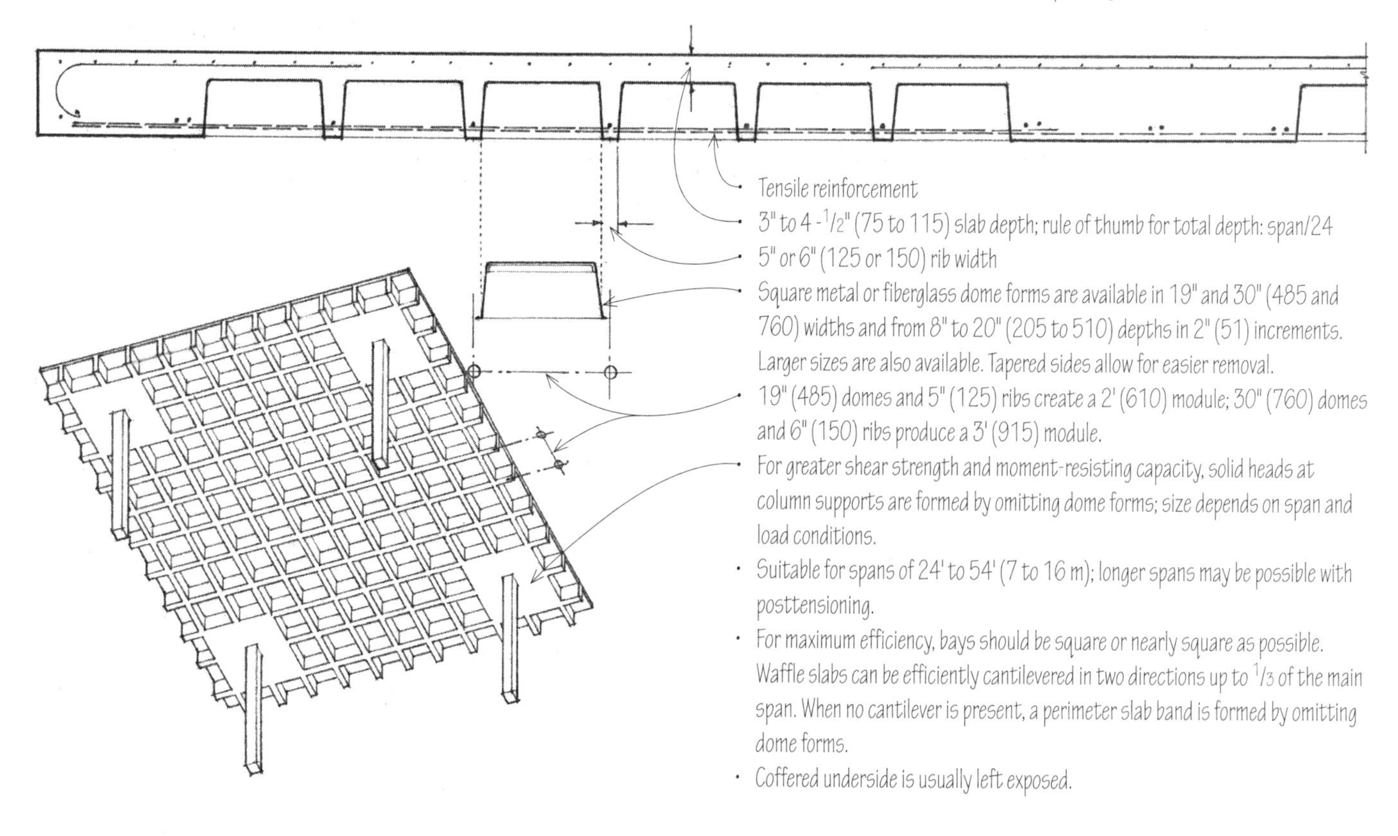

- Tensile reinforcement
- 3" to 4-1/2" (75 to 115) slab depth; rule of thumb for total depth: span/24
- 5" or 6" (125 or 150) rib width
- Square metal or fiberglass dome forms are available in 19" and 30" (485 and 760) widths and from 8" to 20" (205 to 510) depths in 2" (51) increments. Larger sizes are also available. Tapered sides allow for easier removal.
- 19" (485) domes and 5" (125) ribs create a 2' (610) module; 30" (760) domes and 6" (150) ribs produce a 3' (915) module.
- For greater shear strength and moment-resisting capacity, solid heads at column supports are formed by omitting dome forms; size depends on span and load conditions.
- Suitable for spans of 24' to 54' (7 to 16 m); longer spans may be possible with posttensioning.
- For maximum efficiency, bays should be square or nearly square as possible. Waffle slabs can be efficiently cantilevered in two directions up to 1/3 of the main span. When no cantilever is present, a perimeter slab band is formed by omitting dome forms.
- Coffered underside is usually left exposed.

Two-Way Flat Plate

A flat plate is a concrete slab of uniform thickness reinforced in two or more directions and supported directly by columns without beams or girders. Simplicity of forming, lower floor-to-floor heights, and some flexibility in column placement make flat plates practical for apartment and hotel construction.

- Tensile reinforcement
- 5" to 12" (125 to 305) slab depth; rule of thumb for slab depth: span/33
- Suitable for light live to moderate loads over relatively short spans of 12' to 24' (3.6 to 7 m)
- While a regular column grid is most appropriate, some flexibility in column placement is possible.
- Shear at column locations governs the thickness of a flat plate.
- Punching shear is the potentially high shearing stress developed by the reactive force of a column on a reinforced concrete slab.

Two-Way Flat Slab

A flat slab is a flat plate thickened at its column supports to increase its shear strength and moment-resisting capacity.

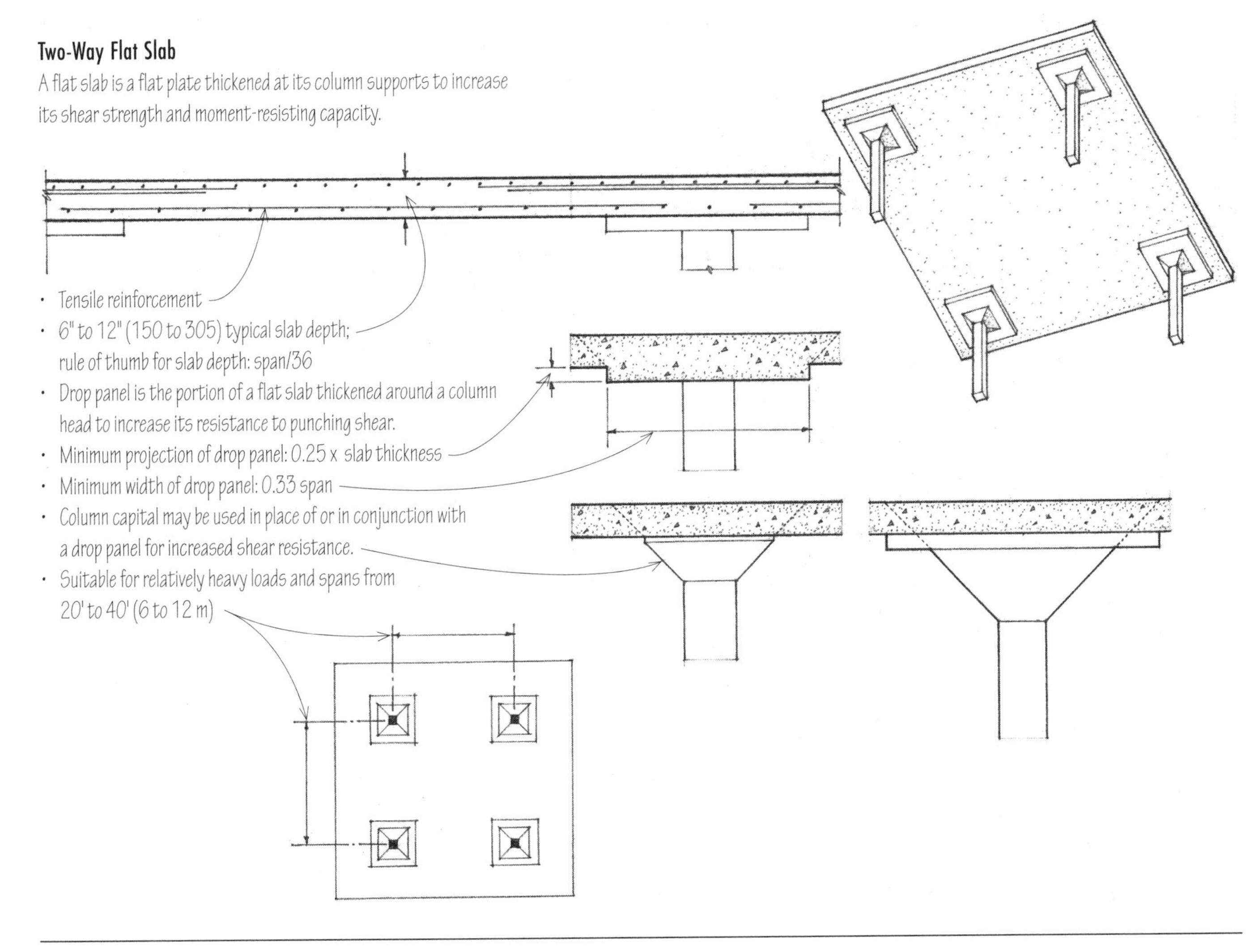

- Tensile reinforcement
- 6" to 12" (150 to 305) typical slab depth; rule of thumb for slab depth: span/36
- Drop panel is the portion of a flat slab thickened around a column head to increase its resistance to punching shear.
- Minimum projection of drop panel: 0.25 x slab thickness
- Minimum width of drop panel: 0.33 span
- Column capital may be used in place of or in conjunction with a drop panel for increased shear resistance.
- Suitable for relatively heavy loads and spans from 20' to 40' (6 to 12 m)

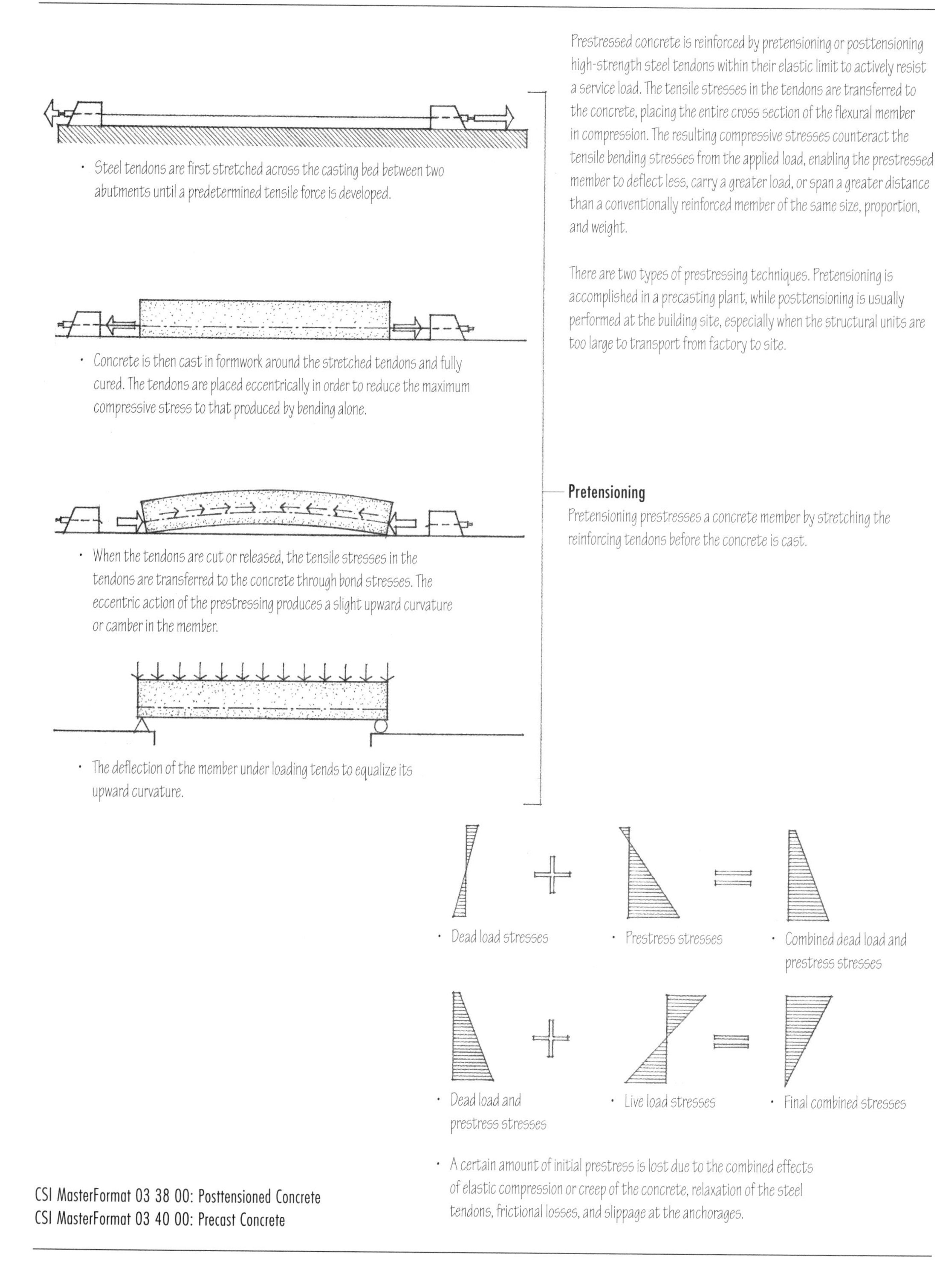

Prestressed concrete is reinforced by pretensioning or posttensioning high-strength steel tendons within their elastic limit to actively resist a service load. The tensile stresses in the tendons are transferred to the concrete, placing the entire cross section of the flexural member in compression. The resulting compressive stresses counteract the tensile bending stresses from the applied load, enabling the prestressed member to deflect less, carry a greater load, or span a greater distance than a conventionally reinforced member of the same size, proportion, and weight.

There are two types of prestressing techniques. Pretensioning is accomplished in a precasting plant, while posttensioning is usually performed at the building site, especially when the structural units are too large to transport from factory to site.

- Steel tendons are first stretched across the casting bed between two abutments until a predetermined tensile force is developed.
- Concrete is then cast in formwork around the stretched tendons and fully cured. The tendons are placed eccentrically in order to reduce the maximum compressive stress to that produced by bending alone.

Pretensioning

Pretensioning prestresses a concrete member by stretching the reinforcing tendons before the concrete is cast.

- When the tendons are cut or released, the tensile stresses in the tendons are transferred to the concrete through bond stresses. The eccentric action of the prestressing produces a slight upward curvature or camber in the member.
- The deflection of the member under loading tends to equalize its upward curvature.

- Dead load stresses
- Prestress stresses
- Combined dead load and prestress stresses
- Dead load and prestress stresses
- Live load stresses
- Final combined stresses

- A certain amount of initial prestress is lost due to the combined effects of elastic compression or creep of the concrete, relaxation of the steel tendons, frictional losses, and slippage at the anchorages.

CSI MasterFormat 03 38 00: Posttensioned Concrete
CSI MasterFormat 03 40 00: Precast Concrete

- The extremely high-strength steel tendons may be in the form of wire cables, bundled strands, or bars.

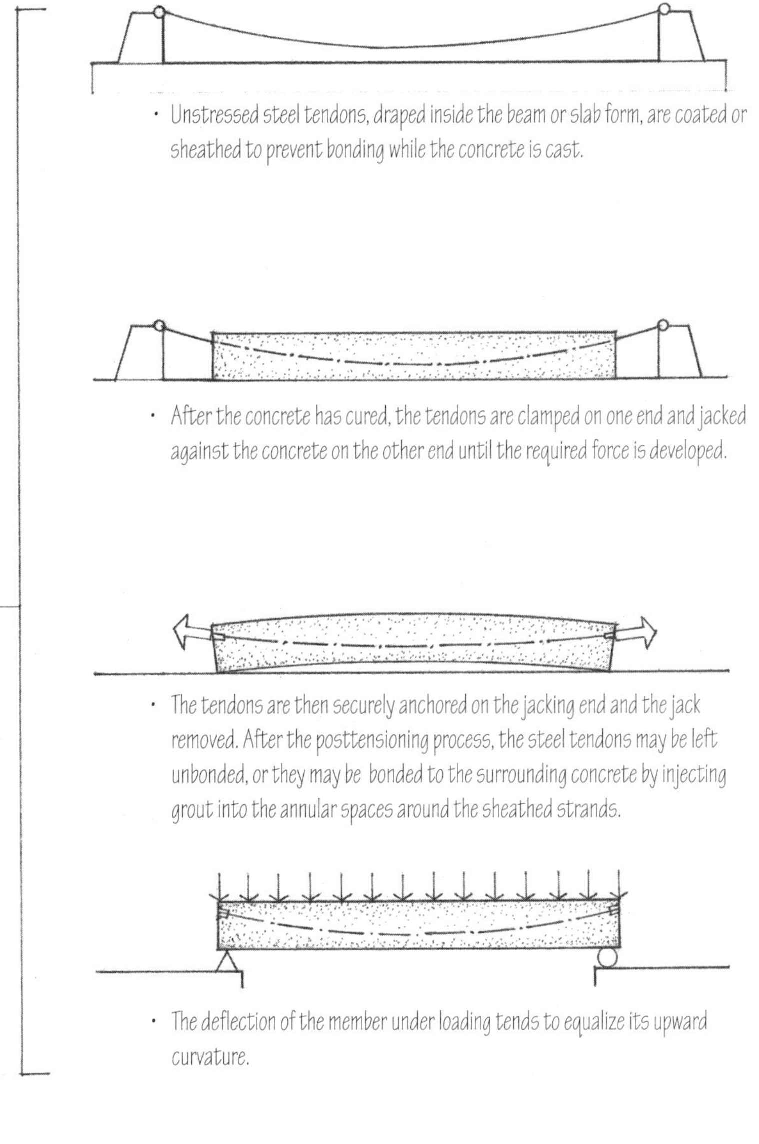

- Unstressed steel tendons, draped inside the beam or slab form, are coated or sheathed to prevent bonding while the concrete is cast.
- After the concrete has cured, the tendons are clamped on one end and jacked against the concrete on the other end until the required force is developed.
- The tendons are then securely anchored on the jacking end and the jack removed. After the posttensioning process, the steel tendons may be left unbonded, or they may be bonded to the surrounding concrete by injecting grout into the annular spaces around the sheathed strands.
- The deflection of the member under loading tends to equalize its upward curvature.

Posttensioning

Posttensioning is the prestressing of a concrete member by tensioning the reinforcing tendons after the concrete has set.

- Posttensioned members tend to shorten over time due to elastic compression, shrinkage, and creep. Adjoining elements that would be affected by this movement should be constructed after the posttensioning process is completed and be isolated from the posttensioned members with expansion joints.

- Load balancing is the concept of prestressing a concrete member with draped tendons, theoretically resulting in a state of zero deflection under a given loading condition.
- Draped tendons have a parabolic trajectory that mirrors the moment diagram of a uniformly distributed gravity load. When tensioned, the tendons produce a variable eccentricity that responds to the variation in applied bending moment along the length of the member.
- Depressed tendons approximate the curve of a draped tendon with straight-line segments. They are used in the pretensioning process because the prestressing force does not allow for draping the tendons. Harped tendons are a series of depressed tendons having varying slopes.

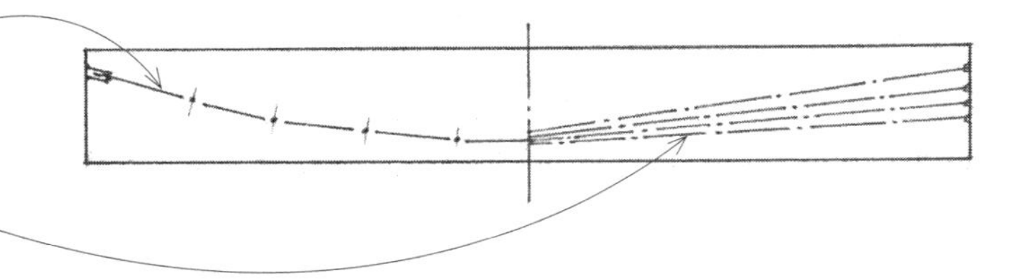

Fresh concrete must be shaped and supported by formwork until it cures and can support itself. This formwork is often designed as a separate structural system by an engineer because of the considerable weight and fluid pressure a concrete mass can exert on it.

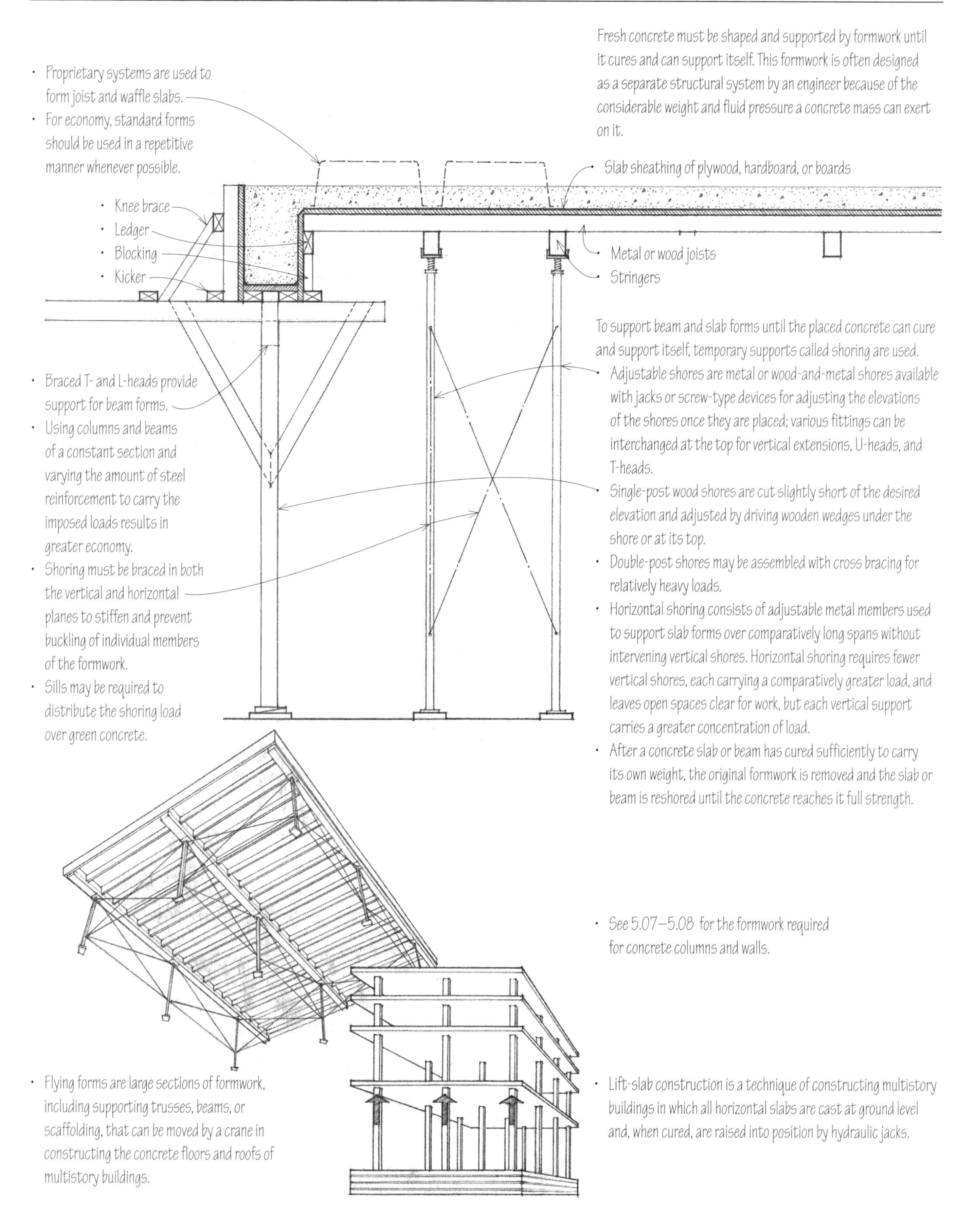

- Proprietary systems are used to form joist and waffle slabs.
- For economy, standard forms should be used in a repetitive manner whenever possible.

- Braced T- and L-heads provide support for beam forms.
- Using columns and beams of a constant section and varying the amount of steel reinforcement to carry the imposed loads results in greater economy.
- Shoring must be braced in both the vertical and horizontal planes to stiffen and prevent buckling of individual members of the formwork.
- Sills may be required to distribute the shoring load over green concrete.

To support beam and slab forms until the placed concrete can cure and support itself, temporary supports called shoring are used.

- Adjustable shores are metal or wood-and-metal shores available with jacks or screw-type devices for adjusting the elevations of the shores once they are placed; various fittings can be interchanged at the top for vertical extensions, U-heads, and T-heads.
- Single-post wood shores are cut slightly short of the desired elevation and adjusted by driving wooden wedges under the shore or at its top.
- Double-post shores may be assembled with cross bracing for relatively heavy loads.
- Horizontal shoring consists of adjustable metal members used to support slab forms over comparatively long spans without intervening vertical shores. Horizontal shoring requires fewer vertical shores, each carrying a comparatively greater load, and leaves open spaces clear for work, but each vertical support carries a greater concentration of load.
- After a concrete slab or beam has cured sufficiently to carry its own weight, the original formwork is removed and the slab or beam is reshored until the concrete reaches it full strength.

- See 5.07–5.08 for the formwork required for concrete columns and walls.

- Flying forms are large sections of formwork, including supporting trusses, beams, or scaffolding, that can be moved by a crane in constructing the concrete floors and roofs of multistory buildings.

- Lift-slab construction is a technique of constructing multistory buildings in which all horizontal slabs are cast at ground level and, when cured, are raised into position by hydraulic jacks.

Precast concrete slabs, beams, and structural tees are one-way spanning units that may be supported by sitecast concrete, precast concrete, or masonry bearing walls, or by steel, sitecast concrete, or precast concrete frames. The precast units are manufactured with normal-density or structural lightweight concrete and prestressed for greater structural efficiency, which results in less depth, reduced weight, and longer spans.

The units are cast and steam-cured in a plant off-site, transported to the construction site, and set in place as rigid components with cranes. The size and proportion of the units may be limited by the means of transportation. Fabrication in a factory environment enables the units to have a consistent quality of strength, durability, and finish, and eliminates the need for on-site formwork. The modular nature of the standard-sized units, however, may not be suitable for irregular building shapes.

Span of precast slab

- Small openings may be cut in the field.
- Narrow openings parallel to slab span are preferred.
- Engineering analysis is required for wide openings.

- A 2" to 3-1/2" (51 to 90) concrete topping reinforced with steel fabric or reinforcing bars bonds with the precast units to form a composite structural unit.
- Grout key

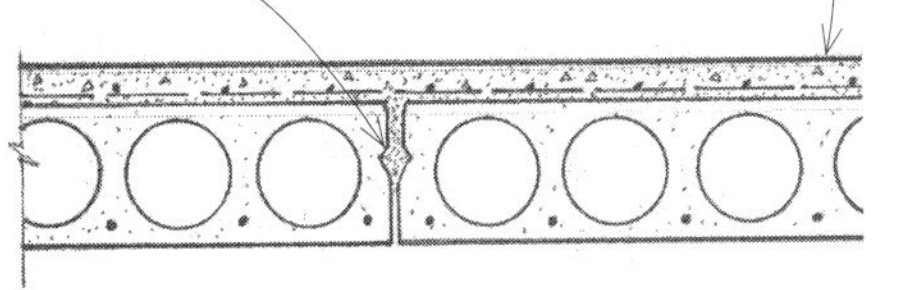

- The topping also conceals any surface irregularities, increases the fire-resistance rating of the slab, and accommodates underfloor conduit for wiring.
- When the flooring is to be pad and carpet, the topping may be omitted if smooth-surface units are used.

Precast slabs may be supported by a structural frame of sitecast or precast concrete girders and columns, or by a loadbearing wall of masonry, sitecast concrete, or precast concrete.

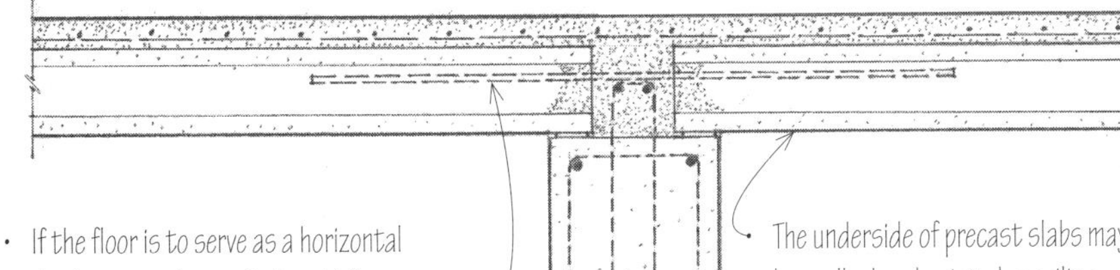

- If the floor is to serve as a horizontal diaphragm and transfer lateral forces to shear walls, steel reinforcement must tie the precast slab units to each other over their supports and at their end bearings.

The underside of precast slabs may be caulked and painted; a ceiling finish may also be applied to or be suspended from the slabs.

CSI MasterFormat 03 40 00: Precast Concrete
CSI MasterFormat 03 50 00: Cast Decks and Underlayment

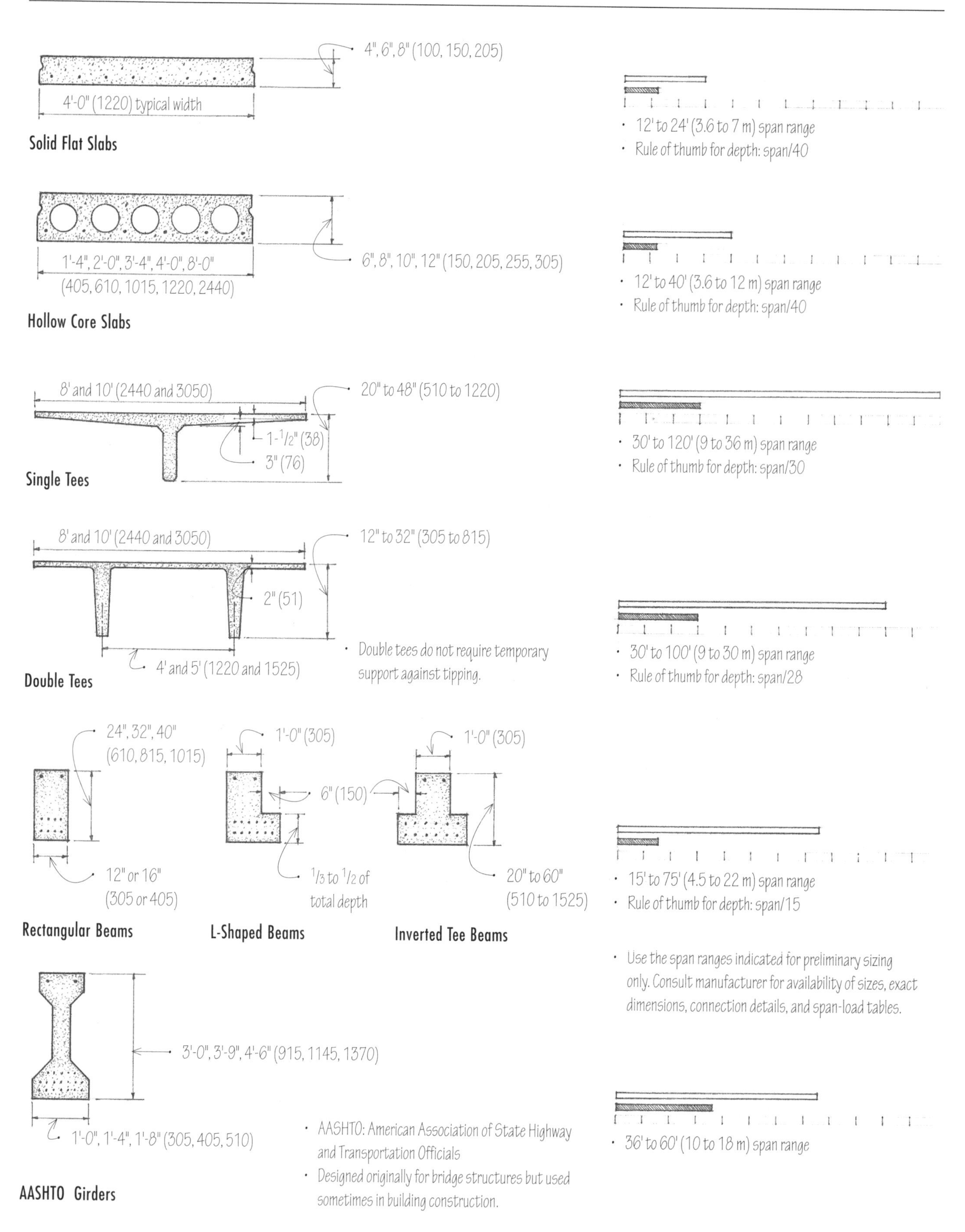
4", 6", 8" (100, 150, 205)
4'-0" (1220) typical width
Solid Flat Slabs
• 12' to 24' (3.6 to 7 m) span range
• Rule of thumb for depth: span/40
1'-4", 2'-0", 3'-4", 4'-0", 8'-0"
(405, 610, 1015, 1220, 2440)
6", 8", 10", 12" (150, 205, 255, 305)
Hollow Core Slabs
• 12' to 40' (3.6 to 12 m) span range
• Rule of thumb for depth: span/40
8' and 10' (2440 and 3050)
20" to 48" (510 to 1220)
1-1/2" (38)
3" (76)
Single Tees
• 30' to 120' (9 to 36 m) span range
• Rule of thumb for depth: span/30
8' and 10' (2440 and 3050)
12" to 32" (305 to 815)
2" (51)
4' and 5' (1220 and 1525)
Double Tees
• Double tees do not require temporary support against tipping.
• 30' to 100' (9 to 30 m) span range
• Rule of thumb for depth: span/28
24", 32", 40"
(610, 815, 1015)
12" or 16"
(305 or 405)
Rectangular Beams
1'-0" (305)
6" (150)
1/3 to 1/2 of total depth
L-Shaped Beams
1'-0" (305)
20" to 60"
(510 to 1525)
Inverted Tee Beams
• 15' to 75' (4.5 to 22 m) span range
• Rule of thumb for depth: span/15
• Use the span ranges indicated for preliminary sizing only. Consult manufacturer for availability of sizes, exact dimensions, connection details, and span-load tables.
3'-0", 3'-9", 4'-6" (915, 1145, 1370)
1'-0", 1'-4", 1'-8" (305, 405, 510)
AASHTO Girders
• AASHTO: American Association of State Highway and Transportation Officials
• Designed originally for bridge structures but used sometimes in building construction.
• 36' to 60' (10 to 18 m) span range

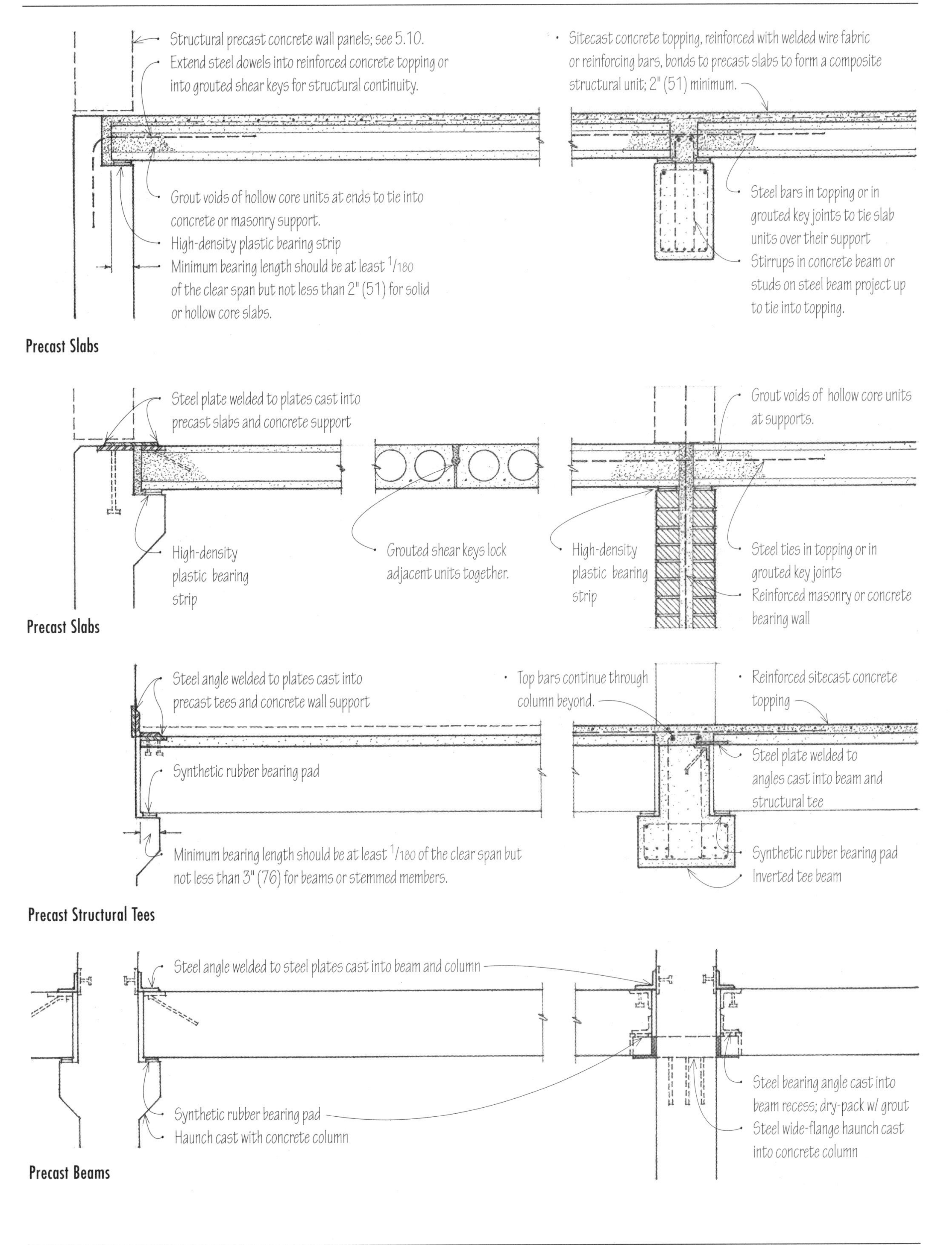

Structural precast concrete wall panels; see 5.10.
Extend steel dowels into reinforced concrete topping or into grouted shear keys for structural continuity.
Grout voids of hollow core units at ends to tie into concrete or masonry support.
High-density plastic bearing strip
Minimum bearing length should be at least 1/180 of the clear span but not less than 2" (51) for solid or hollow core slabs.
Sitecast concrete topping, reinforced with welded wire fabric or reinforcing bars, bonds to precast slabs to form a composite structural unit; 2" (51) minimum.
Steel bars in topping or in grouted key joints to tie slab units over their support
Stirrups in concrete beam or studs on steel beam project up to tie into topping.
Precast Slabs
Steel plate welded to plates cast into precast slabs and concrete support
High-density plastic bearing strip
Grouted shear keys lock adjacent units together.
Grout voids of hollow core units at supports.
High-density plastic bearing strip
Steel ties in topping or in grouted key joints
Reinforced masonry or concrete bearing wall
Precast Slabs
Steel angle welded to plates cast into precast tees and concrete wall support
Synthetic rubber bearing pad
Minimum bearing length should be at least 1/180 of the clear span but not less than 3" (76) for beams or stemmed members.
Top bars continue through column beyond.
Reinforced sitecast concrete topping
Steel plate welded to angles cast into beam and structural tee
Synthetic rubber bearing pad
Inverted tee beam
Precast Structural Tees
Steel angle welded to steel plates cast into beam and column
Synthetic rubber bearing pad
Haunch cast with concrete column
Steel bearing angle cast into beam recess; dry-pack w/ grout
Steel wide-flange haunch cast into concrete column
Precast Beams

Structural steel girders, beams, and columns are used to construct a skeleton frame for structures ranging in size from one-story buildings to skyscrapers. Because structural steel is difficult to work on-site, it is normally cut, shaped, and drilled in a fabrication shop according to design specifications; this can result in relatively fast, precise construction of a structural frame. Structural steel may be left exposed in unprotected noncombustible construction, but because steel can lose strength rapidly in a fire, fire-rated assemblies or coatings are required to qualify as fire-resistive construction. In exposed conditions, corrosion resistance is also required. See 12.08 for a discussion of steel as a construction material; see the Appendix for fire-rated steel assemblies.

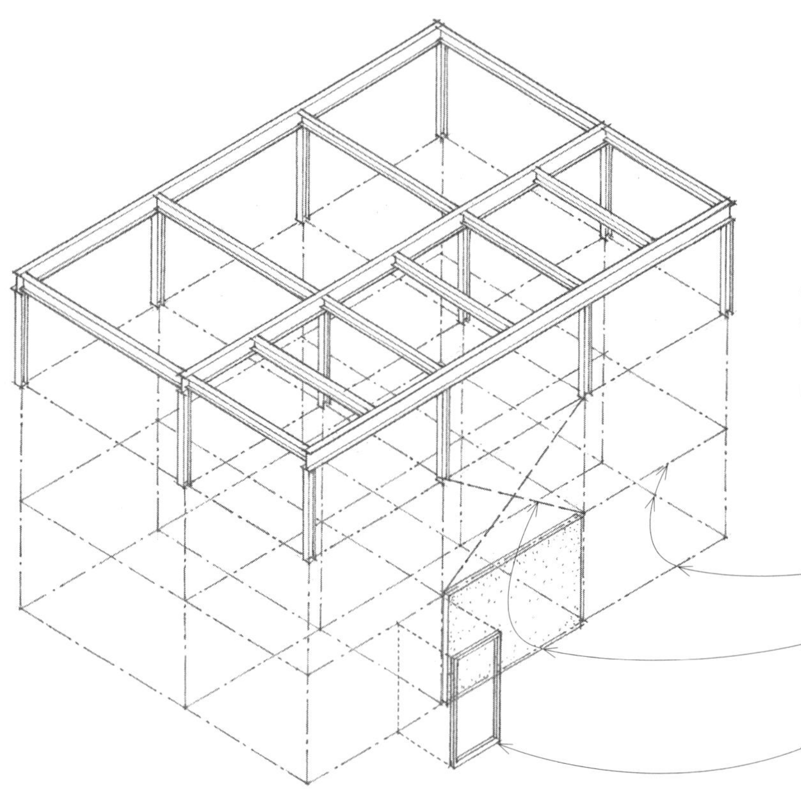

- Steel framing is most efficient when the girder and beam supports are laid out along a regular grid.
- Resistance to lateral wind or earthquake forces requires the use of shear walls, diagonal bracing, or rigid framing with moment-resisting connections.
- For nonbearing or curtain wall options, see 7.24.

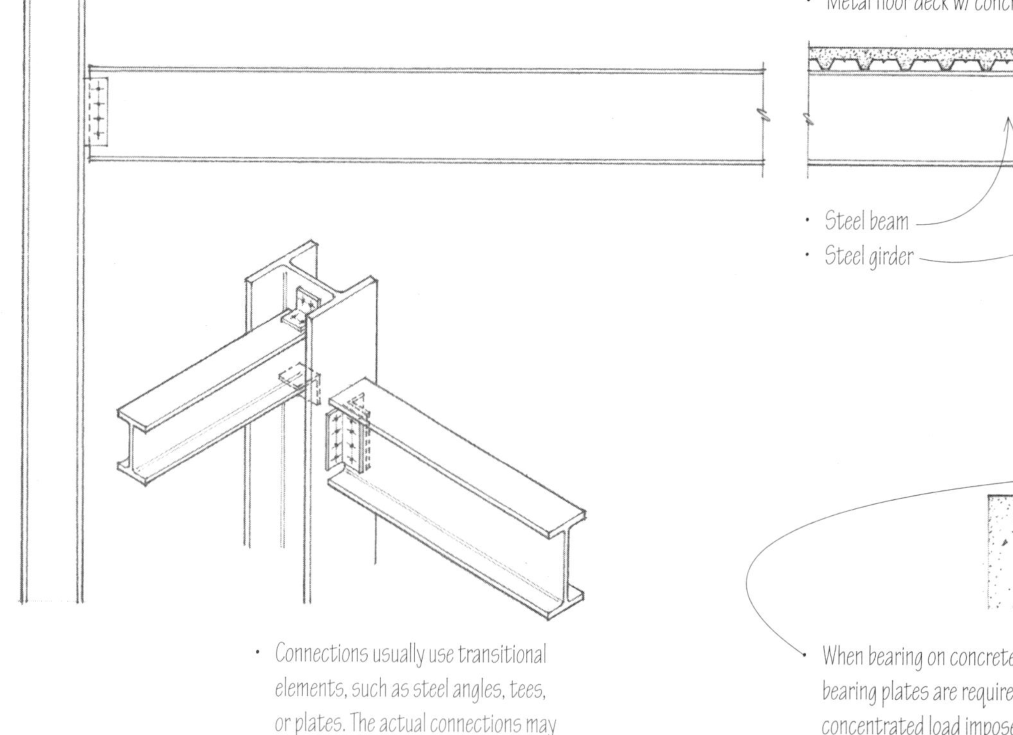

- Connections usually use transitional elements, such as steel angles, tees, or plates. The actual connections may be riveted but are more often bolted or welded.

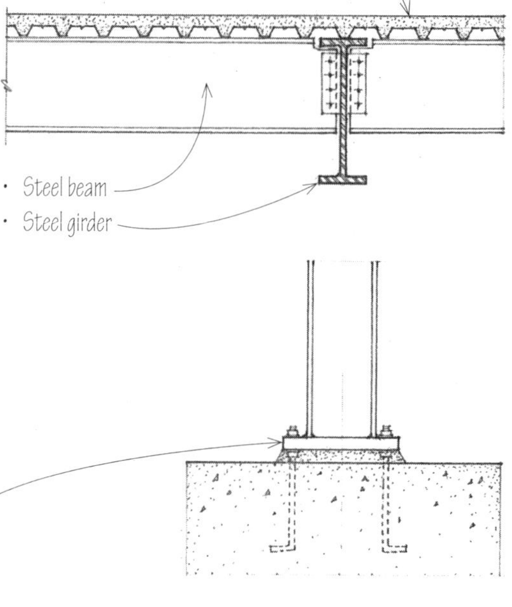

- When bearing on concrete or masonry, steel bearing plates are required to distribute the concentrated load imposed by a column or beam so that the resultant unit bearing pressure does not exceed the allowable unit stress for the supporting material.

CSI MasterFormat 05 12 00: Structural Steel Framing

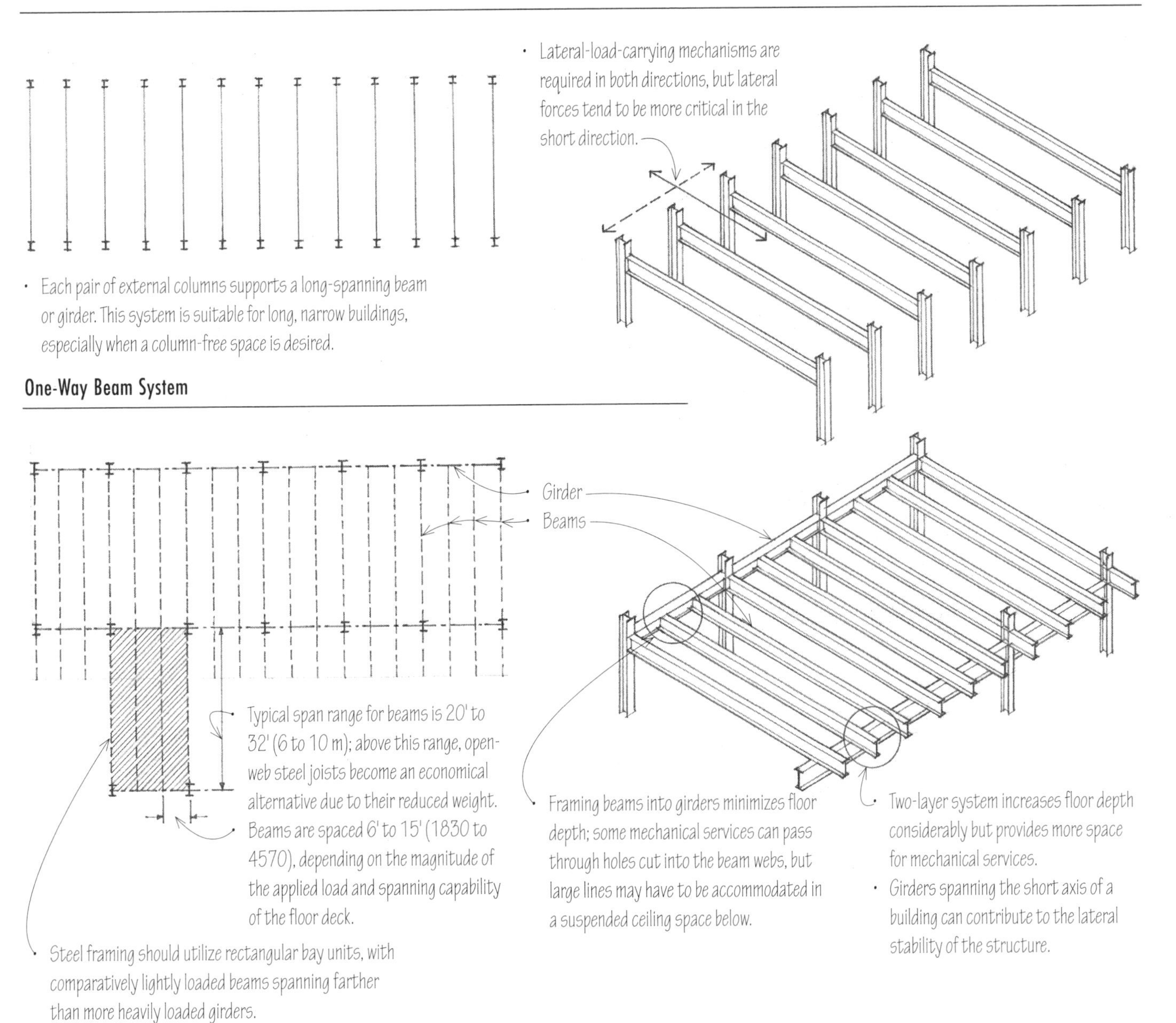

- Lateral-load-carrying mechanisms are required in both directions, but lateral forces tend to be more critical in the short direction.
- Each pair of external columns supports a long-spanning beam or girder. This system is suitable for long, narrow buildings, especially when a column-free space is desired.

One-Way Beam System

- Typical span range for beams is 20' to 32' (6 to 10 m); above this range, open-web steel joists become an economical alternative due to their reduced weight.
- Beams are spaced 6' to 15' (1830 to 4570), depending on the magnitude of the applied load and spanning capability of the floor deck.
- Steel framing should utilize rectangular bay units, with comparatively lightly loaded beams spanning farther than more heavily loaded girders.
- Framing beams into girders minimizes floor depth; some mechanical services can pass through holes cut into the beam webs, but large lines may have to be accommodated in a suspended ceiling space below.
- Two-layer system increases floor depth considerably but provides more space for mechanical services.
- Girders spanning the short axis of a building can contribute to the lateral stability of the structure.

Two-Way Beam System

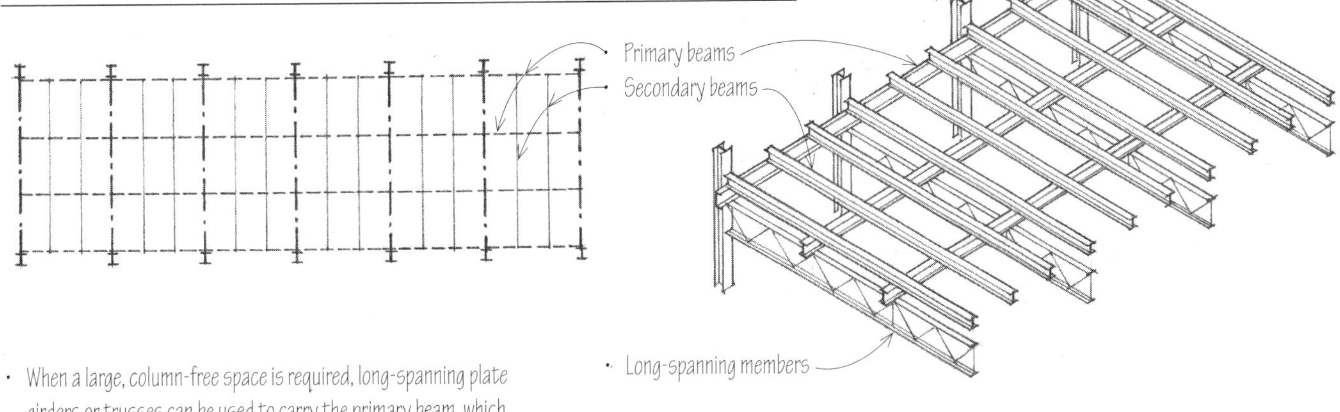

- When a large, column-free space is required, long-spanning plate girders or trusses can be used to carry the primary beam, which in turn support a layer of secondary beams.

Triple Beam System

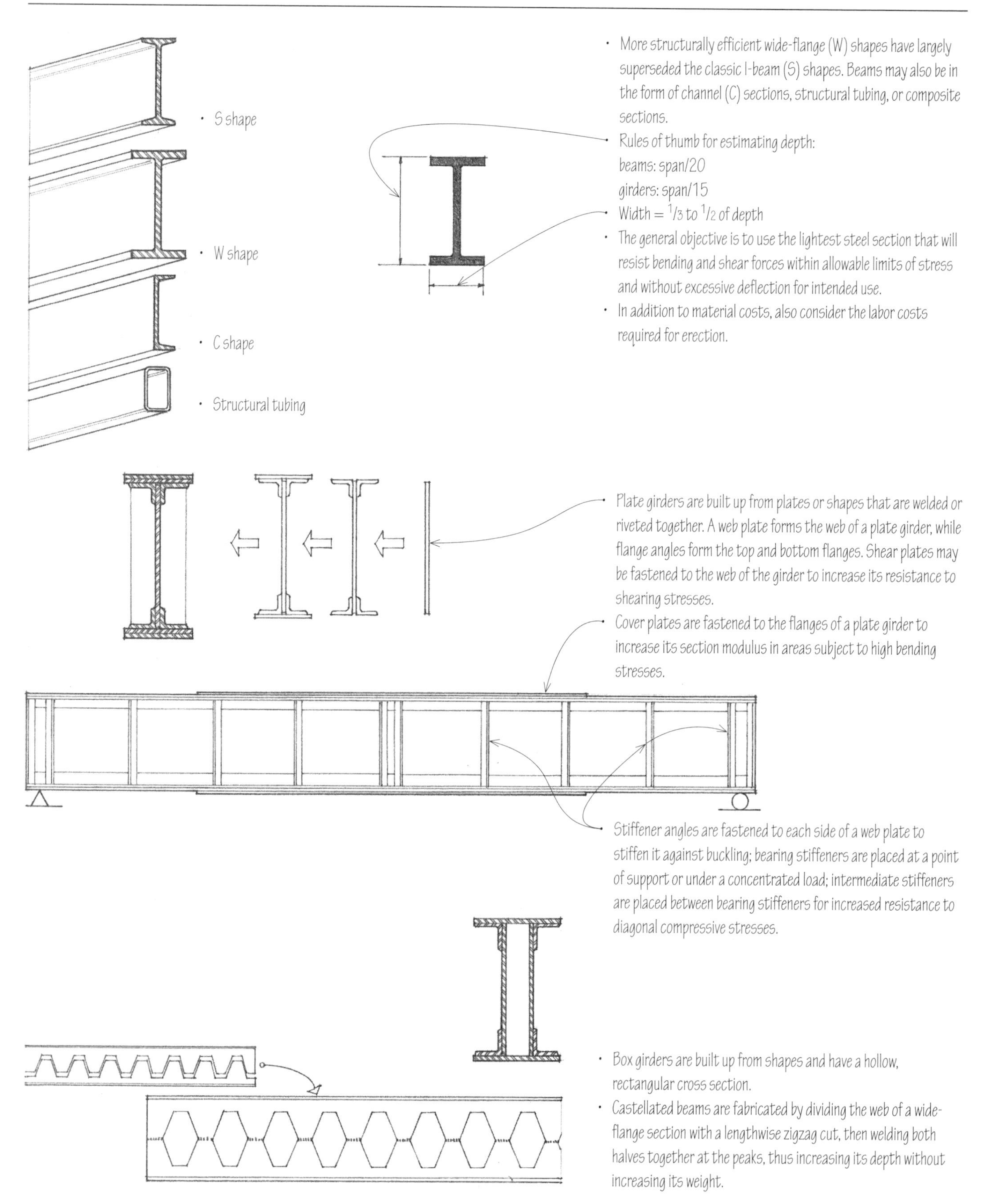

- More structurally efficient wide-flange (W) shapes have largely superseded the classic I-beam (S) shapes. Beams may also be in the form of channel (C) sections, structural tubing, or composite sections.
- Rules of thumb for estimating depth:
  beams: span/20
  girders: span/15
- Width = 1/3 to 1/2 of depth
- The general objective is to use the lightest steel section that will resist bending and shear forces within allowable limits of stress and without excessive deflection for intended use.
- In addition to material costs, also consider the labor costs required for erection.

- Plate girders are built up from plates or shapes that are welded or riveted together. A web plate forms the web of a plate girder, while flange angles form the top and bottom flanges. Shear plates may be fastened to the web of the girder to increase its resistance to shearing stresses.
- Cover plates are fastened to the flanges of a plate girder to increase its section modulus in areas subject to high bending stresses.

- Stiffener angles are fastened to each side of a web plate to stiffen it against buckling; bearing stiffeners are placed at a point of support or under a concentrated load; intermediate stiffeners are placed between bearing stiffeners for increased resistance to diagonal compressive stresses.

- Box girders are built up from shapes and have a hollow, rectangular cross section.
- Castellated beams are fabricated by dividing the web of a wide-flange section with a lengthwise zigzag cut, then welding both halves together at the peaks, thus increasing its depth without increasing its weight.

CSI MasterFormat 05 12 23: Structural Steel for Buildings

There are many ways in which steel connections can be made, using different types of connectors and various combinations of bolts and welds. Refer to the American Institute of Steel Construction's (AISC's) *Manual of Steel Construction* for steel section properties and dimensions, allowable load tables for beams and columns, and requirements for bolted and welded connections. In addition to strength and degree of rigidity, connections should be evaluated for economy of fabrication and erection, and for visual appearance if the structure is exposed to view.

The strength of a connection depends on the sizes of the members and the connecting tees, angles, or plates, as well as the configuration of bolts or welds used. The AISC defines three types of steel framing that govern the sizes of members and the methods for their connections: moment connections, shear connections, and semi-rigid connections.

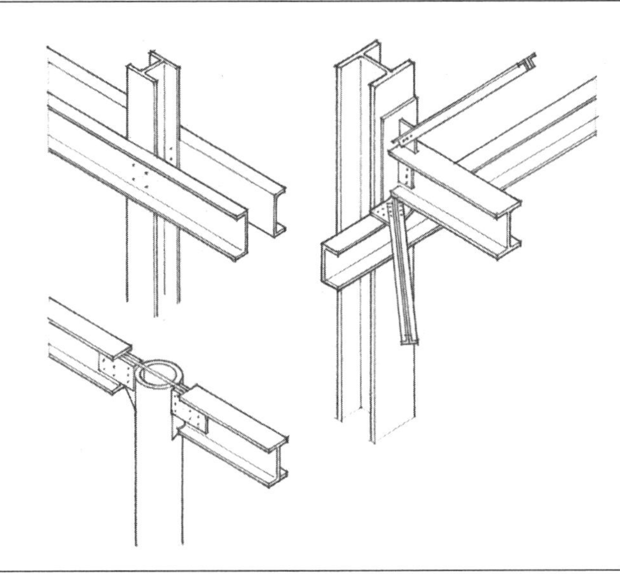

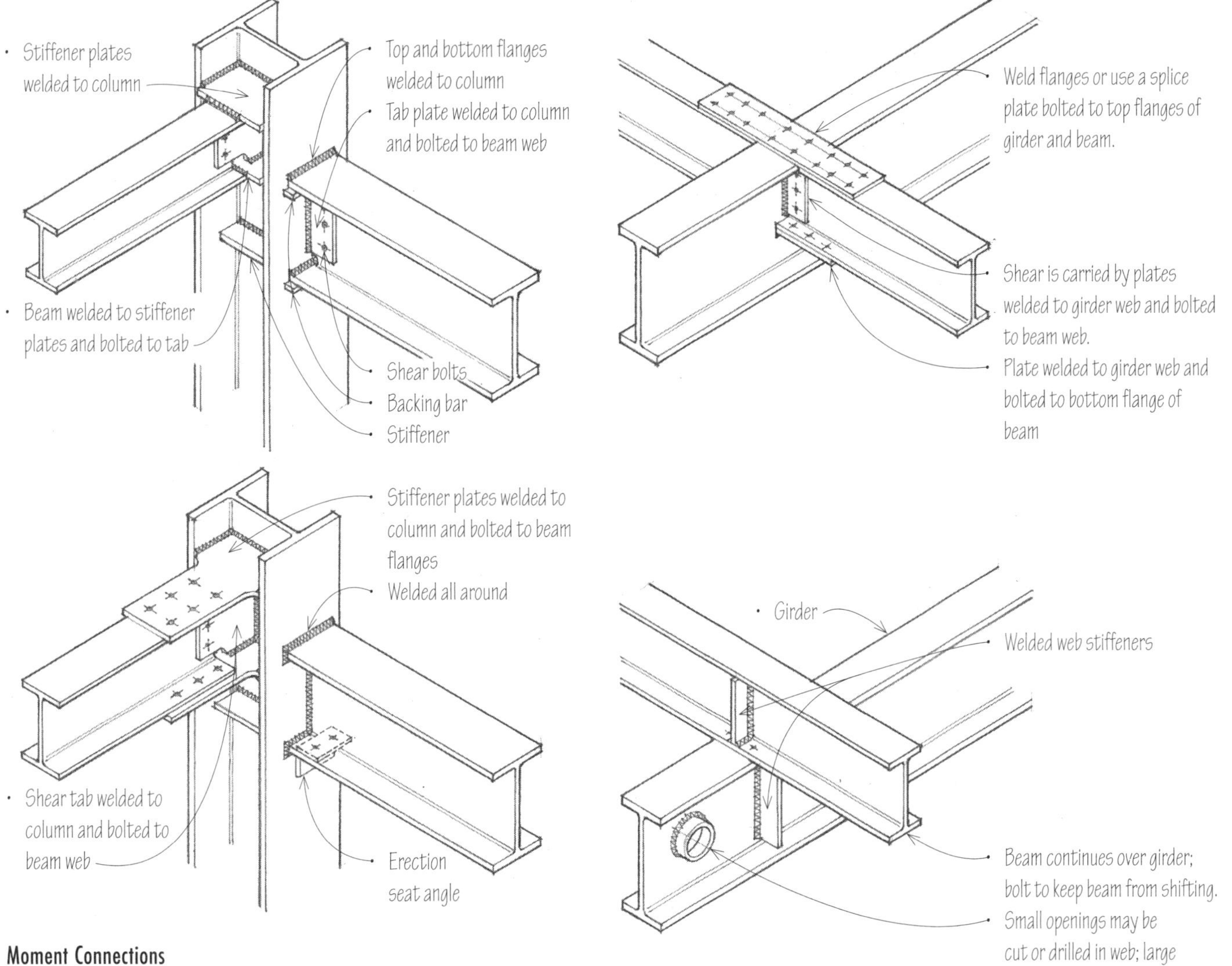

Moment Connections

AISC Type 1—Rigid Frame—connections are able to hold their original angle under loading by developing a specified resisting moment, usually by means of plates welded or bolted to the beam flanges and the supporting column.

- A framed connection is a shear-resisting steel connection made by welding or bolting the web of a beam to the supporting column or girder with two angles or a single tab plate.

- A seated connection is a shear-resisting steel connection made by welding or bolting the flanges of a beam to the supporting column with a seat angle below and a stabilizing angle above.

- A seated connection may be stiffened to resist large beam reactions, usually by means of a vertical plate or pair of angles directly below the horizontal component of the seat angle.

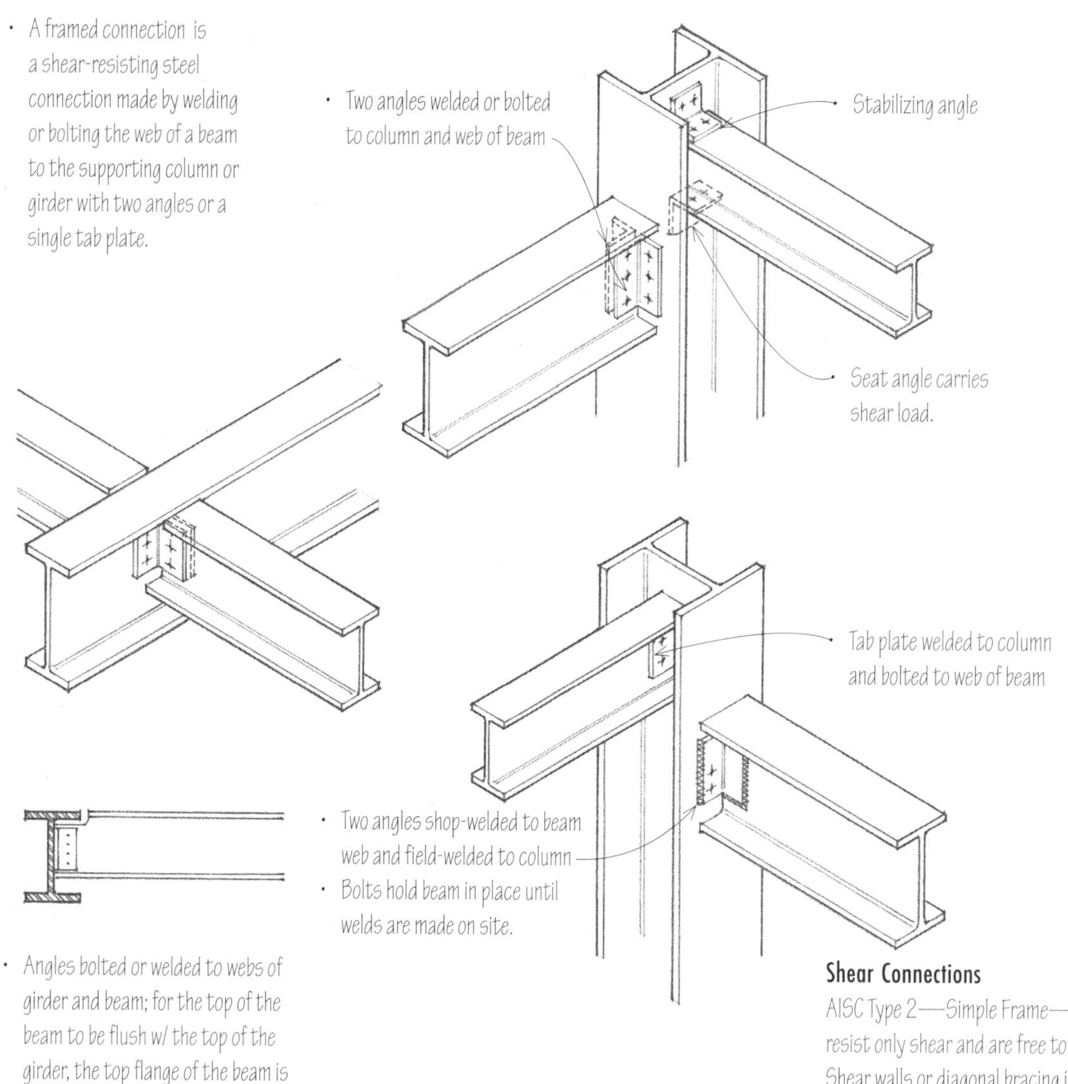

- Angles bolted or welded to webs of girder and beam; for the top of the beam to be flush w/ the top of the girder, the top flange of the beam is coped or cut away.

Shear Connections

AISC Type 2—Simple Frame—connections are made to resist only shear and are free to rotate under gravity loads. Shear walls or diagonal bracing is required for lateral stability of the structure.

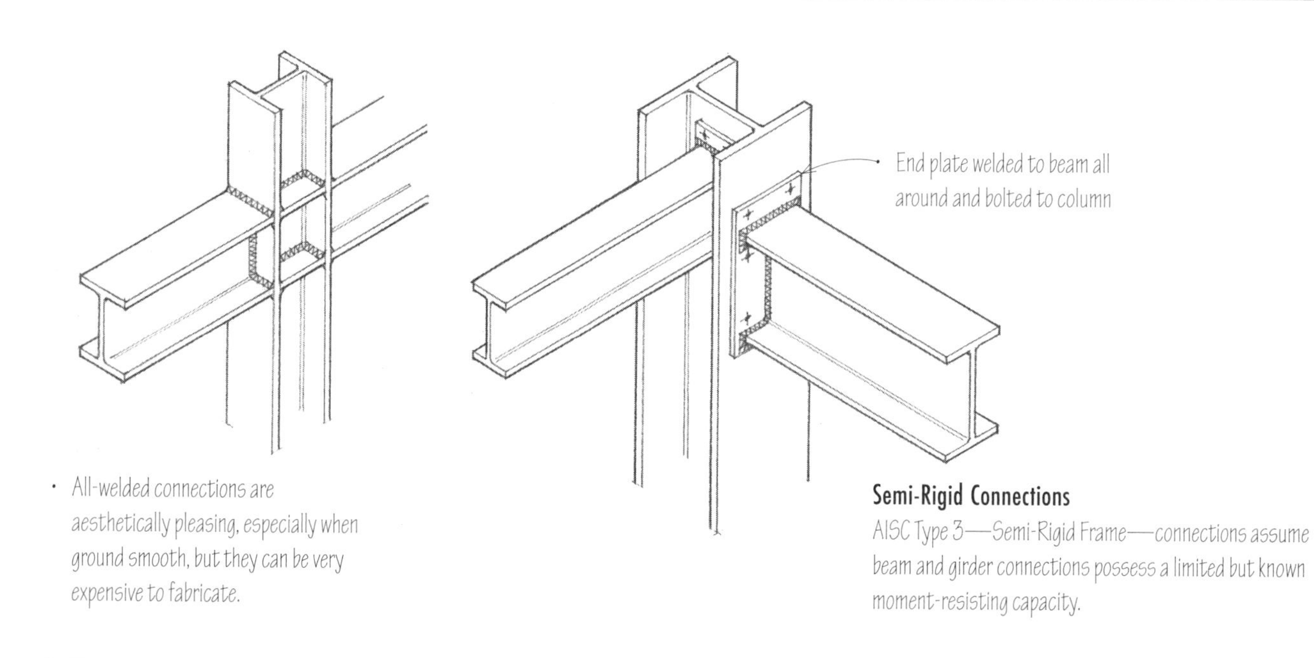

- All-welded connections are aesthetically pleasing, especially when ground smooth, but they can be very expensive to fabricate.

Semi-Rigid Connections

AISC Type 3—Semi-Rigid Frame—connections assume beam and girder connections possess a limited but known moment-resisting capacity.

Open-web joists are lightweight, shop-fabricated steel members having a trussed web. A K series joist has a web consisting of a single bent bar, running in a zigzag pattern between the upper and lower chords. LH and DLH series joists have heavier web and chord members for increased loads and spans.

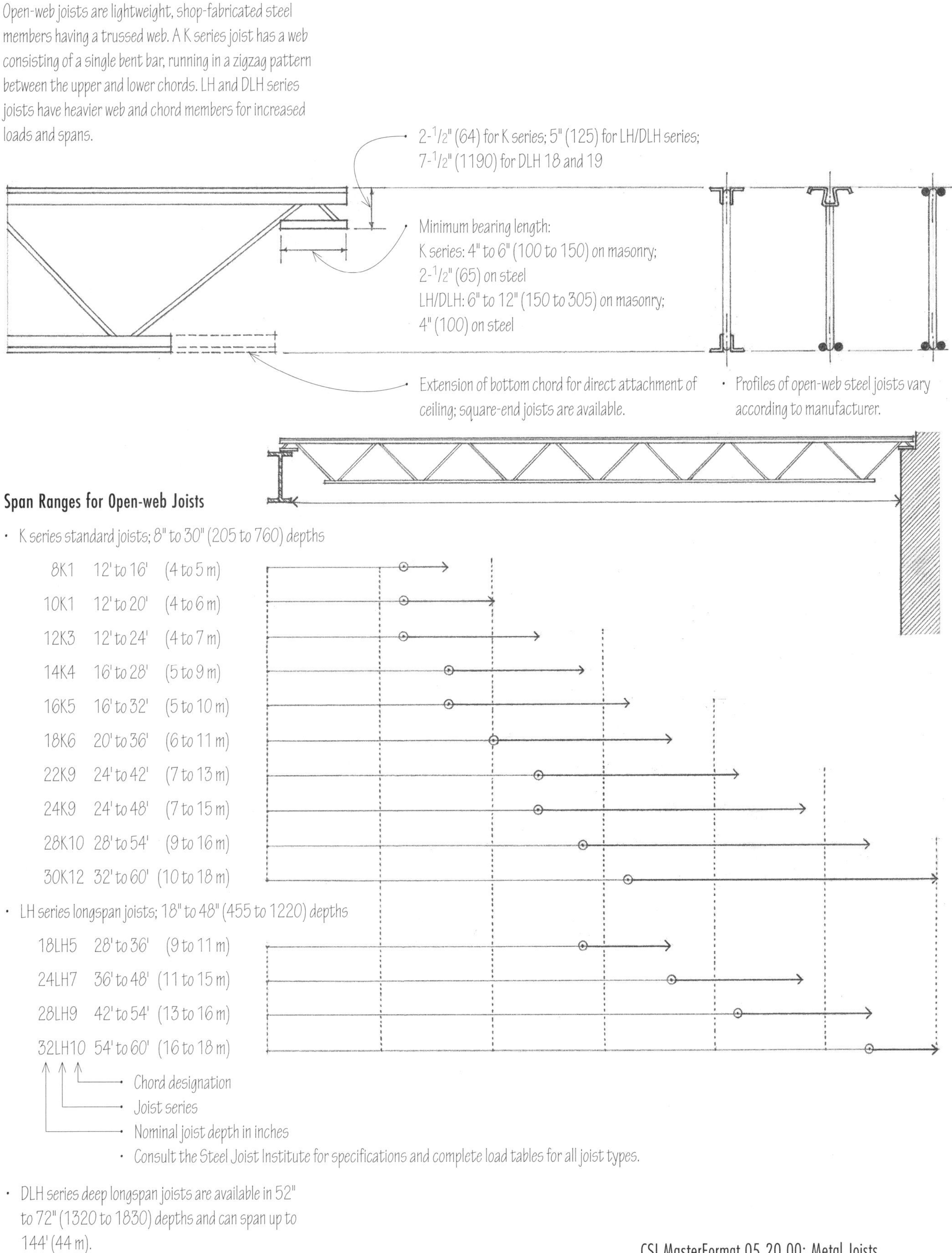

Span Ranges for Open-web Joists

- K series standard joists; 8" to 30" (205 to 760) depths

8K1	12' to 16'	(4 to 5 m)
10K1	12' to 20'	(4 to 6 m)
12K3	12' to 24'	(4 to 7 m)
14K4	16' to 28'	(5 to 9 m)
16K5	16' to 32'	(5 to 10 m)
18K6	20' to 36'	(6 to 11 m)
22K9	24' to 42'	(7 to 13 m)
24K9	24' to 48'	(7 to 15 m)
28K10	28' to 54'	(9 to 16 m)
30K12	32' to 60'	(10 to 18 m)

- LH series longspan joists; 18" to 48" (455 to 1220) depths

18LH5	28' to 36'	(9 to 11 m)
24LH7	36' to 48'	(11 to 15 m)
28LH9	42' to 54'	(13 to 16 m)
32LH10	54' to 60'	(16 to 18 m)

- Chord designation
- Joist series
- Nominal joist depth in inches
- Consult the Steel Joist Institute for specifications and complete load tables for all joist types.

- DLH series deep longspan joists are available in 52" to 72" (1320 to 1830) depths and can span up to 144' (44 m).

CSI MasterFormat 05 20 00: Metal Joists
CSI MasterFormat 05 21 00: Steel Joist Framing

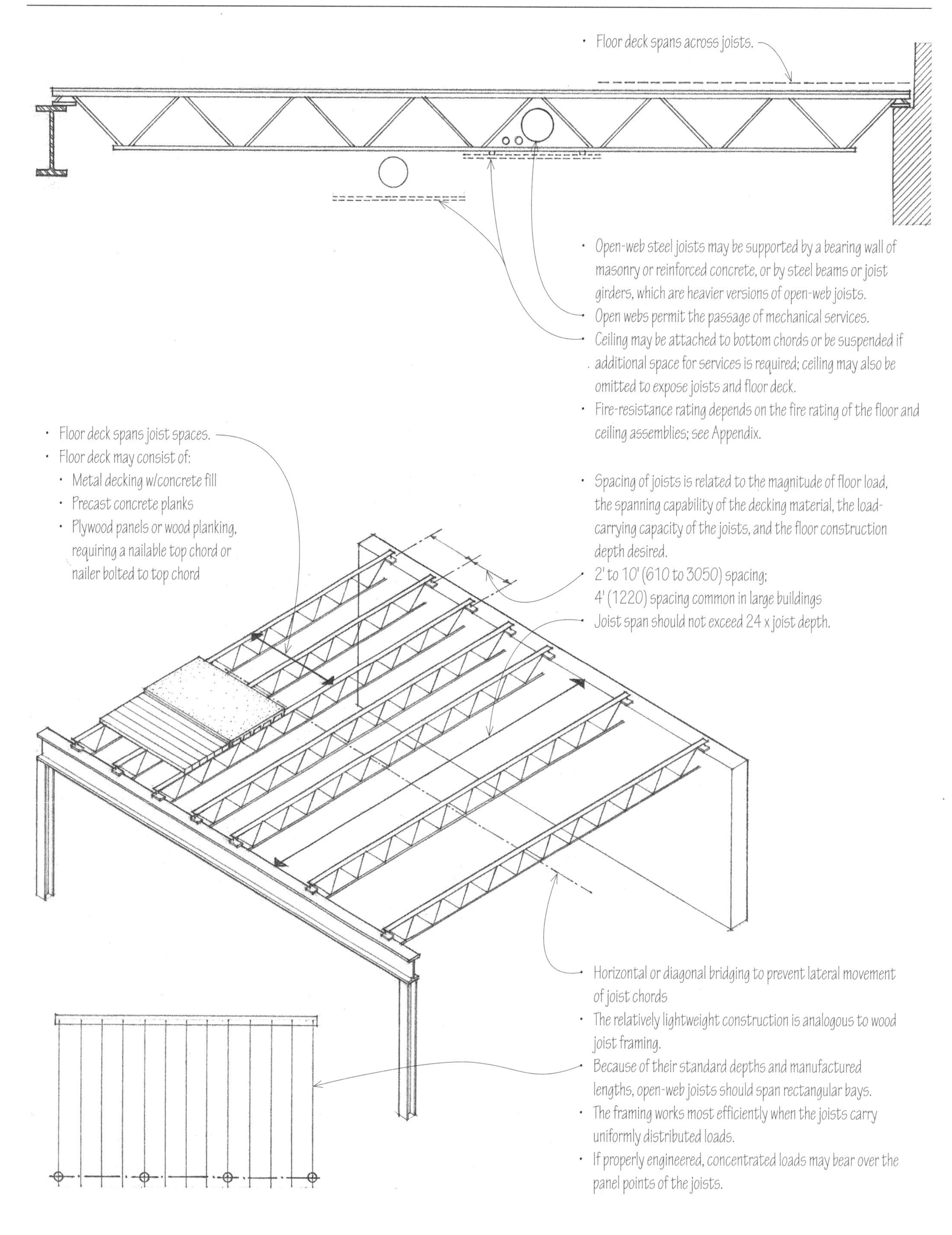

• Floor deck spans across joists.
• Open-web steel joists may be supported by a bearing wall of masonry or reinforced concrete, or by steel beams or joist girders, which are heavier versions of open-web joists.
• Open webs permit the passage of mechanical services.
• Ceiling may be attached to bottom chords or be suspended if additional space for services is required; ceiling may also be omitted to expose joists and floor deck.
• Fire-resistance rating depends on the fire rating of the floor and ceiling assemblies; see Appendix.
• Floor deck spans joist spaces.
• Floor deck may consist of:
• Metal decking w/concrete fill
• Precast concrete planks
• Plywood panels or wood planking, requiring a nailable top chord or nailer bolted to top chord
• Spacing of joists is related to the magnitude of floor load, the spanning capability of the decking material, the load-carrying capacity of the joists, and the floor construction depth desired.
• 2' to 10' (610 to 3050) spacing;
4' (1220) spacing common in large buildings
• Joist span should not exceed 24 x joist depth.
• Horizontal or diagonal bridging to prevent lateral movement of joist chords
• The relatively lightweight construction is analogous to wood joist framing.
• Because of their standard depths and manufactured lengths, open-web joists should span rectangular bays.
• The framing works most efficiently when the joists carry uniformly distributed loads.
• If properly engineered, concentrated loads may bear over the panel points of the joists.

- Reinforced concrete or masonry bearing wall
- Minimum bearing length:
  4" to 6" (100 to 150) for K series joists;
  6" to 12" (150 to 305) for LH/DLH series joists
- Proportion bearing area so that unit bearing stress does not exceed the allowable unit stress for the wall material.
- Secure every joist to a steel bearing plate anchored in the wall.

- Steel beam or joist girder
- Minimum bearing length:
  2-1/2" (65) for K series joists;
  4" (100) for LH/DLH series joists
- Two 1/8" (54) fillet welds 1" (25) long or 1/2" (13) ø bolt
- For LH/DLH series joists, use two 1/4" (57) fillet welds 2" (51) long or two 3/4" (19) ø bolts.

- Horizontal or diagonal bridging is required to prevent lateral movement of joist chords.
- Bridging is spaced from 10' to 20' (3050 to 6095) o.c., depending on joist span and chord size.
- Horizontal bridging angles are welded to top and bottom chords.
- Use diagonal bridging angles for LH/DLH series joists.
- Weld or bolt bridging to clip angles secured to masonry wall or steel edge beam.

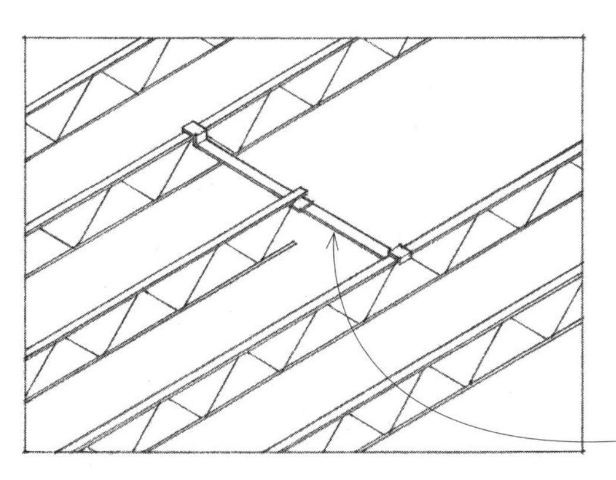

- Small openings may be framed with steel angle headers supported by trimmer joists. Large openings require structural steel framing.
- Header to support tail joist

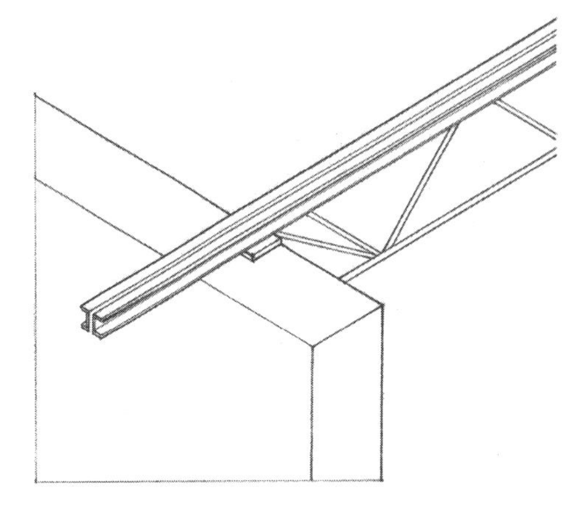

- Limited overhangs are possible by extending the top chords.
- Extended ends of steel angles or channel sections are available for short overhangs. For K series joists, the overhang may extend 5'-6" (1675), with an allowable load of 300 psf (1 psf = 0.479 kPa).

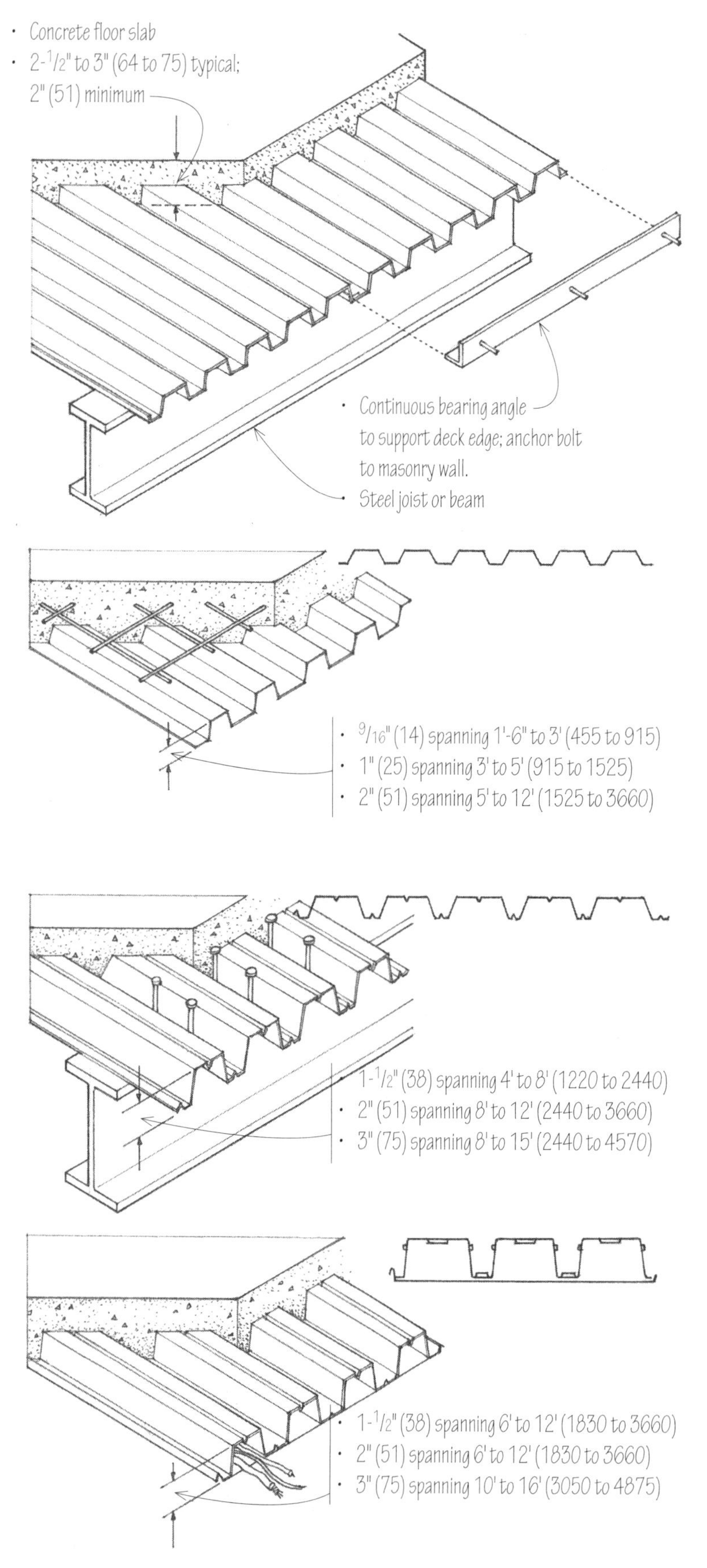

Metal decking is corrugated to increase its stiffness and spanning capability. The floor deck serves as a working platform during construction and as formwork for a sitecast concrete slab.

- The decking panels are secured with puddle-welds or shear studs welded through the decking to the supporting steel joists or beams.
- The panels are fastened to each other along their sides with screws, welds, or button punching standing seams.
- If the deck is to serve as a structural diaphragm and transfer lateral loads to shear walls, its entire perimeter must be welded to steel supports. In addition, more stringent requirements for support and side lap fastening may apply.

There are three major types of metal decking.

Form Decking

- Form decking serves as permanent formwork for a reinforced concrete slab until the slab can support itself and its live load.

Composite Decking

- Composite decking serves as tensile reinforcement for the concrete slab to which it is bonded with embossed rib patterns. Composite action between the concrete slab and the floor beams or joists can be achieved by welding shear studs through the decking to the supporting beam below.

Cellular Decking

- Cellular decking is manufactured by welding a corrugated sheet to a flat steel sheet, forming a series of spaces or raceways for electrical and communications wiring; special cutouts are available for floor outlets. The decking may serve as an acoustic ceiling when the perforated cells are filled with glass fiber.

- Rule of thumb for overall depth: span/24
- Consult the manufacturer for patterns, widths, lengths, gauges, finishes, and allowable spans.

CSI MasterFormat 05 30 00: Metal Decking

Light-gauge steel joists are manufactured by cold-forming sheet or strip steel. The resulting steel joists are lighter, more dimensionally stable, and can span longer distances than their wood counterparts but conduct more heat and require more energy to process and manufacture. The cold-formed steel joists can be easily cut and assembled with simple tools into a floor structure that is lightweight, noncombustible, and dampproof. As in wood light-frame construction, the framing contains cavities for utilities and thermal insulation and accepts a wide range of finishes.

- Nominal depths: 6", 8", 10", 12", 14" (150, 205, 255, 305, 355)
- Flange widths: 1-1/2", 1-3/4", 2", 2-1/2" (38, 45, 51, 64)
- Gauges: 14 through 22

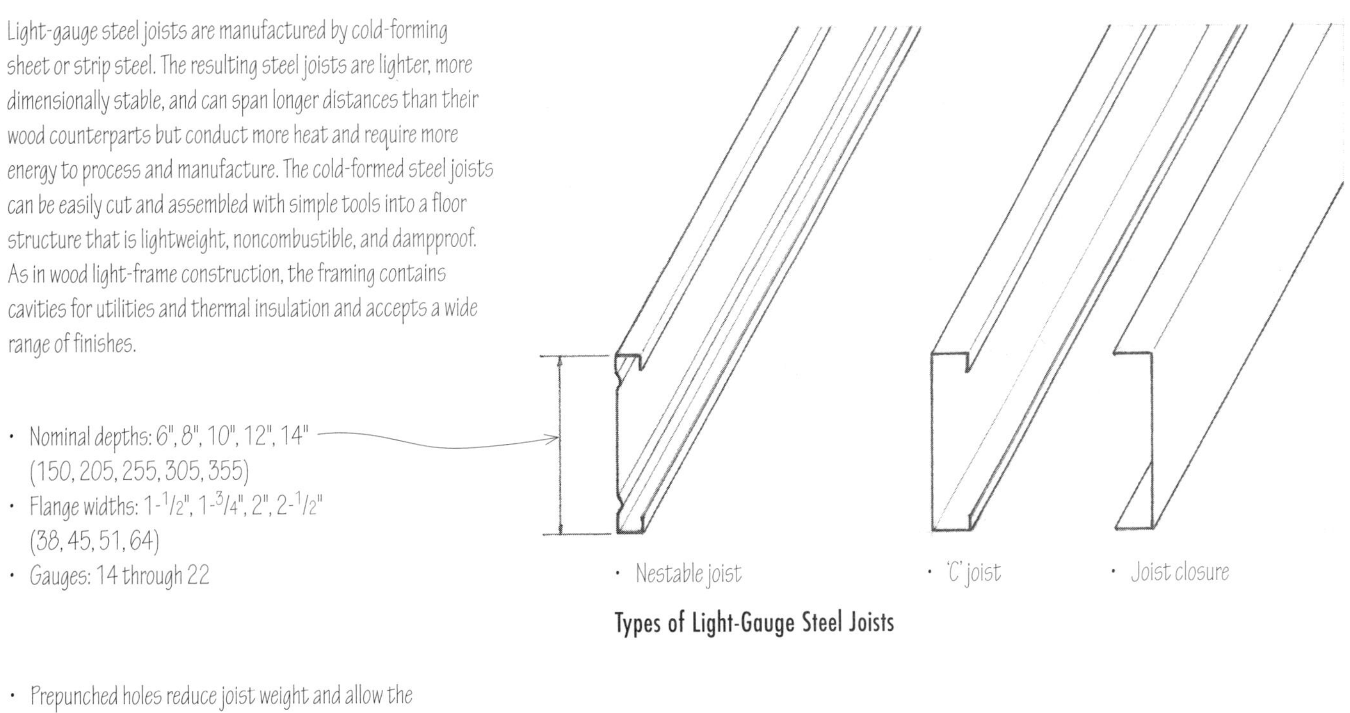

Types of Light-Gauge Steel Joists

- Prepunched holes reduce joist weight and allow the passage of piping, wiring, and bridging straps.

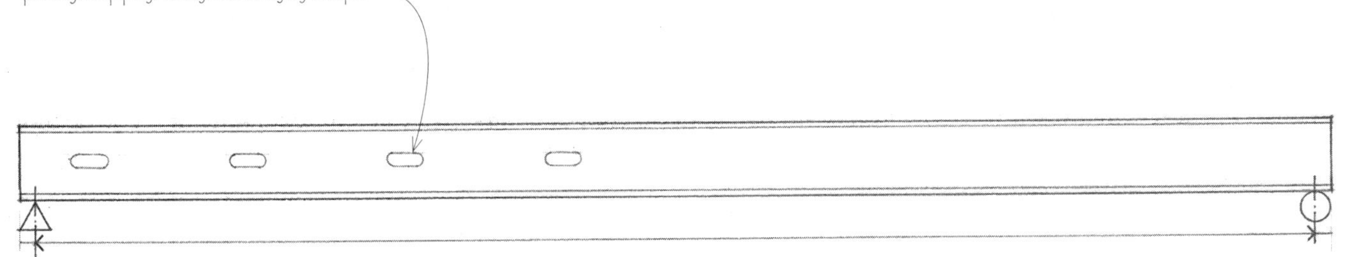

Span Ranges for Light-Gauge Steel Joists

- 6" (150) joists 10' to 14' (3050 to 4265)
- 8" (205) joists 12' to 18' (3660 to 5485)
- 10" (255) joists 14' to 22' (4265 to 6705)
- 12" (305) joists 18' to 26' (5485 to 7925)

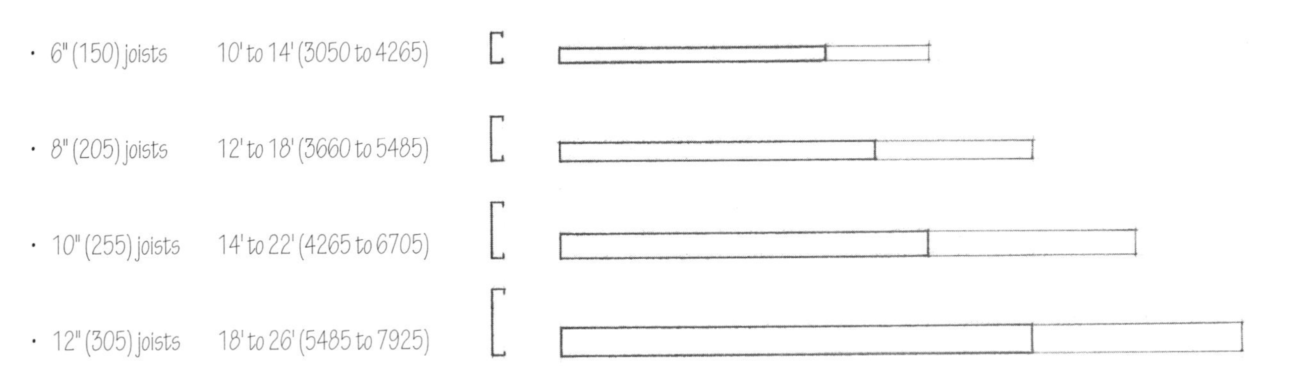

- Rule of thumb for estimating joist depth: span/20
- Consult manufacturer for exact joist dimensions, framing details, and allowable spans and loads.

CSI MasterFormat 05 40 00: Cold-Formed Metal Framing
CSI MasterFormat 05 42 00: Cold-Formed Metal Joist Framing

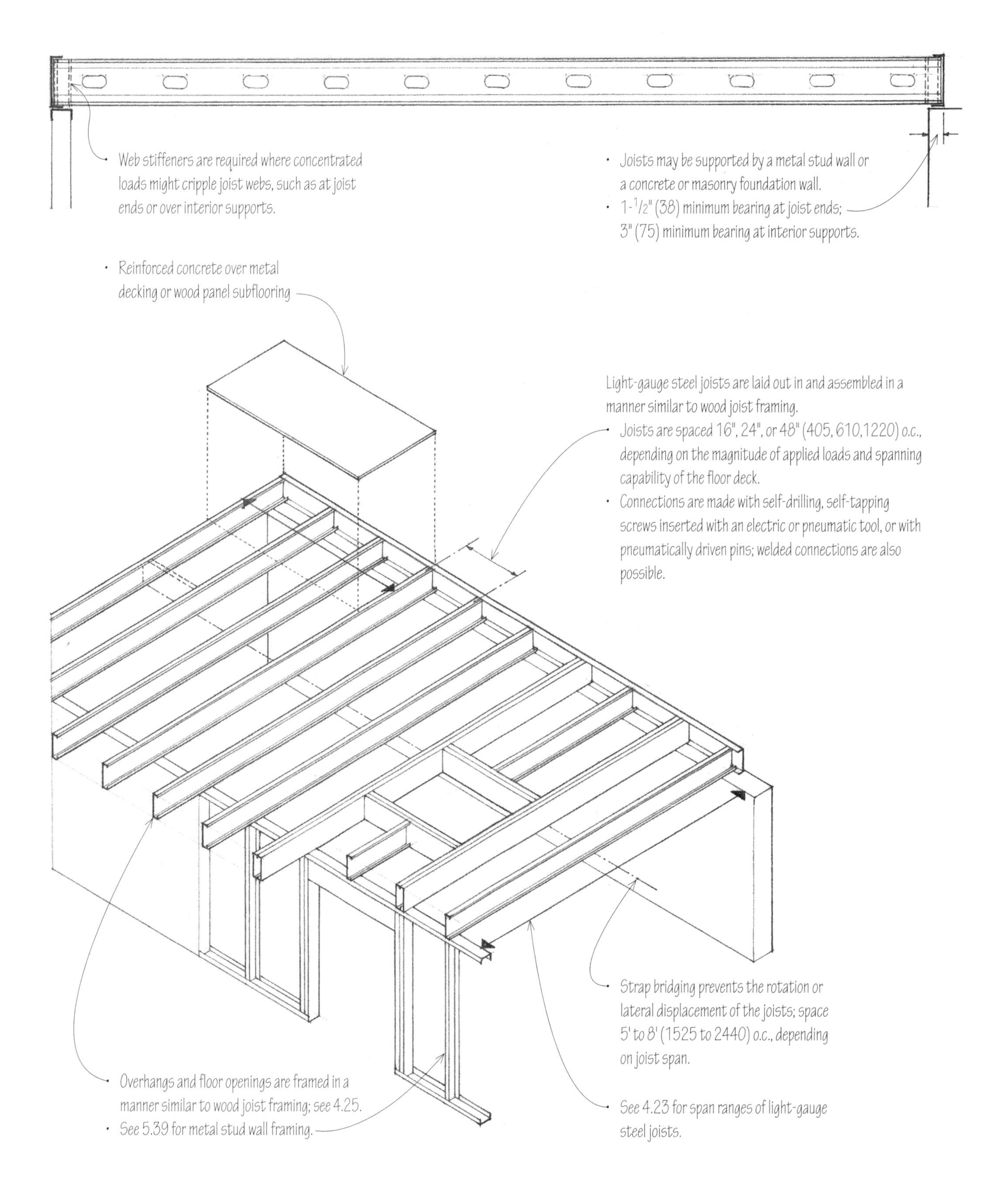
Web stiffeners are required where concentrated loads might cripple joist webs, such as at joist ends or over interior supports.
Joists may be supported by a metal stud wall or a concrete or masonry foundation wall.
1-1/2" (38) minimum bearing at joist ends; 3" (75) minimum bearing at interior supports.
Reinforced concrete over metal decking or wood panel subflooring
Light-gauge steel joists are laid out in and assembled in a manner similar to wood joist framing.
Joists are spaced 16", 24", or 48" (405, 610,1220) o.c., depending on the magnitude of applied loads and spanning capability of the floor deck.
Connections are made with self-drilling, self-tapping screws inserted with an electric or pneumatic tool, or with pneumatically driven pins; welded connections are also possible.
Strap bridging prevents the rotation or lateral displacement of the joists; space 5' to 8' (1525 to 2440) o.c., depending on joist span.
Overhangs and floor openings are framed in a manner similar to wood joist framing; see 4.25.
See 5.39 for metal stud wall framing.
See 4.23 for span ranges of light-gauge steel joists.

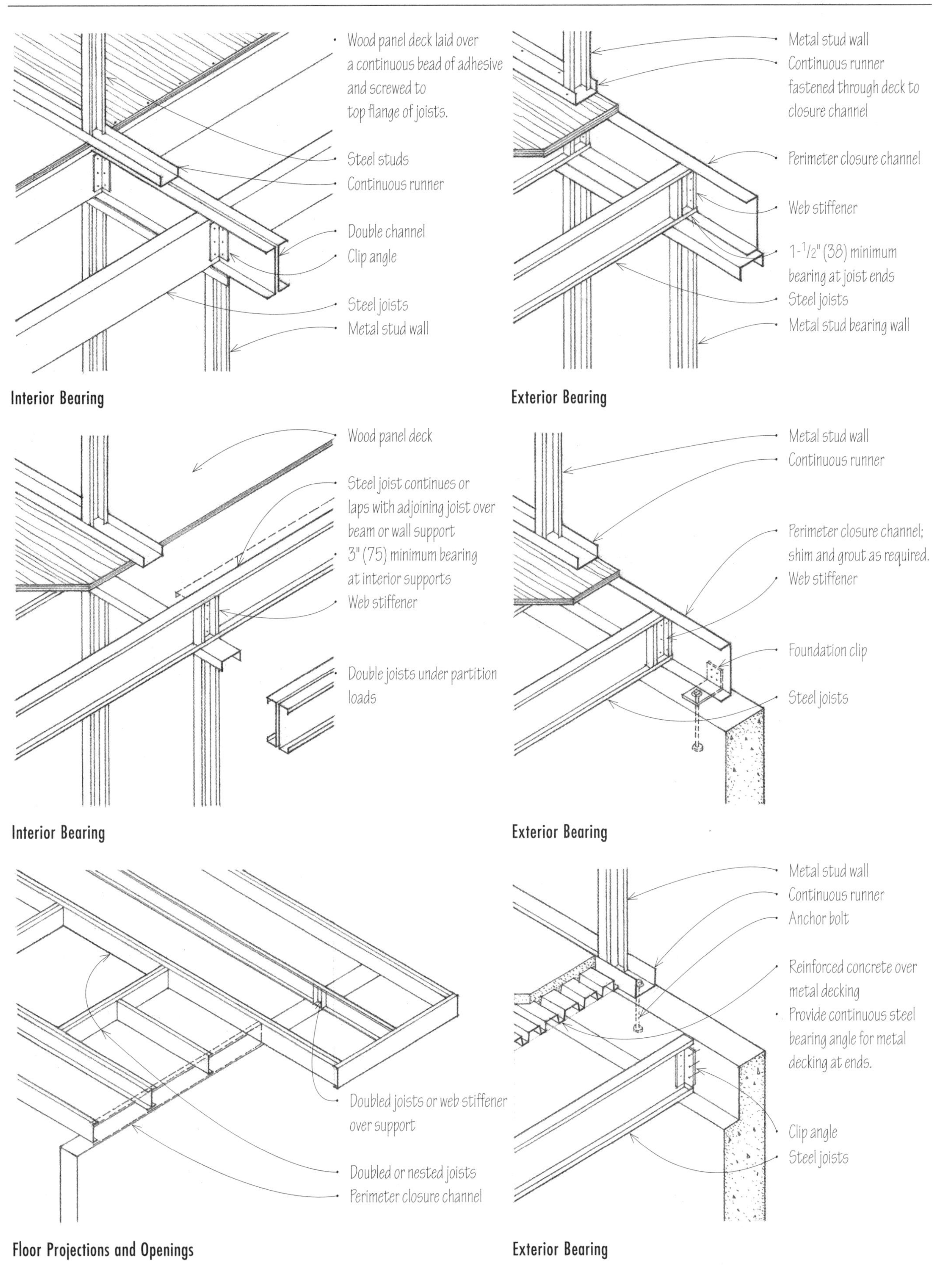
Wood panel deck laid over a continuous bead of adhesive and screwed to top flange of joists.
Steel studs
Continuous runner
Double channel
Clip angle
Steel joists
Metal stud wall
Interior Bearing
Metal stud wall
Continuous runner fastened through deck to closure channel
Perimeter closure channel
Web stiffener
1-1/2" (38) minimum bearing at joist ends
Steel joists
Metal stud bearing wall
Exterior Bearing
Wood panel deck
Steel joist continues or laps with adjoining joist over beam or wall support
3" (75) minimum bearing at interior supports
Web stiffener
Double joists under partition loads
Interior Bearing
Metal stud wall
Continuous runner
Perimeter closure channel; shim and grout as required.
Web stiffener
Foundation clip
Steel joists
Exterior Bearing
Doubled joists or web stiffener over support
Doubled or nested joists
Perimeter closure channel
Floor Projections and Openings
Metal stud wall
Continuous runner
Anchor bolt
Reinforced concrete over metal decking
Provide continuous steel bearing angle for metal decking at ends.
Clip angle
Steel joists
Exterior Bearing

Wood joist floors are an essential subsystem of wood light-frame construction. The dimension lumber used for joists is easily worked and can be quickly assembled on site with simple tools. Together with wood panel sheathing or subflooring, the wood joists form a level working platform for construction. If properly engineered, the resulting floor structure can serve as a structural diaphragm to transfer lateral loads to shear walls; consult the building code for specific requirements.

- Because wood light-framing is combustible, it must rely on finish flooring and ceiling materials for its fire-resistance rating.
- The susceptibility of wood light-framing to decay and insect infestation requires positive site drainage, adequate separation from the ground, appropriate use of pressure-treated lumber, and ventilation to control condensation in enclosed spaces.
- See 12.11–12.12 for discussion of wood as a construction material.

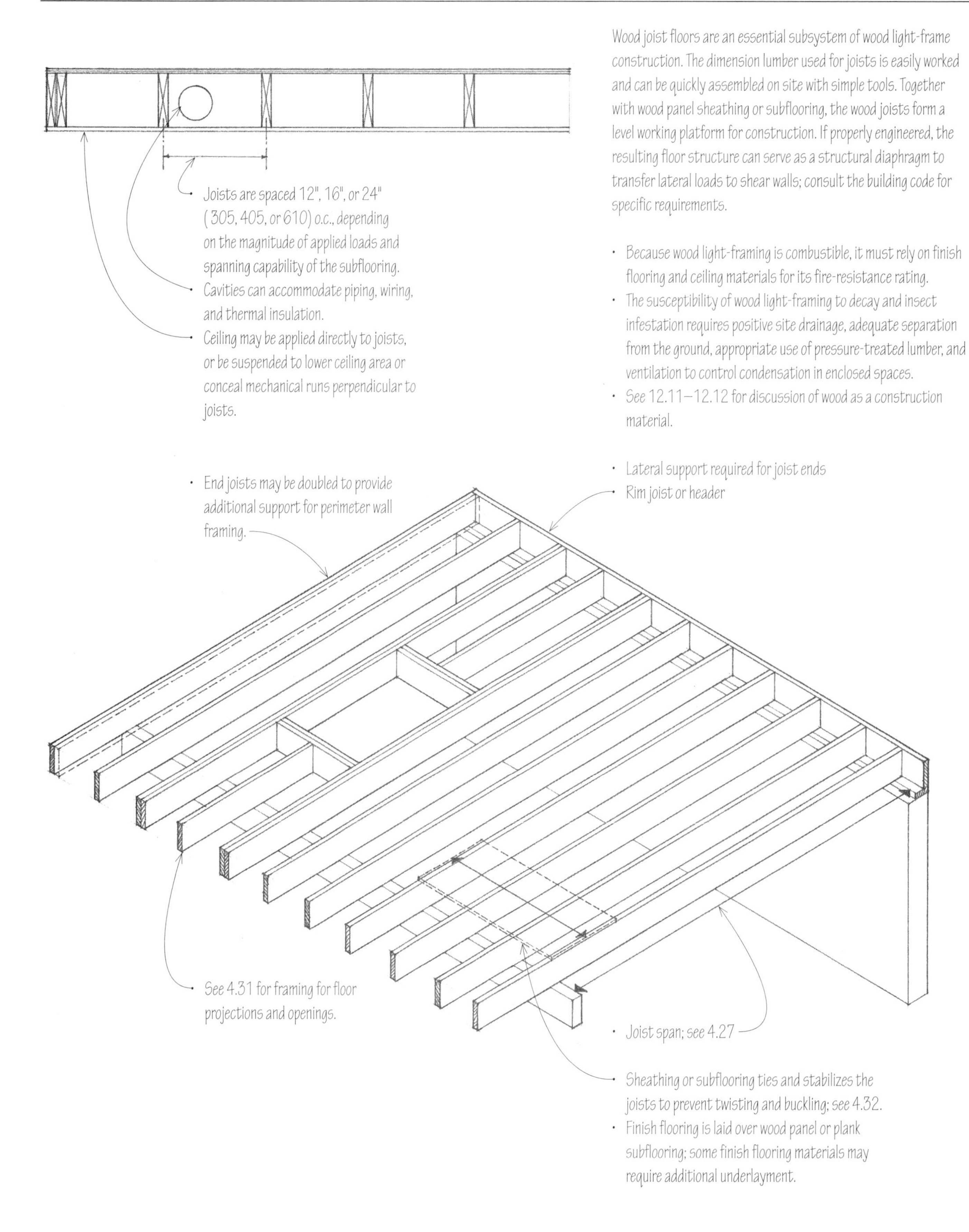

CSI MasterFormat 06 10 00: Rough Carpentry
CSI MasterFormat 06 11 00: Wood Framing

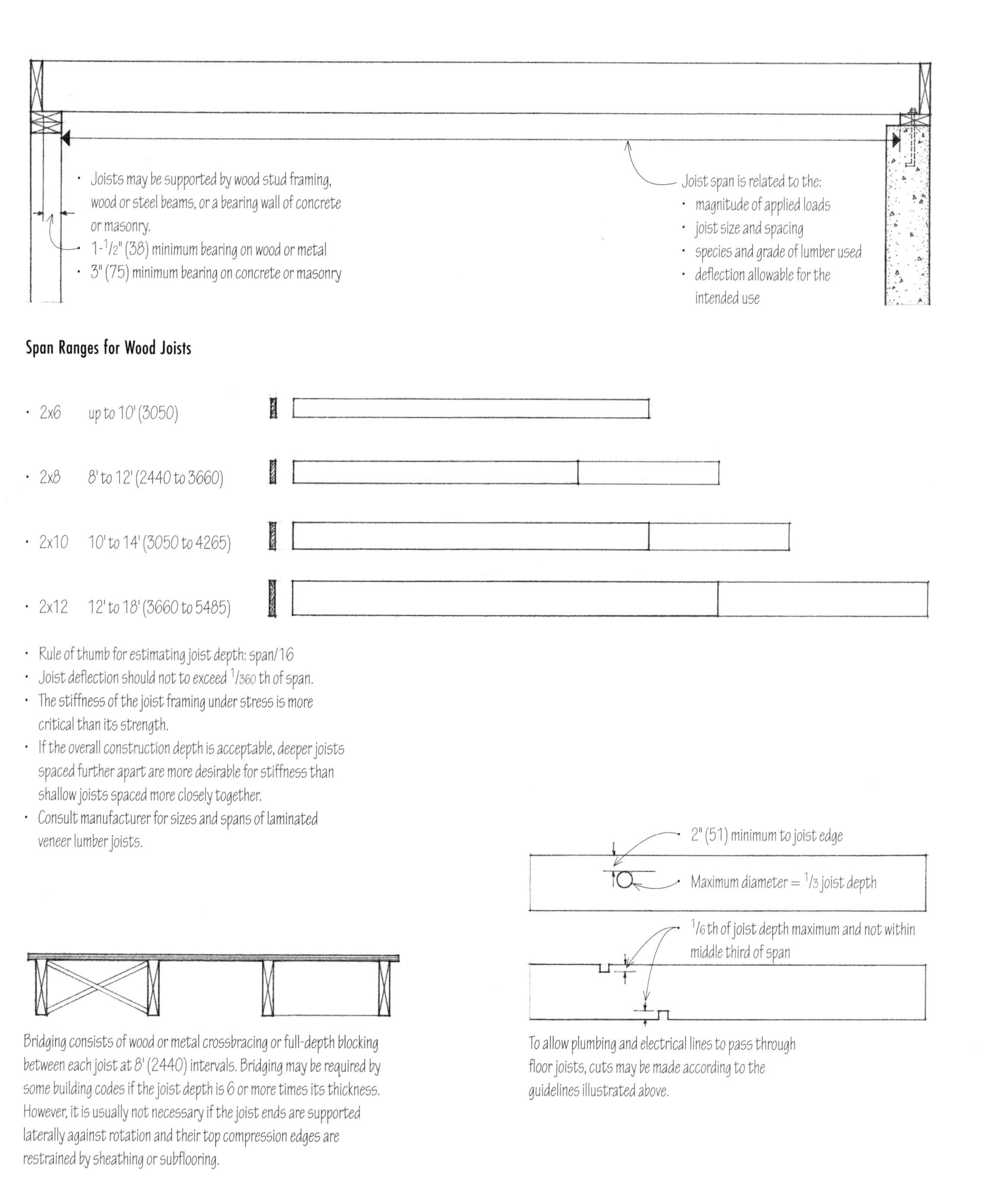
• Joists may be supported by wood stud framing, wood or steel beams, or a bearing wall of concrete or masonry.
• 1-1/2" (38) minimum bearing on wood or metal
• 3" (75) minimum bearing on concrete or masonry
Joist span is related to the:
• magnitude of applied loads
• joist size and spacing
• species and grade of lumber used
• deflection allowable for the intended use
Span Ranges for Wood Joists
• 2x6 up to 10' (3050)
• 2x8 8' to 12' (2440 to 3660)
• 2x10 10' to 14' (3050 to 4265)
• 2x12 12' to 18' (3660 to 5485)
• Rule of thumb for estimating joist depth: span/16
• Joist deflection should not to exceed 1/360 th of span.
• The stiffness of the joist framing under stress is more critical than its strength.
• If the overall construction depth is acceptable, deeper joists spaced further apart are more desirable for stiffness than shallow joists spaced more closely together.
• Consult manufacturer for sizes and spans of laminated veneer lumber joists.
• 2" (51) minimum to joist edge
• Maximum diameter = 1/3 joist depth
1/6 th of joist depth maximum and not within middle third of span
Bridging consists of wood or metal crossbracing or full-depth blocking between each joist at 8' (2440) intervals. Bridging may be required by some building codes if the joist depth is 6 or more times its thickness. However, it is usually not necessary if the joist ends are supported laterally against rotation and their top compression edges are restrained by sheathing or subflooring.
To allow plumbing and electrical lines to pass through floor joists, cuts may be made according to the guidelines illustrated above.

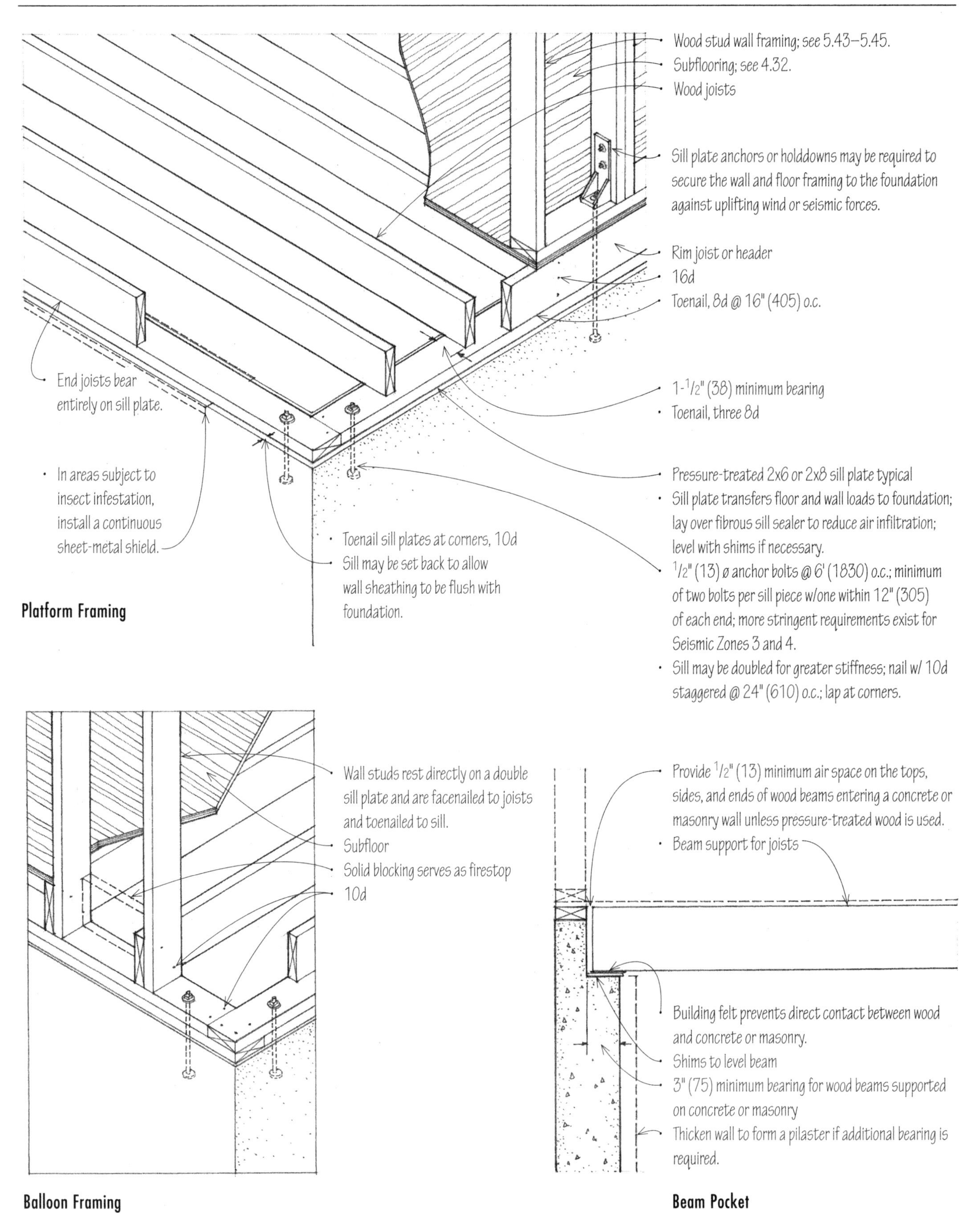

Platform Framing

Balloon Framing

Beam Pocket

• See 5.41–5.42 for discussion of balloon and platform framing.

Wood joists may be supported by wood or steel beams. In either case, the elevation of the beam should be coordinated with the perimeter sill condition and how the beam supports the floor joists. Wood is most susceptible to shrinkage perpendicular to its grain. For this reason, the total depth of wood construction for both the sill condition and the joist–beam connection should be equalized to avoid subsidence of the floor plane.

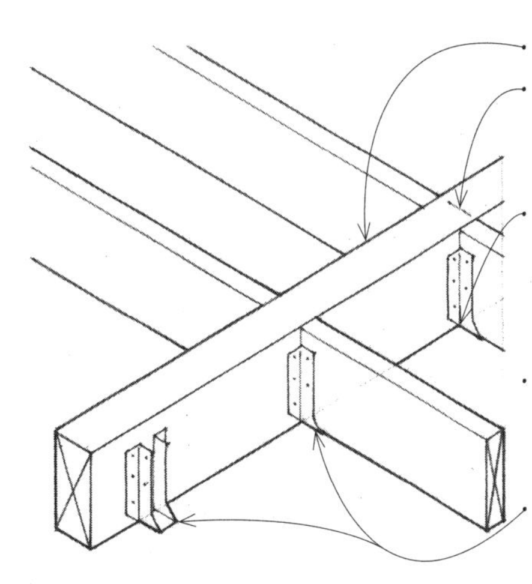

Wood Beam w/Joist Hangers

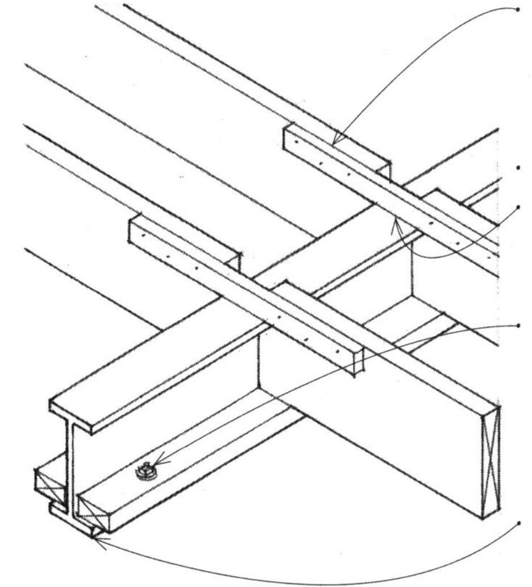

Steel Beam w/Ledger

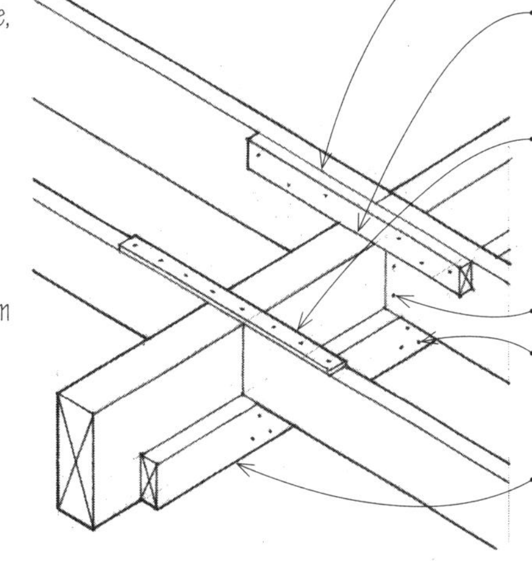

Wood Beam w/Ledger

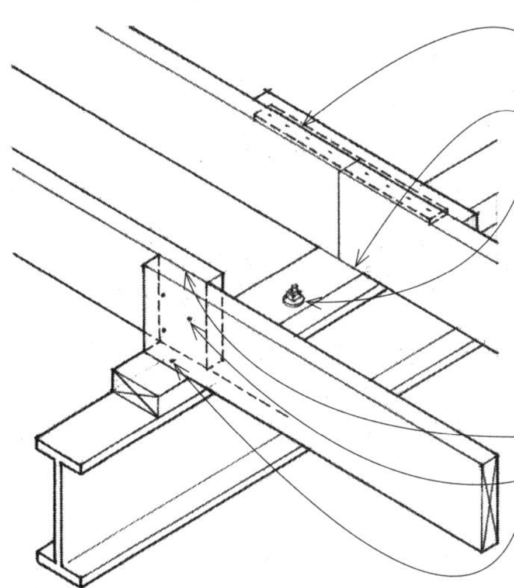

Steel Beam Under Joists

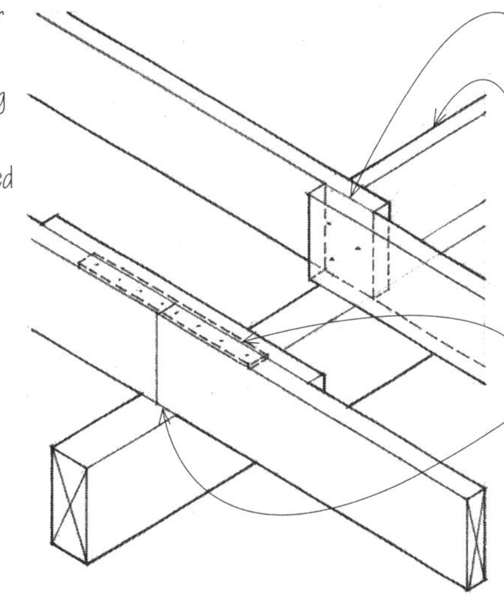

Wood Beam w/Lapped or Spliced Joists

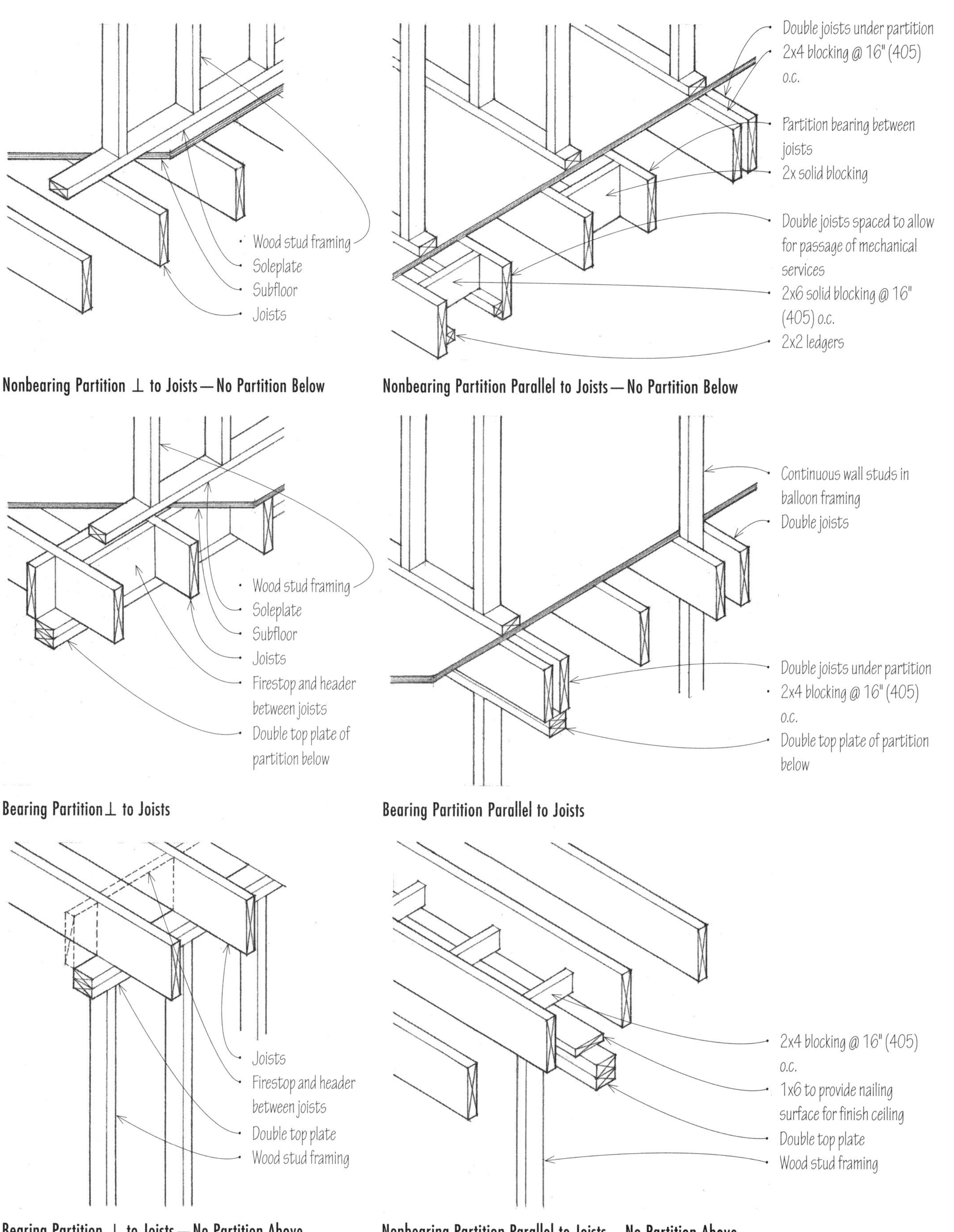

Nonbearing Partition ⊥ to Joists—No Partition Below

Nonbearing Partition Parallel to Joists—No Partition Below

Bearing Partition ⊥ to Joists

Bearing Partition Parallel to Joists

Bearing Partition ⊥ to Joists—No Partition Above

Nonbearing Partition Parallel to Joists—No Partition Above

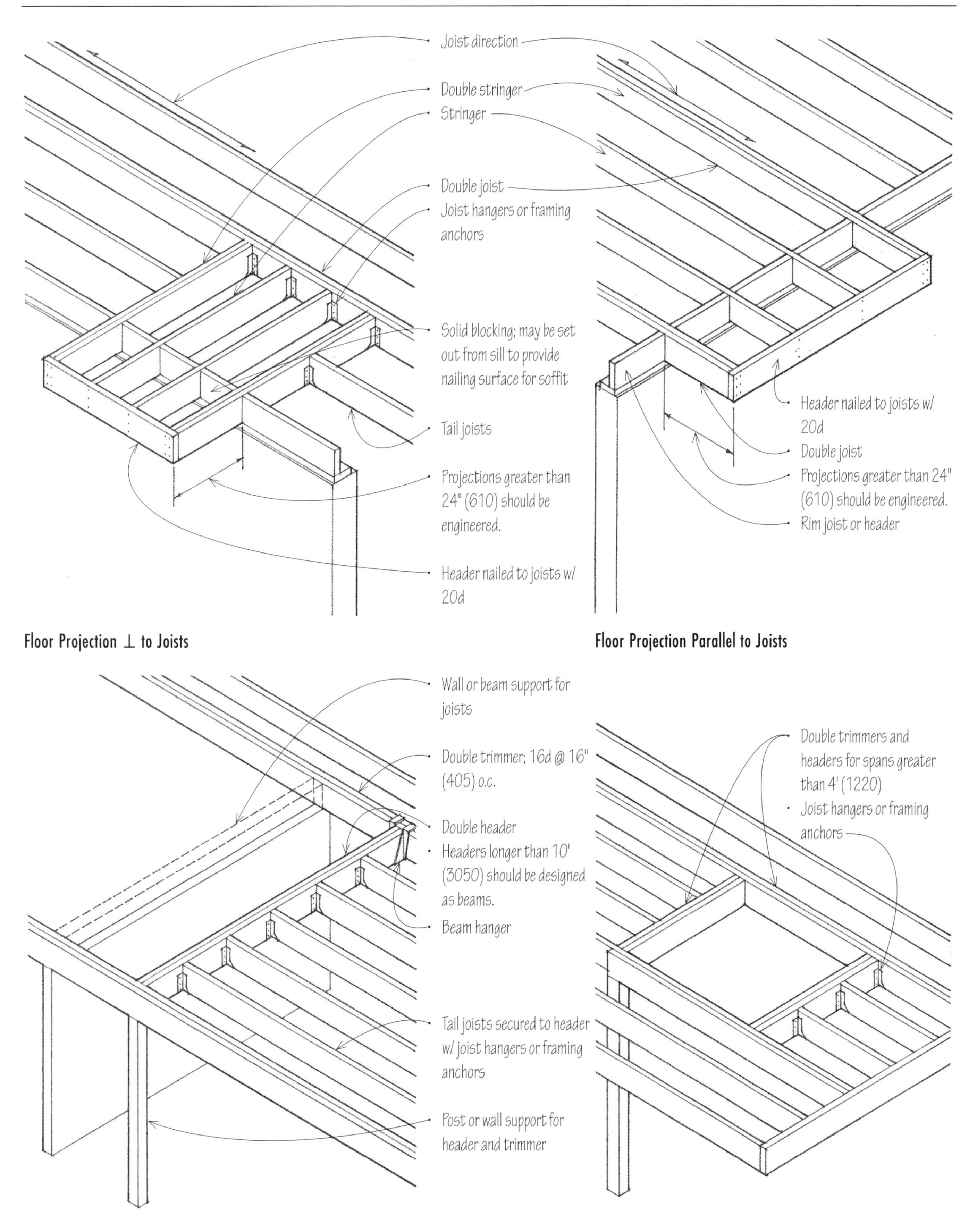

Floor Projection ⊥ to Joists

Floor Projection Parallel to Joists

Floor Opening — Length ⊥ to Joists

Floor Openings — Length Parallel to Joists

Subfloor	Thickness inches*	Span Rating	Span inches (mm)
Panel Subfloor			
For rated	5/8	32/16	16 (405)
sheathing &	1/2, 5/8	36/16	16 (405)
Structural I	5/8, 3/4, 7/8	42/20	20 (510)
& II grades	3/4, 7/8	48/24	24 (610)
Underlayment			
Underlayment	1/4	Over panel subfloor	
or C–C plugged	3/8	Over board subfloor	
ext. grade			
Combined Subfloor–Underlayment			
For APA rated	5/8	16	16 (405)
Sturd-i-floor	5/8, 3/4	20	20 (510)
grades	3/4, 7/8, 1	24	24 (610)
2–4–1	1-1/8	48	48 (1220)

*Metric equivalents:
- 1/2" (13)
- 5/8" (16)
- 3/4" (19)
- 7/8" (22)
- 1" (25)
- 1-1/2" (29)

Subflooring is the structural material that spans across floor joists, serves as a working platform during construction, and provides a base for the finish flooring. The joist and subfloor assembly can also be used as a structural diaphragm to transfer lateral forces to shear walls if constructed according to approved standards. Consult the building code for requirements.

- Subflooring typically consists of plywood, although other nonveneer panel materials such as oriented strand board (OSB), waferboard, and particleboard can be used if manufactured according to approved standards. Consult the American Plywood Association (APA).
- The Span Rating is part of the gradestamp found on the back of each panel. The first number indicates the maximum rafter spacing for roof sheathing and the second number indicates the maximum joist spacing for subflooring.
- Span may be 24" (610) if 25/32" (20) wood strip flooring is laid perpendicular to joists.
- Underlayment provides impact load resistance and a smooth surface for the direct application of nonstructural flooring materials; may be applied as a separate layer over board or panel subflooring, or be combined in a single thickness with the subfloor panel; when floor is subject to unusual moisture conditions, use panels with exterior glue (Exposure 1) or Exterior plywood.

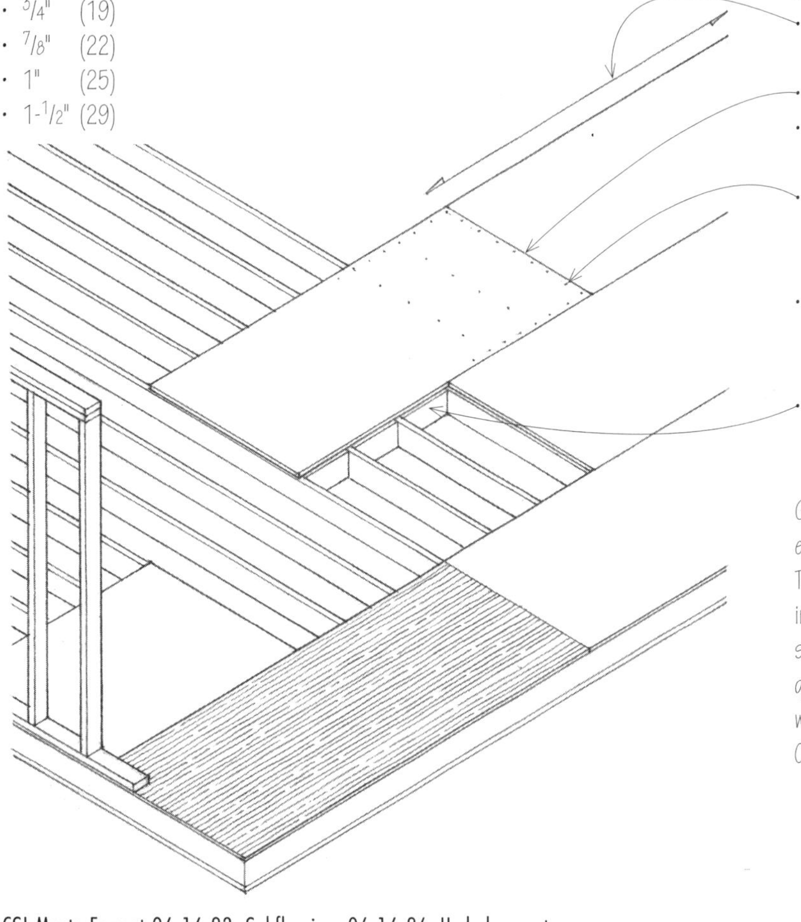

Panel Subfloor and Underlayment

- Indicated spans assume panels are laid continuously over two or more spans with their long dimension perpendicular to the joists.
- Stagger end joints.
- Space joints 1/8" (3) unless otherwise recommended by panel manufacturer; space butt joints in underlayment 1/32" (1).
- Nail @ 6" (150) o.c. along edges and 12" (305) o.c. along intermediate supports; nail @ 6" (150) o.c. along both edges and intermediate supports of 2-4-1 panels.
- Use 6d ring-shank or 8d common nails for thicknesses through 3/4" (19) and 8d ring-shank or common nails for panels 7/8" (22) and thicker.
- Provide blocking under edges or use tongue-and-grove panel edges; not required if underlayment joints are offset from subfloor joints.

Gluing combined subfloor-underlayment panels to the floor joists enables the panels to act together with the joists to form integral T-beam units. This application system lessens floor creep and squeaking, improves floor stiffness and, in some cases, increases the allowable spans for the joists. These benefits, of course, are contingent on the quality of the application. In addition to gluing, the panels are secured with power-driven fasteners or with 6d ring- or screw-shank nails. Consult the APA for detailed recommendations.

CSI MasterFormat 06 16 23: Subflooring; 06 16 26: Underlayment

Prefabricated, pre-engineered wood joists and trusses are increasingly used in the place of dimension lumber to frame floors because they are generally lighter and more dimensionally stable than sawn lumber, are manufactured in greater depths and lengths, and can span longer distances.

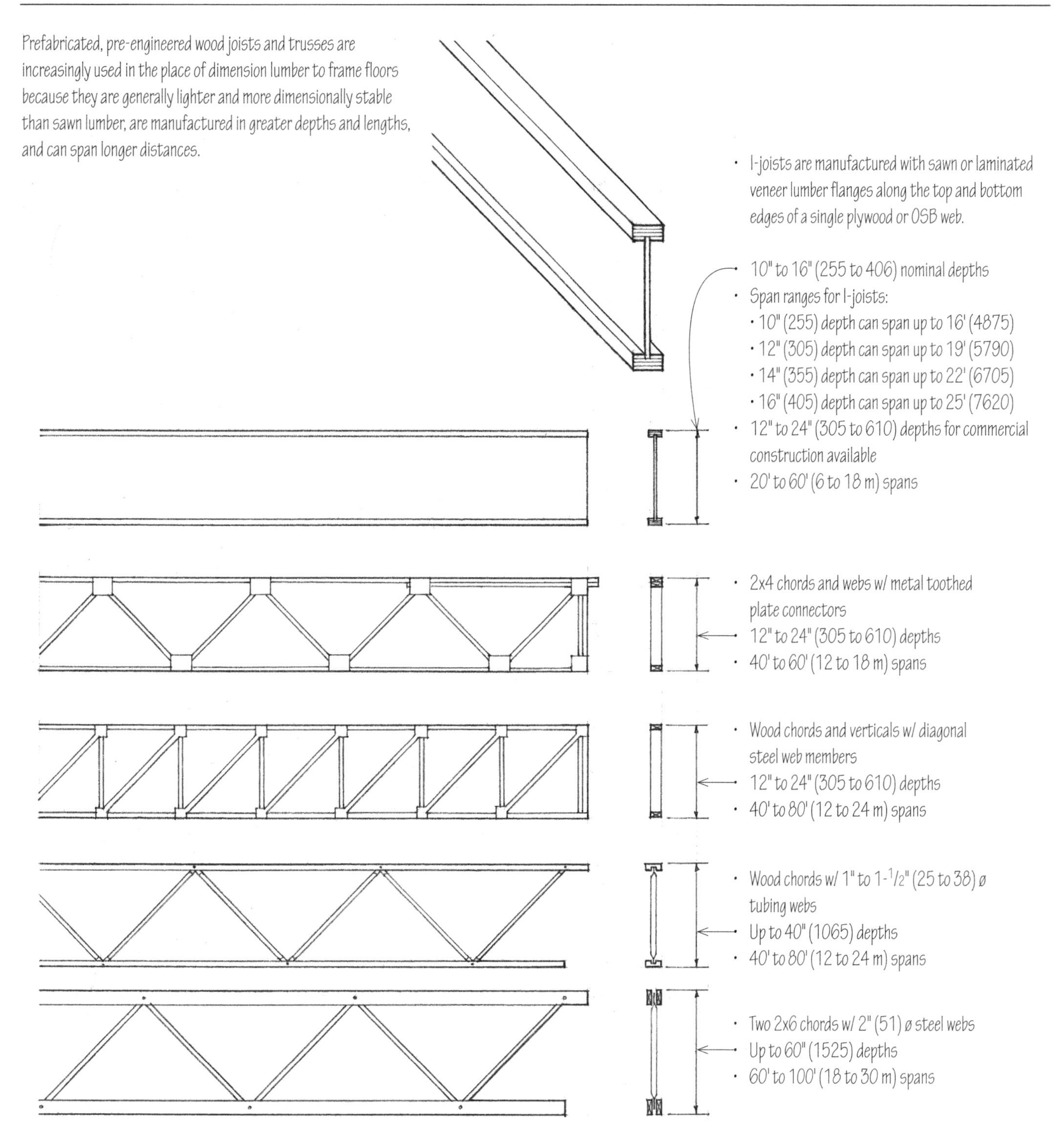

- Rule of thumb for estimating depth of trussed joists: span/18
- Openings in webs allow the passage of electrical and mechanical lines.
- Consult manufacturer for available lengths and depths, recommended spacing and allowable spans, and required bearing conditions.

While the precise form of a prefabricated floor joist or truss varies with the manufacturer, the way they are laid out to frame a floor is similar in principle to conventional wood joist framing. They are most appropriate for long spans and simple floor plans; complex floor layouts may be difficult to frame.

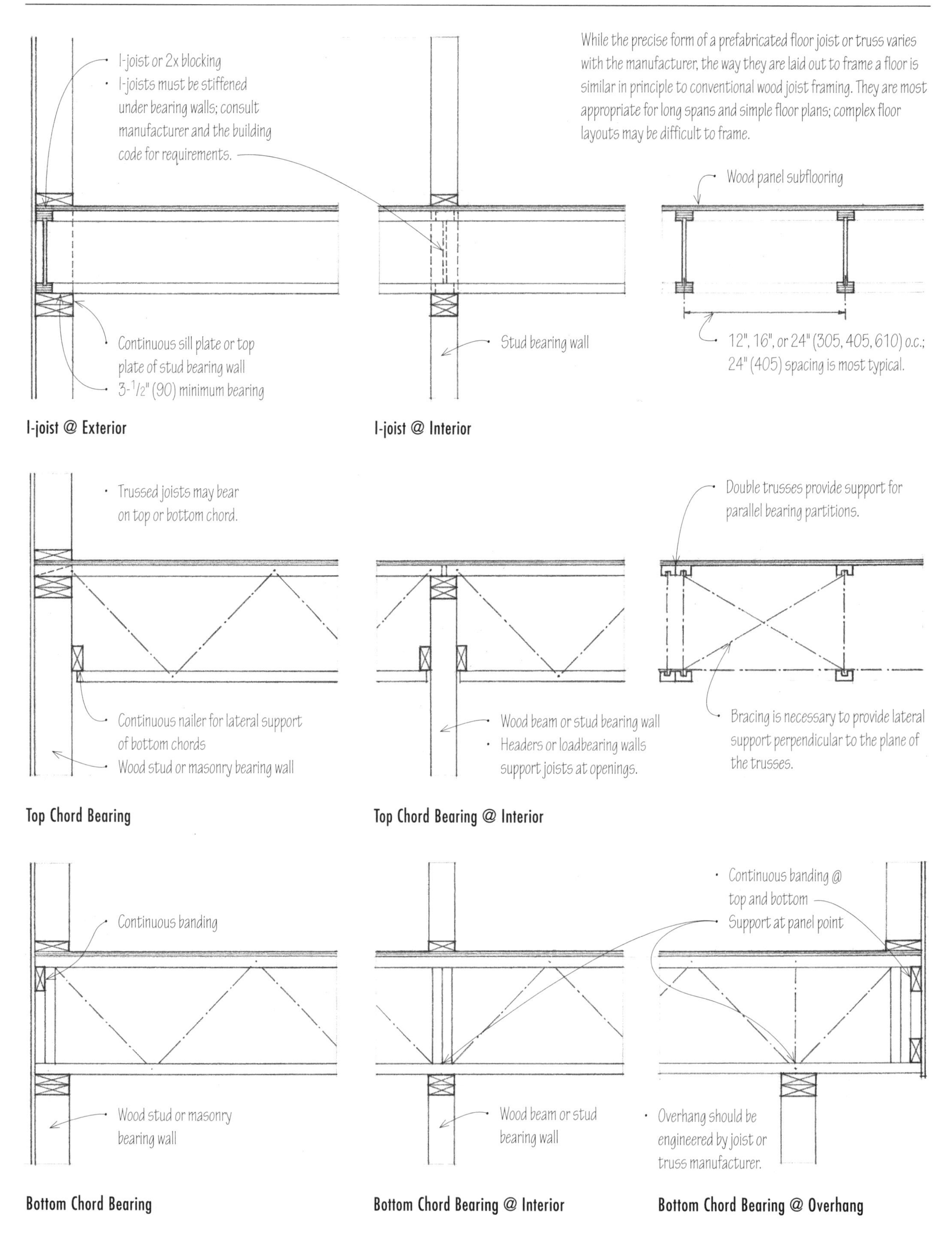

I-joist @ Exterior

I-joist @ Interior

Top Chord Bearing

Top Chord Bearing @ Interior

Bottom Chord Bearing

Bottom Chord Bearing @ Interior

Bottom Chord Bearing @ Overhang

Solid Sawn Lumber

In the selection of a wood beam the following should be considered: lumber species, structural grade, modulus of elasticity, allowable bending and shear stress values, and the minimum deflection permitted for the intended use. In addition, attention should be paid to the precise loading conditions and the types of connections used. See Bibliography for sources of more detailed span and load tables.

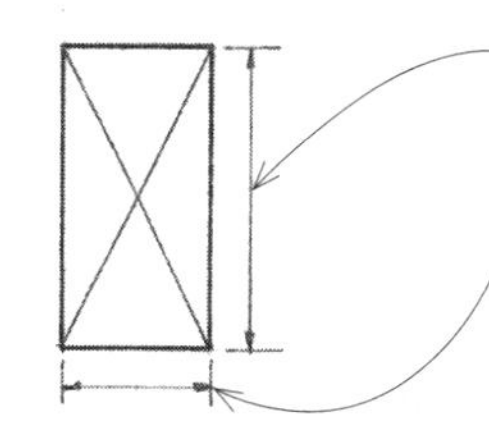

- Rule of thumb for estimating the depth of a wood beam: span/15
- Beam width = $1/3$ to $1/2$ of beam depth
- Limit deflection to $1/360$th of span

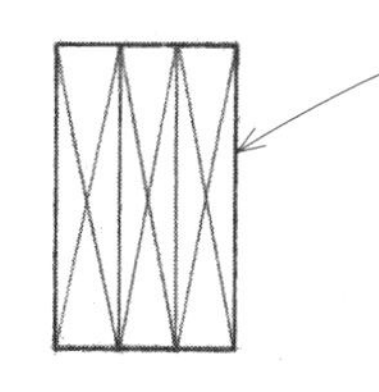

Built-Up Beam

- Equal in strength to the sum of the strengths of the individual pieces if none of the laminations are spliced
- Two members nailed w/ 10d @ 16" (405) o.c. staggered and two 10d @ each end
- Three or more members nailed w/ 20d @ 32" (815) o.c. staggered and two 20d @ each end

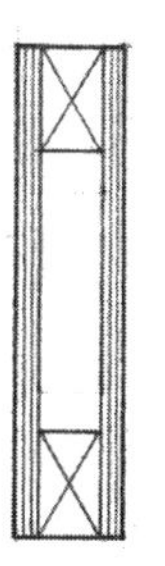

Box Beam

- Made by gluing two or more plywood or OSB webs to sawn or LVL flanges.
- Engineered to span up to 90' (27 m)

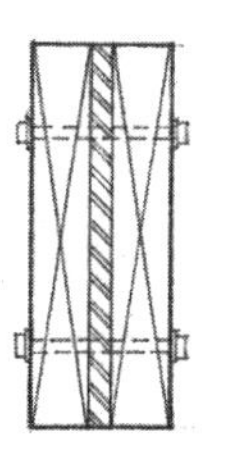

Flitch Beam

- Timbers set on edge and bolted side by side to steel plates or sections
- Engineered design

Spaced Beam

- Blocked and securely nailed at frequent intervals to enable individual member to act as an integral unit

Glue-Laminated Timber

Glue-laminated timber (CSI MasterFormat 06 18 00) is made by laminating stress-grade lumber with adhesive under controlled conditions, usually with the grain of all plies being parallel. The advantages of glued-laminated timber over dimension lumber are generally higher allowable unit stresses, improved appearance, and availability of various sectional shapes. Glue-laminated timbers may be end-joined with scarf or finger joints to any desired length, or edge-glued for greater width or depth.

- Engineered to span up to 80' (24 m)
- Rule of thumb for estimating the depth of glue-laminated beams: span/20
- Beam width = $1/4$ to $1/3$ of beam depth

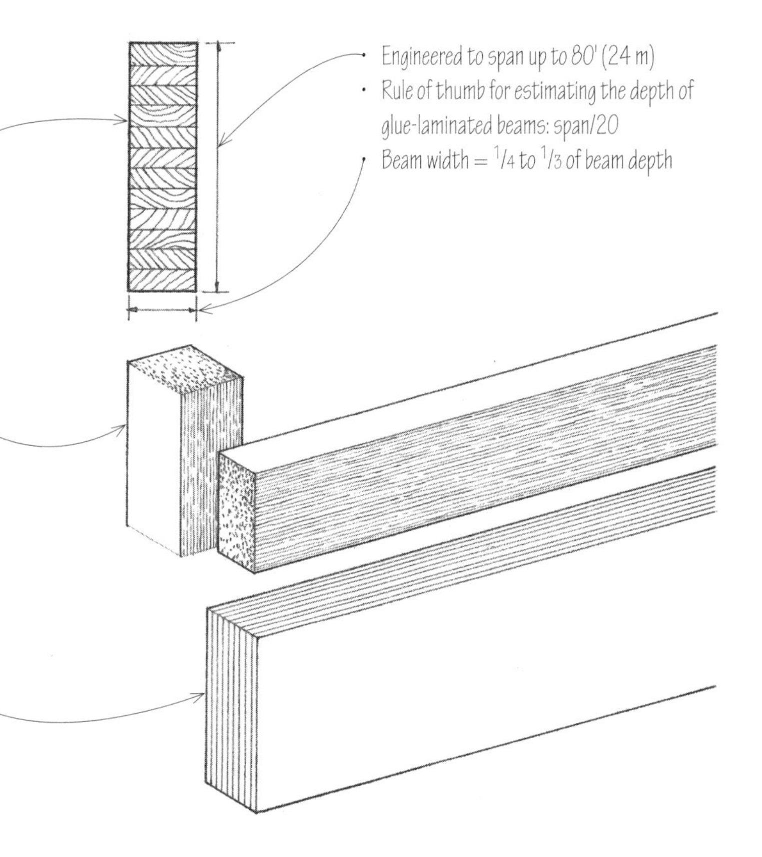

Parallel Strand Lumber

Parallel strand lumber (PSL) is a structural lumber product made by bonding long, narrow wood strands together under heat and pressure using a waterproof adhesive. Parallel strand lumber is a proprietary product marketed under the trademark Parallam, used as beams and columns in post-and-beam construction and for beams, headers, and lintels in light-frame construction.

Laminated Veneer Lumber

Laminated veneer lumber (LVL) is a structural lumber product made by bonding layers of wood veneers together under heat and pressure using a waterproof adhesive. Having the grain of all veneers run in the same longitudinal direction results in a product that is strong when edge-loaded as a beam or face-loaded as a plank. Laminated veneer lumber is marketed under various brand names, such as Microlam, and used as headers and beams or as flanges for prefabricated wood I-joists.

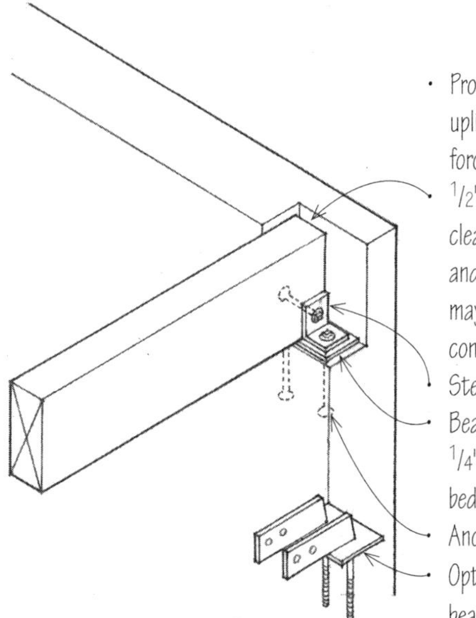

- Provides resistance to uplift and horizontal forces
- 1/2" (13) minimum clearance on top, end, and sides; more space may be required for construction access.
- Steel clip angles
- Bearing plate at least 1/4" (75) thick set on bed of dryset grout
- Anchor bolts
- Optional prefabricated beam seat

Masonry or Concrete Wall Support

A variety of metal attachments are manufactured for wood-to-wood, wood-to-metal, and wood-to-masonry connections. These include joist and beam hangers, post bases and caps, framing angles and anchors, and floor ties and holddowns. Consult manufacturer for specific shapes and sizes, allowable loads, and fastening requirements. Depending on the magnitude of the loads being resisted or transferred, the connectors may be nailed or bolted.

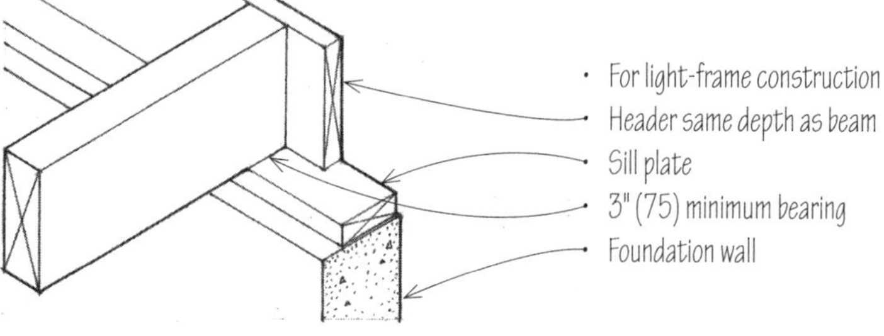

Foundation Wall Support

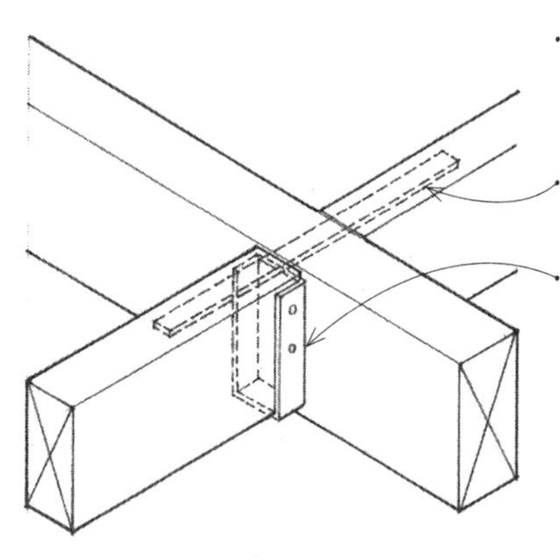

- For well-seasoned or laminated beams and light to moderate loads
- Metal tension tie across girder
- Beam hanger w/ concealed or exposed flanges

Girder Support

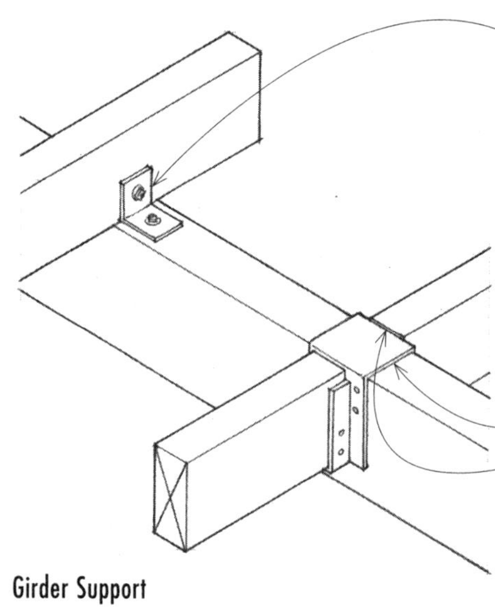

- Clip angles for superimposed beams; provide for lateral stability of supported beam if necessary.
- For moderate to heavy loads
- Exposed beam hanger
- Beams raised above girder for decking to clear saddle

Girder Support

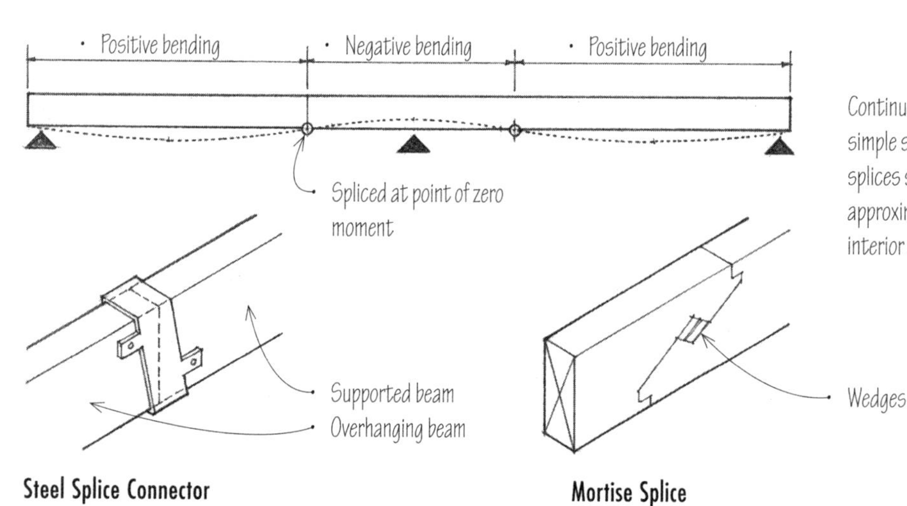

Continuous spans produce more uniform stresses than simple spans, resulting in more efficient use of material. Any splices should occur at points of minimum bending stress, approximately 1/4 to 1/3 of the span on either side of an interior support.

Steel Splice Connector

Mortise Splice

The size and number of bolts required at a connection depend on the thickness of the members, the species of wood, the magnitude and direction of the load relative to the grain of the wood, and the use of metal connectors. Shear plate or split-ring connectors, which can develop greater stresses per unit bearing, can be used when there is insufficient area to accommodate the required number of through bolts. See 5.49 for split-ring and shear plate connectors and bolt spacing guidelines.

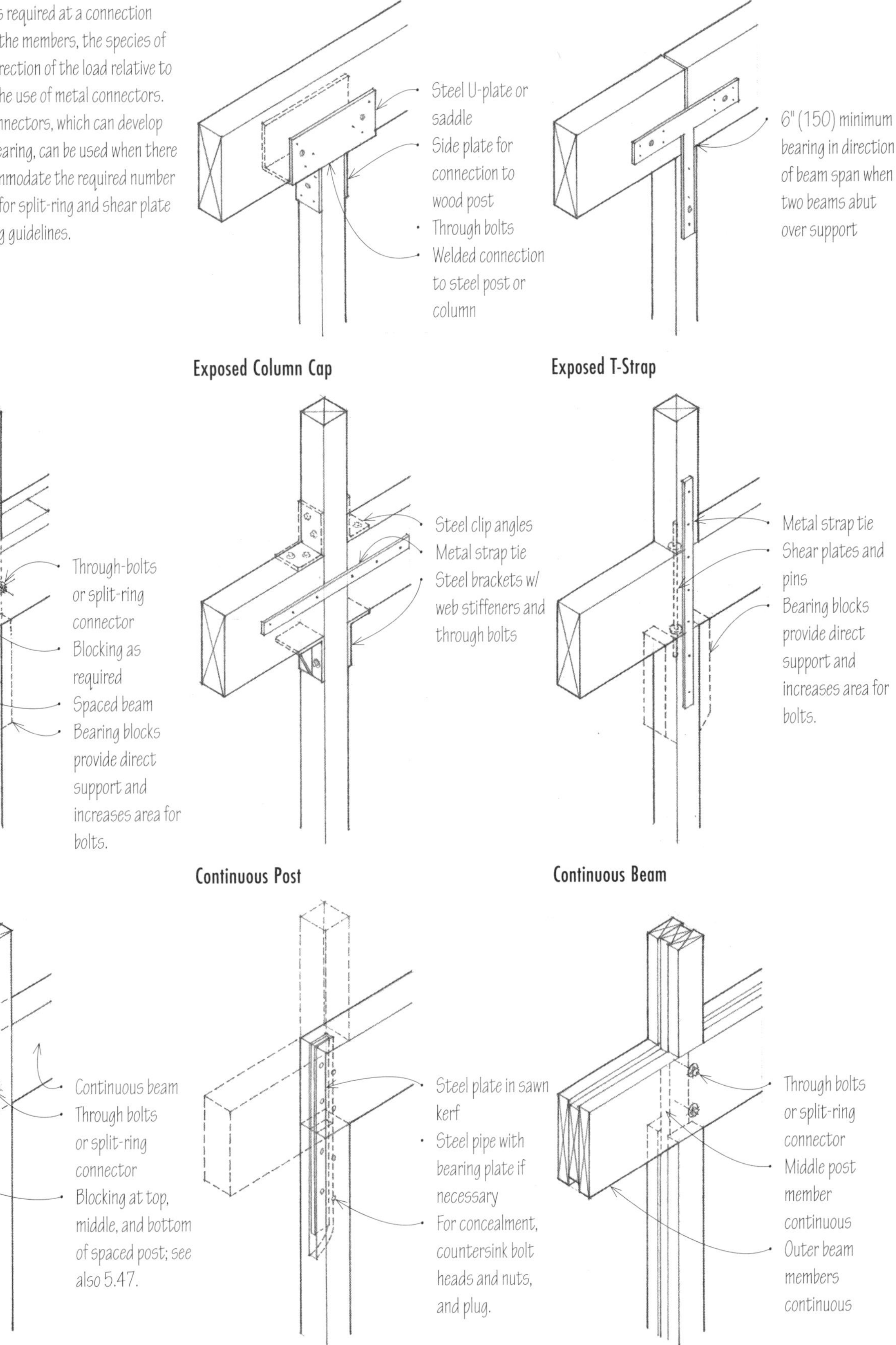

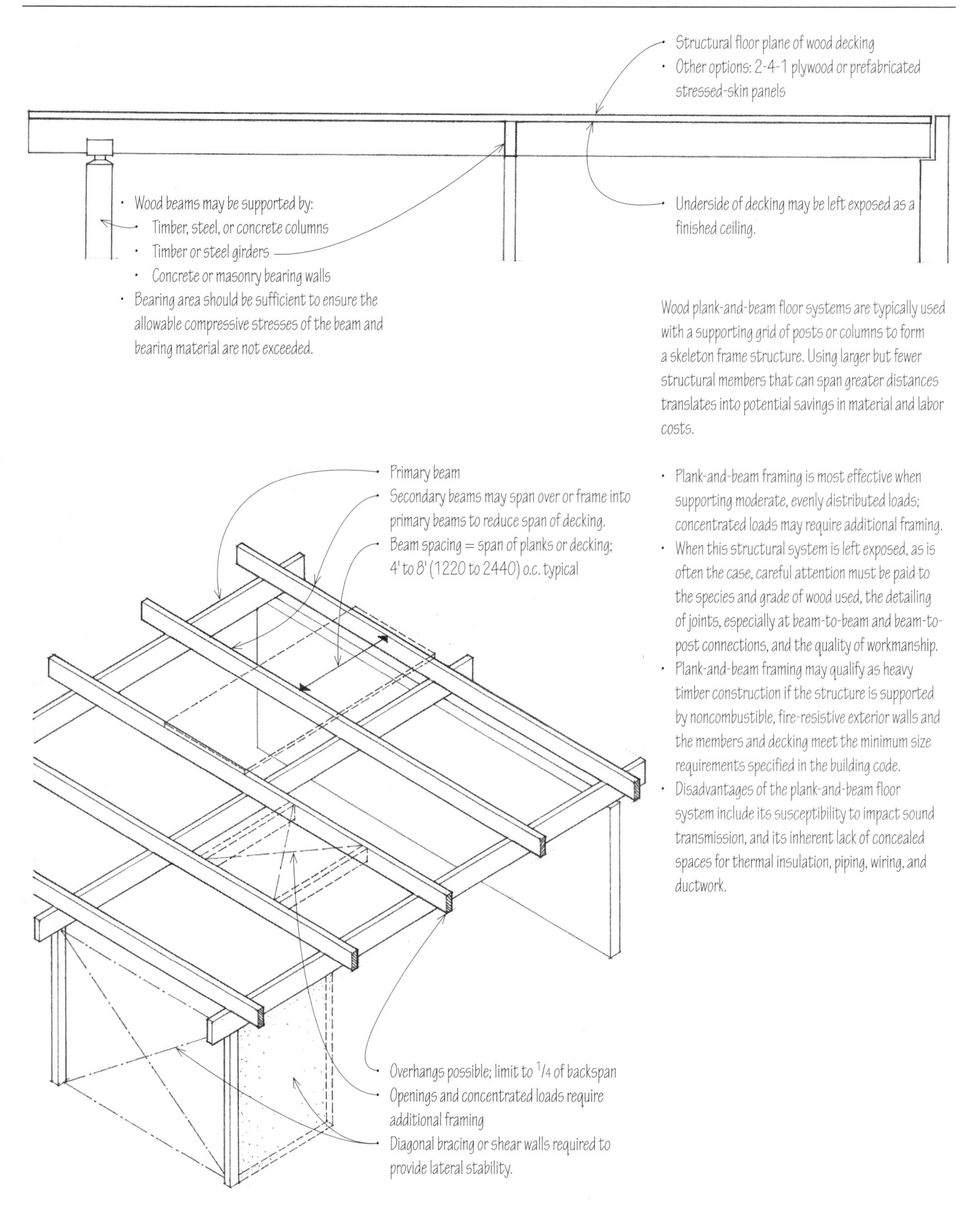

Wood plank-and-beam floor systems are typically used with a supporting grid of posts or columns to form a skeleton frame structure. Using larger but fewer structural members that can span greater distances translates into potential savings in material and labor costs.

- Plank-and-beam framing is most effective when supporting moderate, evenly distributed loads; concentrated loads may require additional framing.
- When this structural system is left exposed, as is often the case, careful attention must be paid to the species and grade of wood used, the detailing of joints, especially at beam-to-beam and beam-to-post connections, and the quality of workmanship.
- Plank-and-beam framing may qualify as heavy timber construction if the structure is supported by noncombustible, fire-resistive exterior walls and the members and decking meet the minimum size requirements specified in the building code.
- Disadvantages of the plank-and-beam floor system include its susceptibility to impact sound transmission, and its inherent lack of concealed spaces for thermal insulation, piping, wiring, and ductwork.

CSI MasterFormat 06 13 00: Heavy Timber Construction
CSI MasterFormat 06 15 00: Wood Decking

In the plank-and-beam framing system, the supporting grid of beams should be carefully integrated with the required placement of interior partitions for both structural and visual reasons. Normally, most partitions in this system are nonloadbearing and may be placed as shown. If bearing partitions are required, however, they should continue down to a foundation wall or be placed directly over beams large enough to carry the imposed load.

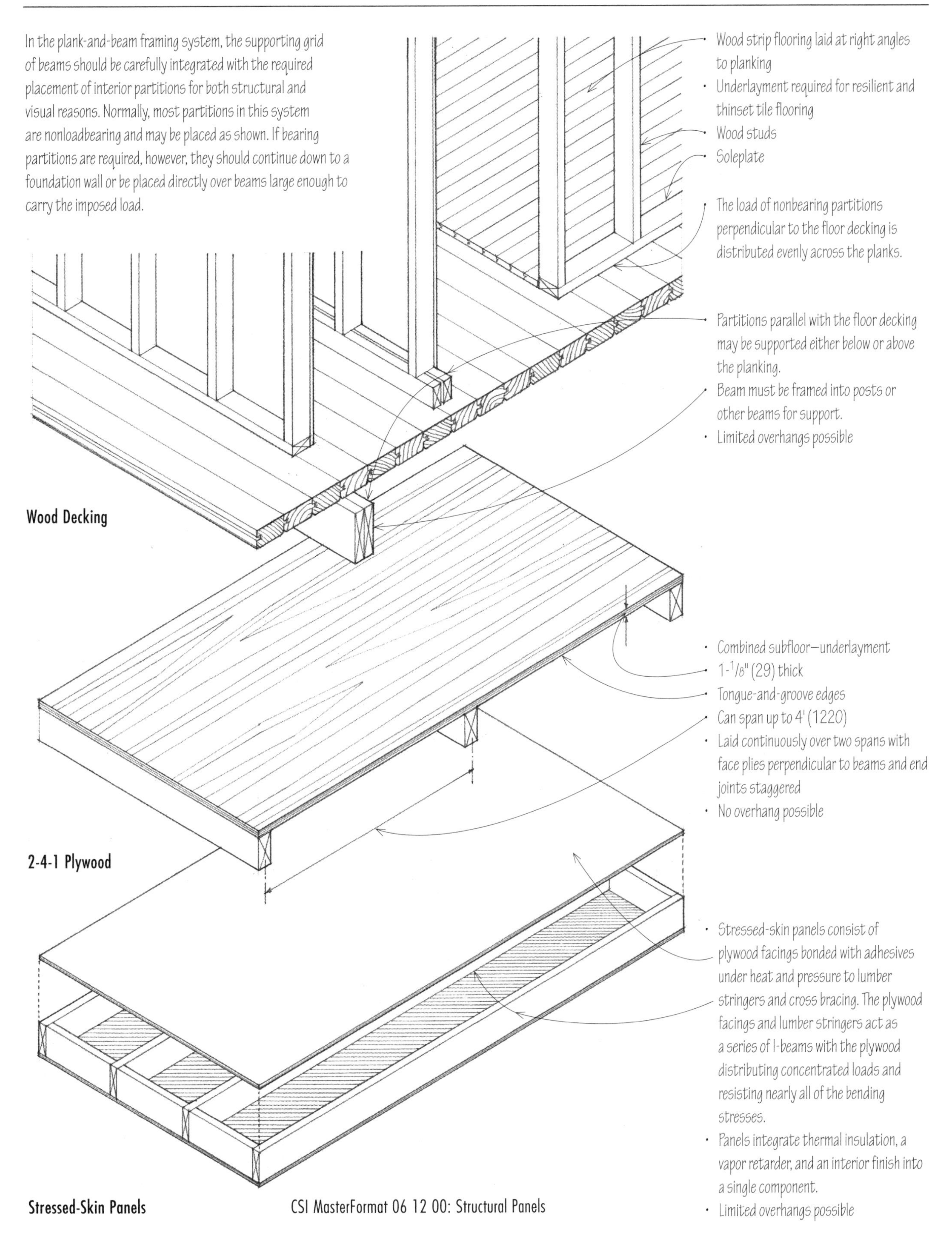

Wood Decking

2-4-1 Plywood

Stressed-Skin Panels

CSI MasterFormat 06 12 00: Structural Panels

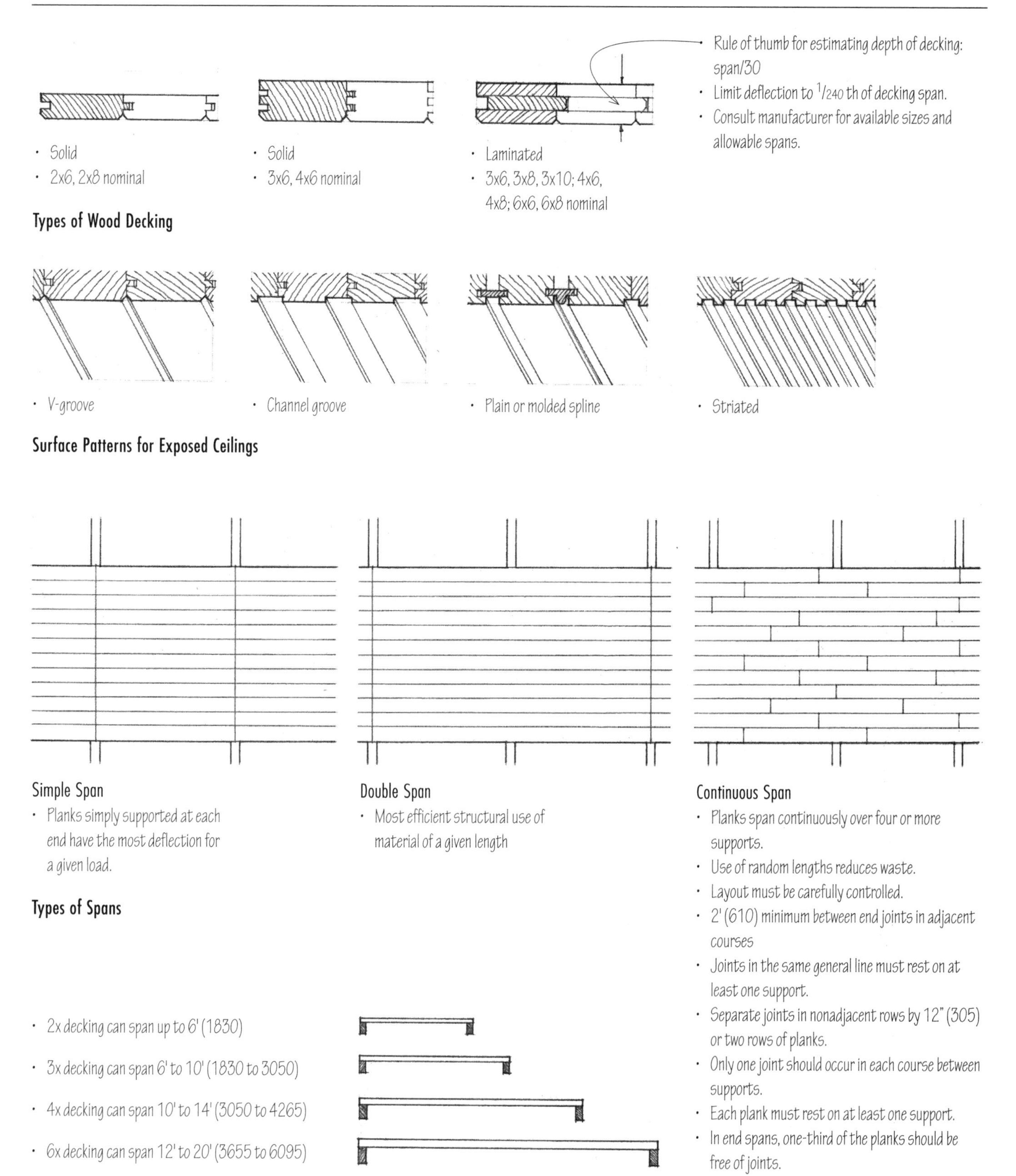

Types of Wood Decking

Surface Patterns for Exposed Ceilings

Simple Span
- Planks simply supported at each end have the most deflection for a given load.

Double Span
- Most efficient structural use of material of a given length

Continuous Span
- Planks span continuously over four or more supports.
- Use of random lengths reduces waste.
- Layout must be carefully controlled.
- 2' (610) minimum between end joints in adjacent courses
- Joints in the same general line must rest on at least one support.
- Separate joints in nonadjacent rows by 12" (305) or two rows of planks.
- Only one joint should occur in each course between supports.
- Each plank must rest on at least one support.
- In end spans, one-third of the planks should be free of joints.

Types of Spans

- 2x decking can span up to 6' (1830)
- 3x decking can span 6' to 10' (1830 to 3050)
- 4x decking can span 10' to 14' (3050 to 4265)
- 6x decking can span 12' to 20' (3655 to 6095)

Span Ranges

CSI MasterFormat 06 15 00: Wood Decking

5

WALL SYSTEMS

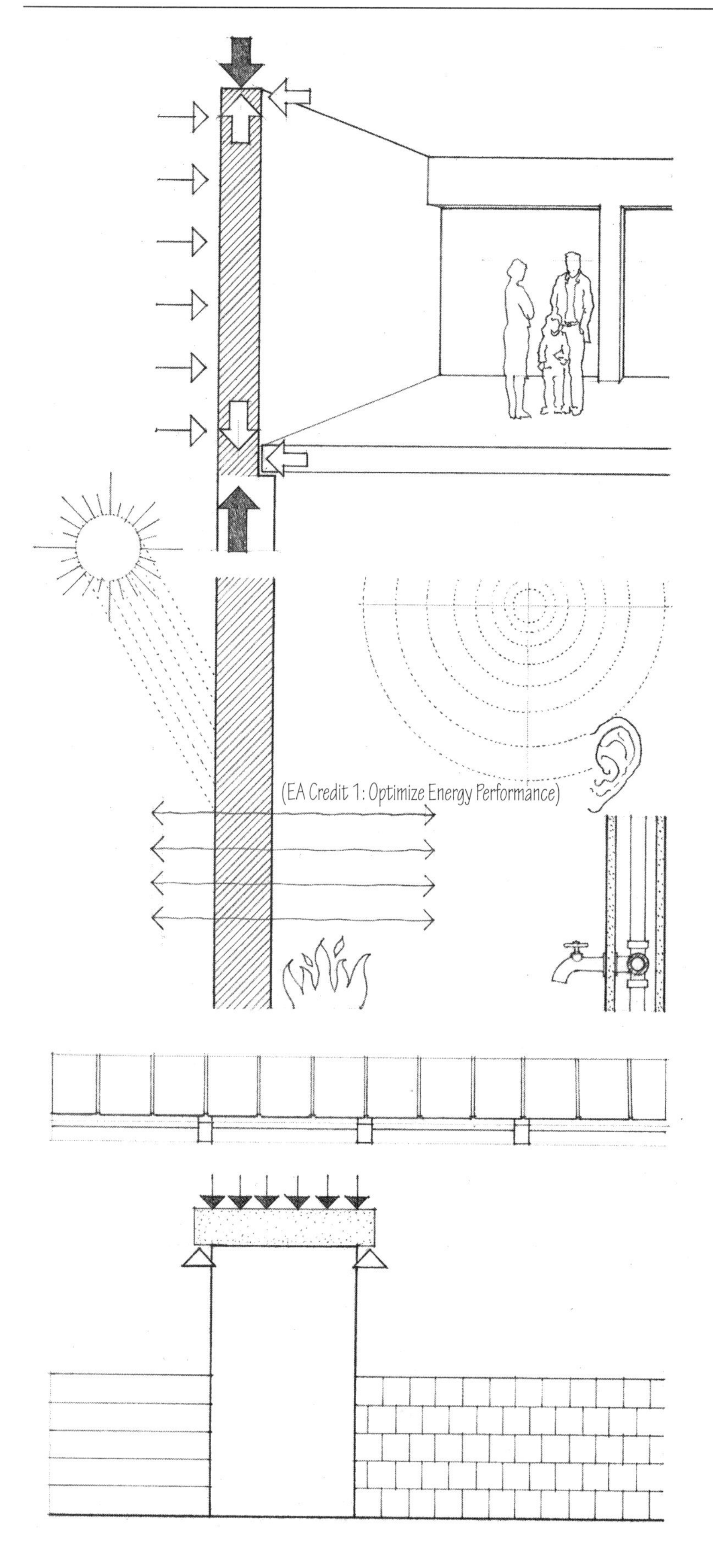

Walls are the vertical constructions of a building that enclose, separate, and protect its interior spaces. They may be loadbearing structures of homogeneous or composite construction designed to support imposed loads from floors and roofs, or consist of a framework of columns and beams with nonstructural panels attached to or filling in between them. The pattern of these loadbearing walls and columns should be coordinated with the layout of the interior spaces of a building.

In addition to supporting vertical loads, exterior wall constructions must be able to withstand horizontal wind loading. If rigid enough, they can serve as shear walls and transfer lateral wind and seismic forces to the ground foundation.

Because exterior walls serve as a protective shield against the weather for the interior spaces of a building, their construction should control the passage of heat, infiltrating air, sound, moisture, and water vapor. The exterior skin, which may be either applied to or integral with the wall structure, should be durable and resistant to the weathering effects of sun, wind, and rain. Building codes specify the fire-resistance rating of exterior walls, loadbearing walls, and interior partitions.

The interior walls or partitions, which subdivide the space within a building, may be either structural or nonloadbearing. Their construction should be able to support the desired finish materials, provide the required degree of acoustical separation, and accommodate when necessary the distribution and outlets of mechanical and electrical services.

Openings for doors and windows must be constructed so that any vertical loads from above are distributed around the openings and not transferred to the door and window units themselves. Their size and location are determined by the requirements for natural light, ventilation, view, and physical access, as well as the constraints of the structural system and modular wall materials.

Structural Frames

- Concrete frames are typically rigid frames and qualify as noncombustible, fire-resistive construction.
- Noncombustible steel frames may utilize moment connections and require fireproofing to qualify as fire-resistive construction.
- Timber frames require diagonal bracing or shear planes for lateral stability and may qualify as heavy timber construction if used with noncombustible, fire-resistive exterior walls and if the members meet the minimum size requirements specified in the building code.
- Steel and concrete frames are able to span greater distances and carry heavier loads than timber structures.
- Structural frames can support and accept a variety of nonbearing or curtain wall systems.
- The detailing of connections is critical for structural and visual reasons when the frame is left exposed.

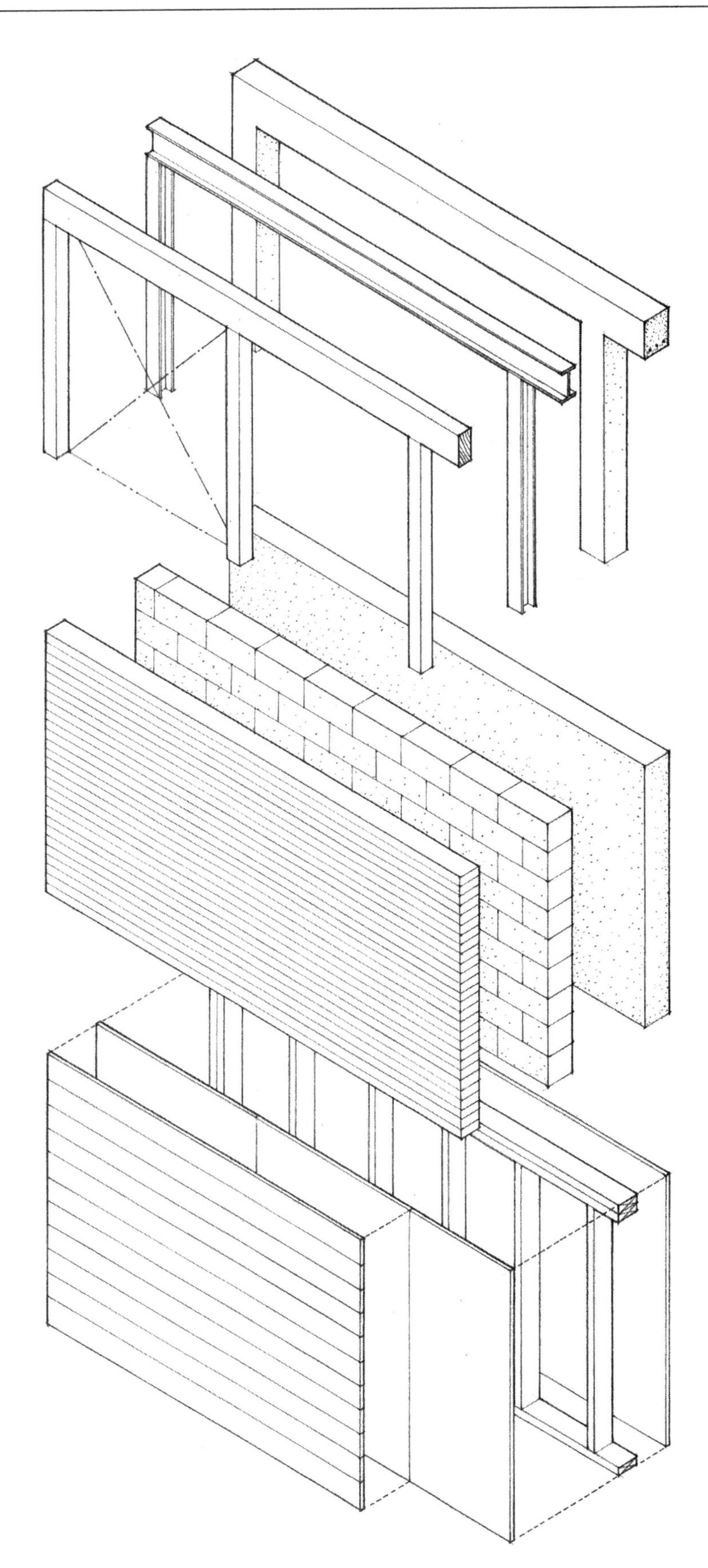

Concrete and Masonry Bearing Walls

- Concrete and masonry walls qualify as noncombustible construction and rely on their mass for their load-carrying capability.
- While strong in compression, concrete and masonry require reinforcing to handle tensile stresses.
- Height-to-width ratio, provisions for lateral stability, and proper placement of expansion joints are critical factors in wall design and construction.
- Wall surfaces may be left exposed.

Metal and Wood Stud Walls

- Studs of cold-formed metal or wood are normally spaced @ 16" or 24" (406 or 610) o.c.; this spacing is related to the width and length of common sheathing materials.
- Studs carry vertical loads while sheathing or diagonal bracing stiffens the plane of the wall.
- Cavities in the wall frame can accommodate thermal insulation, vapor retarders, and mechanical distribution and outlets of mechanical and electrical services.
- Stud framing can accept a variety of interior and exterior wall finishes; some finishes require a nail-base sheathing.
- The finish materials determine the fire-resistance rating of the wall assembly.
- Stud wall frames may be assembled on site or panelized off site.
- Stud walls are flexible in form due to the workability of relatively small pieces and the various means of fastening available.

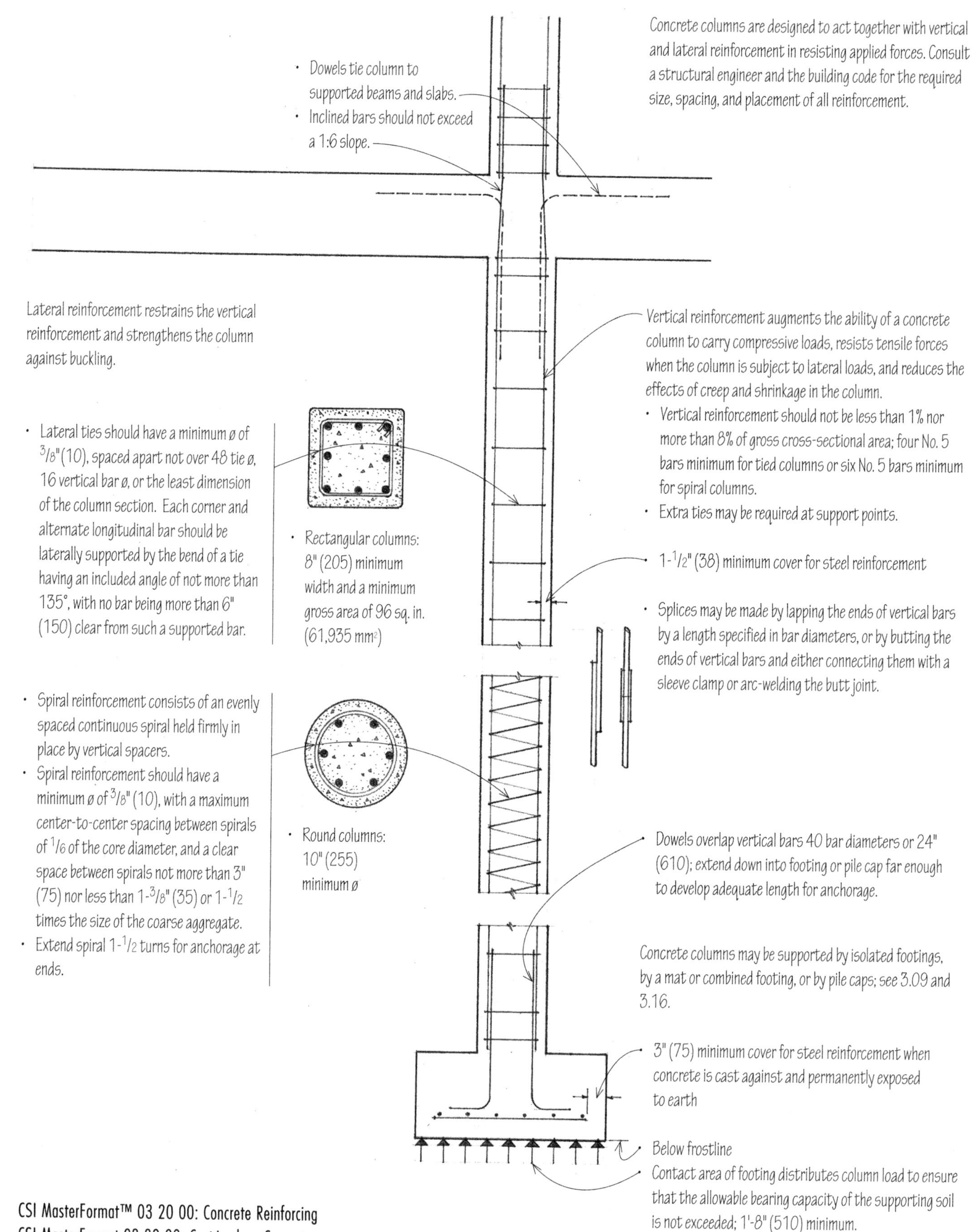

CSI MasterFormat™ 03 20 00: Concrete Reinforcing
CSI MasterFormat 03 30 00: Cast-in-place Concrete
CSI MasterFormat 03 31 00: Structural Concrete

Reinforced concrete columns are usually cast with concrete beams and slabs to form a monolithic structure.

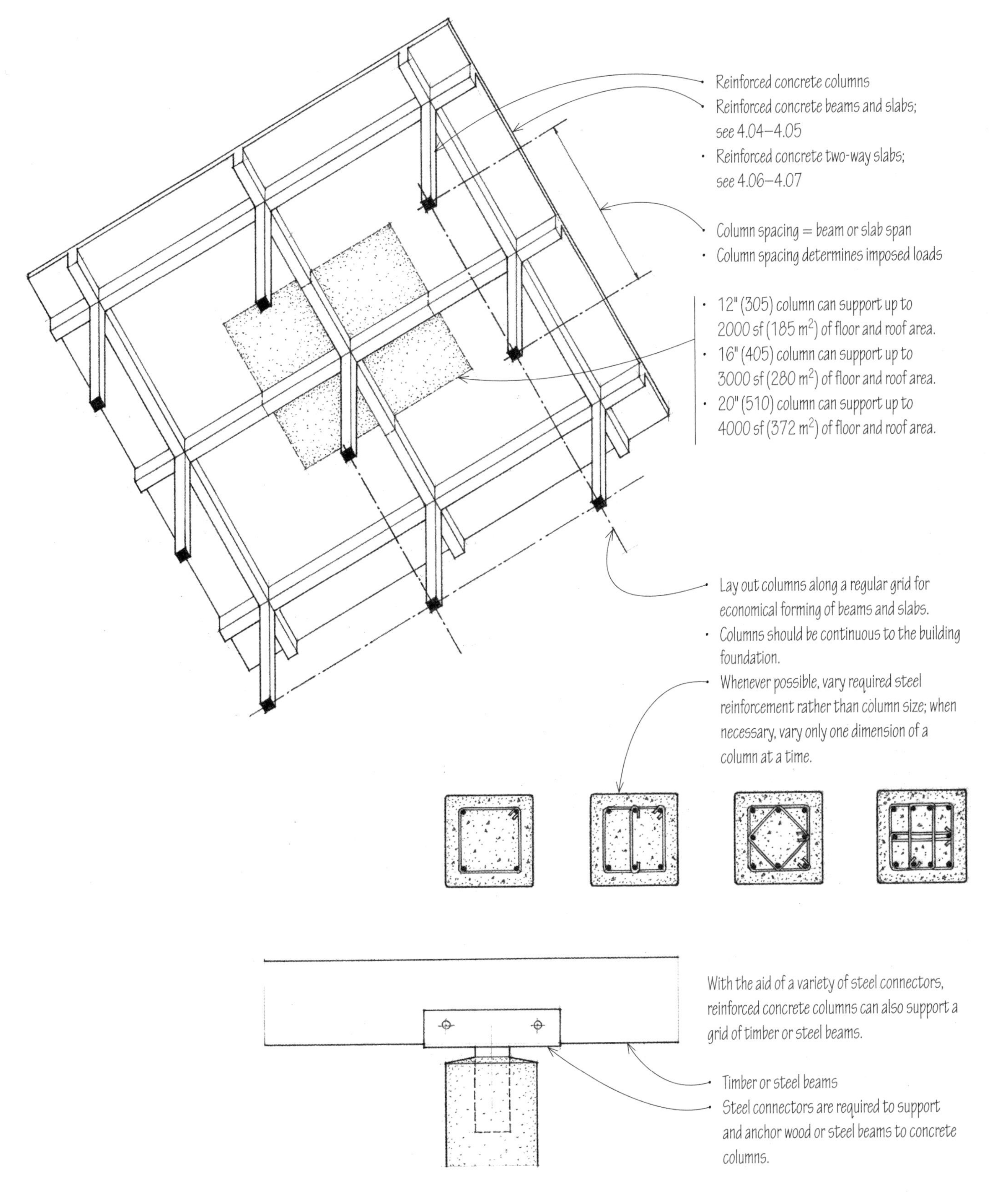

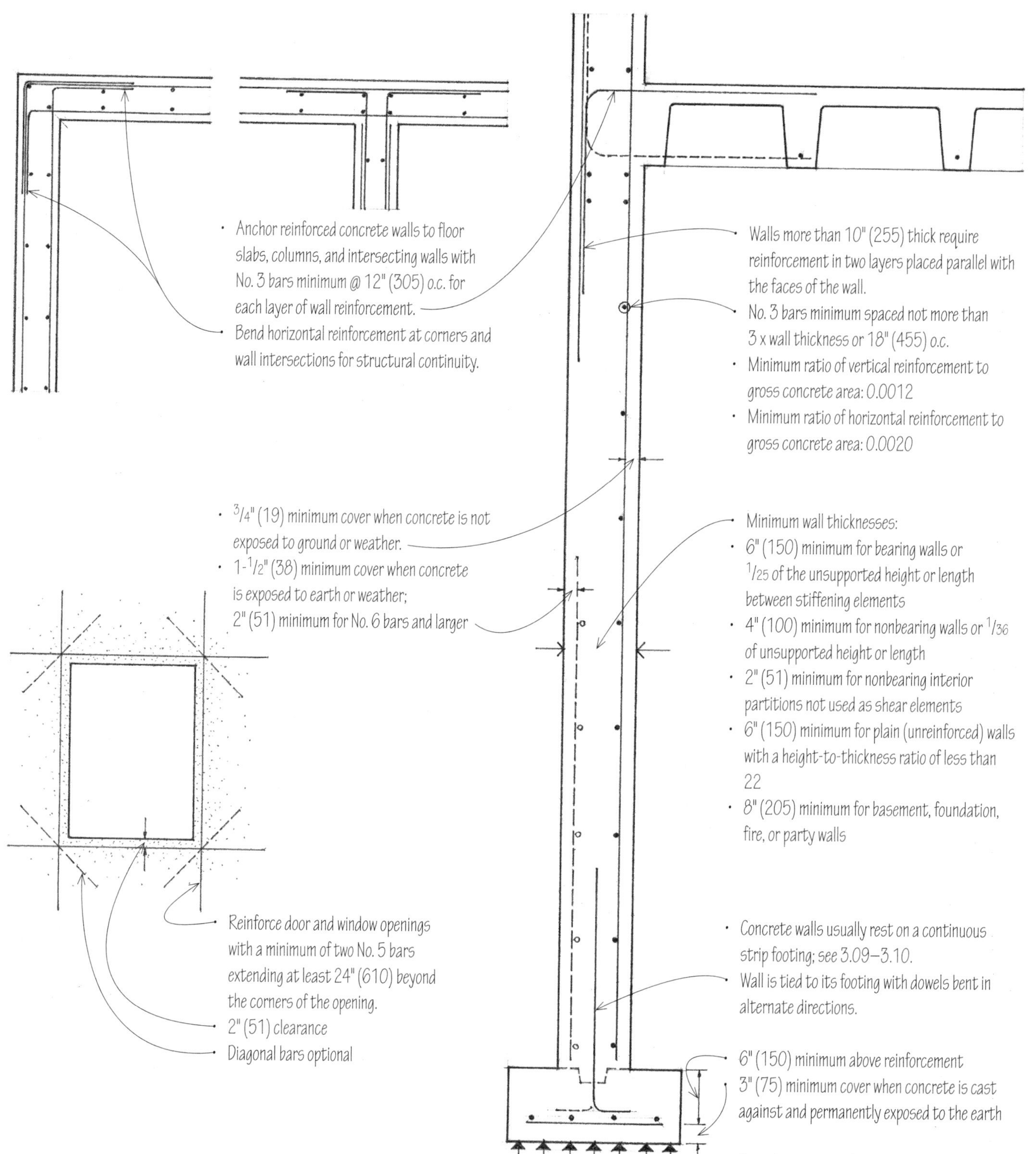

- Consult a structural engineer and the building code for the required size, spacing, and placement of all reinforcement.
- See 12.04–12.05 for a discussion of concrete as a construction material.

CSI MasterFormat 03 20 00: Concrete Reinforcing
CSI MasterFormat 03 30 00: Cast-in-place Concrete
CSI MasterFormat 03 31 00: Structural Concrete

Concrete formwork for columns and walls may be custom-built for a specific job, but prefabricated, reusable panels are used whenever possible. The framework and bracing must be able to maintain the position and shape of the forms until the concrete sets.

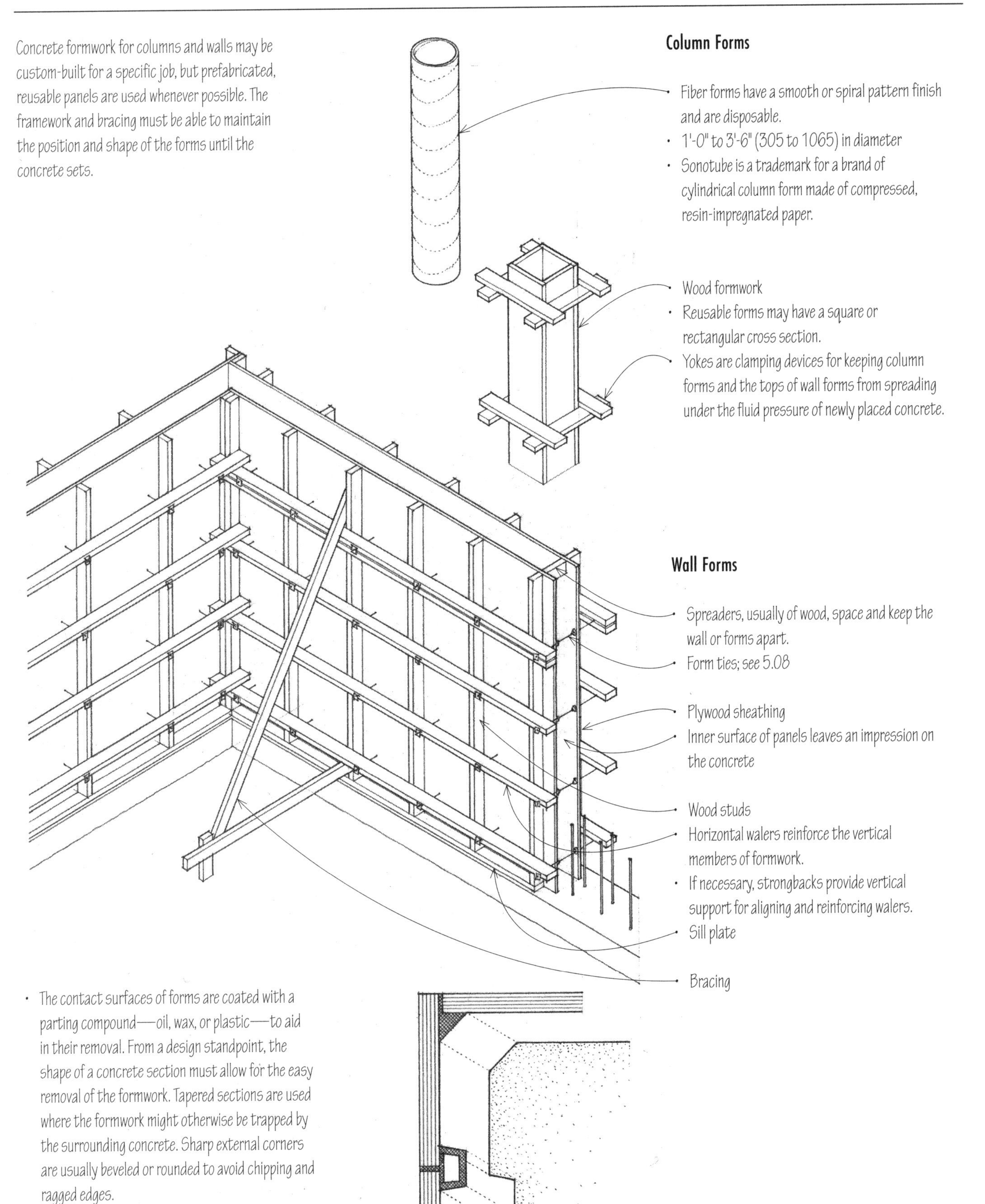

Column Forms

- Fiber forms have a smooth or spiral pattern finish and are disposable.
- 1'-0" to 3'-6" (305 to 1065) in diameter
- Sonotube is a trademark for a brand of cylindrical column form made of compressed, resin-impregnated paper.

- Wood formwork
- Reusable forms may have a square or rectangular cross section.
- Yokes are clamping devices for keeping column forms and the tops of wall forms from spreading under the fluid pressure of newly placed concrete.

Wall Forms

- Spreaders, usually of wood, space and keep the wall or forms apart.
- Form ties; see 5.08
- Plywood sheathing
- Inner surface of panels leaves an impression on the concrete
- Wood studs
- Horizontal walers reinforce the vertical members of formwork.
- If necessary, strongbacks provide vertical support for aligning and reinforcing walers.
- Sill plate
- Bracing

- The contact surfaces of forms are coated with a parting compound—oil, wax, or plastic—to aid in their removal. From a design standpoint, the shape of a concrete section must allow for the easy removal of the formwork. Tapered sections are used where the formwork might otherwise be trapped by the surrounding concrete. Sharp external corners are usually beveled or rounded to avoid chipping and ragged edges.

CSI MasterFormat 03 10 00: Concrete Forming and Accessories

Form ties are required to keep wall forms from spreading under the fluid pressure of newly placed concrete. While various proprietary forms are available, there are two basic types: snap ties and she bolts.

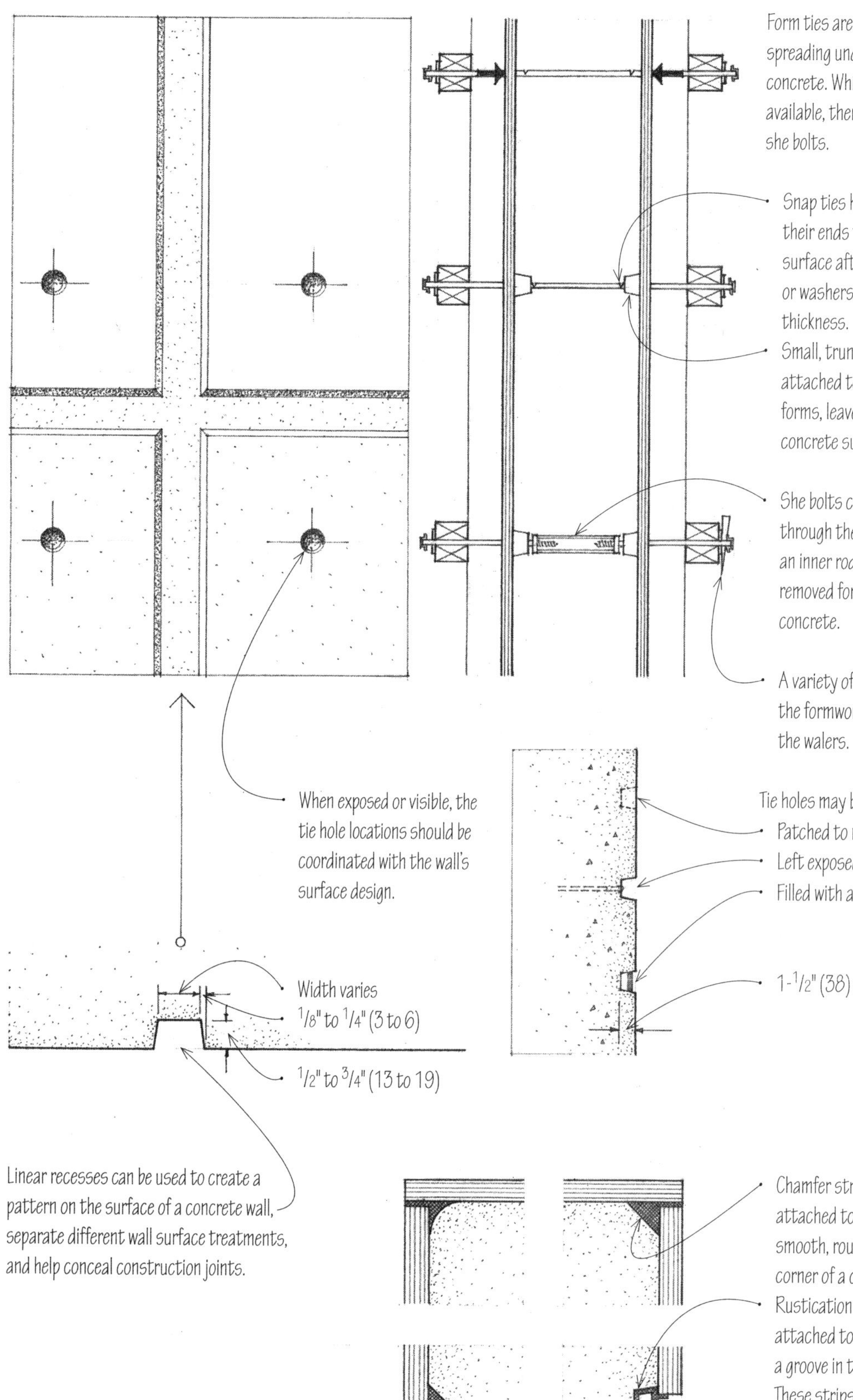

- Snap ties have notches or crimps that allow their ends to be snapped off below the concrete surface after stripping of the forms. Either cones or washers are used to maintain the correct wall thickness.
- Small, truncated cones of wood, steel, or plastic, attached to form ties to space and spread wall forms, leave a neatly finished depression in the concrete surface to be filled or left exposed.
- She bolts consist of waler rods that are inserted through the form and threaded onto the ends of an inner rod. After stripping, the waler rods are removed for reuse while the inner rod remains in the concrete.
- A variety of wedges and slotted devices tighten the formwork and transfer the force in a form tie to the walers.

Tie holes may be:

- Patched to match the surrounding finish
- Left exposed with the exposed tie end epoxied
- Filled with a plastic cap
- 1-1/2" (38)

- Chamfer strips of wood or other material are attached to the inside of a form to produce a smooth, rounded, or beveled edge on the outside corner of a concrete member.
- Rustication strips of wood or other material are attached to the inside face of a form to produce a groove in the surface of a concrete member. These strips are also available as parts of plastic formliner systems.

CSI MasterFormat 03 11 16: Architectural Cast-in-Place Concrete Forming
CSI MasterFormat 03 11 16.13: Concrete Form Liners

A variety of surface patterns and textures can be produced by the following methods.

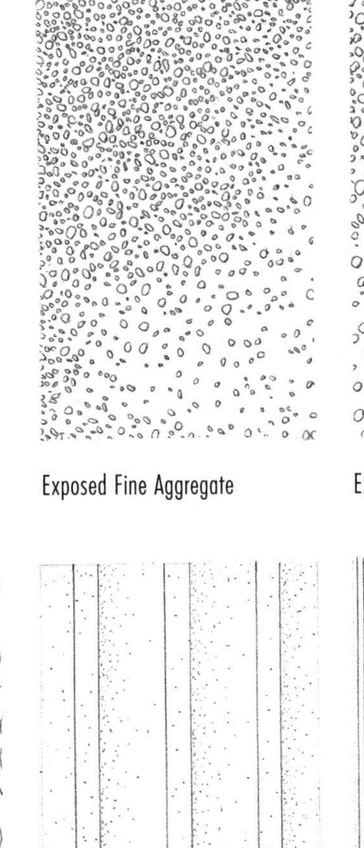

Exposed Fine Aggregate

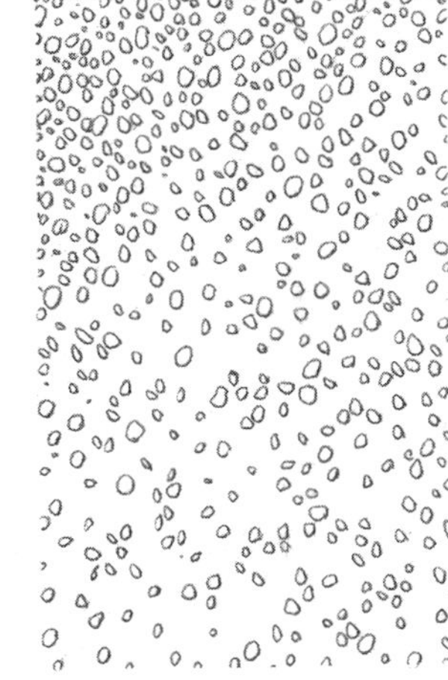

Exposed Coarse Aggregate

Selection of the Concrete Ingredients

- The color of concrete can be controlled with the use of colored cement and aggregates.
- Exposed aggregate finishes are produced by sandblasting, etching with an acid, or scrubbing a concrete surface after the initial set in order to remove the outer layer of cement paste and expose the aggregate. Chemicals can be sprayed on the forms to help retard the setting of the cement paste.

Sandblasted Plywood

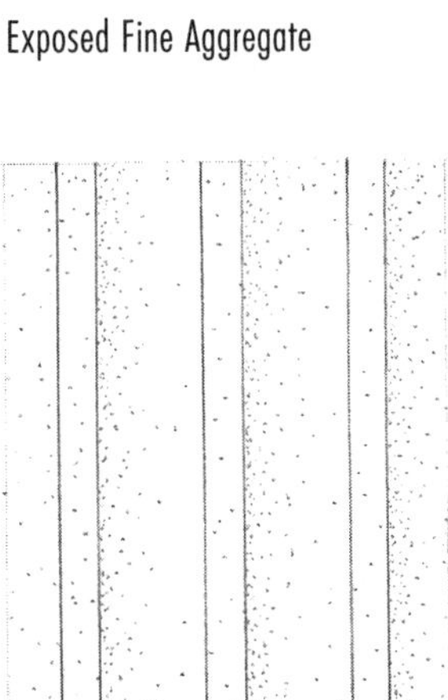

Board-and-Batten Pattern

Ribbed Texture Formliner

The Impressions Left by the Forms

- Béton brut refers to concrete that is left in its natural state after formwork is removed, especially when the concrete surface reflects the texture, joints, and fasteners of a board form.
- Plywood forms can be smooth, or be sandblasted or wirebrushed to accentuate the grain pattern of the face ply.
- Sheathing lumber produces a board texture.
- Metal or plastic formliners can produce a variety of textures and patterns.

Bushhammered Surface

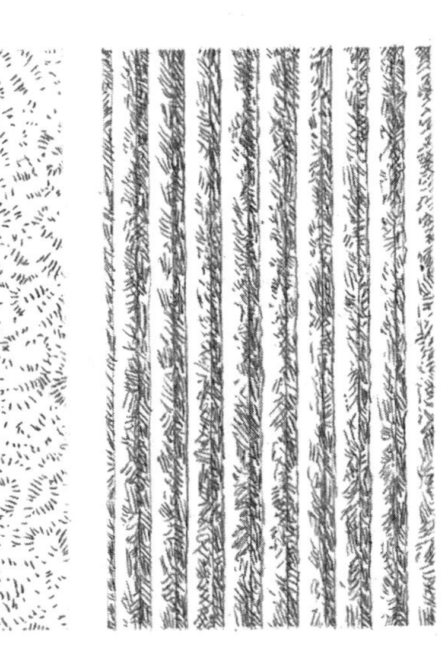

Ribbed Surface Bushhammered

Treatment after the Concrete Sets

- Concrete can be painted or dyed after it has set.
- The concrete surface can be sandblasted, rubbed, or ground smooth.
- Both smooth and textured surfaces can be bush- or jackhammered to produce coarser textures.
- Bushhammered finishes are coarse-textured finishes obtained by fracturing a concrete or stone surface with a power-driven hammer having a rectangular head with a corrugated, serrated, or toothed face.

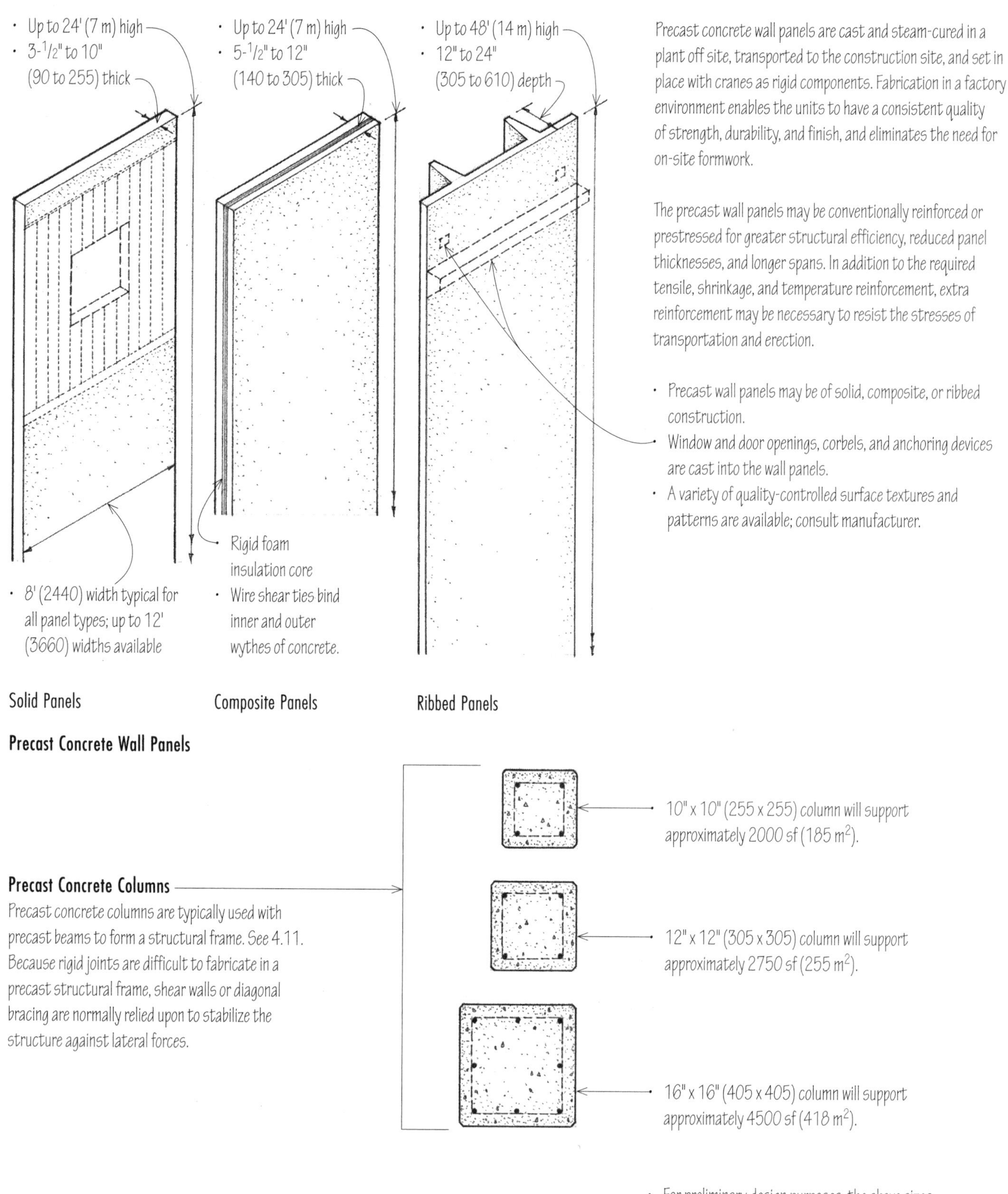

Precast concrete wall panels are cast and steam-cured in a plant off site, transported to the construction site, and set in place with cranes as rigid components. Fabrication in a factory environment enables the units to have a consistent quality of strength, durability, and finish, and eliminates the need for on-site formwork.

The precast wall panels may be conventionally reinforced or prestressed for greater structural efficiency, reduced panel thicknesses, and longer spans. In addition to the required tensile, shrinkage, and temperature reinforcement, extra reinforcement may be necessary to resist the stresses of transportation and erection.

- Precast wall panels may be of solid, composite, or ribbed construction.
- Window and door openings, corbels, and anchoring devices are cast into the wall panels.
- A variety of quality-controlled surface textures and patterns are available; consult manufacturer.

Precast Concrete Wall Panels

Precast Concrete Columns

Precast concrete columns are typically used with precast beams to form a structural frame. See 4.11. Because rigid joints are difficult to fabricate in a precast structural frame, shear walls or diagonal bracing are normally relied upon to stabilize the structure against lateral forces.

- For preliminary design purposes, the above sizes of precast columns can be assumed to support the indicated floor and roof areas.

CSI MasterFormat 03 40 00: Precast Concrete
CSI MasterFormat 03 41 00: Precast Structural Concrete

Precast concrete wall panels may serve as bearing walls capable of supporting sitecast concrete or steel floor and roof systems. Together with precast concrete columns, beams, and slabs, the wall panels form an entirely precast structural system that is inherently modular and fire-resistive. See also 4.11 and 4.13. For nonstructural precast concrete panels, see 7.27.

The lateral stability of a precast concrete structure requires that those floors and roofs that serve as horizontal diaphragms be able to transfer their lateral forces to shear-resisting wall panels. The wall panels, in turn, must be stabilized by columns or cross walls as they transfer the lateral forces to the ground foundation. All forces are transferred by a combination of grouted joints, shear keys, mechanical connectors, steel reinforcement, and reinforced concrete toppings.

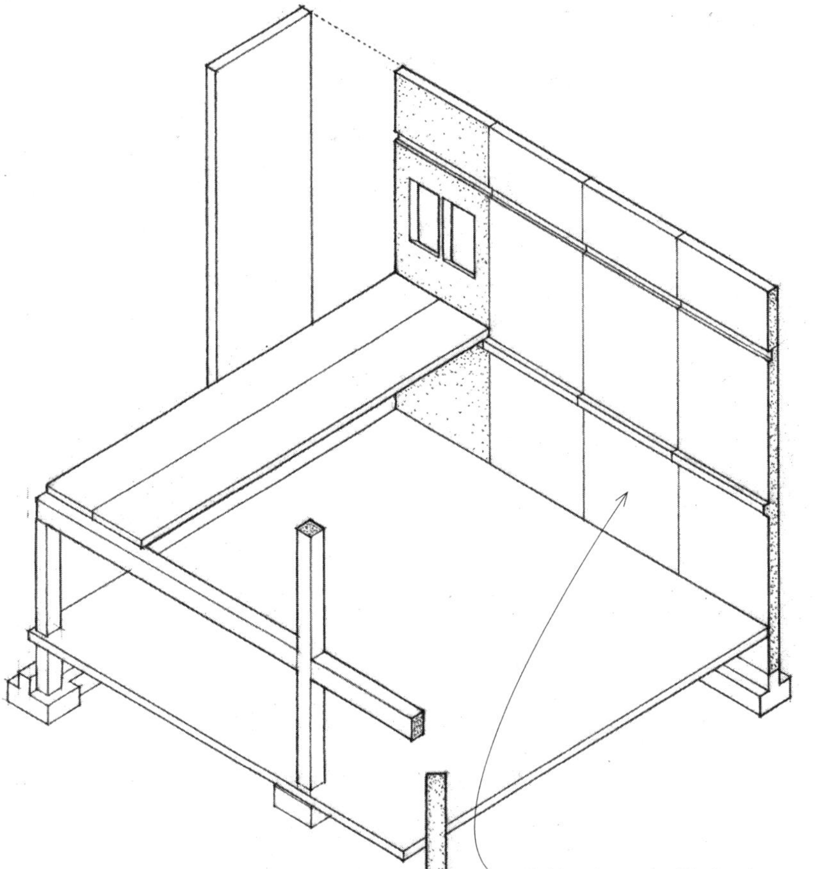

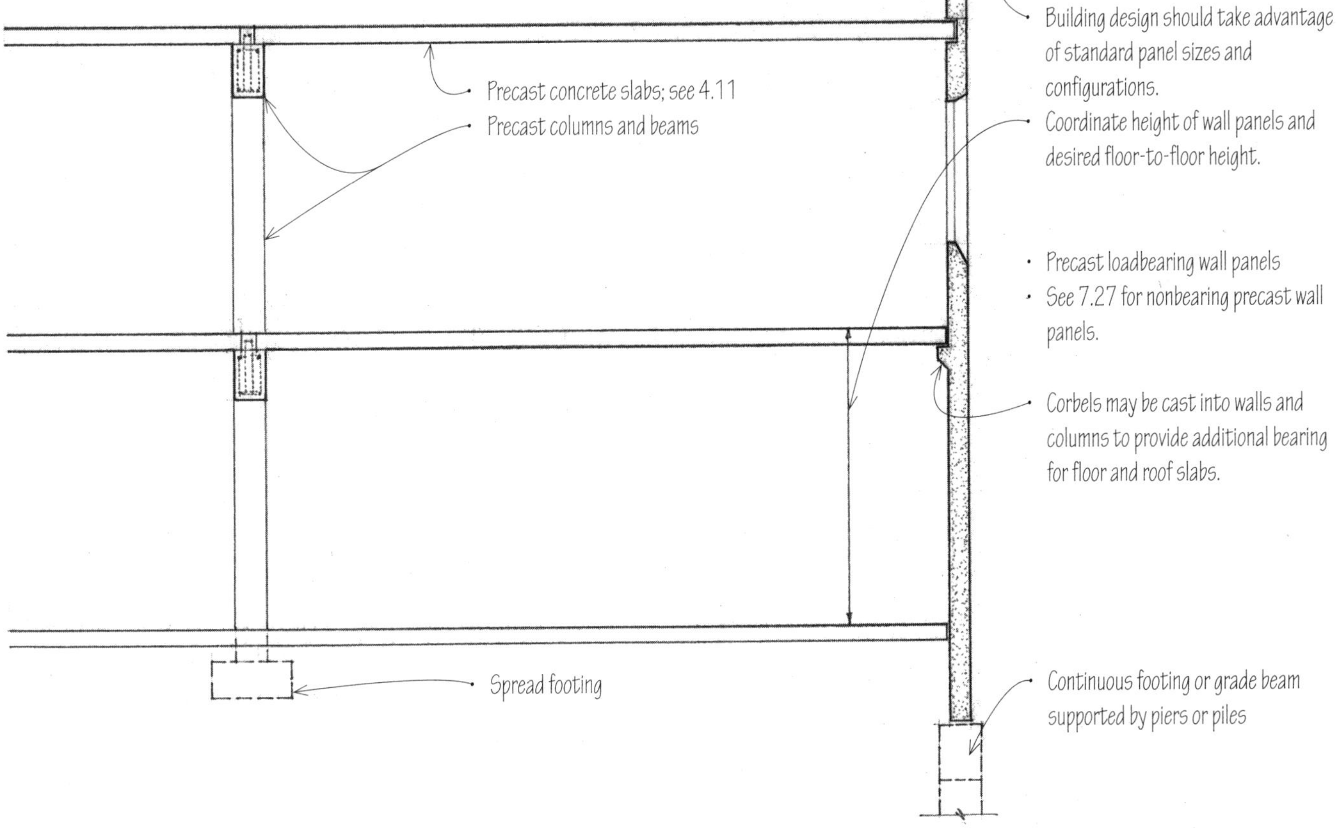

- Building design should take advantage of standard panel sizes and configurations.
- Coordinate height of wall panels and desired floor-to-floor height.
- Precast concrete slabs; see 4.11
- Precast columns and beams
- Precast loadbearing wall panels
- See 7.27 for nonbearing precast wall panels.
- Corbels may be cast into walls and columns to provide additional bearing for floor and roof slabs.
- Spread footing
- Continuous footing or grade beam supported by piers or piles

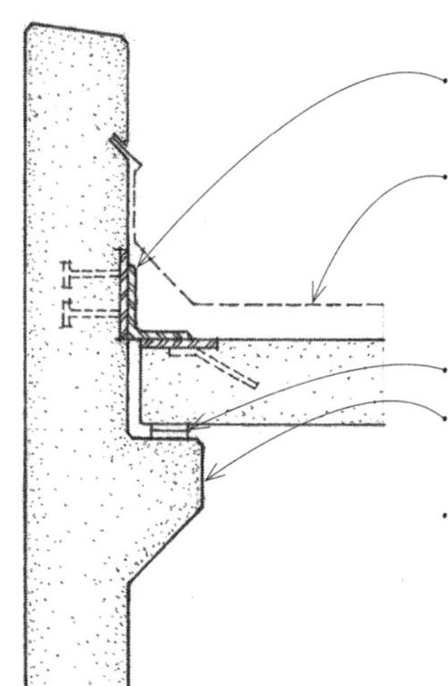

- Steel angle welded to plates anchored in wall and slab units
- Built-up or membrane roofing system; see 7.13–7.14.
- Neoprene bearing strip
- Corbel provides required bearing area.
- Minimum bearing length should be at least 1/180 of the clear span; 2" (51) minimum for solid or hollow core slabs; 3" (75) minimum for beams or stemmed members.

Precast Concrete Slab

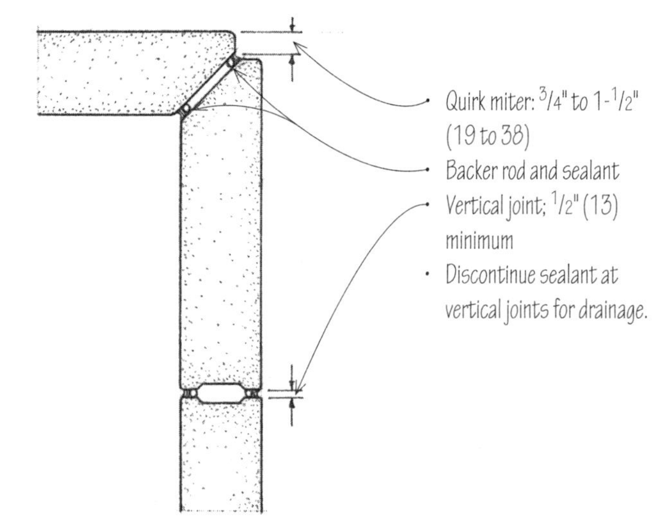

Panel Joints

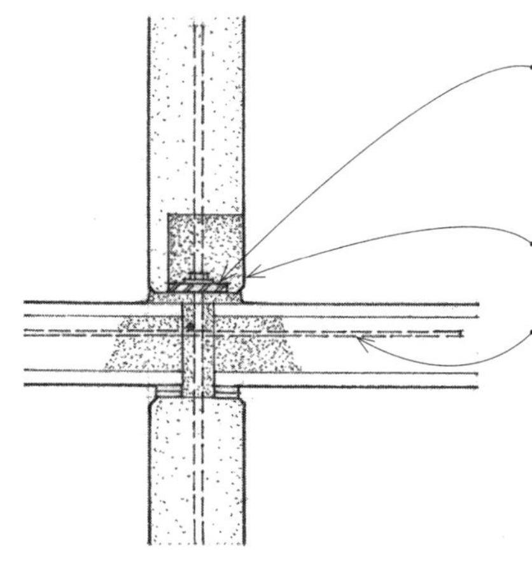

- Steel base plate cast with upper wall panels and secured with anchor bolt cast into lower wall panels.
- Grout pocket after bolted connection is made.
- Dowels in reinforced concrete topping or grouted shear keys tie slabs across supporting wall panels.

Precast Concrete Slab

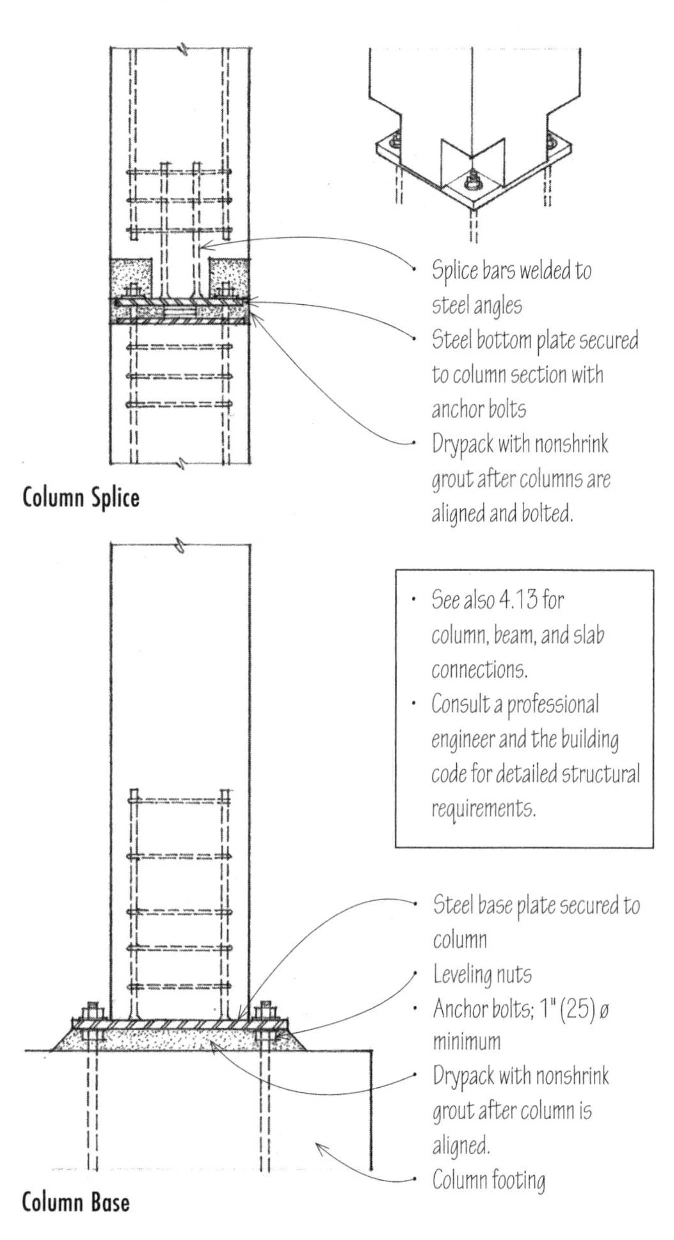

Column Splice

- See also 4.13 for column, beam, and slab connections.
- Consult a professional engineer and the building code for detailed structural requirements.

Column Base

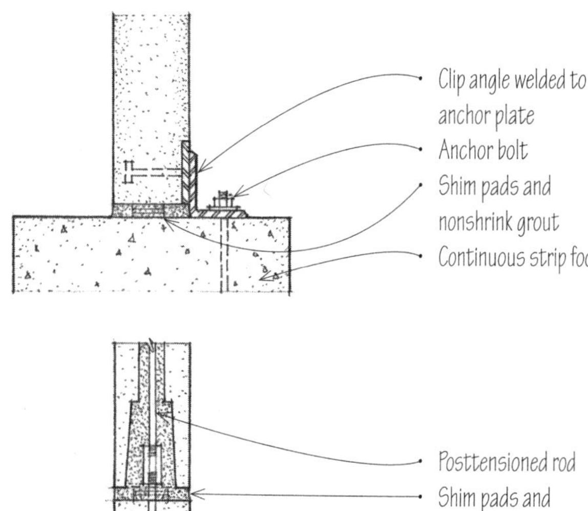

- Clip angle welded to anchor plate
- Anchor bolt
- Shim pads and nonshrink grout
- Continuous strip footing
- Posttensioned rod
- Shim pads and nonshrink grout
- Posttensioned anchor cast into footing

Concrete Footings

Tilt-up construction is a method of casting reinforced concrete wall panels on site in a horizontal position, then tilting them up into their final position. The principal advantage of tilt-up construction is the elimination of the costs associated with constructing and stripping vertical wall forms; this cost savings is offset by the cost of the crane required to lift the completed wall panels into place.

- Once the wall panels are cured to sufficient strength, they are lifted with a crane and set on their footings or piers. They are then temporarily braced until connections can be made to the remaining part of the structure.
- The wall panels must be designed to withstand the stresses of being lifted and moved, which can exceed the in-place loads.

- Full-size panels may be up to 15' (4570) wide.
- 5-1/2" to 11-1/2" (140 to 290) thick

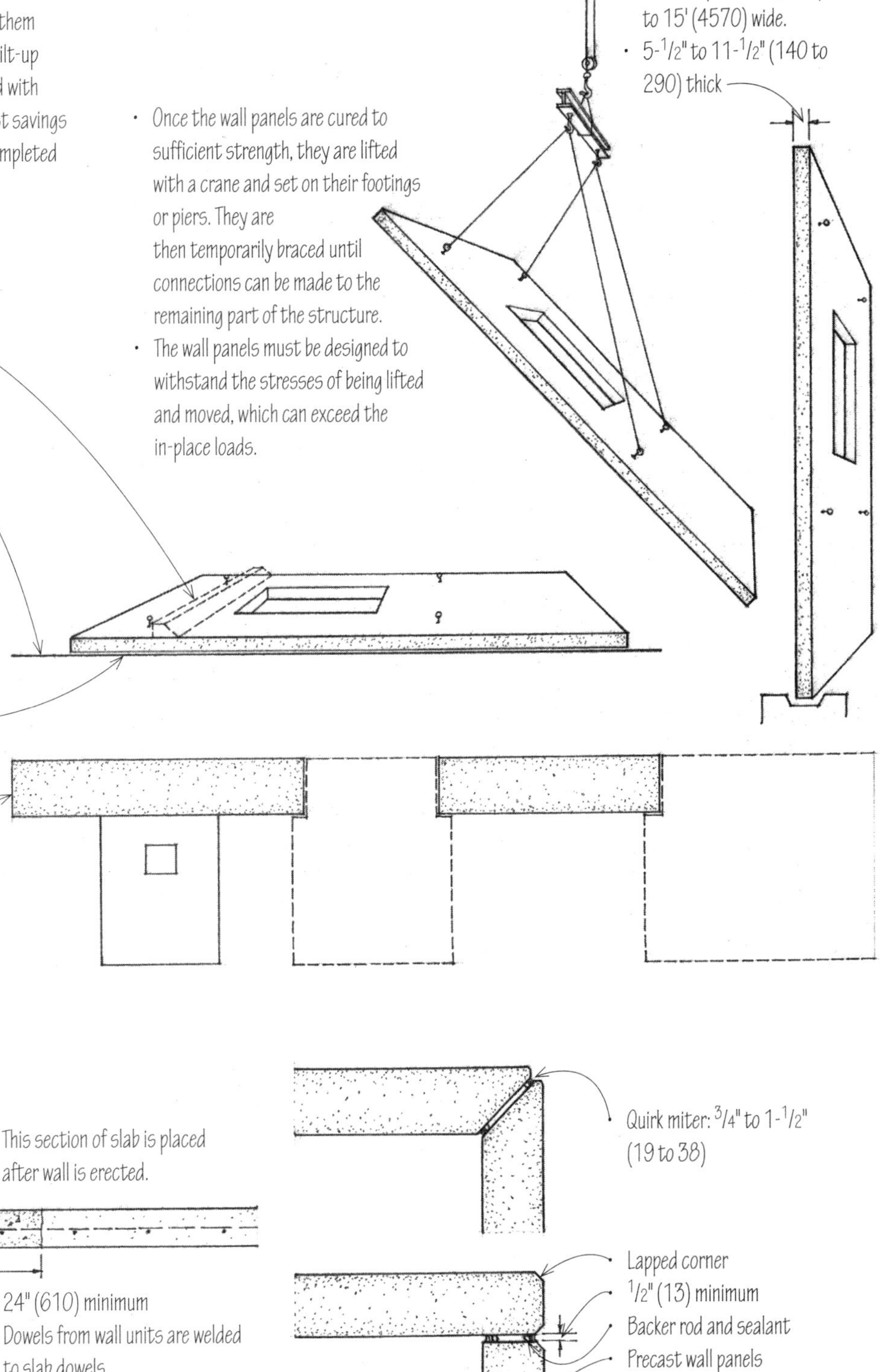

- Projections and the pickup devices are cast into the upper face.
- The concrete ground slab for the building under construction usually serves as the casting platform, although earth, plywood, or steel molds can also be used. The slab must be designed to withstand the truck crane load if the lifting operation requires the presence of the crane on the slab.
- The casting platform should be level and smoothly troweled; a bond breaking agent is used to ensure a clean lift.
- Reveals and recessed steel plates may be cast into the underside of the panels.

- Spandrel units can overhang and span openings up to 30' (9145) wide.
- The floor and roof connections are similar to those shown in 4.13 and 5.12. Shown are typical wall panel connections to adjacent panels and footings.

- This section of slab is placed after wall is erected.
- 24" (610) minimum
- Dowels from wall units are welded to slab dowels.
- Grouted after wall is set
- Groove in continuous footing allows wall panels to be shimmed to a level position.
- Precast wall panels may be supported by isolated spread footings, strip footings, or piers.

Foundations

- Quirk miter: 3/4" to 1-1/2" (19 to 38)
- Lapped corner
- 1/2" (13) minimum
- Backer rod and sealant
- Precast wall panels
- 3/4" (19) chamfers at panel joints
- 1/2" (13) minimum
- Backer rod and sealant

Panel Connections

CSI MasterFormat 03 47 13: Tilt-Up Concrete

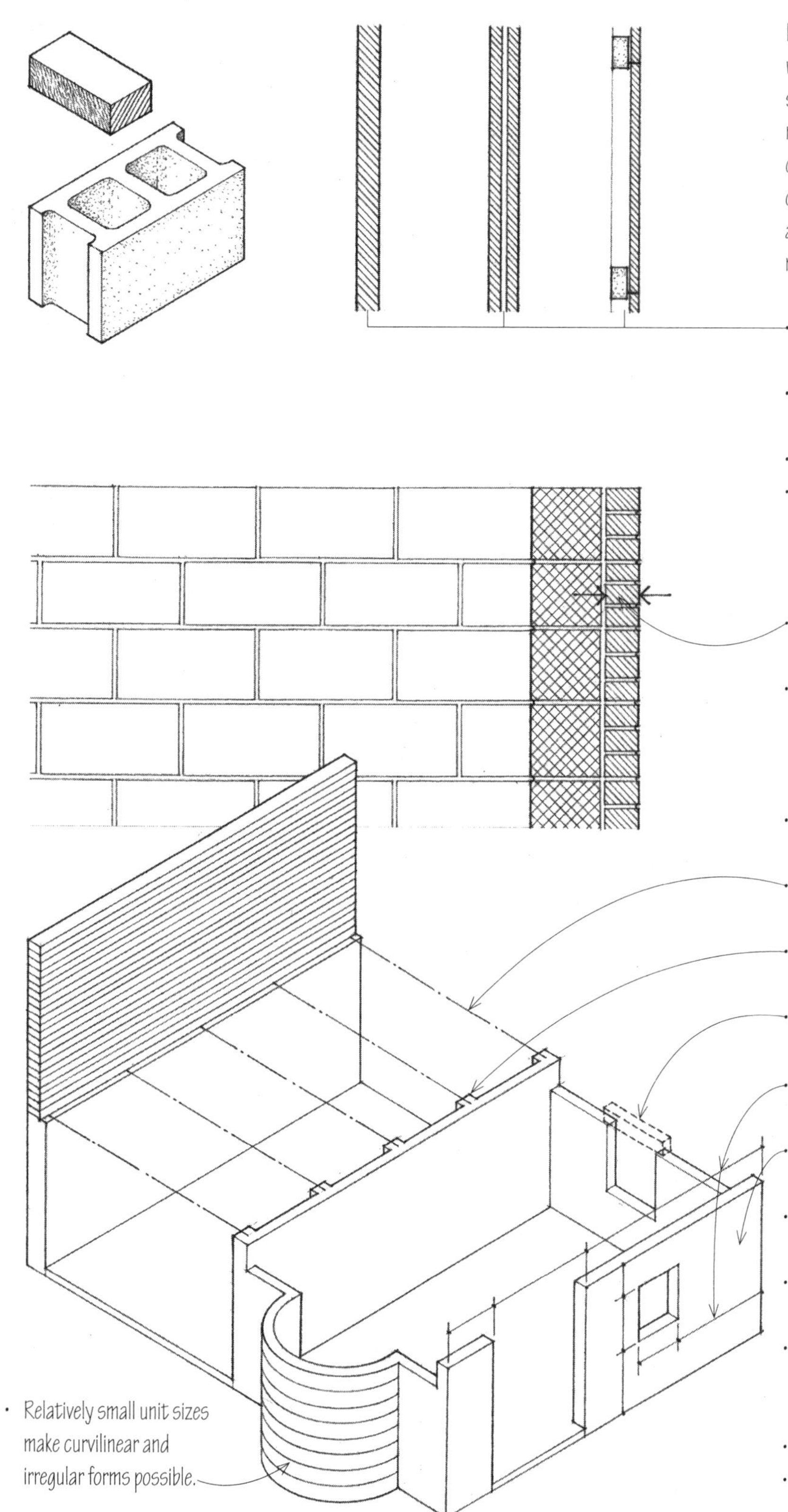

Masonry walls consist of modular building blocks bonded together with mortar to form walls that are durable, fire-resistant, and structurally efficient in compression. The most common types of masonry units are bricks, which are heat-hardened clay units, and concrete blocks, which are chemically hardened units. Other types of masonry units include structural clay tile, structural glass block, and natural or cast stone. See 12.06–12.07 for a discussion of masonry as a construction material.

- Masonry walls may be constructed as solid walls, cavity walls, or veneered walls.
- See 7.28–7.29 for masonry veneer systems.

- Masonry walls may be unreinforced or reinforced.
- Unreinforced masonry walls, also called plain masonry, incorporate horizontal joint reinforcement and metal wall ties to bond the wythes of a solid or cavity wall; see 5.16–5.17 for types of unreinforced masonry walls.
- A wythe refers to a continuous vertical section of a wall that is one masonry unit in thickness.
- Reinforced masonry walls utilize steel reinforcing bars embedded in grout-filled joints and cavities to aid the masonry in resisting stresses; see 5.18 for reinforced masonry walls.

- Masonry bearing walls are typically arranged in parallel sets to support steel, wood, or concrete spanning systems.
- Common spanning elements include open-web steel joists, timber or steel beams, and sitecast or precast concrete slabs.
- Pilasters stiffen masonry walls against lateral forces and buckling, and provide support for large concentrated loads.
- Openings may be arched or spanned with lintels.

- Modular dimensions

- Exterior masonry walls must be weather-resistant and control heat flow.
- Water penetration must be controlled through the use of tooled joints, cavity spaces, flashing, and caulking.
- Cavity walls are preferred for their increased resistance to water penetration and improved thermal performance.
- Differential movements in masonry walls due to changes in temperature or moisture content, or to stress concentrations, require the use of expansion and control joints.
- For installation of thermal insulation, see 7.44.
- For fire-resistance ratings of noncombustible masonry walls, see A.12–A.13.

- Relatively small unit sizes make curvilinear and irregular forms possible.

CSI MasterFormat 04 20 00: Unit Masonry
CSI MasterFormat 04 21 00: Clay Unit Masonry
CSI MasterFormat 04 22 00: Concrete Unit Masonry

Lateral Support for Masonry Walls

Type of Masonry	Maximum L/t or H/t
Bearing Walls	
Solid or grouted	20
All other	18
Nonbearing Walls	
Exterior	18
Interior	36

- L/t = ratio of wall length to thickness; lateral support may be provided by cross walls, columns, or pilasters.
- H/t = ratio of wall height to thickness; lateral support may be provided by floors, beams, or roofs.
- More stringent requirements exist for Seismic Zones 3 and 4.
- Consult a professional engineer and the building code for the structural requirements of all masonry walls.

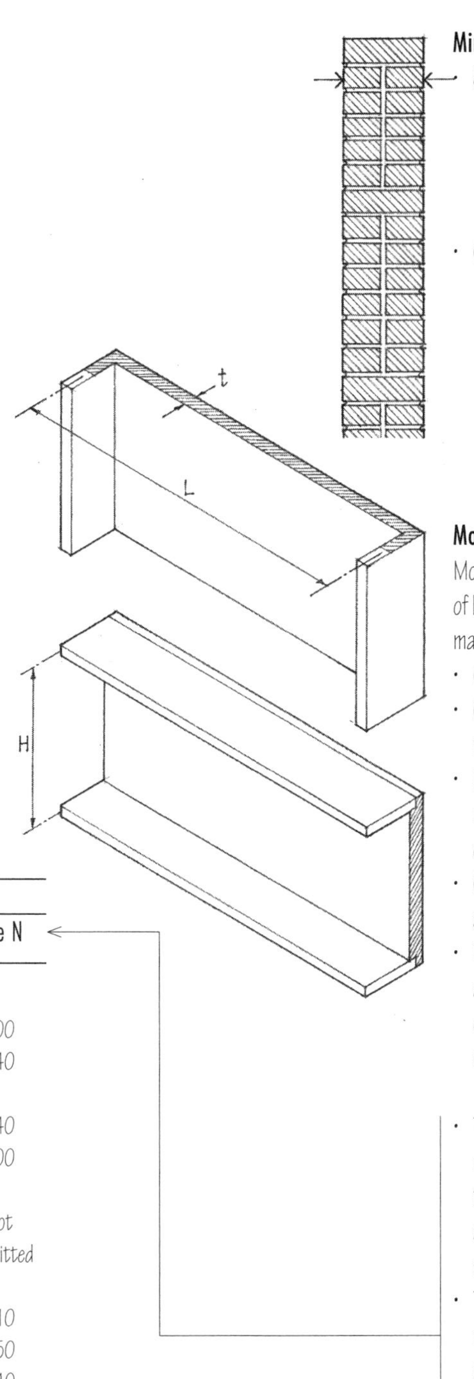

Minimum Wall Thickness

- 8" (205) minimum nominal thickness for:
  - Masonry bearing walls
  - Masonry shear walls
  - Masonry parapets; height of parapet not to exceed 3 x parapet thickness
- 6" (150) minimum nominal thickness for:
  - Reinforced masonry bearing walls
  - Solid masonry walls in one-story buildings not more than 9' (2745) high
  - Masonry walls relied upon for resistance to lateral loading are limited to 35' (10 m) in height.

Allowable Compressive Stresses (psi)* in Unreinforced Masonry Walls

	Mortar Type		
	Type M	Type S	Type N
Solid brick masonry			
4500+ psi	250	225	200
2500–4500 psi	175	160	140
Solid concrete masonry			
Grade N	175	160	140
Grade S	125	115	100
Grouted masonry			
4500+ psi	350	275	Not
2500–4500 psi	275	215	permitted
Cavity walls			
Solid units†	140	130	110
Hollow units‡	70	60	50
Hollow unit masonry	170	150	140
Natural stone	140	120	100

* 1 psi = 6.89 kPa.

† Solid masonry units have a net surface area at least 75% of the gross cross-sectional area parallel to the bedding plane.

‡ Hollow masonry units have a net surface area less than 75% of the gross cross-sectional area parallel to the bedding plane.

Mortar

Mortar is a plastic mixture of cement or lime, or a combination of both, with sand and water, used as a bonding agent in masonry construction.

- CSI MasterFormat 04 05 13: Masonry Mortaring
- Cement mortar is made by mixing portland cement, sand, and water.
- Lime mortar is a mixture of lime, sand, and water that is rarely used because of its slow rate of hardening and low compressive strength.
- Cement-lime mortar is a cement mortar to which lime is added to increase its plasticity and water-retentivity.
- Masonry cement is a proprietary mix of portland cement and other ingredients, as hydrated lime, plasticizers, air-entraining agents, and gypsum, requiring only the addition of sand and water to make cement mortar.

- Type M mortar is a high-strength mortar recommended for use in reinforced masonry below grade or in contact with the earth, as foundation and retaining walls subject to frost action or to high lateral or compressive loads; compressive strength of 2500 psi (17,238 kPa).
- Type S mortar is a medium-high-strength mortar recommended for use in masonry where bond and lateral strength are more important than compressive strength; compressive strength of 1800 psi (12,411 kPa).
- Type N mortar is a medium-strength mortar recommended for general use in exposed masonry above grade where high compressive and lateral strength are not required; compressive strength of 750 psi (5171 kPa).
- Type O mortar is a low-strength mortar suitable for use in interior nonloadbearing walls and partitions.
- Type K mortar is a very-low-strength mortar suitable only for use in interior nonloadbearing walls where permitted by the building code.

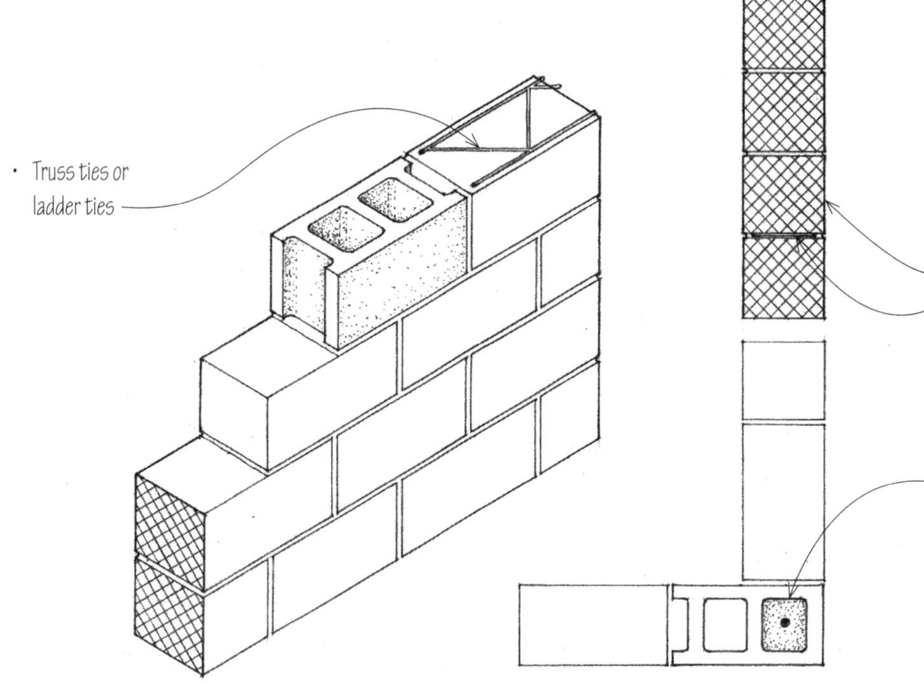

Solid Masonry

Solid masonry walls may be constructed of either solid or hollow masonry units laid contiguously with all joints solidly filled with mortar.

Single wythe walls are typically of concrete masonry units (CMU) bonded with horizontal joint reinforcement.

- Concrete masonry units (CMUs)
- Horizontal joint reinforcement is fully embedded in bed joint mortar.
- Vertical spacing of joint reinforcement not to exceed 16" (405)
- Vertical cells of the concrete blocks may be reinforced with steel bars embedded in grout.

Solid masonry walls may also consist of multiple wythes bonded by grout, corrosion-resistant metal wall ties, or horizontal joint reinforcement.

- Wythes may be bonded by masonry headers or by metal ties.
- Masonry headers should compose at least 4% of exposed face area, with a vertical and horizontal spacing of not less than 24" (610).
- Metal ties should conform to requirements for cavity walls.

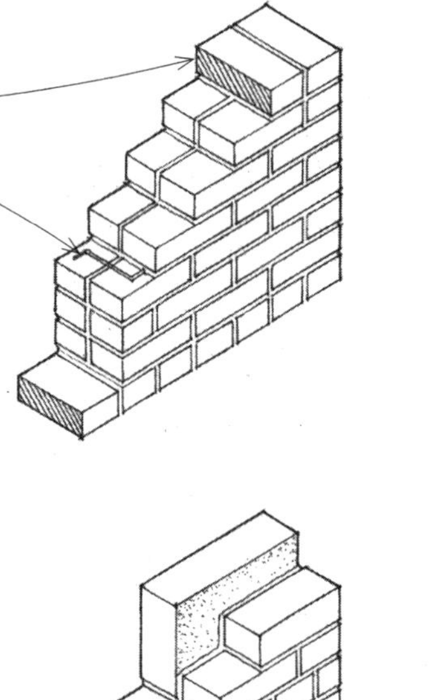

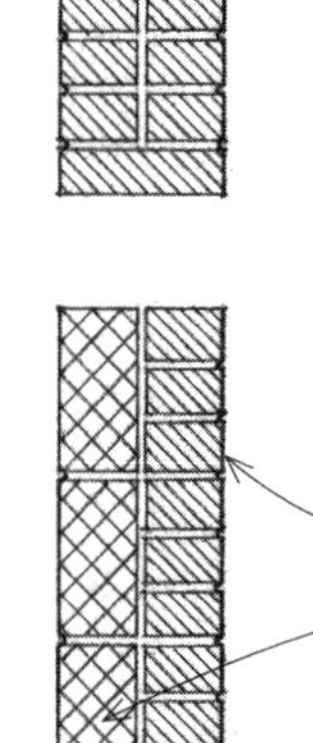

- Composite walls are solid masonry walls having a facing wythe and a backup wythe of different solid or hollow masonry units.
- Facing wythe
- Backup wythe

- 5/8" (16) minimum mortar cover between ties or joint reinforcement and any exposed face
- 1/4" (6) minimum mortar thickness between masonry and ties or joint reinforcement

CSI MasterFormat 04 20 00: Unit Masonry
CSI MasterFormat 04 05 23: Masonry Accessories

Grouted Masonry

Grouted masonry walls have all interior joints filled entirely with grout as the work progresses. The grout used to consolidate the adjoining materials into a solid mass is a fluid portland cement mortar that will flow easily without segregation of the ingredients.

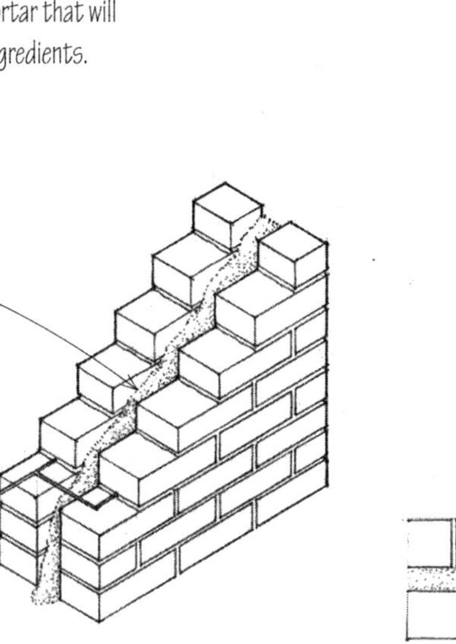

- All interior joints are filled entirely w/grout.

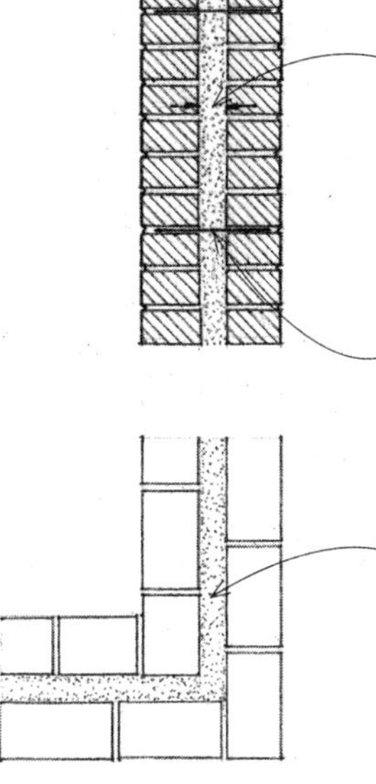

- 3/4" (19) minimum for low-lift grout construction
- Low-lift grouting is executed in lifts not exceeding 6 x the width of the grout space or a maximum of 8" (203 mm) as the wall is constructed.
- 3" (75) minimum for high-lift grouting
- High-lift grouting is completed a story at a time in lifts not exceeding 6' (1830). High-lift grouting requires a wider grout space and rigid metal ties to bond the two tiers together.
- 3/16" (5) minimum ø tie of corrosion-resistant metal for each 2 sf (0.19 m^2) of wall area
- 16" (405) maximum vertical spacing of ties
- Only Type M or Type S mortar is permitted.

CSI MasterFormat 04 05 16: Masonry Grouting

Cavity Walls

Cavity walls are constructed of a facing and a backing wythe of either solid or hollow masonry units, completely separated by a continuous air space and bonded with metal wall ties or horizontal joint reinforcement. Cavity walls have two advantages over other types of masonry walls:

1. The cavity enhances the thermal insulation value of the wall and permits the installation of additional thermal insulation material.
2. The air space acts as a barrier against water penetration if the cavity is kept clear, and if adequate weep holes and flashing are provided.

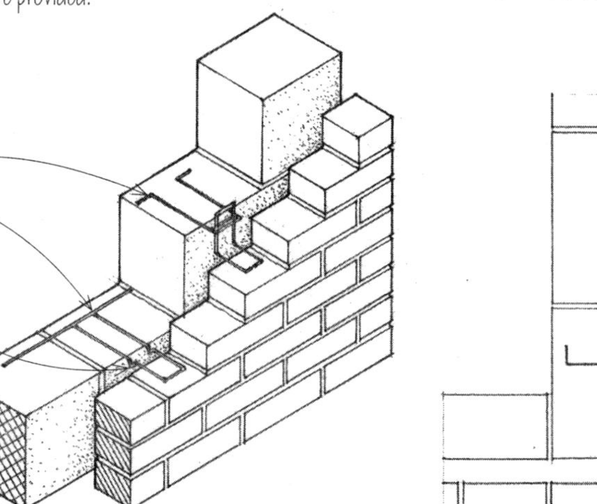

- Adjustable loop tie
- Ladder loop tie
- Drip to prevent water from running across tie to inner wythe

- Cavity to be not less than 2" (51) nor more than 4-1/2" (115) wide
- Solid or hollow masonry units
- Both facing and backing wythes to have a 4" (100) minimum nominal thickness. When computing the ratio of unsupported height or length to thickness, the value for thickness is equal to the sum of the nominal thicknesses of the inner and outer wythes.
- 3/16" (5) minimum ø tie of corrosion-resistant metal for each 4-1/2 sf (0.42 m^2) of wall area for cavities up to 3" (75) wide; for wider cavities, provide a metal tie for each 3 sf (0.28 m^2) of wall area.
- Stagger ties in alternate courses w/ a maximum vertical distance between ties of 16" (405) and a maximum horizontal spacing of 36" (915).
- Place additional ties at 3' (915) o.c. maximum around openings within 12" (305) of the edges of the openings.
- 5/8" (16) minimum mortar cover for joint reinforcement

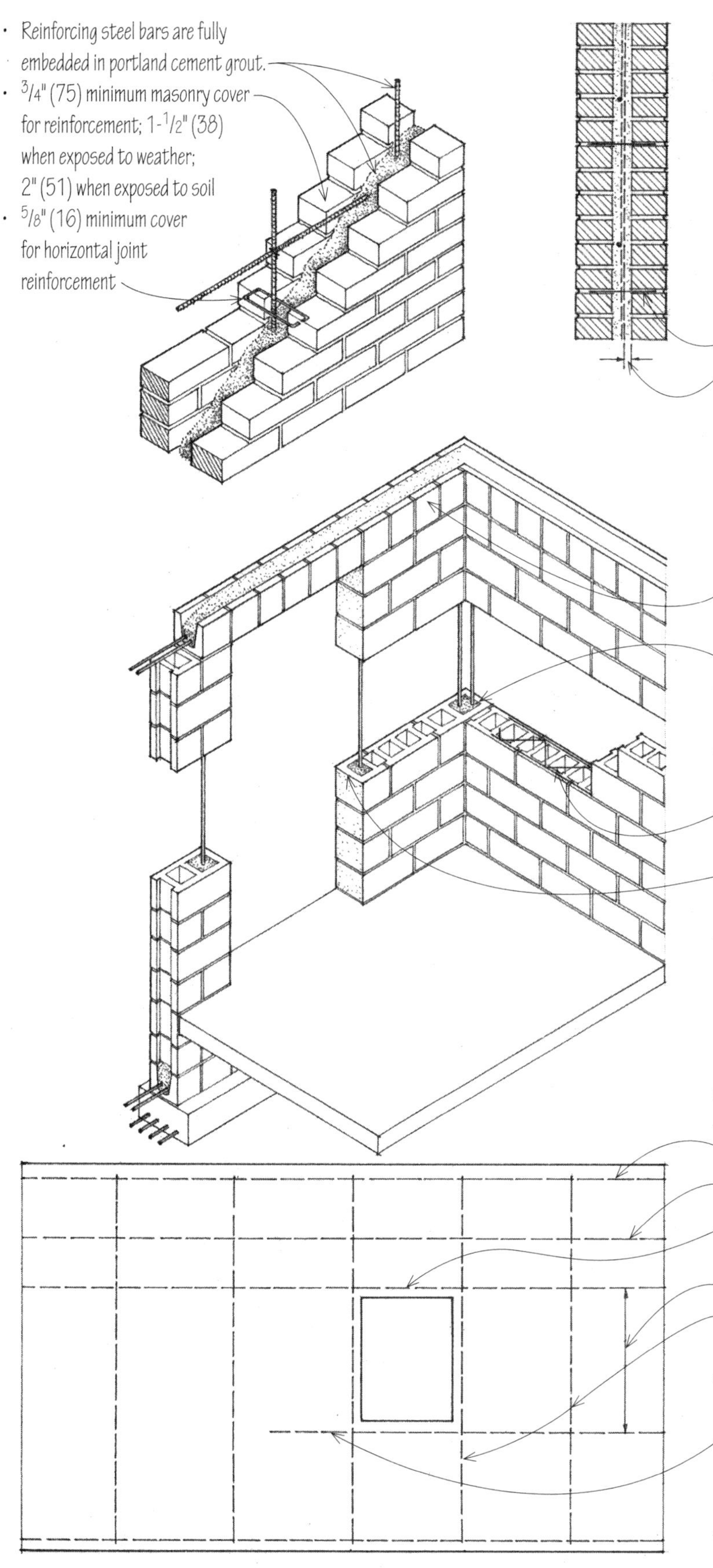

Reinforced masonry walls use steel reinforcing bars placed in thickened joints or cavities with a fluid grout mix of portland cement, aggregate, and water for greater strength in carrying vertical loads and increased resistance to buckling and lateral forces. It is essential that a strong bond develop between the reinforcing steel, grout, and masonry units.

Reinforced Grouted Masonry

- Reinforced grouted masonry should conform to the requirements for plain grouted masonry. See 5.17.
- Metal wall ties
- 1/4" (6) minimum between reinforcement and masonry for fine grout; 1/2" (13) minimum cover for coarse grout

Reinforced Concrete Unit Masonry

- Horizontal bond beam
- All cells containing reinforcement are filled solidly with grout.
- Cells are aligned vertically to form a 2" x 3" (51 x 75) minimum clear, continuous vertical space.
- Reinforcement continues down to a reinforced concrete footing.
- Horizontal joint reinforcement
- Fullbed mortar @ end walls and cross webs form grouted cells; only Type M or Type S mortar is permitted.
- Where grout lift exceeds 4' (1220), provide cleanouts at bottom course of cells to be grouted; inspect and seal before grouting.

General Requirements

- Provide horizontal reinforcement
  - At top of parapet walls
  - At structurally connected floors and roofs
  - At top of wall openings
  - At the top of foundations
- 10' (3050) maximum spacing
- Vertical reinforcement to be 3/8" (10) ø minimum, with a maximum spacing of 4' (1220) o.c.
- The sum of vertical and horizontal reinforcement to be at least 0.002 x the gross cross-sectional area of the wall.
- One No. 4 or two No. 3 bars around openings, extending at least 24" (610) beyond corners
- Consult a structural engineer and the building code for the structural requirements of reinforced masonry walls.

- Minimum nominal width = 12" (305)
- Minimum nominal length = 12" (305); maximum = 3 x column width
- Lateral support for columns = 30 x column width
- The least dimension of reinforced masonry columns to be 12" (305) with a maximum unsupported height of 20 x the least dimension.

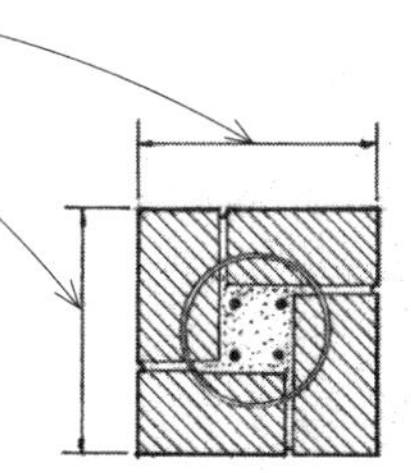

- Minimum of four No. 3 bars with lateral ties at least 18" (455) o.c. or 0.005 of effective area of masonry
- Maximum reinforcement = 0.03 of effective area of masonry
- Minimum lateral reinforcement = 0.0018 of effective area of masonry

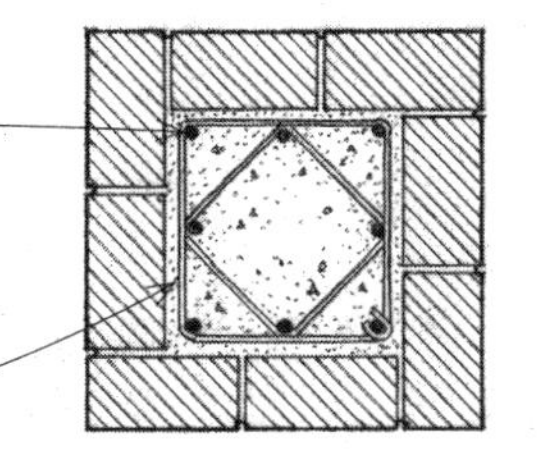

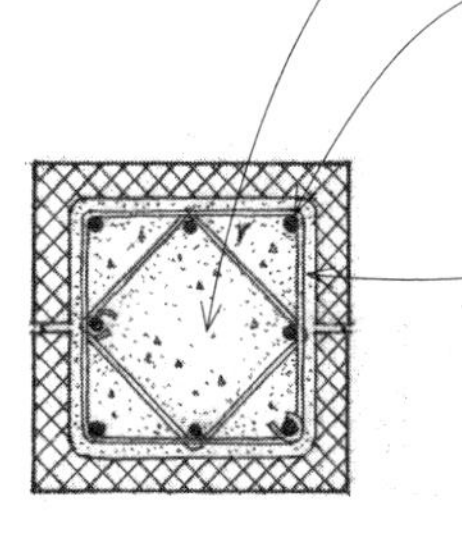

- Vertical core of portland cement grout
- Vertical reinforcement bars extend down and are tied to dowels embedded in column footing.
- Lateral ties
- Embed extra ties or a portion of required lateral reinforcement in mortar joints.

Masonry Columns

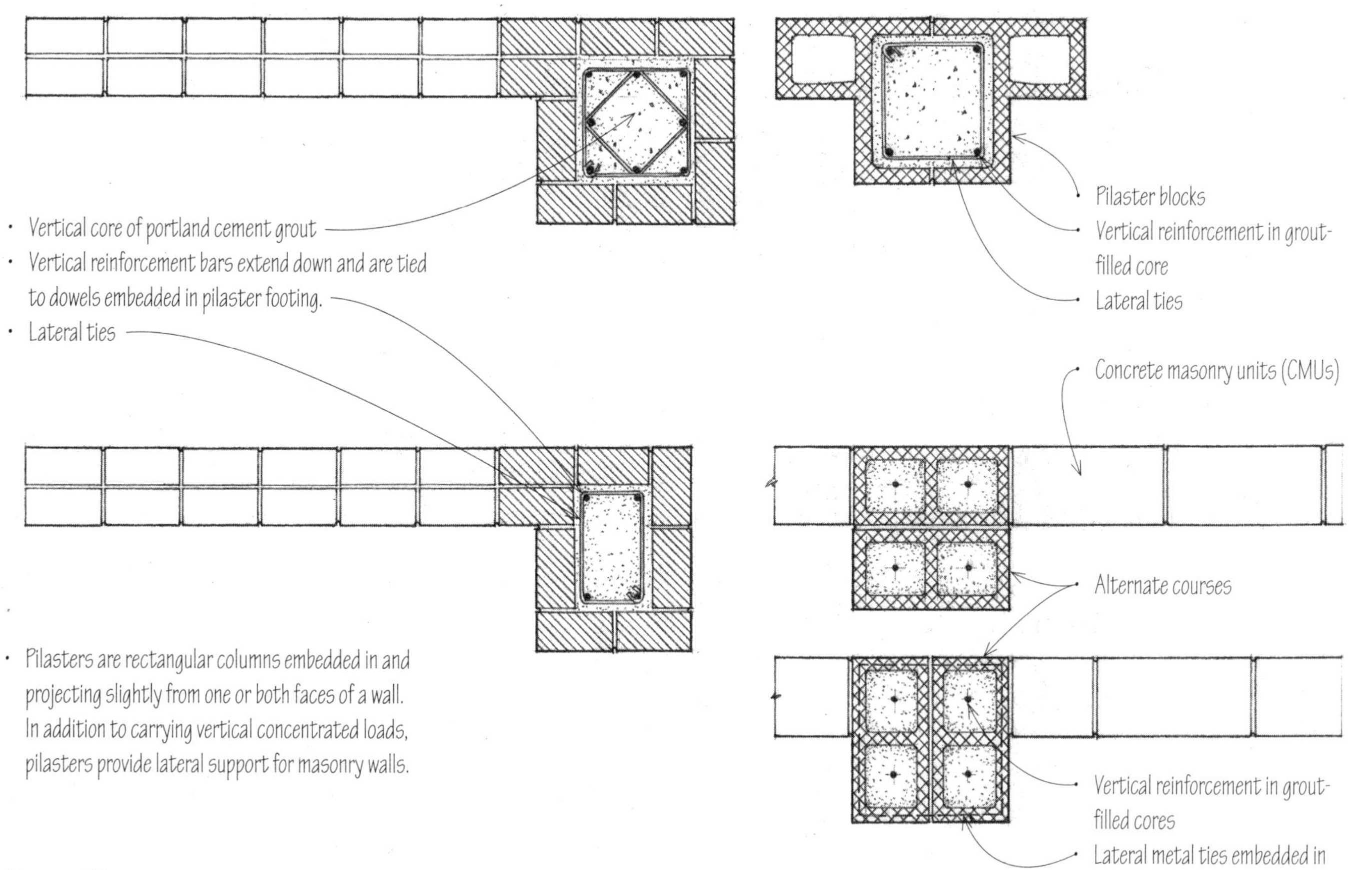

- Vertical core of portland cement grout
- Vertical reinforcement bars extend down and are tied to dowels embedded in pilaster footing.
- Lateral ties

- Pilasters are rectangular columns embedded in and projecting slightly from one or both faces of a wall. In addition to carrying vertical concentrated loads, pilasters provide lateral support for masonry walls.

Masonry Pilasters

- A segmental arch is struck from a center below the springing line.

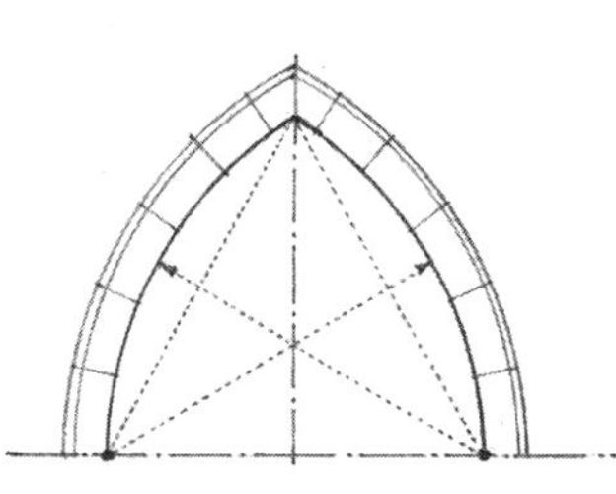

- A Gothic arch is a pointed arch having two centers and usually equal radii.
- A lancet arch is a pointed arch having two centers and radii greater than the span.
- A drop arch is a pointed arch having two centers and radii less than the span.

- A Roman arch has a semicircular intrados.
- Spandrel refers to the triangular-shaped area between the extrados of two adjoining arches, or between the left or right extrados of an arch and the rectangular framework surrounding it.

- A basket-handle arch is a three-centered arch having a crown with a radius much greater than that of the outer pair of curves.
- A Tudor arch is a four-centered arch having an inner pair of curves with radii much greater than that of the outer pair.

- A jack arch has a horizontal soffit with voussoirs radiating from a center below, often built with a slight camber to allow for settling.

Masonry arches utilize the compressive strength of brick and stone to span openings by transforming the vertical forces of a supported load into inclined components. These outward thrusts of the arching action, which are proportional to the total load and span, and inversely proportional to the rise, must be resisted by abutments adjacent to the opening or by equal but opposite thrusts from adjoining arches. For bending to be eliminated throughout an arch, the lines of thrust must coincide with the arch axis.

- A masonry arch may consist of brick coursework or individual stone voussoirs.
- Alternating soldier and rowlock courses
- Two or three rowlock courses
- Skewback is a stone or course of masonry having a sloping face against which the end of a segmental arch rests.

- Keystone is the wedge-shaped, often embellished voussoir at the crown of an arch, serving to lock the other voussoirs in place.
- Voussoirs are any of the wedge-shaped units in a masonry arch, having side cuts converging at one of the arch centers.

- Crown
- Extrados is the exterior curve or boundary of the visible face of an arch.
- Arch axis
- Intrados is the inner curve of the visible face of an arch; soffit refers to the inner surface of an arch forming the concave underside.
- Spring is the point at which an arch, vault, or dome rises from its support.

Span

Rise is the height of an arch from the springing line to the highest point of the soffit; minimum of 1" per foot of span (1:12).

- Skewback 1/2" per foot of span (1:24) for each 4" (100) of arch depth
- Camber = 1/8" per foot of span (1:100)

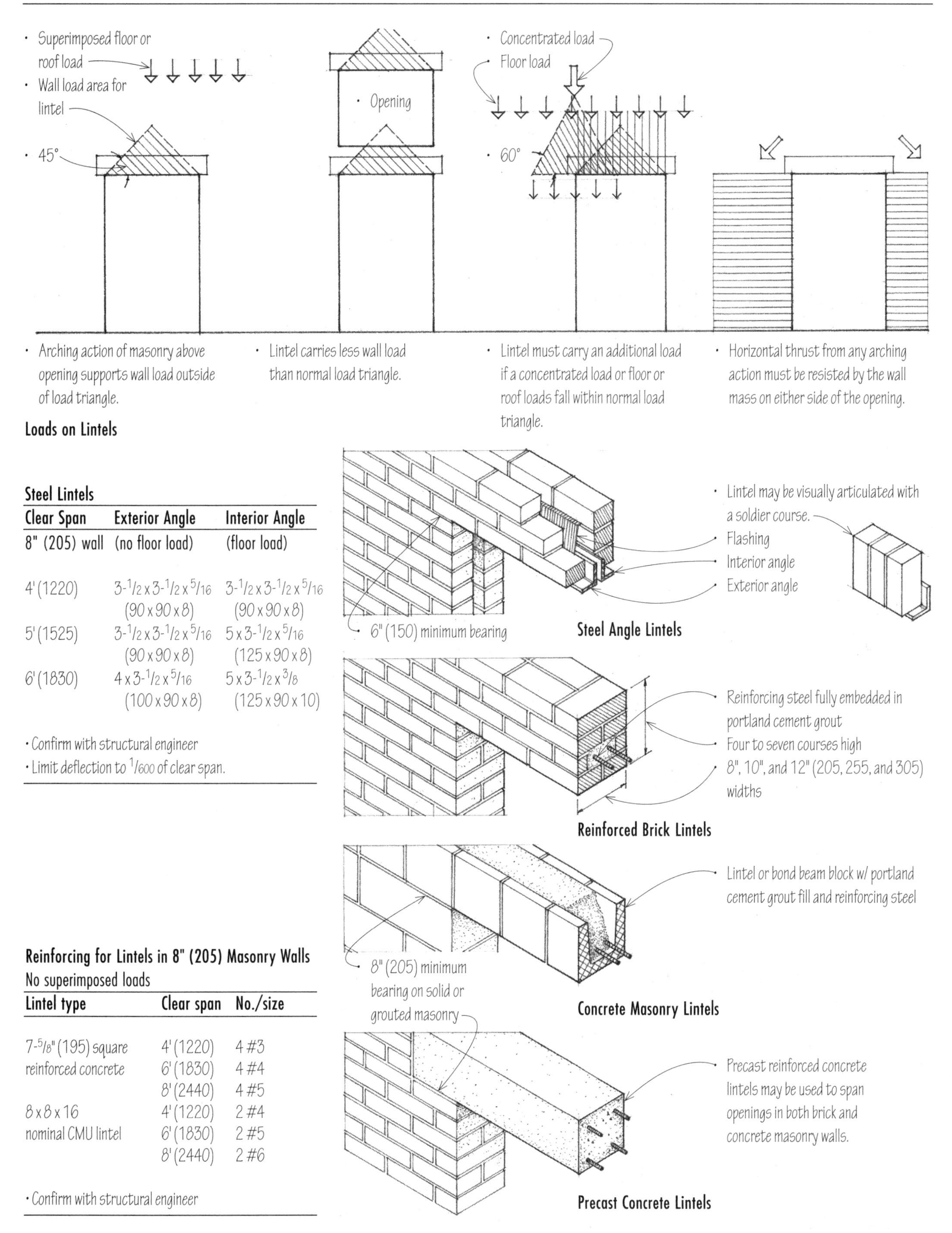

Loads on Lintels

Steel Angle Lintels

Reinforced Brick Lintels

Concrete Masonry Lintels

Precast Concrete Lintels

Steel Lintels

Clear Span 8" (205) wall	Exterior Angle (no floor load)	Interior Angle (floor load)
4' (1220)	3-1/2 x 3-1/2 x 5/16 (90 x 90 x 8)	3-1/2 x 3-1/2 x 5/16 (90 x 90 x 8)
5' (1525)	3-1/2 x 3-1/2 x 5/16 (90 x 90 x 8)	5 x 3-1/2 x 5/16 (125 x 90 x 8)
6' (1830)	4 x 3-1/2 x 5/16 (100 x 90 x 8)	5 x 3-1/2 x 3/8 (125 x 90 x 10)

- Confirm with structural engineer
- Limit deflection to 1/600 of clear span.

Reinforcing for Lintels in 8" (205) Masonry Walls
No superimposed loads

Lintel type	Clear span	No./size
7-5/8" (195) square reinforced concrete	4' (1220)	4 #3
	6' (1830)	4 #4
	8' (2440)	4 #5
8 x 8 x 16 nominal CMU lintel	4' (1220)	2 #4
	6' (1830)	2 #5
	8' (2440)	2 #6

- Confirm with structural engineer

Movement joints should be spaced every 100' to 125' (30 to 38 m) along unbroken wall lengths, and:

(1) At changes in wall height or thickness
(2) At columns, pilasters, and wall intersections
(3) Near corners
(4) On both sides of openings >6' (>1830)
(5) On one side of openings <6' (<1830)

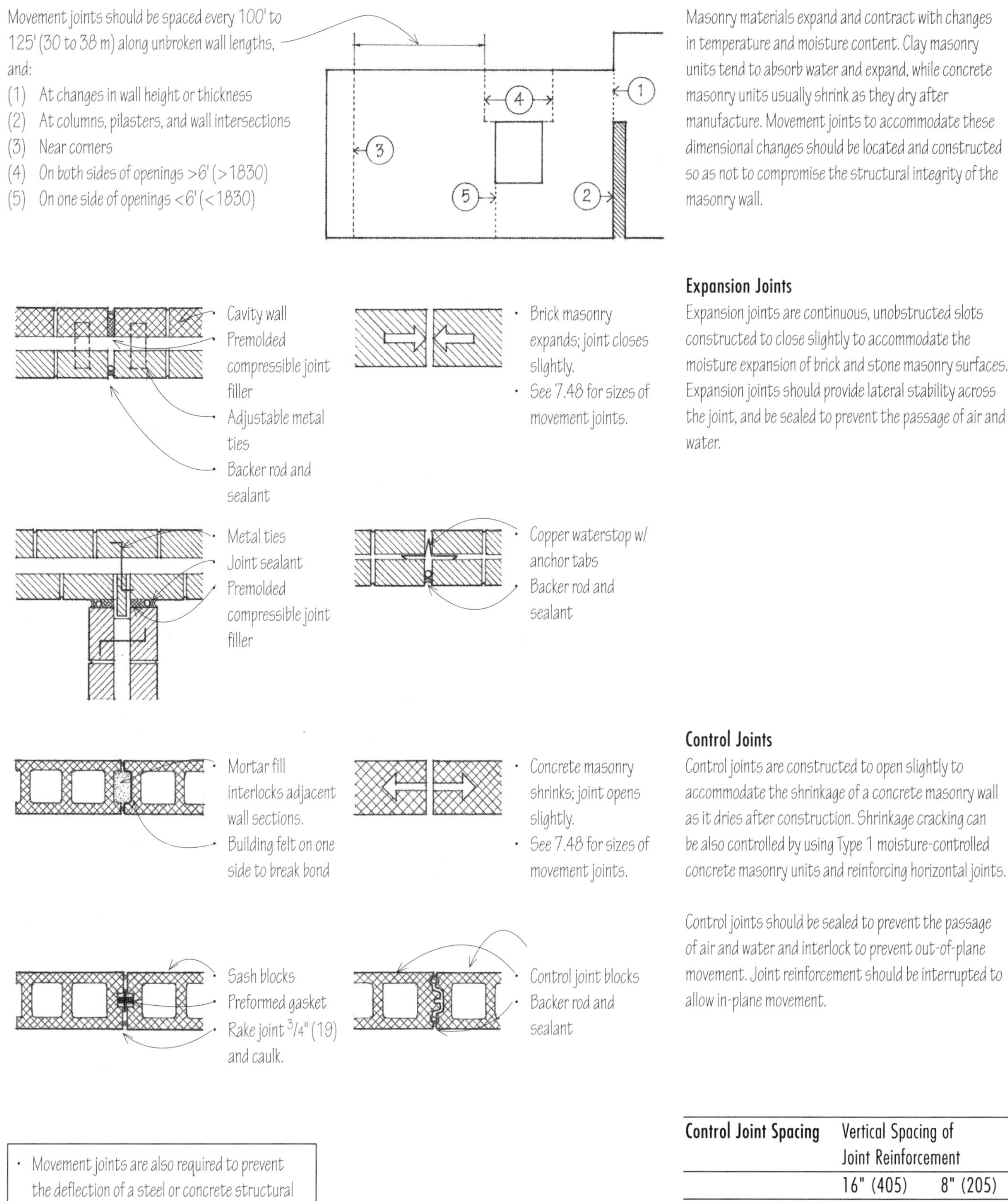

Masonry materials expand and contract with changes in temperature and moisture content. Clay masonry units tend to absorb water and expand, while concrete masonry units usually shrink as they dry after manufacture. Movement joints to accommodate these dimensional changes should be located and constructed so as not to compromise the structural integrity of the masonry wall.

Expansion Joints

Expansion joints are continuous, unobstructed slots constructed to close slightly to accommodate the moisture expansion of brick and stone masonry surfaces. Expansion joints should provide lateral stability across the joint, and be sealed to prevent the passage of air and water.

Control Joints

Control joints are constructed to open slightly to accommodate the shrinkage of a concrete masonry wall as it dries after construction. Shrinkage cracking can be also controlled by using Type 1 moisture-controlled concrete masonry units and reinforcing horizontal joints.

Control joints should be sealed to prevent the passage of air and water and interlock to prevent out-of-plane movement. Joint reinforcement should be interrupted to allow in-plane movement.

Control Joint Spacing	Vertical Spacing of Joint Reinforcement	
	16" (405)	8" (205)
Wall length (L)	50' (15 m)	60' (18 m)
L/H ratio	3	4

- Movement joints are also required to prevent the deflection of a steel or concrete structural frame from placing stress on a supported masonry wall or panel. See 7.28.

The wall sections on this and the following two pages illustrate how concrete, steel, and wood floor and roof systems are supported by and tied to various types of masonry bearing walls. The bearing area of masonry should be proportioned so as not to exceed the allowable compressive stress of the masonry material. All floors and roofs that provide lateral support for a masonry wall must be secured at least 6' (1830) o.c. with anchors embedded in a reinforced, grouted structural element of the wall.

- Wood rafters or joists
- Pressure-treated top plates anchored w/ $^{1}/_{2}$" (13) ø anchor bolts @ no more than 6' (1830) o.c.
- Embed anchor bolts at least 15" (380) into grout-filled cells or weld to bond-beam reinforcement.
- Metal lath to support grout in CMU cells

- Concrete masonry units (CMU): CSI MasterFormat 04 22 00: Concrete Unit Masonry
- Horizontal joint reinforcement vertically spaced @ 16" (405) o.c. typical
- Vertical reinforcing bars in grout-filled cores
- See 7.44 for thermal insulation options.
- See 5.21 for lintel options.

- Wood floor joists
- Metal joist anchors
- Minimum 3 x 8 ledger secured w/ bolts as required
- Fill CMU cells w/ grout.

- Surface bonding consists of laying concrete masonry units dry, without mortar, and plastering each side with a surface bonding compound, a cementitious compound containing short glass fibers.
- CSI MasterFormat 04 22 00.16: Surface-Bonded Concrete Unit Masonry

- Flashing and weep holes @ 32" (815) o.c.
- 2x pressure-treated sill anchor-bolted to grout-filled cells
- Wood floor joists
- Widened CMU foundation wall

Concrete Masonry Bearing Wall

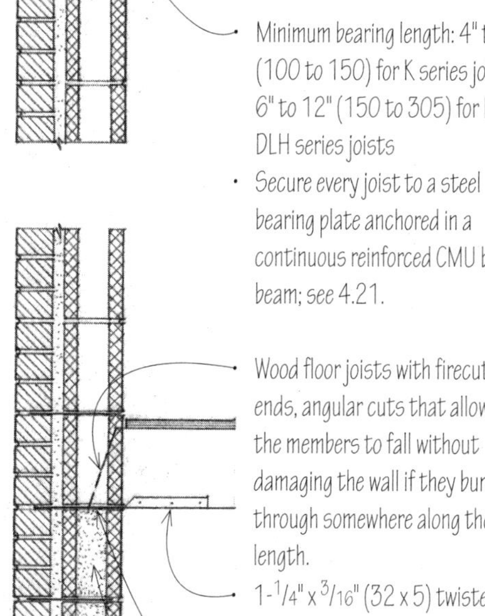

- Flat roof assembly; see 7.12.
- Open-web steel joists
- Minimum bearing length: 4" to 6" (100 to 150) for K series joists; 6" to 12" (150 to 305) for LH/DLH series joists
- Secure every joist to a steel bearing plate anchored in a continuous reinforced CMU bond beam; see 4.21.

- Wood floor joists with firecut ends, angular cuts that allow the members to fall without damaging the wall if they burn through somewhere along their length.
- 1-$^{1}/_{4}$" x $^{3}/_{16}$" (32 x 5) twisted steel strap ties @ 6' (1830) o.c. maximum
- 3" (75) minimum bearing
- Grouted CMU or reinforced CMU bond beam
- Secure joists parallel to wall with steel strap ties @ 6' (1830) o.c. maximum; extend to at least three joists and provide blocking between joists at each strap anchor.

- Rowlock sill
- Continuous flashing and weep holes

- Composite masonry
- Facing wythe; CSI 04 21 00: Clay Unit Masonry
- Backup wythe; CSI 04 22 00: Concrete Unit Masonry
- $^{3}/_{4}$" (19) collar joints typical; fill with mortar or grout.
- Metal wall ties @ 16" (405) o.c. typical
- See 7.44 for thermal insulation options.

Composite Masonry Bearing Wall

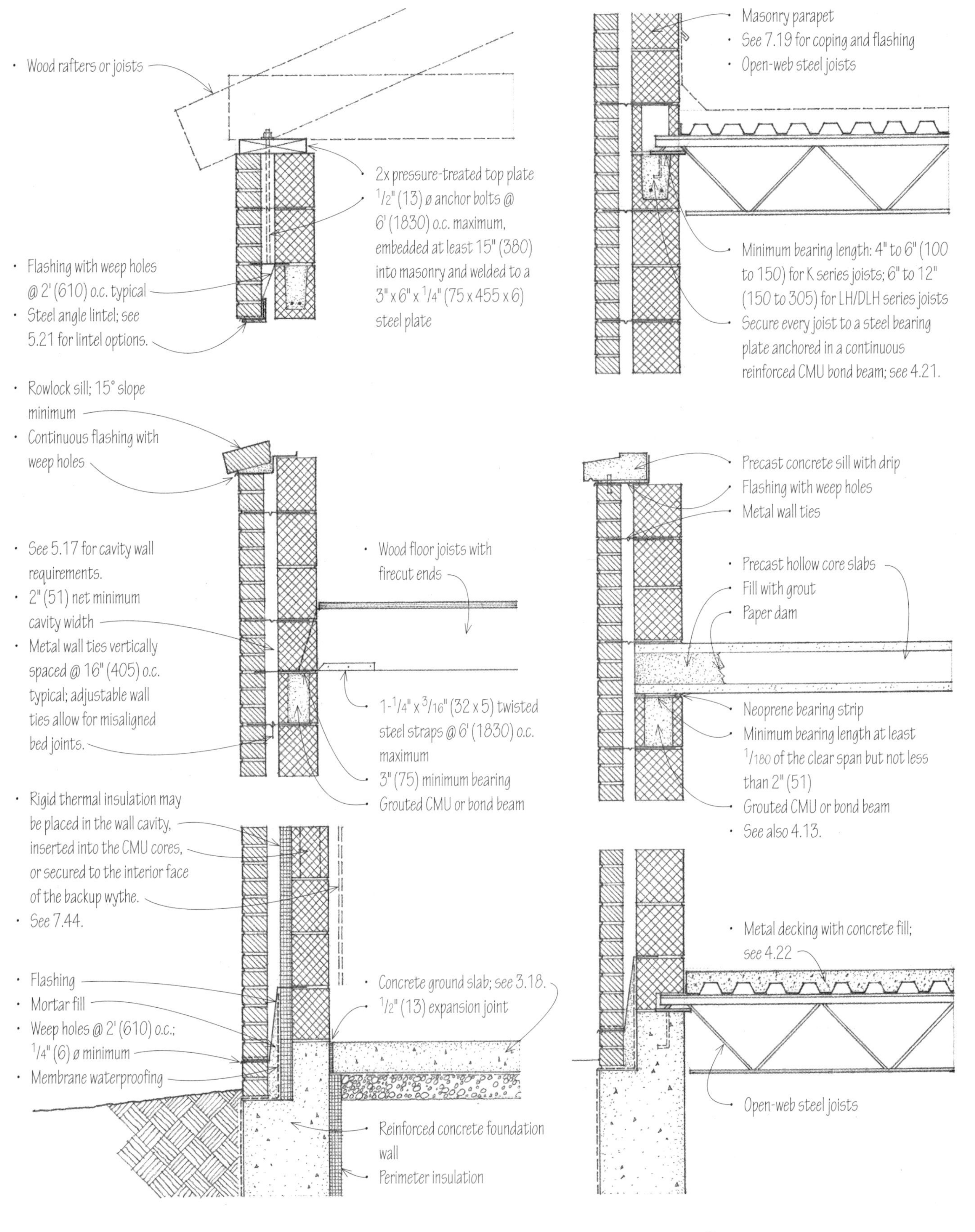

Cavity Bearing Wall

Cavity Bearing Wall

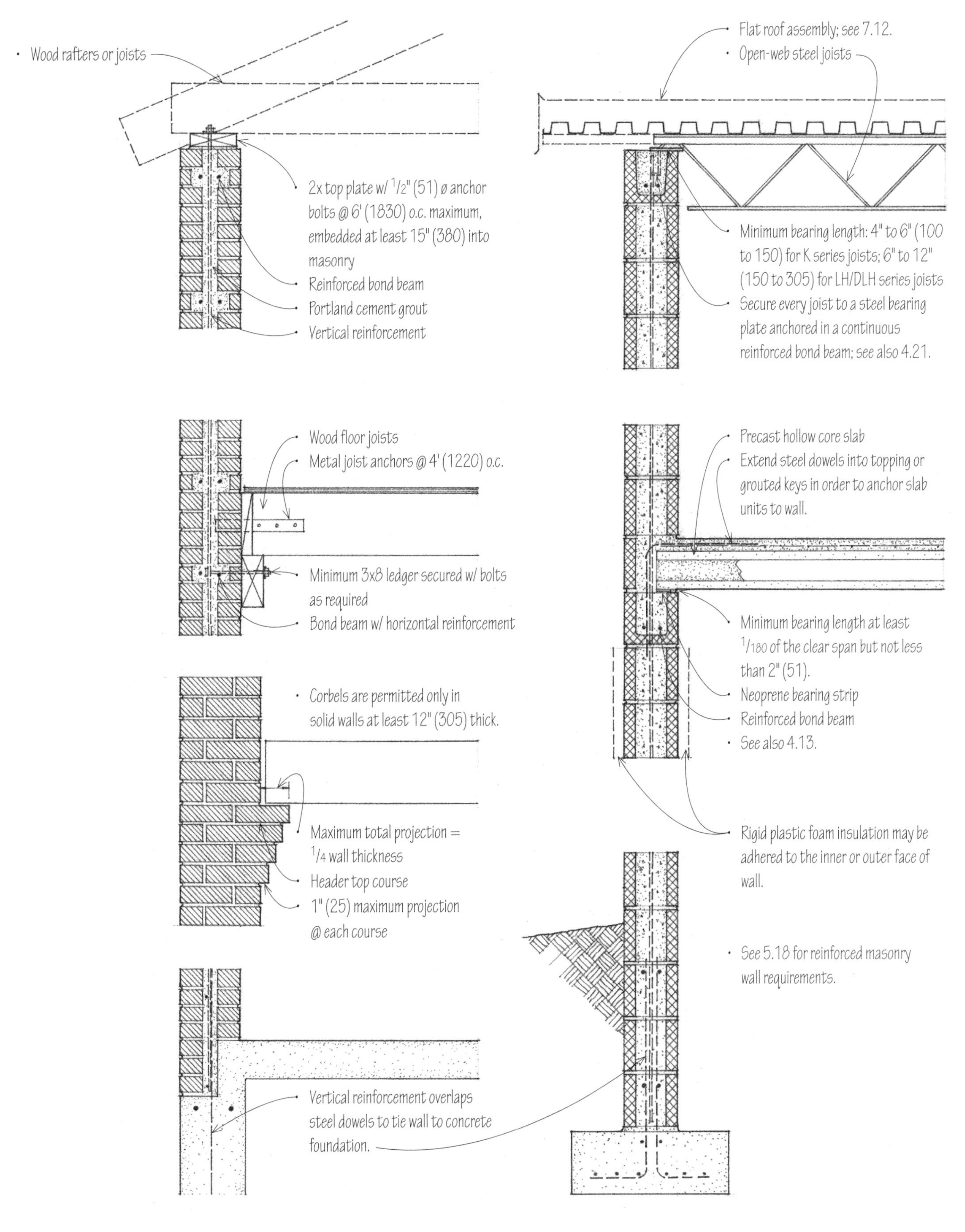

Reinforced Brick Masonry Wall

Reinforced Concrete Masonry Wall

- Wythe is a continuous vertical section of a masonry wall one unit in thickness.
- Course is a continuous horizontal range of masonry units.
- Collar joint is the vertical joint between two wythes of masonry.
- Bed joint is the horizontal joint between two masonry courses. The term bed may refer to the underside of a masonry unit, or to the layer of mortar in which a masonry unit is laid.
- Head joint is the vertical joint between two masonry units, perpendicular to the face of a wall.

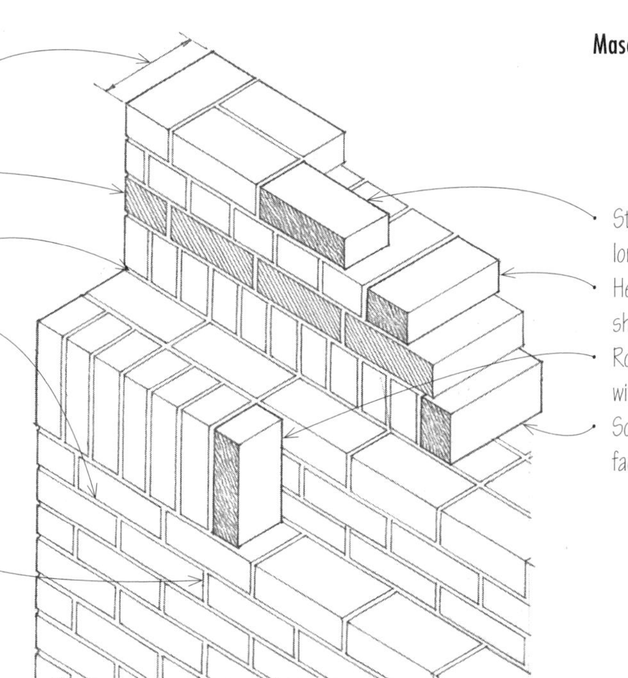

Masonry Terminology

- Stretcher is a masonry unit laid horizontally with the longer edge exposed or parallel to the surface.
- Header is a masonry unit laid horizontally with the shorter end exposed or parallel to the surface.
- Rowlock is a brick laid horizontally on the longer edge with the shorter end exposed.
- Soldier is a brick laid vertically with the longer edge face exposed.

- Concave Joint
- V-joint

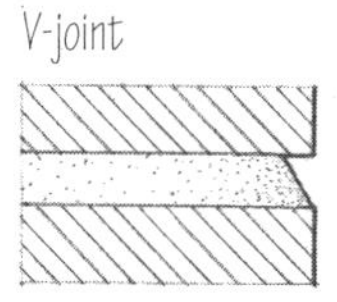

- Weathered Joint
- Struck Joint

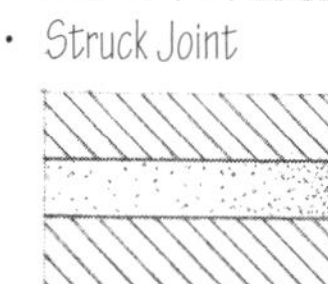

- Flush joint

- Mortar joints vary in thickness from 1/4" to 1/2" (6 to 13) but are typically 3/8" (10) thick.
- Tooled joints are mortar joints compressed and shaped with any tool other than a trowel. Tooling compresses the mortar and forces it tightly against the brick surfaces, providing maximum protection against water penetration in areas subject to high winds or heavy rains.
- Troweled joints are finished by striking off excess mortar with a trowel. In troweled joints, the mortar is cut or struck off with a trowel. The most effective of these is the weathered joint because it sheds water.
- Raked joint is made by removing mortar to a given depth with a square-edged tool before hardening. Raked joints are for interior use only.
- For mortar types, see 5.15.

Mortar Joints

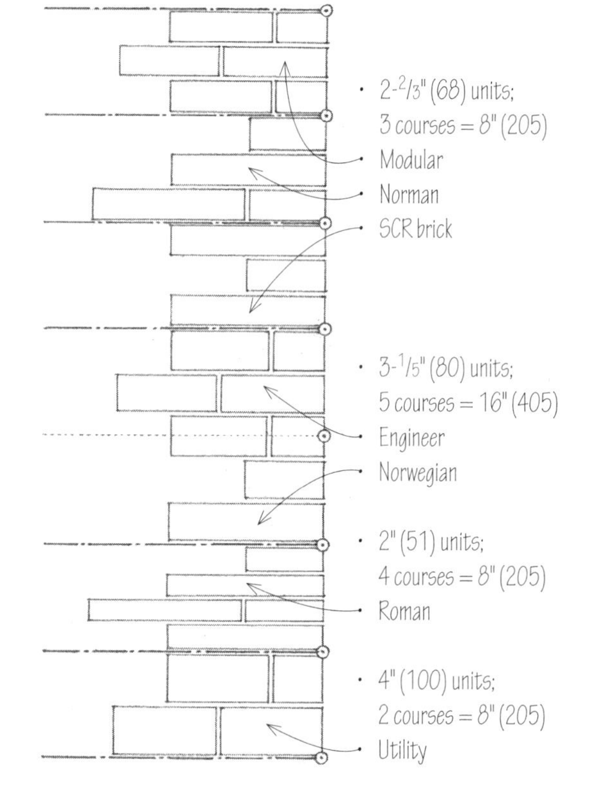

Course Heights

- Relative course heights are nominal dimensions that include the thickness of the mortar joints.
- For lengths, use multiples of 4", 8", or 12" (100, 205, or 305)
- For brick types and sizes, see 12.06.
- Wall thicknesses vary with the type of masonry wall; see 5.14–5.15.

- Running bond, commonly used for cavity and veneer walls, is composed of overlapping stretchers.

- Common bond has a course of headers between every five or six courses of stretchers; also known as American bond.

- Stack bond has successive courses of stretchers with all head joints aligned vertically. Because units do not overlap, horizontal joint reinforcement is required @ 16" (405) o.c. in unreinforced walls.

- Flemish bond has alternating headers and stretchers in each course, each header being centered above and below a stretcher. Flare headers with darker ends are often exposed in patterned brickwork.

- Flemish cross bond is a modified Flemish bond in which courses of alternate headers and stretchers alternate with stretching courses.

- Flemish diagonal bond is a form of Flemish cross bond in which the courses are offset to form a diamond pattern.

- Garden-wall bond, used for lightly loaded boundary walls, has a sequence of a header and three stretchers in each course, with each header being centered over a header in alternate courses.

- English bond has alternate courses of headers and stretchers in which the headers are centered on stretchers and the joints between stretchers line up vertically in all courses.

- To minimize the cutting of brick and enhancing the appearance of bonding patterns, the major dimensions of masonry walls should be based on the size of the modular units used.

Structural clay tile is hollow tile of fired clay having parallel cells or cores and used typically in constructing walls and partitions.

- LB Grade: loadbearing structural clay tile suitable for masonry walls not exposed to frost action, or in exposed masonry where protected by a facing of 3" (75 mm) or more of stone, brick, terra cotta, or other masonry.
- LBX Grade: loadbearing structural clay tile suitable for masonry walls exposed to weathering or frost action.

Structural facing tile is structural clay tile having a glazed surface and used for facing walls and partitions, especially in areas subject to heavy wear, moisture problems, and strict sanitation requirements.

- FTS Grade: structural facing tile suitable for exposed exterior and interior masonry walls and partitions where moderate absorption, slight variation in face dimensions, minor defects in surface finish, and medium color range are acceptable.
- FTX Grade: smooth structural facing tile suitable for exposed exterior and interior masonry walls and partitions where low absorption and stain resistance are required, and where a high degree of mechanical perfection, minimum variation in face dimensions, and narrow color range are desired.

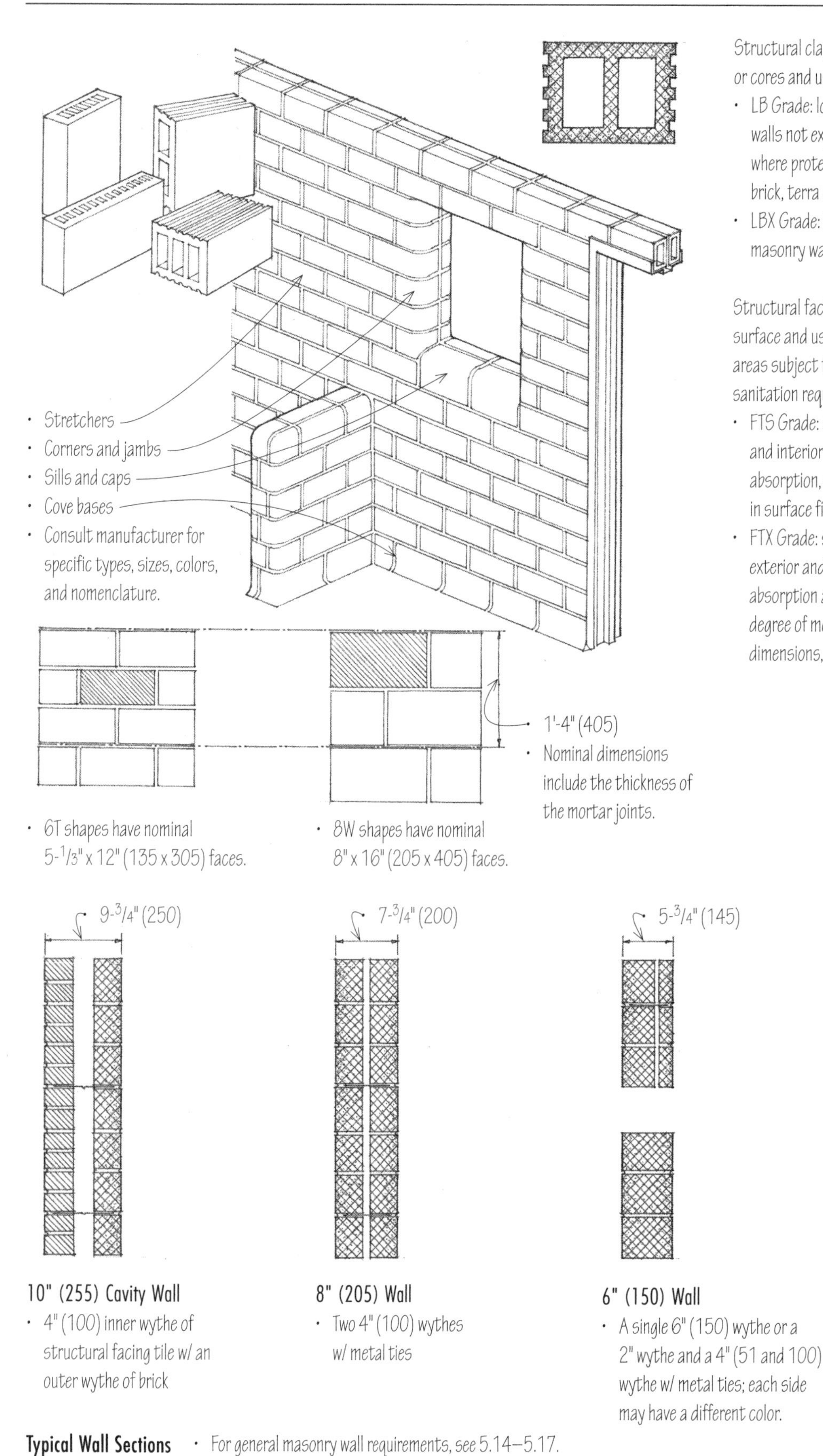

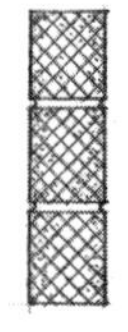

10" (255) Cavity Wall
- 4" (100) inner wythe of structural facing tile w/ an outer wythe of brick

8" (205) Wall
- Two 4" (100) wythes w/ metal ties

6" (150) Wall
- A single 6" (150) wythe or a 2" wythe and a 4" (51 and 100) wythe w/ metal ties; each side may have a different color.

4" (100) Wall
- A single 4" (100) wythe or two 2" (two 51) wythes w/ metal ties

Typical Wall Sections · For general masonry wall requirements, see 5.14–5.17.

CSI MasterFormat 04 21 23: Structural Clay Tile Masonry

Glass block is a translucent, hollow block of glass with clear, textured, or patterned faces, made by fusing two halves together with a partial vacuum inside. Glass block may be used in nonloadbearing exterior and interior walls, and in conventionally framed window openings. The glass block units are laid in Type S or Type N mortar with joints at least 1/4" (6) but not more than 3/8" (10) thick. Typically, a wall panel is mortared at the sill support and provided with expansion joints along the top and sides to allow for movement and settling.

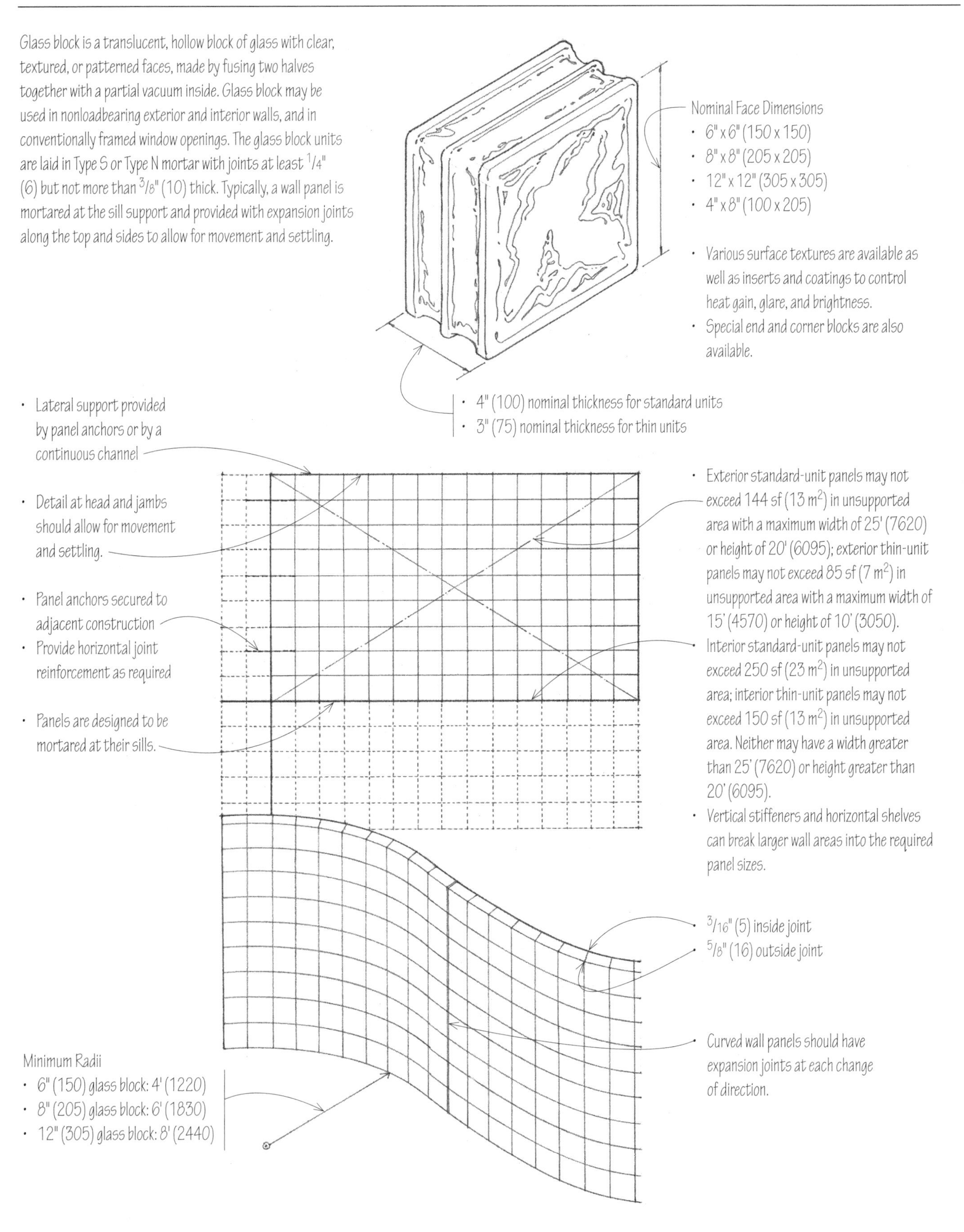

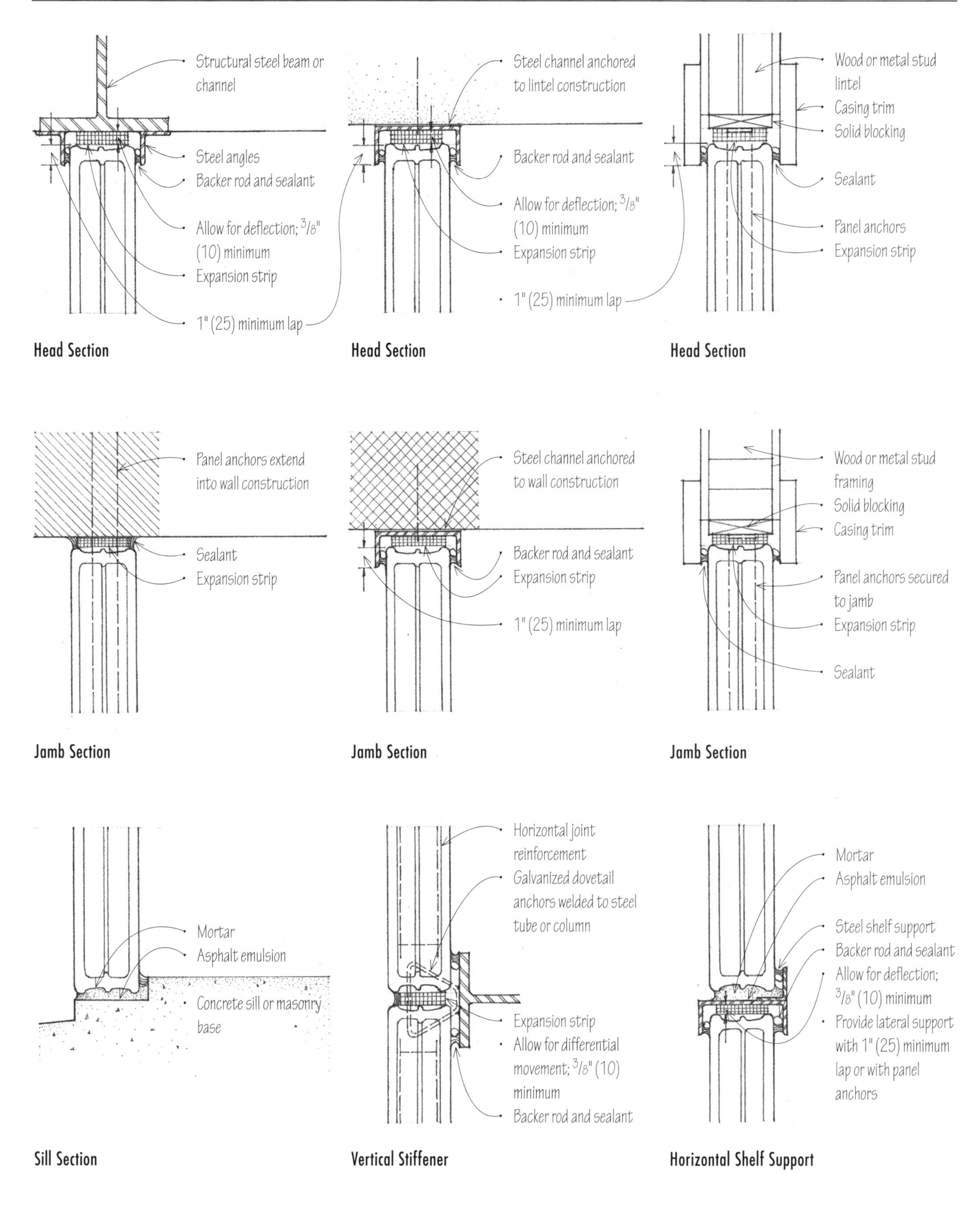

Typical Glass Block Details

Adobe and rammed-earth construction both use unfired, stabilized earth as the primary building material. Current building codes vary in their acceptance of and requirements for adobe and rammed-earth construction. However, the use of earth as a building material is an economic necessity in many areas of the world, and both adobe and rammed-earth remain low-cost alternative building systems.

Adobe is sun-dried clay masonry, traditionally used in countries with little rainfall. Almost any soil having a 15% to 25% clay content may be used for the mud mixture; soils with a higher clay content may require tempering with sand or straw to make satisfactory bricks. Gravel or other coarse aggregate may make up 50% of the volume of the mix. The mixing water should not contain dissolved salts, which can recrystallize and damage the brick upon drying.

Adobe brick is typically made near the point of use with soil obtained from the excavation of basements or from surplus soil from site grading. The mud is mixed by hand or by mechanical means and cast in wood or metal forms, which are set on level ground and wetted with water to aid separation of the units. After initial drying, the units are stacked on edge until fully cured. The brick units are extremely fragile until completely dry.

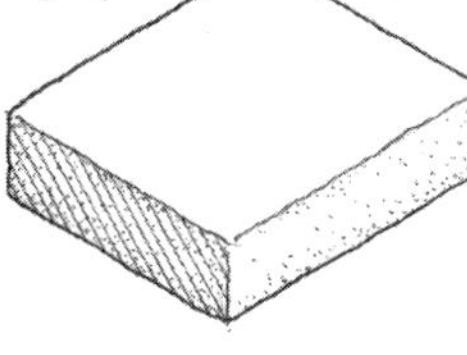

- The dimensions of adobe brick vary according to locale, but a common size is 10" (255) x 14" (355) x 2" to 4" (51 to 100) thick. Thinner bricks dry and cure faster than thicker bricks. Each brick can weigh 25 to 30 pounds (11 to 14 kg).
- Stabilized or treated adobe contains an admixture of portland cement, asphalt emulsion, and other chemical compounds to limit the water absorption of the bricks.
- Allowable loads for top-supported adobe columns 10' (3050) high:
  - 10" x 28" (255 x 710) 12,000 lb. (5400 kg)
  - 14" x 20" (355 x 510) 13,000 lb. (5900 kg)
  - 24" x 24" (610 x 610) 28,000 lb. (12,700 kg)

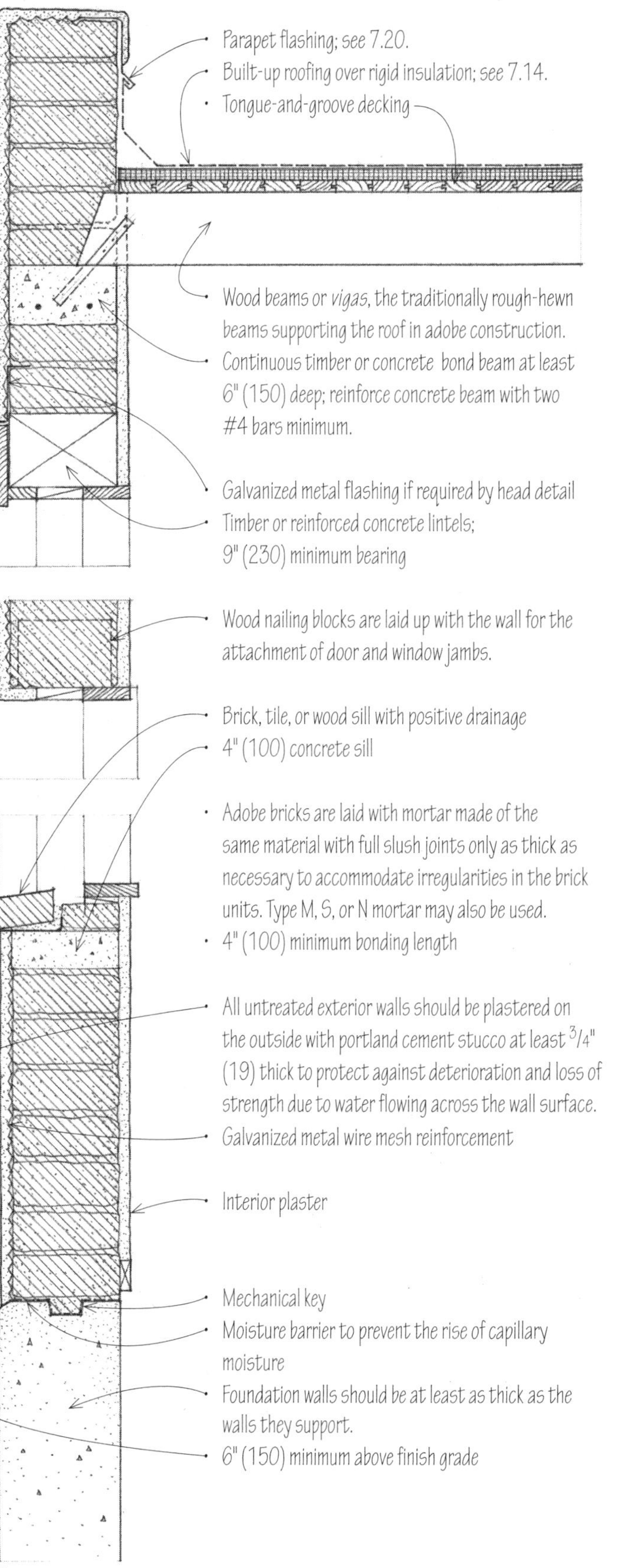

- See 5.32 for general requirements governing both adobe and rammed-earth construction.
- LEED® MR Credit 5: Regional Materials

CSI MasterFormat 04 24 00: Adobe Unit Masonry

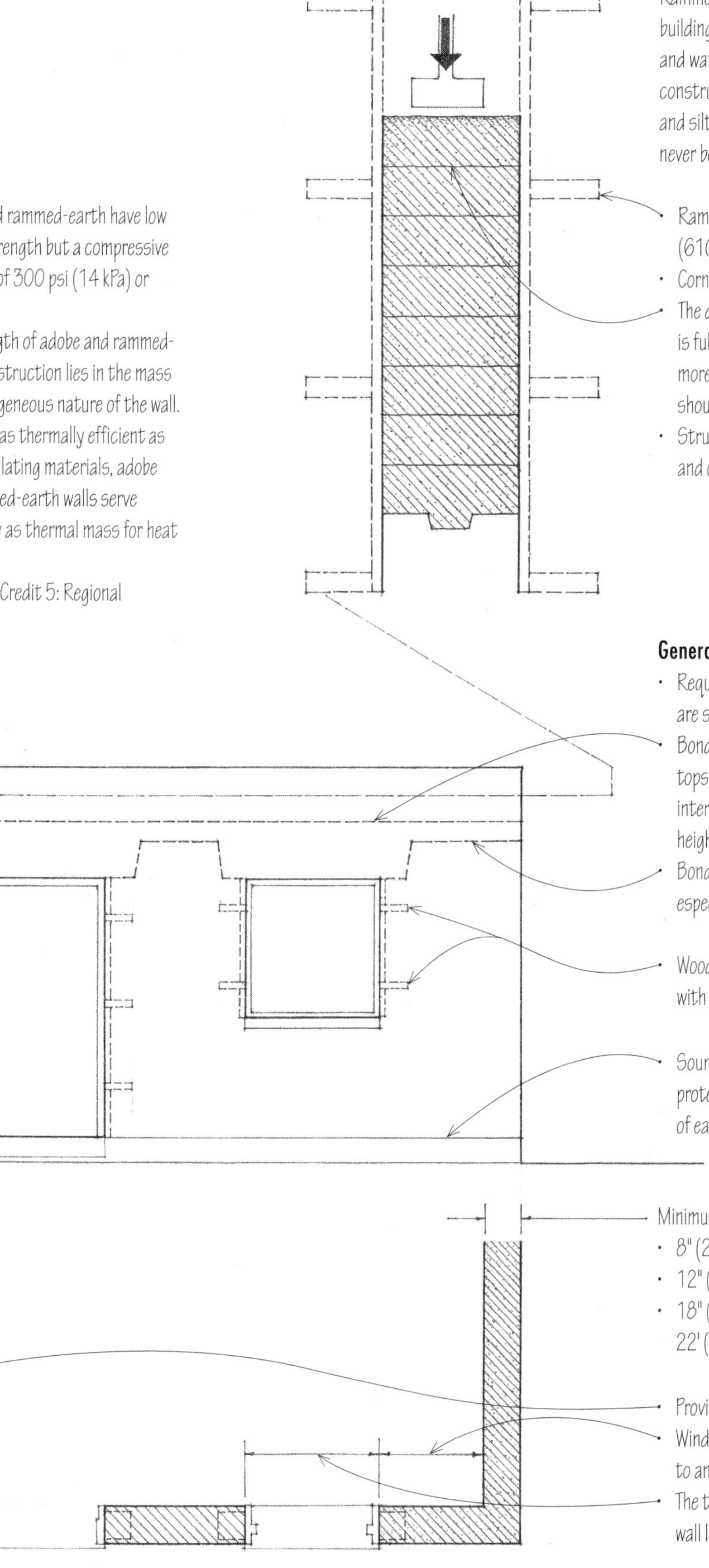

Rammed earth, also called *pisé de terre*, is another traditional building material. It is essentially a stiff mixture of clay, silt, sand, and water that is compressed and dried within forms as a wall construction. The soil mixture should contain less than 50% clay and silt, and a maximum aggregate size of 1/4" (6). Saltwater should never be used in the mix.

- Adobe and rammed-earth have low tensile strength but a compressive strength of 300 psi (14 kPa) or more.
- The strength of adobe and rammed-earth construction lies in the mass and homogeneous nature of the wall.
- While not as thermally efficient as other insulating materials, adobe and rammed-earth walls serve effectively as thermal mass for heat storage.
- LEED MR Credit 5: Regional Materials

- Rammed-earth walls are constructed with slip forms 24" to 36" (610 to 915) high and 10' to 12' (3050 to 3660) long.
- Corners are formed first with special forms.
- The damp soil mixture (approximately 10% moisture content) is fully compacted by hand or mechanically in lifts or layers not more than 6" (150) high before the next lift is placed. Each lift should bond securely with the preceding lift.
- Structural loads cannot be placed until the earth has fully dried and cured.

General Requirements

- Requirements for adobe and rammed-earth construction are similar.
- Bond beams are required to distribute roof loads and stabilize the tops of bearing walls, as well as at each floor level and at regular intervals to maintain the required thickness-to-unsupported wall height ratio.
- Bond beams should be reinforced against tension, especially at corners.
- Wood anchors for attaching door and window frames are cast with the earth wall.
- Sound foundations and ample roof overhangs that protect exterior walls from rain improve the durability of earthen structures.

Minimum Wall Thickness:
- 8" (205) for interior nonbearing walls
- 12" (305) for one-story bearing walls up to 12' (3660) high
- 18" (455) for the first story of two-story bearing walls up to 22' (6705) high and 12" (305) for the second

- Provide cross-wall supports @ 24' (7315) o.c. maximum.
- Window and door openings should be no closer than 2'-4" (710) to any corner.
- The total length of wall openings should be limited to 1/3 of the wall length.

Natural stone is a durable, weather-resistant construction material that may be laid in mortar much like clay and concrete masonry units to make both bearing and nonbearing walls. Some differences result, however, from the irregular shapes and sizes of rubble, the uneven coursing of ashlar masonry, and the varying physical properties of the different types of stone that may be used in the wall construction.

Natural stone may be bonded with mortar and laid up in the traditional manner as a double-faced loadbearing wall. More often, however, stone is used as a facing veneer tied to a concrete or masonry backup wall. To prevent discoloration of the stone, only nonstaining cement and noncorrosive ties, anchors, and flashing should be used. Copper, brass, and bronze may stain under certain conditions.

- See 7.30 for stone veneer walls.
- See 12.10 for a discussion of stone as a construction material.

- Random rubble is a masonry wall of broken stones having discontinuous but approximately level beds or courses. The mortar joints are usually held back of the stone faces to emphasize the natural stone shapes.

- Coursed rubble is a masonry wall of broken stones having approximately level bed joints and brought at intervals to continuous level courses.
- 1/2" to 1-1/2" (13 to 38) face joints

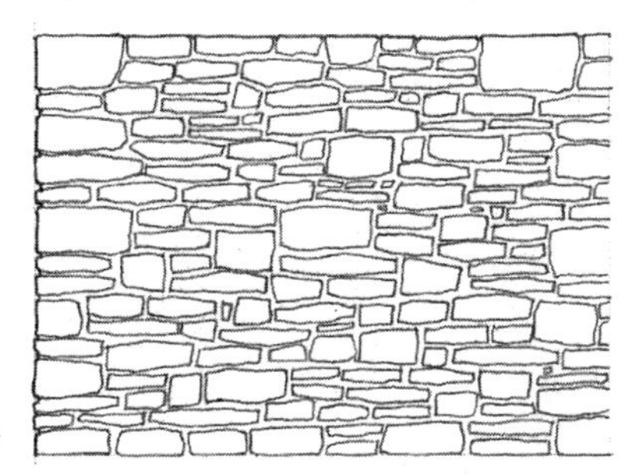

- Squared rubble is a masonry wall built of squared stones of varying sizes and coursed at every third or fourth stone.

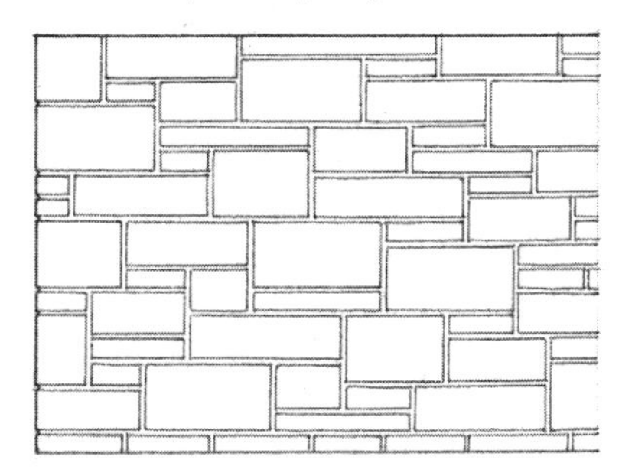

- Random ashlar is built with stones in discontinuous courses.

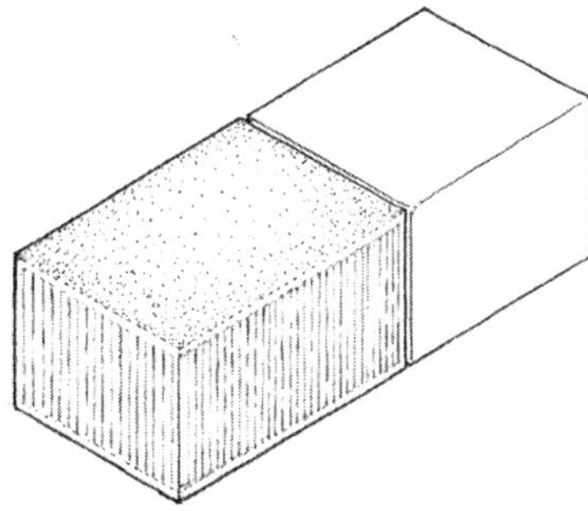

- Ashlar refers to squared building stone finely dressed on all faces adjacent to those of other stones so as to permit very thin mortar joints.

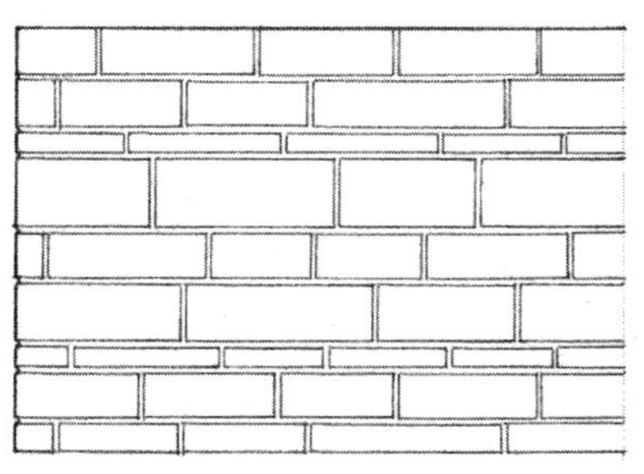

- Coursed ashlar is built of stones having the same height within each course, but with each course varying in height.

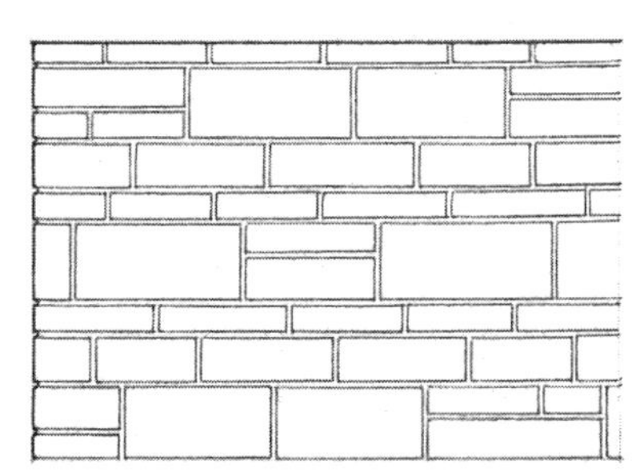

- Broken rangework is ashlar masonry laid in horizontal courses of varying heights, any one of which may be broken at intervals into two or more courses.

- 3/8" to 3/4" (10 to 19) face joints

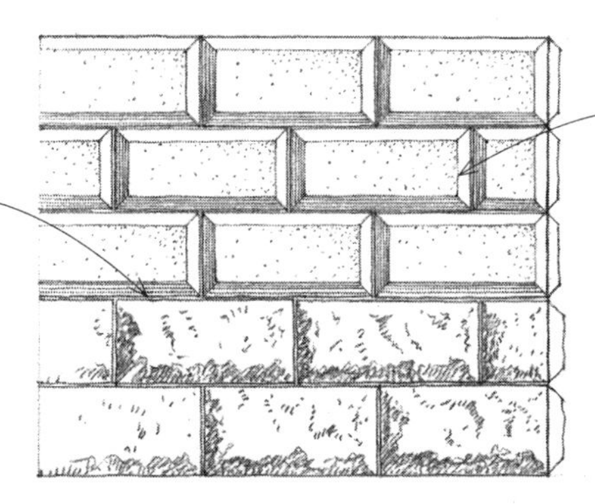

- Rustication is masonry having the visible faces of the dressed stones raised or otherwise contrasted with the horizontal and usually the vertical joints, which may be rabbeted, chamfered, or beveled.

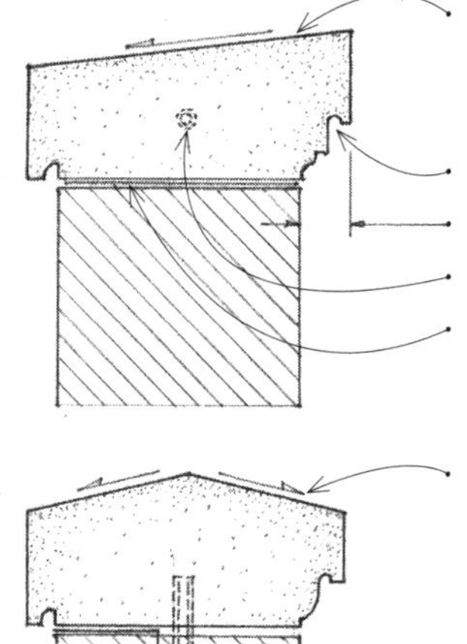

- Splayed copings slope only in one direction.
- Drip on each side
- 1-1/2" (38) minimum
- Center dowel
- Flashing

- Saddle copings slope to either side of a center ridge.
- Two vertical dowels per stone
- Step flashing

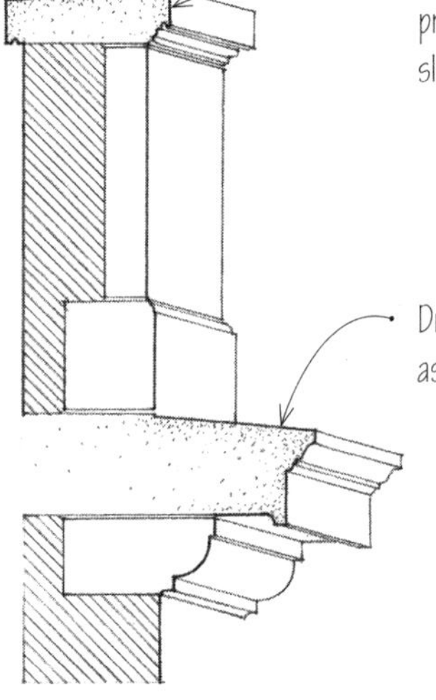

- Copestones form a coping, the finishing or protective course to an exterior wall, usually sloped or curved to shed water.
- Dripstone is a stone molding used as a drip, as on a cornice over a window or doorway.

- Stone facing; 4" (100) minimum
- CMU or reinforced concrete backup wythe
- Slush fill voids for solid wall or create a continuous air space for cavity wall
- Concrete may be waterproofed to prevent staining.
- Noncorrosive metal ties or anchors in dovetail slots
- See 7.30 for stone veneer walls.

Stone Masonry over a Backup Wall

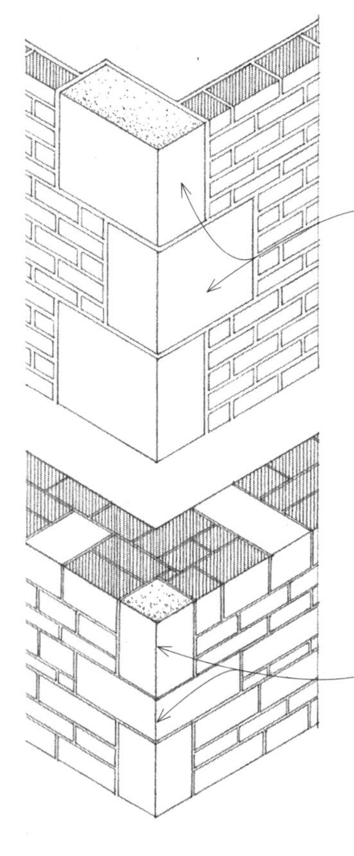

- Quoin refers to an exterior angle of a masonry wall, or one of the stones or bricks forming such an angle, usually differentiated from adjoining surfaces by material, texture, color, size, or projection.
- Long-and-short work is an arrangement of rectangular quoins or jambstones set alternately horizontally and vertically.

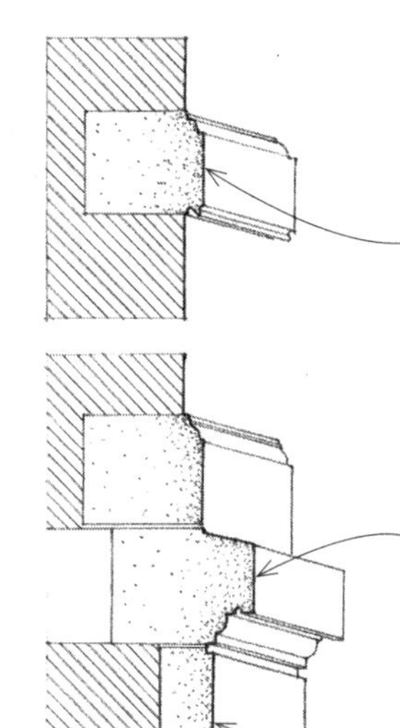

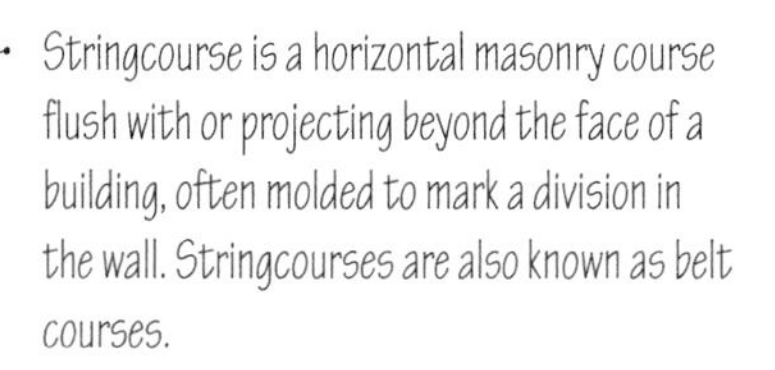

- Stringcourse is a horizontal masonry course flush with or projecting beyond the face of a building, often molded to mark a division in the wall. Stringcourses are also known as belt courses.
- Water table is a projecting stringcourse, molding, or ledge placed so as to divert rainwater from a building.
- Plinth is a continuous, usually projecting course of stones forming the base or foundation of a wall.

Conventional steel-framed structures are constructed of hot-rolled beams and columns, open-web joists, and metal decking. Since structural steel is difficult to work on site, it is normally cut, shaped, and drilled in a fabrication shop according to design specifications; this can result in relatively fast, precise construction.

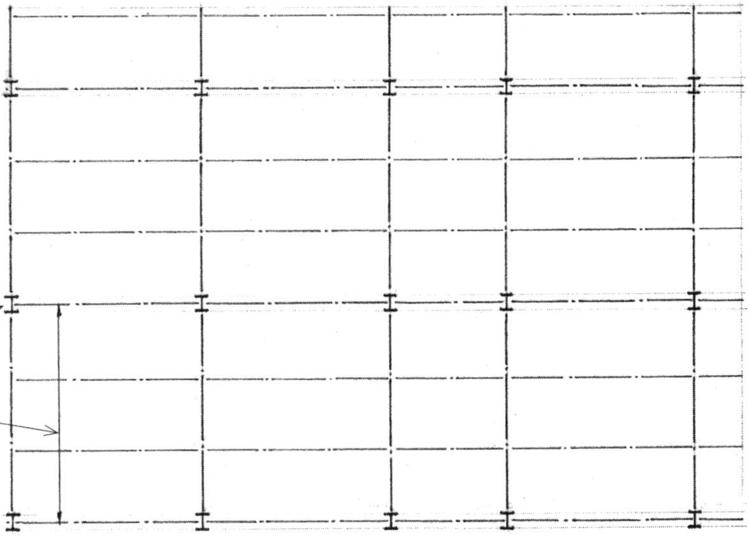

- Steel framing is most efficient when columns are laid out to support a regular grid of girders, beams, and joists.
- Column spacing = beam or girder spans
- Orient the webs of columns parallel to the short axis of the structural frame or the direction along which the structure is most susceptible to lateral forces.
- Orient the flanges on perimeter columns to the outside to facilitate attachment of curtain walls to the structural frame.

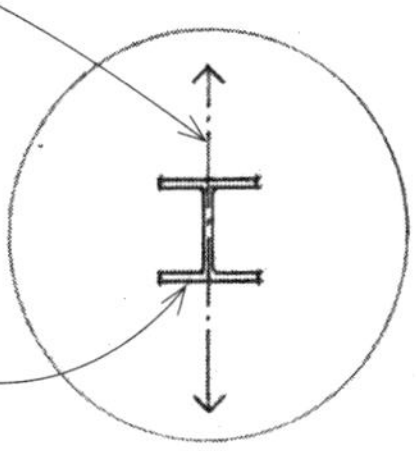

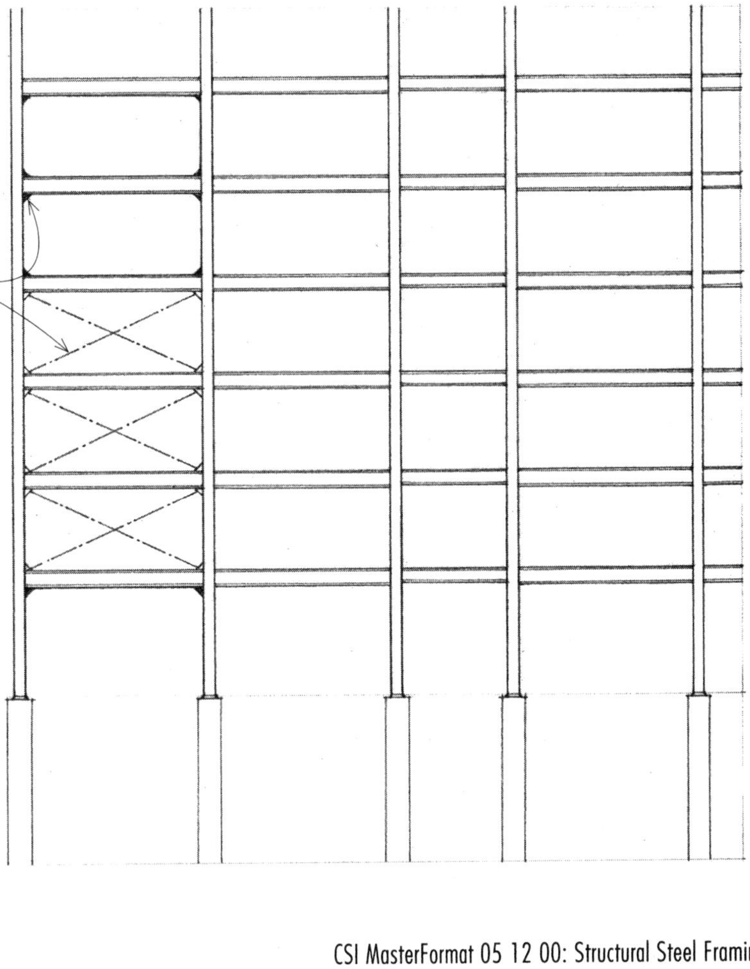

- Resistance to lateral wind and seismic forces requires the use of shear planes, diagonal bracing, or rigid framing with moment-resisting connections.

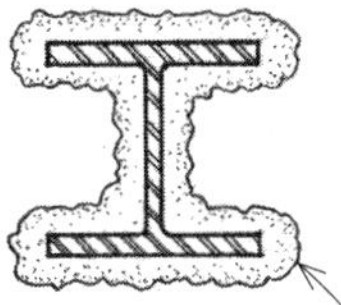

- Because steel can lose strength rapidly in a fire, fire-resistive assemblies or coatings are required; see A.12. In unprotected noncombustible construction, steel framing may be left exposed.
- See 4.14 for steel beams and floor framing systems.
- See 12.08 for a discussion of steel as a construction material.

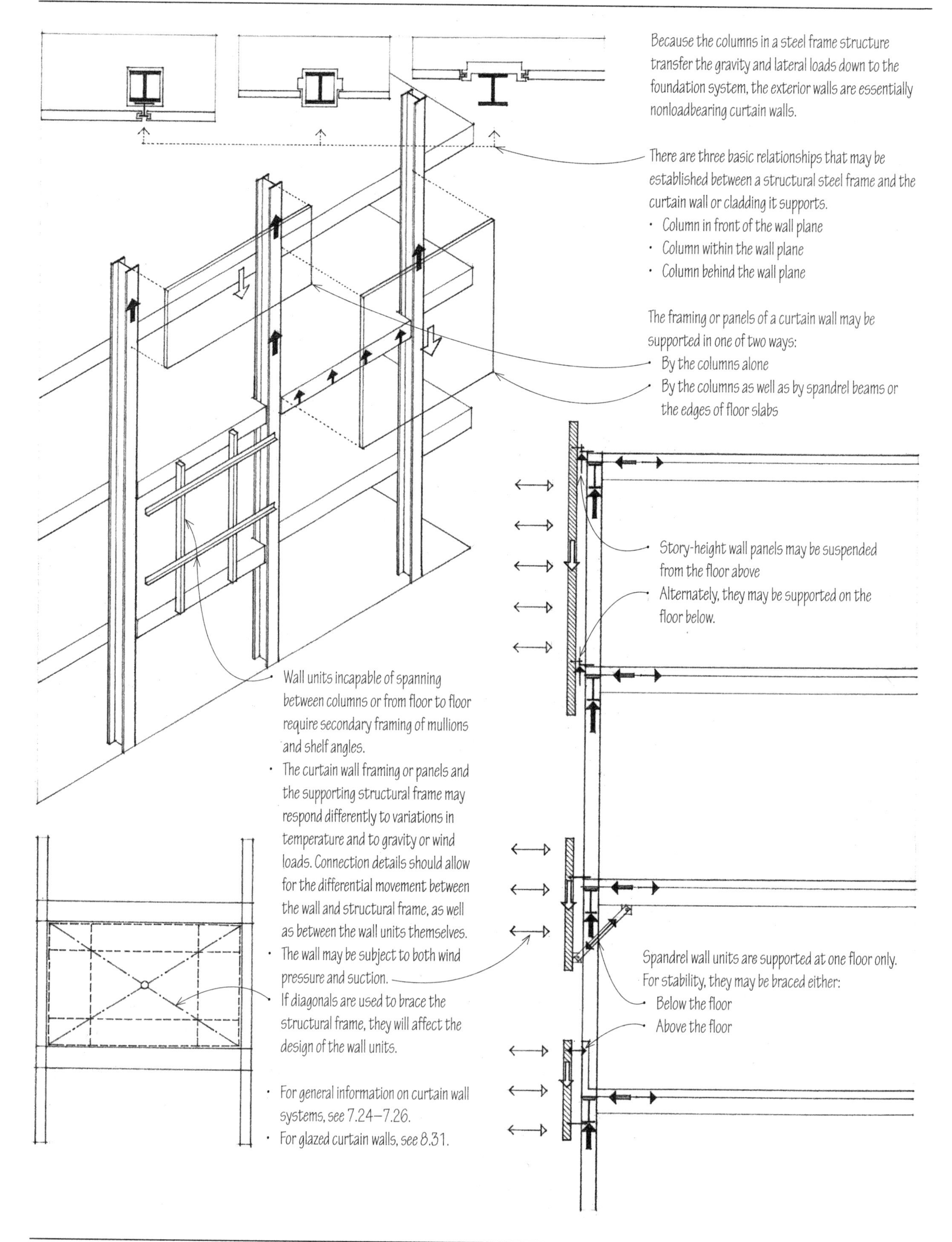

Because the columns in a steel frame structure transfer the gravity and lateral loads down to the foundation system, the exterior walls are essentially nonloadbearing curtain walls.
There are three basic relationships that may be established between a structural steel frame and the curtain wall or cladding it supports.
• Column in front of the wall plane
• Column within the wall plane
• Column behind the wall plane
The framing or panels of a curtain wall may be supported in one of two ways:
• By the columns alone
• By the columns as well as by spandrel beams or the edges of floor slabs
• Story-height wall panels may be suspended from the floor above
• Alternately, they may be supported on the floor below.
• Wall units incapable of spanning between columns or from floor to floor require secondary framing of mullions and shelf angles.
• The curtain wall framing or panels and the supporting structural frame may respond differently to variations in temperature and to gravity or wind loads. Connection details should allow for the differential movement between the wall and structural frame, as well as between the wall units themselves.
• The wall may be subject to both wind pressure and suction.
• If diagonals are used to brace the structural frame, they will affect the design of the wall units.
Spandrel wall units are supported at one floor only. For stability, they may be braced either:
• Below the floor
• Above the floor
• For general information on curtain wall systems, see 7.24–7.26.
• For glazed curtain walls, see 8.31.

The most frequently used section for columns is the wide-flange (W) shape. It is suitable for connections to beams in two directions, and all of its surfaces are accessible for making bolted or welded connections. Other steel shapes used for columns are round pipes and square or rectangular tubing. Column sections may also be fabricated from a number of shapes or plates to fit the desired end-use of a column.

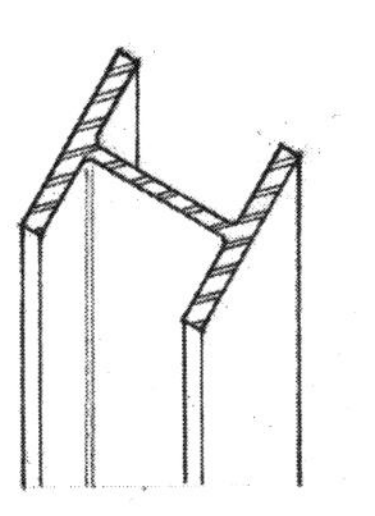

- W shape
- Round pipe
- Rectangular or square tubing

- Compound columns are structural steel columns encased in concrete at least 2-1/2" (64 mm) thick, reinforced with wire mesh.
- Composite columns are structural steel sections thoroughly encased in concrete reinforced with both vertical and spiral reinforcement.

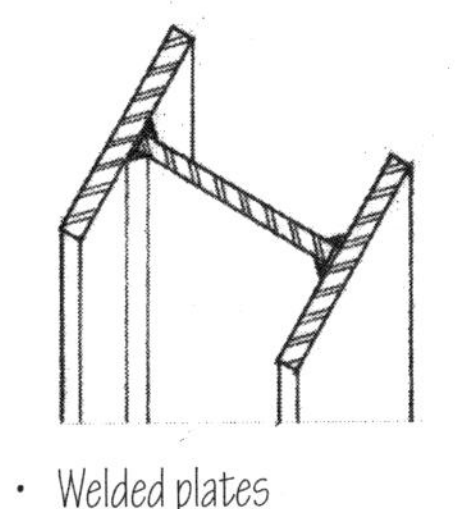

- Welded plates

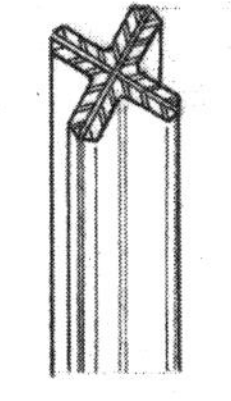

- Cruciform (4 angles)

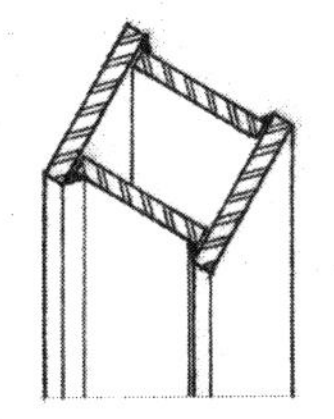

- Welded plates

Column Shapes

The allowable load on a steel column depends on its cross-sectional area and its slenderness ratio (L/r), where (L) is the unsupported length of the column in inches and (r) is the least radius of gyration for the cross section of the column.

Estimating Guidelines for Steel Columns

- 4x4 steel tube column may support up to 750 sf (70 m^2) of floor and roof area.
- 6x6 steel tube column may support up to 2400 sf (223 m^2) of floor and roof area.
- W6x6 may support up to 750 sf (70 m^2) of floor and roof area.
- W8x8 may support up to 3000 sf (279 m^2) of floor and roof area.
- W10x10 may support up to 4500 sf (418 m^2) of floor and roof area.
- W12x12 may support up to 6000 sf (557 m^2) of floor and roof area.
- W14x14 may support up to 12,000 sf (1115 m^2) of floor and roof area.

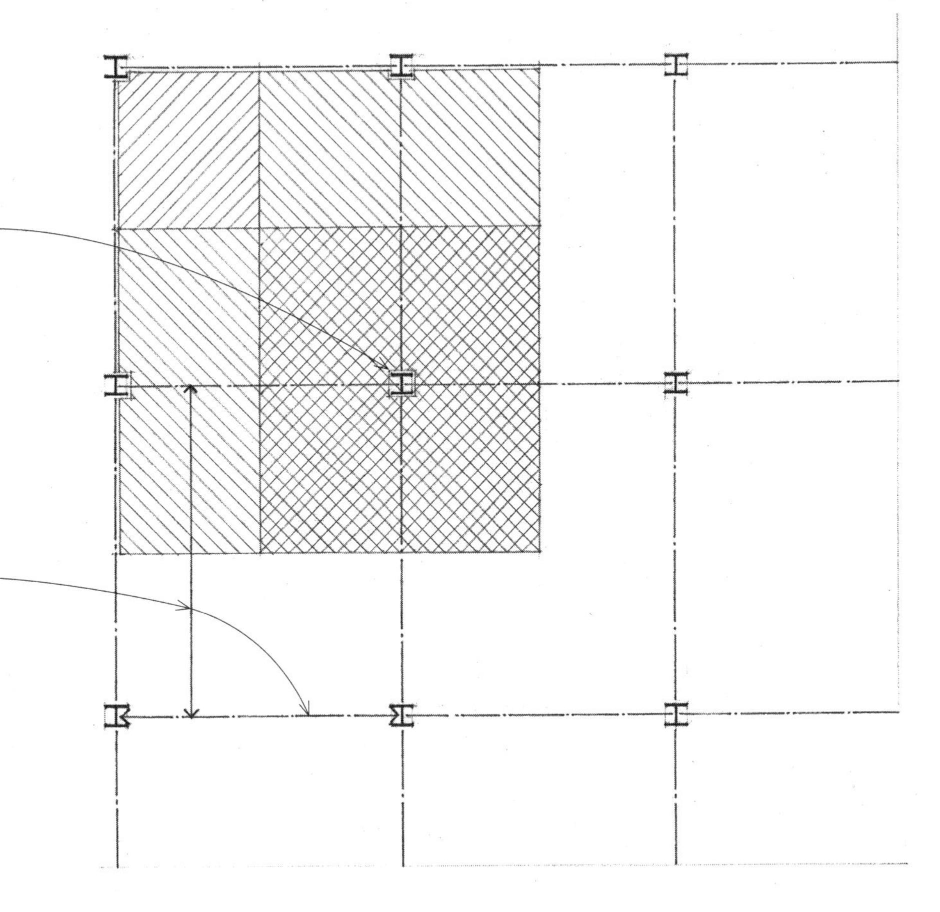

- Column spacing = beam span; see 4.16.
- Columns are assumed to have an effective length of 12' (3660).
- Increased sizes or weights are required for columns supporting heavy loads, rising to greater heights, or contributing to the lateral stability of a structure.
- Consult a structural engineer for final design requirements.

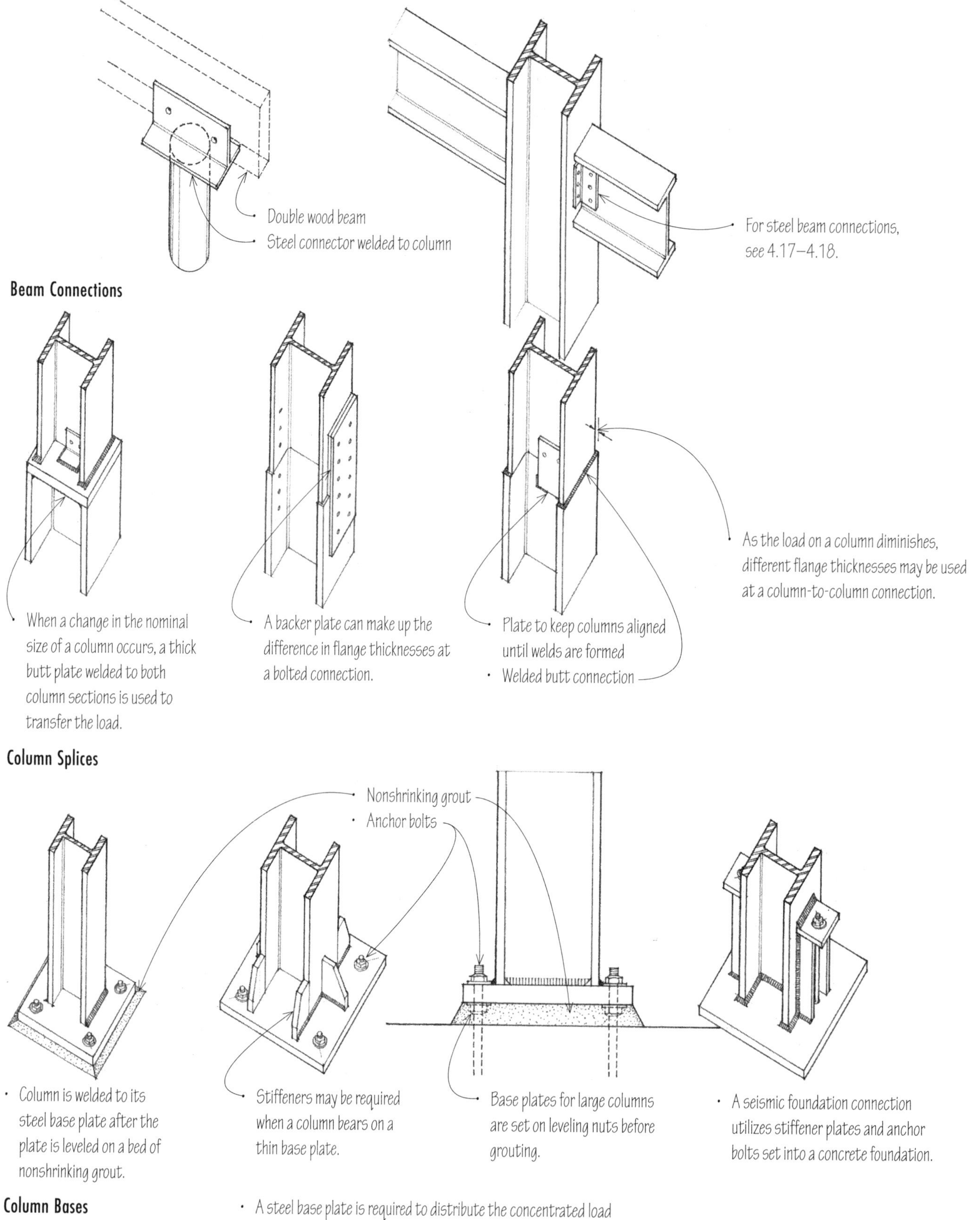

Column Bases

- A steel base plate is required to distribute the concentrated load from a column to the concrete foundation to ensure that the allowable stresses in the concrete are not exceeded.

Light-gauge steel studs are manufactured by cold-forming sheet or strip steel. The cold-formed steel studs can be easily cut and assembled with simple tools into a wall structure that is lightweight, noncombustible, and dampproof. Metal stud walls may be used as nonloadbearing partitions or as bearing walls supporting light-gauge steel joists. As in wood light-frame construction, the stud framing contains cavities for utilities and thermal insulation and accepts a wide range of finishes.

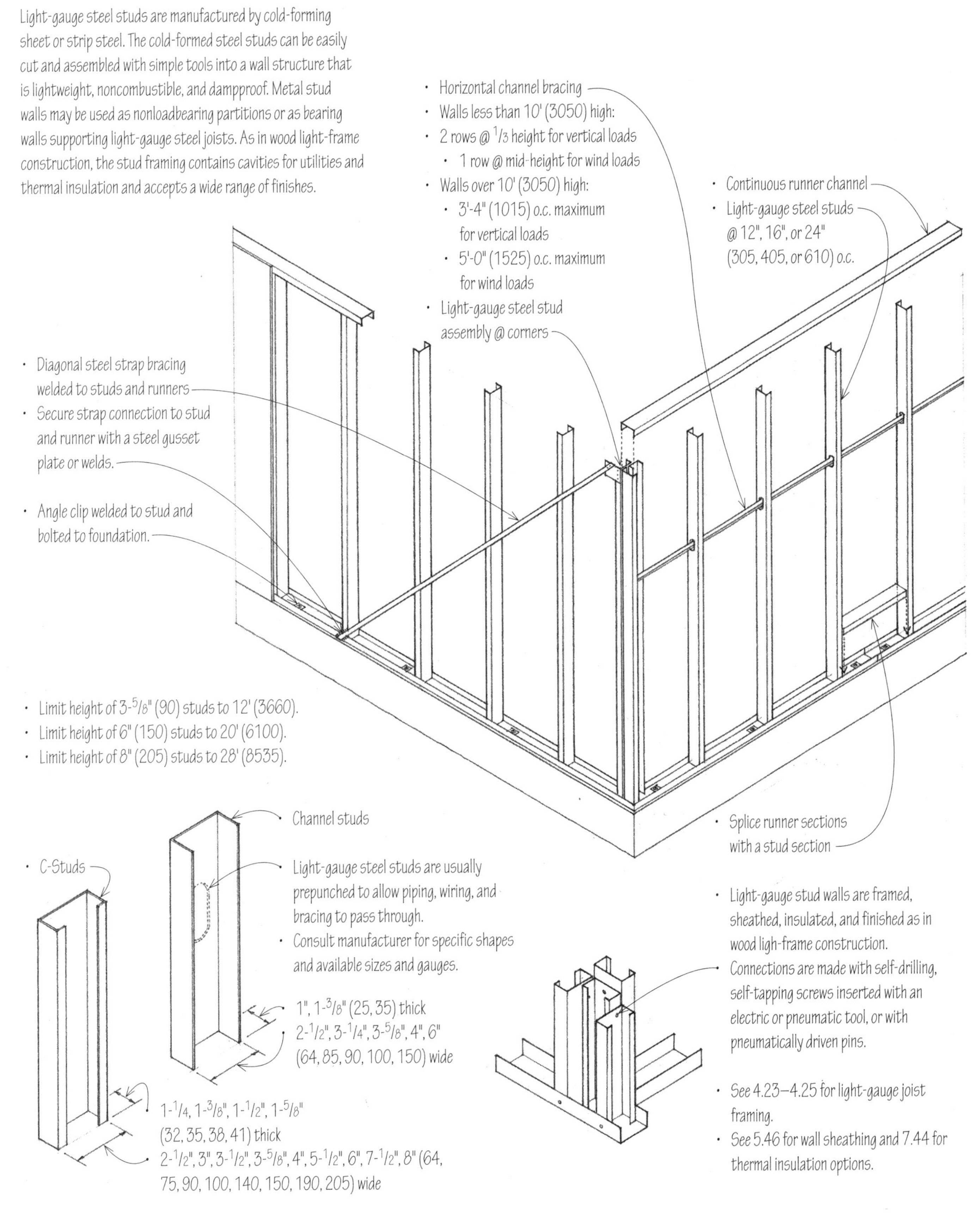

CSI MasterFormat 05 40 00: Cold-Formed Metal Framing
CSI MasterFormat 05 41 00: Structural Metal Stud Framing

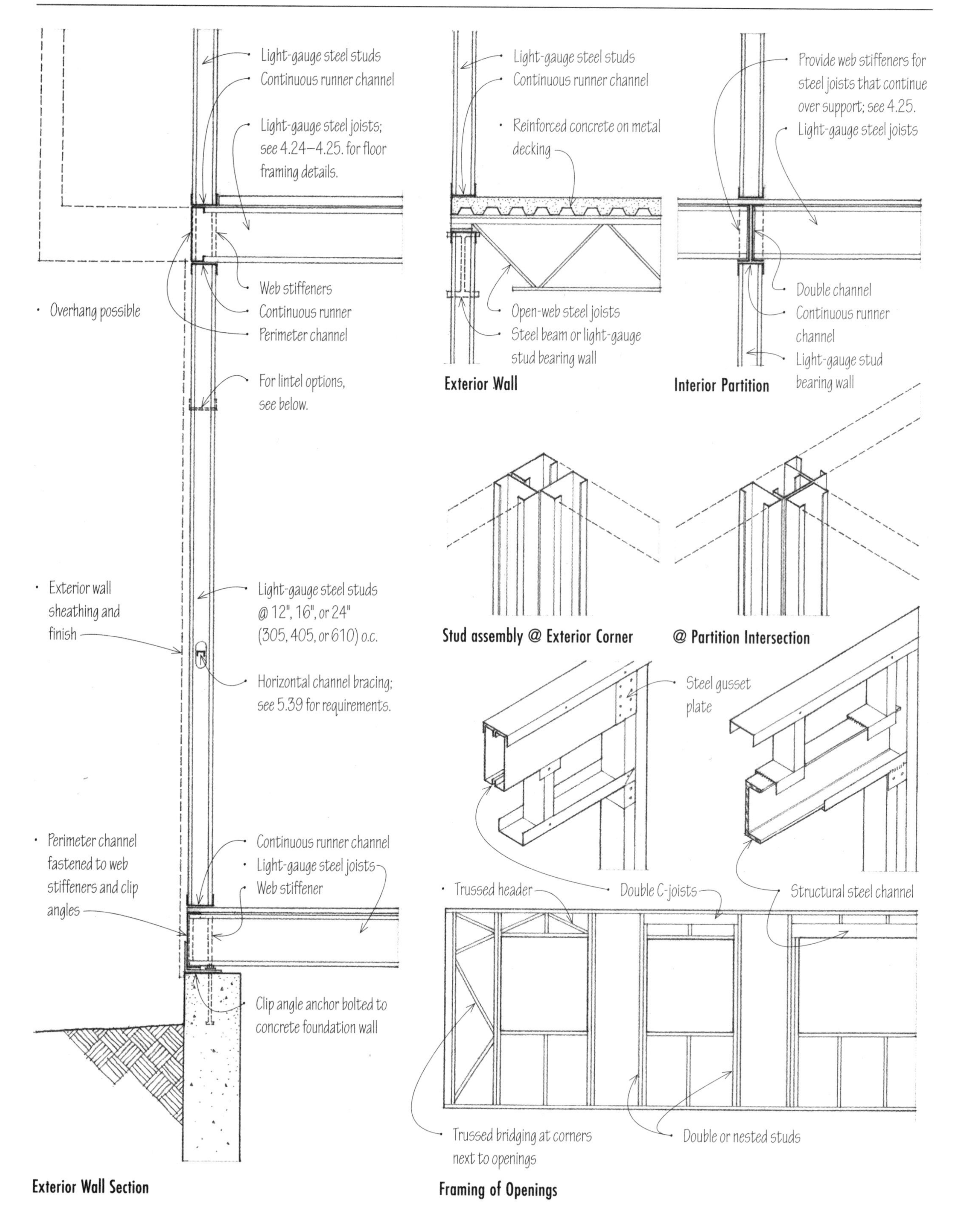

Exterior Wall Section

Exterior Wall

Interior Partition

Stud assembly @ Exterior Corner

@ Partition Intersection

Framing of Openings

Balloon framing utilizes studs that rise the full height of the frame from the sill plate to the roof plate, with joists nailed to the studs and supported by sills or by ribbons let into the studs. Balloon framing is rarely used today, but the minimal vertical shrinkage it affords may be desirable for brick veneer and stucco finishes.

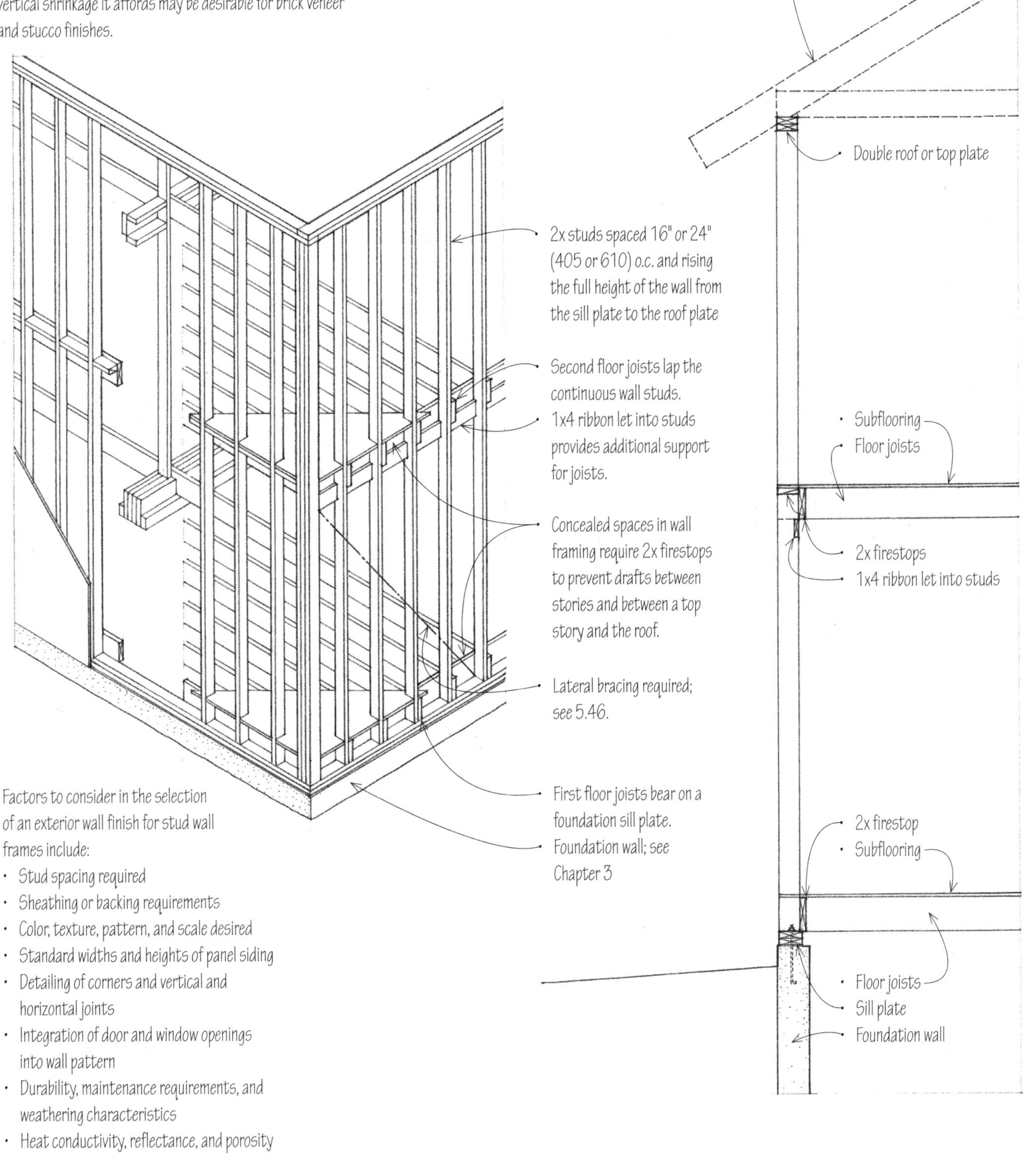

Factors to consider in the selection of an exterior wall finish for stud wall frames include:

- Stud spacing required
- Sheathing or backing requirements
- Color, texture, pattern, and scale desired
- Standard widths and heights of panel siding
- Detailing of corners and vertical and horizontal joints
- Integration of door and window openings into wall pattern
- Durability, maintenance requirements, and weathering characteristics
- Heat conductivity, reflectance, and porosity of the material
- Expansion joints, if required

CSI MasterFormat 06 10 00: Rough Carpentry
CSI MasterFormat 06 11 00: Wood Framing

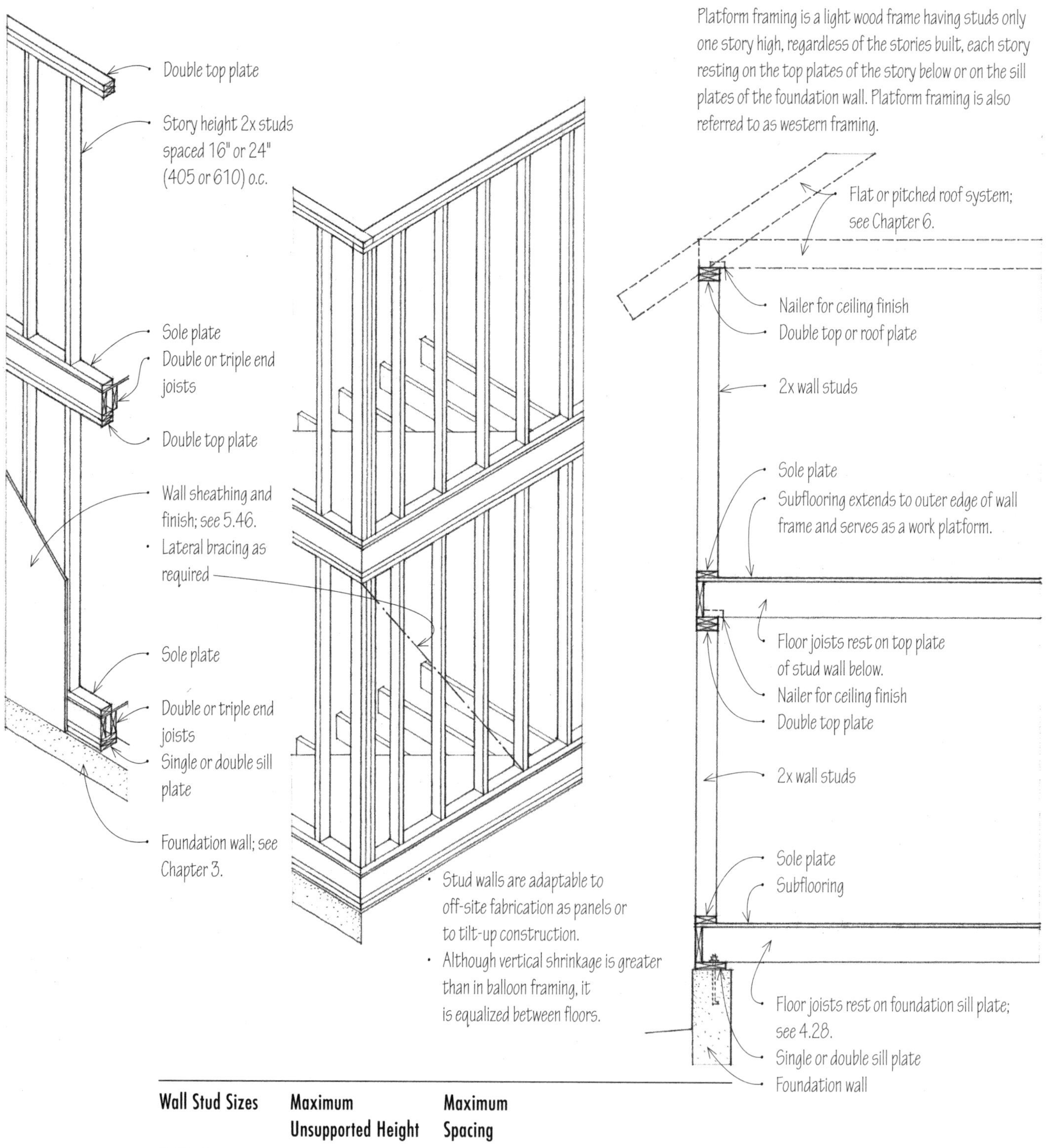

Platform framing is a light wood frame having studs only one story high, regardless of the stories built, each story resting on the top plates of the story below or on the sill plates of the foundation wall. Platform framing is also referred to as western framing.

Wall Stud Sizes	Maximum Unsupported Height	Maximum Spacing
2x4 studs	14' (4265)	16" (405) o.c., except when supporting only a ceiling and roof, 2x4 studs not more than 10' (3050) high may be spaced 24" (610) o.c.
2x6 studs	20' (6100)	24" (610) o.c., except when supporting 2 stories and a roof, space 2x6 studs not more than 16" (405) o.c.

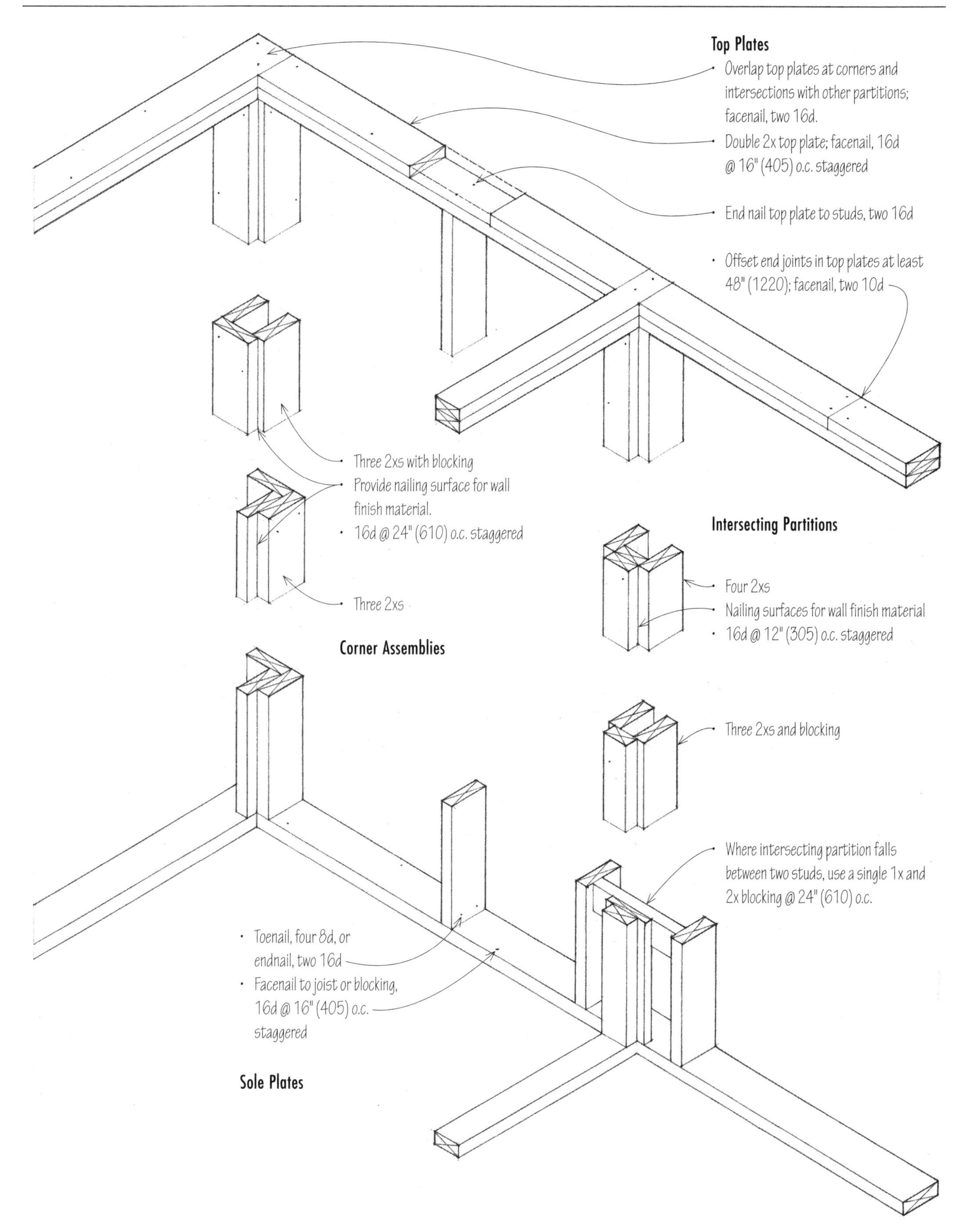
Top Plates
Overlap top plates at corners and intersections with other partitions; facenail, two 16d.
Double 2x top plate; facenail, 16d @ 16" (405) o.c. staggered
End nail top plate to studs, two 16d
Offset end joints in top plates at least 48" (1220); facenail, two 10d
Three 2xs with blocking
Provide nailing surface for wall finish material.
16d @ 24" (610) o.c. staggered
Three 2xs
Corner Assemblies
Intersecting Partitions
Four 2xs
Nailing surfaces for wall finish material
16d @ 12" (305) o.c. staggered
Three 2xs and blocking
Where intersecting partition falls between two studs, use a single 1x and 2x blocking @ 24" (610) o.c.
Toenail, four 8d, or endnail, two 16d
Facenail to joist or blocking, 16d @ 16" (405) o.c. staggered
Sole Plates

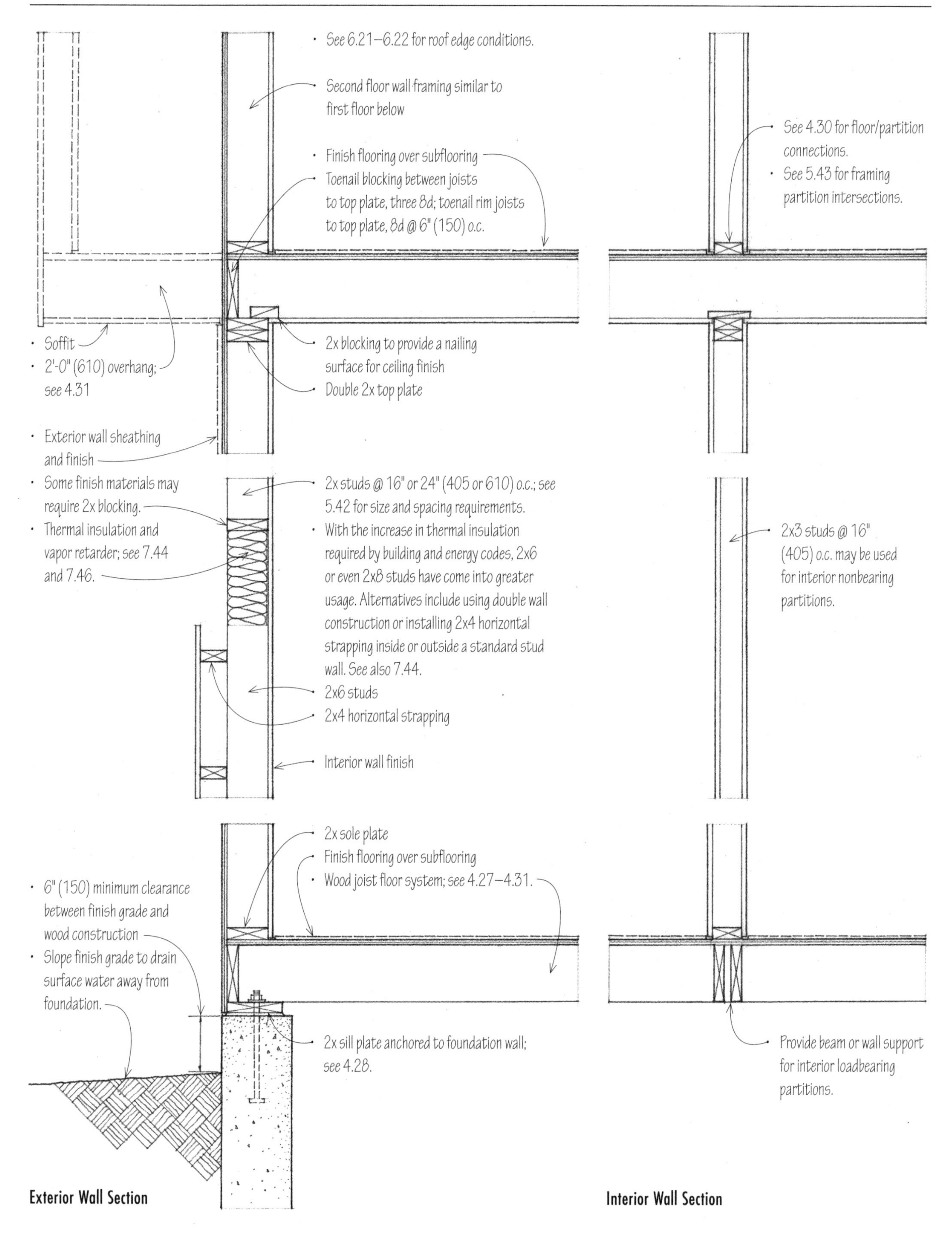
• See 6.21–6.22 for roof edge conditions.
• Second floor wall framing similar to first floor below
• Finish flooring over subflooring
• Toenail blocking between joists to top plate, three 8d; toenail rim joists to top plate, 8d @ 6" (150) o.c.
• Soffit
• 2'-0" (610) overhang; see 4.31
• 2x blocking to provide a nailing surface for ceiling finish
• Double 2x top plate
• See 4.30 for floor/partition connections.
• See 5.43 for framing partition intersections.
• Exterior wall sheathing and finish
• Some finish materials may require 2x blocking.
• Thermal insulation and vapor retarder; see 7.44 and 7.46.
• 2x studs @ 16" or 24" (405 or 610) o.c.; see 5.42 for size and spacing requirements.
• With the increase in thermal insulation required by building and energy codes, 2x6 or even 2x8 studs have come into greater usage. Alternatives include using double wall construction or installing 2x4 horizontal strapping inside or outside a standard stud wall. See also 7.44.
• 2x6 studs
• 2x4 horizontal strapping
• Interior wall finish
• 2x3 studs @ 16" (405) o.c. may be used for interior nonbearing partitions.
• 2x sole plate
• Finish flooring over subflooring
• Wood joist floor system; see 4.27–4.31.
• 6" (150) minimum clearance between finish grade and wood construction
• Slope finish grade to drain surface water away from foundation.
• 2x sill plate anchored to foundation wall; see 4.28.
• Provide beam or wall support for interior loadbearing partitions.
Exterior Wall Section
Interior Wall Section

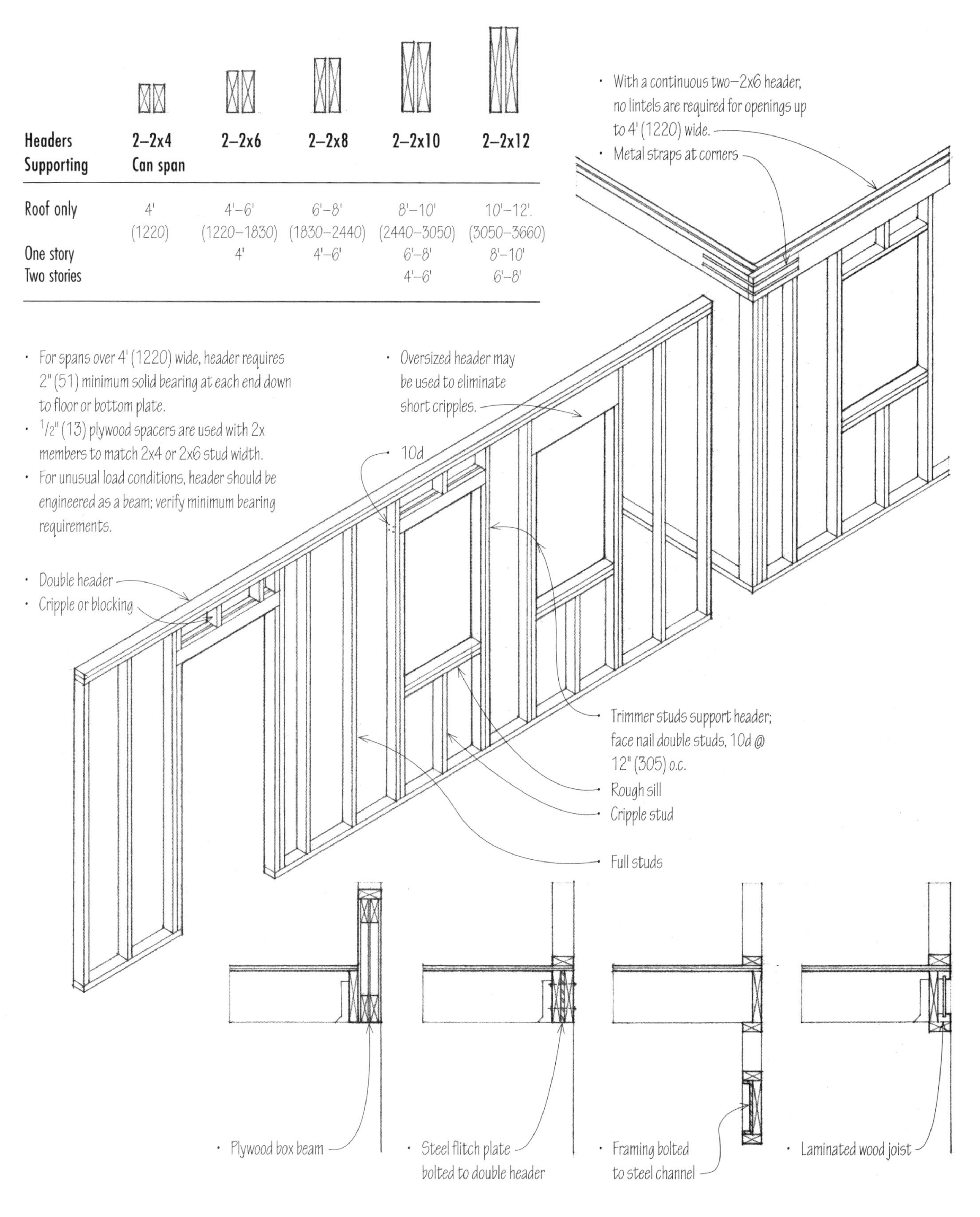

Lintel Options for Wide Openings • These lintels should be engineered as beams; verify minimum bearing requirements.

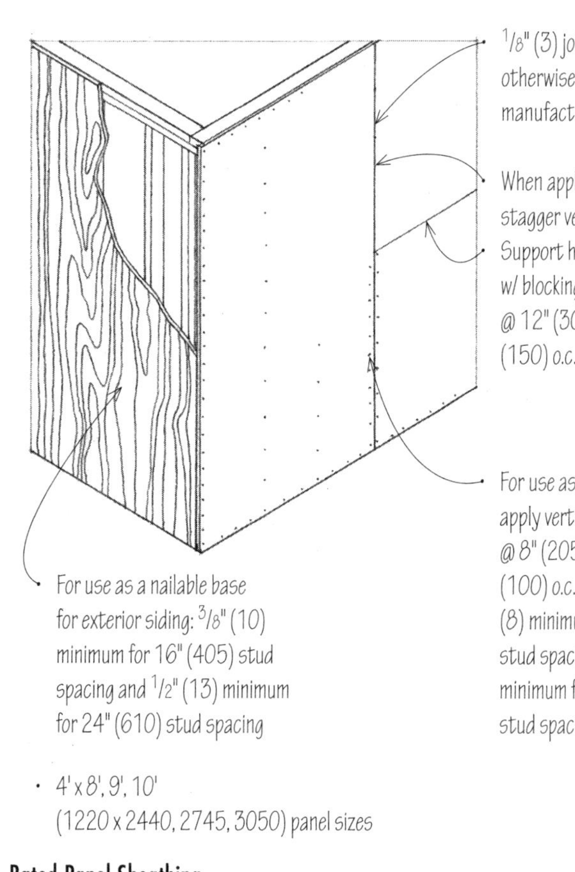

- $^1/_8$" (3) joint spacing unless otherwise recommended by manufacturer
- When applied horizontally, stagger vertical joints.
- Support horizontal edges w/ blocking or plyclips; nail @ 12" (305) o.c. and 6" (150) o.c. along edges.
- For use as a nailable base for exterior siding: $^3/_8$" (10) minimum for 16" (405) stud spacing and $^1/_2$" (13) minimum for 24" (610) stud spacing
- For use as corner bracing, apply vertically and nail @ 8" (205) o.c., and 4" (100) o.c. along edges; $^5/_{16}$" (8) minimum for 16" (405) stud spacing and $^3/_8$" (10) minimum for 24" (610) stud spacing.

- 4' x 8', 9', 10' (1220 x 2440, 2745, 3050) panel sizes

Rated Panel Sheathing

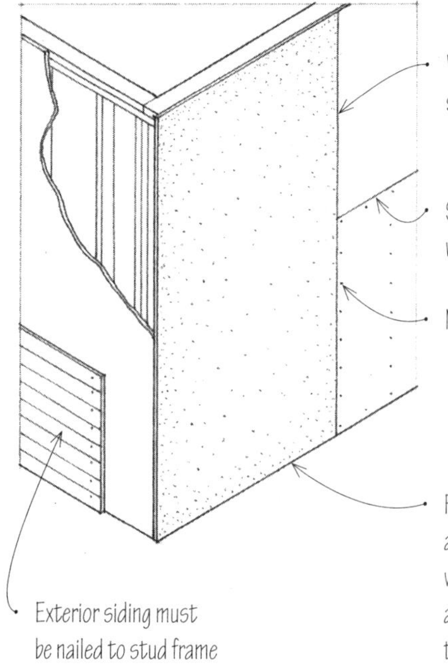

- When applied horizontally, stagger vertical joints.
- Support horizontal edges w/ blocking.
- Nail @ 8" (205) o.c.
- Exterior siding must be nailed to stud frame because gypsum board is not a nailable base.
- For use as corner bracing, apply $^1/_2$" (13) panels vertically and nail or use adhesives according to manufacturer's recommendations.

- 4' x 8', 10', 12', 14' (1220 x 2440, 3050, 3660, 4265) panel sizes

Gypsum Sheathing

- When applied horizontally, stagger vertical joints.
- Solid blocking or V-groove joints along horizontal edges
- Nail @ 8" (205) o.c. and 4" (100) o.c. along edges
- For use as corner bracing, use $^1/_2$" (13) high-density panels applied vertically; nail @ 6" (150) o.c. and 3" (75) o.c. along edges.

- High-density panels may be used as a nailable base for exterior siding.

- 4' x 8', 9', 10', 12' (1220 x 2440, 2745, 3050, 3660) panel sizes

Fiberboard Sheathing

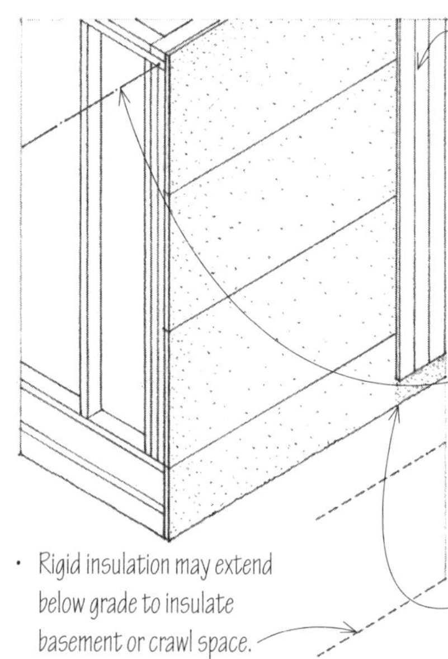

- Exterior siding must be nailed directly to stud frame.
- Because the foam plastic is an effective vapor barrier, the wall assembly must be properly vented.
- Rigid insulation cannot be used as corner bracing; use steel strap or 1x4 let into studs.
- Protect exposed surfaces w/ treated plywood or stucco.

- Rigid insulation may extend below grade to insulate basement or crawl space.
- See also 7.44.

- 2' x 4', 8' and 4' x 8', 9' (610 x 1220, 2440 and 1220 x 2440, 2745) panel sizes

Rigid Foam Plastic Sheathing

Wood columns may be solid, built-up, or spaced. In selecting a wood column, the following should be considered: lumber species; structural grade; modulus of elasticity; and allowable compressive, bending, and shear stress values permitted for the intended use. In addition, attention should be paid to the precise loading conditions and the types of connections used.

Wood columns and posts are loaded axially in compression. Failure can result from crushing of the wood fibers if the maximum unit stress exceeds the allowable unit stress in compression parallel to the grain. The load capacity of a column is also determined by its slenderness ratio. As the slenderness ratio of a column increases, a column can fail from buckling.
See 2.13.

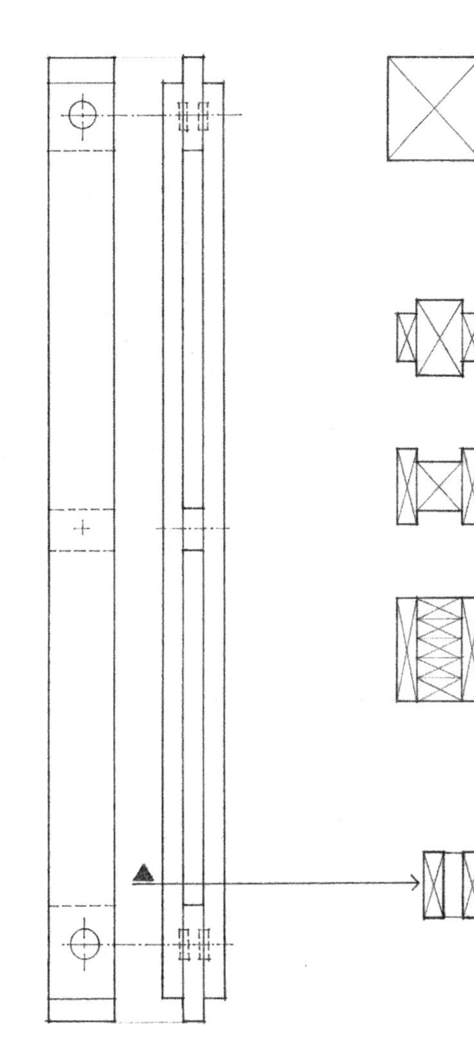

- Solid sawn columns should be of well-seasoned wood.

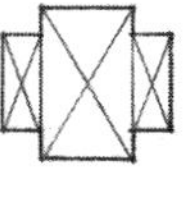

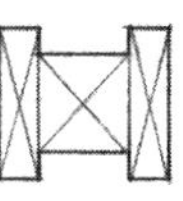

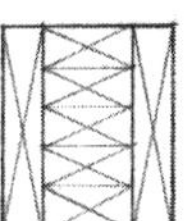

- Built-up columns may be glue-laminated or mechanically fastened. Glue-laminated columns may have a higher allowable compressive stress than solid sawn columns, while mechanically fastened columns cannot equal the strength of a solid column of the same dimensions and material.

- Spaced columns consist of two or more members separated at their ends and middle points by blocking and joined at their ends by timber connectors and bolts.

- $l/d < 50$ for solid or built-up columns
- $l/d < 80$ for individual member of a spaced column
- l = unsupported length in inches
- d = the least dimension of the compression member in inches

Estimating Guidelines for Wood Columns

- 6x6 may support up to 500 sf (46 m^2) of floor and roof area.
- 8x8 may support up to 1000 sf (93 m^2) of floor and roof area.
- 10x10 may support up to 2500 sf (232 m^2) of floor and roof area.
- Columns are assumed to have an unsupported height of 12' (3660).
- Increased sizes are required for columns supporting heavy loads, rising to greater heights, or resisting lateral forces.
- See the Bibliography for sources of more detailed load tables.
- Consult a structural engineer for final design requirements.

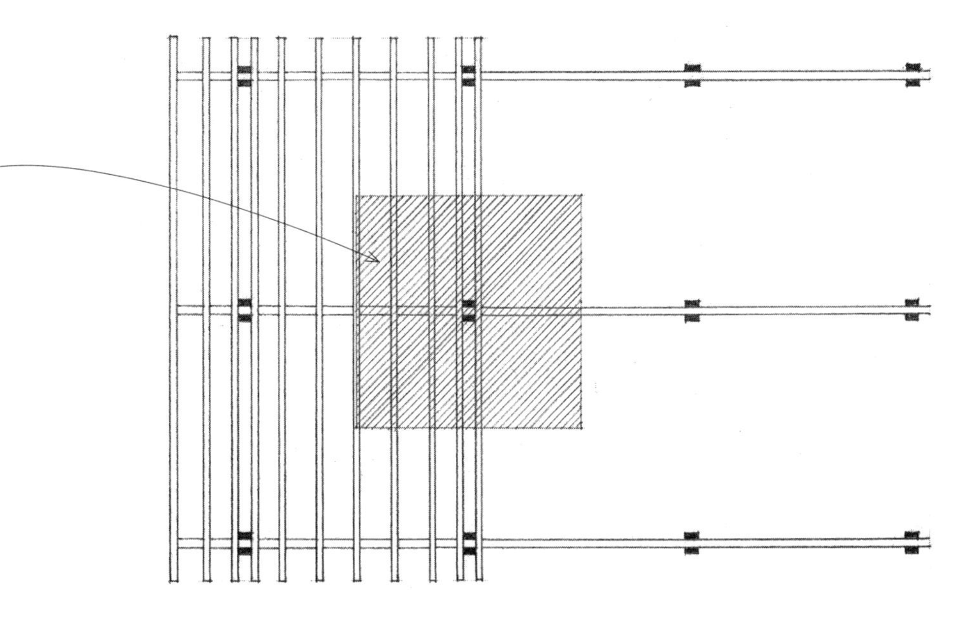

- Roof system: conventional wood rafters or plank-and-beam framing; see Chapter 6.

Post-and-beam construction uses a framework of vertical posts and horizontal beams to carry both floor and roof loads. The beams supporting the floor and roof systems transmit their loads to posts or columns that, in turn, carry the loads down to the foundation system.

- Together with plank-and-beam floor and roof systems, the post-and-beam wall system forms a three-dimensional structural grid, which may be expanded vertically or horizontally.
- The skeleton frame of posts and beams is often left exposed to form a visible framework within which nonbearing wall panels, doors, and windows are integrated.
- When the post-and-beam frame is left exposed, as is often the case, careful attention must be paid to the species and grade of wood used, the detailing of joints, especially at beam-to-beam and beam-to-post connections, and the quality of workmanship.

- Floor system: conventional joists or plank-and-beam framing; see Chapter 4.
- Resistance to lateral wind and seismic forces requires the use of rigid shear walls or diagonal bracing.
- Posts or columns may be supported by individual piers or by a wall foundation.
- Column spacing is related to the desired size and proportion of the bays, and the spanning capability of the beams, joists, and decking.

Heavy Timber Construction

- Post-and-beam framing may qualify as heavy timber construction if the plank-and-beam floor and roof structures are supported by noncombustible, fire-resistive exterior walls and the wood members and decking meet the minimum size requirements specified in the building code.

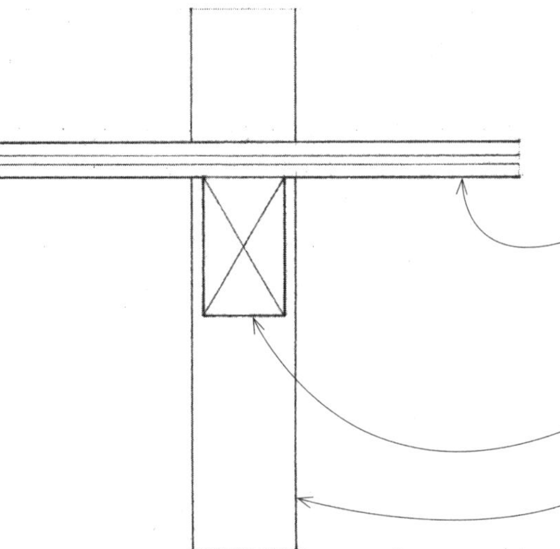

- Floor decking: not less than 3" (75) nominal tongue-and-groove (t&g) or splined planks with 1" (25) nominal t&g flooring or 1/2" (13) wood structural panel subflooring
- Roof decking: not less than 2" (51) nominal t&g or splined planks or 1-1/8" (32) thick wood structural panel
- Beams and girders: not less than 6" (150) nominal in thickness and 10" (255) nominal in depth
- Columns: not less than 8x8 nominal when supporting floors; not less than 8x6 nominal when supporting roofs only.

CSI MasterFormat 06130: Heavy Timber Construction

The strength of a post-and-beam connection depends on:

- The species and grade of lumber used
- The thickness of the wood members
- The angle of the resisting force relative to the grain of the wood
- The size and number of bolts or timber connectors used

The size and number of bolts required for a connection depend on the magnitude of the loads being transferred. Generally, greater efficiency is achieved with a few large bolts rather than with more smaller ones. The drawings to the right illustrate general guidelines for the placement of bolts.

Load Parallel to Grain

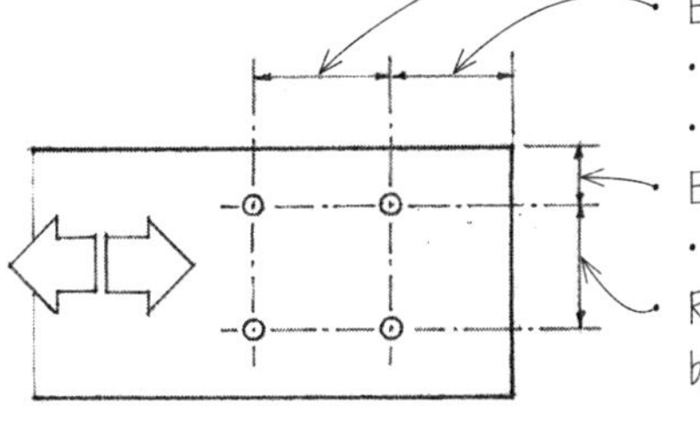

- 4d
- End distance:
  - 4d in compression; d = bolt diameter
  - 7d in tension
- Edge distance:
  - 1-1/2 d or 1/2 row spacing for l/d ratios <6
- Row spacing parallel to grain is determined by net section requirements.

Timber Connectors

If there is insufficient surface contact area to accommodate the required number of bolts, timber connectors can be used. Timber connectors are metal rings, plates, or grids for transferring shear between the faces of two timber members, used with a single bolt that serves to restrain and clamp the assembly together. Timber connectors are more efficient than bolts or lag screws used alone because they enlarge the area of wood over which a load is distributed and develop higher stresses per unit of bearing.

Load Perpendicular to Grain

- Row spacing perpendicular to grain:
  - 2-1/2 d for l/d of 2
  - 5d for l/d of 6
- Edge distance for edge toward which load is acting > or = 4d
- 4d

Split-ring connectors consist of a metal ring inserted into corresponding grooves cut into the faces of the joining members and held in place by a single bolt. The tongue-and-groove split in the ring permits it to deform slightly under loading and maintain bearing at all surfaces, while the beveled cross section eases insertion and ensures a tight-fitting joint after the ring is fully seated in the grooves.

- Available in 2-1/2" and 4" (64 and 100) diameters
- 3-5/8" (90) minimum face width for 2-1/2" (64) split rings; 5-1/2" (140) minimum for 4" (100) split rings
- 1/2" (13) ø bolt for 2-1/2" (64) split rings; 3/4" (19) ø for 4" (100) split rings

Shear plates consist of a round plate of malleable iron inserted into a corresponding groove, flush with the face of a timber, and held in place by a single bolt. Shear plates are used in back-to-back pairs to develop shear resistance in demountable wood-to-wood connections, or singly in a wood-to-metal connection.

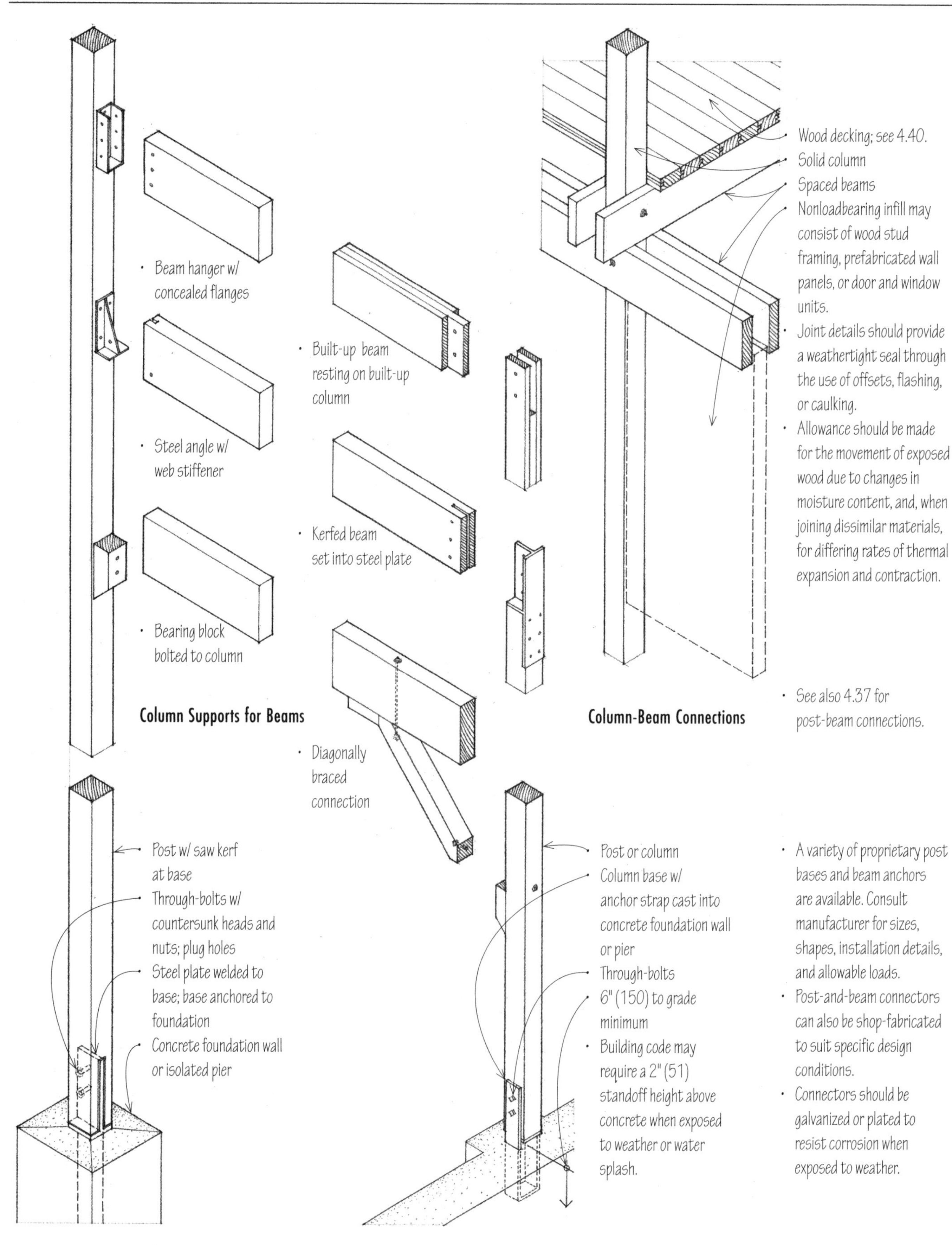

Beam hanger w/ concealed flanges
Steel angle w/ web stiffener
Bearing block bolted to column
Column Supports for Beams
Built-up beam resting on built-up column
Kerfed beam set into steel plate
Diagonally braced connection
Wood decking; see 4.40.
Solid column
Spaced beams
Nonloadbearing infill may consist of wood stud framing, prefabricated wall panels, or door and window units.
Joint details should provide a weathertight seal through the use of offsets, flashing, or caulking.
Allowance should be made for the movement of exposed wood due to changes in moisture content, and, when joining dissimilar materials, for differing rates of thermal expansion and contraction.
Column-Beam Connections
See also 4.37 for post-beam connections.
Post w/ saw kerf at base
Through-bolts w/ countersunk heads and nuts; plug holes
Steel plate welded to base; base anchored to foundation
Concrete foundation wall or isolated pier
Post or column
Column base w/ anchor strap cast into concrete foundation wall or pier
Through-bolts
6" (150) to grade minimum
Building code may require a 2" (51) standoff height above concrete when exposed to weather or water splash.
A variety of proprietary post bases and beam anchors are available. Consult manufacturer for sizes, shapes, installation details, and allowable loads.
Post-and-beam connectors can also be shop-fabricated to suit specific design conditions.
Connectors should be galvanized or plated to resist corrosion when exposed to weather.
Column Base Supports

6

ROOF SYSTEMS

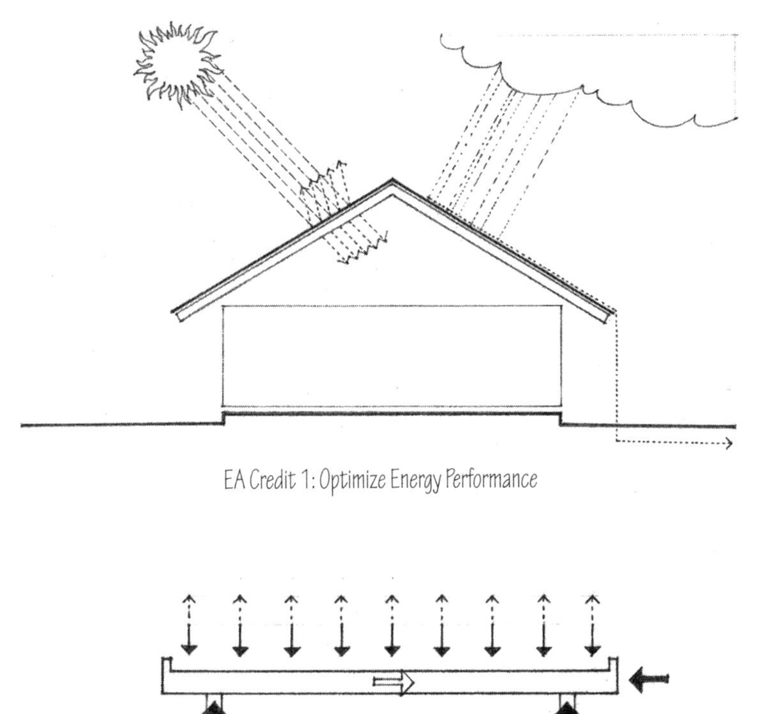

EA Credit 1: Optimize Energy Performance

The roof system functions as the primary sheltering element for the interior spaces of a building. The form and slope of a roof must be compatible with the type of roofing—shingles, tiles, or a continuous membrane—used to shed rainwater and melting snow to a system of drains, gutters, and downspouts. The construction of a roof should also control the passage of moisture vapor, the infiltration of air, and the flow of heat and solar radiation. And depending on the type of construction required by the building code, the roof structure and assembly may have to resist the spread of fire.

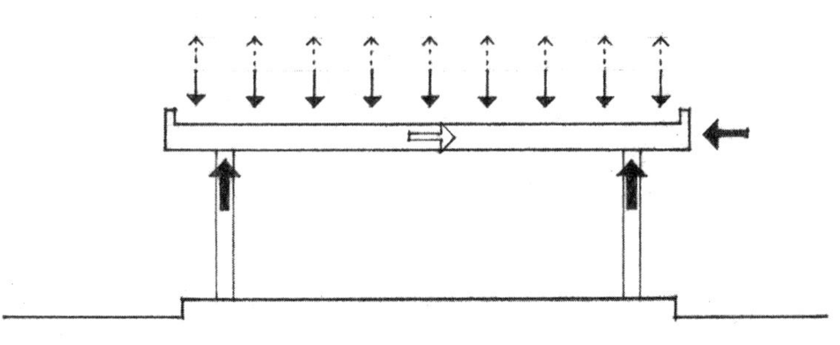

Like floor systems, a roof must be structured to span across space and carry its own weight as well as the weight of any attached equipment and accumulated rain and snow. Flat roofs used as decks are also subject to live occupancy loads. In addition to these gravity loads, the planes of the roof may be required to resist lateral wind and seismic forces, as well as uplifting wind forces, and transfer these forces to the supporting structure.

Because the gravity loads for a building originate with the roof system, its structural layout must correspond to that of the column and bearing wall systems through which its loads are transferred down to the foundation system. This pattern of roof supports and the extent of the roof spans, in turn, influences the layout of interior spaces and the type of ceiling that the roof structure may support. Long roof spans would open up a more flexible interior space while shorter roof spans might suggest more precisely defined spaces.

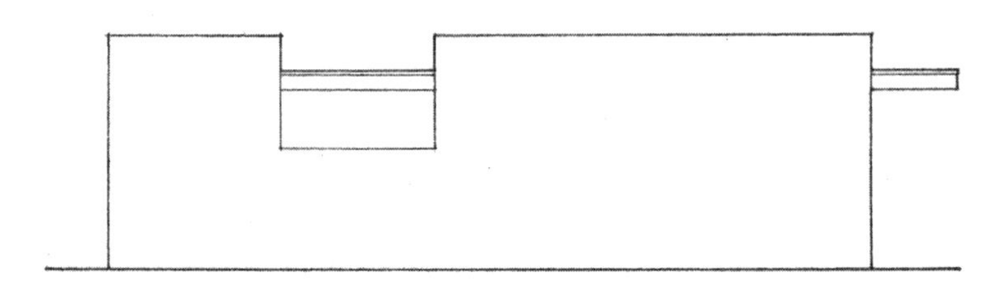

The form of a roof structure—whether flat or pitched, gabled or hipped, broad and sheltering, or rhythmically articulated—has a major impact on the image of a building. The roof may be exposed with its edges flush with or overhanging the exterior walls, or it may be concealed from view, hidden behind a parapet. If its underside remains exposed, the roof also transmits its form to the upper boundaries of the interior spaces below.

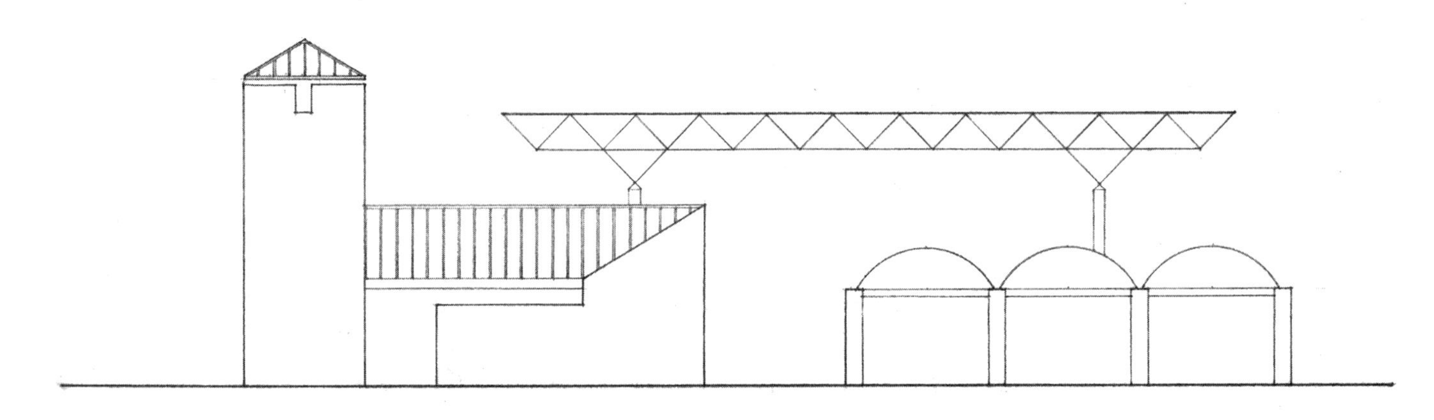

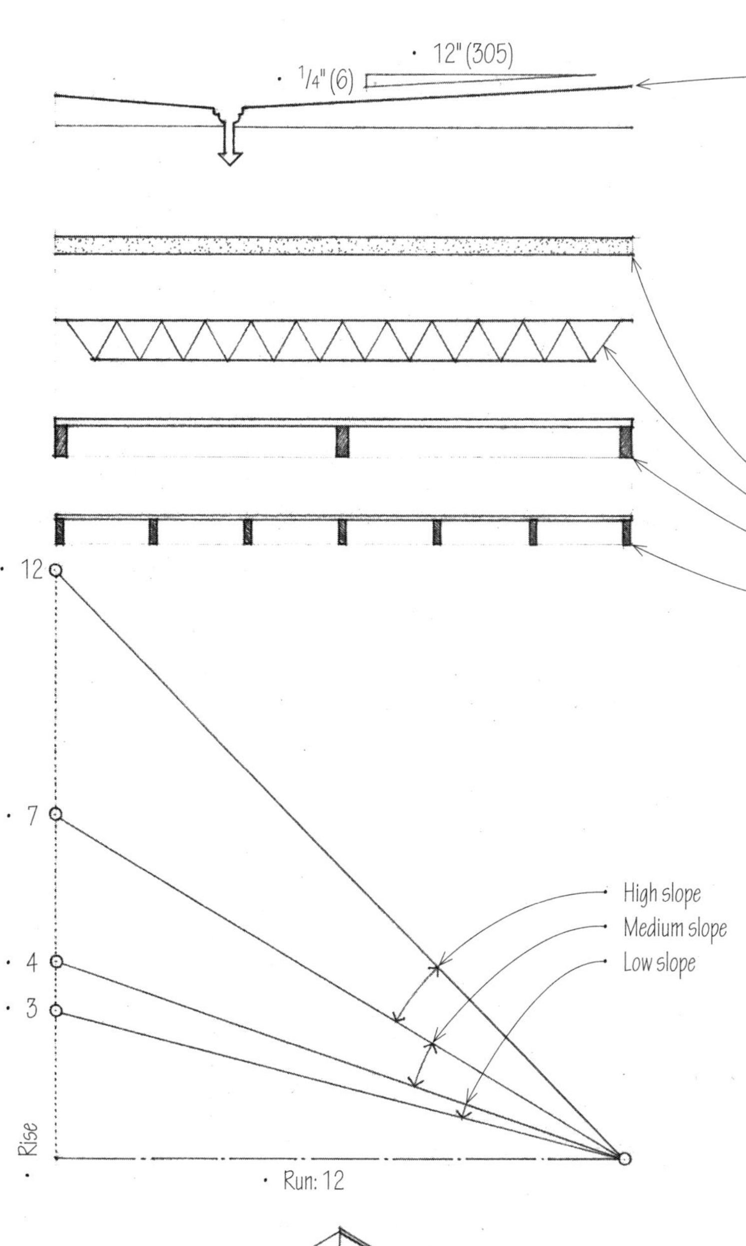

Flat Roofs

- Flat roofs require a continuous membrane roofing material.
- Minimum recommended slope: 1/4" per foot (1:50)
- The roof slope may be achieved by inclining the structural members or roof deck, or by tapering the layer of thermal insulation.
- The slope usually leads to interior drains. Secondary, emergency overflow roof drains or scuppers are required in cases where water might be trapped if the primary roof drains are blocked.

- Flat roofs can efficiently cover a building of any horizontal dimension, and may be structured and designed to serve as an outdoor space.
- The structure of a flat roof may consist of:
  - Reinforced concrete slabs
  - Flat timber or steel trusses
  - Timber or steel beams and decking
  - Wood or steel joists and sheathing

Sloping Roofs

- Sloping roofs may be categorized into
  - Low-slope roofs—up to 3:12
  - Medium- to high-slope roofs—4:12 to 12:12

- The roof slope affects the choice of roofing material, the requirements for underlayment and eave flashing, and design wind loads.
- Low-slope roofs require roll or continuous membrane roofing; some shingles and sheet materials may be used on 3:12 pitches.
- Medium- and high-slope roofs may be covered with shingles, tiles, or sheet materials.
- Sloping roofs shed rainwater easily to eave gutters.

- The height and area of a sloping roof increase with its horizontal dimensions.
- The space under a sloping roof may be usable.
- Sloping roof planes may be combined to form a variety of roof forms.

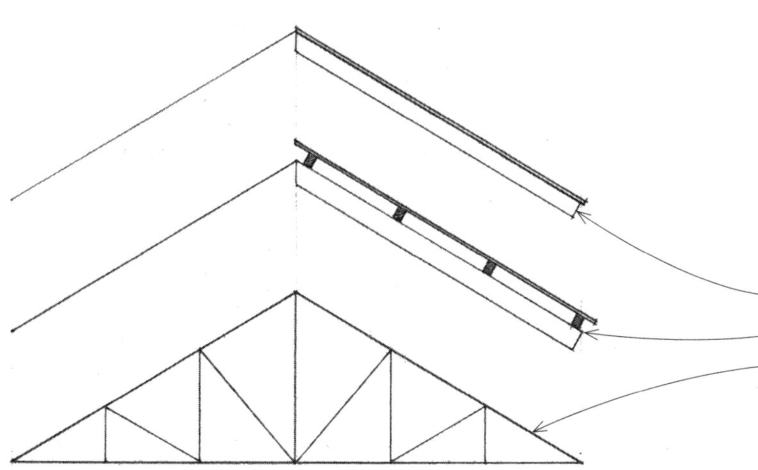

- Sloping roofs may have a structure of:
  - Wood or steel rafters and sheathing
  - Timber or steel beams, purlins, and decking
  - Timber or steel trusses

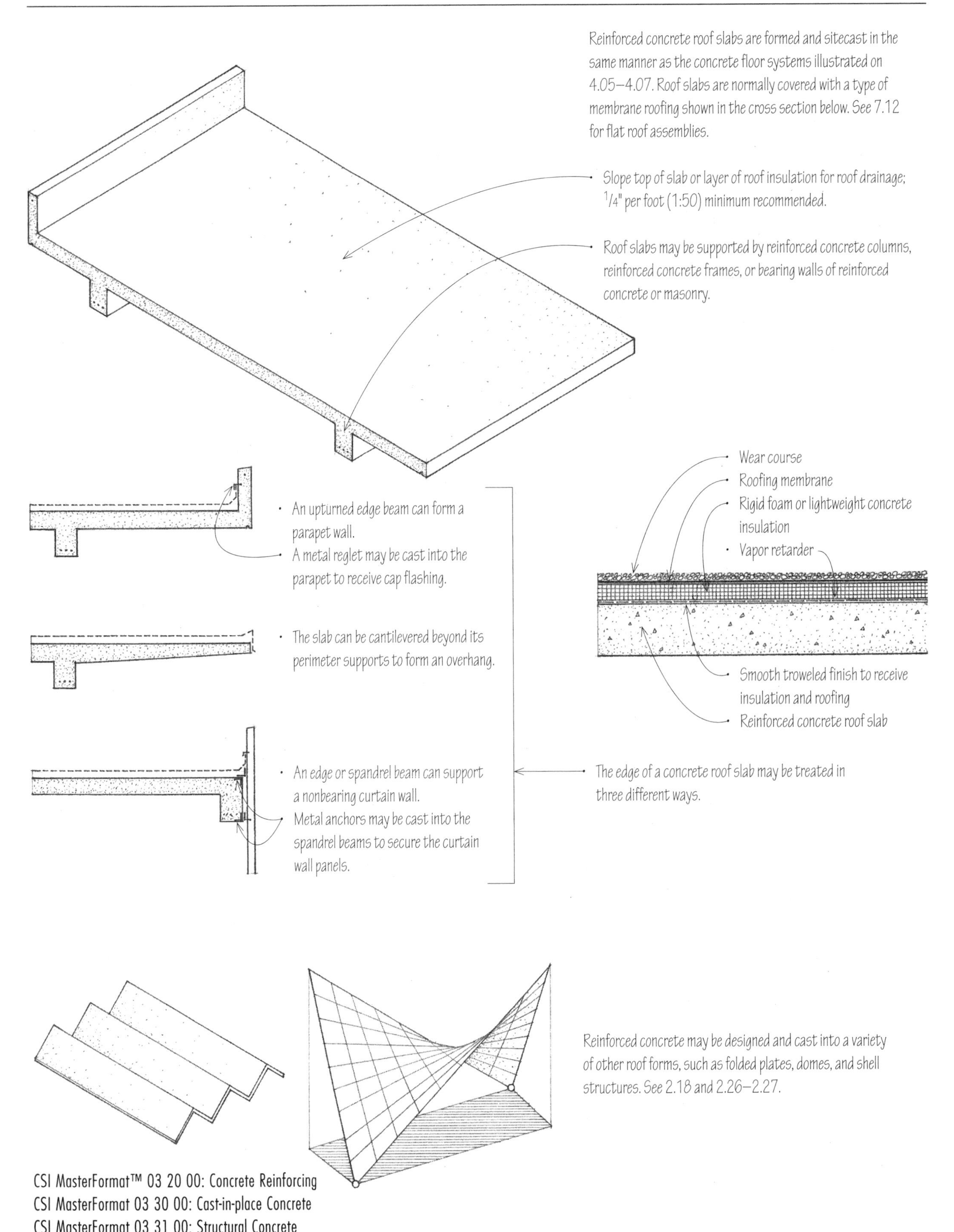
Reinforced concrete roof slabs are formed and sitecast in the same manner as the concrete floor systems illustrated on 4.05–4.07. Roof slabs are normally covered with a type of membrane roofing shown in the cross section below. See 7.12 for flat roof assemblies.
Slope top of slab or layer of roof insulation for roof drainage; 1/4" per foot (1:50) minimum recommended.
Roof slabs may be supported by reinforced concrete columns, reinforced concrete frames, or bearing walls of reinforced concrete or masonry.
An upturned edge beam can form a parapet wall.
A metal reglet may be cast into the parapet to receive cap flashing.
The slab can be cantilevered beyond its perimeter supports to form an overhang.
An edge or spandrel beam can support a nonbearing curtain wall.
Metal anchors may be cast into the spandrel beams to secure the curtain wall panels.
The edge of a concrete roof slab may be treated in three different ways.
Wear course
Roofing membrane
Rigid foam or lightweight concrete insulation
Vapor retarder
Smooth troweled finish to receive insulation and roofing
Reinforced concrete roof slab
Reinforced concrete may be designed and cast into a variety of other roof forms, such as folded plates, domes, and shell structures. See 2.18 and 2.26–2.27.
CSI MasterFormat™ 03 20 00: Concrete Reinforcing
CSI MasterFormat 03 30 00: Cast-in-place Concrete
CSI MasterFormat 03 31 00: Structural Concrete

Precast concrete roof systems are similar in form and construction to precast floor systems and use the same types of slab units. See 4.11–4.13 for general conditions and requirements.

- See 7.19 for parapet flashing.
- Roofing membrane over rigid insulation; see 7.12.
- Sitecast concrete topping, reinforced with welded wire fabric or reinforcing bars, bonds to precast slabs to form a composite structural unit; 2" (51) minimum. The topping may be omitted if rigid foam insulation is laid over smooth-surfaced precast units.
- Slope concrete topping or precast units to drain; 1/4" per foot (1:50) minimum recommended.
- Bearing connections should allow for limited horizontal movement due to creep, shrinkage, and temperature changes.
- To serve as a horizontal diaphragm and transfer lateral forces to shear walls, steel reinforcement must tie the precast slab units to each other over their supports and at their end bearings.

- Precast hollow core slab unit
- Extend steel dowels into reinforced concrete topping or into grouted shear keys for structural continuity.
- Grout voids at ends of hollow core units.
- High-density plastic bearing strip
- Minimum bearing length should be at least 1/180 of the clear span but not less than 2" (51) for solid or hollow core slabs.

Bearing Wall

- Reinforced concrete topping; 2" (51) minimum
- Hooked bars cast into slab edge slots @ 4' (1220) o.c.
- Underside of precast slabs may be caulked and painted; a ceiling finish may also be applied to or be suspended from slab.

End Wall

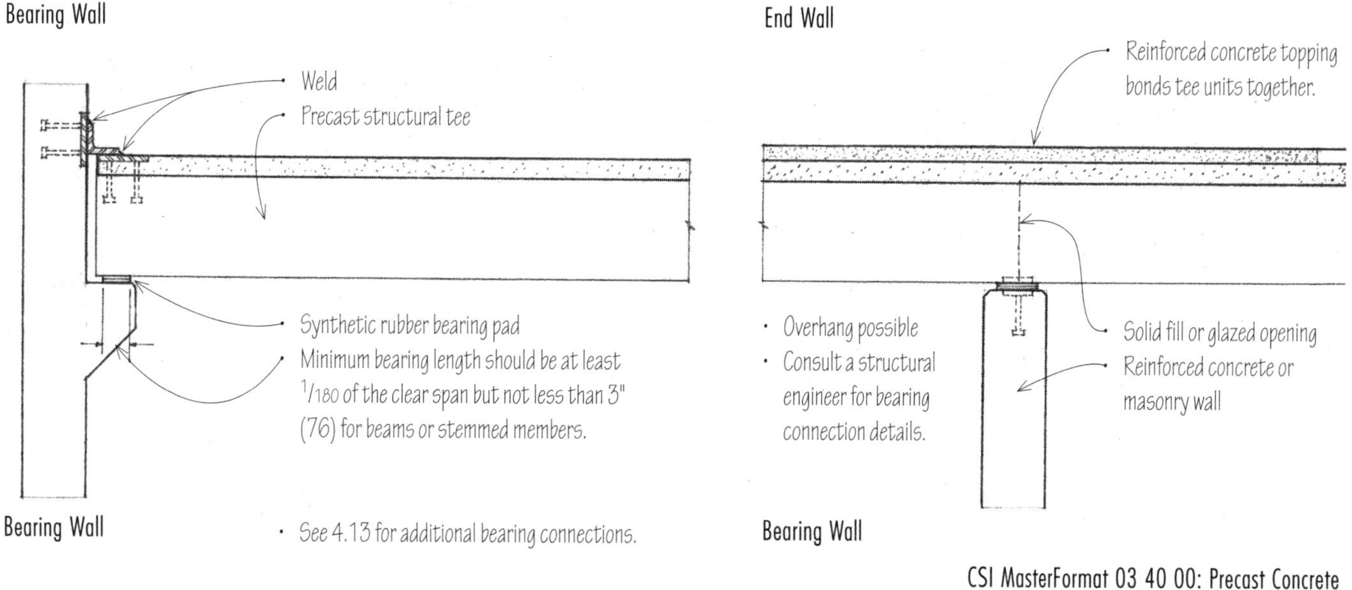

- See 4.13 for additional bearing connections.

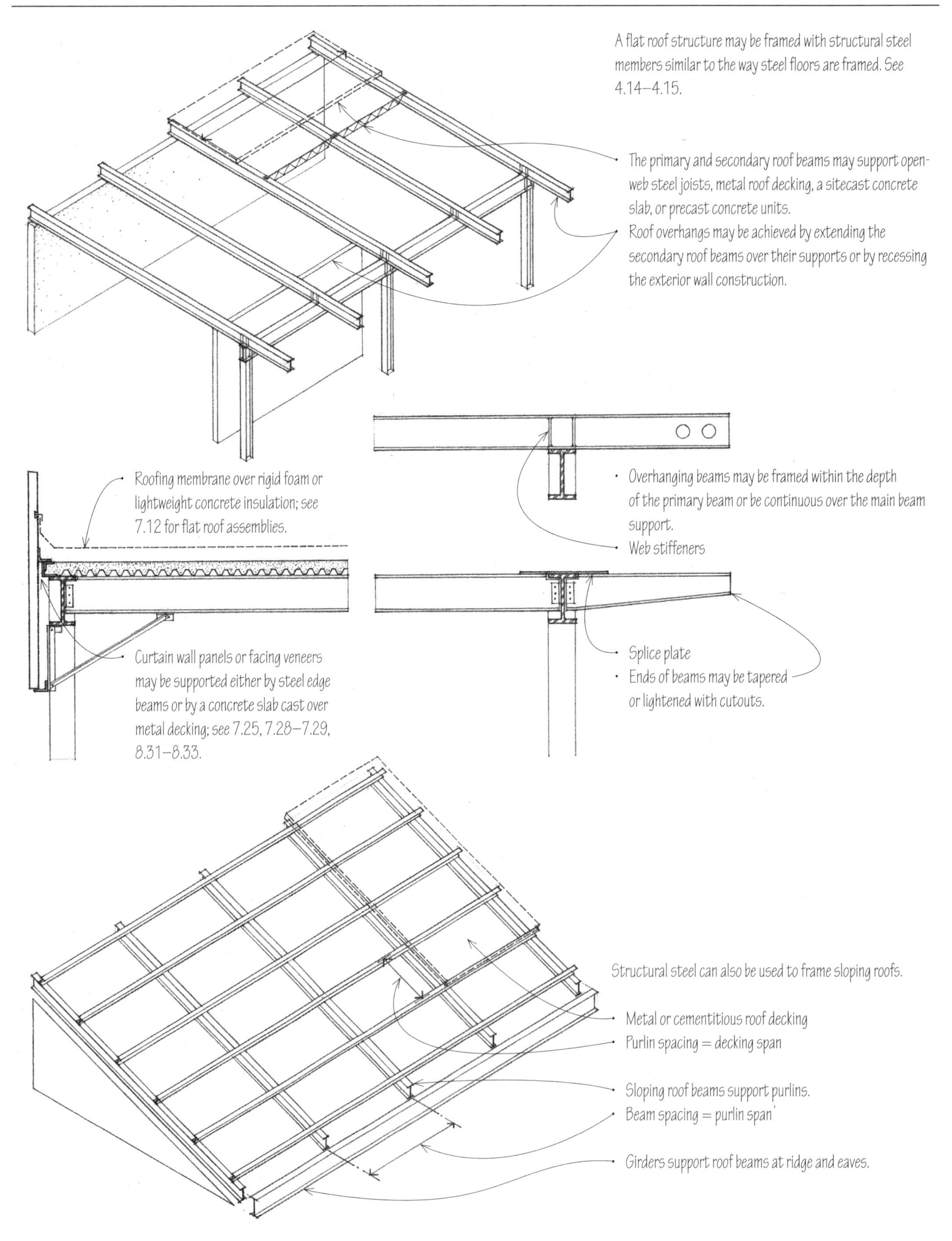
A flat roof structure may be framed with structural steel members similar to the way steel floors are framed. See 4.14–4.15.
The primary and secondary roof beams may support open-web steel joists, metal roof decking, a sitecast concrete slab, or precast concrete units.
Roof overhangs may be achieved by extending the secondary roof beams over their supports or by recessing the exterior wall construction.
Roofing membrane over rigid foam or lightweight concrete insulation; see 7.12 for flat roof assemblies.
Curtain wall panels or facing veneers may be supported either by steel edge beams or by a concrete slab cast over metal decking; see 7.25, 7.28–7.29, 8.31–8.33.
Overhanging beams may be framed within the depth of the primary beam or be continuous over the main beam support.
Web stiffeners
Splice plate
Ends of beams may be tapered or lightened with cutouts.
Structural steel can also be used to frame sloping roofs.
Metal or cementitious roof decking
Purlin spacing = decking span
Sloping roof beams support purlins.
Beam spacing = purlin span
Girders support roof beams at ridge and eaves.

Rigid frames consist of two columns and a beam or girder that are rigidly connected at their joints. Applied loads produce axial, bending, and shear forces in all members of the frame since the rigid joints restrain the ends of the members from rotating freely. In addition, vertical loads cause a rigid frame to develop horizontal thrusts at its base. A rigid frame is statically indeterminate and rigid only in its plane.

- Various shapes of rigid frames can be fabricated of steel to span from 30' to 120' (9 to 36 m).
- Rigid frames typically form one-story structures used for light-industrial buildings, warehouses, and recreational facilities.

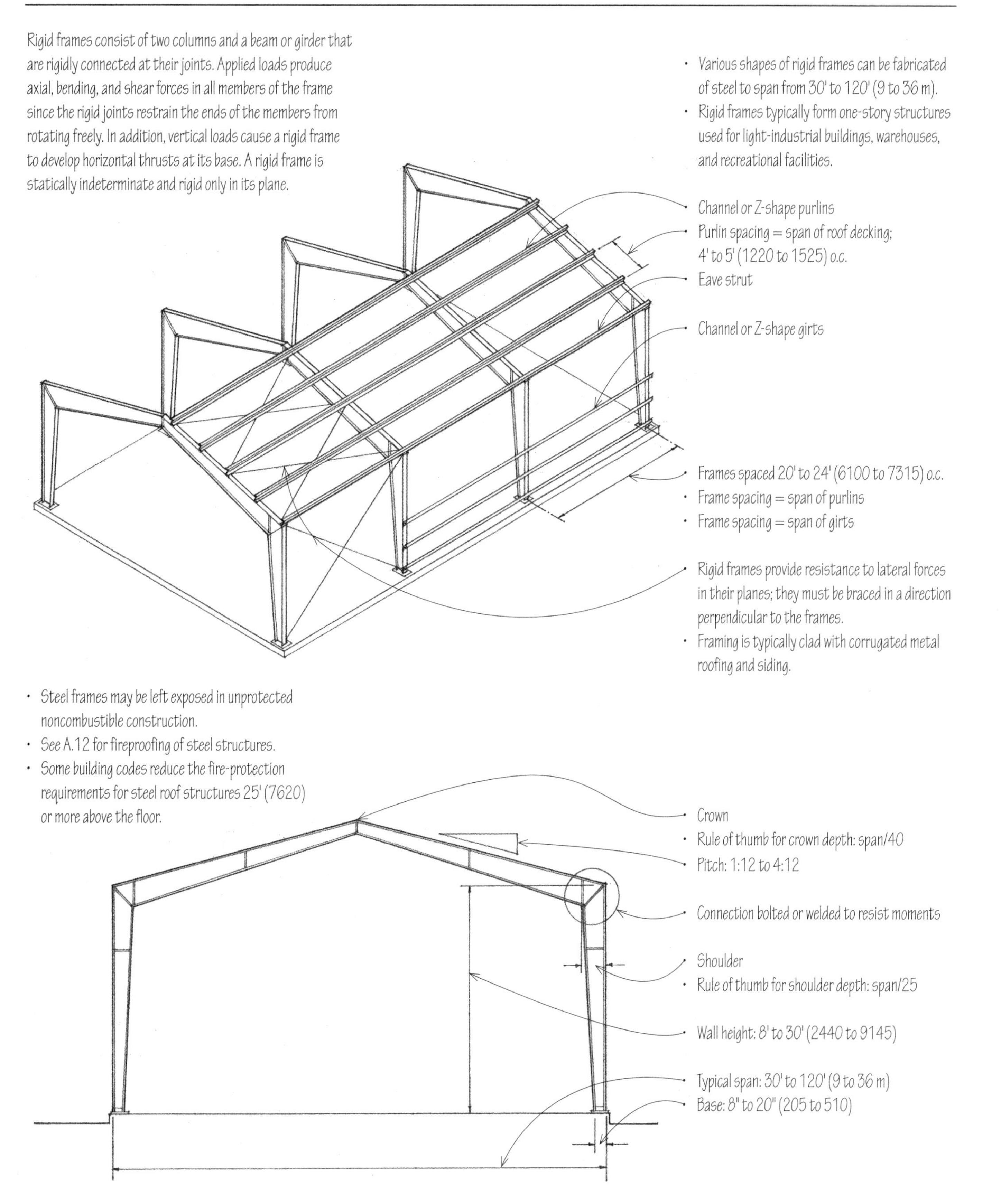

- Steel frames may be left exposed in unprotected noncombustible construction.
- See A.12 for fireproofing of steel structures.
- Some building codes reduce the fire-protection requirements for steel roof structures 25' (7620) or more above the floor.

6.08 STEEL TRUSSES

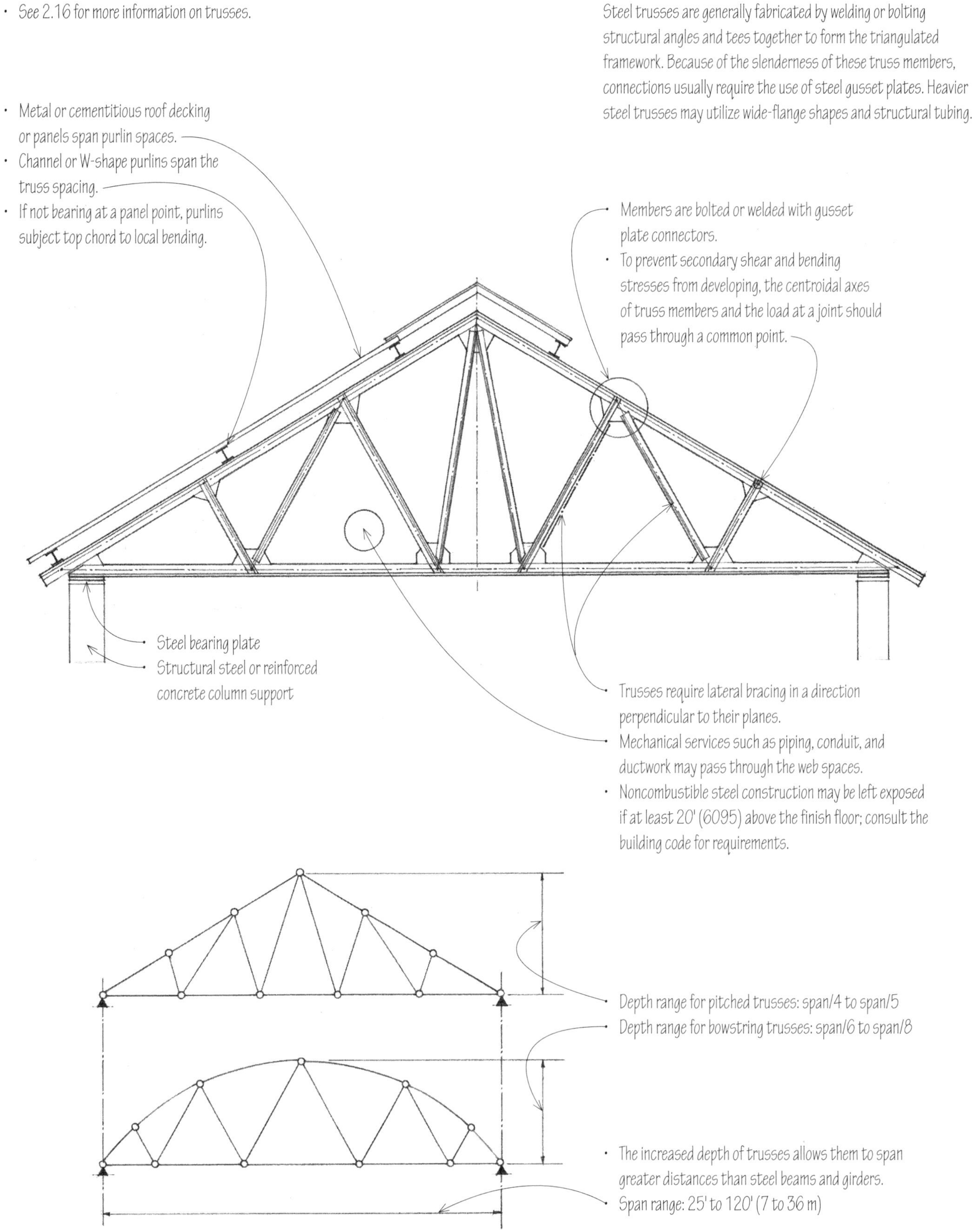

CSI MasterFormat 05 12 00: Structural Steel Framing

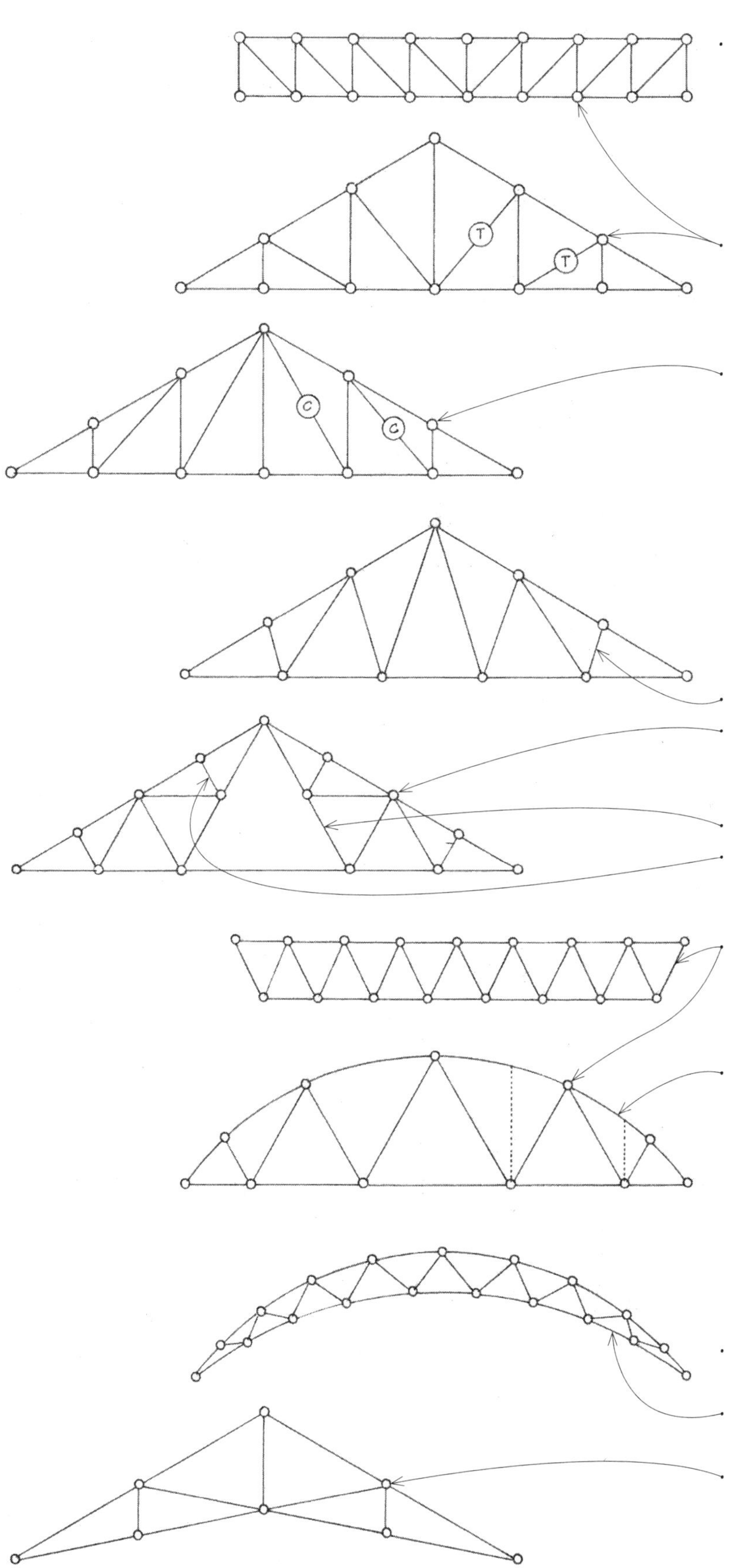

- Flat trusses have parallel top and bottom chords. Flat trusses are generally not as efficient as pitched or bowstring trusses.
- Pratt trusses have vertical web members in compression and diagonal web members in tension. It is generally more efficient to use a truss type in which the longer web members are loaded in tension.
- Howe trusses have vertical web members in tension and diagonal web members in compression.
- Belgian trusses have only inclined web members.
- Fink trusses are Belgian trusses having subdiagonals to reduce the length of compression web members toward the centerline of the span.
- Diagonals connect a top to a bottom chord.
- Subdiagonals join a chord with a main diagonal.
- Warren trusses have inclined web members forming a series of equilateral triangles. Vertical web members are sometimes introduced to reduce the panel lengths of the top chord, which is in compression.
- Bowstring trusses have a curved top chord meeting a straight bottom chord at each end.
- Raised-chord trusses have a bottom chord raised substantially above the level of the supports.
- Crescent trusses have both top and bottom chords curving upward from a common point at each side.
- Scissors trusses have tension members extending from the foot of each top chord to an intermediate point on the opposite top chord.

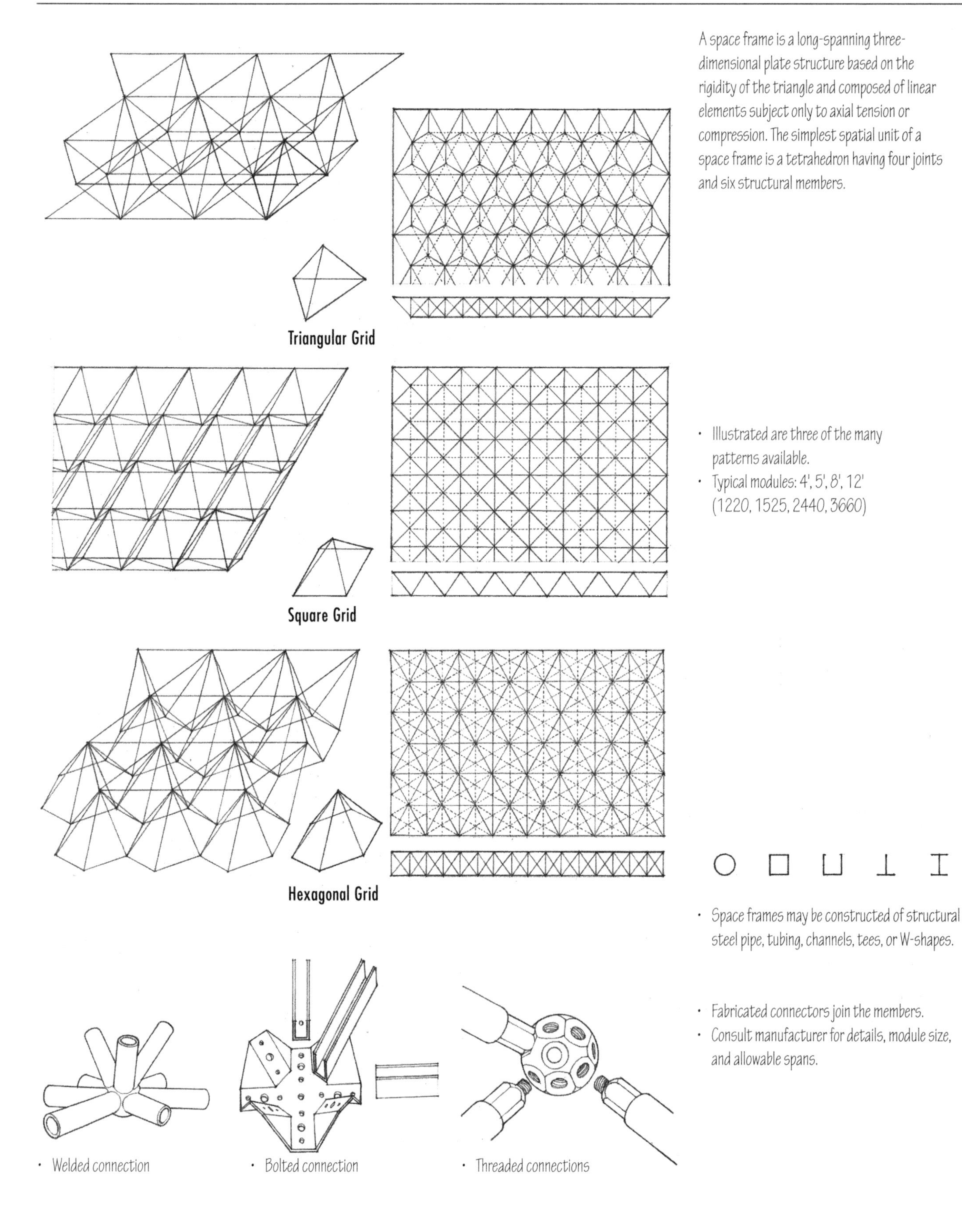

A space frame is a long-spanning three-dimensional plate structure based on the rigidity of the triangle and composed of linear elements subject only to axial tension or compression. The simplest spatial unit of a space frame is a tetrahedron having four joints and six structural members.

Triangular Grid

Square Grid

Hexagonal Grid

- Illustrated are three of the many patterns available.
- Typical modules: 4', 5', 8', 12' (1220, 1525, 2440, 3660)

- Space frames may be constructed of structural steel pipe, tubing, channels, tees, or W-shapes.

- Fabricated connectors join the members.
- Consult manufacturer for details, module size, and allowable spans.

- Welded connection
- Bolted connection
- Threaded connections

- As with other constant-depth plate structures, the supporting bay for a space frame should be square or nearly square to ensure that it acts as a two-way structure.

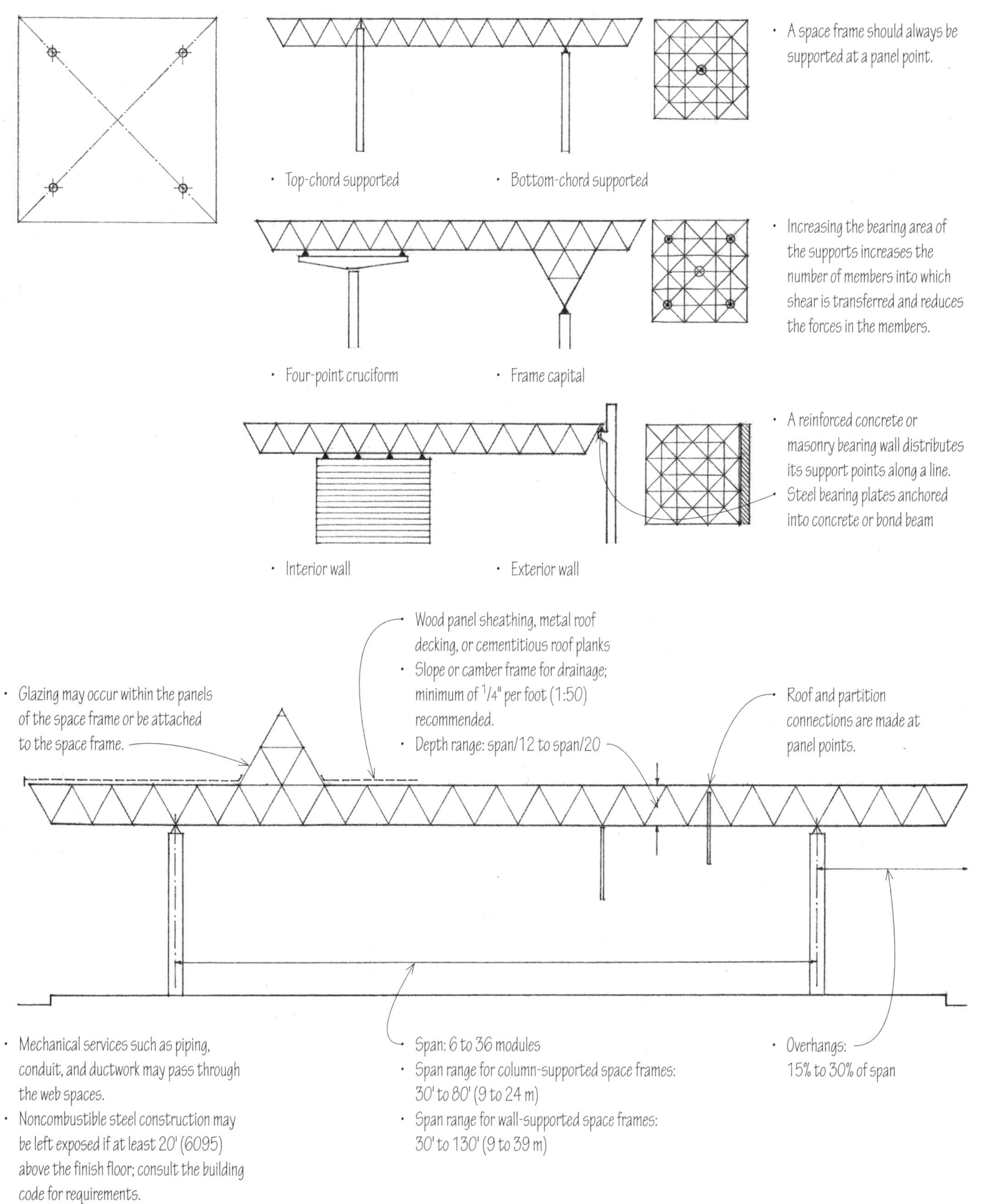

6.12 OPEN-WEB STEEL JOISTS

Roof systems using open-web steel joists are similar in layout and construction to steel joist floor systems. For joist sizes and span ranges, refer to 4.19–4.21.

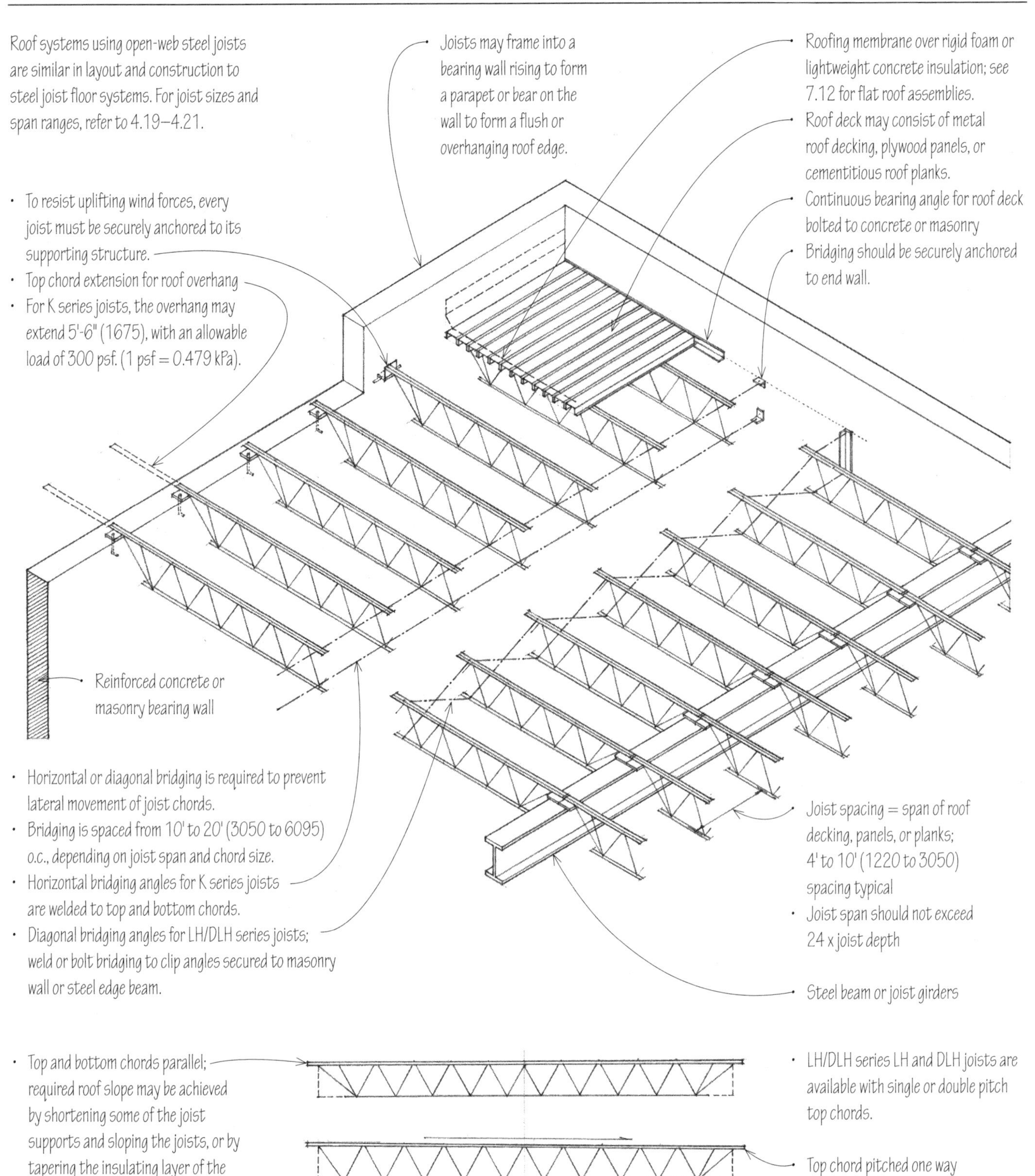

• Top and bottom chords parallel; required roof slope may be achieved by shortening some of the joist supports and sloping the joists, or by tapering the insulating layer of the roof deck.

• LH/DLH series LH and DLH joists are available with single or double pitch top chords.

Top chord pitched one way

Top chord pitched two ways

• Standard slope is 1/8" per foot (1:100).

CSI MasterFormat 05 21 00: Steel Joist Framing

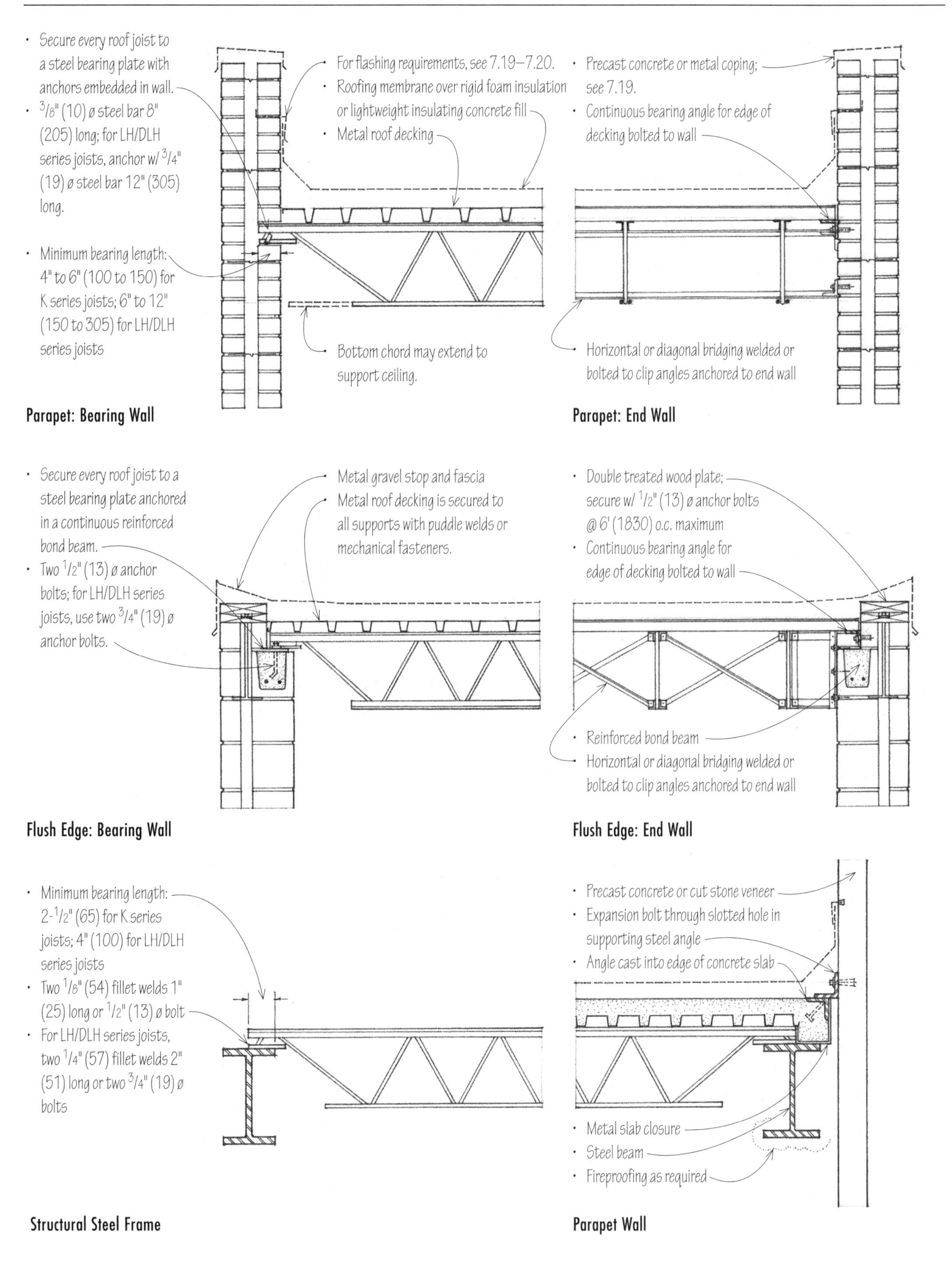

• Secure every roof joist to a steel bearing plate with anchors embedded in wall.
• 3/8" (10) ø steel bar 8" (205) long; for LH/DLH series joists, anchor w/ 3/4" (19) ø steel bar 12" (305) long.
• Minimum bearing length: 4" to 6" (100 to 150) for K series joists; 6" to 12" (150 to 305) for LH/DLH series joists
• For flashing requirements, see 7.19–7.20.
• Roofing membrane over rigid foam insulation or lightweight insulating concrete fill
• Metal roof decking
• Bottom chord may extend to support ceiling.
Parapet: Bearing Wall
• Precast concrete or metal coping; see 7.19.
• Continuous bearing angle for edge of decking bolted to wall
• Horizontal or diagonal bridging welded or bolted to clip angles anchored to end wall
Parapet: End Wall
• Secure every roof joist to a steel bearing plate anchored in a continuous reinforced bond beam.
• Two 1/2" (13) ø anchor bolts; for LH/DLH series joists, use two 3/4" (19) ø anchor bolts.
• Metal gravel stop and fascia
• Metal roof decking is secured to all supports with puddle welds or mechanical fasteners.
Flush Edge: Bearing Wall
• Double treated wood plate; secure w/ 1/2" (13) ø anchor bolts @ 6' (1830) o.c. maximum
• Continuous bearing angle for edge of decking bolted to wall
• Reinforced bond beam
• Horizontal or diagonal bridging welded or bolted to clip angles anchored to end wall
Flush Edge: End Wall
• Minimum bearing length: 2-1/2" (65) for K series joists; 4" (100) for LH/DLH series joists
• Two 1/8" (54) fillet welds 1" (25) long or 1/2" (13) ø bolt
• For LH/DLH series joists, two 1/4" (57) fillet welds 2" (51) long or two 3/4" (19) ø bolts
Structural Steel Frame
• Precast concrete or cut stone veneer
• Expansion bolt through slotted hole in supporting steel angle
• Angle cast into edge of concrete slab
• Metal slab closure
• Steel beam
• Fireproofing as required
Parapet Wall

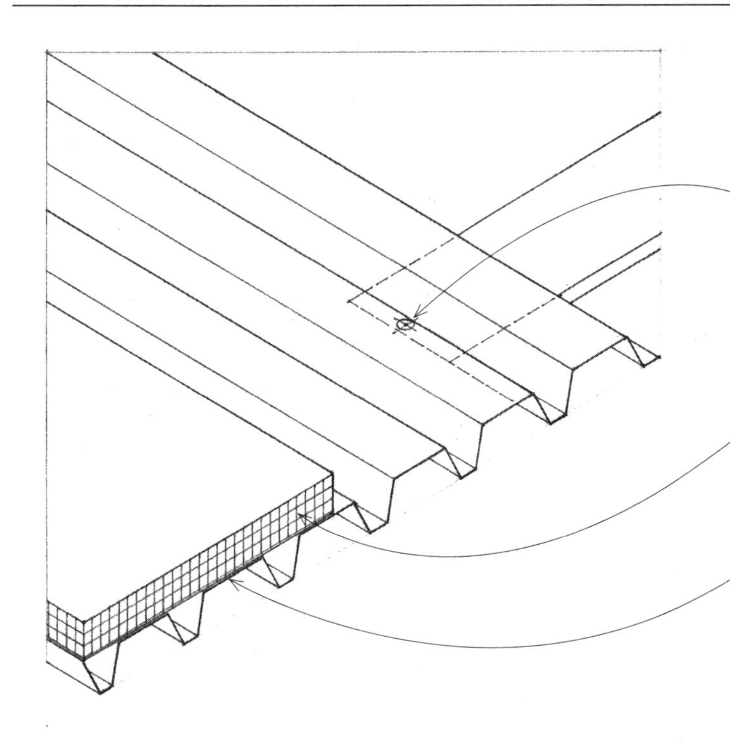

Metal roof decking is corrugated to increase its stiffness and ability to span across open-web steel joists or more widely spaced steel beams and to serve as a base for thermal insulation and membrane roofing.

- The decking panels are puddle-welded or mechanically fastened to the supporting steel joists or beams.
- The panels are fastened to each other along their sides with screws, welds, or button-punching standing seams.
- If the deck is to serve as a structural diaphragm and transfer lateral loads to shear walls, its entire perimeter must be welded to steel supports. In addition, more stringent requirements for support and side lap fastening may apply.
- Metal roof decking is commonly used without a concrete topping, requiring structural wood or cementitious panels or rigid foam insulation panels to bridge the gaps in the corrugation and provide a smooth, firm surface for the thermal insulation and membrane roofing.
- To provide maximum surface area for the effective adhesion of rigid foam insulation, the top flange should be wide and flat. If the decking has stiffening grooves, the insulation layer may have to be mechanically fastened.
- Metal decking has low-vapor permeance but because of the many discontinuities between the panels, it is not airtight. If an air barrier is required to prevent the migration of moisture vapor into the roofing assembly, a concrete topping can be used. When a lightweight insulating concrete fill is used, the decking may have perforated vents for the release of latent moisture and vapor pressure.

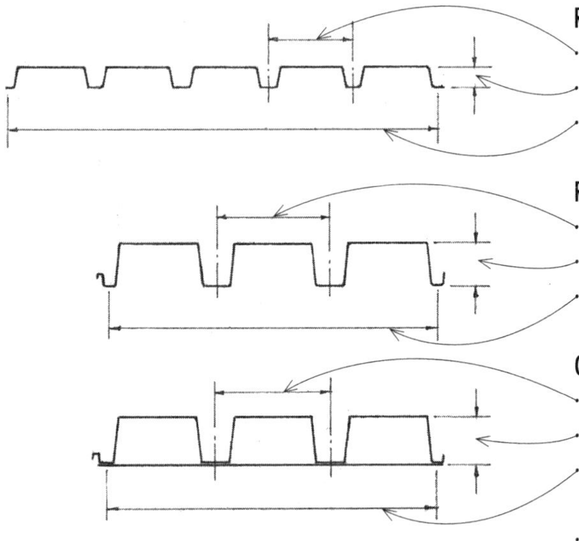

Ribbed Roof Decking

- 6" (150)
- 1-$\frac{1}{2}$" (38) depth spans 4' to 9' (1220 to 2745)
- 24", 30", 36" (610, 760, 915)

Ribbed Roof Decking

- 8" (205)
- 3" and 4-$\frac{1}{2}$" (75 and 115) depths span 8' to 16' (2440 to 4875)
- 12", 24" (305, 610)

Cellular Roof Decking

- 6" (150)
- 3" (75) depth spans 10' to 20' (3050 to 6095)
- 24" (610)

- Acoustic roof decking used as a sound-absorbing ceiling contains glass fiber between the perforated webs of ribbed decking or in the perforated cells of cellular decking.
- Decking profiles vary. Consult manufacturer for available profiles, lengths, gauges, allowable spans, and installation details.

Cementitious roof planks are manufactured with portland cement, lightweight aggregate, an aerating compound, and galvanized welded wire fabric reinforcement.

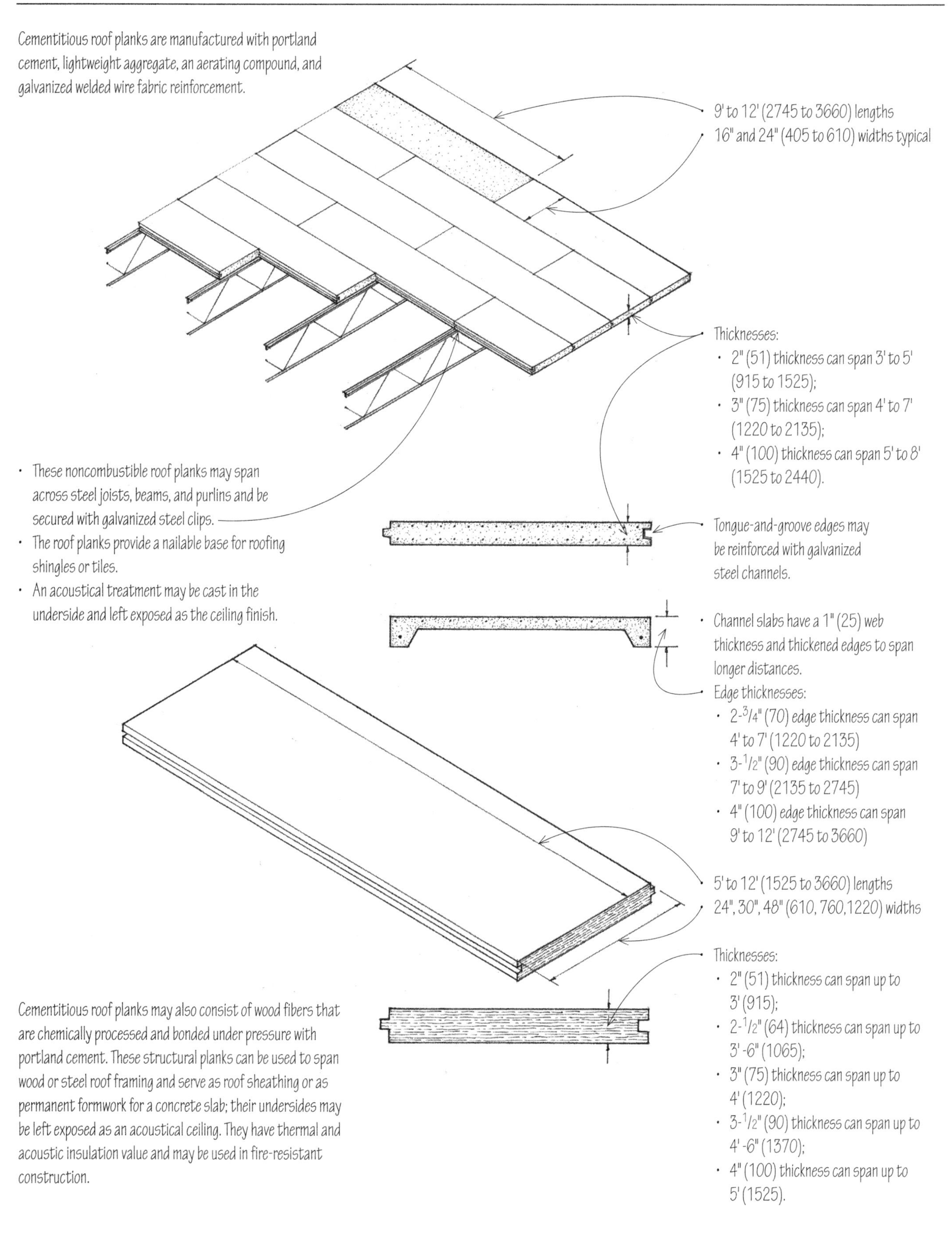

- These noncombustible roof planks may span across steel joists, beams, and purlins and be secured with galvanized steel clips.
- The roof planks provide a nailable base for roofing shingles or tiles.
- An acoustical treatment may be cast in the underside and left exposed as the ceiling finish.

Cementitious roof planks may also consist of wood fibers that are chemically processed and bonded under pressure with portland cement. These structural planks can be used to span wood or steel roof framing and serve as roof sheathing or as permanent formwork for a concrete slab; their undersides may be left exposed as an acoustical ceiling. They have thermal and acoustic insulation value and may be used in fire-resistant construction.

CSI MasterFormat 03 51 13: Cementitious Wood Fiber Decks

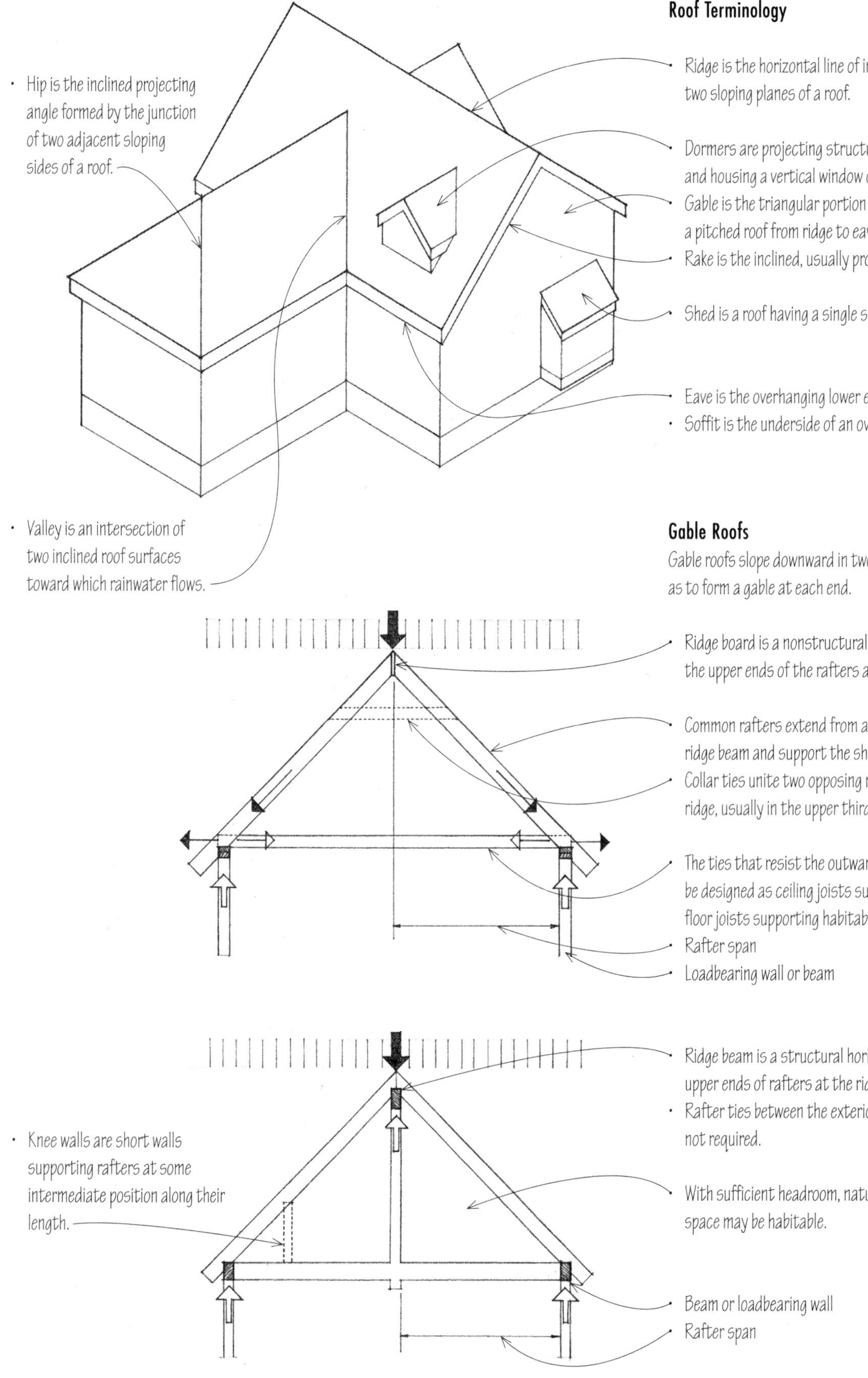

Roof Terminology

- Ridge is the horizontal line of intersection at the top between two sloping planes of a roof.
- Dormers are projecting structures built out from a sloping roof and housing a vertical window or ventilating louver.
- Gable is the triangular portion of wall enclosing the end of a pitched roof from ridge to eaves.
- Rake is the inclined, usually projecting edge of a sloping roof.
- Shed is a roof having a single slope.
- Eave is the overhanging lower edge of a roof.
- Soffit is the underside of an overhanging roof eave.

Gable Roofs

Gable roofs slope downward in two parts from a central ridge, so as to form a gable at each end.

- Ridge board is a nonstructural horizontal member to which the upper ends of the rafters are aligned and fastened.
- Common rafters extend from a wall plate to a ridge board or ridge beam and support the sheathing and covering of a roof.
- Collar ties unite two opposing rafters at a point below the ridge, usually in the upper third of the rafter length.
- The ties that resist the outward thrust of the rafters may be designed as ceiling joists supporting only attic loads or as floor joists supporting habitable space.
- Rafter span
- Loadbearing wall or beam

- Ridge beam is a structural horizontal member supporting the upper ends of rafters at the ridge of a roof.
- Rafter ties between the exterior wall or beam supports are not required.
- With sufficient headroom, natural light, and ventilation, attic space may be habitable.
- Beam or loadbearing wall
- Rafter span

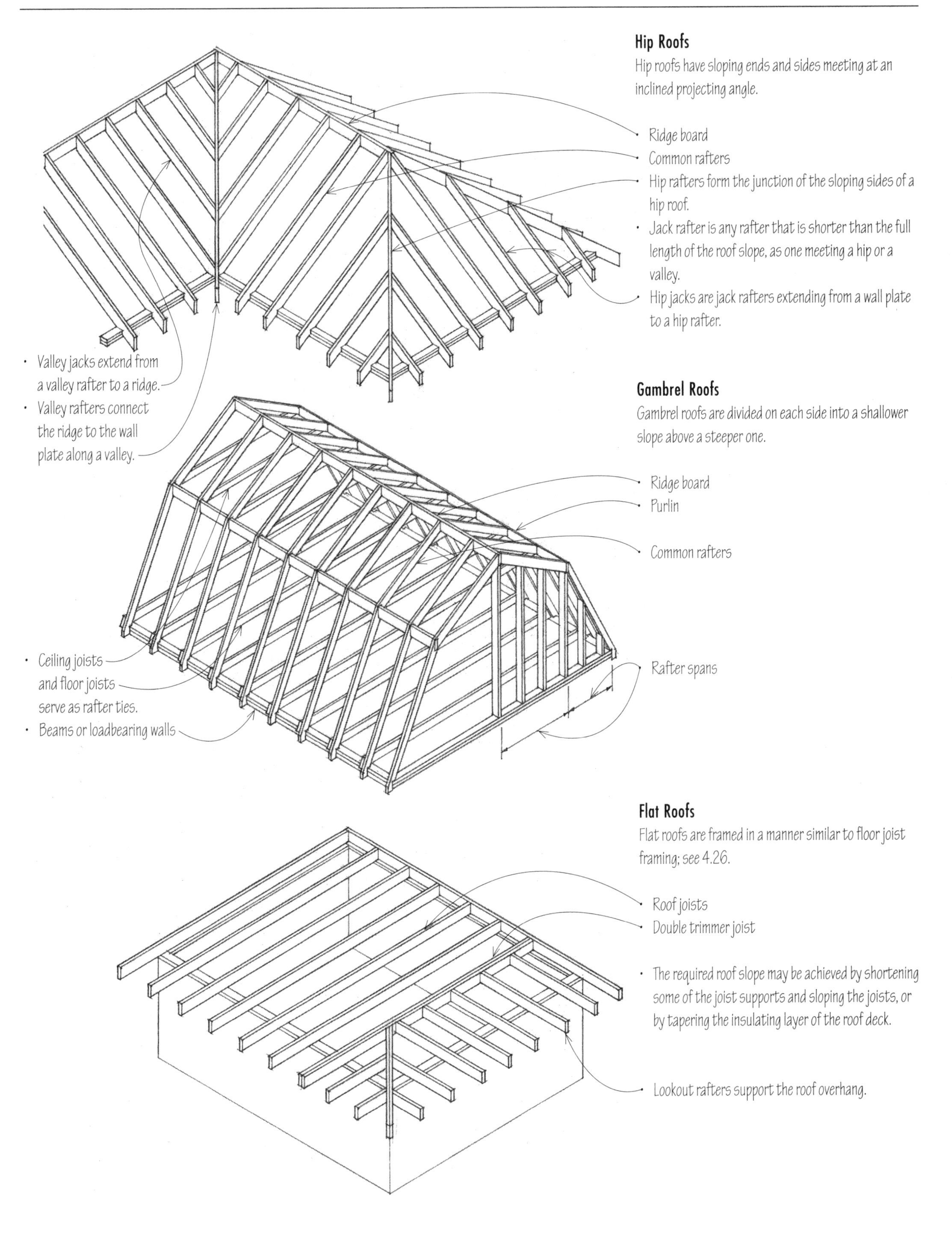
Hip Roofs
Hip roofs have sloping ends and sides meeting at an inclined projecting angle.
Ridge board
Common rafters
Hip rafters form the junction of the sloping sides of a hip roof.
Jack rafter is any rafter that is shorter than the full length of the roof slope, as one meeting a hip or a valley.
Hip jacks are jack rafters extending from a wall plate to a hip rafter.
Valley jacks extend from a valley rafter to a ridge.
Valley rafters connect the ridge to the wall plate along a valley.
Gambrel Roofs
Gambrel roofs are divided on each side into a shallower slope above a steeper one.
Ridge board
Purlin
Common rafters
Rafter spans
Ceiling joists and floor joists serve as rafter ties.
Beams or loadbearing walls
Flat Roofs
Flat roofs are framed in a manner similar to floor joist framing; see 4.26.
Roof joists
Double trimmer joist
The required roof slope may be achieved by shortening some of the joist supports and sloping the joists, or by tapering the insulating layer of the roof deck.
Lookout rafters support the roof overhang.

Roofs and ceilings may be constructed with light-gauge steel members in a manner similar to wood light-frame construction; see 6.19–6.22. The light-gauge steel members may also be screwed or welded together to form roof trusses similar to those described in 6.29.

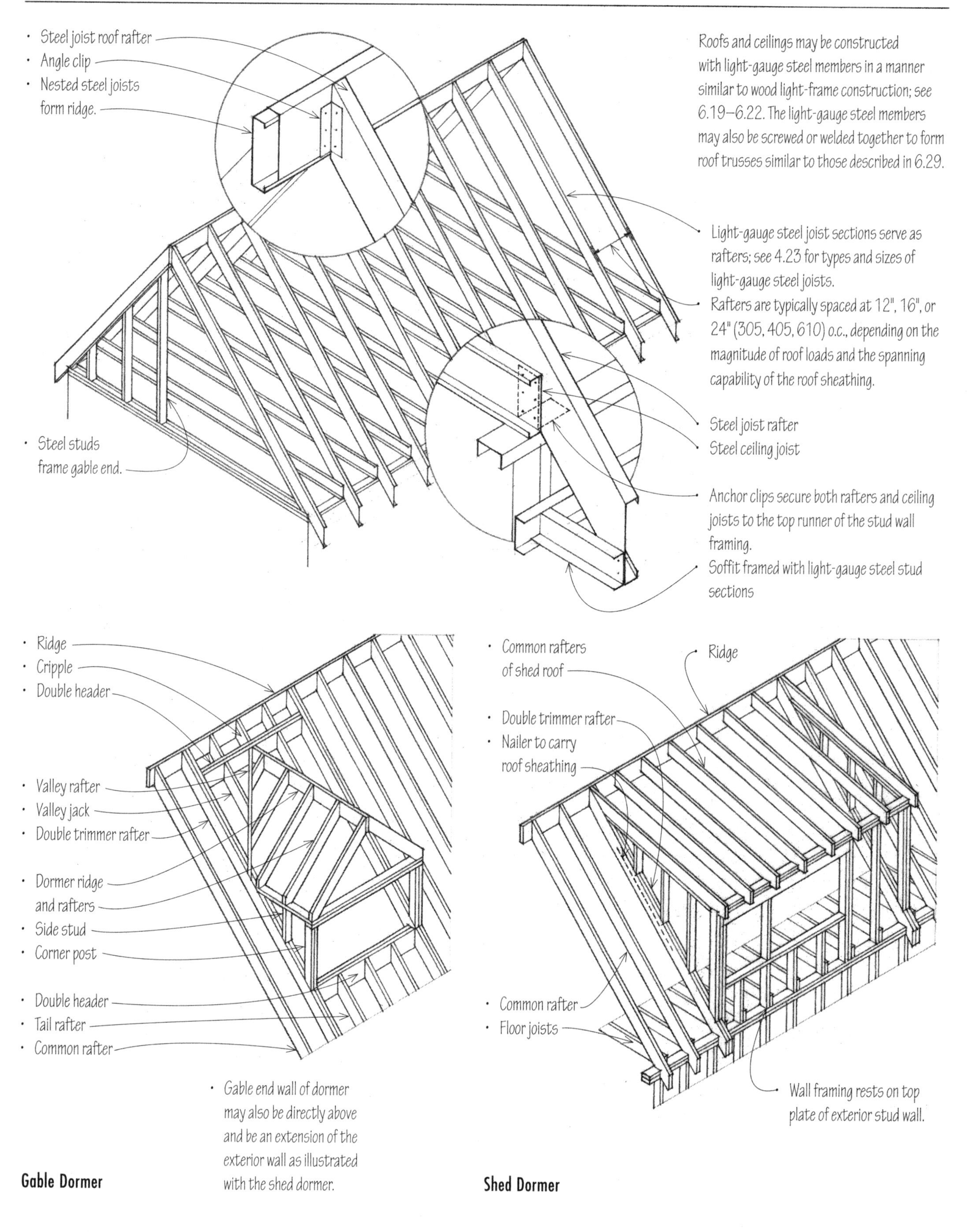

Gable Dormer

Shed Dormer

CSI MasterFormat 05 40 00: Cold-Formed Metal Framing

• See 6.16–6.17 for light-frame roof forms and terminology.

Roof structures framed with wood rafters are an essential subsystem of wood light-frame construction. The dimension lumber used for roof joists and rafters is easily worked and can be quickly assembled on site with simple tools.

- Ridge members supporting rafters having a slope of less than 3:12 must be designed as beams.
- Rake overhangs are constructed with lookouts framed into a double common rafter and bearing on the top plate of the gable end wall.
- Barge or fly rafters are the end rafters in the part of a gable roof that projects beyond the gable wall.
- Roof openings are framed in a manner similar to floor joist openings; see 4.31.
- Double header
- Double rafters for large openings 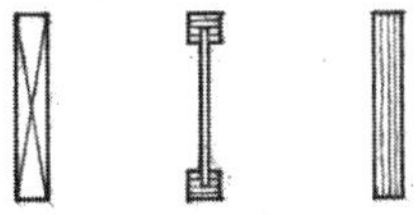
- Sloping rafters and flat roof joists are typically of solid-sawn 2x lumber, but I-joists and laminated veneer lumber may also be used.
- Rafters and roof joists are typically spaced at 12", 16", or 24" (305, 405, 610) o.c., depending on the magnitude of roof loads and the spanning capability of the roof sheathing.

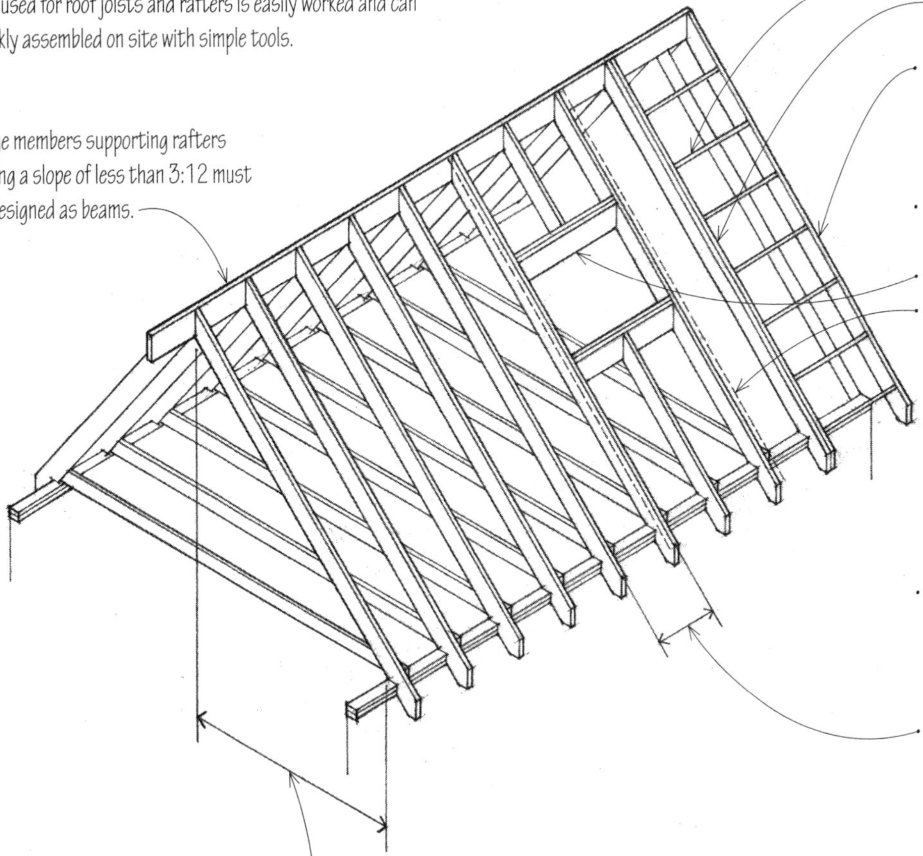

Rafter span ranges:

- 2x6 can span up to 10' (3050);
- 2x8 can span up to 14' (4265);
- 2x10 can span up to 16' (4875);
- 2x12 can span up to 22' (6705).
- Rafter spans are related to the magnitude of applied loads, the rafter size and spacing, and the species and grade of lumber used.
- Rafters may be oversized to accommodate the required thermal insulation and provide space for ventilating the concealed roof spaces.
- Consult manufacturer for sizes and spans of laminated veneer lumber joists.

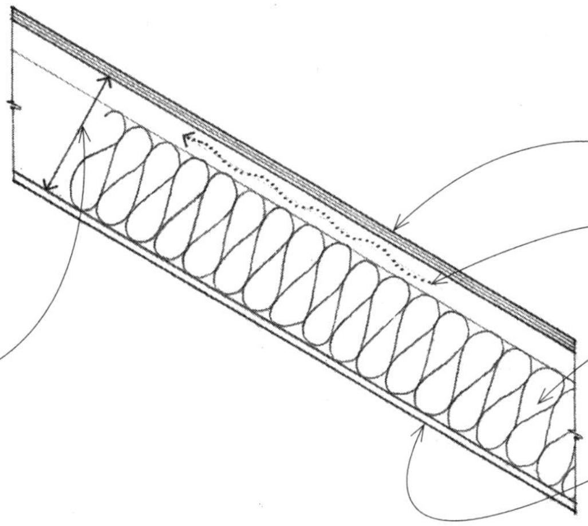

- Because wood light-framing is combustible, it must rely on roofing and ceiling materials for its fire-resistance rating.
- Roof sheathing; see 6.23.
- The susceptibility of wood light-framing to decay requires ventilation to control condensation in enclosed roof spaces.
- See 7.43 for thermal insulation of roofs.
- A ceiling finish is usually applied directly to the underside of roof rafters or ceiling joists.
- If ceiling joists are used, attic space may accommodate mechanical equipment.

CSI MasterFormat 06 10 00: Rough Carpentry
CSI MasterFormat 06 11 00: Wood Framing

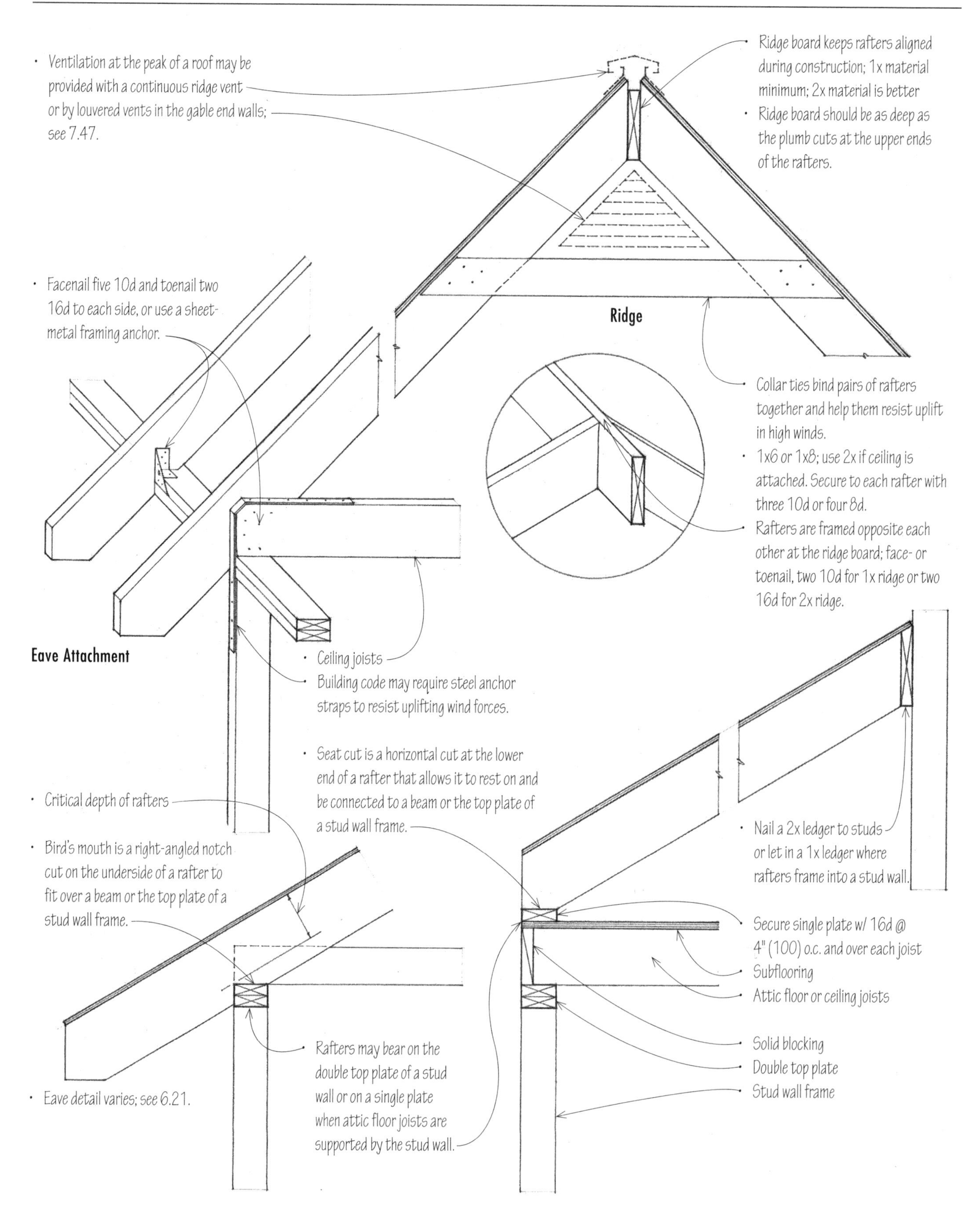

• Ventilation at the peak of a roof may be provided with a continuous ridge vent or by louvered vents in the gable end walls; see 7.47.
• Ridge board keeps rafters aligned during construction; 1x material minimum; 2x material is better
• Ridge board should be as deep as the plumb cuts at the upper ends of the rafters.
• Facenail five 10d and toenail two 16d to each side, or use a sheet-metal framing anchor.
Ridge
• Collar ties bind pairs of rafters together and help them resist uplift in high winds.
• 1x6 or 1x8; use 2x if ceiling is attached. Secure to each rafter with three 10d or four 8d.
• Rafters are framed opposite each other at the ridge board; face- or toenail, two 10d for 1x ridge or two 16d for 2x ridge.
Eave Attachment
• Ceiling joists
• Building code may require steel anchor straps to resist uplifting wind forces.
• Seat cut is a horizontal cut at the lower end of a rafter that allows it to rest on and be connected to a beam or the top plate of a stud wall frame.
• Critical depth of rafters
• Bird's mouth is a right-angled notch cut on the underside of a rafter to fit over a beam or the top plate of a stud wall frame.
• Nail a 2x ledger to studs or let in a 1x ledger where rafters frame into a stud wall.
• Secure single plate w/ 16d @ 4" (100) o.c. and over each joist
• Subflooring
• Attic floor or ceiling joists
• Solid blocking
• Double top plate
• Stud wall frame
• Rafters may bear on the double top plate of a stud wall or on a single plate when attic floor joists are supported by the stud wall.
• Eave detail varies; see 6.21.
Eave Support Conditions

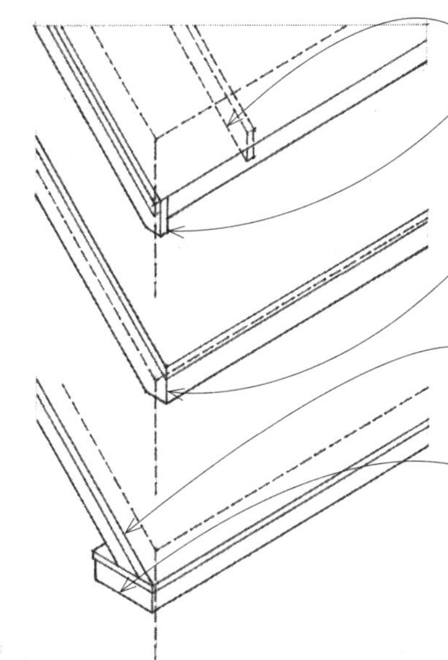

- Exposed rafter tails or sloping soffit
- Rake trim and bargeboard may extend beyond eave fascia to terminate the end of the eave fascia and gutter.
- Closed rake with a narrow eave soffit
- Rake trim and bargeboard may be terminated by a cornice return.
- A cornice return extends the eave fascia and soffit around the corner and turns into the gable end wall.

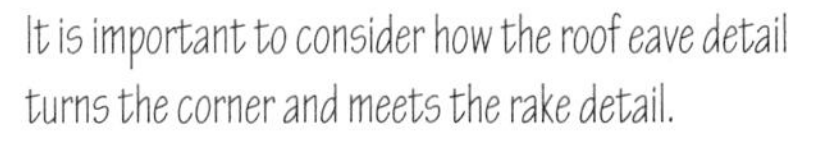

It is important to consider how the roof eave detail turns the corner and meets the rake detail.

- Roof sheathing
- Blocking w/ screened vent holes
- Exposed rafters
- Rafter tails may extend and be cut in an ornamental manner.
- Fascia is the broad, flat surface of the outer edge of a roof.
- Header

Exposed Rafters

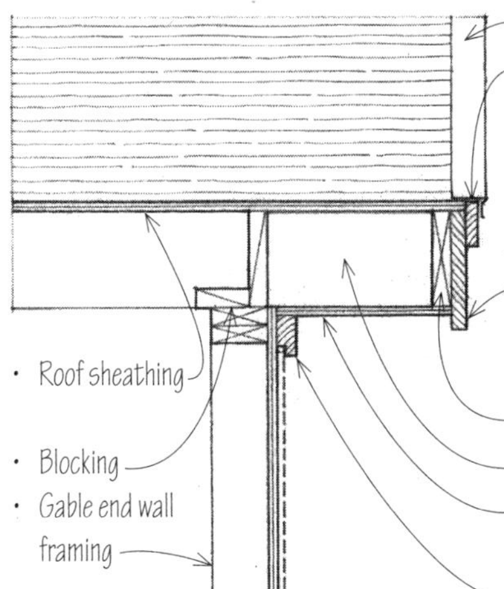

- Metal drip edge
- Rake trim
- Coordinate rake trim and bargeboard with eave fascia and gutter detail.
- Bargeboard extended to form drip; sometimes carved for ornamental effect.
- Fly rafter
- Lookout rafter
- Soffit of plywood or t&g boards
- Frieze board
- Roof sheathing
- Blocking
- Gable end wall framing

Closed Rake

- Roof sheathing
- Common rafters
- Top plate of stud wall frame
- 2x ledger
- Frieze board
- Continuous screened or slotted vent strip
- Fascia
- Header
- Lookout
- Soffit of plywood or t&g boards

Wide Vented Soffit

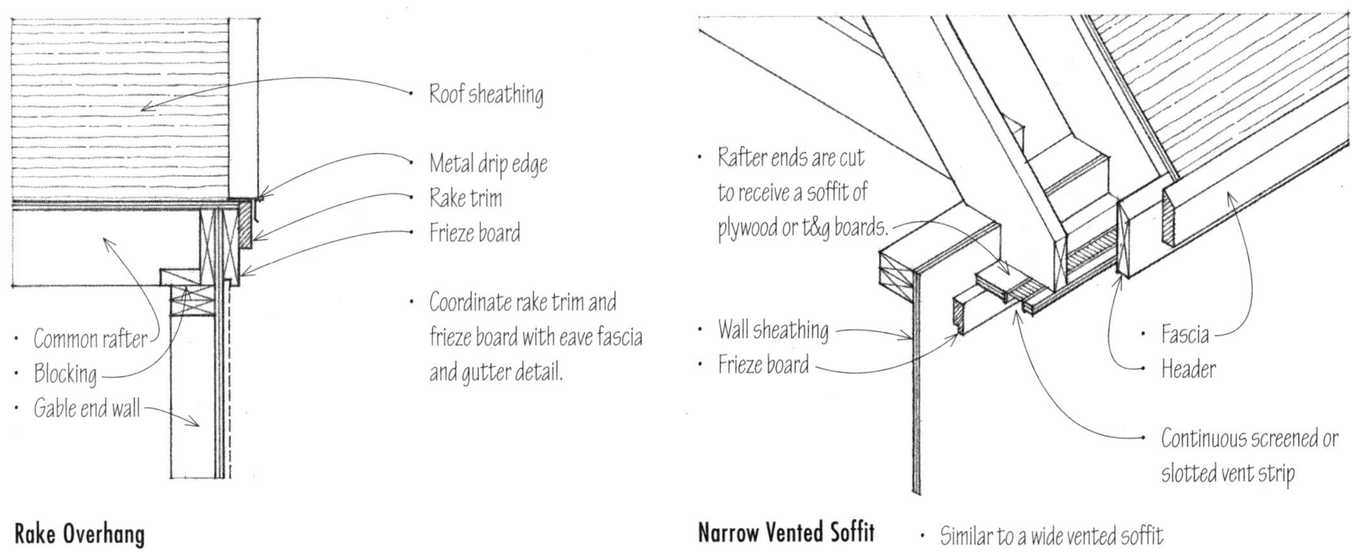

Rake Overhang

Narrow Vented Soffit

- Similar to a wide vented soffit

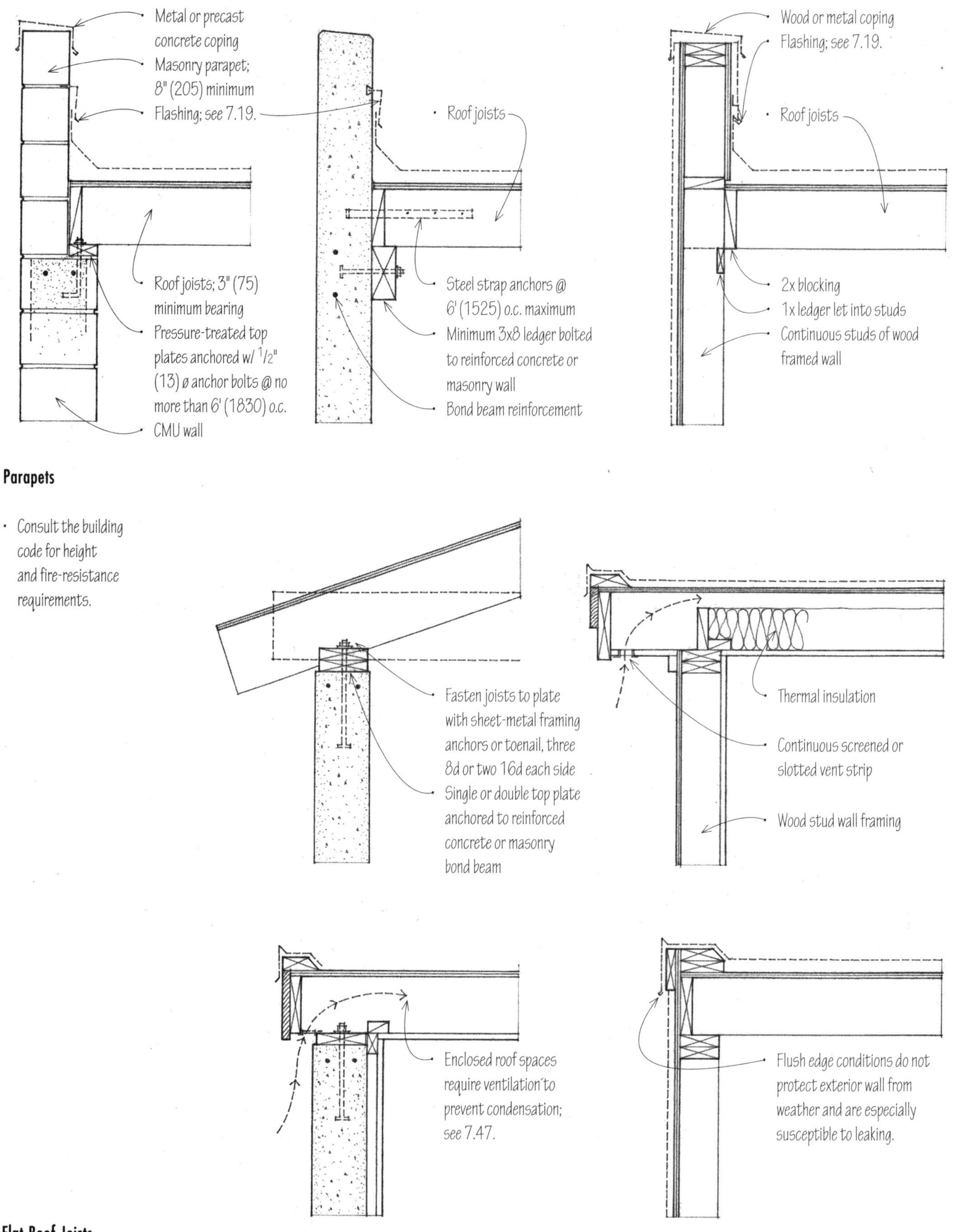

Parapets

- Consult the building code for height and fire-resistance requirements.

Flat Roof Joists

Sheathing over wood or light-gauge metal rafters typically consists of APA-rated plywood or nonveneered wood panels. The panels enhance the stiffness of the rafter framing and provide a solid base for the application of various roofing materials. Sheathing and underlayment requirements should be in accordance with the recommendations of the roofing manufacturer. In damp climates not subject to blizzard conditions, spaced sheathing of 1x4 or 1x6 boards may be used with wood shingle or shake roofing. See 7.04–7.05.

Panel Roof Sheathing

Panel Span Rating	Panel Thickness inch (mm)	Maximum Span in inches (mm)	
		w/ edge support	w/o edge support
12/0	5/16 (8)	12 (305)	
16/0	5/16, 3/8 (8, 10)	16 (405)	
20/0	5/16, 3/8 (8, 10)	20 (510)	
24/0	3/8 (10)	24 (610)	16 (405)
24/0	1/2 (13)	24 (610)	24 (610)
32/16	1/2, 5/8 (13, 16)	32 (815)	28 (710)
40/20	5/8, 3/4, 7/8 (16, 19, 22)	40 (1015)	32 (815)
48/24	3/4, 7/8 (19, 22)	48 (1220)	36 (915)

- The span rating of a panel can be determined from its identifying grade stamp.
- The table above assumes that the panels are laid continuously over two or more spans with their long dimension perpendicular to the supports, and capable of carrying 30 psf live load and 10 psf dead load; 1 psf = 0.479 kPa.

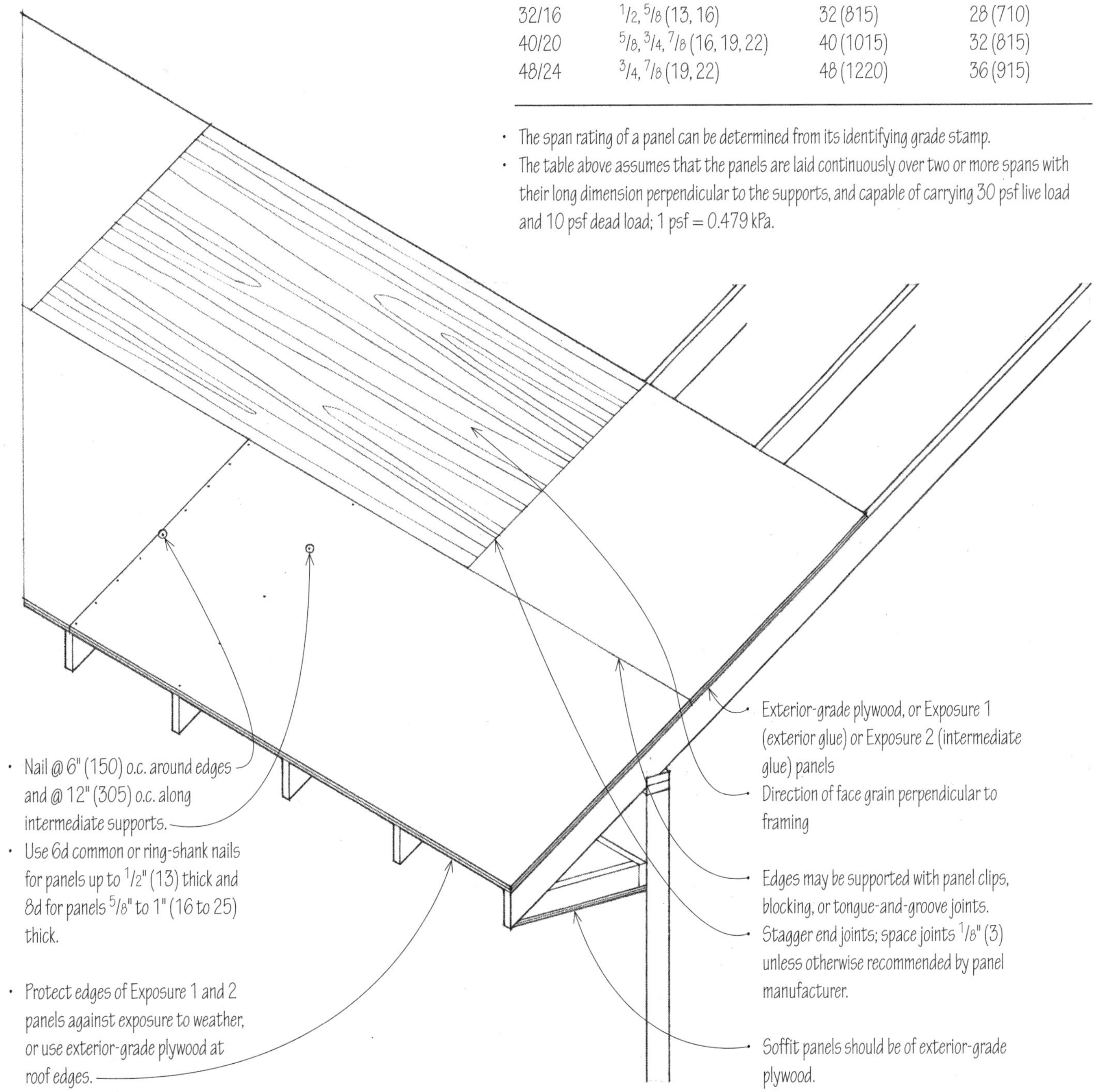

- Nail @ 6" (150) o.c. around edges and @ 12" (305) o.c. along intermediate supports.
- Use 6d common or ring-shank nails for panels up to 1/2" (13) thick and 8d for panels 5/8" to 1" (16 to 25) thick.
- Protect edges of Exposure 1 and 2 panels against exposure to weather, or use exterior-grade plywood at roof edges.

- Exterior-grade plywood, or Exposure 1 (exterior glue) or Exposure 2 (intermediate glue) panels
- Direction of face grain perpendicular to framing
- Edges may be supported with panel clips, blocking, or tongue-and-groove joints.
- Stagger end joints; space joints 1/8" (3) unless otherwise recommended by panel manufacturer.
- Soffit panels should be of exterior-grade plywood.

Wood plank-and-beam roof systems typically use the same supporting grid of posts or columns as do plank-and-beam floor systems. See 4.38 and 5.50.

- Beam spacing = span of wood decking; 4' to 8' (1220 to 2440) typical
- The guidelines in 4.40 may be used to estimate the span of wood decking.
- Underside of decking may be left exposed as a finished ceiling.
- Other options:
  - 2-4-1 plywood
  - Prefabricated composite or stressed-skin panels
  - Cementitious roof planks

- Roof beams may be supported by:
  - Timber, steel, or concrete columns
  - Timber or steel girders
  - A reinforced concrete or masonry bearing wall
- Bearing area should be sufficient to ensure the allowable compressive stresses of the beam and bearing material are not exceeded.

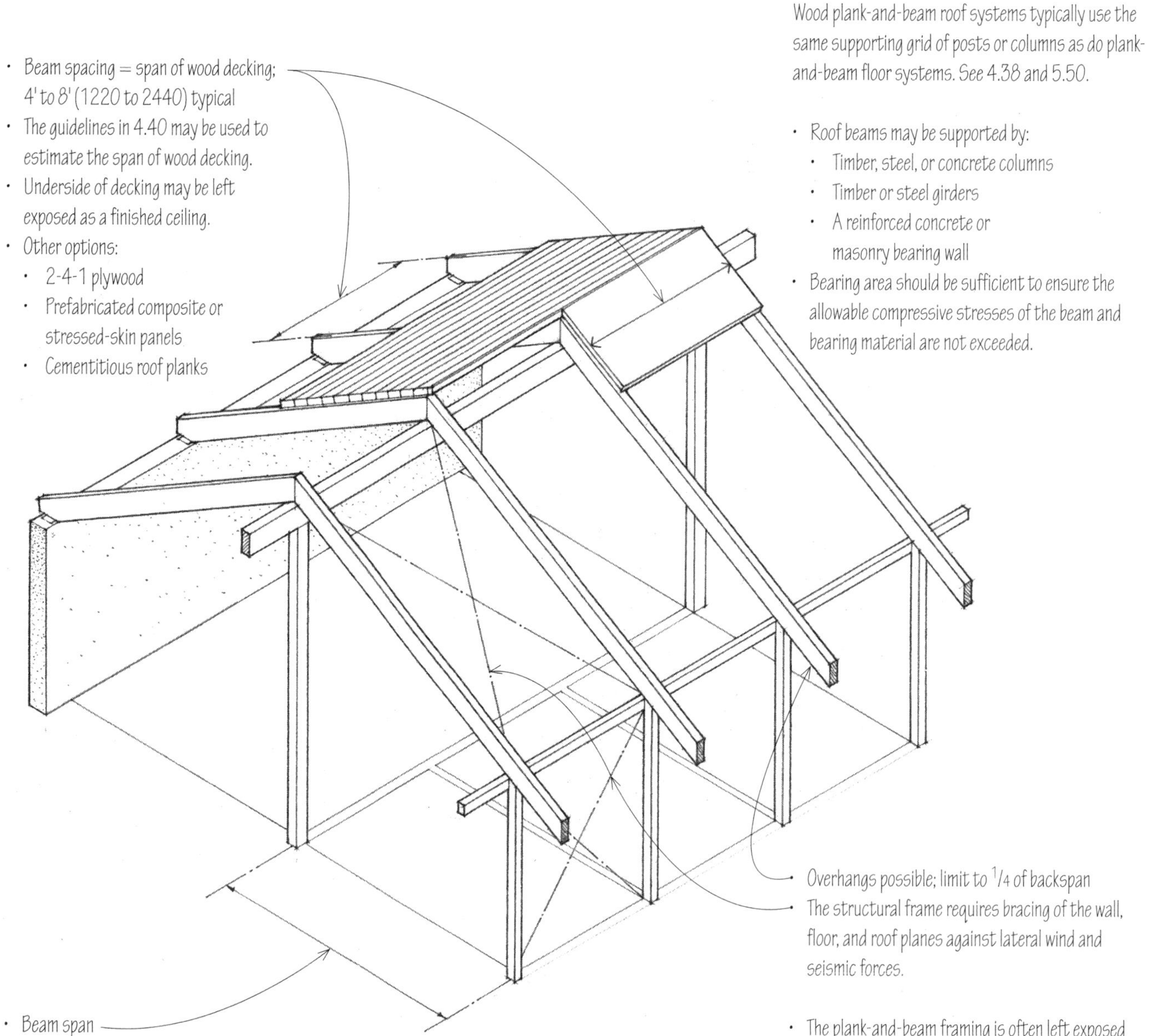

- Overhangs possible; limit to 1/4 of backspan
- The structural frame requires bracing of the wall, floor, and roof planes against lateral wind and seismic forces.

- Beam span
- Rule of thumb for estimating the depth of a beam:
- Solid sawn wood beams: span/15; beam width = 1/3 to 1/2 of beam depth
- Glue-laminated beams: span/20; beam width = 1/4 to 1/3 of beam depth
- The required size of a wood beam is directly related to the magnitude of the roof load, the species and grade of lumber used, and the beam spacing and span.

- The plank-and-beam framing is often left exposed to the interior with rigid thermal insulation being applied over the roof deck and a vapor retarder. Exposed structures require thoughtful detailing of connections, the use of quality materials, and careful workmanship.
- Plank-and-beam framing offers no concealed spaces for overhead ductwork, pipes, or wiring, except when a layered structure or spaced structural members are used.
- Plank-and-beam framing may qualify as heavy timber construction if the structure is supported by noncombustible, fire-resistive exterior walls and the wood members and decking meet the minimum size requirements specified in the building code.

There are alternatives for how a plank-and-beam roof structure can be framed, depending on the direction and spacing of the roof beams, the elements used to span the beam spacing, and the overall depth of the construction assembly.

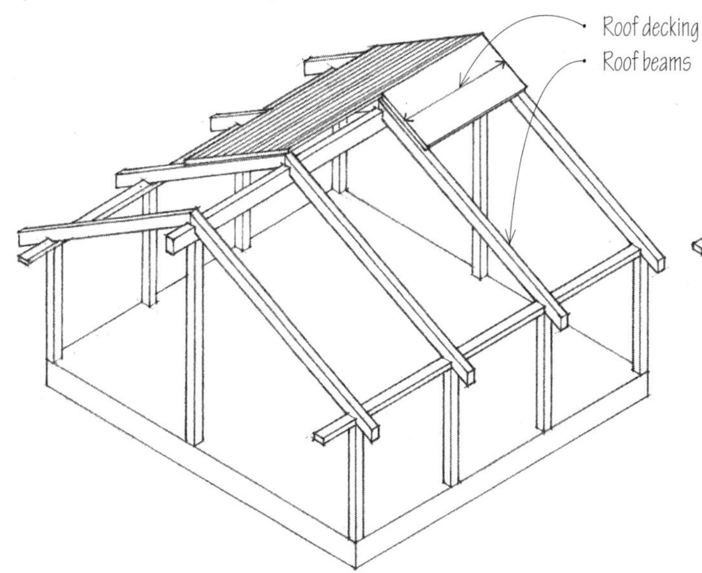

The roof beams may be spaced 4' to 8' (1220 to 2440) o.c. and spanned with solid or glue-laminated wood decking. The beams may be supported by girders, columns, or a reinforced concrete or masonry bearing wall.

In this two-layer system, the roof beams may be spaced farther apart and support a series of purlins. These purlins, in turn, are spanned with wood decking or a rigid, sheet roofing material.

Roof Beams Parallel with Slope

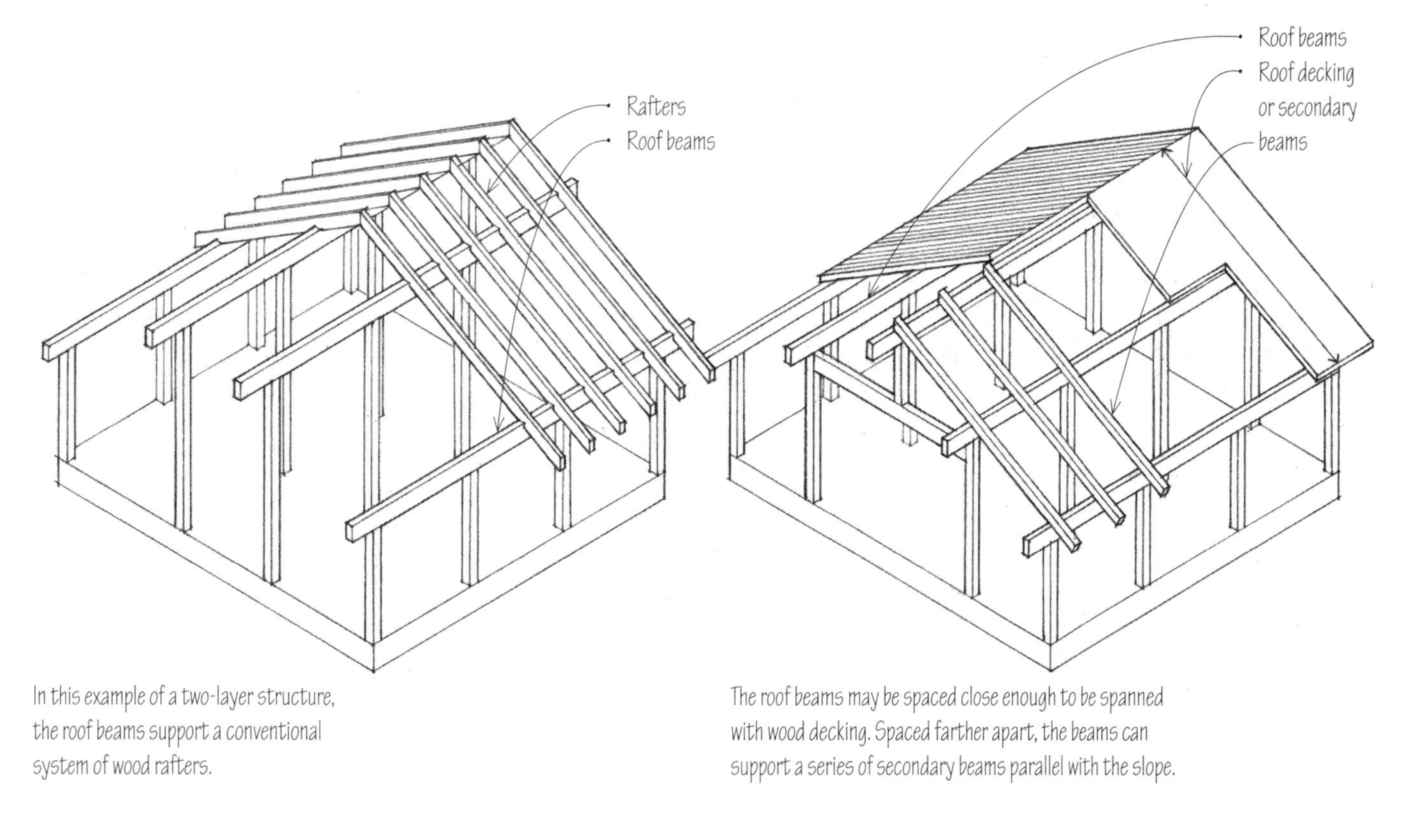

In this example of a two-layer structure, the roof beams support a conventional system of wood rafters.

The roof beams may be spaced close enough to be spanned with wood decking. Spaced farther apart, the beams can support a series of secondary beams parallel with the slope.

Roof Beams Perpendicular to Slope

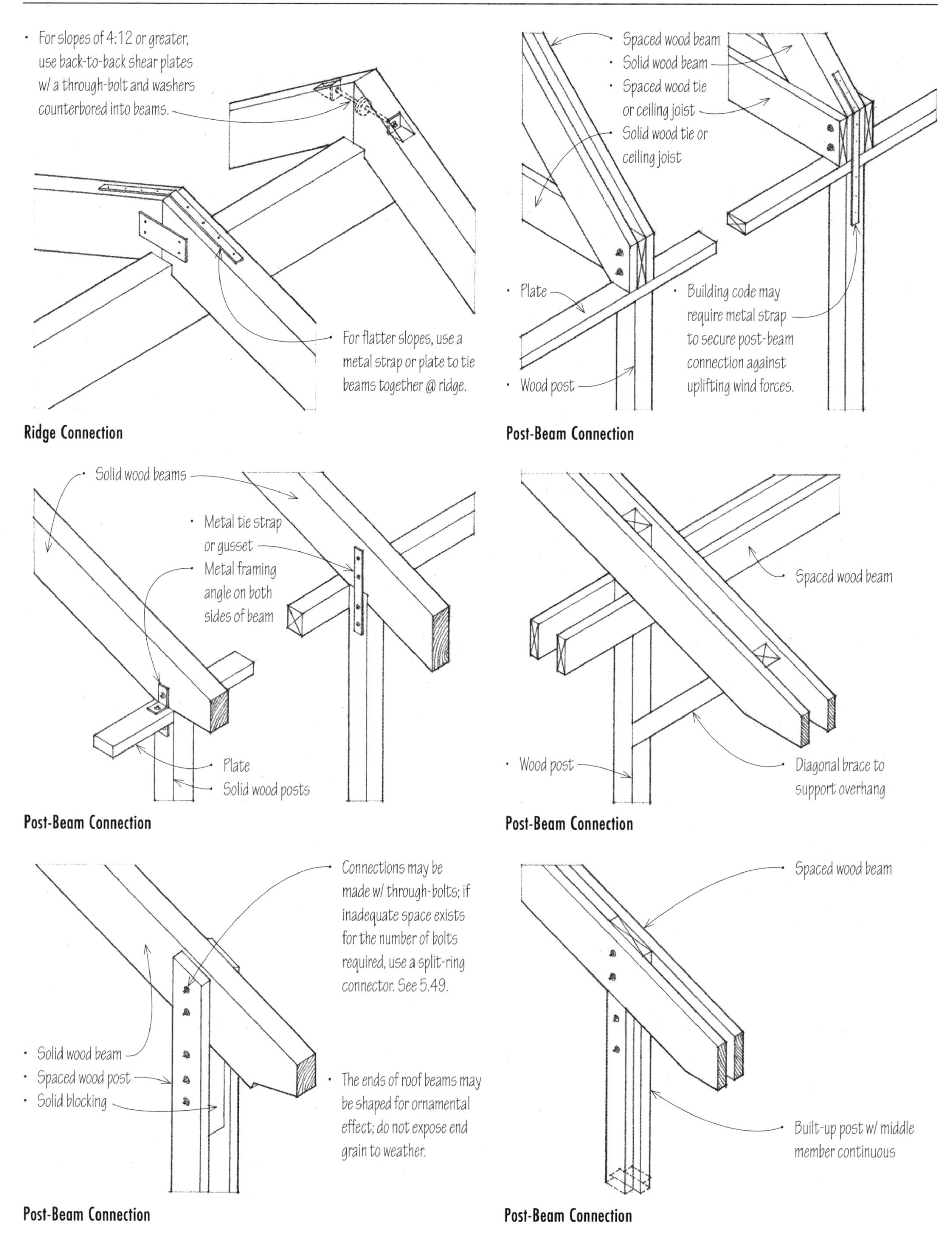

• For slopes of 4:12 or greater, use back-to-back shear plates w/ a through-bolt and washers counterbored into beams.
• For flatter slopes, use a metal strap or plate to tie beams together @ ridge.
Ridge Connection
• Spaced wood beam
• Solid wood beam
• Spaced wood tie or ceiling joist
• Solid wood tie or ceiling joist
• Plate
• Wood post
• Building code may require metal strap to secure post-beam connection against uplifting wind forces.
Post-Beam Connection
• Solid wood beams
• Metal tie strap or gusset
• Metal framing angle on both sides of beam
• Plate
• Solid wood posts
Post-Beam Connection
• Spaced wood beam
• Wood post
• Diagonal brace to support overhang
Post-Beam Connection
• Connections may be made w/ through-bolts; if inadequate space exists for the number of bolts required, use a split-ring connector. See 5.49.
• Solid wood beam
• Spaced wood post
• Solid blocking
• The ends of roof beams may be shaped for ornamental effect; do not expose end grain to weather.
Post-Beam Connection
• Spaced wood beam
• Built-up post w/ middle member continuous
Post-Beam Connection

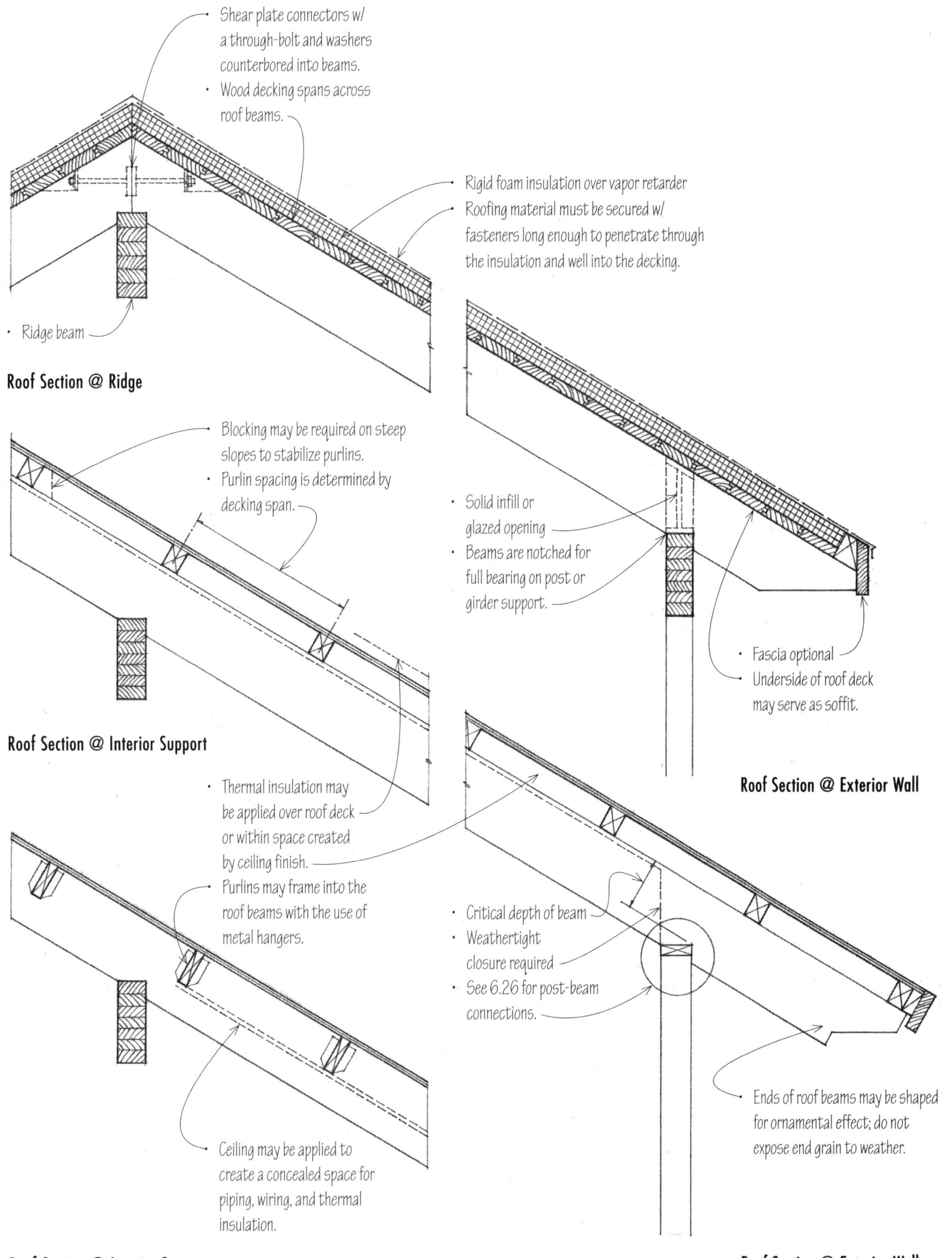

Roof Section @ Ridge

Roof Section @ Interior Support

Roof Section @ Exterior Wall

Roof Section @ Interior Support

Roof Section @ Exterior Wall

In contrast to monoplanar trussed rafters, heavier wood trusses can be assembled by layering multiple members and joining them at the panel points with split-ring connectors. These wood trusses are capable of carrying greater loads than trussed rafters and are spaced farther apart. Consult a structural engineer for design, bracing, and anchorage requirements.

- To avoid additional bending stresses in truss members, loads should be applied at panel points.
- Vertical sway bracing may be required between the top and bottom chords of adjacent trusses to provide resistance against lateral wind and seismic forces.
- Horizontal cross-bracing may be required in the plane of the top or bottom chord if the diaphragm action of the roof framing is not adequate for end-wall forces.
- Any knee bracing should connect to the top or bottom chord at a panel point.

- Wood trusses may be spaced up to 8' (2440) o.c., depending on the spanning capability of the roof decking or planking. When purlins span across the trusses, the truss spacing may be increased up to 20' (6095).

- Span range for shaped trusses: 40' to 150' (12 to 45 m)
- Depth range for shaped trusses: span/2 to span/6

- See 6.09 for a description of truss configurations.

- Span range for flat trusses: 40' to 110' (12 to 33 m)
- Depth range for flat trusses: span/10 to span/15

- Composite trusses have timber compression members and steel tension members.
- Truss rods are metal tie rods that serve as tension members in a truss or trussed beam.

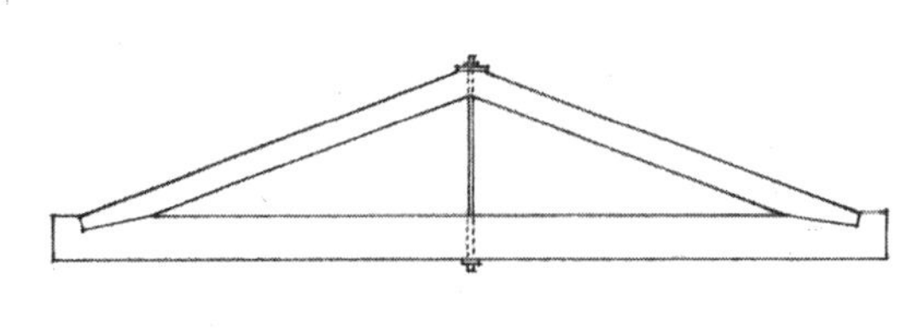

- Trussed beams are timber beams stiffened by a combination of diagonal truss rods and either compression struts or suspension rods.

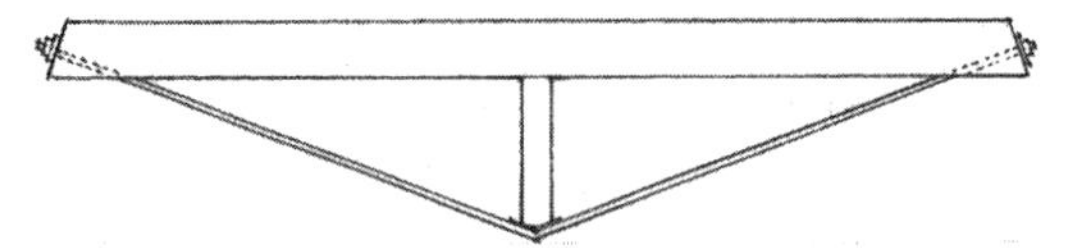

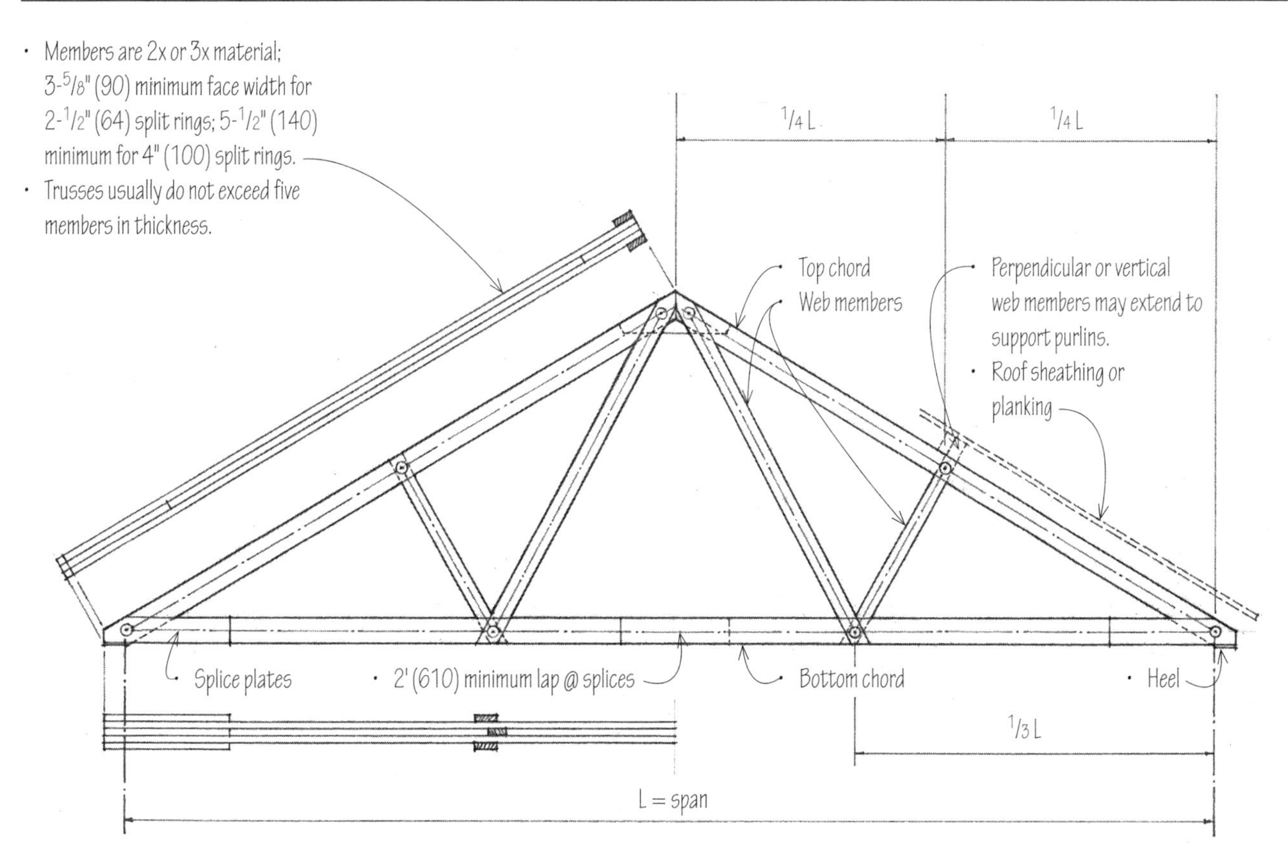

Example of a Belgian Truss

- Member sizes and joint details are determined by engineering calculations based on truss type, load pattern, span, and grade and species of lumber used.
- The size of compression members is generally governed by buckling while the size of tension members is controlled by tensile stresses at connections.
- Consult building code for minimum member thicknesses if trusses are to qualify as heavy timber construction.

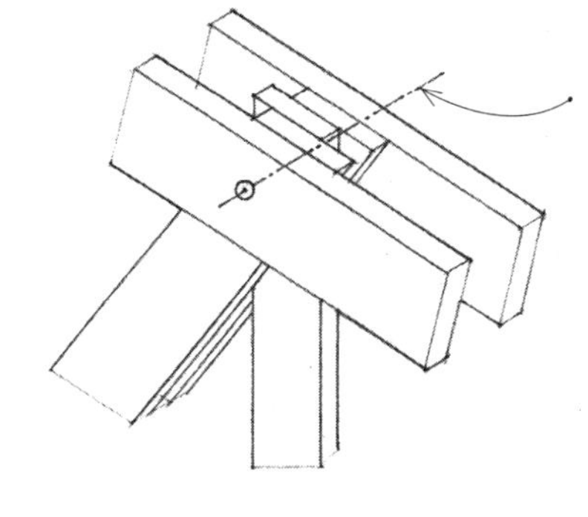

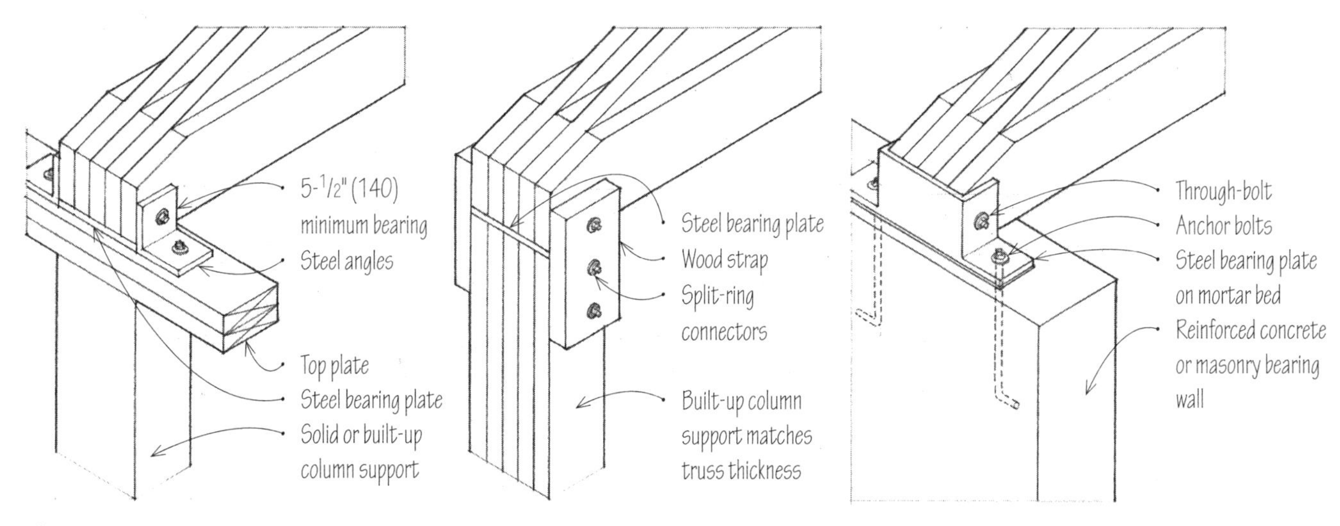

Heel Joints

Wood trussed rafters are pre-engineered and shop-fabricated monoplanar trusses. Because the individual members are subject primarily to compressive and tensile forces, they are typically 2x4s; 2x6s are sometimes used for the top chord. Trussed rafters are best used when a rectangular plan requires a quantity of a single truss type and clear spans over 18' (5485). Consult the truss manufacturer for configurations, allowable spans and loads, and construction details.

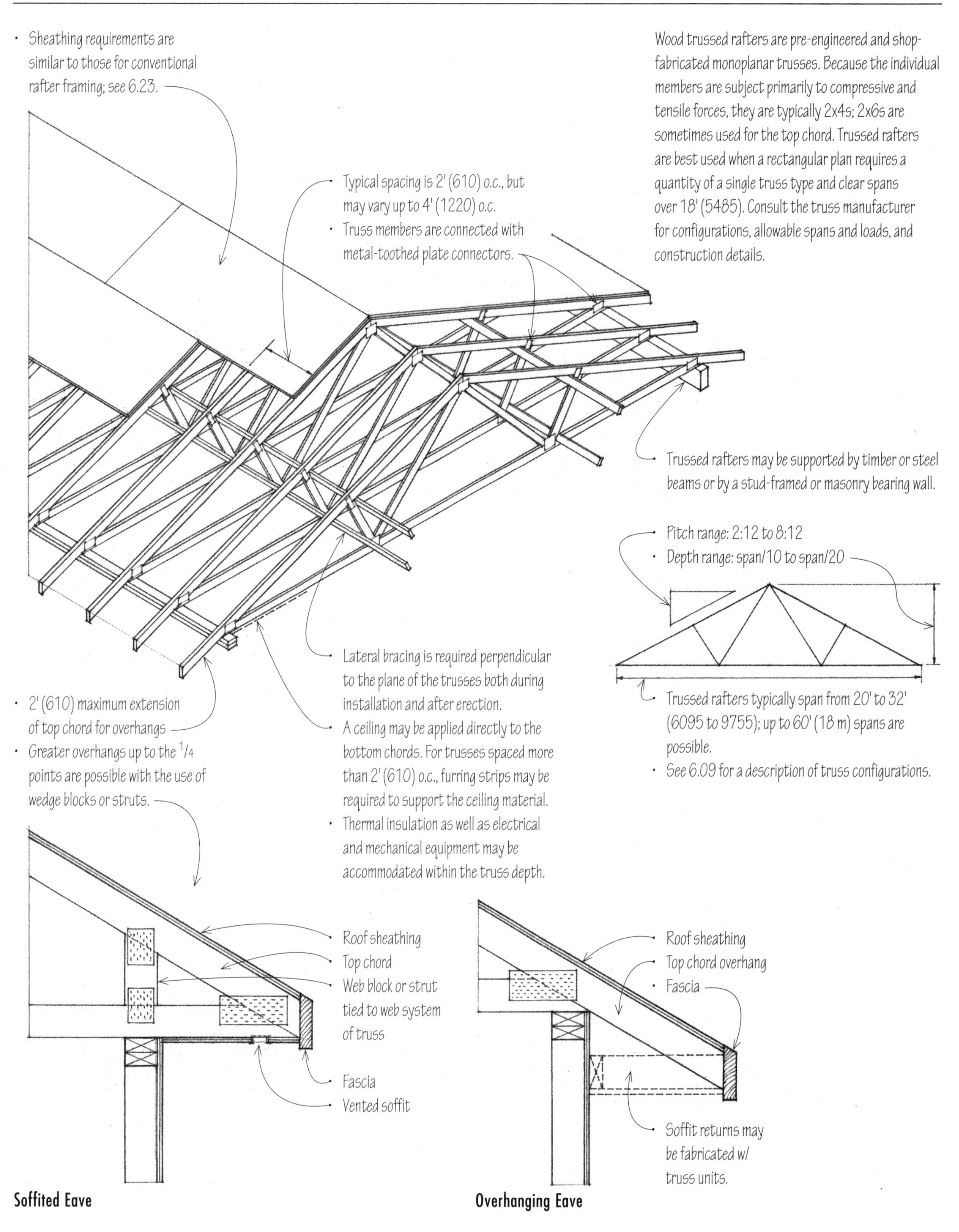

CSI MasterFormat 06 17 53: Shop-Fabricated Wood Trusses

7

MOISTURE & THERMAL PROTECTION

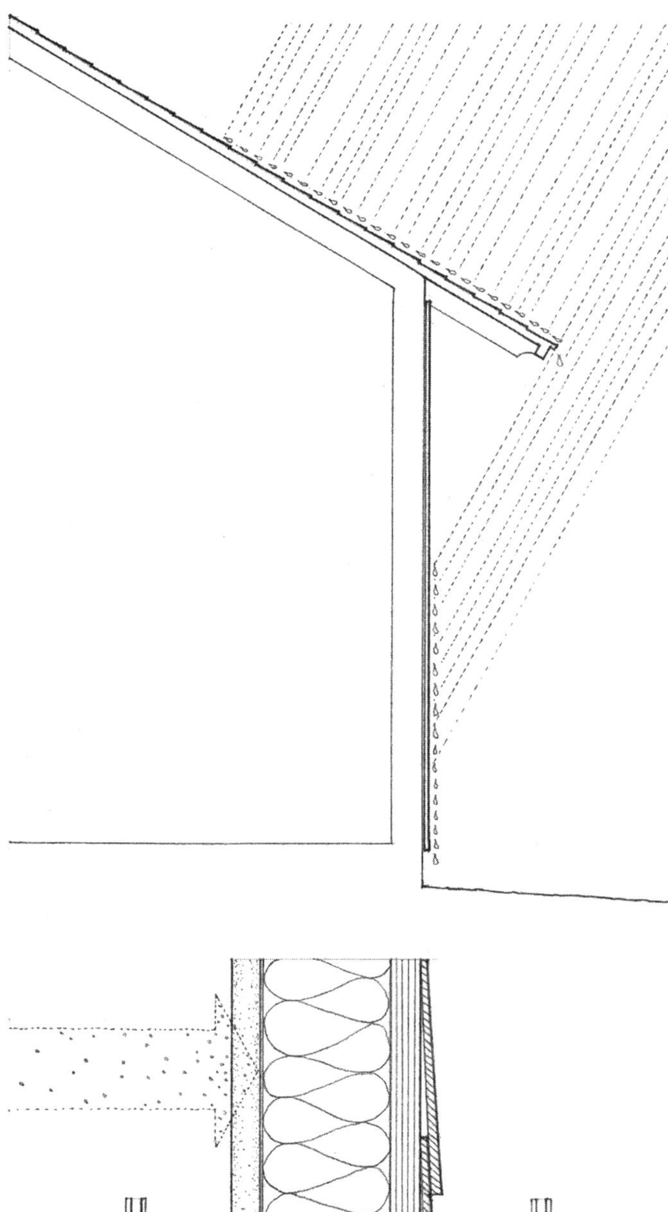

LEED® EA Credit 1: Optimize Energy Performance

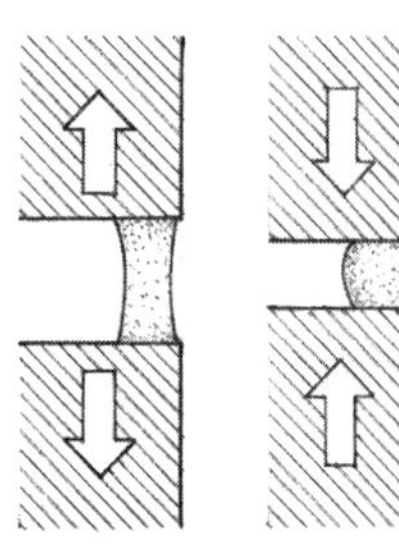

Roofing materials provide the water-resistant covering for a roof system. They range in form from virtually continuous, impervious membranes to overlapping or interlocking pieces of shingles and tiles. The type of roofing that may be used depends on the pitch of the roof structure. While a sloping roof easily sheds water, a flat roof must depend on a continuous waterproof membrane to contain the water while it drains or evaporates. A flat roof as well as any well-insulated sloping roof capable of retaining snow may therefore have to be designed to support a greater live load than a moderately or high-pitched roof. Additional factors to consider in the selection of a roofing material include requirements for installation, maintenance, and durability, resistance to wind and fire, and, if visible, the roofing pattern, texture, and color.

To prevent water from leaking into a roof assembly and eventually the interior of a building, flashing must be installed along roof edges, where roofs change slope or abut vertical planes, and where roofs are penetrated by chimneys, vent pipes, and skylights. Exterior walls must also be flashed where leakage might occur — at door and window openings and along joints where materials meet in the plane of the wall.

Exterior walls also must provide protection from the weather. While some exterior wall systems, such as solid masonry and concrete loadbearing walls, use their mass as barriers against the penetration of water into the interior of a building, other wall systems, such as cavity walls and curtain walls, use an interior drainage system to carry away any moisture that finds its way through the facing or cladding.

Moisture is normally present in the interior spaces of a building in the form of water vapor. When this water vapor reaches a surface cooled by heat loss to the colder outside air, condensation may occur. This condensation may be visible, as on an uninsulated window pane, or it can collect in concealed roof, wall, or floor spaces. Means of combating condensation include the correct placement of thermal insulation and vapor retarders, and the ventilation of concealed spaces, such as attics and crawl spaces.

Potential heat loss or gain through the exterior enclosure of a building is an important factor when estimating the amount of mechanical equipment and energy required to maintain the desired level of environmental comfort in the interior spaces. The proper selection of building materials, the correct construction and insulation of the building enclosure, and the orientation of a building on its site are the basic means of controlling heat loss and gain.

Building materials expand and contract due to variations within the normal temperature range, as well as exposure to solar radiation and wind. To allow for this movement and help relieve the stresses caused by thermal expansion and contraction, expansion joints should be flexible, weathertight, durable, and correctly placed to be effective.

Underlayment protects the roof sheathing from moisture until the roofing shingles are applied. Once the roofing is applied, the underlayment provides the sheathing with additional protection from wind-driven rain. The underlayment material should have low vapor resistance so that moisture does not accumulate between the underlayment and the roof sheathing. Only enough nails are used to hold the underlayment in place until the roofing shingles are applied.

Eave Flashing

Eave flashing is required whenever there is a possibility that ice might form along the eave and cause melting ice and snow to back up under the roofing shingles.

- On normal slope roofs, eave flashing consists of two layers of 15 lb. felt or a single layer of 50 lb. smooth roll roofing extending from the eave up the roof to a point 24" (610) inside the interior wall line.
- On low slope roofs, an additional course of underlayment is cemented in place, and extended to a point 36" (915) inside the interior wall line.

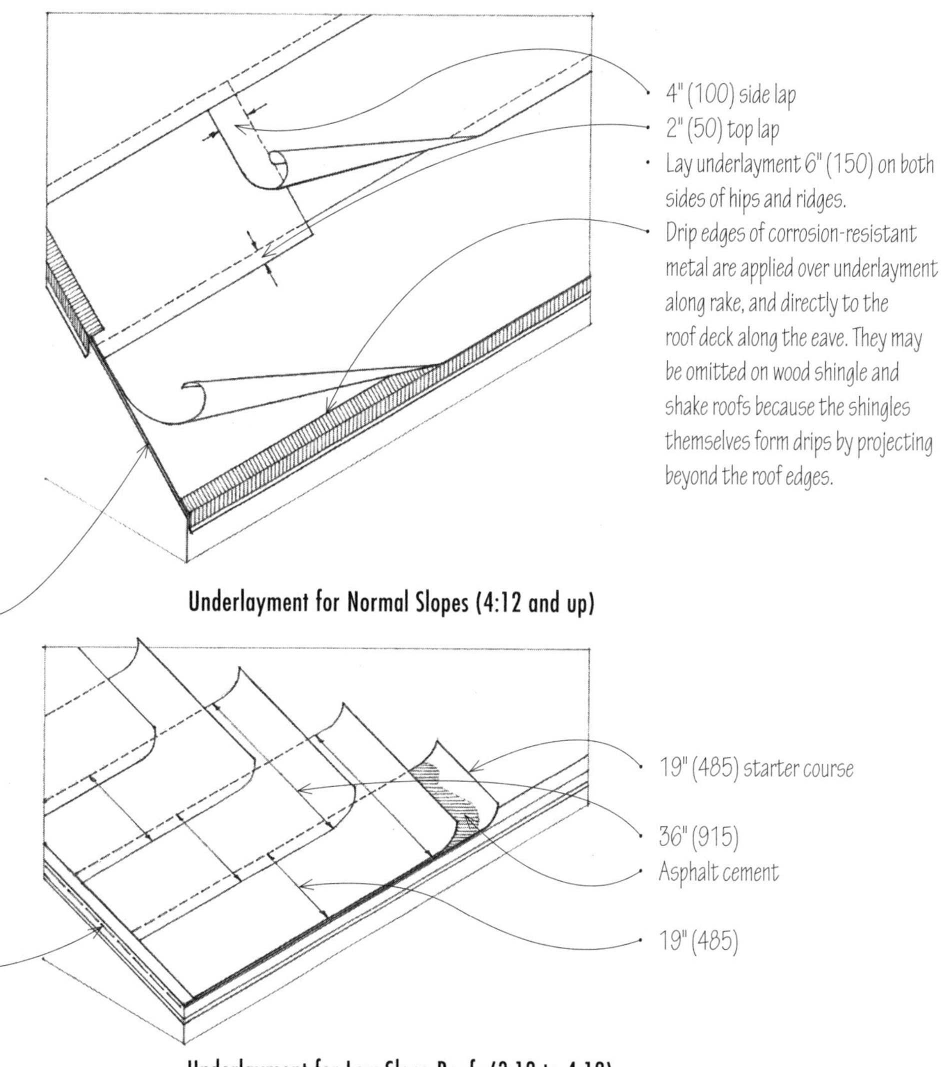

Underlayment for Normal Slopes (4:12 and up)

Underlayment for Low Slope Roofs (3:12 to 4:12)

Underlayment and Sheathing for Shingle Roofs

Roofing Type	Sheathing	Underlayment	Normal Slope		Low Slope	
Fiberglass shingles	Solid	15 lb. asphalt-saturated felt	4:12 and up	Single layer	3:12 to 4:12	Double layer
Asphalt shingles	Solid	15 lb. asphalt-saturated felt	4:12 and up	Single layer	2:12 to 4:12	Double layer
Wood shingles	Spaced	15 lb. asphalt-saturated felt	4:12 and up	Optional	3:12 to 4:12	Reduce weather exposure
	Solid	15 lb. asphalt-saturated felt	4:12 and up	Optional; eave flashing required in snow areas	3:12 to 4:12	Optional; eave flashing required in snow areas
Wood shakes	Spaced	30 lb. asphalt-saturated felt (interlayment)	4:12 and up		Not recommended	
	Solid	30 lb. asphalt-saturated felt (interlayment)	4:12 and up		3:12 to 4:12	Single layer underlayment and interlayment over entire roof

Maximum Recommended Exposure in Inches (mm)

Shingle Grade	Length Inches (mm)	Roof Slope 4:12 and up	3:12 to 4:12
No. 1	16" (405)	5" (125)	3-3/4" (95)
	18" (455)	5-1/2" (140)	4-1/4" (110)
	24" (610)	7-1/2" (190)	5-3/4" (145)
No. 2	16" (405)	4" (100)	3-1/2" (90)
	18" (455)	4-1/2" (115)	4" (100)
	24" (610)	6-1/2" (165)	5-1/2" (140)
No. 3	16" (405)	3-1/2" (90)	3" (75)
	18" (455)	4" (100)	3-1/2" (90)
	24" (610)	5-1/2" (140)	5" (125)
Wood Shakes	18" (455)	7-1/2" (190)	Not
	24" (610)	10" (255)	recommended

Wood shingles and shakes are normally cut from red cedar, although white cedar, redwood, and red cypress shingles may be available. Red cedar has a fine, even grain and is naturally resistant to water, rot, and sunlight.

Red cedar shingles are available in 16" (405), 18" (455), and 24" (610) lengths, and in the following grades:

- No. 1 Premium Grade (Blue Label):
  - 100% heartwood, 100% clear, 100% edge grain
- No. 2 Intermediate Grade (Red Label):
  - 10" (255) clear on 16" (405) shingles
  - 11" (280) clear on 18" (455) shingles
  - 16" (405) clear on 24" (610) shingles
  - Some flat grain permitted
- No. 3 Utility Grade (Black Label):
  - 6" (150) clear on 16" (405) and 18" (455) shingles
  - 10" (255) clear on 24" (610) shingles

- Use only corrosion-resistant nails, such as hot-dipped galvanized steel or aluminum-alloy nails. Two fasteners per shingle are required. Nails should be driven flush with, but not into the surface of the shingles.
- Spaced 1x4 or 1x6 sheathing provides ventilation for the shingles. Board spacing is equal to the shingle exposure.

1/4" to 3/8" (6 to 10) joints to allow for expansion

1-1/2" (38) minimum offset between adjacent courses; do not align joints in alternate courses.

26-gauge galvanized steel or 0.019" (0.5 mm) corrosion-resistant flashing; provide underlayment under flashing in severe climates.

11" (280) minimum to each side of valley centerline with 36" (915) wide underlayment for roof slopes of 3:12 (25%) and over

Lap 4" (100)

1/2" (13) edge crimps

1" (25) high center crimp

4" (100)

Open Valley

- Shingle exposure depends on shingle length and roof slope; see table above.
- In areas subject to wind-driven snow and roof ice buildup, eave flashing over solid sheathing is required; see 7.03.

Doubled first course; extend 1" to 1-1/2" (25 to 38) to form drip.

Shingles may project 1" (25) beyond rake trim to form drip, or be canted with a beveled strip to eliminate drip.

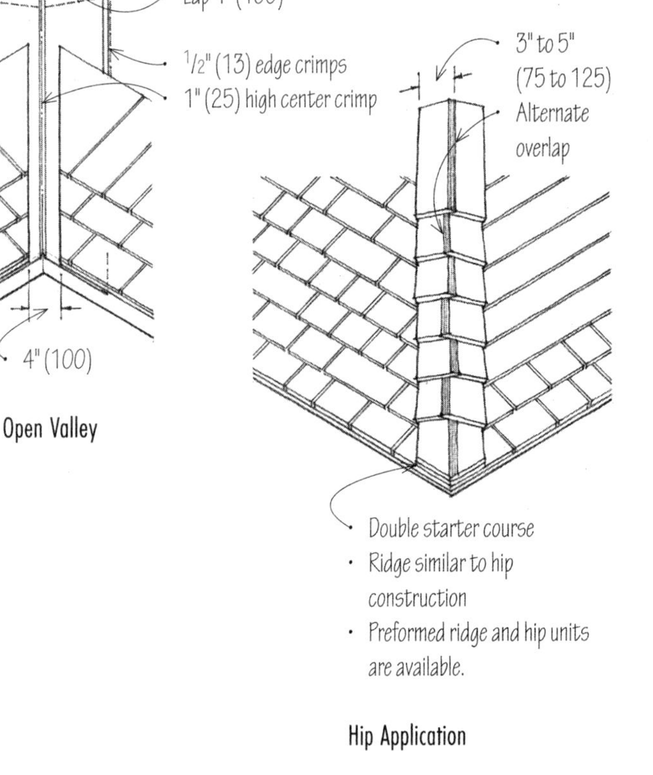

- Ridge similar to hip construction
- Preformed ridge and hip units are available.

Hip Application

CSI MasterFormat™ 07 31 29: Wood Shingles and Shakes

While wood shingles are sawn, wood shakes are formed by splitting a short log into a number of tapered radial sections, resulting in at least one textured face. Shakes are normally 100% clear heart wood, and available in 18" (150) and 24" (610) lengths. Tapersplit and straightsplit shakes have 100% edge grain, while handsplit and resawn shakes have at least 90% edge grain.

Both wood shingles and shakes are flammable unless chemically treated to receive a UL Class C rating. A Class B rating may be possible if Class C shingles or shakes are used over a solid roof deck of 5/8" (16) plywood w/ exterior glue covered with a plastic-coated sheet foil.

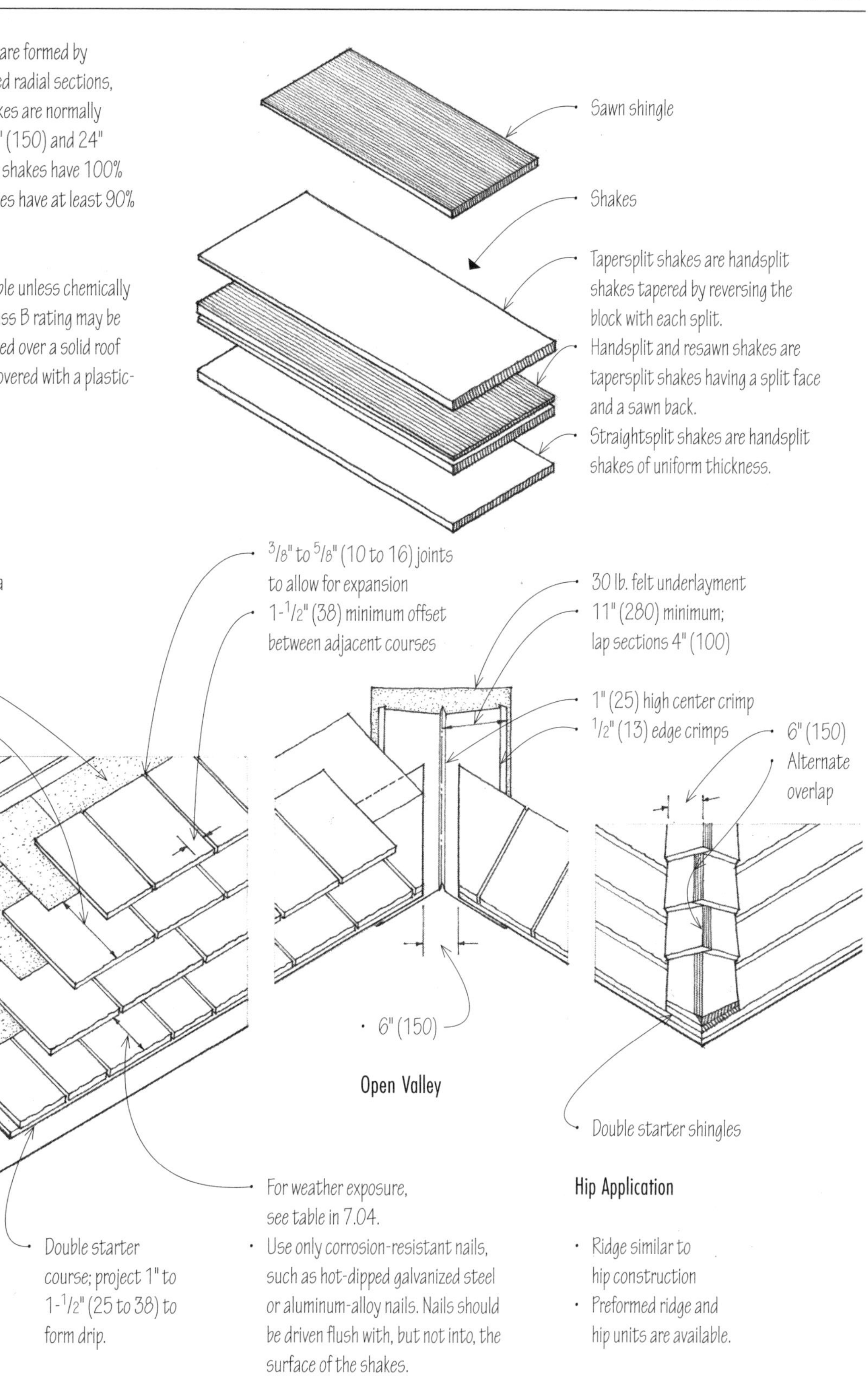

Composition shingles have either an inorganic fiberglass base or an organic felt base surfaced on the weather side with colored mineral or ceramic granules. Inorganic fiberglass base shingles have excellent fire resistance (UL Class A); organic felt base shingles possess only a moderate resistance to fire (UL Class C). Most composition shingles have tabs with a self-sealing adhesive or locking tabs that make them wind-resistant. Wind resistance is important when shingles are used on low-slope roofs and in areas subject to high winds.

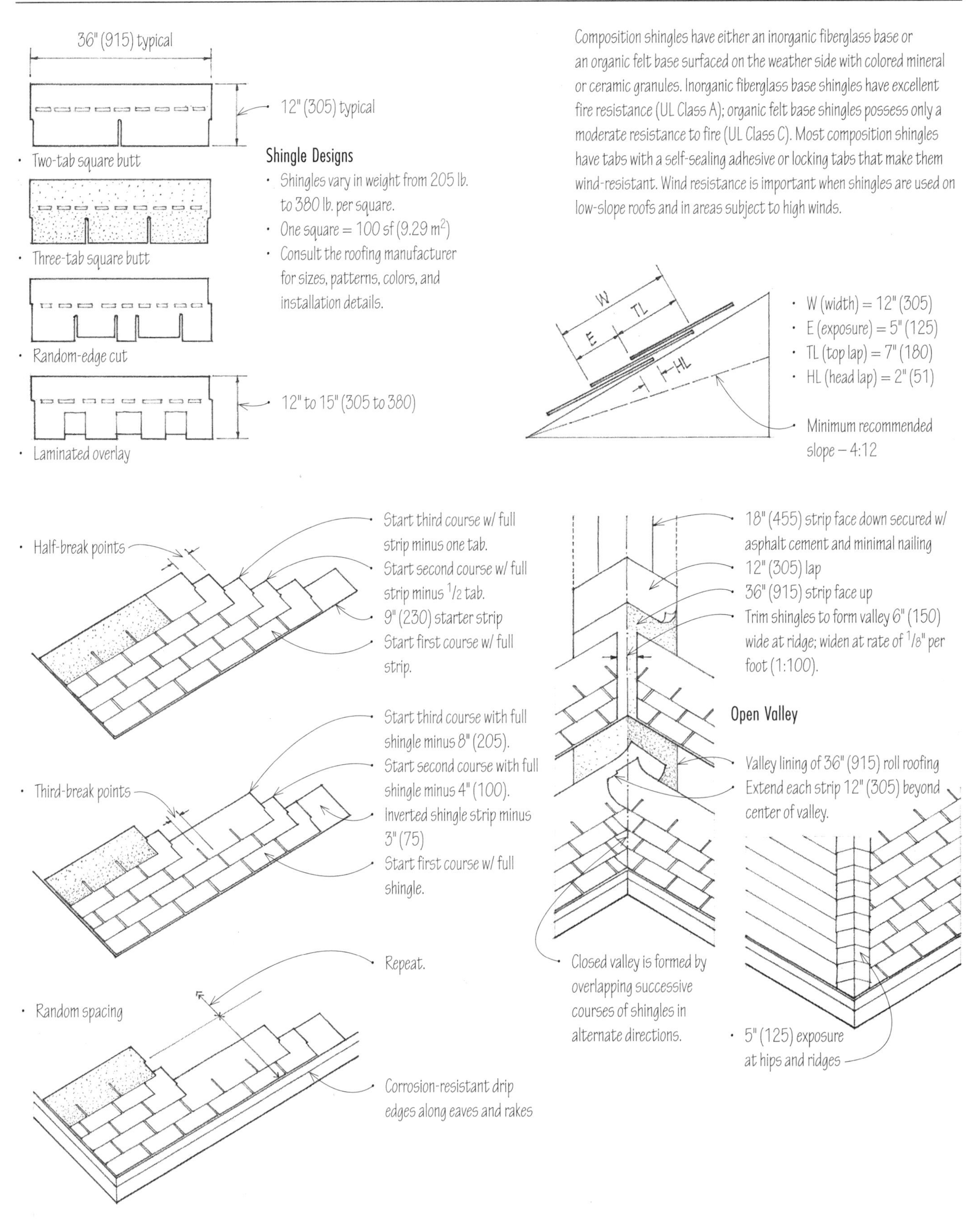

CSI MasterFormat 07 31 13: Asphalt Shingles
CSI MasterFormat 07 31 13.13: Fiberglass-Reinforced Asphalt Shingles

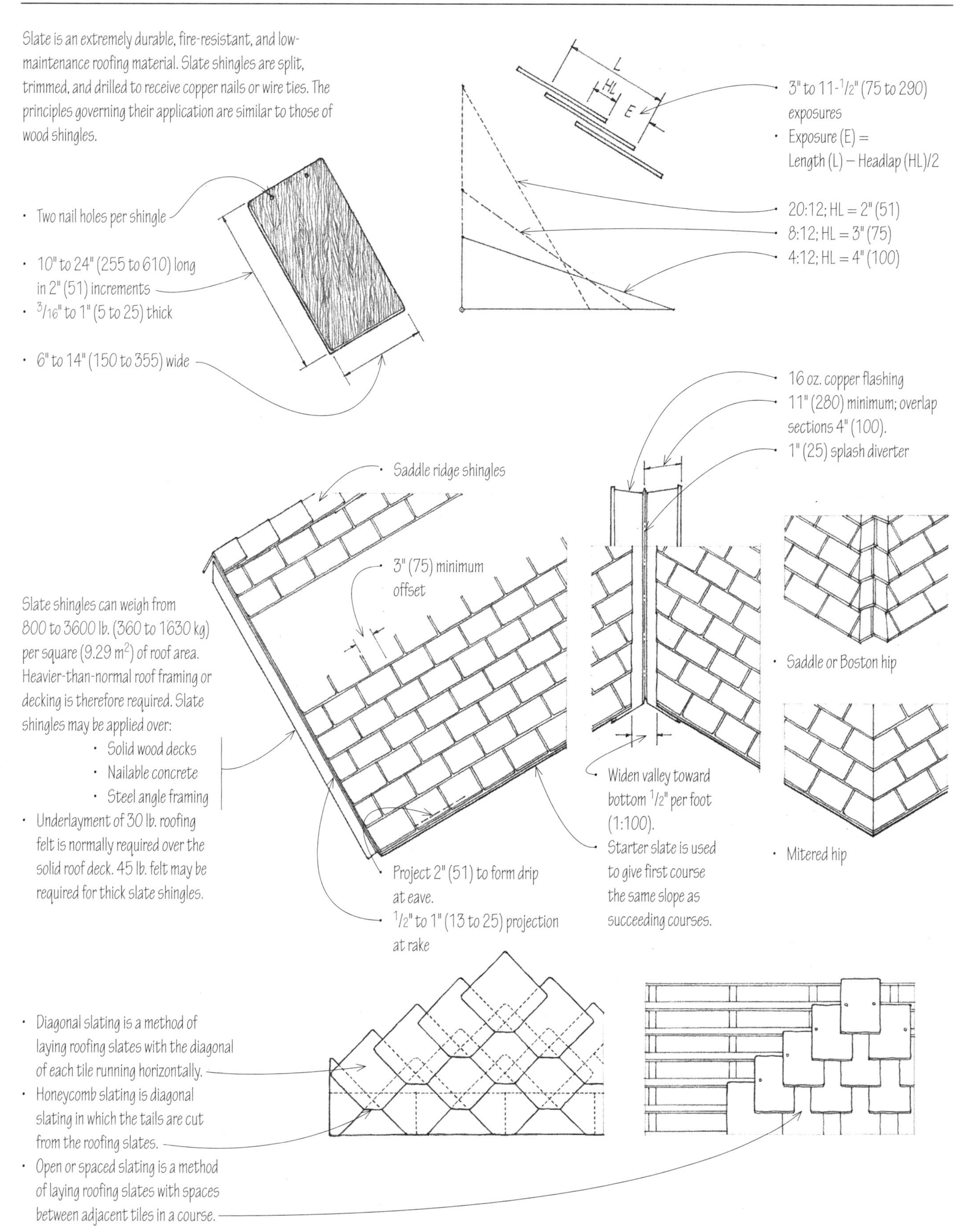
Slate is an extremely durable, fire-resistant, and low-maintenance roofing material. Slate shingles are split, trimmed, and drilled to receive copper nails or wire ties. The principles governing their application are similar to those of wood shingles.
• Two nail holes per shingle
• 10" to 24" (255 to 610) long in 2" (51) increments
• 3/16" to 1" (5 to 25) thick
• 6" to 14" (150 to 355) wide
L
HL
E
• 3" to 11-1/2" (75 to 290) exposures
• Exposure (E) = Length (L) – Headlap (HL)/2
20:12; HL = 2" (51)
8:12; HL = 3" (75)
4:12; HL = 4" (100)
• 16 oz. copper flashing
• 11" (280) minimum; overlap sections 4" (100).
• 1" (25) splash diverter
• Saddle ridge shingles
• 3" (75) minimum offset
Slate shingles can weigh from 800 to 3600 lb. (360 to 1630 kg) per square (9.29 m2) of roof area. Heavier-than-normal roof framing or decking is therefore required. Slate shingles may be applied over:
• Solid wood decks
• Nailable concrete
• Steel angle framing
• Underlayment of 30 lb. roofing felt is normally required over the solid roof deck. 45 lb. felt may be required for thick slate shingles.
• Project 2" (51) to form drip at eave.
• 1/2" to 1" (13 to 25) projection at rake
• Widen valley toward bottom 1/2" per foot (1:100).
• Starter slate is used to give first course the same slope as succeeding courses.
• Saddle or Boston hip
• Mitered hip
• Diagonal slating is a method of laying roofing slates with the diagonal of each tile running horizontally.
• Honeycomb slating is diagonal slating in which the tails are cut from the roofing slates.
• Open or spaced slating is a method of laying roofing slates with spaces between adjacent tiles in a course.

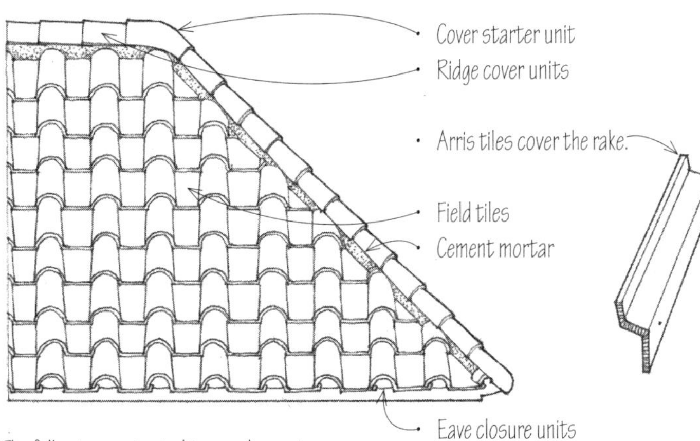

Tile roofing consists of clay or concrete units that overlap or interlock to create a strong textural pattern. Like slate, roofing tiles are fire-resistant, durable, and require little maintenance. They are also heavy (800 to 1000 lb. per square; 363 to 454 kg per 9.29 m^2) and require roof framing that is strong enough to carry the weight of the tiles. Roofing tiles are normally installed over a solid plywood deck with an underlayment of 30 lb. or 45 lb. roofing felt. Special tile units are used at ridges, hips, rakes, and eaves.

The following are typical types, dimensions, and weights of clay tiles. Confirm sizes, weights, and installation details with tile manufacturer.

- Mission or Spanish tiles are tapered, semicylindrical roofing tiles laid convex side up to overlap flanking, similar tiles laid concave side up.
- Imbrex laid convex side up; tegula laid concave side up.
- Taper allows tiles to nest into the overlapping tiles.
- Minimum recommended slope – 4:12
- 10" (255) wide; 19" (485) long
- Exposure: 16" (405)

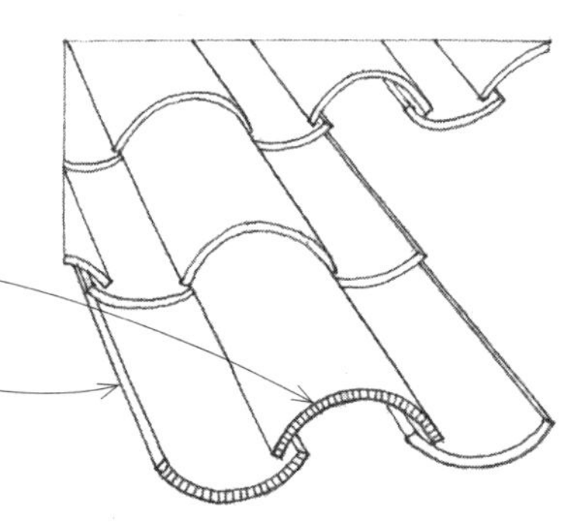

- Pantiles have an S-shaped cross section, laid so the downturn of one overlaps the upturn of the next in the same course.
- Minimum recommended slope – 4:12
- 14" (355) wide; 19" (485) long
- Exposure: 16" (405)

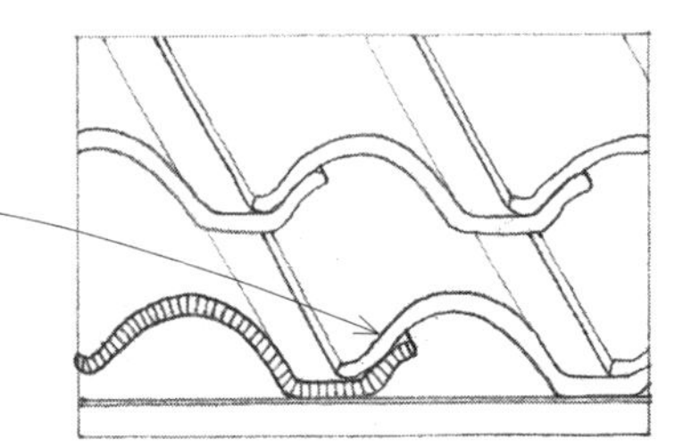

- Interlocking tiles are flat, rectangular roofing tiles having a groove along one edge that fits over a flange in the next tile in the same course.
- Minimum recommended slope – 3:12
- 9" (230) wide; 12" (305) long
- Exposure: 9" (230)

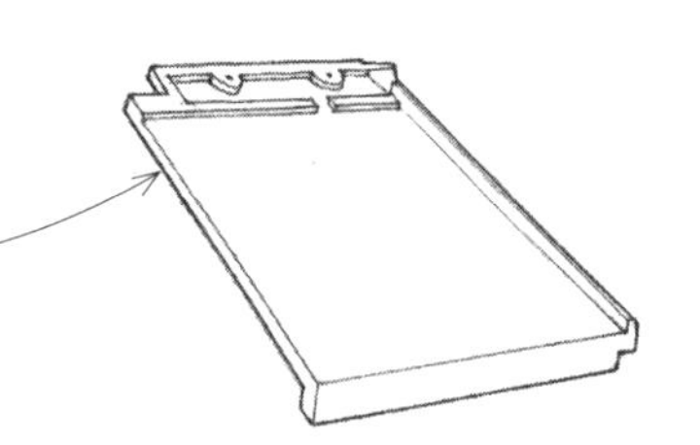

- Shingle tiles are flat, rectangular roofing tiles laid in an overlapping pattern.
- Minimum recommended slope – 3:12
- 10" (255) wide; 16" (405) long
- Exposure: 13" (330)

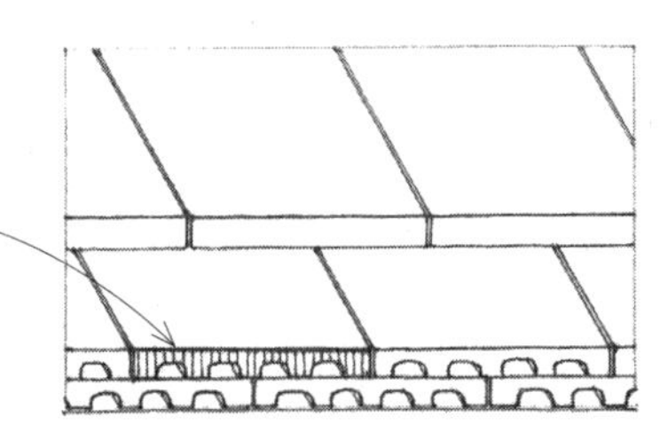

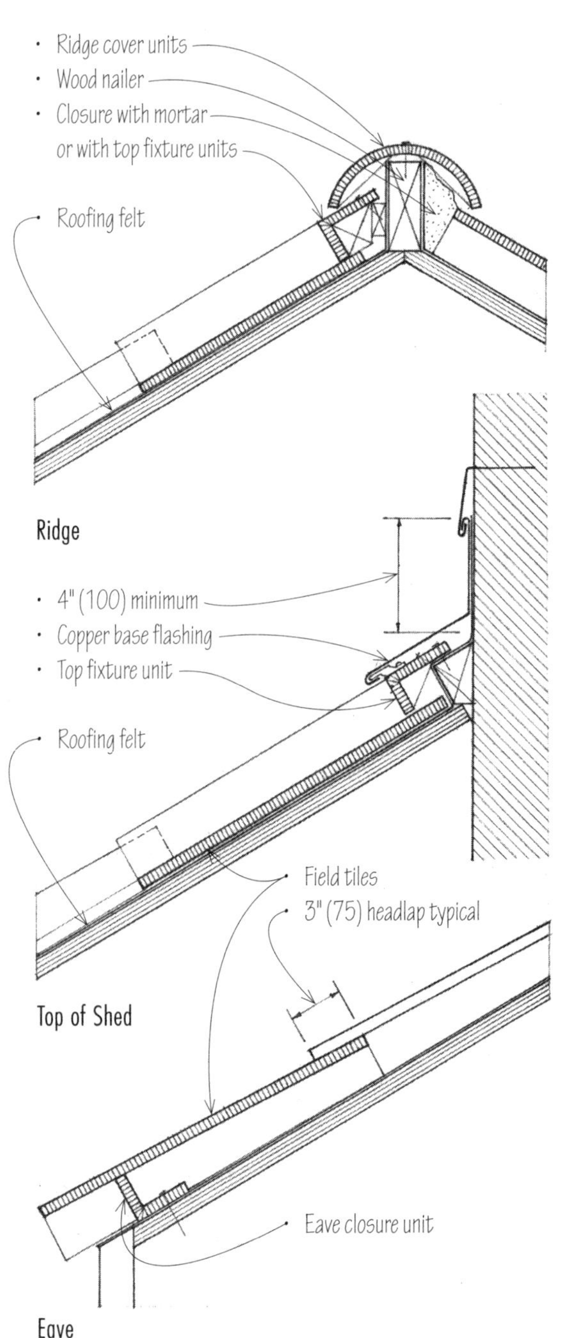

Vegetated roofing, also known as "green roofing," refers to a natural roof covering typically consisting of vegetation planted in engineered soil or growing medium over a waterproof membrane. While vegetated roofing typically requires a greater initial investment, the natural covering protects the waterproof membrane from daily temperature fluctuations and the ultraviolet radiation of the sun that breaks down conventional roofing systems. Vegetated roofing also offers environmental benefits, including conserving a pervious area otherwise replaced by a building's footprint, controlling the volume of stormwater runoff, and improving air and water quality.

The surface temperature of traditional roofing can be up to 90°F (32°C) warmer than the air temperature on a hot summer day. A vegetated roof, having a much lower surface temperature, helps reduce the heat island effect in urban areas. The increased insulation value of a vegetated roofing system can also help stabilize indoor air temperatures and humidity and reduce the heating and cooling costs for a building.

There are three types of vegetated roofing systems: intensive, extensive, and modular block.

- Intensive vegetative roofing systems require a minimum of one foot (305) of soil depth to create an accessible roof garden with larger trees, shrubs, grasses, and other landscapes. They require irrigation and drainage systems to maintain the plant materials, which can add 80 to 150 psf (2870 to 4310 Pa) to the load on the roof structure. Concrete is usually the best choice for a roof deck.
- Extensive vegetative roofing systems are low maintenance and built primarily for their environmental benefit. The lightweight growing medium they use is typically 4" to 6" (100 to 150) in depth and contain small, hardy plants and thick grasses that are accessed for maintenance only. Extensive vegetative roof systems can add between 15 and 50 psf (715 to 2395 Pa) to the load on the roof structure and can be installed over any properly designed concrete, steel, or wood roof deck.
- Modular block systems consist of anodized aluminum containers or recycled polystyrene trays with 3" to 4" (75 to100) of engineered soil supporting low-growing plant species. A pad fastened to the bottom of each block protects the roof surface and allows controlled drainage through the unit. The typical weight of the system is 12 to 18 psf (575 to 860 Pa).

LEED SS Credit 6: Stormwater Design
LEED SS Credit 7: Reduce Heat Island Effect

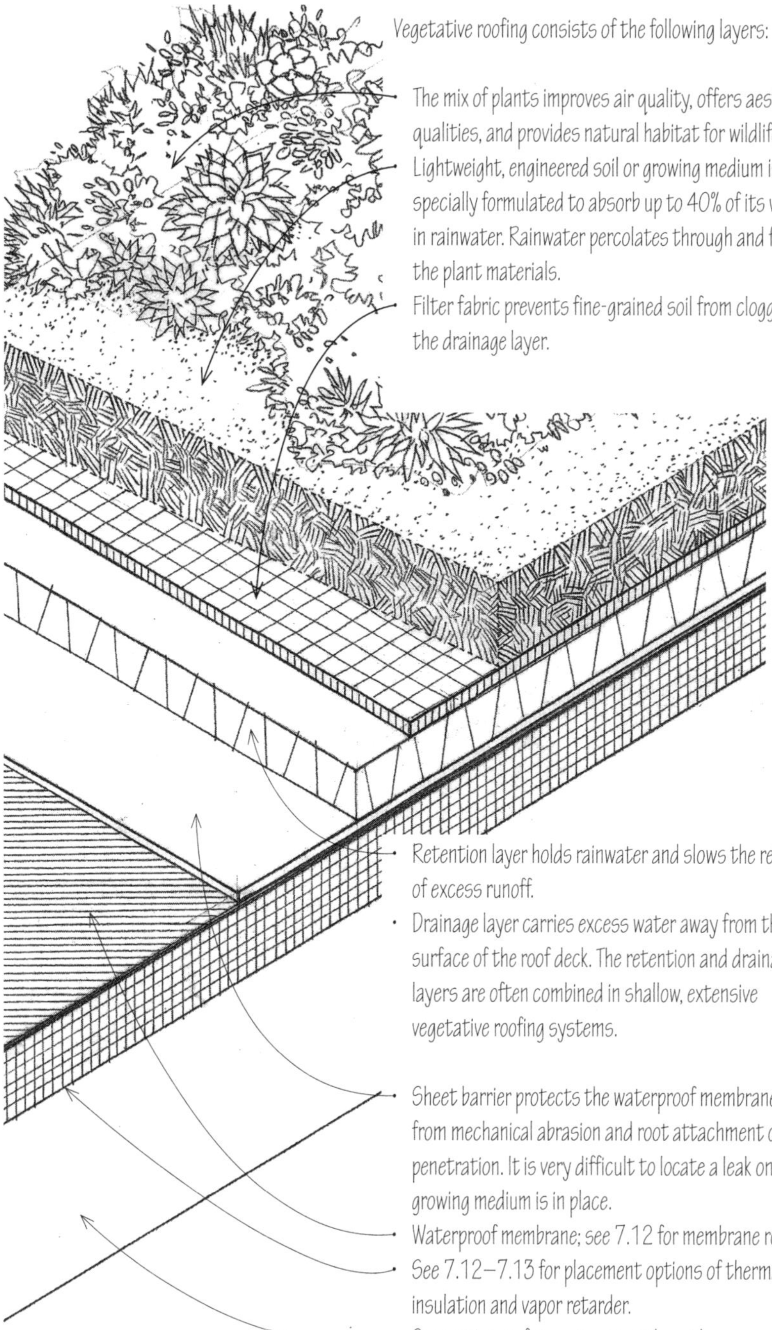

Vegetative roofing consists of the following layers:

- The mix of plants improves air quality, offers aesthetic qualities, and provides natural habitat for wildlife.
- Lightweight, engineered soil or growing medium is specially formulated to absorb up to 40% of its volume in rainwater. Rainwater percolates through and feeds the plant materials.
- Filter fabric prevents fine-grained soil from clogging the drainage layer.
- Retention layer holds rainwater and slows the release of excess runoff.
- Drainage layer carries excess water away from the surface of the roof deck. The retention and drainage layers are often combined in shallow, extensive vegetative roofing systems.
- Sheet barrier protects the waterproof membrane from mechanical abrasion and root attachment or penetration. It is very difficult to locate a leak once the growing medium is in place.
- Waterproof membrane; see 7.12 for membrane roofing.
- See 7.12–7.13 for placement options of thermal insulation and vapor retarder.
- Supporting roof structure must have the necessary load-bearing capacity to support wet densities of 60 to 90 pcf (960 to 1440 kg/m^3).

- Vegetated roofing is easiest to create on low-sloped roofs, but it can also be installed on roof slopes up to 12:12 (100%) if a suitable system for stabilizing the soil or growing medium is in place.

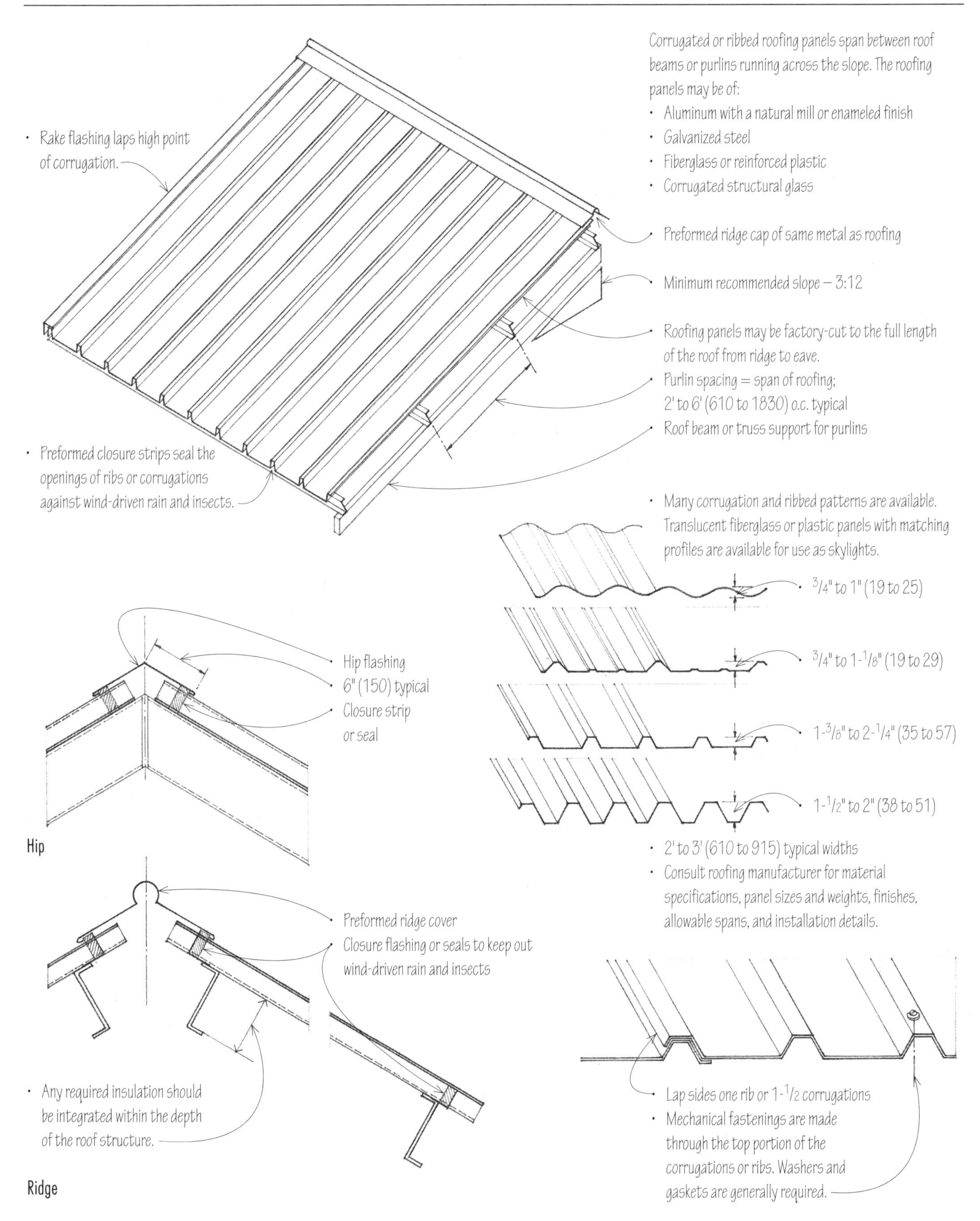

CSI MasterFormat 07 41 13: Metal Roof Panels

A sheet metal roof is characterized by a strong visual pattern of interlocking seams and articulated ridges and roof edges. The metal sheets may be of copper, zinc alloy, galvanized steel, or terne metal, a stainless steel plated with an alloy of tin and lead. To avoid possible galvanic action in the presence of rain water, flashing, fastenings, and metal accessories should be of the same metal as the roofing material. Other factors to consider in the use of metal roofing are the weathering characteristics and coefficient of expansion of the metal.

- Metal roofing is installed over an underlayment of roofing felt. Rosin paper is used to prevent bonding between the felt and terne metal roofing.

- Standing or batten seams
- Horizontal and valley seams are flat and usually soldered.
- Provide expansion joints on runs exceeding 30' (9 m).
- Vertical seams are spaced from 12" to 24" (305 to 610) o.c., depending on the starting width of the metal sheets and the size of the standing or batten seams.
- The seams on prefabricated batten roofs are spaced from 24" to 36" (610 to 915) o.c.
- Metal pan may continue down to form a deep fascia.
- Minimum slope — 3:12
- May be less if locked and soldered seams are used.
- Nailing strips must be provided if roofing is laid over a non-nailable roof deck.
- Interlocking gutter and lining of same metal as roofing
- Batten ridge seam
- Standing ridge seam
- Roll seams are joints between two pieces of sheet metal in the direction of fall of a curved or sloping roof, made by turning up the adjoining edges against each other, then bending them around to form a cylindrical roll.
- Cleats and holddowns restrain the edge or joint of sheet metal roofing.
- Eave seam

- Standing seams are made by folding up the adjoining edges against each other, then folding their upper portion over in the same direction a number of times.
- Batten seams are made by turning up the adjoining edges against a batten and locking them in place with a metal strip placed over the batten.
- Taper batten to allow for expansion of roofing.

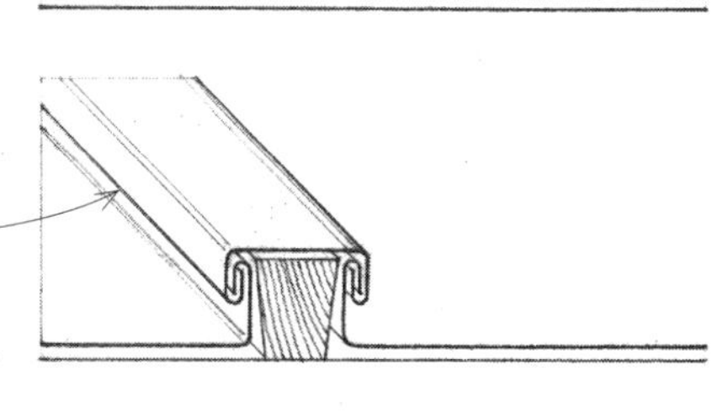

- Various prefabricated standing and batten seams are available from metal roofing manufacturers.

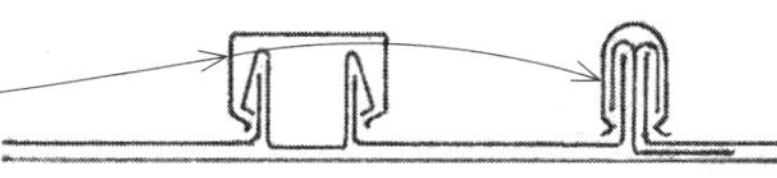

- Lock seams are made by folding up the adjoining edges against each other, folding them over, and flattening the interlock.

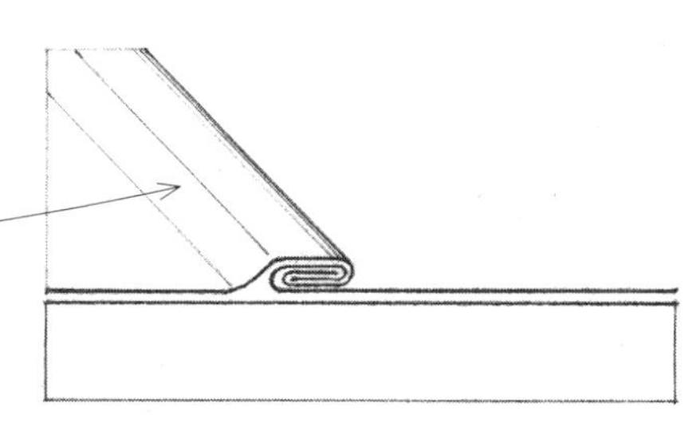

Types of Seams

The construction of a flat roof requires the following elements:

1. The wear course protects the roofing from uplifting wind forces and mechanical abrasion. It may be provided by built-up roofing aggregate, ballast aggregate, or plaza deck pavers.
2. The drainage layer permits the free flow of water to the roof drains. It may consist of the aggregate layer in a built-up roofing system, the ballast layer in a loose-laid, single-ply roofing system, the surface of a fully adhered single-ply roof, or the drainage fabric or space under the pavers in a plaza deck system.
3. The roofing membrane is the waterproofing layer of the roof. It should be sloped at least 1/4" per foot (1:50) to transport rainwater to the roof drains. This slope may be created by adjusting the height of the beams that support the roof deck, placing a tapered fill of insulating lightweight concrete over the roof deck, or installing tapered panels of rigid foam insulation over the roof deck. The two major types of membrane systems are:
   - Built-up roofing systems; see 7.14.
   - Single-ply roofing systems; see 7.15–7.16. 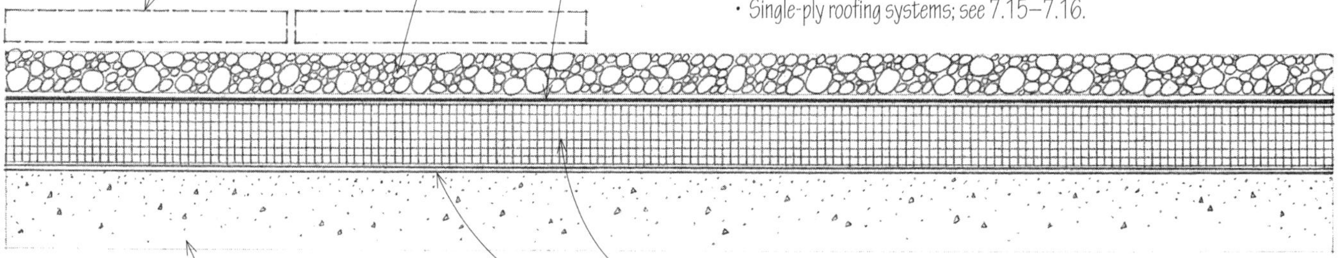
4. Thermal insulation; see 7.13 for placement options.
5. A vapor retarder impedes the passage of water vapor into the roofing assembly. The use of a vapor retarder is generally recommended in geographic locations where the average outdoor temperature in January is below 40°F (4°C) and the interior relative humidity in winter is 45% or greater at 68°F (20°C). Consisting of conventional built-up roofing or a proprietary material of low permeance, the vapor retarder is placed on the predominantly warm side of an assembly. The temperature at the vapor retarder should be warmer than the dew-point temperature to prevent condensation from occurring, which can damage the thermal insulation, roofing membrane, and structural materials. It is also important that the vapor retarder be continuous, sealed at all roof penetrations, and tied into the wall assembly around the perimeter of the roof. When a vapor retarder is present, topside vents may be required to allow any trapped moisture to escape from between the vapor retarder and the roofing membrane. For more information on moisture control, see 7.45.
6. The roof deck must be stiff enough to maintain the desired slope under expected loading conditions, and be smooth, clean, and dry enough for the rigid insulation or roofing membrane to adhere properly. See 7.14 for a list of types of roof decks and substrates. Large roof areas may require expansion joints or area dividers. For these and other flashing details, see 7.19–7.20.

CSI MasterFormat 07 50 00: Membrane Roofing

Thermal insulation provides the required resistance to heat flow through the roof assembly. It may be installed in three positions: below the structural roof deck, between the roof deck and the roofing membrane, or above the roofing membrane.

- When located below the roof deck, the thermal insulation typically consists of batt insulation installed over a vapor retarder. A ventilated air space between the insulation and the roof deck is required to dissipate any water vapor that migrates into the construction assembly.

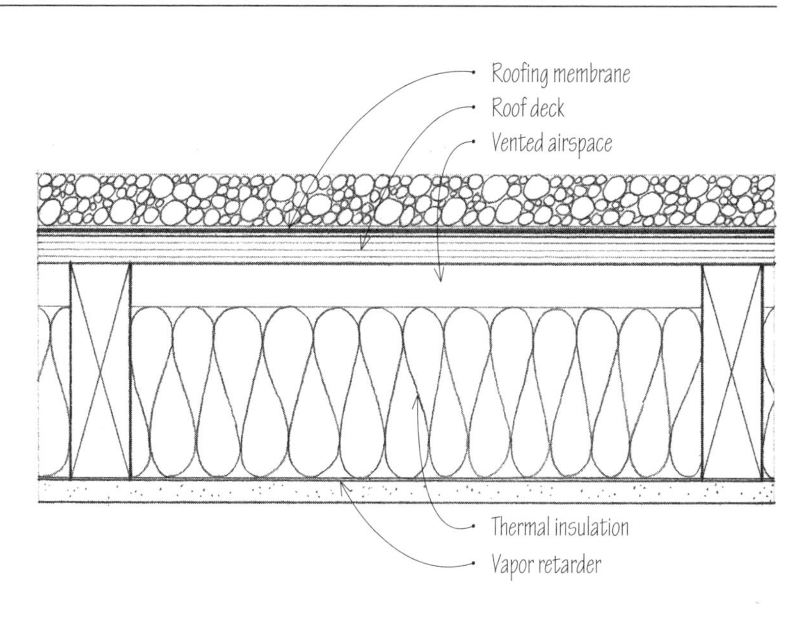

- When located between the roof deck and the roofing membrane, the thermal insulation may be in the form of a lightweight insulating concrete fill or rigid foam insulation boards capable of supporting the roofing membrane. Rigid insulation should be installed in at least two staggered layers to minimize heat loss through the joints. The first layer should be mechanically fastened to resist wind uplift; the upper layers are fully adhered with shot steep asphalt. When rigid polyurethane, polystyrene, or polyisocyanurate insulation is used, the top layer should be perlite or gypsum board to provide a stable underlayment for the roofing membrane and to comply with building code requirements.

- In the protected membrane system, the thermal insulation is placed over the roofing membrane. In this position, the insulation protects the roofing membrane from temperature extremes but not from almost continual dampness. The thermal insulation consists of moisture-resistant extruded polystyrene boards laid loosely or adhered to the roofing membrane with hot asphalt. The insulation is protected from sunlight and held in place by stone ballast laid over a filtration fabric, an integrally bonded concrete facing, or interlocking concrete blocks.

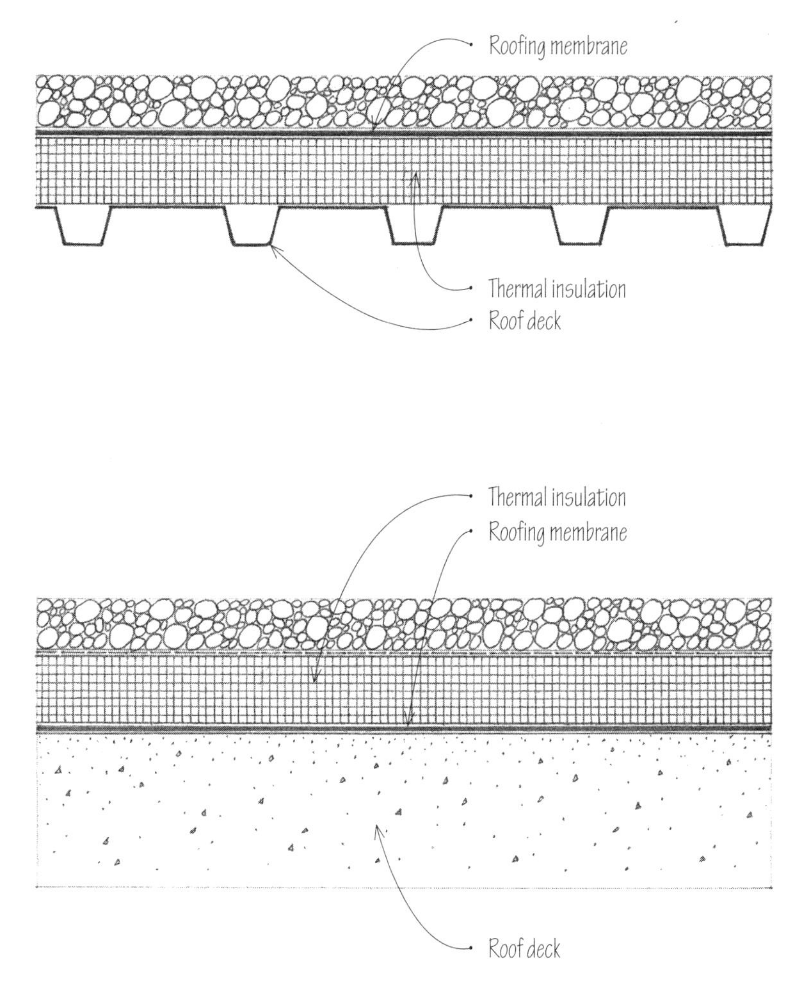

CSI MasterFormat 07 22 00: Roof and Deck Insulation
CSI MasterFormat 07 55 00: Protected Membrane Roofing

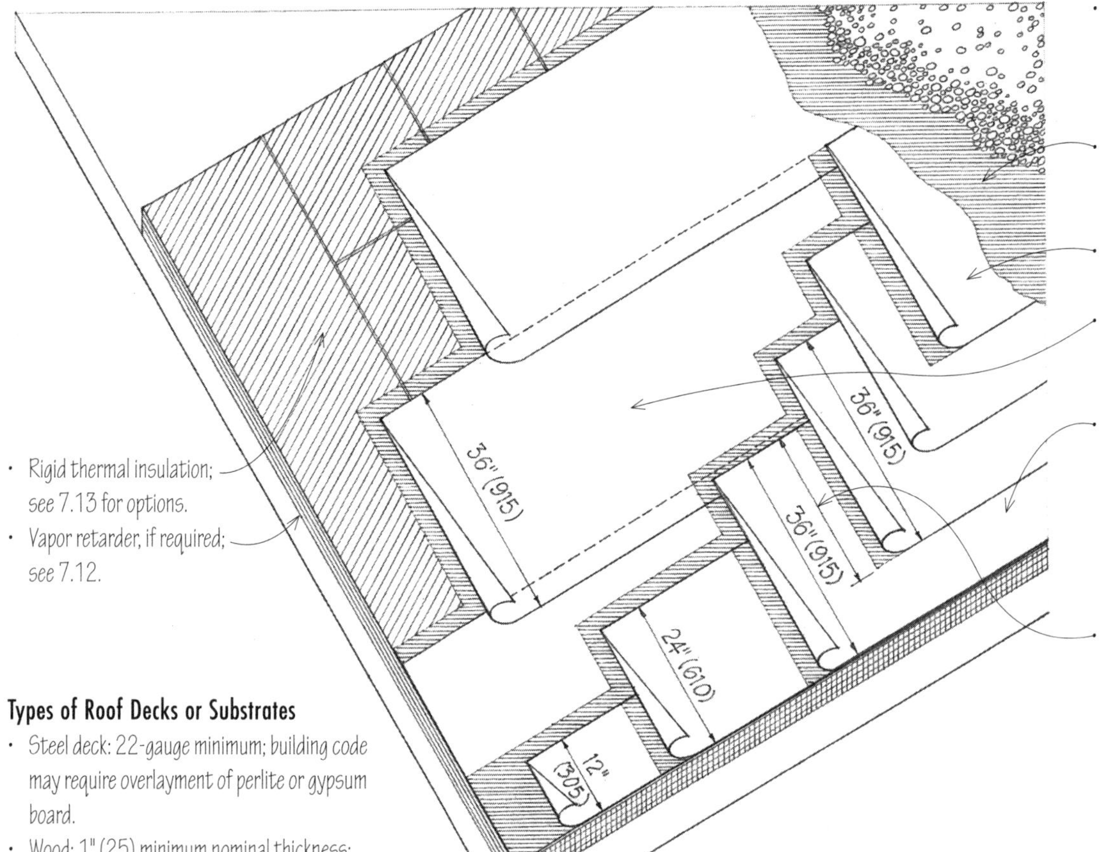

- Wear course of gravel, slag, or marble chips aids in stiffening membrane and resisting wind blowoff.
- Surfacing bitumen of coal tar or asphalt
- Cap sheet of coated, mineral-surfaced felt
- Base sheet of glass fiber with 2" (51) laps, or organic base roofing felt with 4" (100) laps
- Plysheets of fiberglass, asphalt-saturated felt, or coat tar-saturated felt placed with hot steep asphalt or coal tar bitumen
- Felt overlap
  - 2-ply roofing: 19" (485)
  - 3-ply roofing: 24-2/3" (625)
  - 4-ply roofing: 27-1/2" (700)

- Rigid thermal insulation; see 7.13 for options.
- Vapor retarder, if required; see 7.12.

Types of Roof Decks or Substrates

- Steel deck: 22-gauge minimum; building code may require overlayment of perlite or gypsum board.
- Wood: 1" (25) minimum nominal thickness; should be of well-seasoned lumber with tongue-and-groove or splined joints; knotholes and large cracks should be covered with sheet metal.
- Plywood: 1/2" (13) minimum thickness, laid with face plies perpendicular to supports spaced no more than 24" (610) o.c.; tongue-and-groove joints or blocking under joints.
- Structural wood-fiber deck must be dense enough to hold mechanical fasteners.
- Cast-in-place concrete should be well-cured, dry, frost-free, smooth, and sloped to drain.
- Precast concrete must have all joints grouted; any unevenness between units must be leveled with a vented topping or fill.
- Lightweight insulating concrete: must be fully cured and air dried; consult roofing manufacturer for acceptability as a roof deck.
- Consult roofing manufacturer for approved types of deck, insulation, and fastenings, installation details, requirements for vapor retarder and venting, and the Underwriters' Laboratories (UL) Fire-Hazard Classification of the roofing assembly.

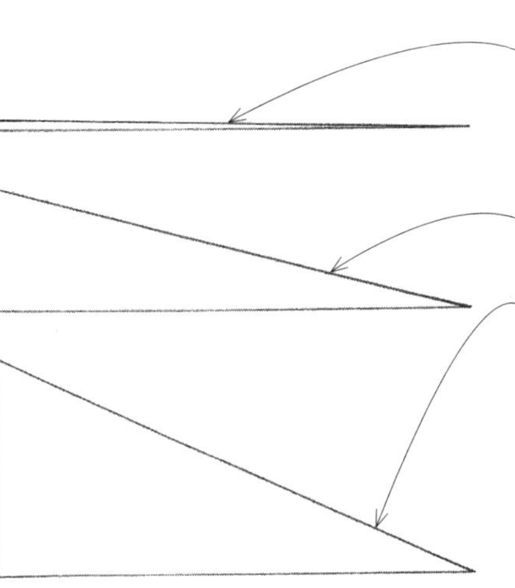

- Minimum recommended slope – 1:48
- Maximum slope for aggregate surfaces – 3:12
- Maximum slope for smooth-surface roofs – 6:12 to 9:12
- For slopes over 1:12, lay plies parallel to slope and back nail to prevent slippage; provide non-nailable decks with treated wood nailers for back- nailing.

CSI MasterFormat 07 51 00: Built-Up Bituminous Roofing

Single-ply membrane roofing may be applied in liquid or sheet form. Large domed, vaulted, or complex roof forms require that the roofing membrane be rolled or sprayed on in liquid form. Materials used for liquid-applied membranes include silicone, neoprene, butyl rubber, and polyurethane. On planar roof forms, the roofing membrane may be applied in sheet form. Sheet materials used for single-ply roofing include:

- Thermoplastic membranes, which may be heat- or chemically welded.
- PVC (polyvinyl chloride) and PVC alloys
- Polymer-modified bitumens, asphaltic materials to which polymers have been added for increased flexibility, cohesion, and toughness; often reinforced with glass fibers or plastic films
- Thermosetting membranes, which can be bonded only by adhesives.
- EPDM (ethylene propylene diene monomer), a vulcanized elastomeric material
- CSPE (chlorosulfonated polyethylene), a synthetic rubber
- Neoprene (polychloroprene), a synthetic rubber

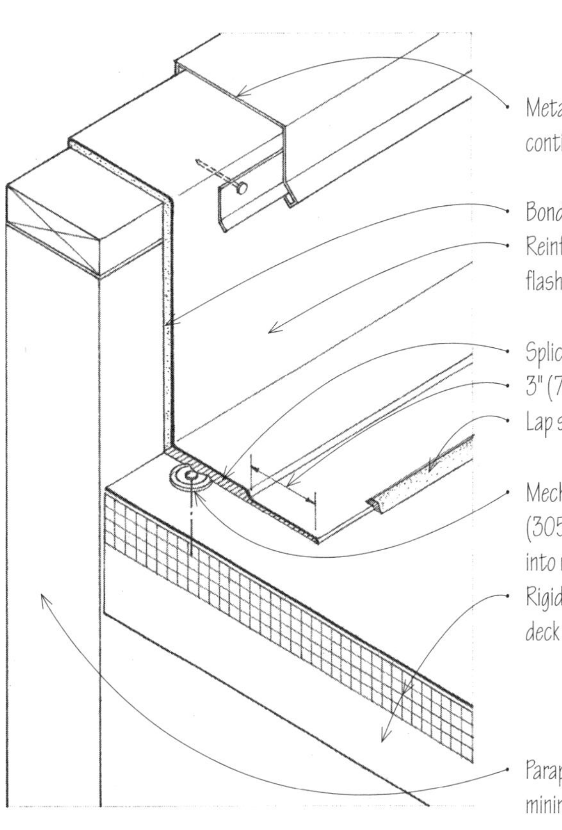

These materials are very thin – from 0.03" to 0.10" (0.8 to 2.5) thick – flexible, and strong. They vary in their resistance to flame propagation, abrasion, and degradation from ultraviolet rays, pollutants, oils, and chemicals. Some are reinforced with glass fiber or polyester; others have coatings for greater heat-reflectance or resistance to flame spread. Consult the roofing manufacturer for:

- Material specifications
- Approved types of roof deck, insulation, and fastenings
- Installation and flashing details
- Underwriters' Laboratories (UL) Fire-Hazard Classification of the roofing assembly

The details on this and the following page refer to EPDM roofing. Details for other single-ply membranes are similar in principle. There are three generic systems for the application of EPDM roofing:

- Fully adhered system
- Mechanically fastened system
- Loose laid, ballasted system

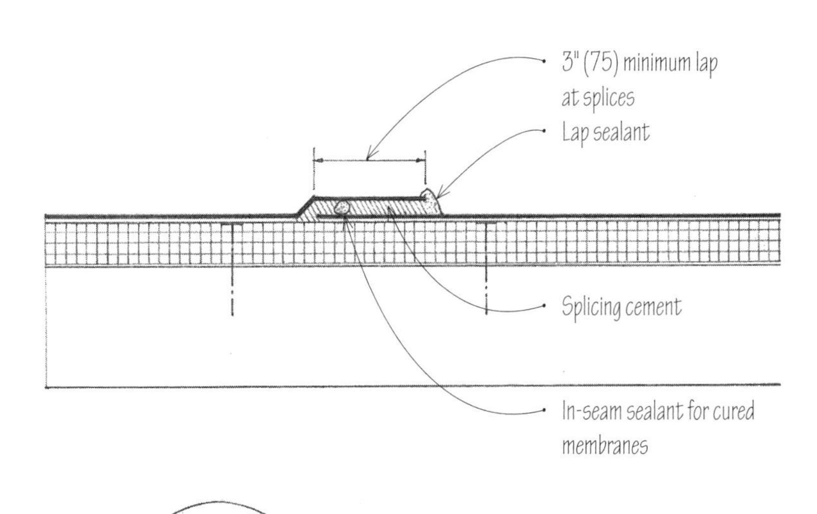

Fully Adhered System

The membrane is fully adhered with bonding adhesive to a smooth-surfaced concrete or wood deck, or to rigid insulation boards that are mechanically fastened to the roof deck. The membrane is mechanically fastened along the perimeter and at roof penetrations.

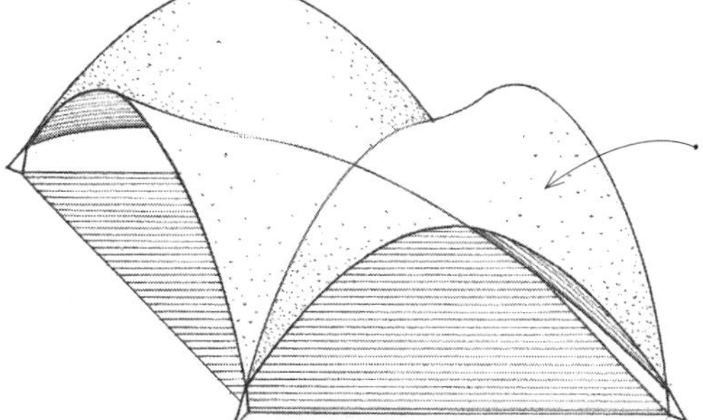

CSI MasterFormat 07 52 00: Modified Bituminous Membrane Roofing
CSI MasterFormat 07 53 00: Elastomeric Membrane Roofing
CSI MasterFormat 07 54 00: Thermoplastic Membrane Roofing

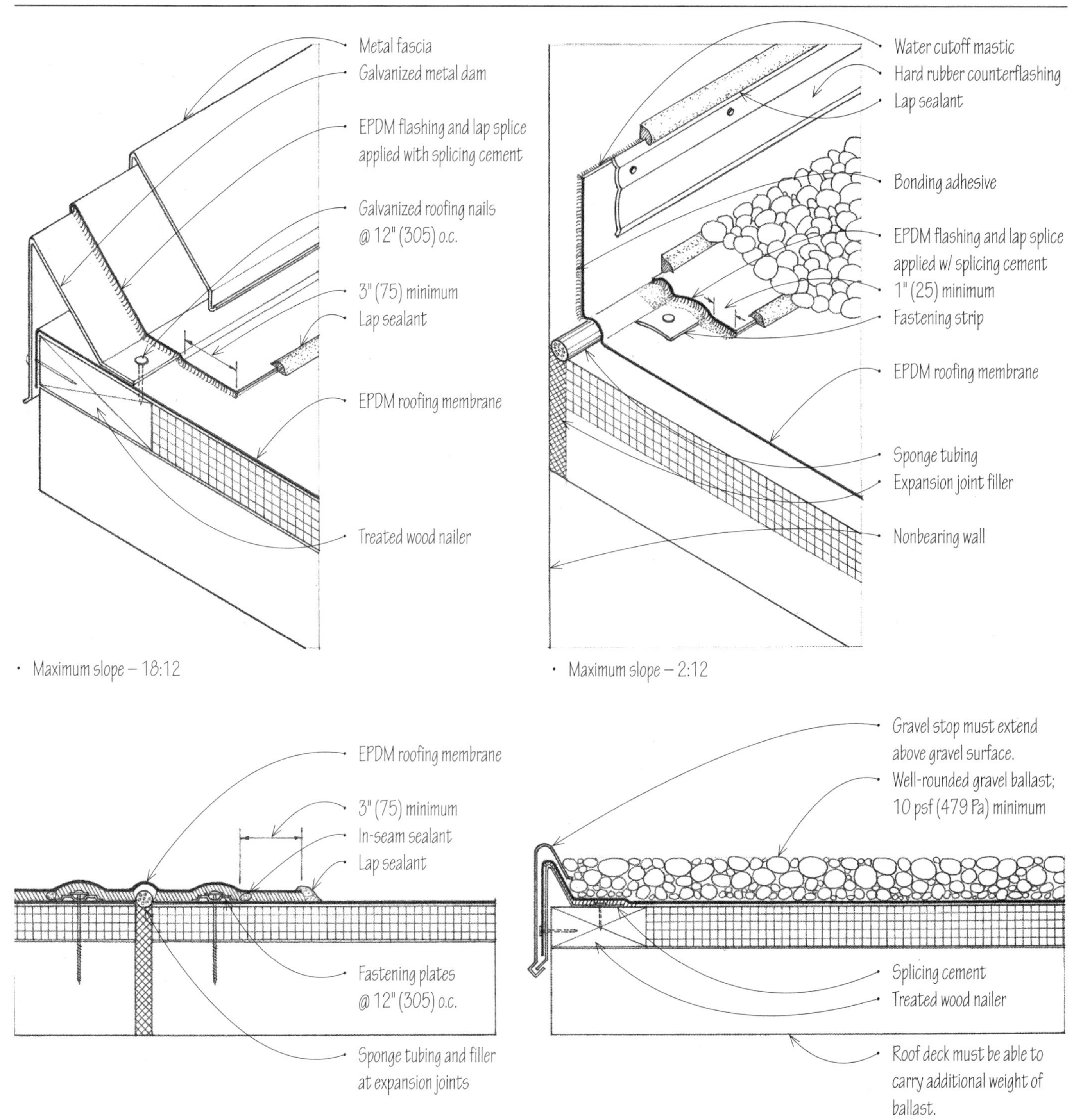

Mechanically Fastened System

After the thermal insulation boards have been mechanically fastened to the roof deck, the membrane is also secured to the deck with plates and fasteners in the membrane splices.

Loose Laid, Ballasted System

Both the insulation and the membrane are laid loosely over the roof deck and covered with a layer of river-washed gravel or a roof paver system. The membrane is mechanically fastened to the roof deck only along the perimeter and at roof penetrations.

The amount of rain or melting snow a roof and its drainage system must handle is a function of:

- The roof area leading to the roof drains or gutters
- The frequency and intensity of the rainfall for the region

Flat roofs should be pitched to roof drains that are located at the low points and that connect to the storm drain system of the building. A system of scuppers or overflow drains may also be required with the inlet flow 2" (51) above the low points of the roof.

Rainwater shed by sloping roofs should be caught by gutters along the eave to prevent ground erosion. Gutters empty into vertical downspouts or leaders that, in turn, discharge into a dry well or storm sewer system. In dry climates or for small roof areas with adequate overhangs, gutters may be omitted and a bed of gravel or a masonry strip set in the ground under the eave line.

Gutters are typically of vinyl, galvanized steel, or aluminum, although copper, stainless steel, terne metal, and wood ones are also available. Aluminum gutters can be cold-formed on-site in continuous runs without joints.

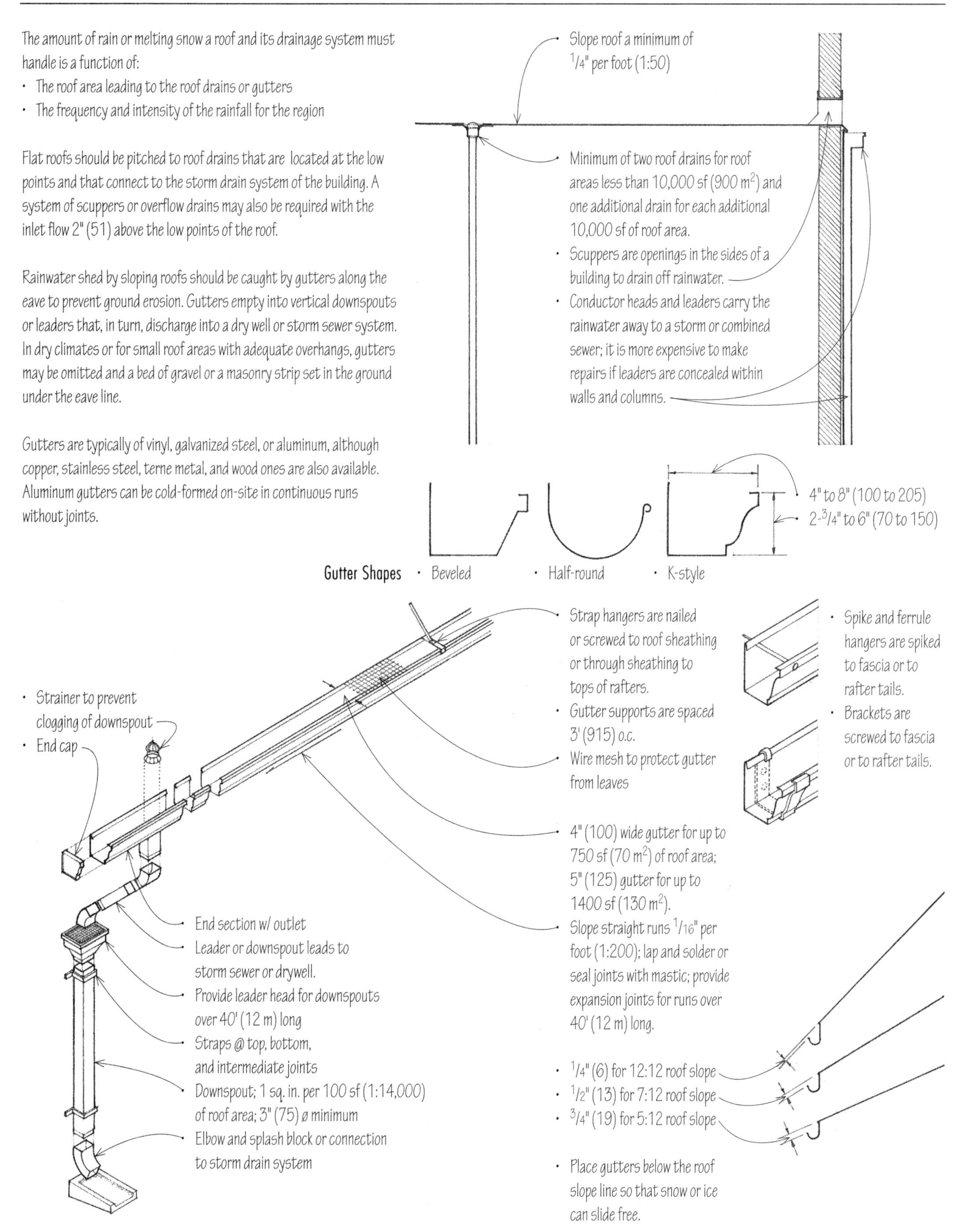

CSI MasterFormat 07 63 00: Sheet Metal Roofing Specialties

Flashing refers to thin continuous pieces of sheet metal or other impervious material installed to prevent the passage of water into a structure from an angle or joint. Flashing generally operates on the principle that, for water to penetrate a joint, it must work itself upward against the force of gravity, or, in the case of wind-driven rain, it would have to follow a tortuous path during which the driving force would be dissipated. See also 7.23 for a discussion of pressure-equalized rainscreen wall design.

Flashing may be exposed or concealed. Exposed flashing is usually of a sheet metal, such as aluminum, copper, painted galvanized steel, stainless steel, zinc alloy, terne metal, or copper-clad lead. Metal flashing should be provided with expansion joints on long runs to prevent deformation of the metal sheets. The selected metal should not stain or be stained by adjacent materials or react chemically with them. See 12.09.

Flashing concealed within a construction assembly may be of sheet metal or a waterproofing membrane such as bituminous fabric or plastic sheet material, depending on climate and structural requirements.

- Aluminum and lead react chemically with cement mortar.
- Some flashing materials can deteriorate with exposure to sunlight.

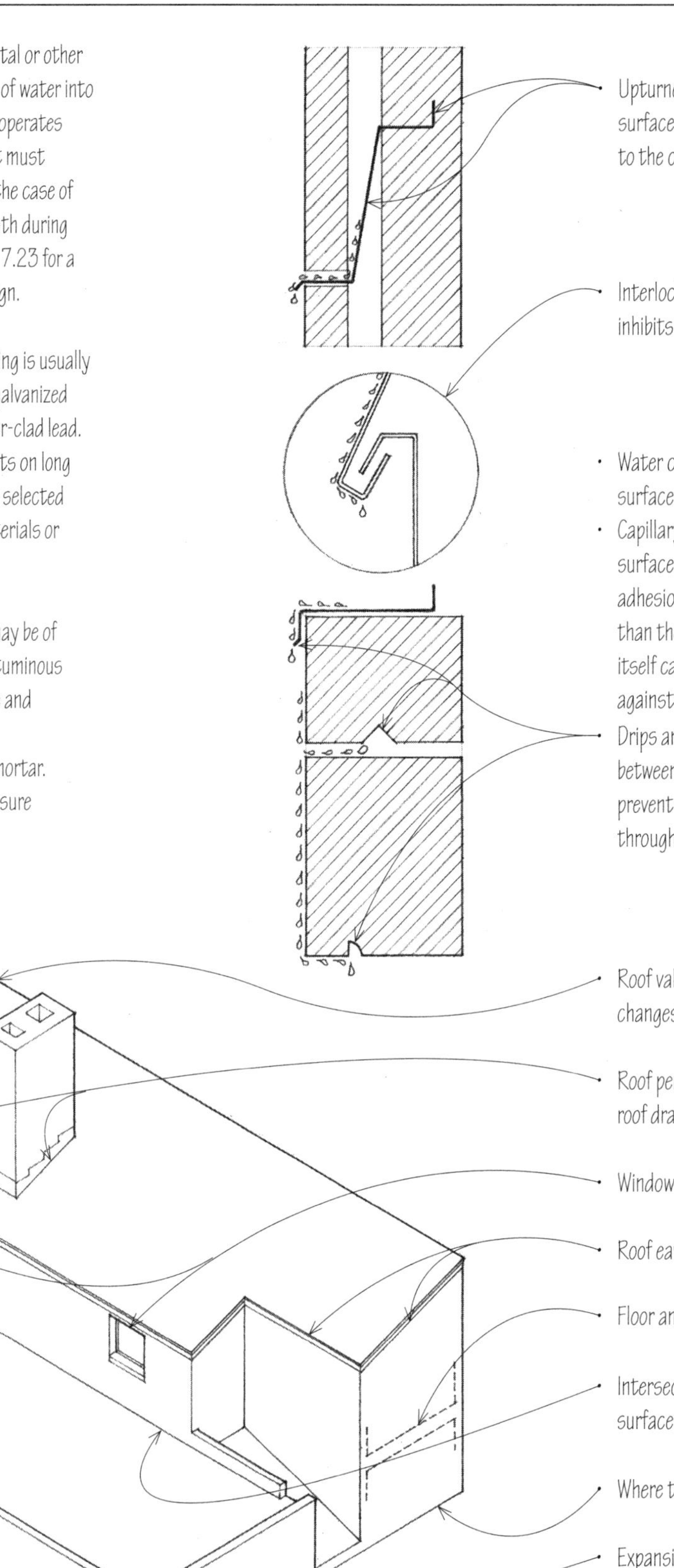

Flashing Locations

CSI 07600: Flashing
CSI 07620: Sheet Metal Flashing
CSI 07650: Flexible Flashing

The flashing details on this and the following pages illustrate general conditions and can be adapted for use with various building materials and assemblies. All dimensions are minimum. Weather conditions and roof slope may dictate greater overlaps. Consult the manufacturer for details of flashing and flashing accessories.

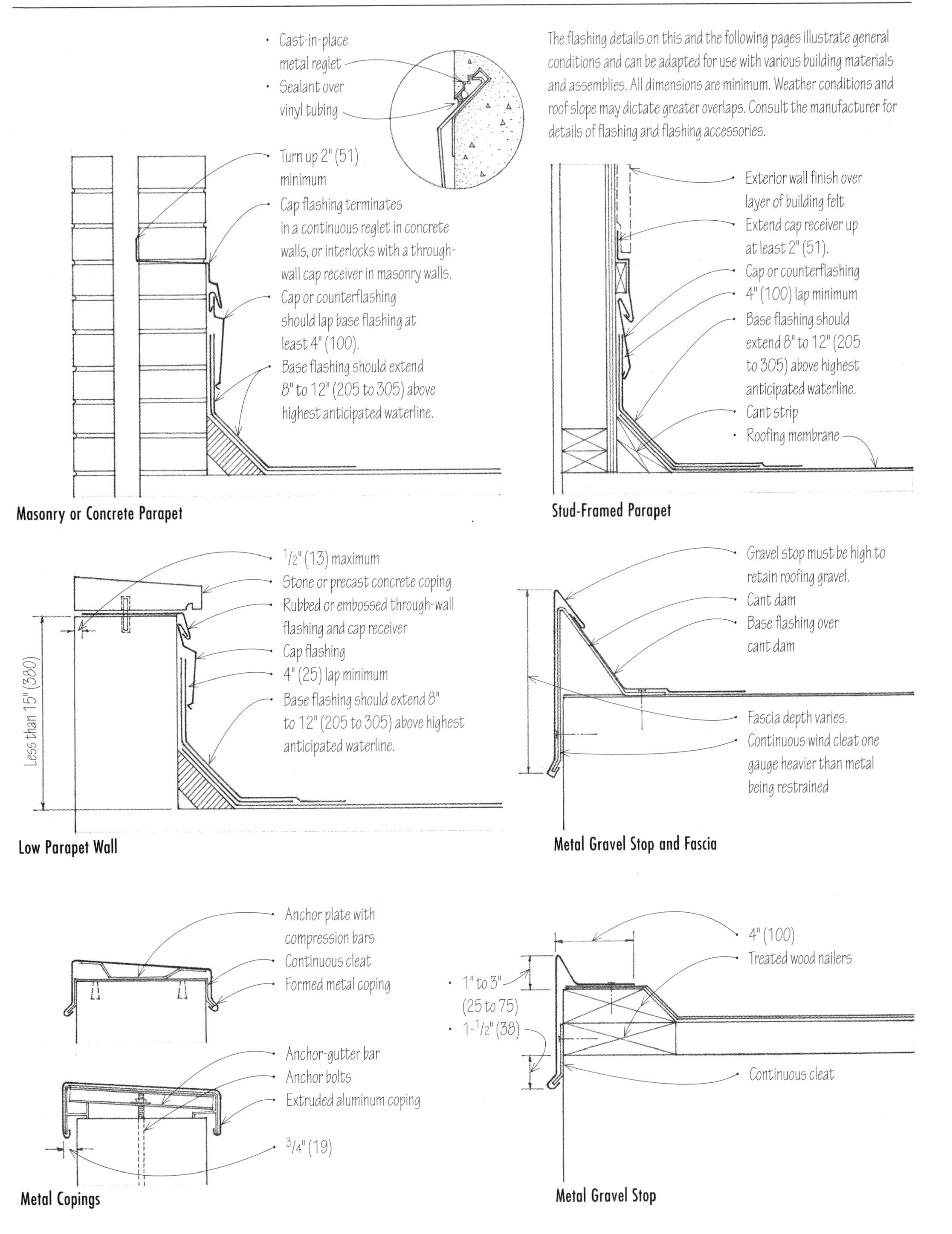

Masonry or Concrete Parapet

Stud-Framed Parapet

Low Parapet Wall

Metal Gravel Stop and Fascia

Metal Copings

Metal Gravel Stop

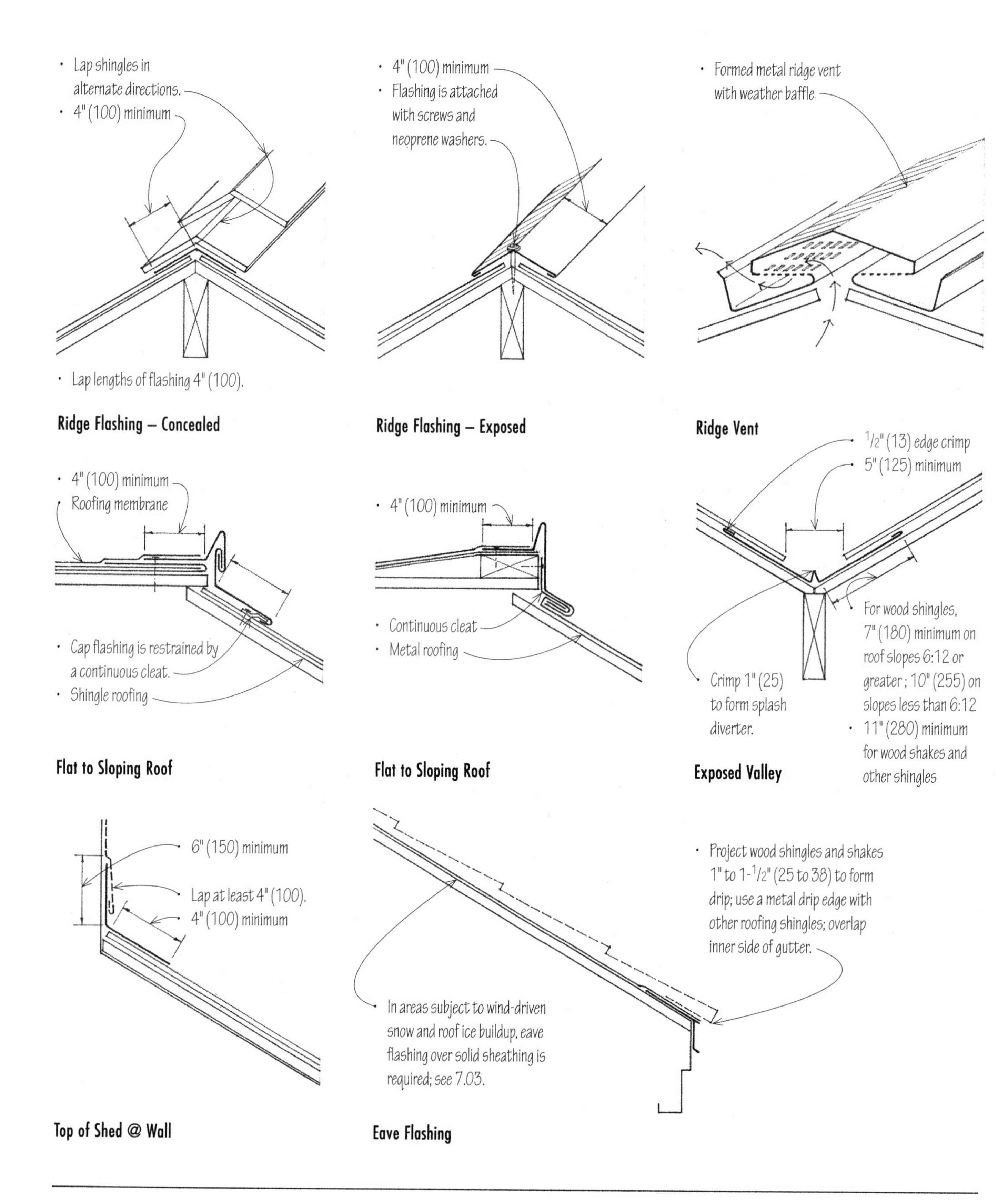
Lap shingles in alternate directions.
4" (100) minimum
Lap lengths of flashing 4" (100).
Ridge Flashing – Concealed
4" (100) minimum
Flashing is attached with screws and neoprene washers.
Ridge Flashing – Exposed
Formed metal ridge vent with weather baffle.
Ridge Vent
4" (100) minimum
Roofing membrane
Cap flashing is restrained by a continuous cleat.
Shingle roofing
Flat to Sloping Roof
4" (100) minimum
Continuous cleat
Metal roofing
Flat to Sloping Roof
1/2" (13) edge crimp
5" (125) minimum
Crimp 1" (25) to form splash diverter.
For wood shingles, 7" (180) minimum on roof slopes 6:12 or greater; 10" (255) on slopes less than 6:12
11" (280) minimum for wood shakes and other shingles
Exposed Valley
6" (150) minimum
Lap at least 4" (100).
4" (100) minimum
Top of Shed @ Wall
In areas subject to wind-driven snow and roof ice buildup, eave flashing over solid sheathing is required; see 7.03.
Project wood shingles and shakes 1" to 1-1/2" (25 to 38) to form drip; use a metal drip edge with other roofing shingles; overlap inner side of gutter.
Eave Flashing

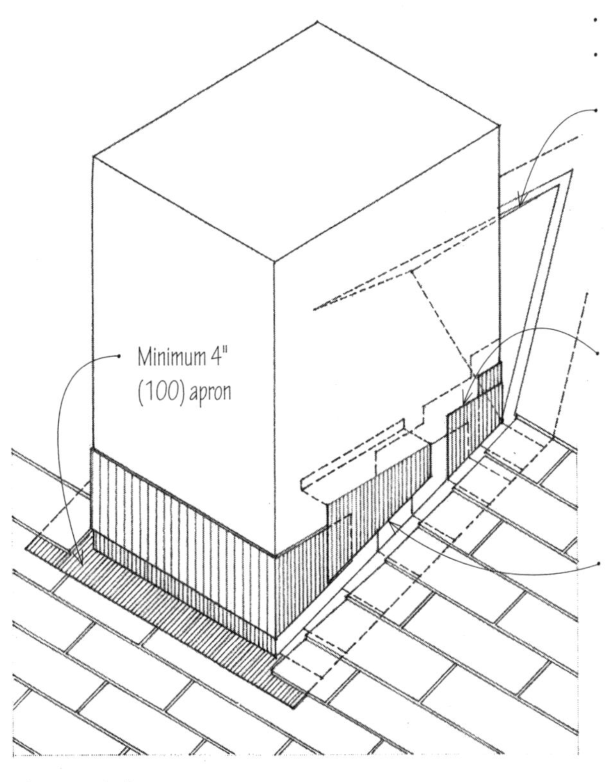

Chimney Flashing

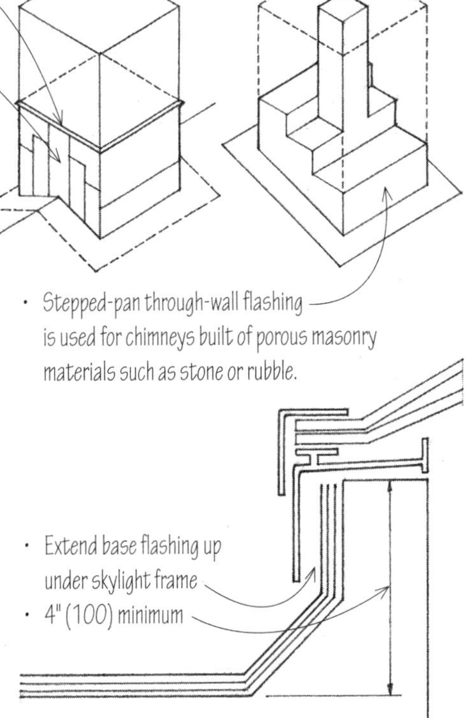

Skylight

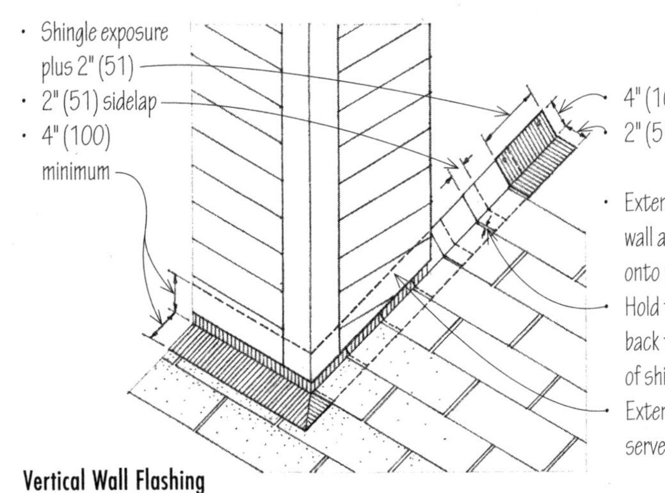

Vertical Wall Flashing

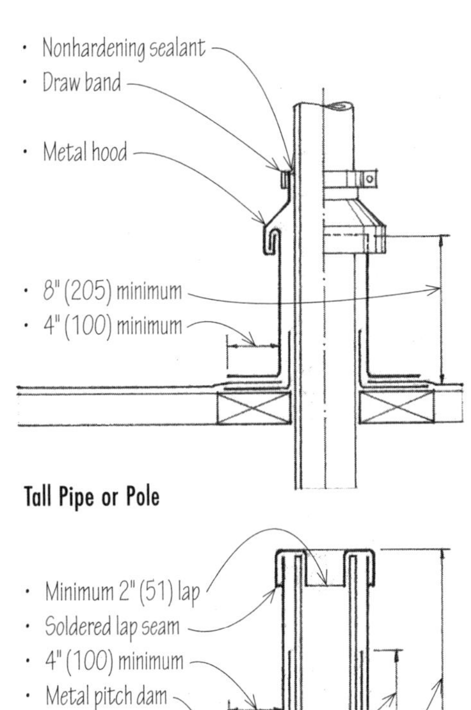

Tall Pipe or Pole

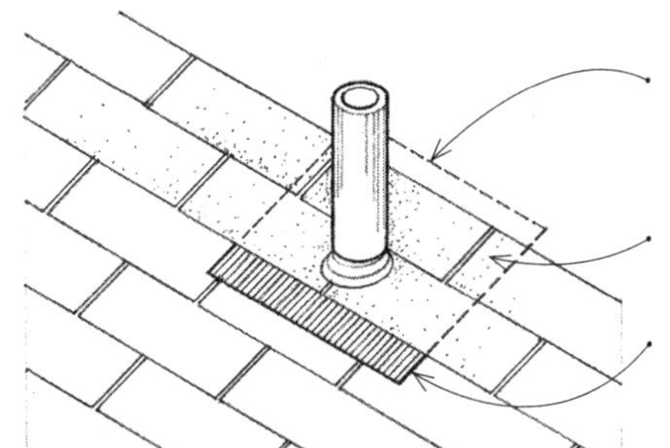

Stack Flashing

Vent Pipe

Wall flashing is installed to collect any moisture that may penetrate a wall and divert it to the outside through weep holes. The drawings on this page illustrate where wall flashing is usually required. Masonry walls are especially susceptible to water penetration. Rain penetration can be controlled by properly tooling mortar joints, sealing joints such as those around window and door openings, and sloping the horizontal surfaces of sills and copings. Cavity walls are especially effective in resisting the penetration of water.

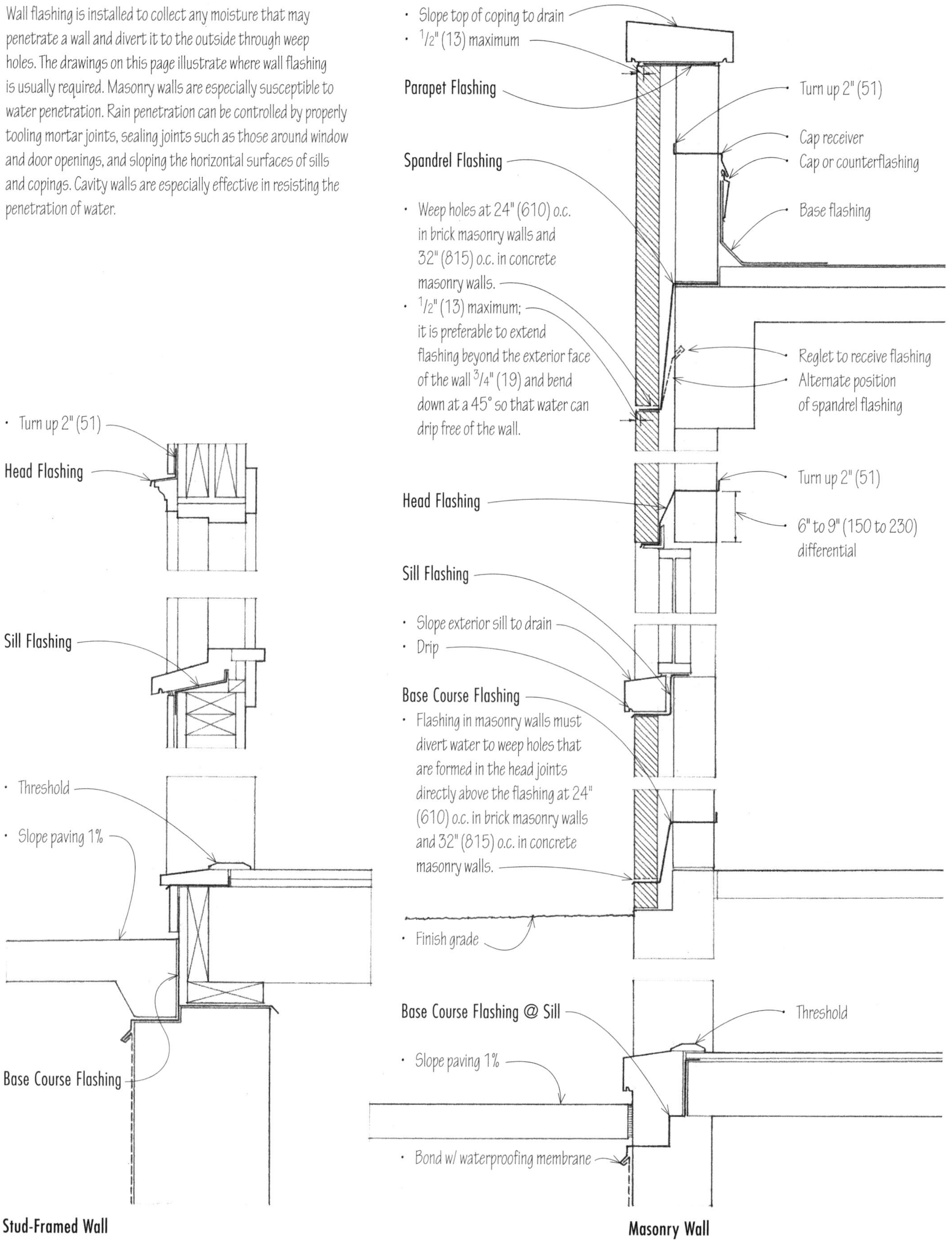

Water can penetrate exterior wall joints and assemblies by the kinetic energy of raindrops, gravity flow, surface tension, capillary action, and pressure differential. According to how exterior walls deter water penetration, they can be categorized as follows:

- Mass wall systems, such as concrete and solid masonry walls, shed most rain at the exterior face, absorb the remainder, and dry by releasing the absorbed moisture as vapor.
- Barrier wall systems, such as EIFS walls, rely on a continuous seal at the exterior face, which requires ongoing maintenance to be effective in resisting solar radiation, thermal movement, and cracking.
- Drainage walls, such as traditional stucco and clapboard walls, use a drainage plane or moisture barrier between the exterior cladding and supporting wall for additional moisture resistance.
- Rainscreen walls consist of an outer layer of cladding (the rainscreen), an air cavity, and a drainage plane on a rigid, water-resistant, and airtight support wall.

Simple rainscreen walls, such as brick cavity walls and furred-out clapboard walls, rely on cladding to shed most of the rain while the air cavity serves as a drainage layer to remove any water that may penetrate the outer layer. The cavity should be wide enough to prevent the capillary movement of this water from bridging the cavity and reaching the support wall.

Pressure differential can drive water through an opening in a wall assembly, no matter how small, when water is present on one side of the opening, and the air pressure on that side is greater than that on the other side. Pressure-equalized rainscreen (PER) walls utilize vented cladding and an air cavity, often divided into drainable compartments, to facilitate pressure equalization with the outside atmosphere and limit water penetration through joints in the cladding assembly. The primary seals against air and vapor are located on the indoor side of the air cavity, where they are exposed to little if any water.

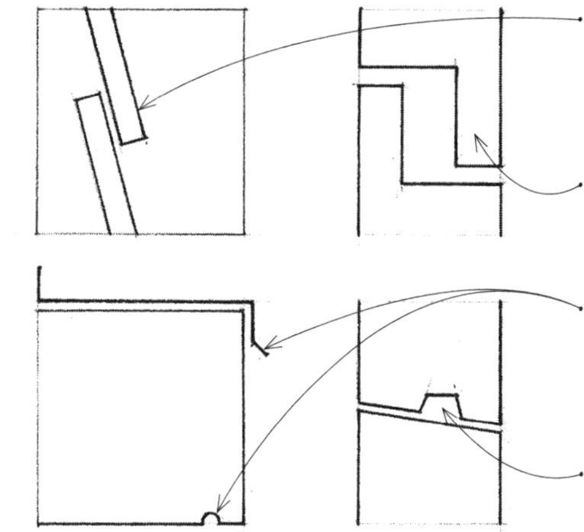

- Lapping horizontal joints in shingle fashion, sealing vertical joints, and sloping horizontal surfaces away from the interior can stem gravity flow.
- Overlapping materials or internal baffles deflect the kinetic energy of wind-driven raindrops.
- Drips break the surface tension that causes water to cling to and flow along the underside of horizontal, or nearly horizontal, surfaces.
- Discontinuities or air gaps disrupt the capillary movement of water.

- Lapped or shingled siding serves as a rainscreen.
- Furring strips space the siding material away from the wall framing, creating a vented cavity that is drained and back-ventilated to promote evaporation of any collected water.
- Sheathing and a weather barrier behind the furring strips create a drainage plane.
- Insect screening
- Metal flashing and drip

Interior Side

- An air-barrier system contains the primary joint seals, controls the flow of air and noise through the wall, and is airtight and rigid enough to withstand wind pressures.
- Thermal insulation is situated on the indoor side of the air cavity. The air barrier itself may be a continuous membrane placed on either side of the insulation or on either side of the interior wall layer.

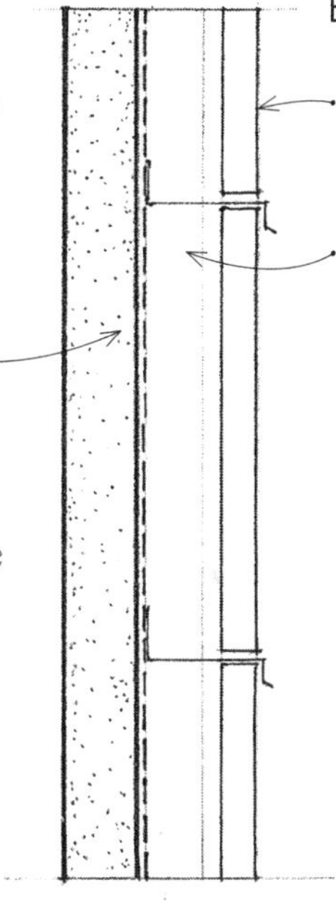

Exterior Side

- Vented cladding (the rainscreen) deflects the kinetic force of rain and deters water penetration at the exterior face of a wall.
- An air cavity provides a place for the equalization of air pressure to occur, is wide enough to prevent the capillary movement of water, and serves as a drainage layer for any water that manages to penetrate the rainscreen.

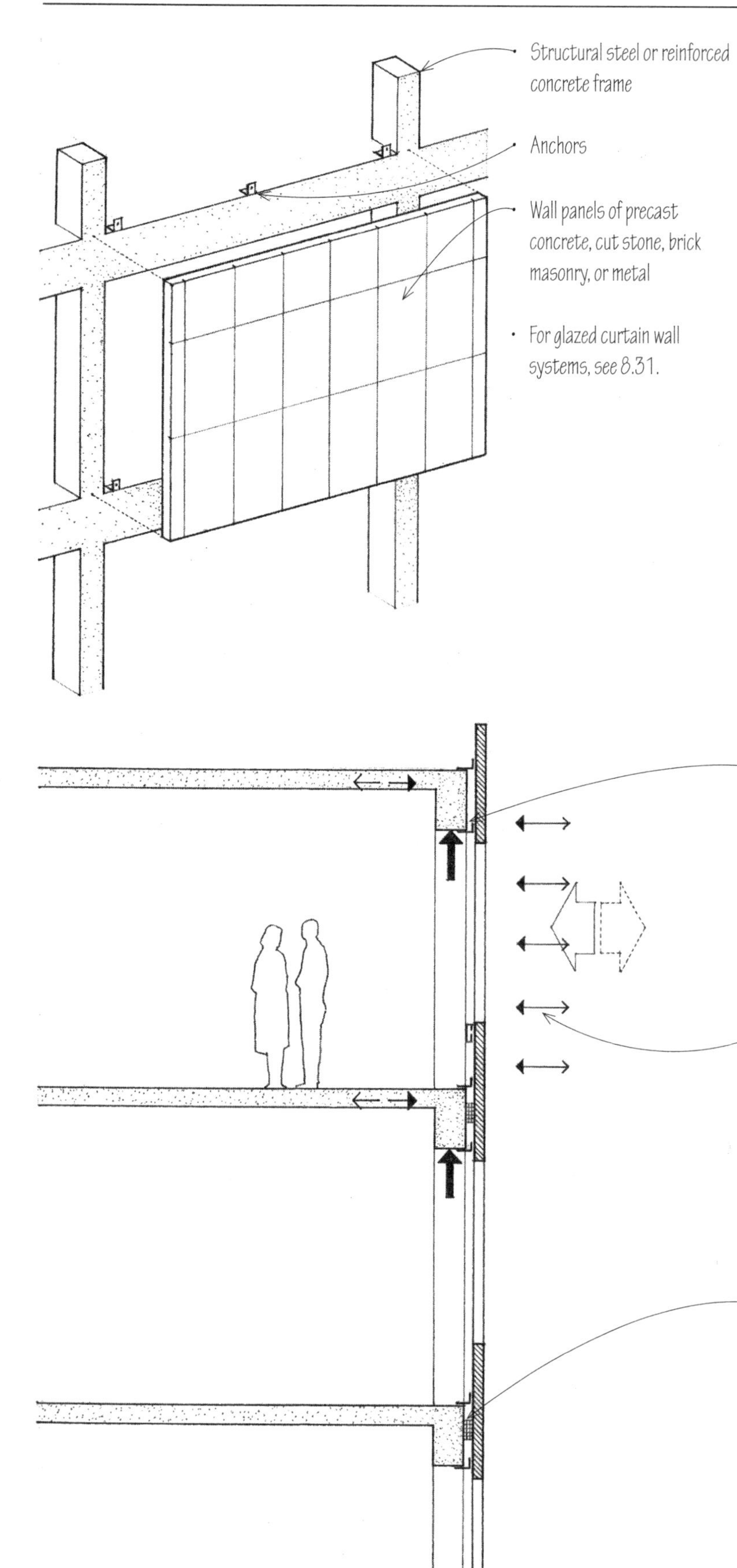

- For glazed curtain wall systems, see 8.31.

A curtain wall is an exterior wall supported wholly by the steel or concrete structural frame of a building and carrying no loads other than its own weight and wind loads. A curtain wall may consist of metal framing holding either vision glass or opaque spandrel units, or of thin veneer panels of concrete, stone, masonry, or metal.

Panel systems consist entirely of precast concrete, masonry, or cut stone units. The wall units may be one, two, or three stories in height, and may be preglazed or glazed after installation. Panel systems offer controlled shop assembly and rapid erection, but are bulky to ship and handle.

While simple in theory, curtain wall construction is complex and requires careful development, testing, and erection. Close coordination is also required between the architect, structural engineer, contractor, and a fabricator who is experienced in curtain wall construction.

As with other exterior walls, a curtain wall must be able to withstand the following elements:

Loads

- The curtain wall panels must be adequately supported by the structural frame.
- Any deflection or deformation of the structural frame under loading should not be transferred to the curtain wall.
- Seismic design requires the use of energy-absorbing connections.

Wind

- Wind can create both positive and negative pressure on a wall, depending on its direction and the shape and height of the building.
- The wall must be able to transfer any wind loads to the structural frame of the building without excessive deflection. Wind-induced movement of the wall should be anticipated in the design of its joints and connections.

Fire

- A noncombustible material, sometimes referred to as safing, must be installed to prevent the spread of fire at each floor within column covers and between the wall panels and the slab edge or spandrel beam.
- The building code also specifies the fire-resistance requirements for the structural frame and the curtain wall panels themselves.

Sun

- Brightness and glare should be controlled with shading devices or the use of reflective or tinted glass.
- The ultraviolet rays of the sun can also cause deterioration of joint and glazing materials and fading of interior furnishings.

Temperature

- Daily and seasonal variations in temperature cause expansion and contraction of the materials comprising a wall assembly, especially metals. Allowance must be made for differential movement caused by the variable thermal expansion of different materials.
- Joints and sealants must be able to withstand the movement caused by thermal stresses.
- Heat flow through glazed curtain walls should be controlled by using insulating glass, insulating opaque panels, and by incorporating thermal breaks into metal frames.
- Thermal insulation of veneer panels may also be incorporated into the wall units, attached to their backsides, or provided with a backup wall constructed on site.

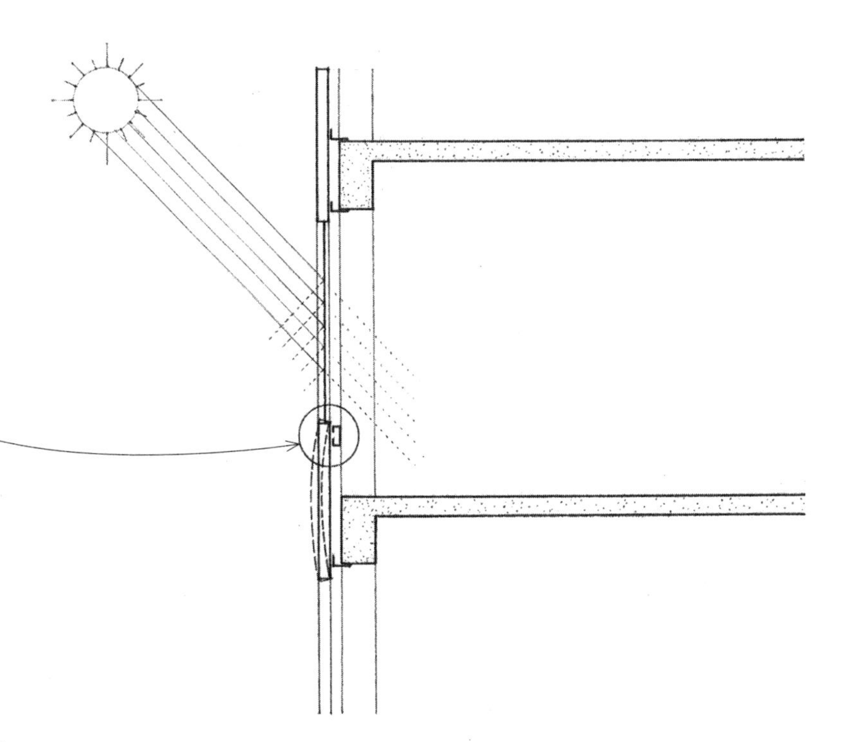

Water

- Rain can collect on the wall surface and be wind-driven under pressure through the smallest openings.
- Water vapor that condenses and collects within the wall must be drained to the outside.

Pressure-Equalized Design

The pressure-equalized design principles outlined in 7.23 become critical in the detailing of curtain walls, especially in larger and taller buildings, where the pressure differential between the outside atmosphere and an interior environment can cause rainwater to migrate through even the smallest openings in wall joints.

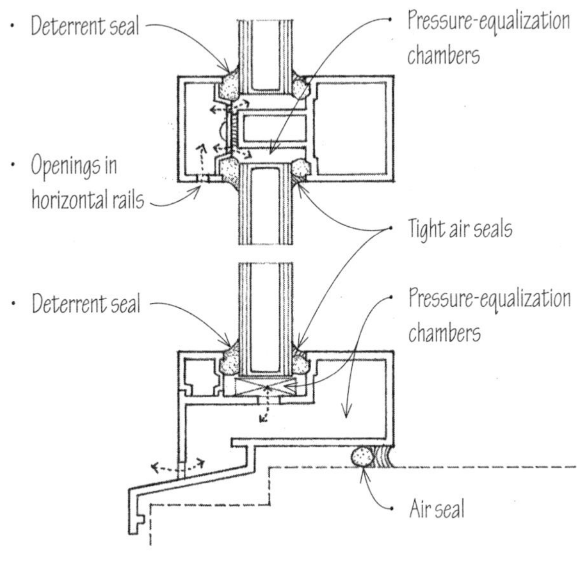

Application of Pressure-Equalization Principle in Glazed Curtain Wall

Any of various metal devices may be used to secure a curtain wall to the structural frame of a building. Some connections are fixed to resist loads applied from any direction. Others are designed to resist only lateral wind loads. These joints typically permit adjustment in three dimensions in order to allow for discrepancies between the dimensions of the curtain wall units and the structural frame, as well as to accommodate the differential movement when the structural frame deflects under loading or when the curtain wall reacts to thermal stresses and changes in temperature.

Shim plates and angles with slotted holes allow adjustments to be made in one direction; a combination of angles and plates allows adjustments to be made in three dimensions. After final adjustments are made, the connections can be permanently secured by welding if a fixed connection is required.

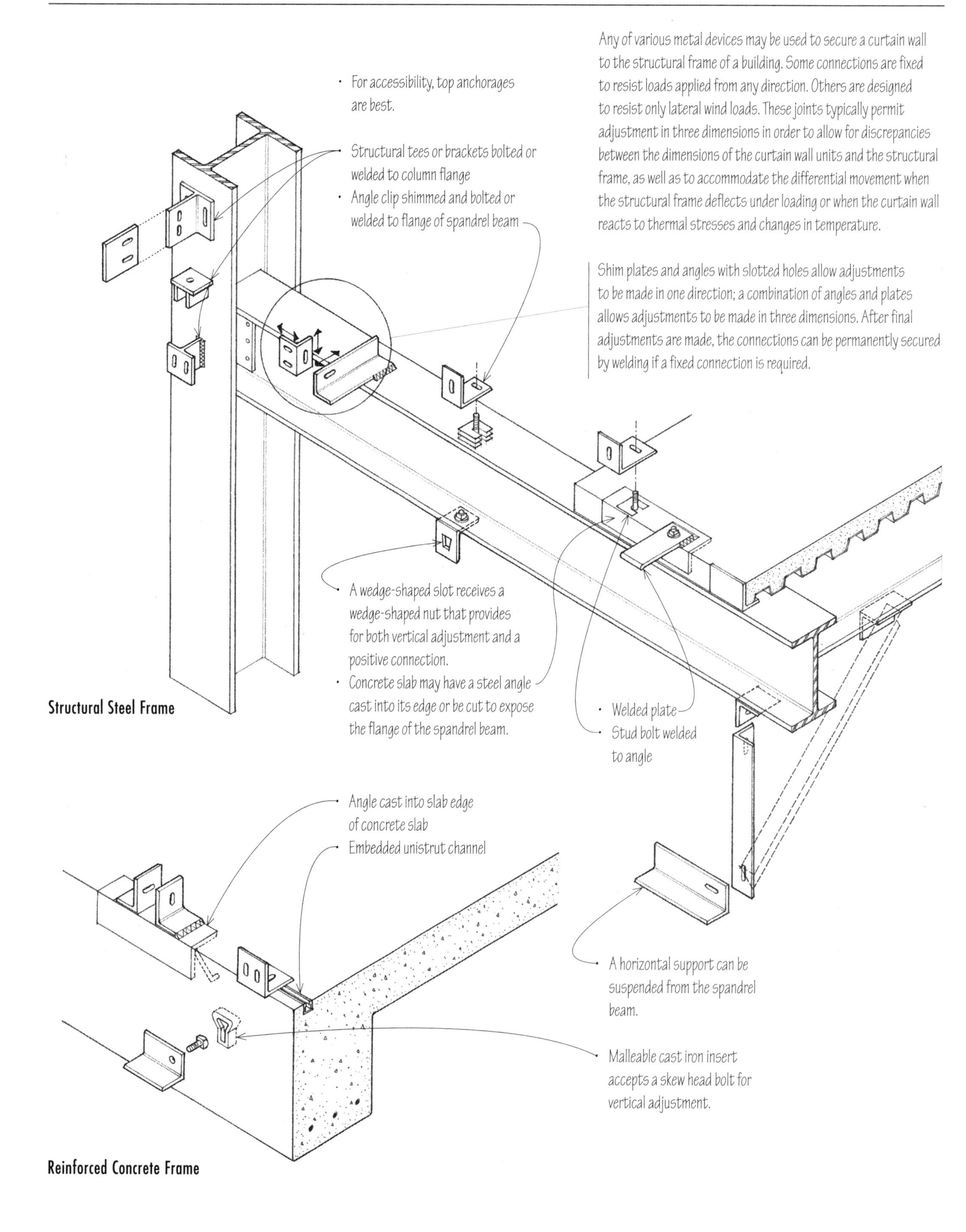

Structural Steel Frame

Reinforced Concrete Frame

Precast concrete wall panels may be used as nonbearing facings supported by a structural steel or reinforced concrete frame. See 5.10 for loadbearing precast wall panels.

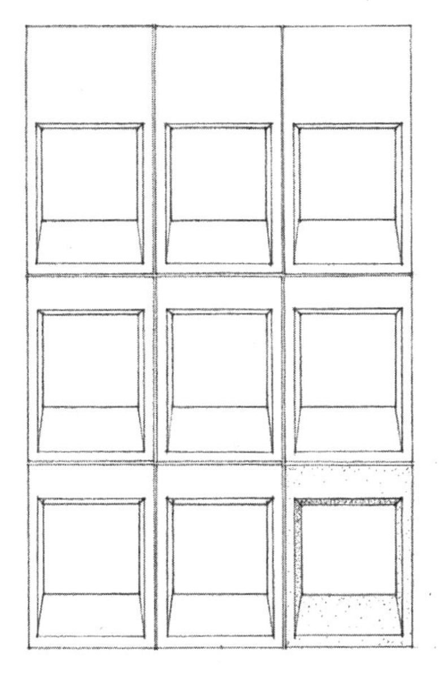

- A variety of quality-controlled smooth and textured finishes are available.
- Ceramic tile and thin brick or stone facings may be fixed to the wall panels.
- Thermal insulation may be sandwiched in the wall panel, attached to its backside, or provided with a backup wall constructed on site.

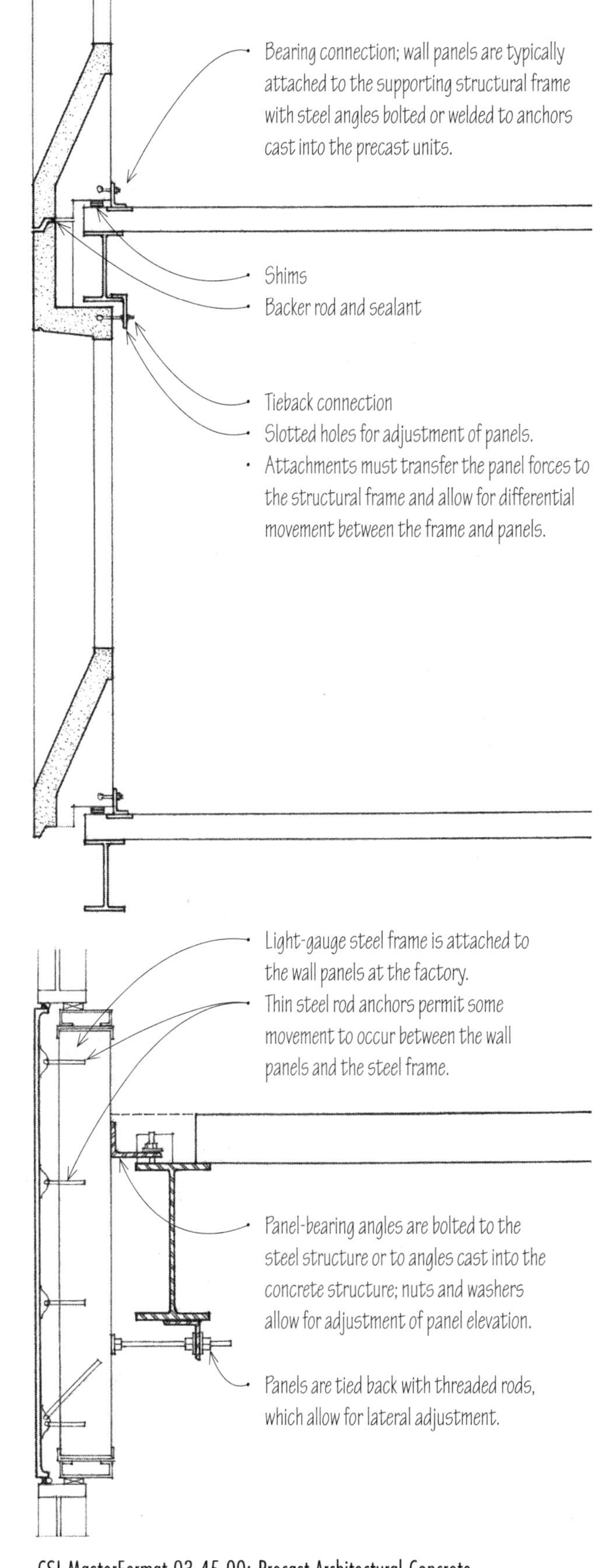

Glass-fiber-reinforced concrete can be used in place of conventionally reinforced concrete to produce much thinner and lighter veneer panels. The panels are produced by spraying short glass fibers onto a mold with a portland cement and sand slurry. A variety of three-dimensional panel designs and finishes are possible.

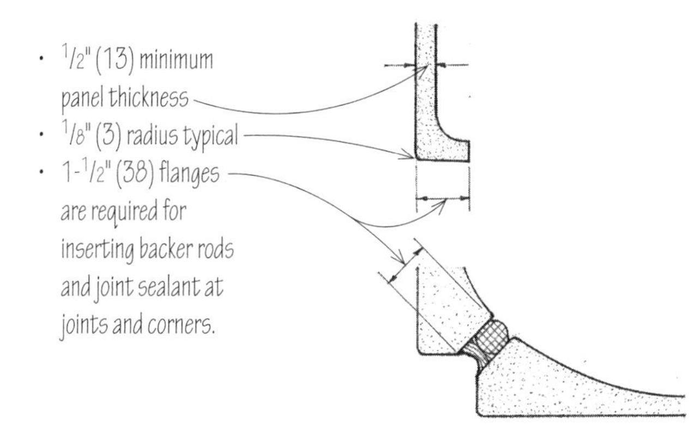

CSI MasterFormat 03 45 00: Precast Architectural Concrete
CSI MasterFormat 07 44 53: Glass-Fiber-Reinforced Cementitious Panels

Masonry veneer construction consists of a single wythe of masonry serving as a weather barrier and anchored but not bonded to a supporting structural frame. In residential construction, wood or metal stud walls are typically faced with a brick veneer.

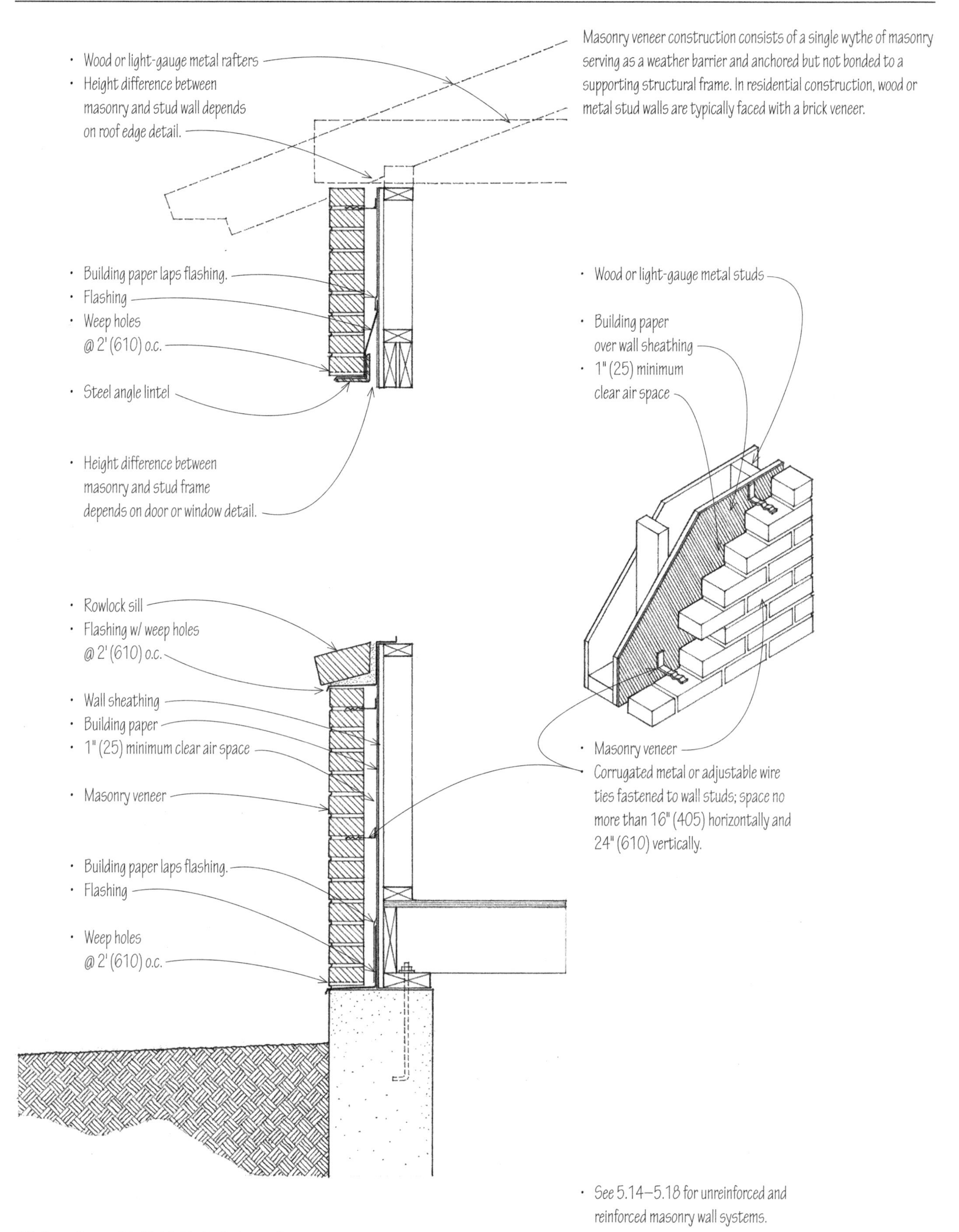

• See 5.14–5.18 for unreinforced and reinforced masonry wall systems.

CSI MasterFormat 04 21 13.13: Brick Veneer Masonry

Masonry veneers can also be used as curtain walls supported by steel or concrete frames.

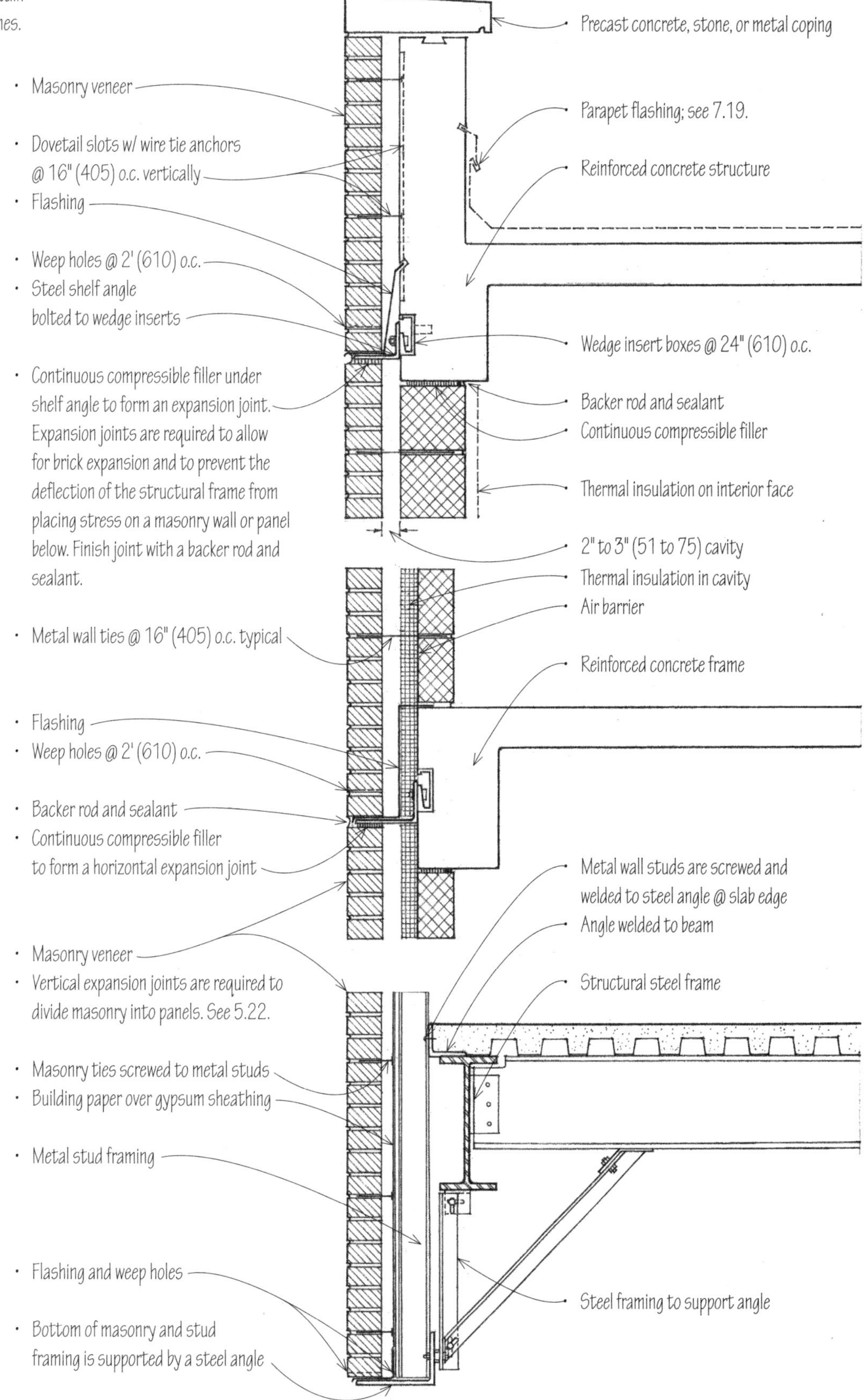

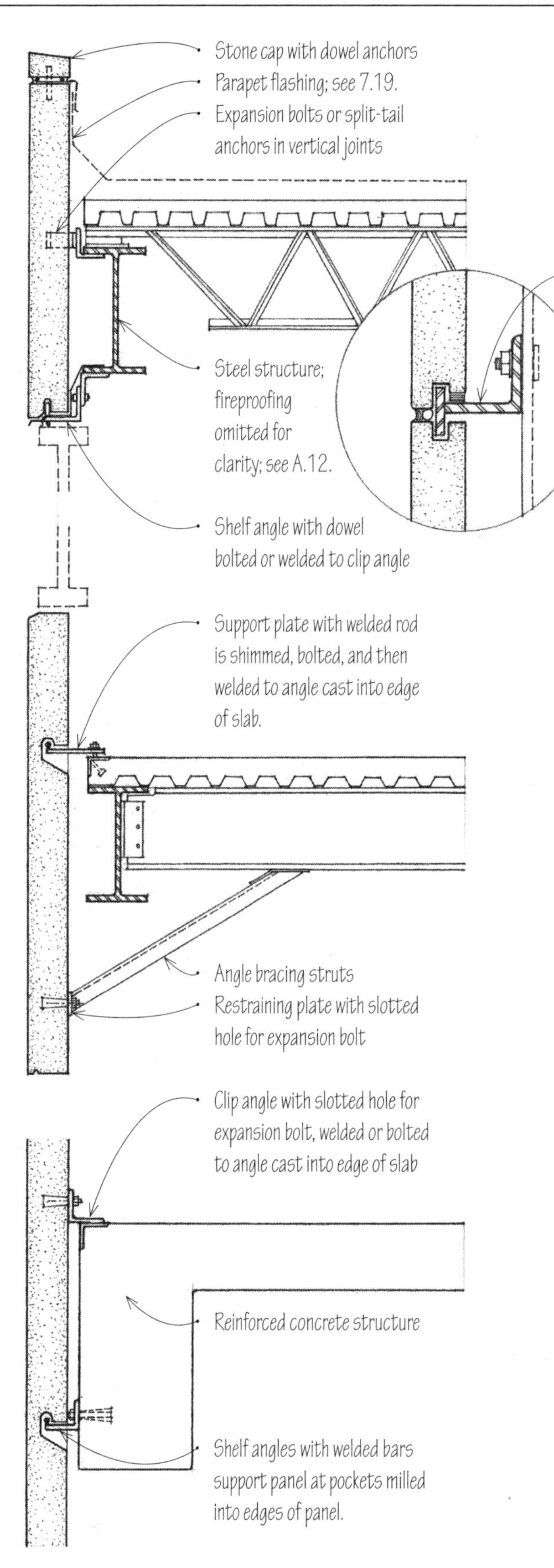

Typical Monolithic Stone Panel Details

Stone facings may be set in mortar and tied to a concrete or masonry backup wall; see 5.33–5.34. Large stone veneer panels from 1-1/2" to 3" (32 to 75) thick may also be supported by the steel or concrete structural frame of a building in a number of ways.

- Monolithic stone panels may be fastened directly to the structural frame of a building.
- Stone panels may be mounted on a steel subframe designed to transmit gravity and lateral loads from the slabs to the structural frame of a building. The subframe consists of vertical steel struts that support horizontal stainless steel or aluminum angles. Bars welded to the angles engage slots in the lower and upper edges of the stone panels.
- Stone veneers may be preassembled into larger panels by mounting the thin slabs on non-corrosive metal framing, or by bonding them to reinforced precast concrete panels with bent stainless steel anchors. A moisture barrier and bonding agent may be applied between the concrete and stone to prevent concrete salts from staining the stonework.

The required anchorages should be carefully engineered and take into account the strength of the stone veneer, especially at anchorage points, the gravity and lateral loads to be sustained, and the anticipated range of structural and thermal movement. Some anchors must carry the weight of the stonework and transfer the load to the supporting structural wall or frame. Others only restrain the stonework from lateral movement. Still others must offer resistance to shear. All connecting hardware should be of stainless steel or nonferrous metal to resist corrosion and prevent staining of the stonework. Adequate tolerances must be built in to allow for proper fitting and shimming, if necessary.

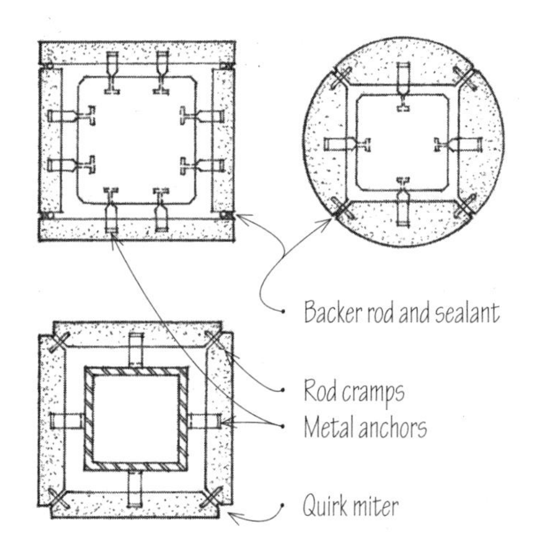

Columns

CSI MasterFormat 04 42 00: Exterior Stone Cladding

Insulated and bonded metal panels are used primarily to clad industrial-type buildings; see 6.07. They may have facings of anodized aluminum or steel with porcelain, vinyl, acrylic, or enamel finishes. The panels are typically 3' (915) wide and span vertically between horizontal steel girts spaced 8' to 24' (2.5 to 7.3 m) apart, depending on the type and profile of panel used. Consult manufacturer for profiles, sizes, allowable spans, thermal and acoustical ratings, and installation details.

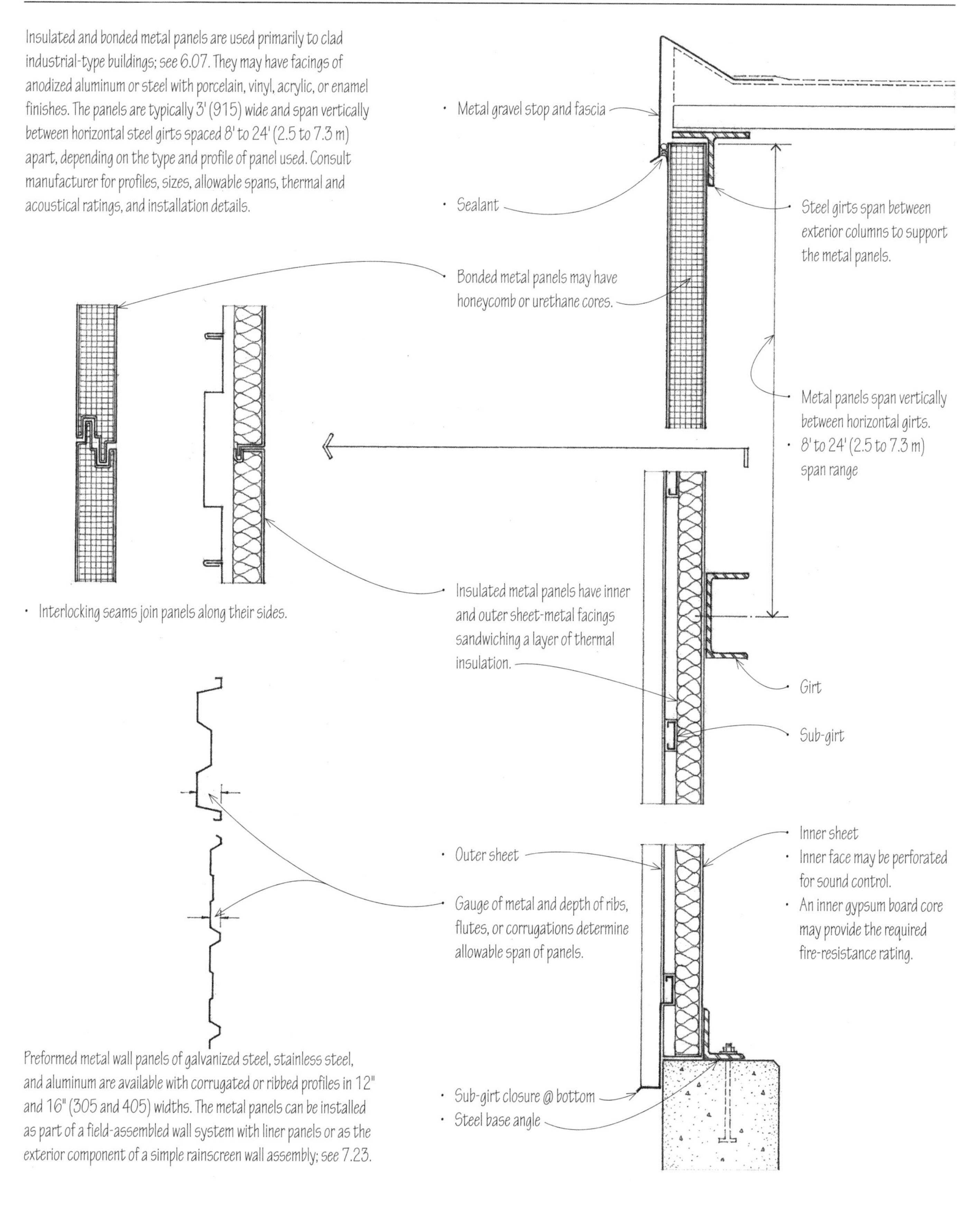

Preformed metal wall panels of galvanized steel, stainless steel, and aluminum are available with corrugated or ribbed profiles in 12" and 16" (305 and 405) widths. The metal panels can be installed as part of a field-assembled wall system with liner panels or as the exterior component of a simple rainscreen wall assembly; see 7.23.

CSI MasterFormat 07 42 13: Metal Wall Panels
CSI MasterFormat 13 34 19: Metal Building Systems

Exterior-grade plywood siding panels are typically 4' x 8' (1220 x 2440), although 9' and 10' (2.8 and 3.0 m) lengths are available. The most common patterns imitate vertical board siding. The panel surface may have a grooved, rough-sawn, brushed, or striated texture, and may be stained or treated with a clear water-repellant finish. Medium-density overlay (MDO) is an exterior plywood panel having a melamine or phenolic resin overlay on one or both sides, providing a smooth base for painting.

Horizontal joints, which must be protected by flashing or other means, are very noticeable. These horizontal lines should therefore be coordinated with other exterior wall elements such as window and door openings.

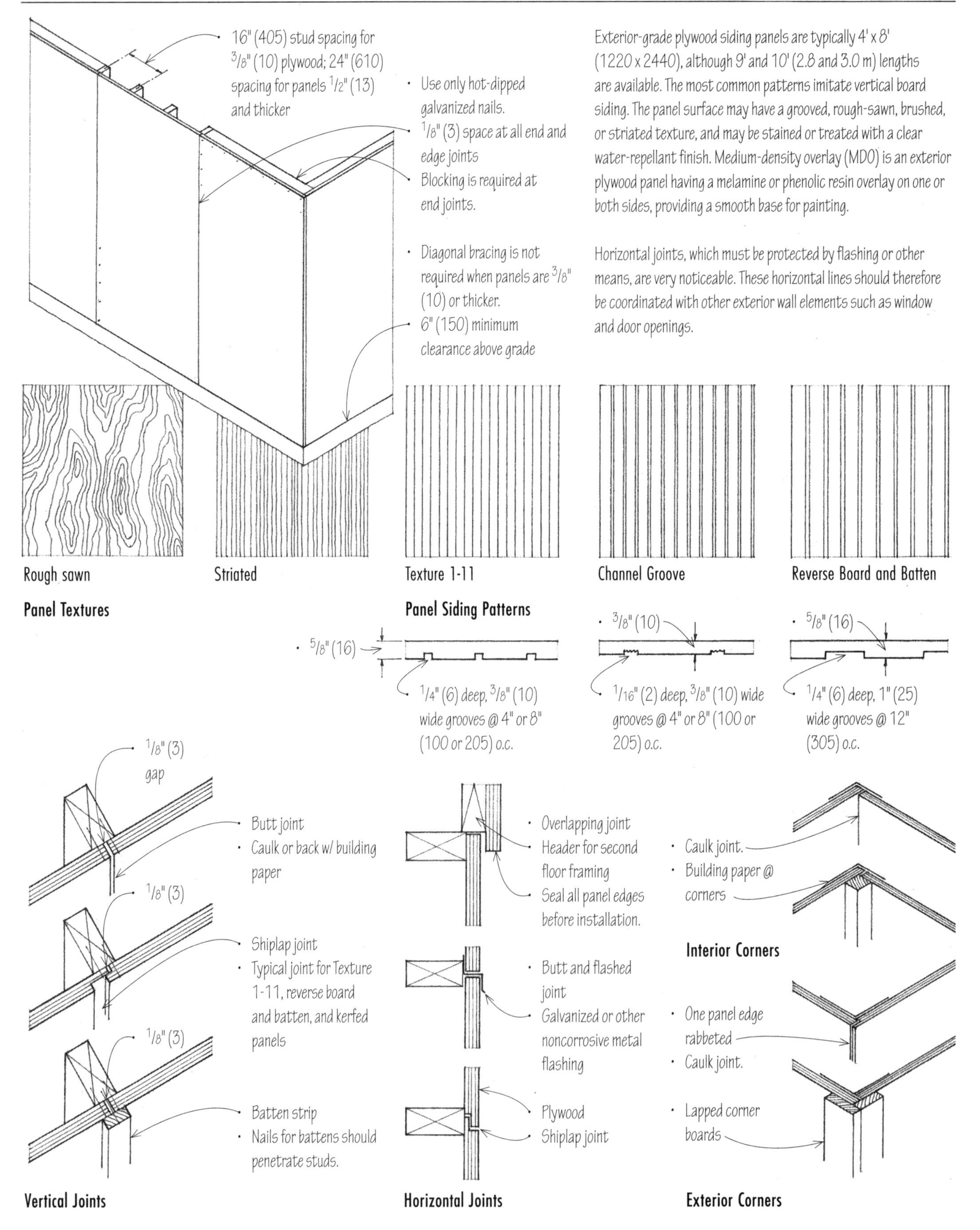

CSI MasterFormat 07 46 29: Plywood Siding

On exterior walls, wood shingles are laid in uniform courses that resemble lap siding. The courses should be adjusted to meet the heads and sills of window openings and other horizontal bands neatly. The shingles may be stained or painted. Premium-grade shingles can be left unpainted to weather naturally.

Wood shingle siding may be applied in single or double coursing, with the following exposures:

Shingle Length	Single Coursing	Double Coursing
16" (405)	6" to 7-1/2" (150 to 190)	8" to 12" (205 to 305)
18" (455)	6" to 8-1/2" (150 to 190)	9" to 14" (230 to 355)
24" (610)	8" to 11-1/2" (205 to 290)	12" to 20" (305 to 510)

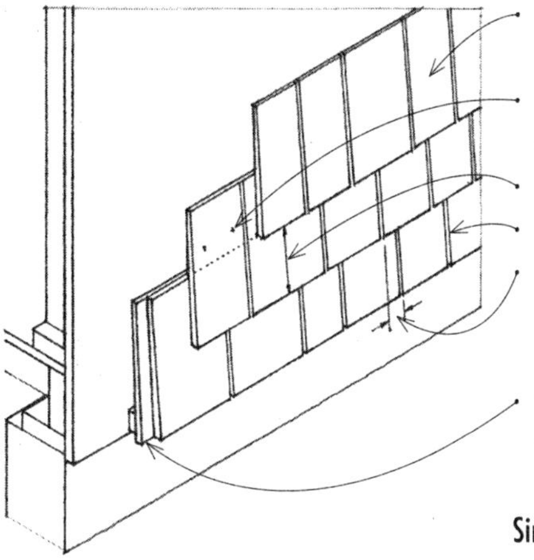

No. 2 red label shingles
Nail 2" (51) above beltline of succeeding course
Exposure (see table)
1/4" (6) joints
1-1/4" (32) minimum offset between joints
Double starting course; lap foundation wall 1" (25).

Single Coursing Application

No. 3 undercourse
No. 1 blue label shingles
Outer course is 1/2" (13) lower than undercourse.
Exposure (see table)
Triple starting course; lap foundation wall 1" (25).

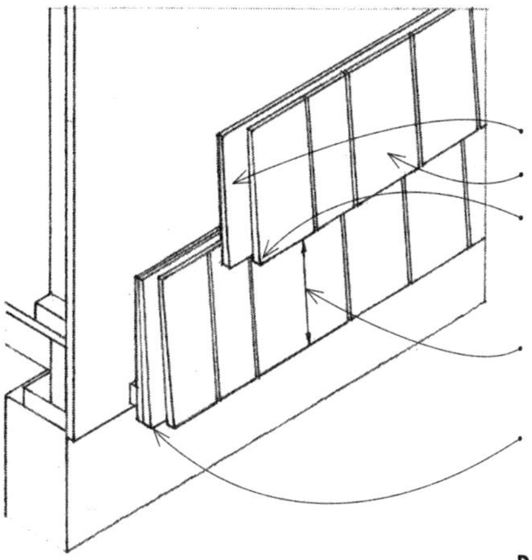

Double Coursing Application

Breathing type building paper
1x3s or 1x4s nailed to wall studs
Spacing equals shingle exposure
• Plywood or nailable fiberboard
• Spaced sheathing is required over non-nailable sheathing or siding

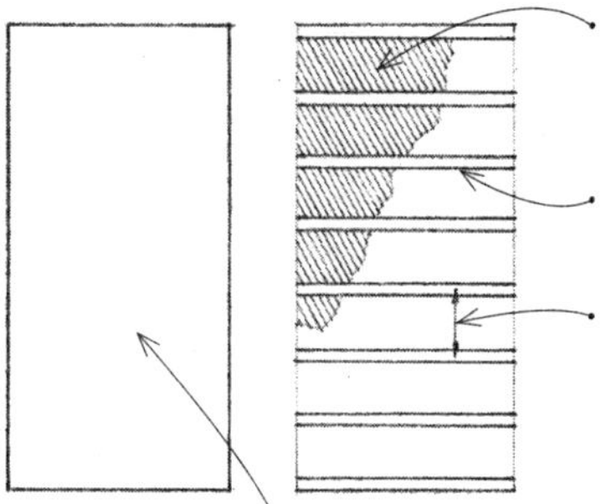

Types of Sheathing

Dimension and fancy butt shingles are cut to uniform widths and shapes. They are used on walls to create certain effects such as scalloped or fish-scale textures.

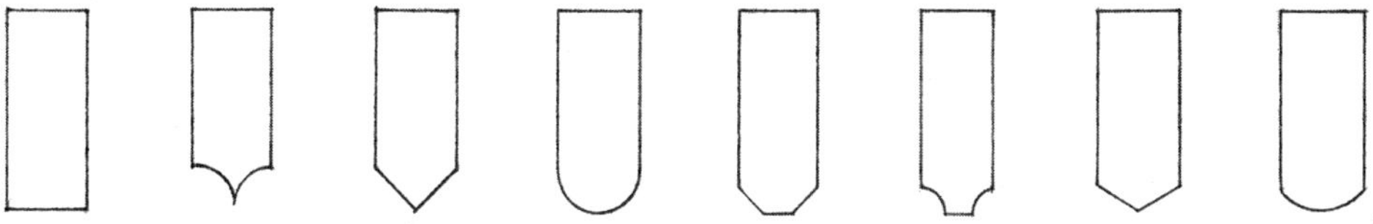

Dimension Shingles

• Square • Arrow • Diamond • Round • Octagonal • Half cove • Hexagonal • Fish scale

At corners, alternating courses are lapped over the adjacent corner shingles on the other side. Exposed edges should be treated. Corner boards can also be used to receive the shingles at both interior and exterior corners. Building paper should be used to flash corners and wherever the shingles abut wood trim.

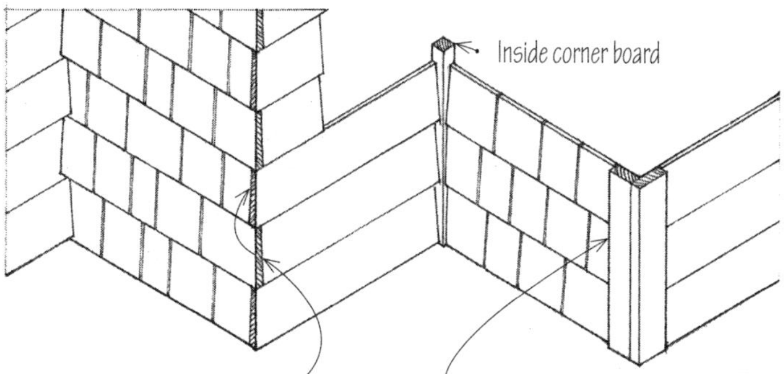

Corners

Inside corner board
• Alternating overlap
Lapping outside corner boards

CSI MasterFormat 07 46 23: Wood Siding

Horizontal board sidings are available in different forms.

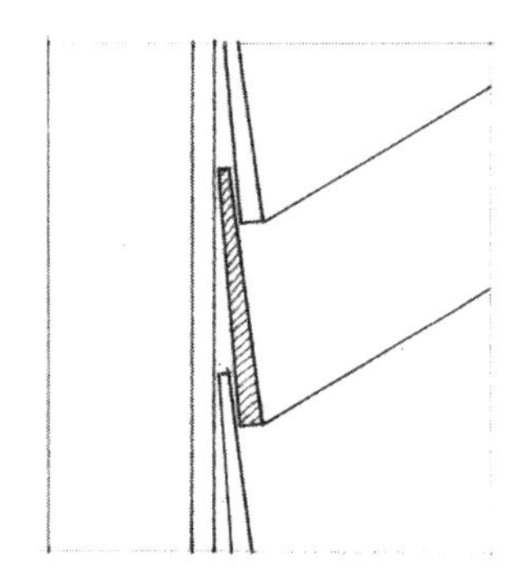

- Bevel siding, also known as lap siding, is made by cutting a board diagonally across its cross section so that the siding has one thin edge and one thick edge. The rough, resawn side can be exposed for stain finishes, while the smooth, planed side can be either painted or stained.

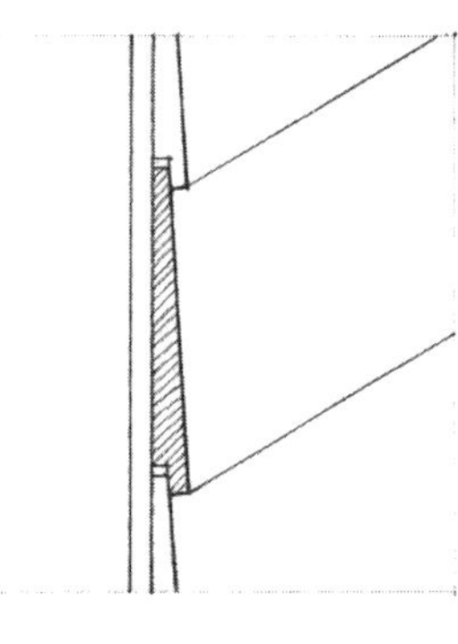

- Dolly Varden siding is bevel siding rabbeted along the lower edge to receive the upper edge of the board below it.

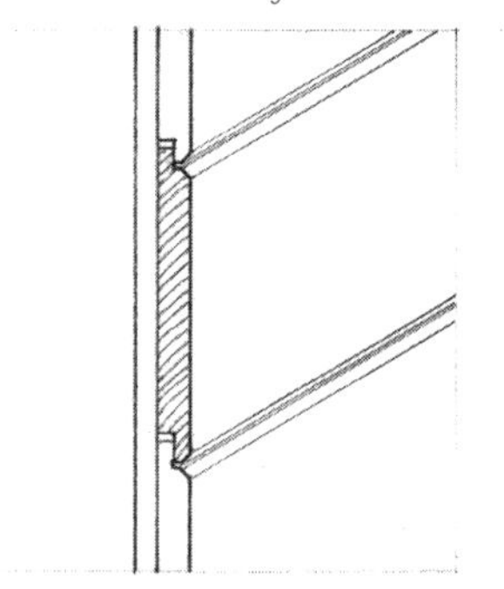

- Shiplap siding consists of boards joined edge to edge with overlapping rabbeted joints.

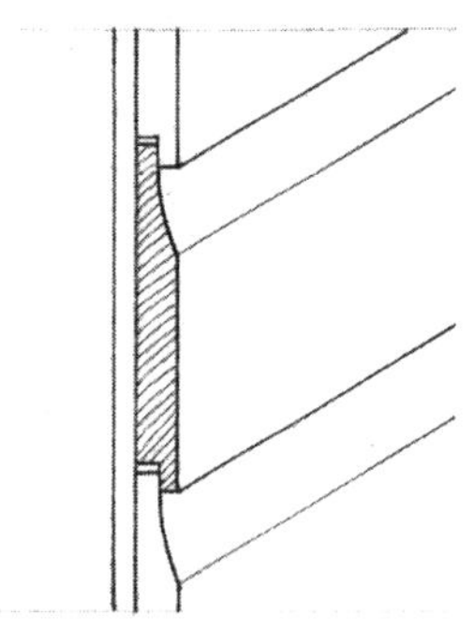

- Drop siding is composed of boards narrowed along the upper edges to fit into rabbets or grooves in the lower edges, laid horizontally with their backs flat against the sheathing or studs of the wall.

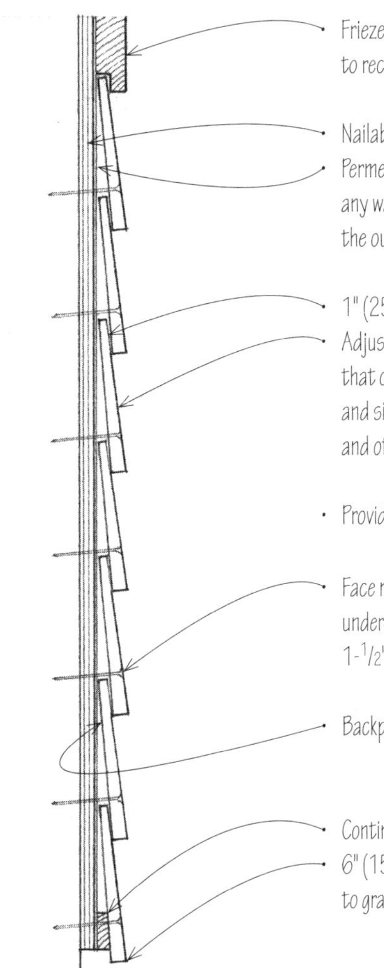

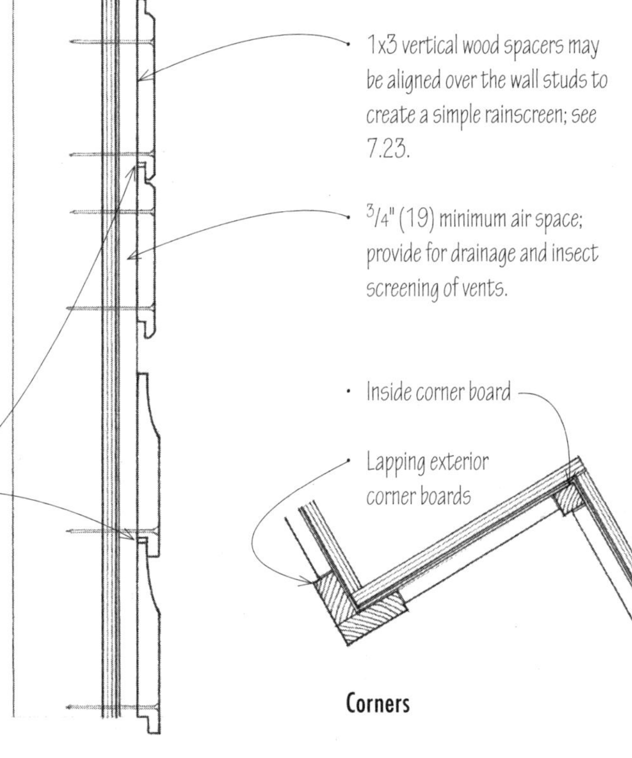

Corners

Siding Application

Horizontal board sidings are fastened through the wall sheathing and into the wall studs with hot-dipped galvanized, aluminum, or stainless steel nails. Nailing is done to allow the individual boards to expand and contract freely with changes in moisture content. Board ends should meet over a stud or butt against corner boards or window and door trim; a sealant is usually applied to the board ends during installation and the joints caulked.

Vertical board siding can be laid in various patterns. Matched boards that interlap or interlock can have flush, V-groove, or beaded joints. Square-edged boards can be used with other boards or battens to protect their vertical joints and form board-and-board or board-and-batten patterns.

While horizontal siding is nailed directly to the wall studs, vertical siding requires solid blocking at 24" (610) o.c., or plywood sheathing at least 5/8" or 3/4" (16 or 19) thick. Over thinner sheathing, 1x4 furring can be used at 24" (610) o.c A permeable building paper that allows water vapor to escape to the outside is used under the siding.

As with other wood siding materials, only hot-dipped galvanized or other corrosion-resistant nails should be used. Treat ends and edges of siding, and the back of batten strips, with a preservative before installation.

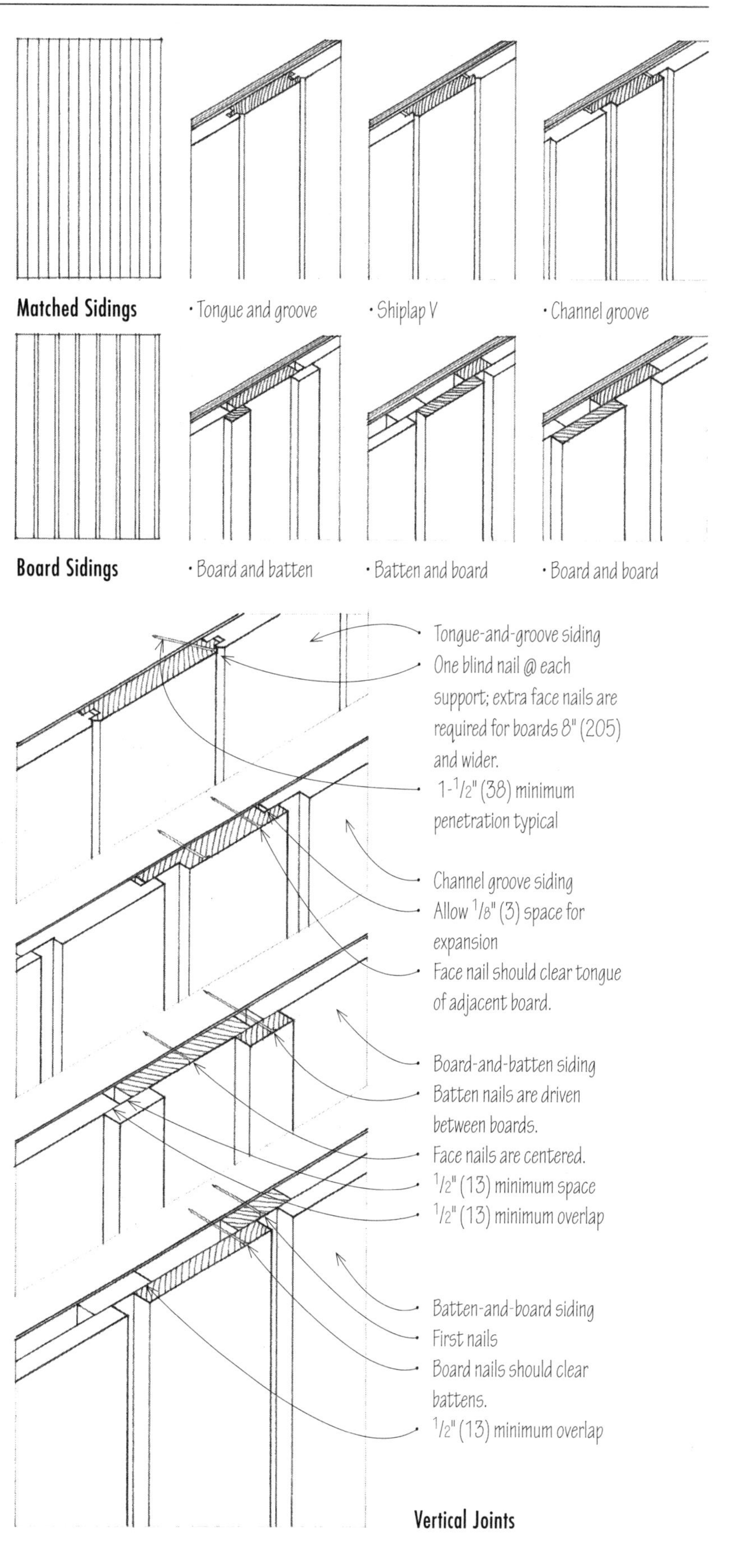

Vertical Joints

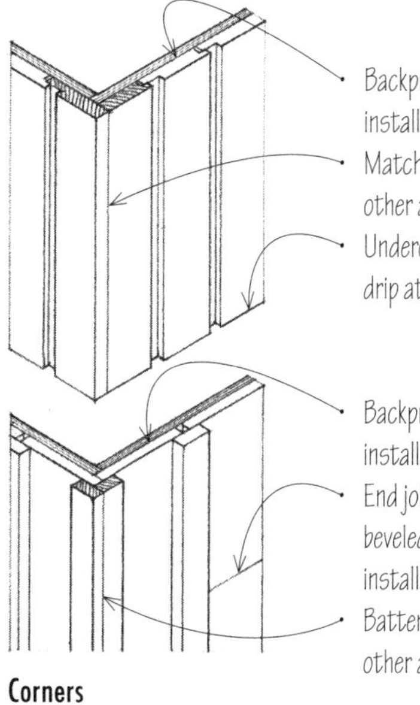

Corners

Alternative Sidings

A variety of siding materials have been designed to mimic the appearance of traditional wood siding, offer improved durability and resistance to weathering, and reduce maintenance costs. These alternatives include aluminum siding, vinyl (PVC) siding, and fiber-cement planks and panels. Consult the following sources for more information on the suitability of these siding alternatives for specific applications.

- American Architectural Manufacturers Association (AAMA) Publication 1402 for aluminum siding
- Vinyl Siding Institute's (VSI) publication on rigid vinyl siding application instructions
- National Evaluation Service, Inc. (NES) Report No. NER-405 for cement-fiber products

Stucco is a coarse plaster composed of portland or masonry cement, sand, and hydrated lime, mixed with water and applied in a plastic state to form a hard covering for exterior walls. This weather- and fire-resistant finish is normally used for exterior walls and soffits, but it can also be used for interior walls and ceilings that are subject to direct wetting or damp conditions.

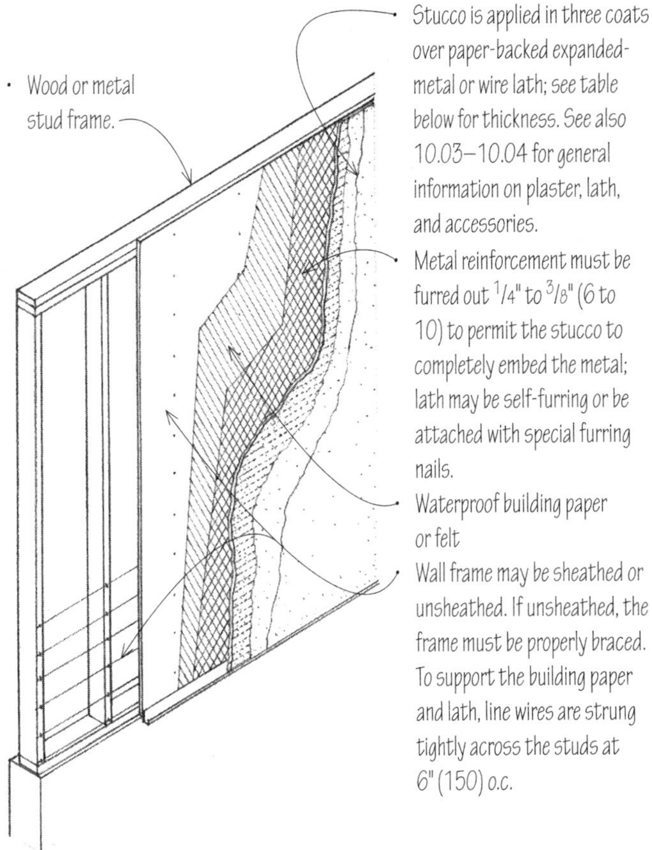

• Stucco is applied in three coats over paper-backed expanded-metal or wire lath; see table below for thickness. See also 10.03–10.04 for general information on plaster, lath, and accessories.
• Metal reinforcement must be furred out 1/4" to 3/8" (6 to 10) to permit the stucco to completely embed the metal; lath may be self-furring or be attached with special furring nails.
• Waterproof building paper or felt
• Wall frame may be sheathed or unsheathed. If unsheathed, the frame must be properly braced. To support the building paper and lath, line wires are strung tightly across the studs at 6" (150) o.c.

Stud Wall Base

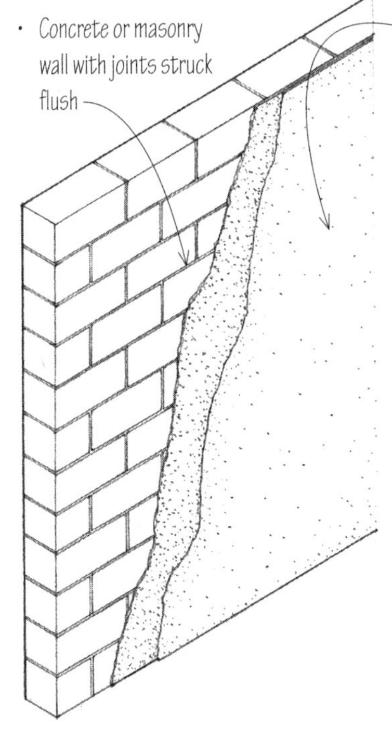

• Stucco is applied in two coats over a suitable masonry or concrete surface; see table below for thickness.
• The masonry or concrete wall should be structurally sound and its surface should be free of dust, grease, or other contaminants that would prevent good suction or chemical bond. In addition, the surface should be rough and porous enough to ensure a good mechanical bond.
• Metal reinforcement, a dash coat of portland cement and sand, or a bonding agent is used if a good bond is doubtful.

Masonry or Concrete Base

Stucco Finishes

The finish coat may have a float, stippled, combed, or pebbled texture. The finish may be natural or be integrally colored through the use of pigment, colored sand, or stone chips.

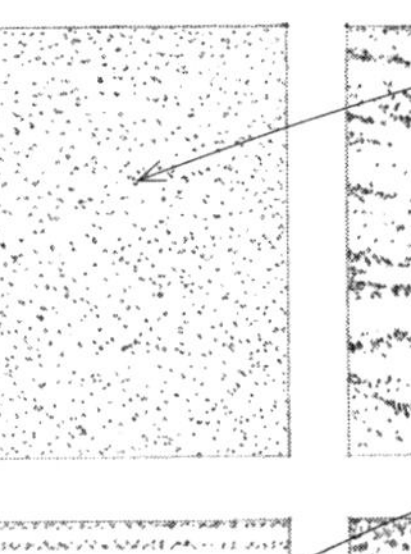
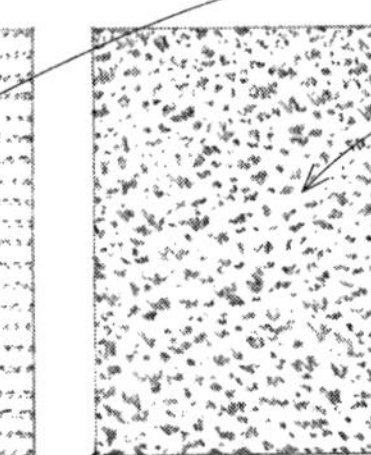

• Float finish is a fine-textured finish produced with a carpet or rubber-faced float.
• Stipple-troweled finish is first stippled with a broom; the high spots are then troweled.
• Combed finish is formed with a notched or serrated tool.
• Rock-dash finish is produced by machine-spraying small pebbles onto unset stucco.

Thickness of Portland Cement Stucco

Base	Minimum Finished Thickness from Face of Base
Expanded metal or wire lath	7/8" (22); exterior 5/8" (16); interior
Masonry walls	1/2" (13)
Concrete walls	7/8" (22) maximum
Concrete ceilings	3/8" (10) maximum

CSI MasterFormat 09 24 23: Portland Cement Stucco

Like gypsum plaster, stucco is a relatively thin, hard, brittle material that requires reinforcement or a sturdy, rigid, unyielding base. Unlike gypsum plaster, which expands slightly as it hardens, portland cement stucco shrinks as it cures. This shrinkage, along with the stresses caused by structural movement of the base support and variations in temperature and humidity, can cause the stucco to crack. Control and relief joints are required to eliminate or minimize any cracking.

Relief Joints

- Relief joints relieve stress by permitting independent movement along the perimeter of a stucco membrane. They are required where two planes of stucco meet at an internal corner, or where a stucco membrane abuts or is penetrated by a structural element, such as a beam, column, or loadbearing wall.

Control Joints

- Control joints relieve stress in the stucco membrane and prealign the cracking that can be caused by structural movement in the supporting construction, drying shrinkage, and variations in temperature. When stucco is applied over metal reinforcement, control joints should be spaced no more than 18' (5.5 m) apart and define panels no larger than 150 sf (14 m^2).
- When stucco is applied directly to a masonry base, control joints should be installed directly over and aligned with any control joints existing in the masonry base.
- Control joints are also required where dissimilar base materials meet and along floor lines in wood frame construction.

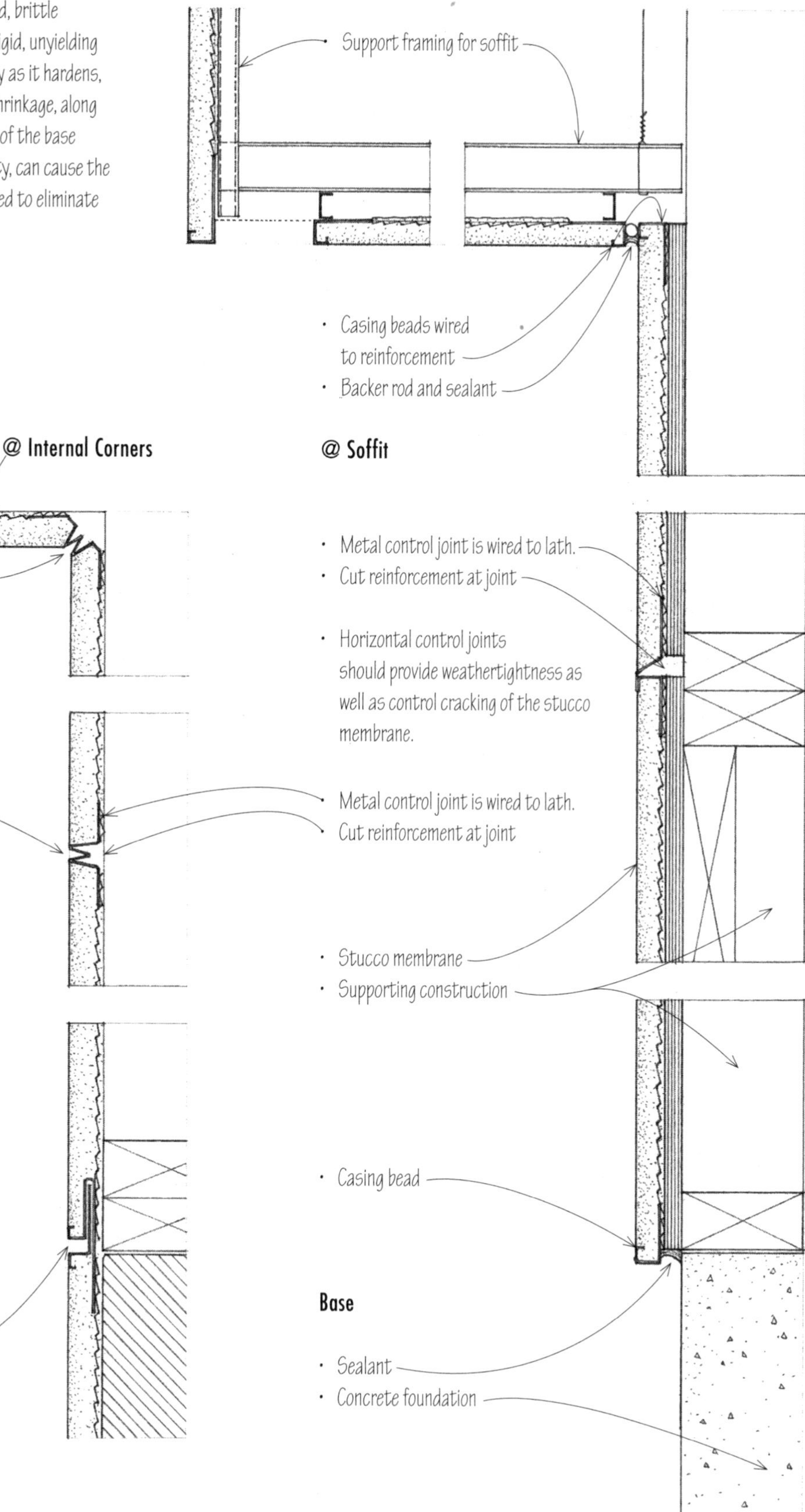

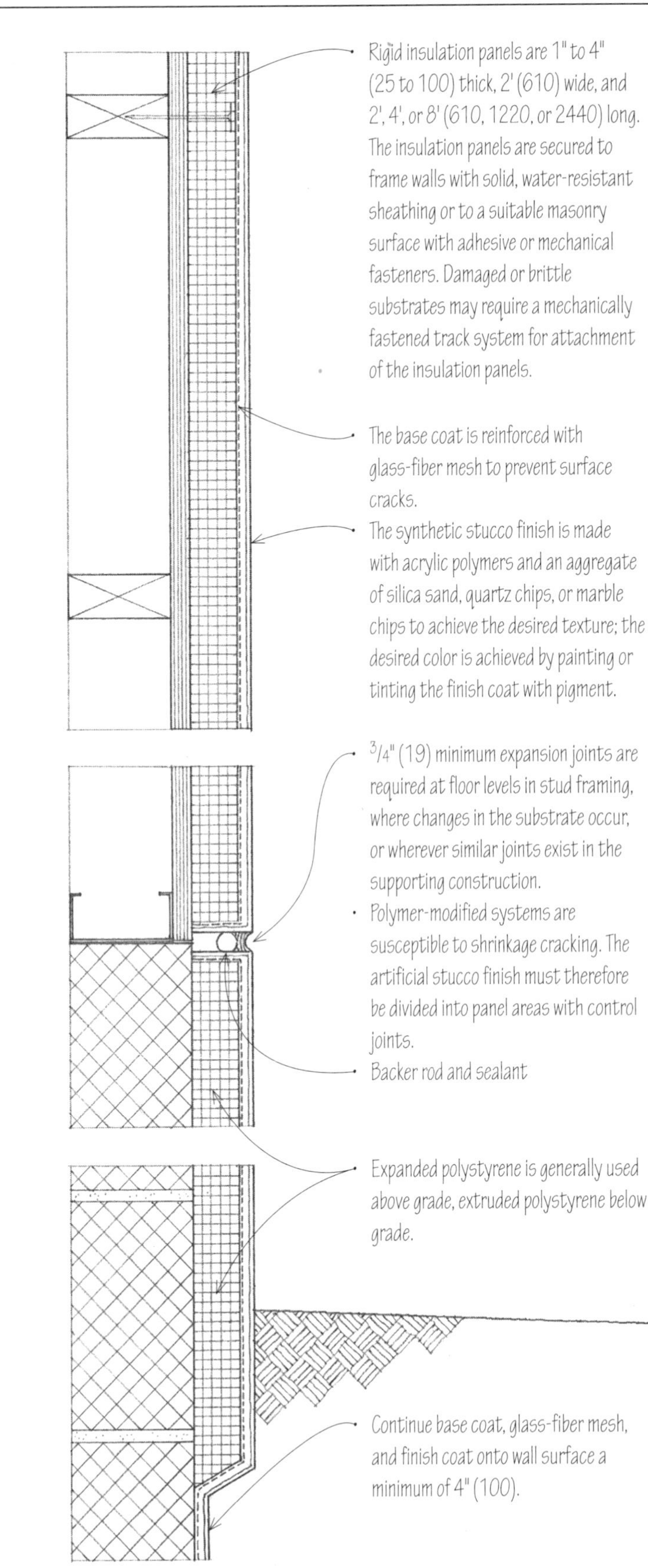

Exterior insulation and finish systems (EIFS) are available for cladding the exterior of new structures as well as insulating and refacing existing buildings. The system consists of a thin layer of synthetic stucco troweled, rolled, or sprayed over a layer of rigid plastic foam insulation.

EIFS systems are susceptible to leaking around windows and doors because of poor detailing or faulty installation. There is no internal drainage system that would allow any water that does penetrate the system to escape. This trapped water can cause the insulation layer to separate from the substrate or the sheathing to deteriorate. To address this problem, a proprietary system uses a drainage mat installed between an air and water barrier and the insulation layer to allow water to drain to plastic flashings above wall openings and at the base of the wall.

There are two generic types of EIFS systems: polymer-modified and polymer-based systems. Polymer-modified systems consist of a portland cement base coat from 1/4" to 3/8" (6.4 to 9.5 mm) thick, reinforced with metal wire lath or glass-fiber mesh fastened to the insulation layer. In areas subject to impact, heavy-duty fiberglass mesh is used in place of, or in addition to, the standard mesh. The finish coat of portland cement is modified with acrylic polymers.

Polymer-based systems consist of a portland cement or acrylic polymer base coat 1/16" to 1/4" (1.6 to 6.4 mm) thick, reinforced with glass-fiber mesh embedded at the time of installation. The finish coat is made with acrylic polymers. Polymer-based systems are more elastic and crack-resistant than polymer-modified systems, but also more susceptible to denting and puncturing.

- Consult standards published by the Exterior Insulation Manufacturers Association (EIMA) for installation details.

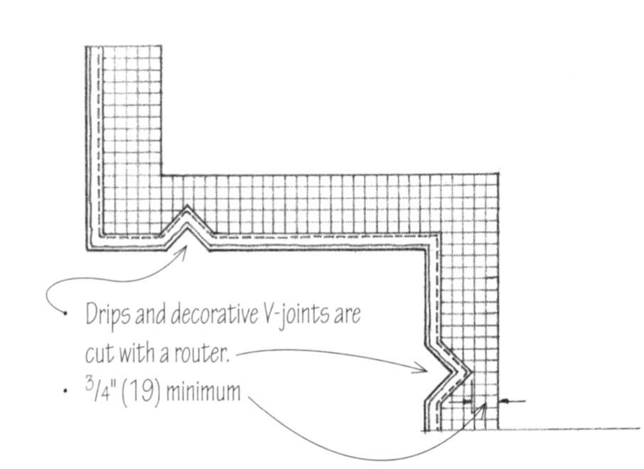

The primary purpose of thermal insulation is to control the flow or transfer of heat through the exterior assemblies of a building and thereby prevent excessive heat loss in cold seasons and heat gain in hot weather. This control can effectively reduce the amount of energy required by heating and cooling equipment to maintain conditions for human comfort in a building.

LEED EA Credit 1: Optimize Energy Performance

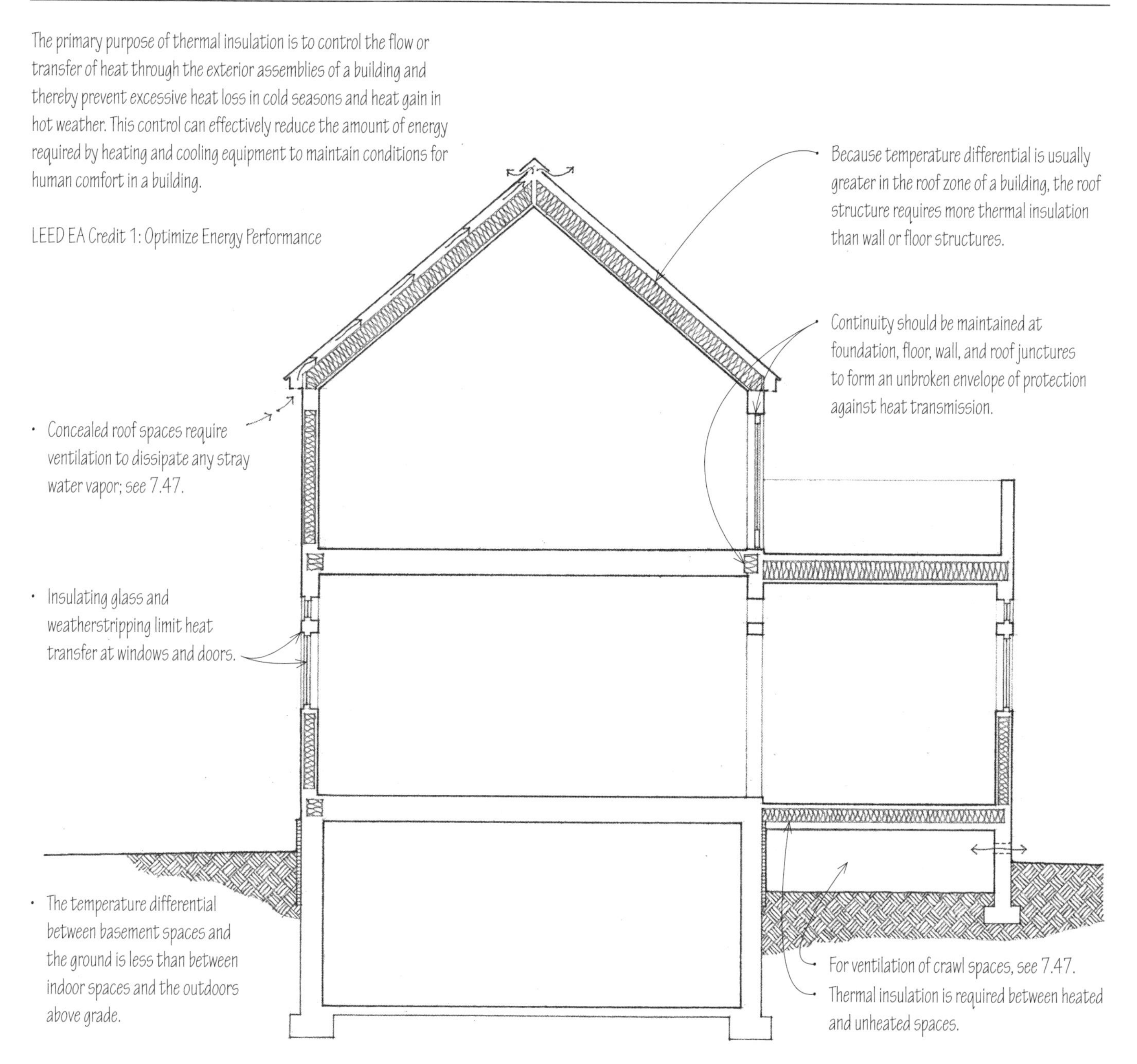

- For a discussion of the factors that affect human comfort, see 11.03.
- For siting factors that also affect potential heat loss or gain, see Chapter 1.

Recommended Minimum Thermal Resistances of Building Insulation*

Zone	Ceiling or Roof	Exterior Wall	Floor over Unheated Space
Minimum recommended	19	11	11
Southern zone	26	13	11
Temperate zone	30	19	19
Northern zone	38	19	22

* Use these R-values only for preliminary design. Consult the local or state energy code for specific requirements.

CSI MasterFormat 07 21 00: Thermal Insulation

Material	1/k*	1/C†
Concrete		
Concrete		
Sand & gravel aggregate	0.08	
Lightweight aggregate	0.60	
Cement mortar	0.20	
Stucco	0.20	
Masonry		
Common brick	0.20	
Face brick	0.11	
Concrete block, 8" (205)		
Sand & gravel aggregate		1.11
Lightweight aggregate		2.00
Granite and marble	0.05	
Sandstone	0.08	
Metal		
Aluminum	0.0007	
Brass	0.0010	
Copper	0.0004	
Lead	0.0041	
Steel	0.0032	
Wood		
Hardwoods	0.91	
Softwoods	1.25	
Plywood	1.25	
Particleboard, 5/8" (16)		0.82
Wood fiberboard	2.00	
Roofing		
Built-up roofing		0.33
Fiberglass shingles		0.44
Slate roofing		0.05
Wood shingles		0.94
Siding		
Aluminum siding		0.61
Wood shingles		0.87
Wood bevel siding		0.81
Vinyl siding		1.00

Material	1/k*	1/C†
Building Paper		
Vapor-permeable felt		0.06
Polyethylene film		0.00
Plaster & Gypsum		
Cement plaster, sand aggregate	0.20	
Gypsum plaster,		
sand aggregate	0.18	
perlite aggregate	0.67	
Gypsum board, 1/2" (13)		0.45
Flooring		
Carpet & pad		1.50
Hardwood, 25/32" (20)		0.71
Terrazzo		0.08
Vinyl tile		0.05
Doors		
Steel, mineral fiber core		1.69
Steel, polystyrene core		2.13
Steel, urethane core		5.56
Wood hollow core, 1-3/4" (45)		2.04
Wood solid core, 1-3/4" (45)		3.13
Glass		
Single, clear, 1/4" (6)		0.88
Double, clear, 3/16" (5) space		1.61
1/4" (6) space		1.72
1/2" (13) space		2.04
Double, blue/clear		2.25
gray/clear		2.40
green/clear		2.50
Double, clear, low-e coating		3.23
Triple, clear		2.56
Glass block, 4" (100)		1.79
Air Space		
3/4" (19), nonreflective		1.01
3/4" (19), reflective		3.48

* 1/k = R per inch of thickness

† 1/C = R for the thickness indicated

The tables to the left can be used to estimate the thermal resistance of a construction assembly. For specific R-values of materials and building components such as windows, consult the product manufacturer.

R = F°/Btu/hr • sf, where:

- R is a measure of thermal resistance of a given material. It is expressed as the temperature difference required to cause heat to flow through a unit area of material at the rate of one heat unit per hour.

$U = 1/R_t$, where:

- R_t is the total thermal resistance for a construction assembly and is simply the sum of the individual R-values of the component materials of an assembly.
- U is a measure of the thermal transmittance of a building component or assembly. It is expressed as the rate of heat transfer through a unit area of a building component or assembly caused by a difference of one degree between the air temperatures on the two sides of the component or assembly. The U-value for a component or assembly is the reciprocal of its R-value.

$Q = U \times A \times (t_i - t_o)$, where:

- Q is the rate of heat flow through a construction assembly and is equal to:
- U = overall coefficient of assembly
- A = exposed area of assembly
- $(t_i - t_o)$ = difference between the inside and outside air temperatures

Almost all building materials offer some resistance to heat flow. To achieve the desired R value, however, wall, floor, and roof assemblies usually require the addition of an insulating material. Below is an outline of the basic materials used to insulate the components and assemblies of a building. Note that all effective insulating materials usually incorporate some form of captured dead air space.

- Batt insulation consists of flexible, fibrous thermal insulation of glass or mineral wool, made in various thicknesses and lengths and in 16" or 24" (406 or 610) widths to fit between studs, joists, and rafters in light wood frame construction, sometimes faced with a vapor retarder of kraft paper, metal foil, or plastic sheet. Batt insulation is also as a component in sound-insulating construction.
- Rigid foam insulation is a preformed, nonstructural insulating board of foamed plastic or cellular glass. Cellular glass insulation is fire-resistant, impervious to moisture, and dimensionally stable, but has a lower thermal-resistance value than foamed plastic insulations, which are flammable and must be protected by a thermal barrier when used on the interior surfaces of a building. Rigid insulations having closed-cell structures, such as extruded polystyrene and cellular glass, are moisture-resistant and may be used in contact with the earth.
- Foamed-in-place insulation consists of a foamed plastic, as polyurethane, that is sprayed or injected into a cavity where it adheres to the surrounding surfaces.
- Loose-fill insulation consists of mineral wool fibers, granular vermiculite or perlite, or treated cellulosic fibers, poured by hand or blown through a nozzle into a cavity or over a supporting membrane.
- Reflective insulation uses a material of high reflectivity and low emissivity, as paper-backed aluminum foil or foil-backed gypsum board, in conjunction with a dead-air space to reduce the transfer of heat by radiation.

Form	Material	R-value per Inch of Thickness	
Batt or blanket	Fiberglass	3.3	Installed between studs, joists, rafters, or furring;
	Rock wool	3.3	considered incombustible except for paper facing
Rigid board	Cellular glass	2.5	Boards may be applied over a roof deck, over wall
	Polystyrene, molded	3.6	framing as sheathing, in cavity walls, or beneath an
	Polystyrene, extruded	5.0	interior finish material; the plastics are combustible
	Polyurethane, expanded	6.2	and give off toxic fumes when burned; extruded
	Polyisocyanurate	7.2	polystyrene can be used in contact with the earth but
	Perlite, expanded	2.6	any exposed surfaces should be protected from sunlight
Foamed in place	Polyurethane	6.2	Used to insulate irregularly shaped spaces
Loose fill	Cellulose	3.7	Used to insulate attic floors and wall cavities; cellulose
	Perlite	2.7	may be combined with adhesives for sprayed application;
	Vermiculite	2.1	cellulose should be treated and UL-listed for fire resistance
Cast	Insulating concrete	1.12	Used primarily as an insulating layer under membrane roofing; insulating value depends on its density

The steady state method for calculating heat loss or gain takes into account primarily the total thermal resistance (R_t) of the construction assembly and the differential in air temperature. Other factors that affect heat loss or gain are:

- The surface color and reflectivity of the materials used; light colors and shiny surfaces tend to reflect more thermal radiation than dark, textured ones.
- The mass of the assembly, which affects the time lag or delay before any absorbed and stored heat is released by the structure; time lag becomes a significant factor with thick, dense materials.
- The orientation of the exterior surfaces of a building, which affects solar heat gain as well as exposure to wind and the attendant potential for air infiltration.
- Latent heat sources and heat gain from the occupants, lighting, and equipment within a building.
- Proper installation of thermal insulation and vapor retarders.

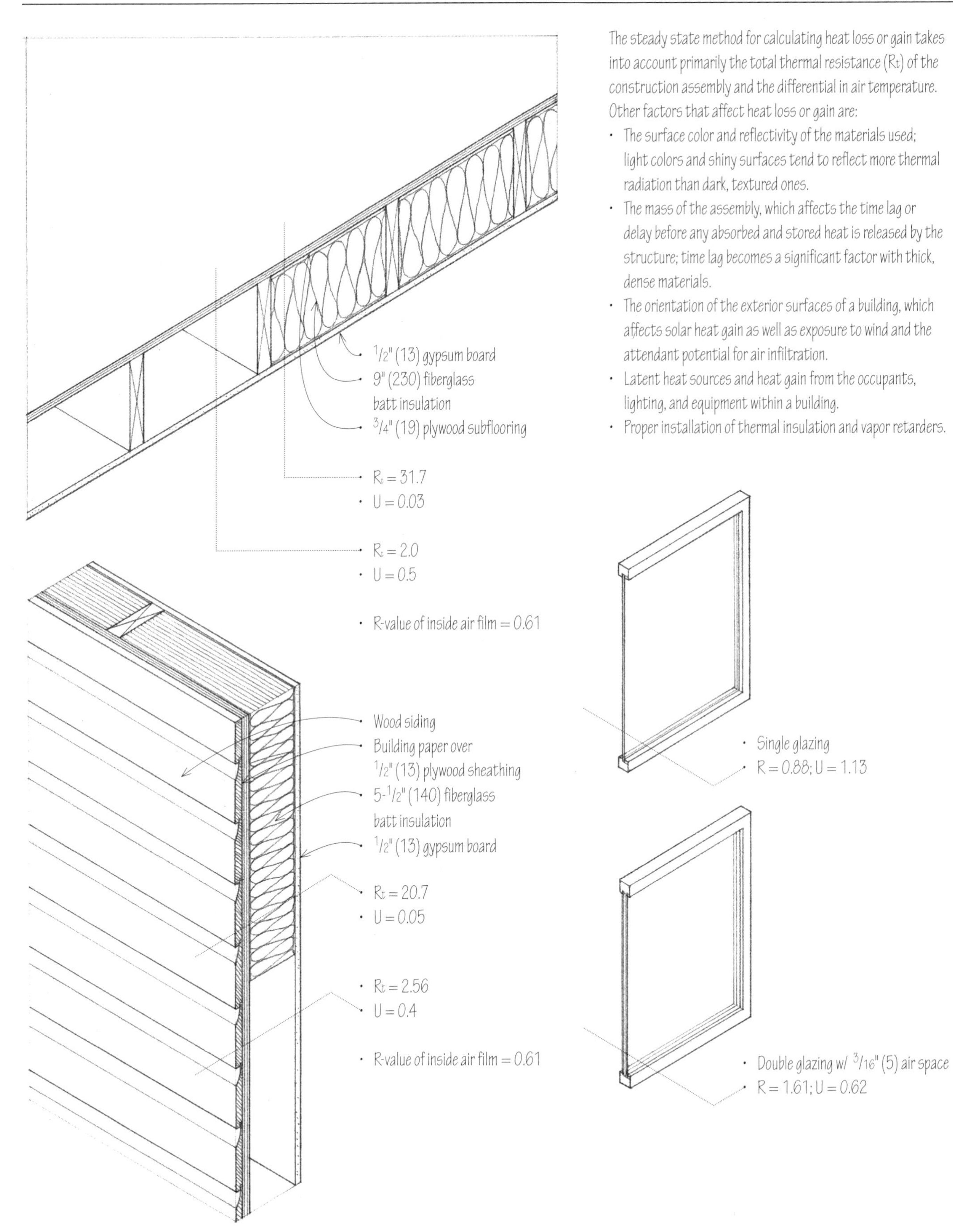

Comparison of R-values for Insulated and Uninsulated Assemblies

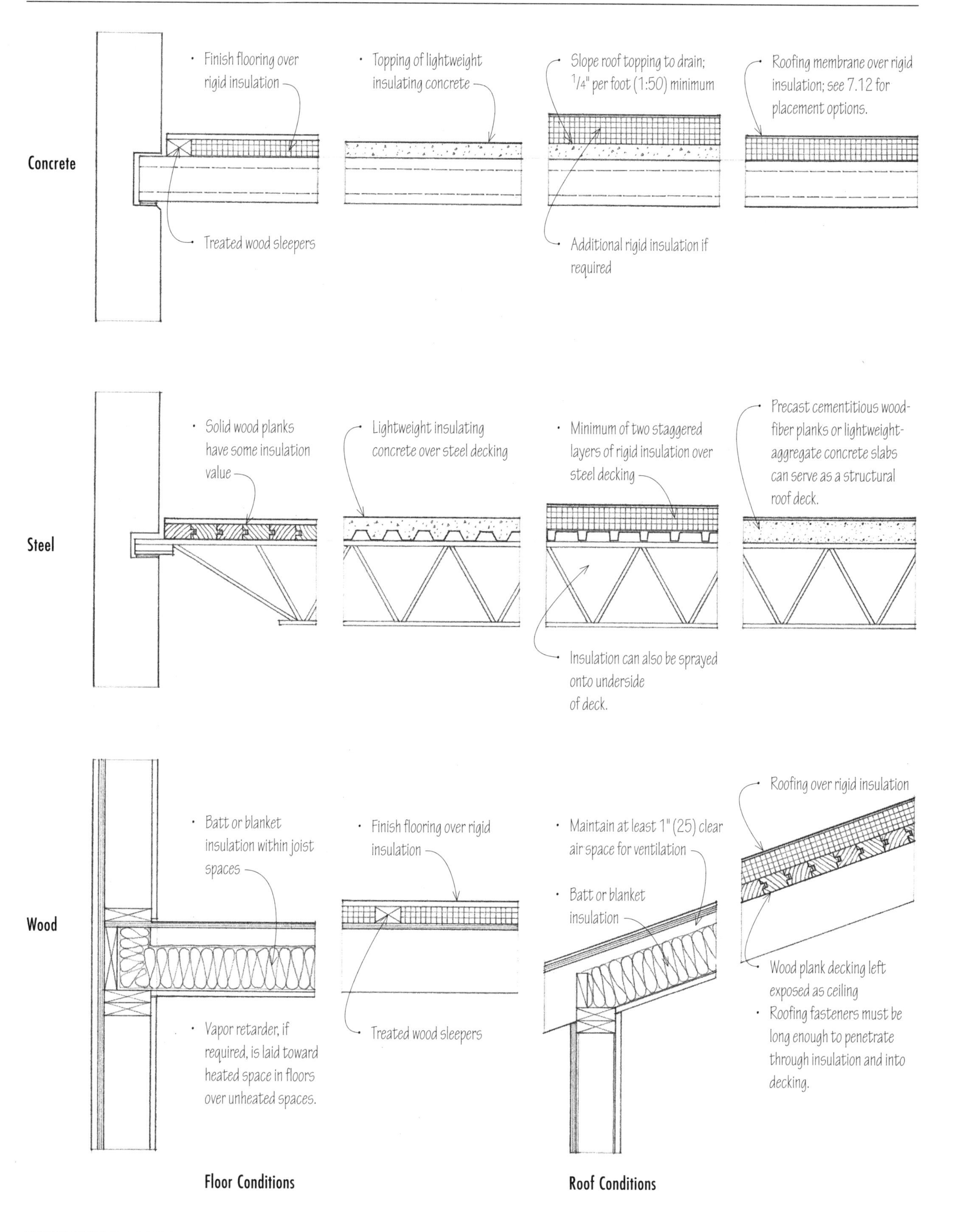
Concrete
Finish flooring over rigid insulation
Treated wood sleepers
Topping of lightweight insulating concrete
Slope roof topping to drain; 1/4" per foot (1:50) minimum
Additional rigid insulation if required
Roofing membrane over rigid insulation; see 7.12 for placement options.
Steel
Solid wood planks have some insulation value
Lightweight insulating concrete over steel decking
Minimum of two staggered layers of rigid insulation over steel decking
Insulation can also be sprayed onto underside of deck.
Precast cementitious wood-fiber planks or lightweight-aggregate concrete slabs can serve as a structural roof deck.
Wood
Batt or blanket insulation within joist spaces
Vapor retarder, if required, is laid toward heated space in floors over unheated spaces.
Finish flooring over rigid insulation
Treated wood sleepers
Maintain at least 1" (25) clear air space for ventilation
Batt or blanket insulation
Roofing over rigid insulation
Wood plank decking left exposed as ceiling
Roofing fasteners must be long enough to penetrate through insulation and into decking.
Floor Conditions
Roof Conditions

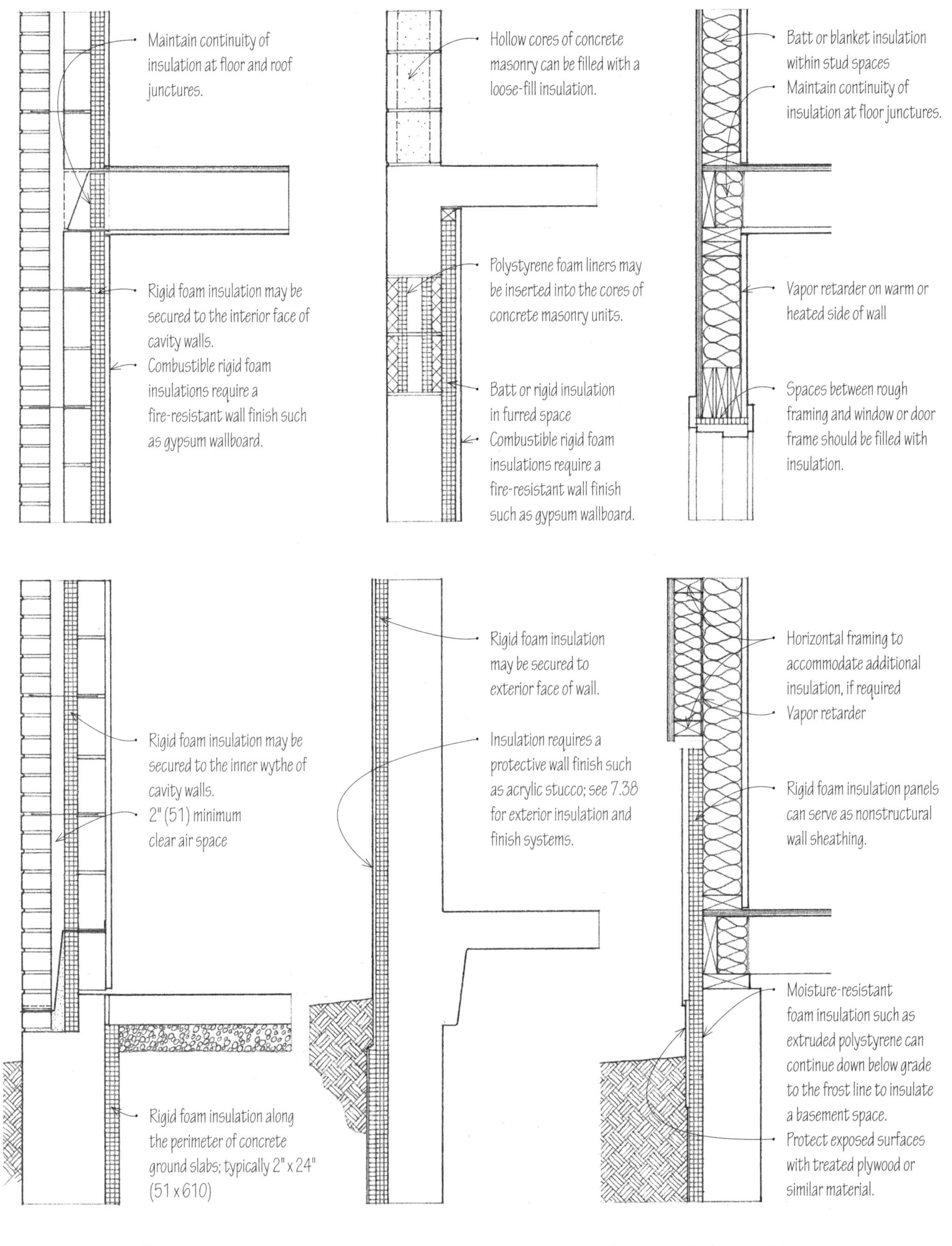

Masonry Cavity Walls | **Cast Concrete or Concrete Masonry Walls** | **Stud Frame Walls**

Moisture is normally present in the air as water vapor. Evaporation from occupants and equipment can raise the humidity of the air in a building. This moisture vapor will transform itself into a liquid state or condense when the air in which it exists becomes completely saturated with all the vapor it can hold and reaches its dew point temperature. Warm air is capable of holding more moisture vapor and has a higher dew point than cooler air.

Because it is a gas, moisture vapor always migrates from high to lower pressure areas. This normally means it tends to diffuse from the higher humidity levels of a building's interior toward the lower humidity levels outside. This flow is reversed when hot, humid conditions exist outdoors and a building's interior spaces are cooler. Most building materials offer little resistance to this passage of moisture vapor. If the moisture vapor comes into contact with a cool surface whose temperature is at or below the dew point of the air, it will condense.

Condensation can lessen the effectiveness of thermal insulation, be absorbed by building materials, and deteriorate finishes. Moisture vapor, therefore, must be:

- Prevented by vapor retarders from penetrating the enclosed spaces of exterior construction;
- Or be allowed to escape, by means of ventilation, before it can condense into a liquid.

- Surface condensation on windows can be controlled by raising the surface temperature with a warm air supply or by using double or triple glazing.

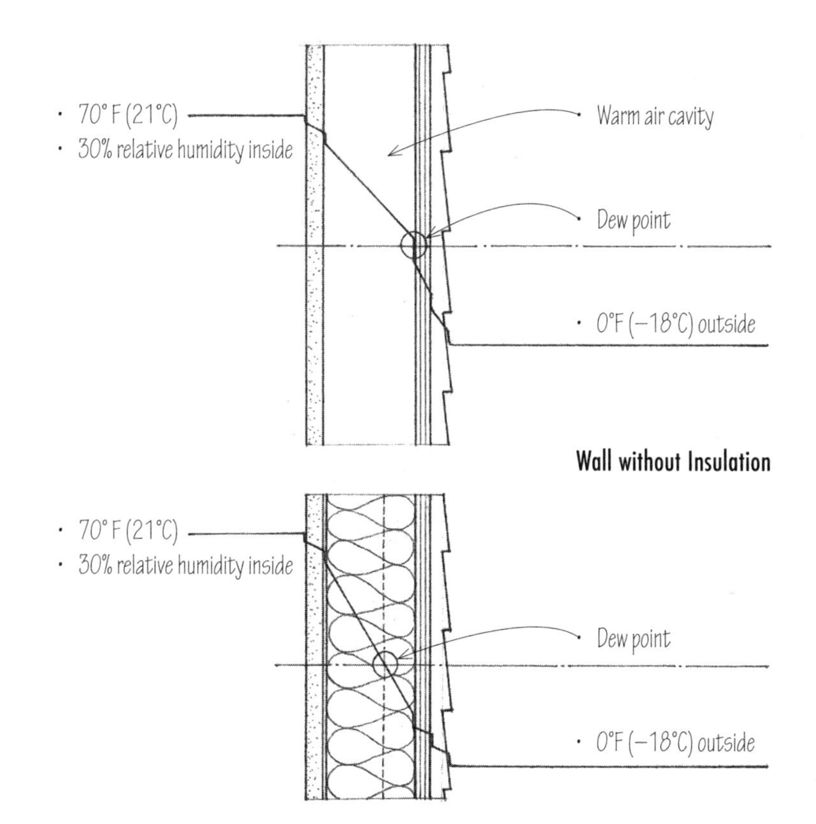

Wall with Insulation

- Wall requires a vapor retarder to prevent water vapor from condensing within the layer of insulation. A vapor retarder becomes more important as the level of thermal insulation increases.

Permeability of Some Building Materials

Material	Permeance (perms)*
Brick, 4" (100)	0.800
Concrete, 1" (25)	3.200
Concrete block, 8" (205)	2.400
Gypsum board, 3/8" (10)	50.000
Plaster, 3/4" (19)	15.000
Plywood, 1/4" (6), exterior glue	0.700
Built-up roofing	0.000
Aluminum foil, 1 mil	0.000
Polyethylene, 4 mil	0.080
Polyethylene, 6 mil	0.060
Duplex sheet, asphalt + foil	0.002
Asphalt-saturated + coated paper	0.200
Kraft paper, foil-faced	0.500
Blanket insulation, faced	0.400
Cellular glass	0.000
Polystyrene, molded	2.000
Polystyrene, extruded	1.200
Paint, two coats, exterior	0.900

* Perm is a unit of water vapor transmission, expressed in grains of vapor per one square foot per hour per inch of mercury pressure difference.

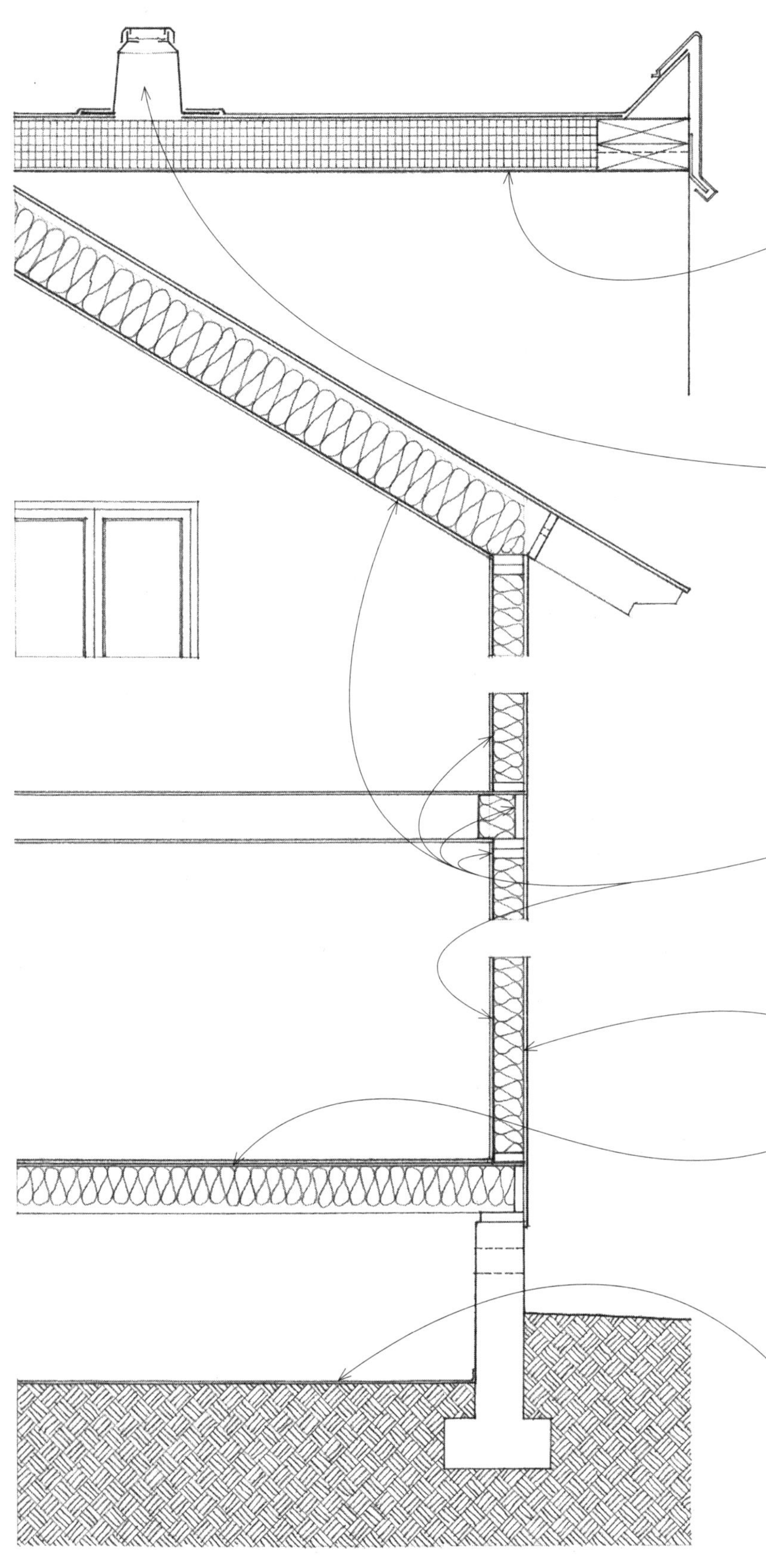

A vapor retarder is a material of low permeance installed in a construction to prevent moisture from entering and reaching a point where it can condense into a liquid. Vapor retarders are normally placed as close as possible to the warm side of insulated construction in temperate and cold climates. In warm, humid climates, the vapor retarder may have to be placed closer to the outer face of the construction.

- The use of a vapor retarder is generally recommended to protect the insulation layer of flat roof assemblies in geographic locations where the average outdoor temperature in January is below 40°F (4°C) and the interior relative humidity in winter is 45% or greater at 68°F (20°C).
- The barrier may be in the form of asphalt-saturated roofing felt or a proprietary material of low permeance.
- When a vapor retarder is present, topside vents may be required to allow any trapped moisture to escape from between the vapor retarder and the roofing membrane. Consult roofing manufacturer for recommendations.

- Some rigid foam insulation boards have inherent vapor resistance, while other insulating materials have a vapor-retarding facing. A vapor retarder is most effective, however, when it is applied as a separate layer of aluminum foil, polyethylene film, or treated paper.

- Vapor retarders should have a flow rating of one perm or less and be installed with all seams at joints and openings lapped and sealed. In this case, a vapor retarder is sometimes referred to as an air barrier.

- Exterior sheathing, building paper, and siding should be permeable to allow any vapor in the wall construction to escape to the outside.

- Over unheated spaces, the vapor retarder is placed on the warm side of the insulated floor. The vapor retarder may be laid on top of the subfloor or be integral with the insulation.

- A moisture barrier, such as polyethylene film, is usually required to retard the migration of ground moisture into a crawl space.

CSI MasterFormat 07 26 00: Vapor Retarders

Whole-House Ventilation

- Whole-house ventilators are motor-driven fans for pulling stale air from the living areas of a house and exhausting it through attic and roof vents.
- Vapor retarders installed in a seamless manner can result in airtight construction, in which case, a forced-air ventilating system with an air-to-air heat exchanger is required to rid interior spaces of moisture, odors, and pollutants.

Energy-Recovery Ventilation

- Energy-recovery ventilation systems are whole-house ventilators that provide a controlled way of ventilating a building while minimizing energy loss through the use of either heat-recovery ventilators or energy-recovery ventilators.
- Heat-recovery ventilators (HRV) use a heat-exchange core to transfer heat from the exhaust air stream to the prefiltered fresh air stream in the winter and, during the summer, cool the prefiltered fresh air stream with the exhaust air stream.
- Energy-recovery ventilators (ERV) have a heat exchanger that transfers both heat and moisture, cooling and dehumidifying the incoming fresh airstream in the summer while heating and humidifying the cold, drier incoming airstream in the winter.

Roof and Attic Ventilation

- Ventilation of concealed roof spaces and attics is provided by eave vents and, on sloping roofs, by vents close to the ridge. The total net free ventilating area should be at least 1/300th of the area being vented, with at least 50% of the required area being at or along the ridge. Openings should be protected against the penetration of rain, snow, and insects.
- Eave or soffit vents may consist of a continuous screened vent slot or a metal vent strip installed in the eave soffit, or comprise a series of evenly distributed circular plug vents in frieze boards.

Crawl Space Ventilation

- Unheated crawl spaces also require ventilation. Openings should have a net area of at least 1-1/4 sf (0.14 m^2) for each 25 lineal feet (7620) of perimeter wall. There should be at least one opening on each side of the crawl space, located as high as possible and near a corner to promote cross ventilation. Openings should be protected against insects and vermin with wire mesh screening.

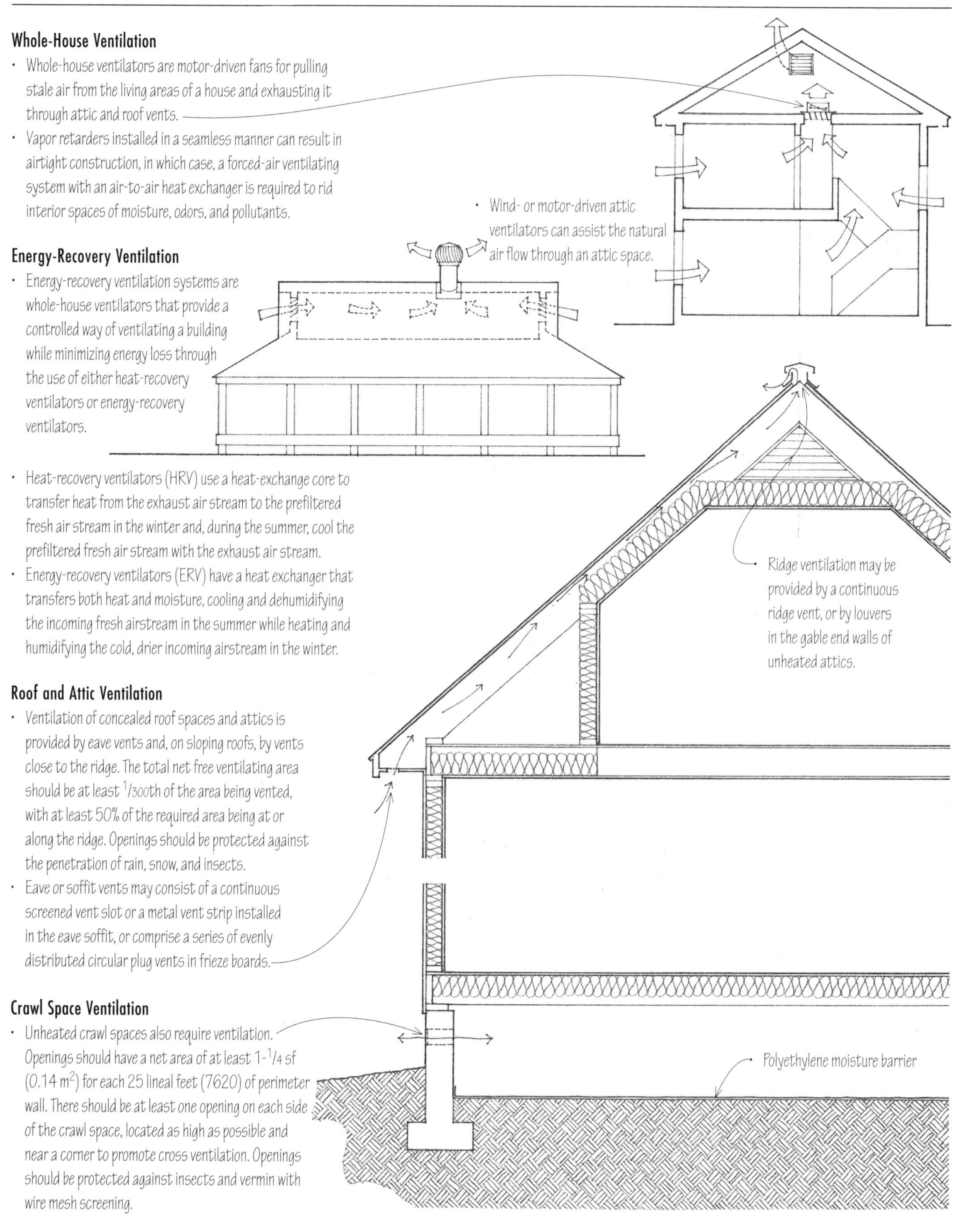

CSI MasterFormat 07 71 00: Roof Specialties
CSI MasterFormat 08 90 00: Louvers and Vents

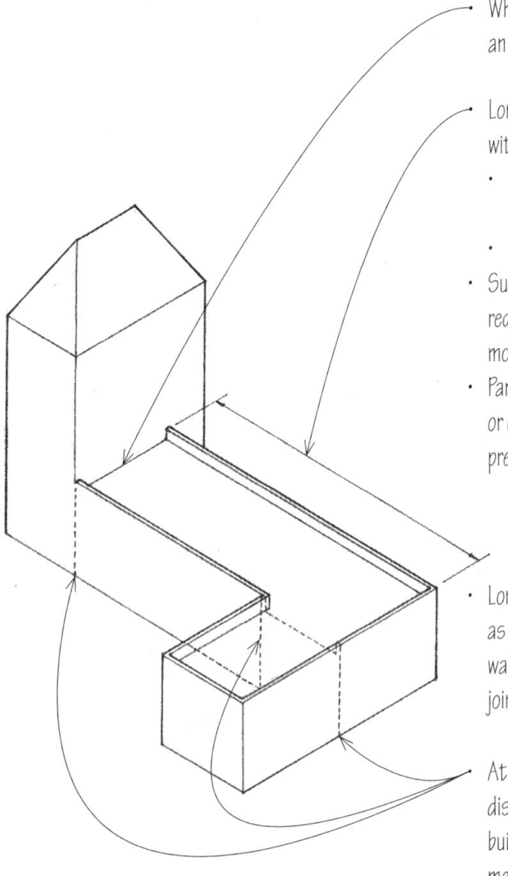

Where a new building adjoins an existing structure

Long surface areas; maximum length without expansion joints:

- Steel, concrete, or built-up roofing – 200' (60 m)
- Masonry – 125' (38 m)

- Surfaces with severe solar exposure require expansion or control joints at more frequent intervals.
- Parapet walls require expansion or control joints near corners to prevent their displacement.

- Long, linear building elements, such as fascias, gravel stops, and curtain wall framing also require expansion joints.

At horizontal and vertical discontinuities in the massing of a building, such as where a low mass meets a taller mass, or at wings and intersections of L-, T-, and U-shaped buildings.

Location of Movement Joints

All building materials expand and contract in response to normal changes in temperature. Some also swell and shrink with changes in moisture content, while others deflect under loading. Joints must be constructed to allow this movement to occur in order to prevent distortion, cracking, or breaks in the building materials. Movement joints should provide a complete separation of material and allow free movement while, at the same time, maintaining the weathertightness of the construction.

Types of Movement Joints

- Expansion joints are continuous, unobstructed slots constructed between two parts of a building or structure permitting thermal or moisture expansion to occur without damage to either part. Expansion joints can often serve as control and isolation joints. See 5.22 for expansion joints in brick masonry walls, 7.29 for horizontal expansion joints in masonry veneer walls, and 10.04 for expansion joints in gypsum plaster.
- Control joints are continuous grooves or separations formed in concrete ground slabs and concrete masonry walls to form a plane of weakness and thus regulate the location and amount of cracking resulting from drying shrinkage, thermal stresses, or structural movement. See 3.19 for control joints in concrete ground slabs and 5.22 for control joints in concrete masonry walls.
- Isolation joints divide a large or geometrically complex structure into sections so that differential movement or settlement can occur between the parts. At a smaller scale, an isolation joint can also protect a nonstructural element from the deflection or movement of an abutting structural member.

Coefficients of Linear Expansion

Per Unit Length Per 1 Degree Change in Temperature (°F)*

	$\times 10^{-7}$		$\times 10^{-7}$		$\times 10^{-7}$
Aluminum	128	Wood parallel to grain:		Brick masonry	34
Brass	104	Fir	21	Concrete masonry	52
Bronze	101	Maple	36	Concrete	55
Copper	93	Oak	27	Granite	47
Iron, cast	59	Pine	36	Limestone	44
Iron, wrought	67	Wood perpendicular to grain:		Marble	73
Lead	159	Fir	320	Plaster	76
Nickel	70	Maple	270	Rubble masonry	35
Steel, carbon	65	Oak	300	Slate	44
Steel, stainless	99	Pine	190	Glass	50

* One degree Fahrenheit is equal to approximately 0.6 degree Celsius or Centigrade. To find degrees Celsius or Centigrade, first subtract 32 from the degrees Fahrenheit and then multiply by 5/9.

The width of an expansion joint depends on the building material and the temperature range involved. It varies from 1/4" (6) to 1" (25) or more, and should be calculated for each specific situation.

- The coefficient of surface expansion is approximately twice the linear coefficient.
- The coefficient of volume expansion is approximately three times the linear coefficient.

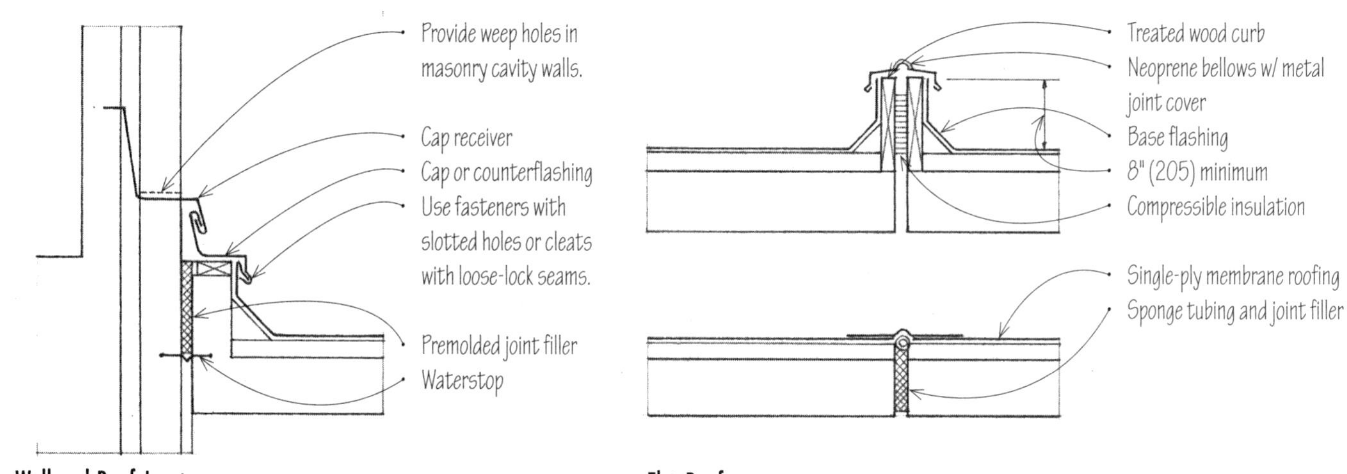

Wall and Roof Juncture

Flat Roof

These expansion joint details, although general in nature, have the following elements in common:

- A joint that creates a complete break through the structure, which is then usually filled with a compressible material
- A weatherstop that may be in the form of an elastic joint sealant, a flexible waterstop embedded within the construction, or a flexible membrane over flat roof joints.

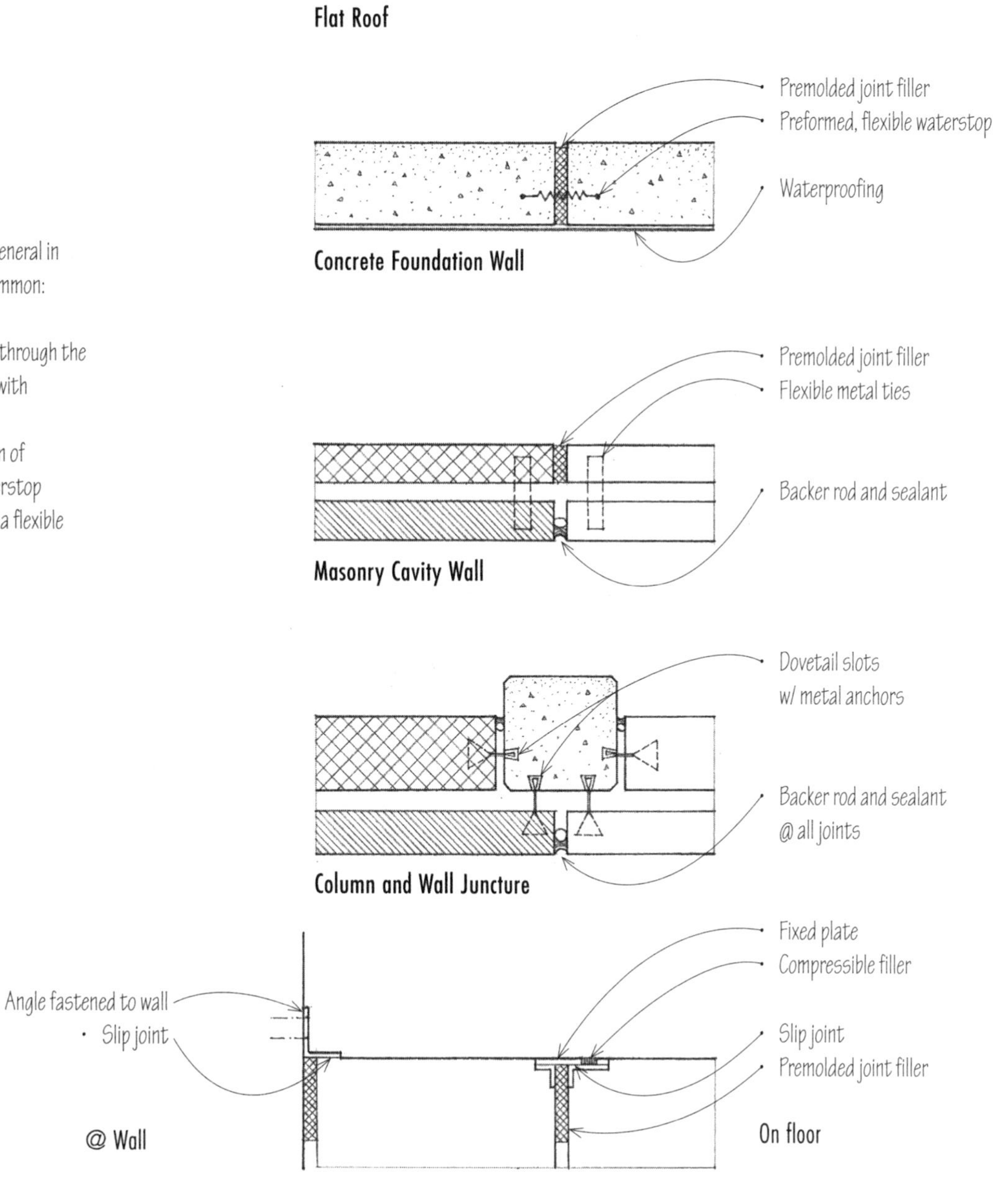

Concrete Foundation Wall

Masonry Cavity Wall

Column and Wall Juncture

Expansion Joint Covers

To provide an effective seal against the passage of water and air, a joint sealant must be durable, resilient, and have both cohesive and adhesive strength. Sealants can be classified according to the amount of extension and compression they can withstand before failure.

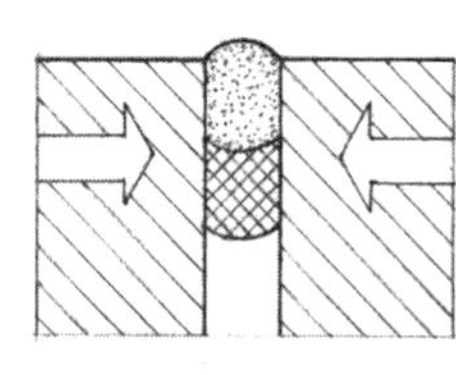

• Compressed

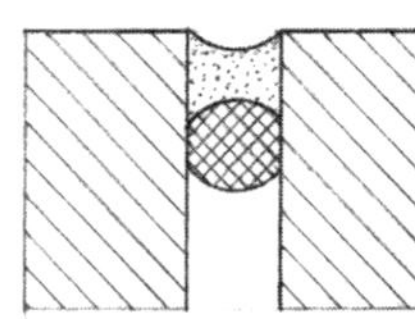

• As installed

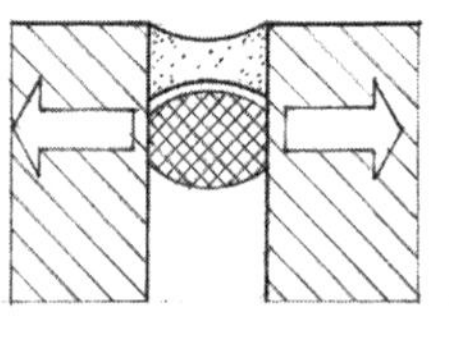

• Elongated

Joint Movement

- Joints should be tooled to ensure full contact with and adhesion to substrate
- Sealant joint depth
- Full contact depth
- Sealant depth
- 1/4" (6) minimum for 1/4" (6) joints
- Equal to joint width for joints up to 1/2" (13)
- Half of joint width for joints 1/2" (13) and wider, but not more than 2" (51)
- Joint width = sealant width
- 1/4" (6) to 1" (25) or more
- Width depends on the joint spacing, expected temperature range, anticipated movement due to wind or structural displacement, and the movement capability of the sealant.

Low Range Sealants
- Movement capability of +/− 5%
- Oil-based or acrylic compounds
- Often referred to as caulking and used for small joints where little movement is expected

Medium Range Sealants
- Movement capability of +/− 5% to +/− 10%
- Butyl rubber, acrylic, or neoprene compounds
- Used for nonworking, mechanically fastened joints

High Range Sealants
- Movement capability of +/− 12% to +/− 25%
- Polymercaptans, polysulfides, polyurethanes, and silicones
- Used for working joints subject to a significant amount of movement, such as those in curtain

- The substrate must be clean, dry, and compatible with the sealant material.
- A primer may be required to improve the adhesion of a sealant to the substrate.
- The joint filler controls the depth of the sealant contact with the joining parts. It should be compressible and be compatible with but not adhere to the sealant. It may be in the form of a rod or tubing of polyethylene foam, polyurethane foam, neoprene, or butyl rubber.

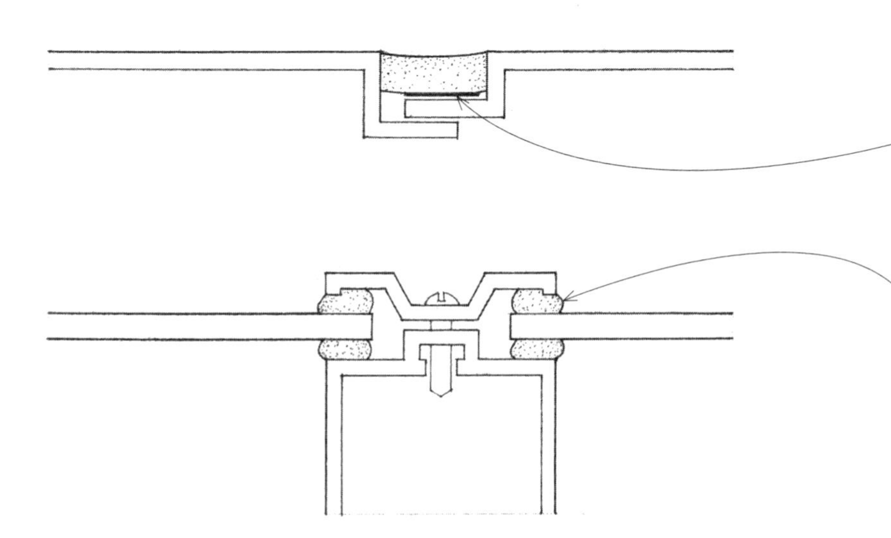

- When there is insufficient depth for a compressible filler, a bond breaker, such as polyethylene tape, is required to prevent adhesion between the sealant and the bottom of the joint recess.
- Most sealants are viscous liquids that cure after being applied with a hand-operated or power gun. These are referred to as gunnable sealants. Some lap joints, however, are difficult to seal with gunnable sealants. These joints may require instead a preformed solid polybutene or polyisobutylene tape sealant that is held in place under compression.

8

DOORS & WINDOWS

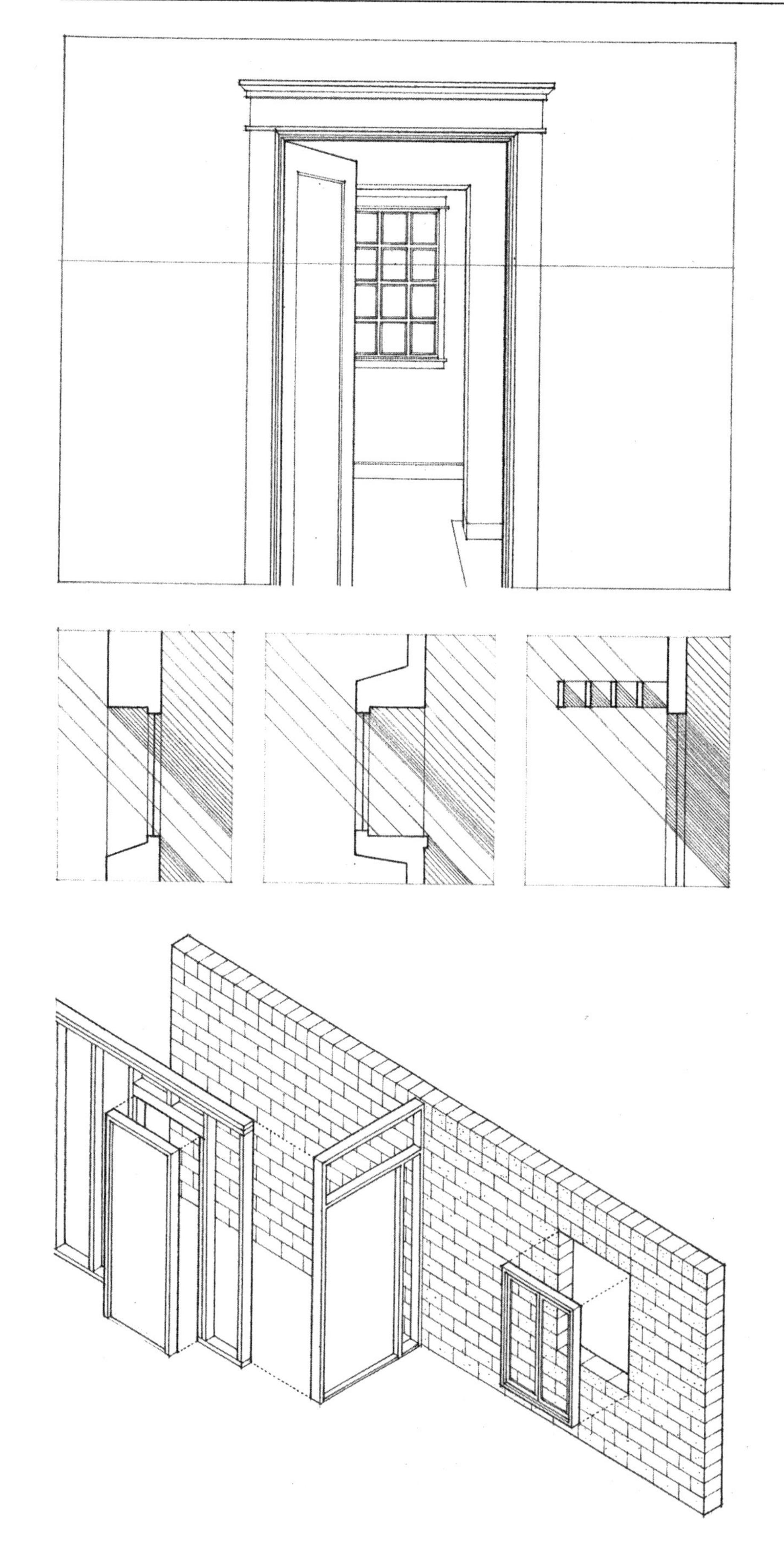

Doors and doorways provide access from the outside into the interior of a building as well as passage between interior spaces. Doorways should therefore be large enough to move through easily and accommodate the moving of furnishings and equipment. They should be located so that the patterns of movement they create between and within spaces are appropriate to the uses and activities housed by the spaces.

Exterior doors should provide weathertight seals when closed and maintain the approximate thermal insulation value of the exterior walls they penetrate. Interior doors should offer the desired degree of visual and acoustical privacy. All doors should be evaluated for their ease of operation, durability under the anticipated frequency of use, security provisions, and the light, ventilation, and view they may offer. Further, there may be building code requirements for fire resistance, emergency egress, and safety glazing that must be satisfied.

There are many types and sizes of windows, the choice of which affects not only the physical appearance of a building, but also the natural lighting, ventilation, view potential, and spatial quality of the building's interior spaces. As with exterior doors, windows should provide a weathertight seal when closed. Window frames should have low thermal conductivity or be constructed to interrupt the flow of heat. Window glazing should retard the transmission of heat and control solar radiation and glare.

Because door and window units are normally factory-built, their manufacturers may have standard sizes and corresponding rough-opening requirements for the various door and window types. The size and location of doors and windows should be carefully planned so that adequate rough openings with properly sized lintels can be built into the wall systems that will receive them.

From an exterior point of view, doors and windows are important compositional elements in the design of building facades. The manner in which they punctuate or divide exterior wall surfaces affects the massing, visual weight, scale, and articulation of the building form.

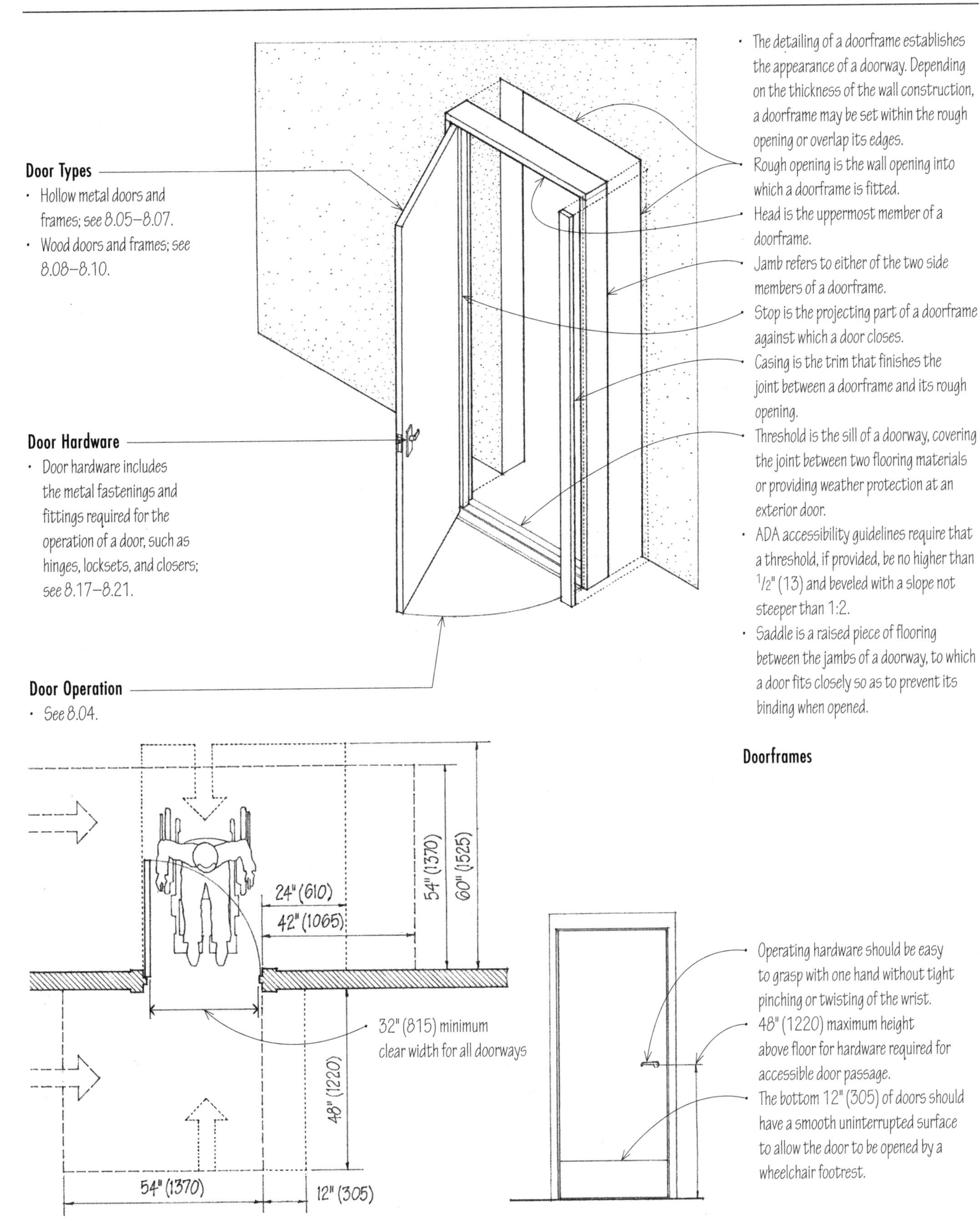

Door Types

- Hollow metal doors and frames; see 8.05–8.07.
- Wood doors and frames; see 8.08–8.10.

Door Hardware

- Door hardware includes the metal fastenings and fittings required for the operation of a door, such as hinges, locksets, and closers; see 8.17–8.21.

Door Operation

- See 8.04.

- The detailing of a doorframe establishes the appearance of a doorway. Depending on the thickness of the wall construction, a doorframe may be set within the rough opening or overlap its edges.
- Rough opening is the wall opening into which a doorframe is fitted.
- Head is the uppermost member of a doorframe.
- Jamb refers to either of the two side members of a doorframe.
- Stop is the projecting part of a doorframe against which a door closes.
- Casing is the trim that finishes the joint between a doorframe and its rough opening.
- Threshold is the sill of a doorway, covering the joint between two flooring materials or providing weather protection at an exterior door.
- ADA accessibility guidelines require that a threshold, if provided, be no higher than 1/2" (13) and beveled with a slope not steeper than 1:2.
- Saddle is a raised piece of flooring between the jambs of a doorway, to which a door fits closely so as to prevent its binding when opened.

Doorframes

- 32" (815) minimum clear width for all doorways

Minimum Maneuvering Clearances at Doorways

- Operating hardware should be easy to grasp with one hand without tight pinching or twisting of the wrist.
- 48" (1220) maximum height above floor for hardware required for accessible door passage.
- The bottom 12" (305) of doors should have a smooth uninterrupted surface to allow the door to be opened by a wheelchair footrest.

ADA Accessibility Guidelines for Doors

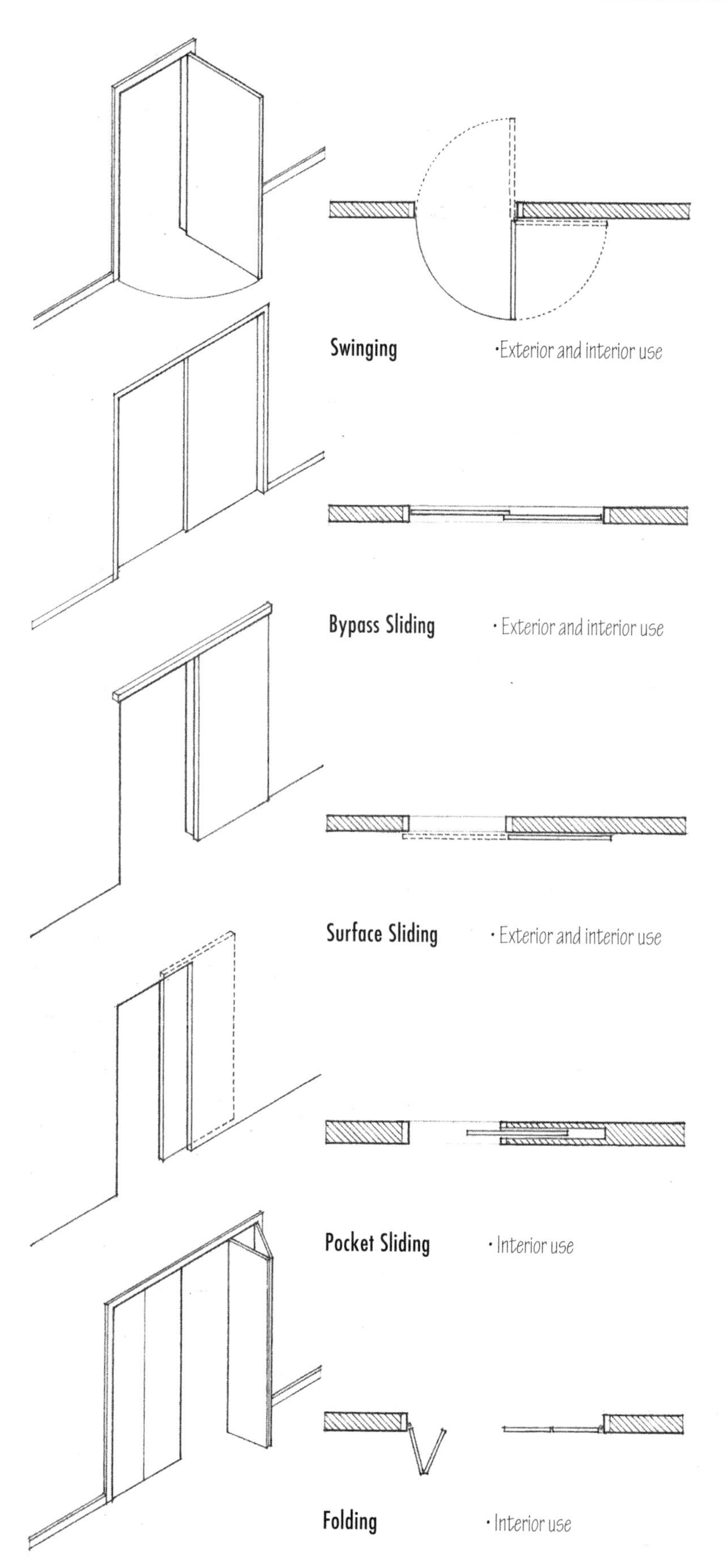

- Door normally turns on hinges about a side jamb when pushed or pulled, but may also be pivoted from head jamb and threshold.
- Requires space around doorway for door swing; check clearance required
- Most convenient operation for entry and passage
- Most effective door type for thermal and acoustic insulation and for weather resistance; can be fire-rated

- Doors slide on overhead track and along guides or a track on the floor.
- Requires no operating space but is difficult to seal against weather and sound
- Offers access only through 50% of doorway width
- Used on exterior as sliding glass doors
- Used in interiors primarily for visual screening

- Similar to a bypass sliding door but provides access through full width of doorway
- No operating space required but is difficult to weatherproof
- Door is surface-hung on an exposed overhead track

- Door slides on an overhead track into and out of a recess within the width of a wall.
- Doorway has a finished appearance when fully open.
- Often used where a normal door swing would interfere with the use of a space

- Hinged door panels fold flat against one another when opened.
- Bifold doors divide into two parts, require little operating space, and are used primarily as a visual screen to enclose closet and storage spaces.
- Accordion doors are multileafed doors that are used primarily to subdivide interior spaces. They are hung from an overhead track and open by folding back in the manner of an accordion.

- See 8.16 for revolving doors.

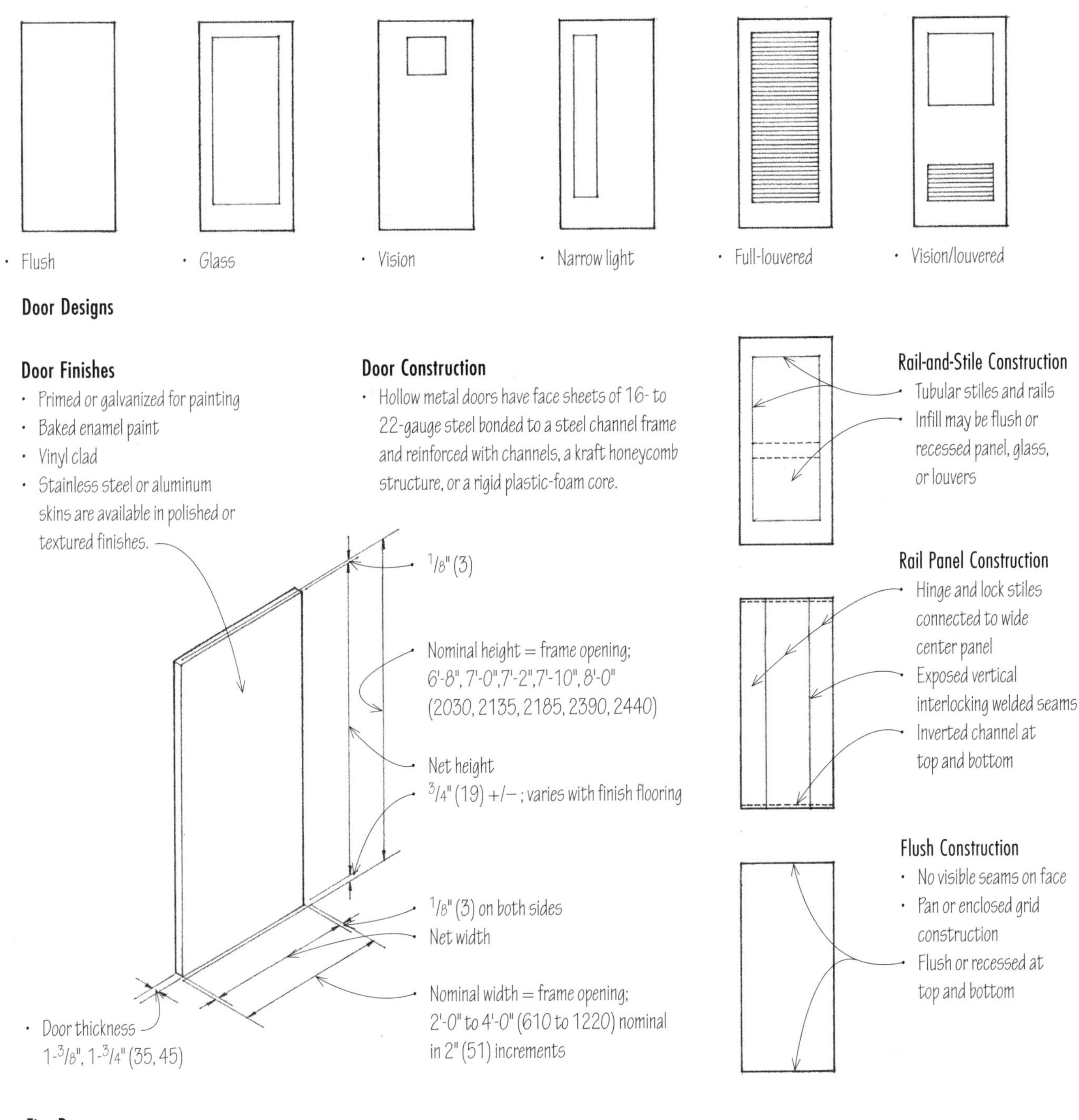

Fire Doors

UL Label	Rating	Glazing Permitted: 1/4" (6) wired glass
A	3 hour	No glass permitted
B	1-1/2 hour	100 sq. in. (0.06 m^2) per leaf
C	3/4 hour	1296 sq. in. (0.84 m^2) per leaf: 54" (1370) max. dimension
D	1-1/2 hour	No glass permitted
E	3/4 hour	720 sq. in. (0.46 m^2) per light; 54" (1370) max. dimension

- Fire door assemblies, consisting of a fire-resistive door, doorframe, and hardware, are required to protect openings in fire-rated walls. See 2.07.
- Maximum door size: 4' x 10' (1220 x 3050)
- Doorframe and hardware must have a fire-resistance rating similar to that of the door.
- Door must be self-latching and be equipped with closers.
- Louvers with fusible links are permitted for B and C label doors; maximum area = 576 sq. in. (0.37 m^2)
- No glass and louver combinations are permitted.

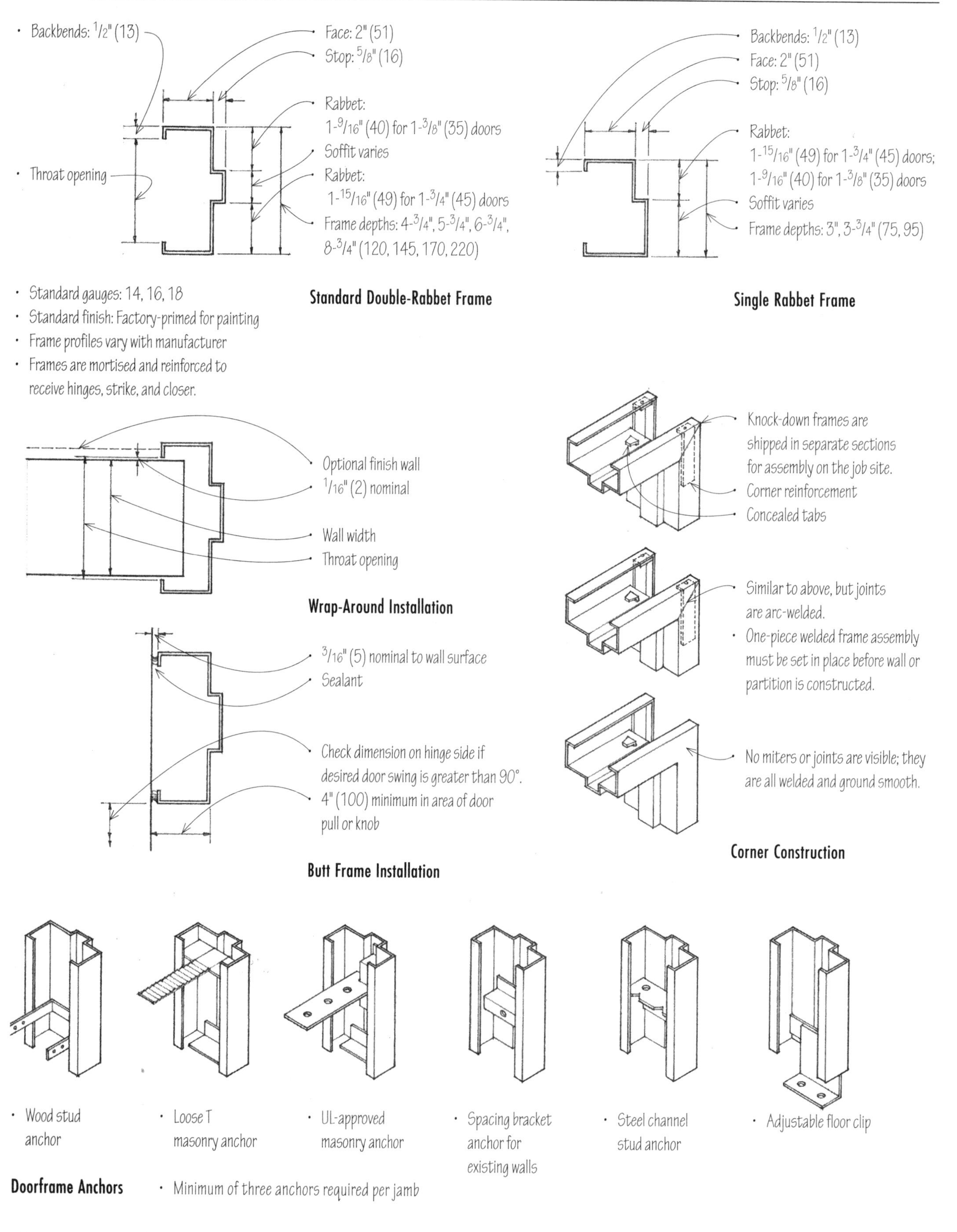

Standard Double-Rabbet Frame

Single Rabbet Frame

- Standard gauges: 14, 16, 18
- Standard finish: Factory-primed for painting
- Frame profiles vary with manufacturer
- Frames are mortised and reinforced to receive hinges, strike, and closer.

Wrap-Around Installation

Butt Frame Installation

Corner Construction

Doorframe Anchors

- Minimum of three anchors required per jamb

CSI MasterFormat 08 12 13: Hollow Metal Frames

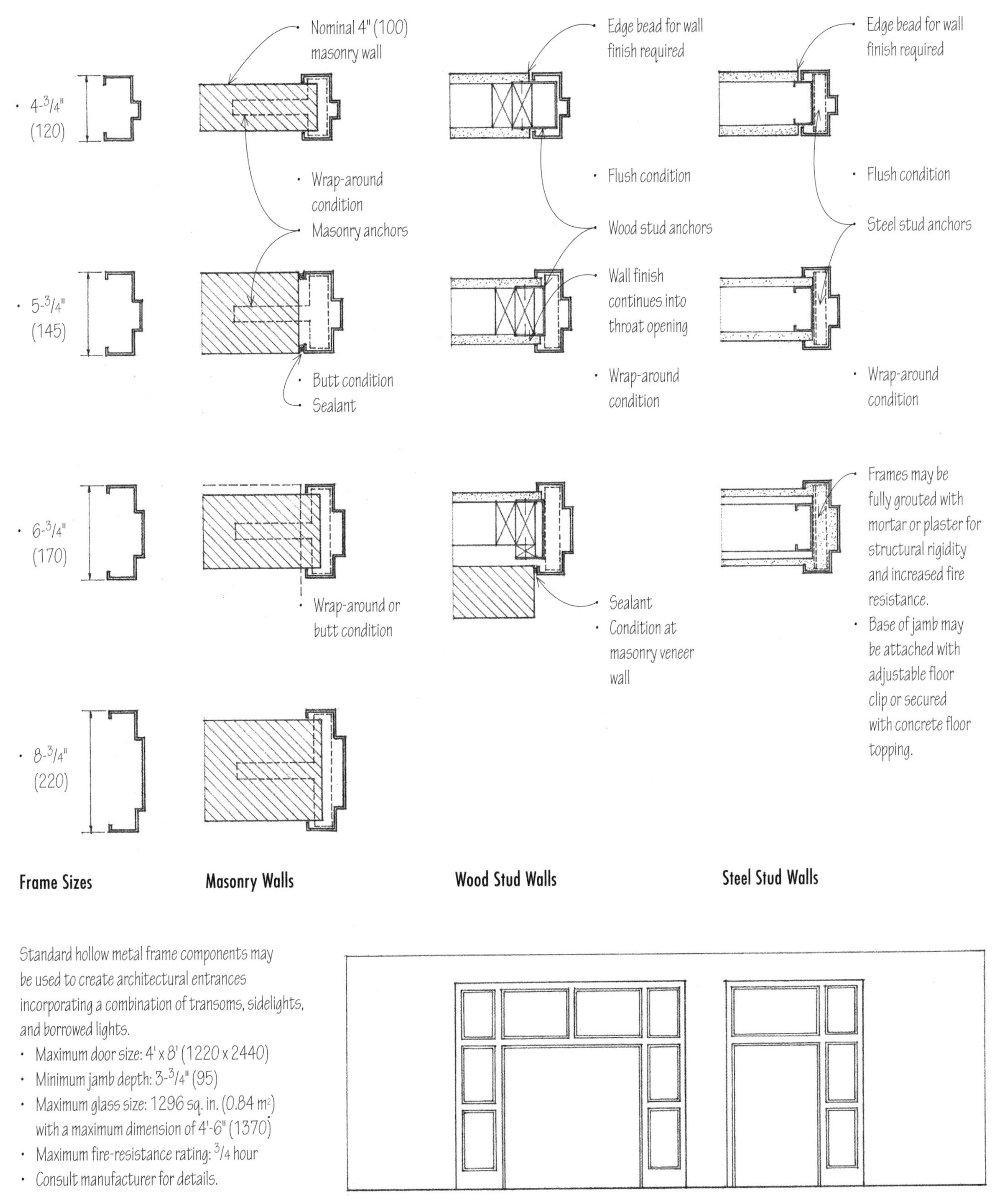

Standard hollow metal frame components may be used to create architectural entrances incorporating a combination of transoms, sidelights, and borrowed lights.

- Maximum door size: 4' x 8' (1220 x 2440)
- Minimum jamb depth: 3-3/4" (95)
- Maximum glass size: 1296 sq. in. (0.84 m^2) with a maximum dimension of 4'-6" (1370)
- Maximum fire-resistance rating: 3/4 hour
- Consult manufacturer for details.

Hollow Metal Stick Systems

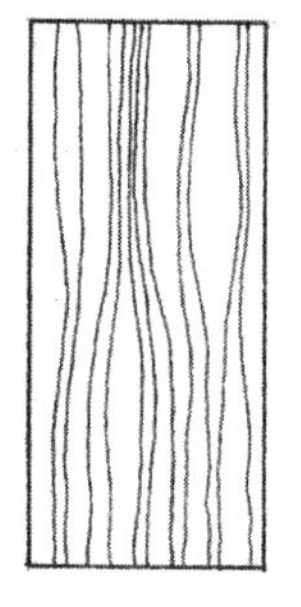

• Flush door

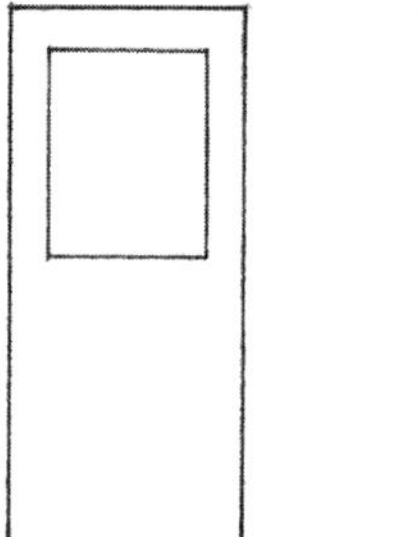

• Flush door w/ glass inserts

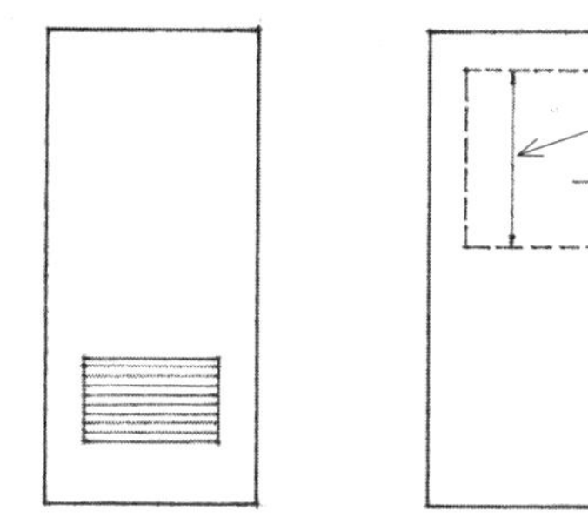

• Flush door w/ louvered insert

Door Designs

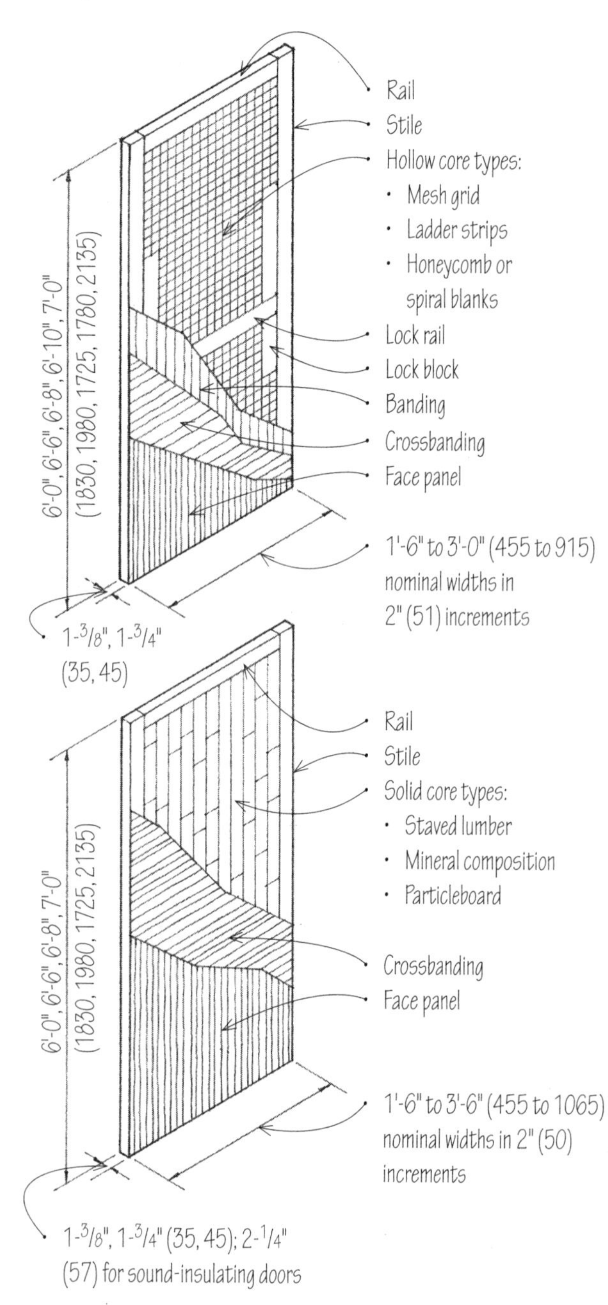

Hollow Core Doors

Hollow core doors have a framework of stiles and rails encasing an expanded honeycomb core of corrugated fiberboard or a grid of interlocking horizontal and vertical wood strips. They are lightweight but have little inherent thermal or acoustic insulation value. While intended primarily for interior use, they may be used for exterior doors if bonded with waterproof adhesives.

Solid Core Doors

Solid core doors have a core of bonded lumber blocks, particleboard, or a mineral composition. Of these, the bonded lumber core is the most economical and widely used. The mineral composition core is lightest but has low screw-holding strength and cutouts are difficult. Solid core doors are used primarily as exterior doors, but they may also be used wherever increased fire resistance, sound insulation, or dimensional stability is desired.

Grades and Finishes

- There are three hardwood veneer grades: premium, good, and sound.
- Premium grade veneers are suitable for natural, transparent finishes.
- Good grade veneers are for transparent or paint finishes.
- Sound grade veneers are for paint finishes only; they require two coats to cover surface defects.
- Hardboard face panels are suitable for paint finishes.
- High-pressure plastic laminates may be bonded to the face panels.
- Flush doors may also be factory-finished partially with a seal coat or completely including prefitting and premachining for hinges and locksets.

Special Doors

- Fire-rated doors have mineral composition cores.
- B-label doors have a 1 hour or 1-1/2 hour UL-approved rating.
- C-label doors have a 3/4 hour UL-approved rating.
- Sound-insulating doors have faces separated by a void or damping compound. Special stops, gaskets, and thresholds are also required.

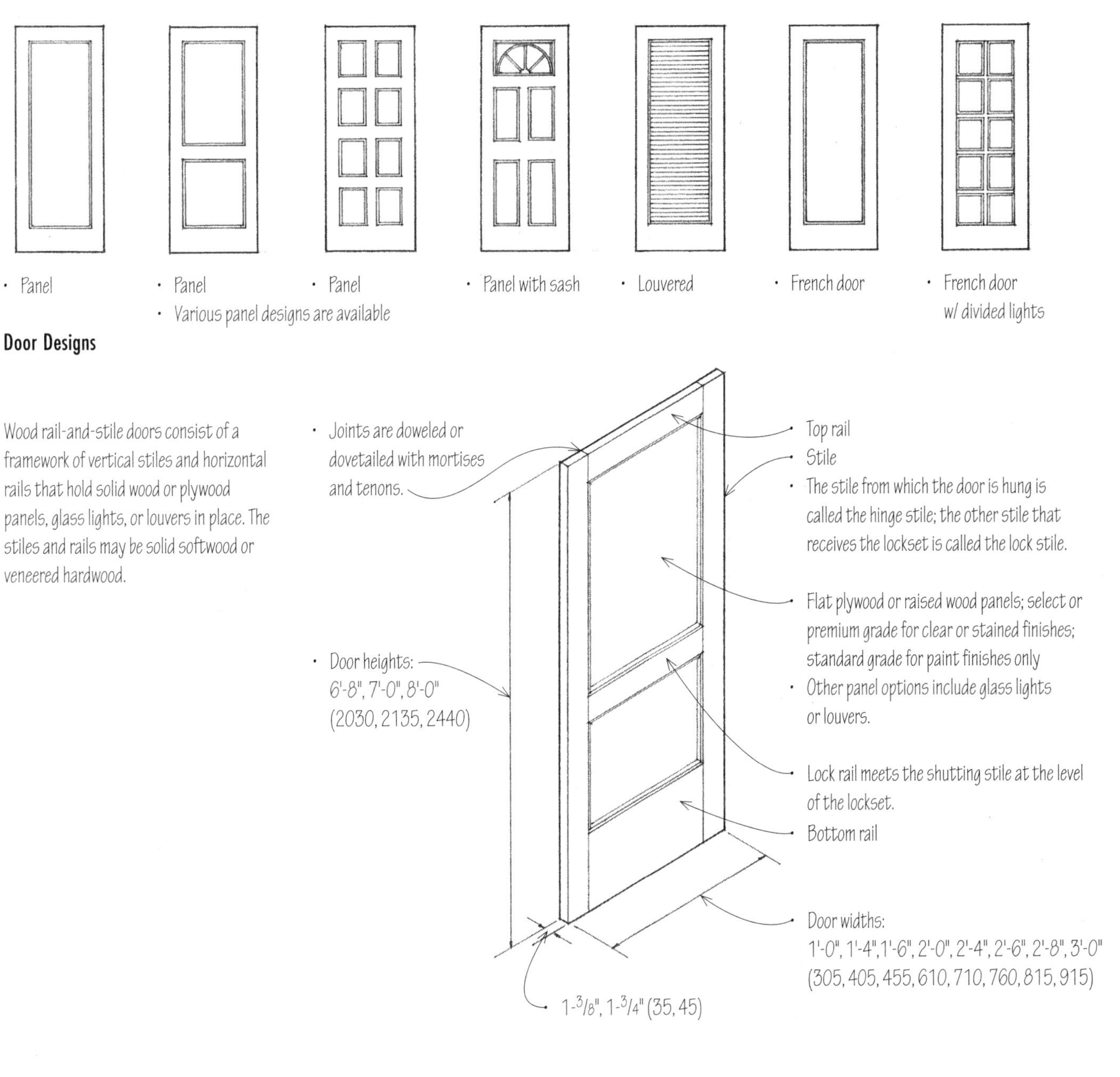

Door Designs

Wood rail-and-stile doors consist of a framework of vertical stiles and horizontal rails that hold solid wood or plywood panels, glass lights, or louvers in place. The stiles and rails may be solid softwood or veneered hardwood.

Batten doors consist of vertical board sheathing nailed at right angles to cross strips or ledgers. Diagonal bracing is nailed between and notched into the ledgers.

- Used primarily for economy in rough construction
- Usually site-fabricated
- Tongue-and-groove sheathing is recommended for weathertightness.
- Subject to expansion and contraction with changes in moisture content

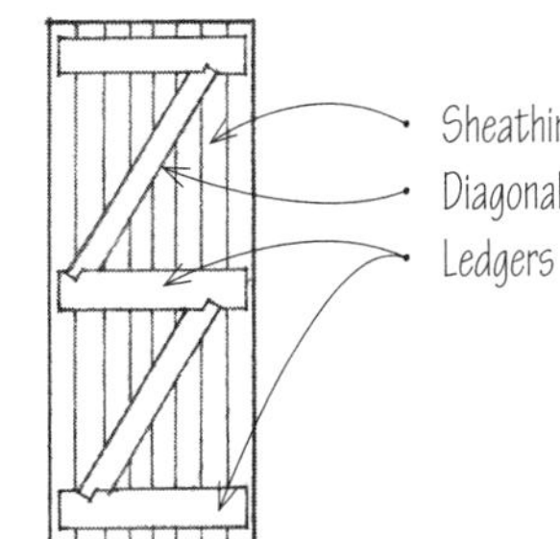

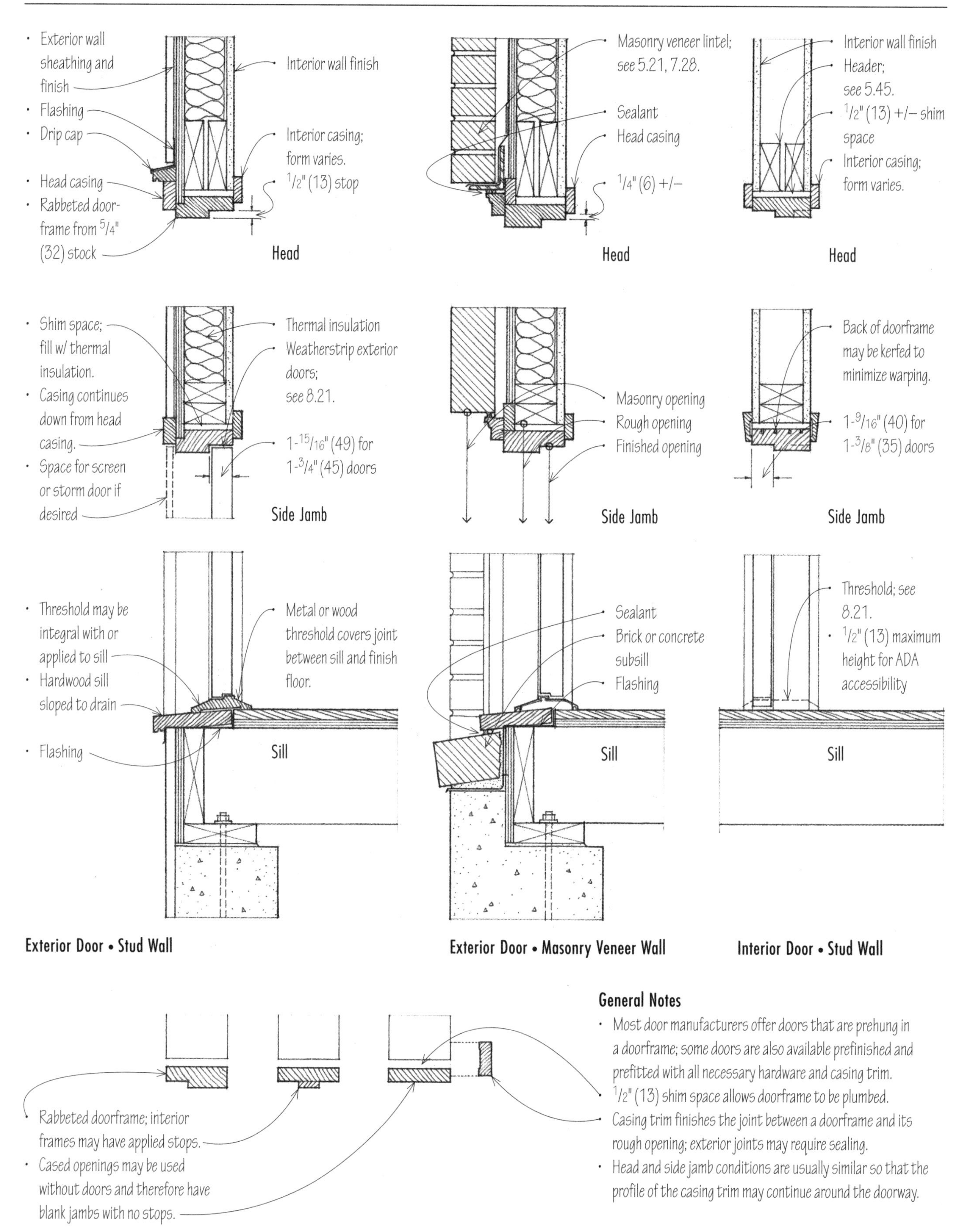

General Notes

- Most door manufacturers offer doors that are prehung in a doorframe; some doors are also available prefinished and prefitted with all necessary hardware and casing trim.
- 1/2" (13) shim space allows doorframe to be plumbed.
- Casing trim finishes the joint between a doorframe and its rough opening; exterior joints may require sealing.
- Head and side jamb conditions are usually similar so that the profile of the casing trim may continue around the doorway.

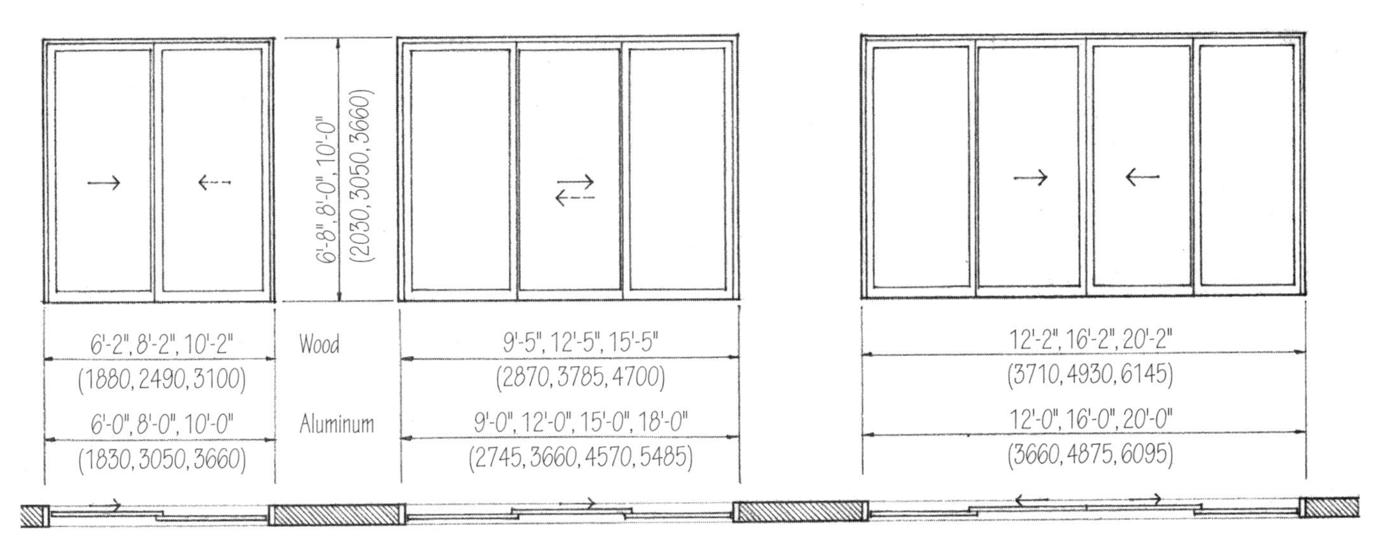

- Dimensions are nominal stock sizes; consult manufacturer for stock sizes, required rough or masonry openings, glazing options, and installation details.
- As a guide, add 1" (25) to nominal width for rough framed openings and 3" (75) for masonry openings.

Typical Sizes

Sliding glass doors are available with wood, aluminum, or steel frames. Wood frames may be treated with preservative, primed for painting, or clad in aluminum or vinyl. Metal frames are available in a variety of finishes, with thermal breaks and integral windproof mounting fins.

- Sliding glass doors are manufactured as standard units complete with operating hardware and weatherstripping. Screen and operating door panels may be on the interior or exterior.

ADA Accessibility Guideline

- Thresholds for exterior residential sliding doors should be no higher than for 3/4" (19).

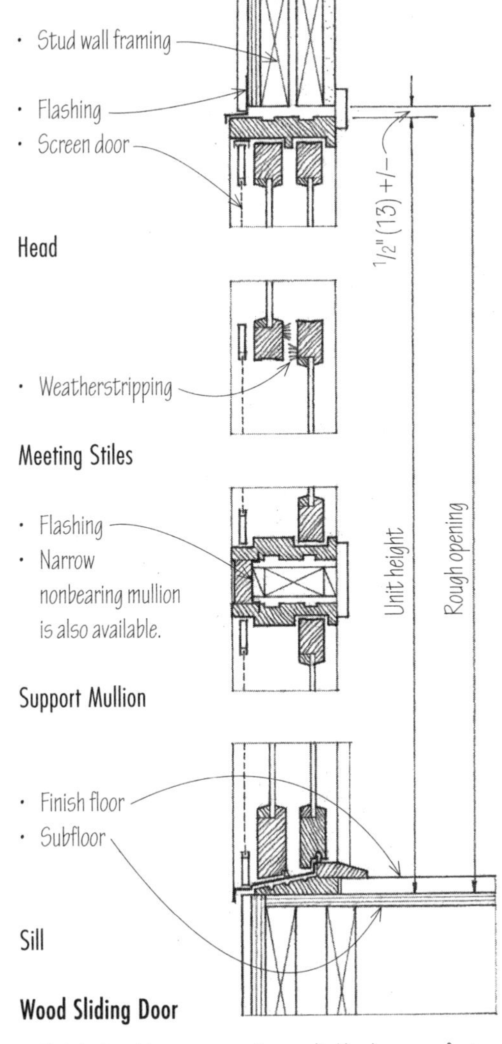

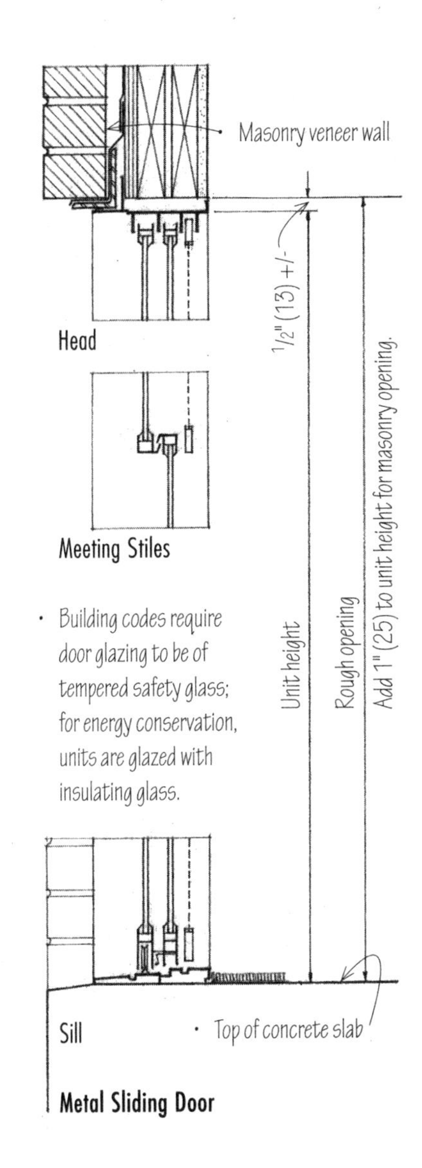

- Building codes require door glazing to be of tempered safety glass; for energy conservation, units are glazed with insulating glass.

- Hatched sections are normally supplied by door manufacturer.

CSI MasterFormat 08 32 00: Sliding Glass Doors

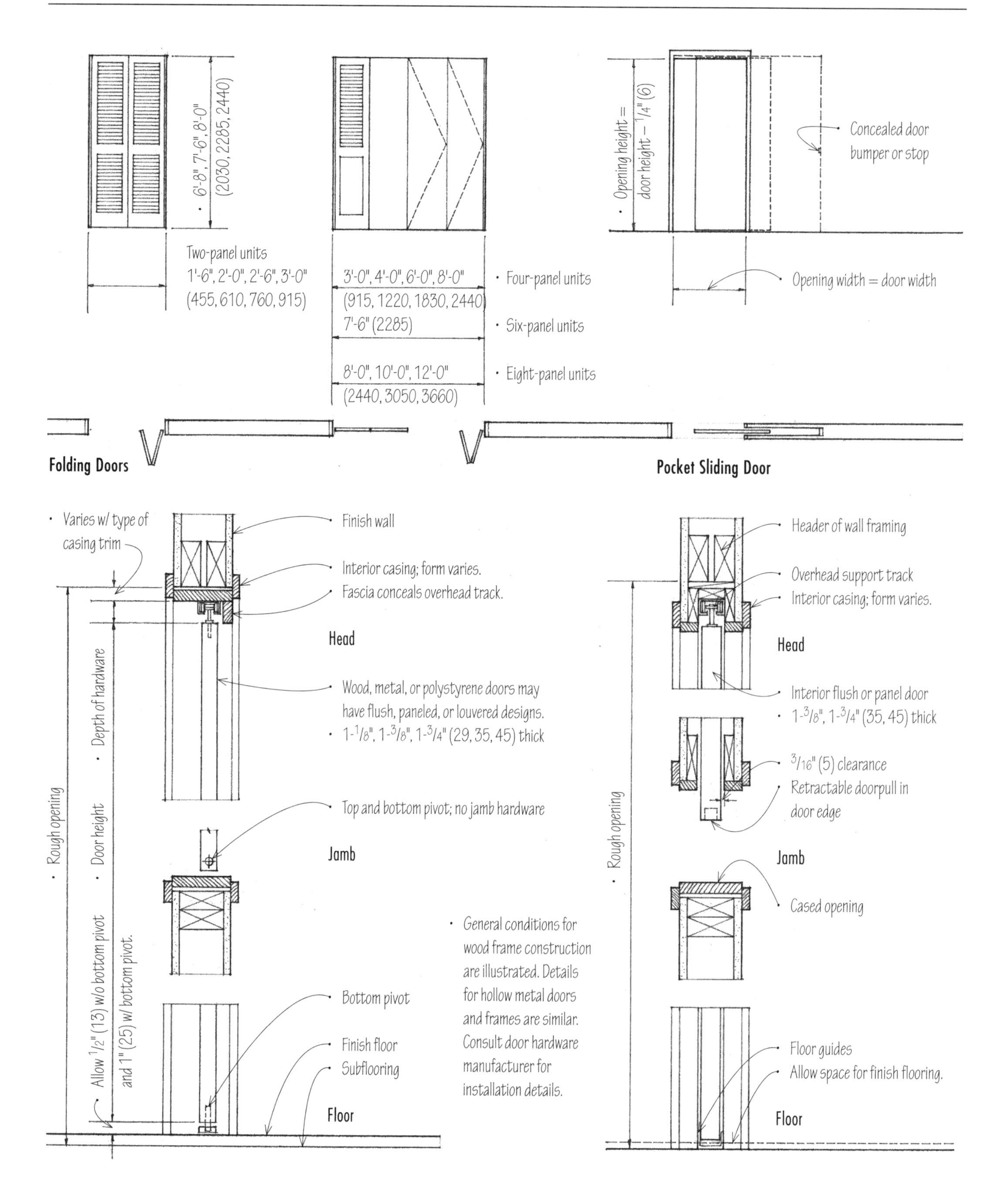

CSI MasterFormat 08 13 76: Bifolding Metal Doors; 08 14 76: Bifolding Wood Doors; 08 15 76: Bifolding Plastic Doors

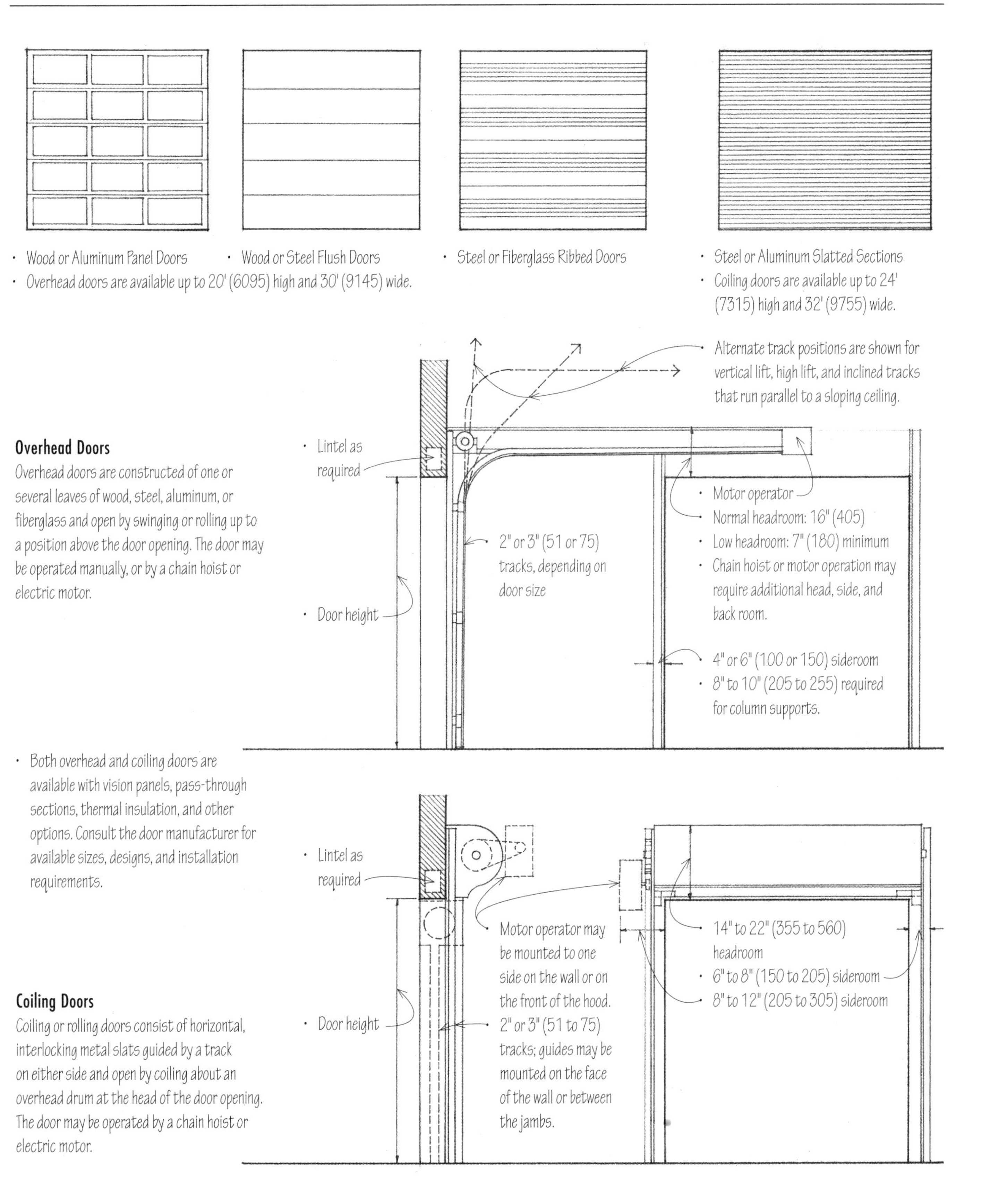

Overhead Doors

Overhead doors are constructed of one or several leaves of wood, steel, aluminum, or fiberglass and open by swinging or rolling up to a position above the door opening. The door may be operated manually, or by a chain hoist or electric motor.

- Both overhead and coiling doors are available with vision panels, pass-through sections, thermal insulation, and other options. Consult the door manufacturer for available sizes, designs, and installation requirements.

Coiling Doors

Coiling or rolling doors consist of horizontal, interlocking metal slats guided by a track on either side and open by coiling about an overhead drum at the head of the door opening. The door may be operated by a chain hoist or electric motor.

CSI MasterFormat 08 33 00: Coiling Doors and Grilles
CSI MasterFormat 08 36 00: Panel Doors

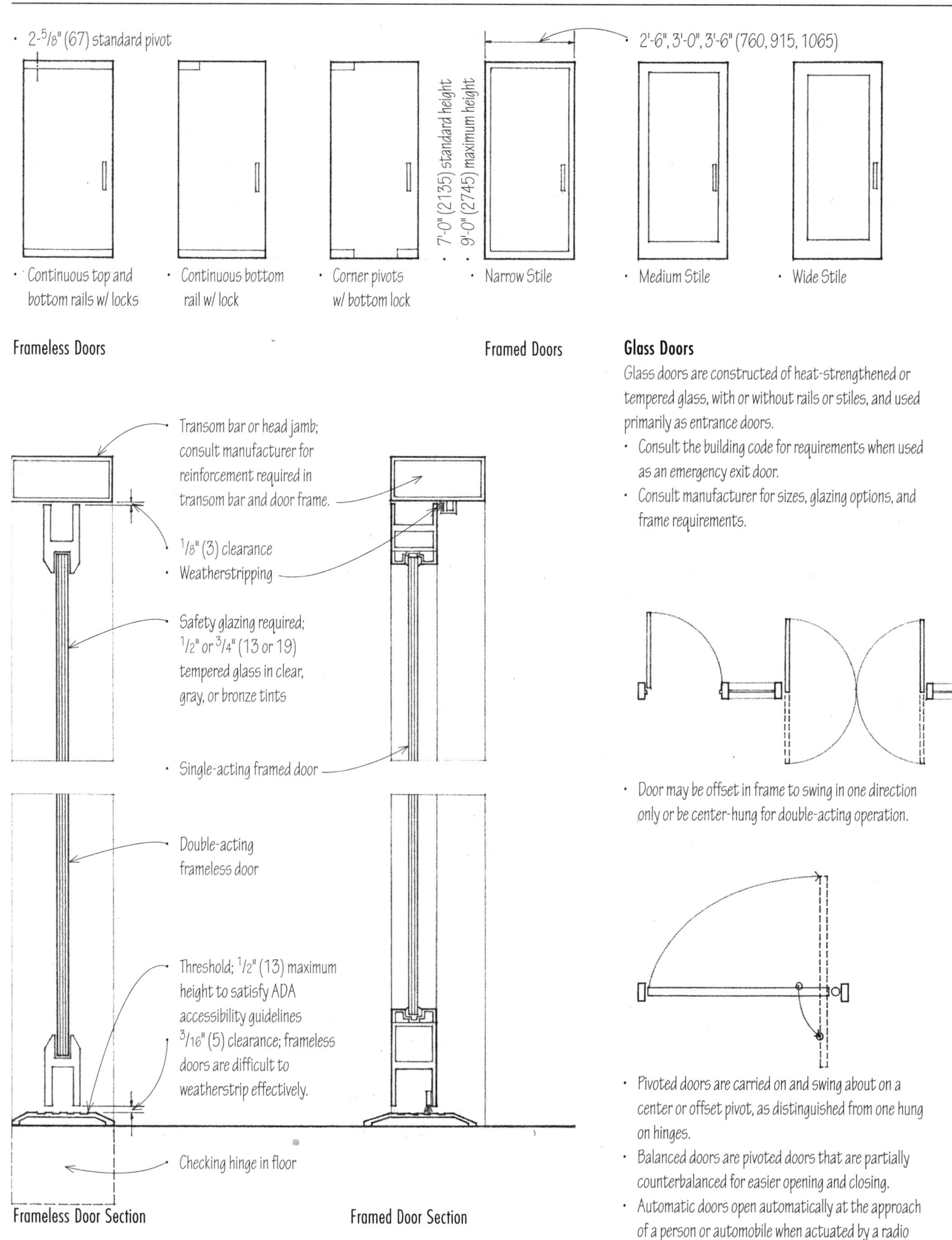

Glass Doors

Glass doors are constructed of heat-strengthened or tempered glass, with or without rails or stiles, and used primarily as entrance doors.

- Consult the building code for requirements when used as an emergency exit door.
- Consult manufacturer for sizes, glazing options, and frame requirements.

- Door may be offset in frame to swing in one direction only or be center-hung for double-acting operation.

- Pivoted doors are carried on and swing about on a center or offset pivot, as distinguished from one hung on hinges.
- Balanced doors are pivoted doors that are partially counterbalanced for easier opening and closing.
- Automatic doors open automatically at the approach of a person or automobile when actuated by a radio transmitter, electric eye, or other device.

CSI MasterFormat 08 41 00: Entrances and Storefronts

Storefronts are coordinated systems of extruded metal frames, glass panels, glass entrance doors, and hardware fittings. The size and spacing of the mullions are determined by the glass strength and thickness and the wind load on the wall plane. The deflection normal to the wall plane should be limited to 1/200 of each component's clear span; the deflection of glass supports should be limited to 1/300 of the support distance.

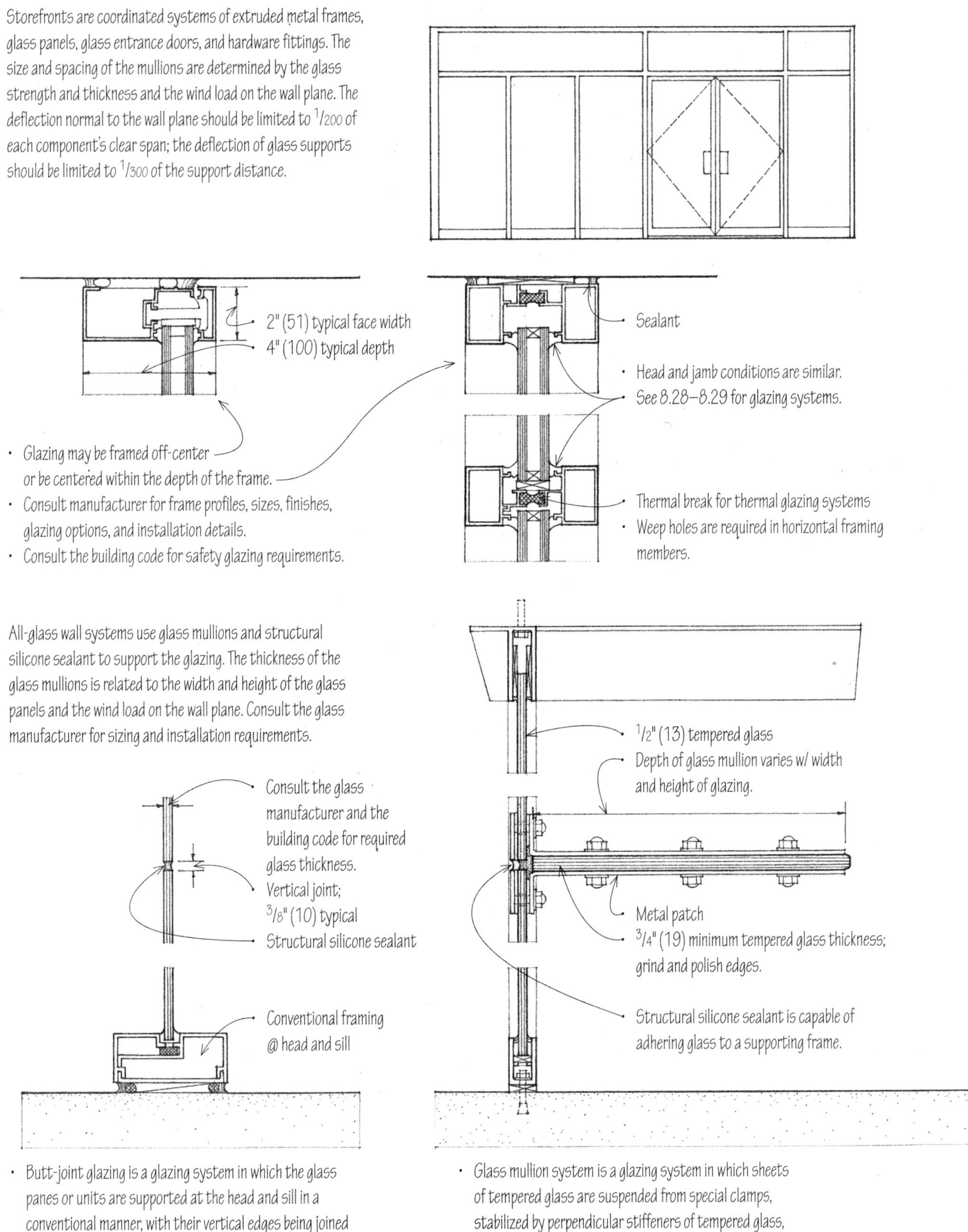

- Glazing may be framed off-center or be centered within the depth of the frame.
- Consult manufacturer for frame profiles, sizes, finishes, glazing options, and installation details.
- Consult the building code for safety glazing requirements.

All-glass wall systems use glass mullions and structural silicone sealant to support the glazing. The thickness of the glass mullions is related to the width and height of the glass panels and the wind load on the wall plane. Consult the glass manufacturer for sizing and installation requirements.

- Butt-joint glazing is a glazing system in which the glass panes or units are supported at the head and sill in a conventional manner, with their vertical edges being joined with a structural silicone sealant without mullions.

- Glass mullion system is a glazing system in which sheets of tempered glass are suspended from special clamps, stabilized by perpendicular stiffeners of tempered glass, and joined by a structural silicone sealant and by metal patch plates at corners and edges.

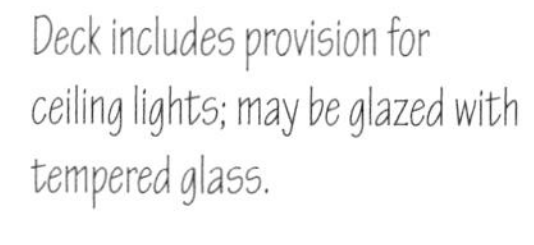

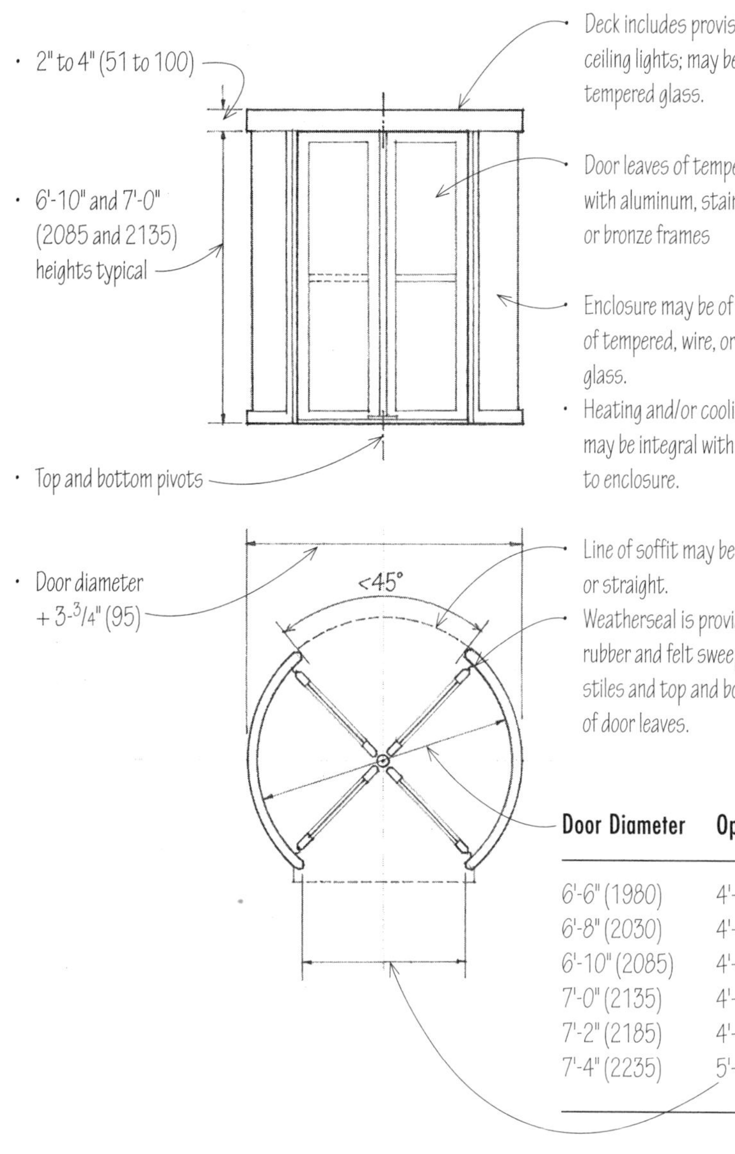

Door Diameter	Opening
6'-6" (1980)	4'-5" (1345)
6'-8" (2030)	4'-6" (1370)
6'-10" (2085)	4'-8" (1420)
7'-0" (2135)	4'-9" (1450)
7'-2" (2185)	4'-11" (1500)
7'-4" (2235)	5'-0" (1525)

Revolving doors consist of three or four leaves that rotate about a central, vertical pivot within a cylindrically shaped vestibule. Used typically as entrance doors in large commercial and institutional buildings, revolving doors provide a continuous weatherseal, eliminate drafts, and hold heating and cooling losses to a minimum while accommodating traffic up to 2000 persons per hour.

- 6'-6" (1980) diameter for general use; 7'-0" (2135) diameter or greater for high traffic areas
- An optional speed control automatically aligns doors at quarter points when not in use and turns wings 3/4 of a revolution at walking speed when activated by slight pressure.
- Some revolving doors have leaves that automatically fold back in the direction of egress when pressure is applied, providing a legal passageway on both sides of the door pivot.
- Some building codes may credit revolving doors with satisfying 50% of the legal exit requirements. Other codes do not credit revolving doors and require adjacent hinged doors for use as emergency exits.

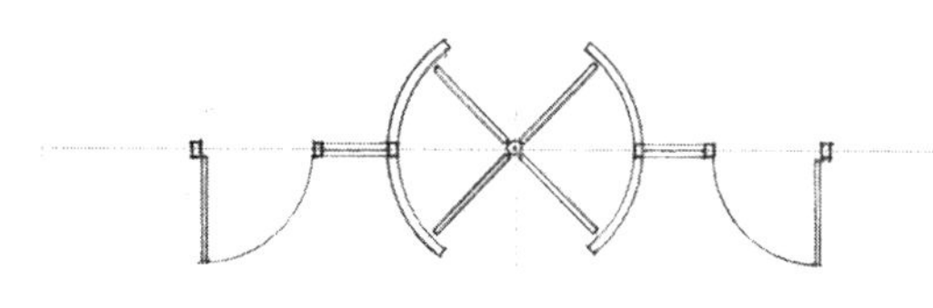

- Enclosure flanked by hinged doors

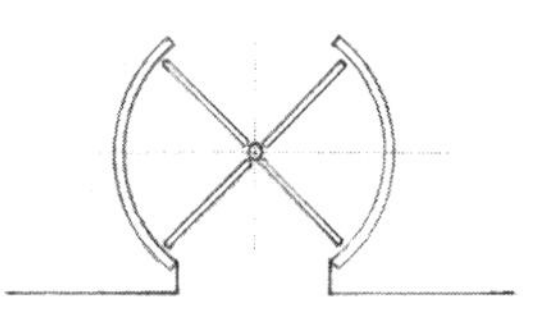

- Enclosure set within a wall plane

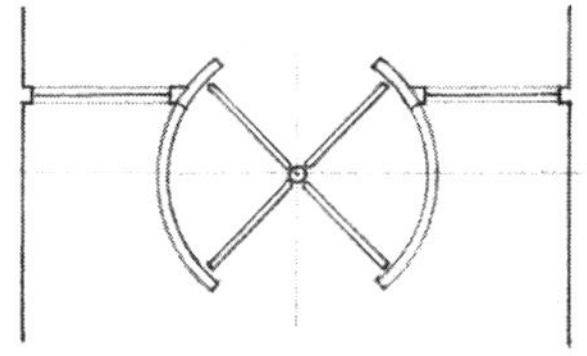

- Enclosure projecting from sidelights

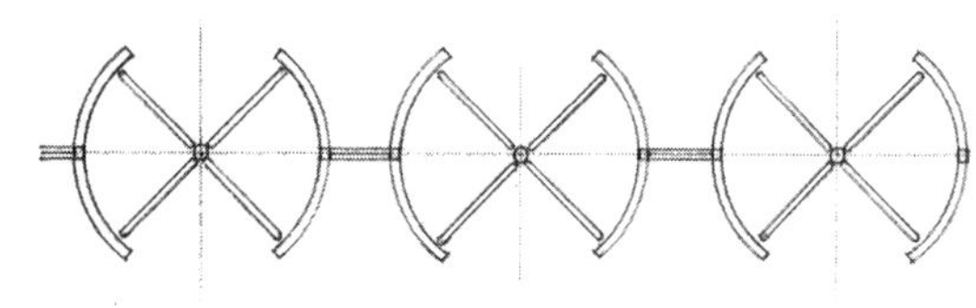

- Bank of enclosures with sidelights between

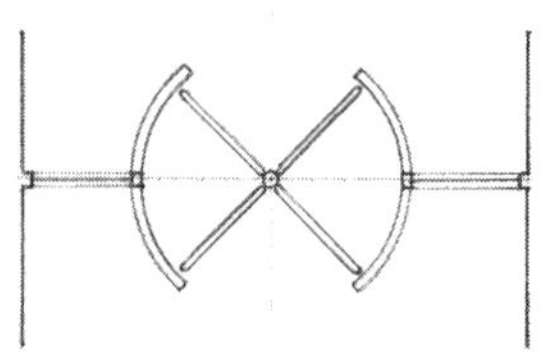

- Sidelights centered on enclosure

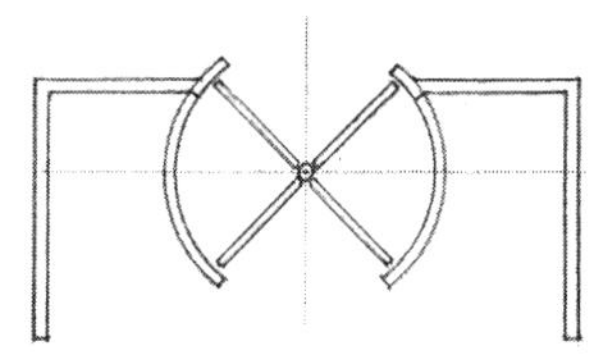

- Enclosure set back within a wall recess

Revolving Door Layouts

Finish door hardware for doors include the following items:

- Locksets incorporating locks, latches, and bolts, a cylinder and stop works, and operating trim
- Hinges
- Closers
- Panic hardware
- Push and pull bars and plates
- Kick plates
- Door stops, holders, and bumpers
- Thresholds
- Weatherstripping
- Door tracks and guides

Hardware selection factors:

- Function and ease of operation
- Recessed or surface-mounted installation
- Material, finish, texture, and color
- Durability in terms of anticipated frequency of use and possible exposure to weather or corrosive conditions

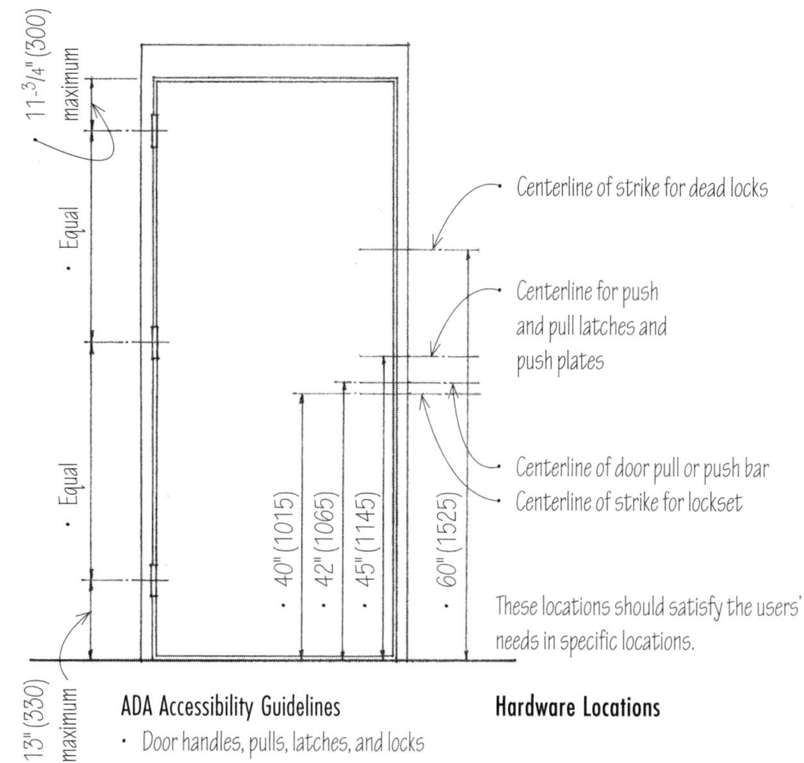

These locations should satisfy the users' needs in specific locations.

ADA Accessibility Guidelines

- Door handles, pulls, latches, and locks should be easy to grasp with one hand without tight grasping, pinching, or twisting of the wrist.
- Hardware should be mounted within the reach ranges specified in A.03.

Hardware Locations

Hardware Finishes

BHMA Code	US No.	Finish
600	USP	Steel primed for painting
603	US2G	Zinc plated steel
605	US3	Bright brass, clear coated
606	US4	Satin brass, clear coated
611	US9	Bright bronze, clear coated
612	US10	Satin bronze, clear coated
613	US10B	Oxidized satin bronze, oil rubbed
618	US14	Bright nickel plated, clear coated brass
619	US15	Satin nickel plated, clear coated brass
622	US19	Flat black coated brass or bronze
623	US20	Light oxidized bright bronze
624	US20A	Dark oxidized statuary bronze
625	US26	Bright chromium plated brass or bronze
626	US26D	Satin chromium plated brass or bronze
628	US28	Satin aluminum, clear anodized
629	US32	Bright stainless steel
630	US32D	Satin stainless steel
684	—	Black chrome, bright brass or bronze
685	—	Black chrome, satin brass or bronze

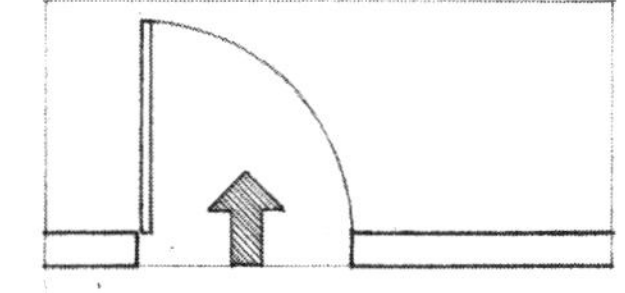

- Left hand (LH)
- Door opens inward; hinges on left

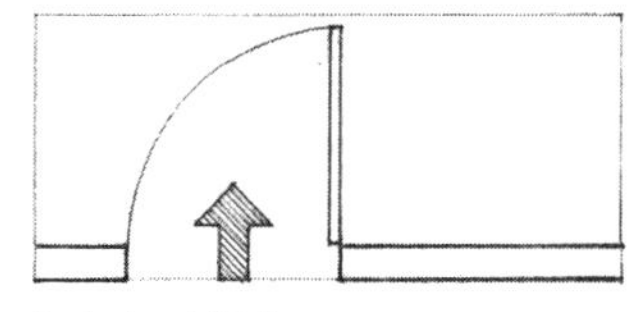

- Right hand (RH)
- Door opens inward; hinges on right

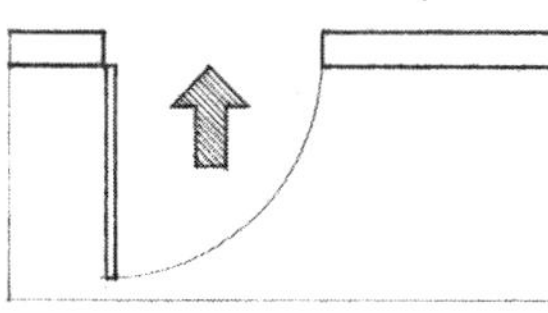

- Left hand reverse (LHR)
- Door opens outward; hinges on left

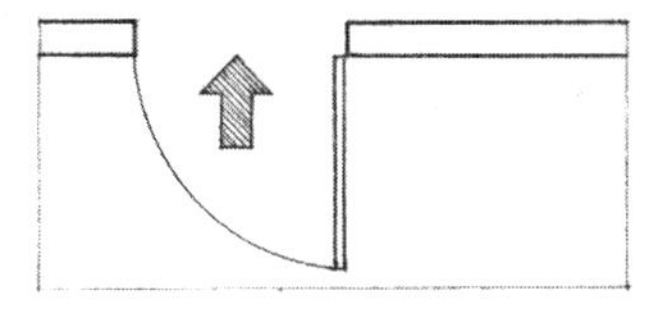

- Right hand reverse (RHR)
- Door opens outward; hinges on right

Door Hand Conventions

Door hand conventions are used in specifying door hardware such as locksets and closers. The terms right and left assume a view from the exterior of the building or room to which the doorway leads.

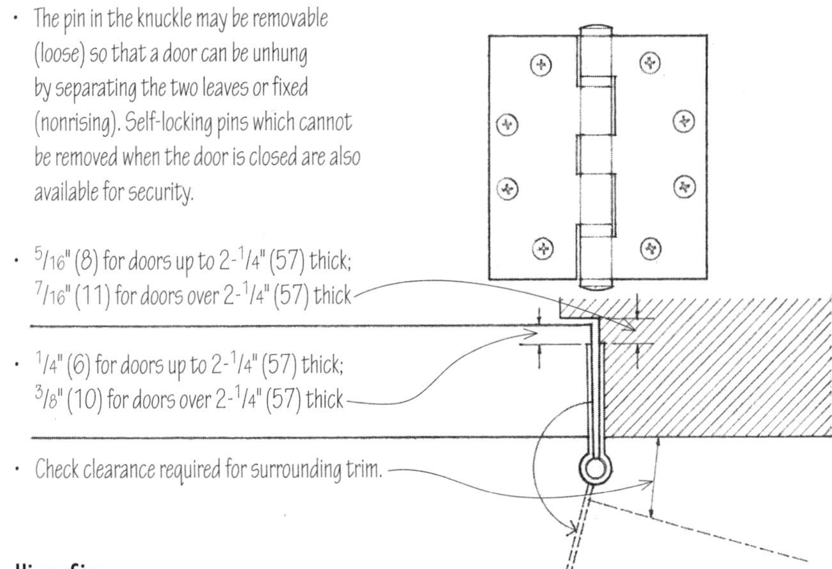

- The pin in the knuckle may be removable (loose) so that a door can be unhung by separating the two leaves or fixed (nonrising). Self-locking pins which cannot be removed when the door is closed are also available for security.
- 5/16" (8) for doors up to 2-1/4" (57) thick; 7/16" (11) for doors over 2-1/4" (57) thick
- 1/4" (6) for doors up to 2-1/4" (57) thick; 3/8" (10) for doors over 2-1/4" (57) thick
- Check clearance required for surrounding trim.

Hinge Size

- Hinge width is determined by door thickness and clearance required.
- Hinge height is determined by the door width and thickness.

Door Thickness	Door Width	Hinge Height	Clearance Required	Hinge Width
3/4" to 1" (19 to 25)	To 24" (610)	2-1/2" (64)		
1-1/8" (29)	To 36" (915)	3" (75)		
1-3/8" (35)	To 36" (915)	3-1/2" (90)	1-1/4" (32)	3-1/2" (90)
	Over 36" (915)	4" (100)	1-3/4" (45)	4" (100)
1-3/4" (45)	To 36" (915)	4-1/2" (115)	1-1/2" (38)	4-1/2" (115)
	36" to 48" (915 to 1220)	5" (125)	2" (51)	5" (125)
2-1/4" (57)	To 42" (1065)	5" (125)	1" (25)	5" (125)
	Over 42" (1065)	6" (150)	2" (51)	6" (150)

Butt Hinges

Butt hinges are composed of two plates or leaves joined by a pin and secured to the abutting surfaces of wood and hollow metal doors and door jambs.

- Full-mortise hinges have both leaves fully mortised into the abutting surfaces of a door and door jamb so that only the knuckle is visible when the door is closed.
- Template hinges are mortise hinges manufactured to fit the recess and match the arrangement of holes of hollow metal doors and frames; non-template hinges are used for wood doors.
- Half-mortise hinges have one leaf mortised into the edge of a door and the other surface-mounted to the doorframe.
- Half-surface hinges have one leaf mortised into a doorframe and the other surface-mounted to the face of the door.
- Full-surface hinges have two leaves surface-mounted to the adjacent faces of a door and doorframe.

Special-Purpose Hinges

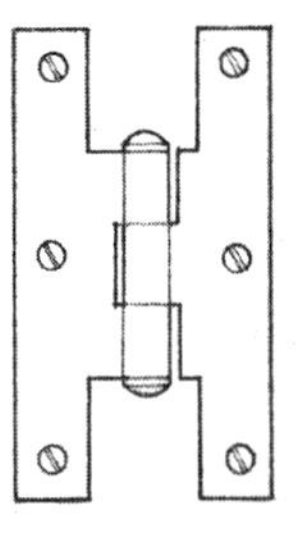

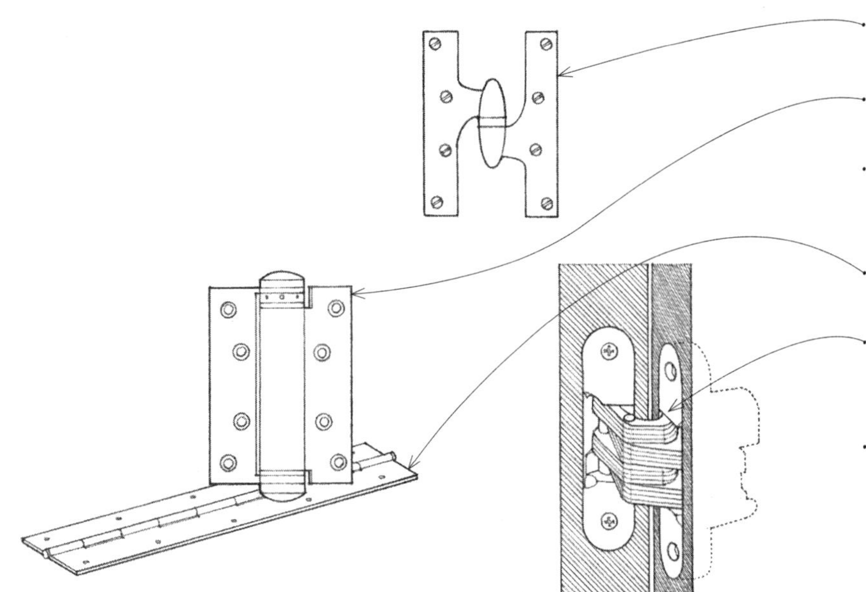

- Parliament hinges have T-shaped leaves and a protruding knuckle so that a door can stand away from the wall when fully opened.
- Olive knuckle hinges have a single, pivoting joint and an oval-shaped knuckle.
- Spring hinges contain coiled springs in their barrels for closing a door automatically.
- Double-acting hinges permit a door to swing in either direction, and are usually fitted with springs to bring the door to a closed position after opening.
- Piano hinges are long, narrow hinges that run the full length of the two surfaces to which their leaves are joined.
- Invisible hinges consist of a number of flat plates rotating about a central pin, with shoulders mortised into the door edge and doorframe so as to be concealed when closed.
- Floor hinges are used with a mortise pivot at door head to enable a door to swing in either direction; may be provided with a closer mechanism.

Locksets are manufactured assemblies of parts making up a complete locking system, including knobs, plates, and a locking mechanism. Described below are the major types of locksets: mortise locks, unit and integral locks, and cylinder locks. Consult hardware manufacturer for lockset functions, installation requirements, trim designs, dimensions, and finishes.

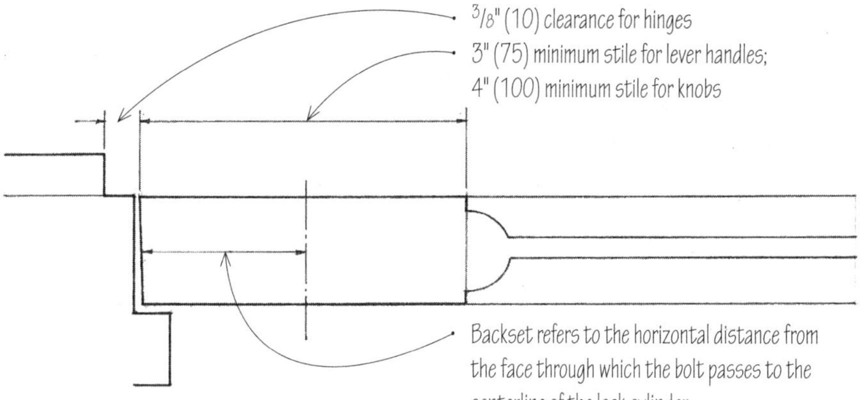

3/8" (10) clearance for hinges

3" (75) minimum stile for lever handles; 4" (100) minimum stile for knobs

Backset refers to the horizontal distance from the face through which the bolt passes to the centerline of the lock cylinder.

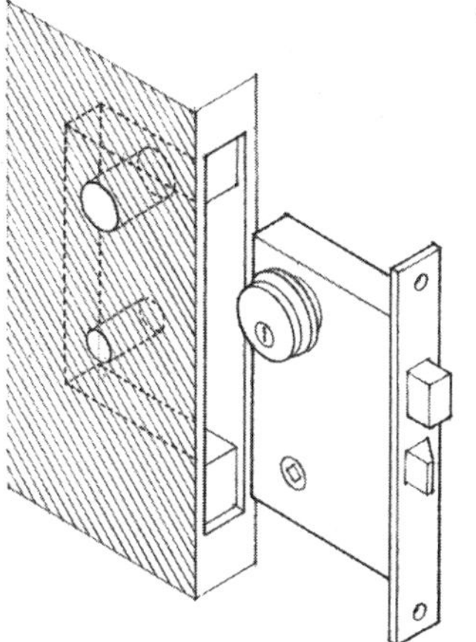

Mortise Lock

- Mortise lock is housed within a mortise cut into a door edge so that the lock mechanism is covered on both sides.
- Lock is concealed except for a faceplate at the door edge, knobs or levers, a cylinder, and operating trim.
- Backset: 2-1/2" (64) for 1-3/8" (35) doors; 2-3/4" (70) for 1-3/4" (45) doors

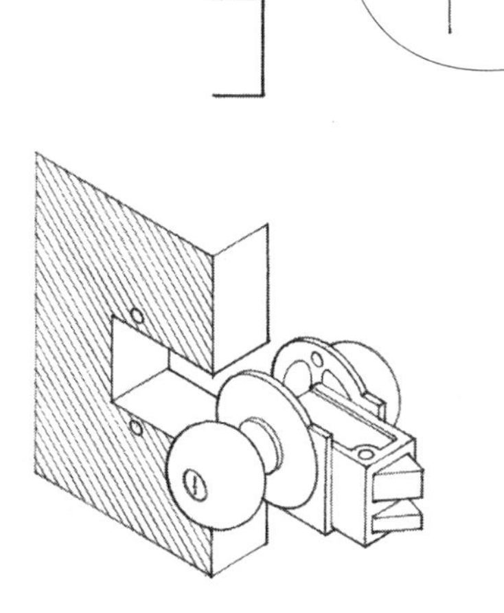

Unit and Integral Locks

- Unit lock is housed within a rectangular notch cut into the edge of a door.
- Integral lock fits into a mortise cut into the edge of a door.
- Unit and integral locks combine the security advantages of a mortise lock with the economy of a cylinder lock.
- Backset: 2-3/4" (70) for unit locks; 2-1/4" (57) for integral locks

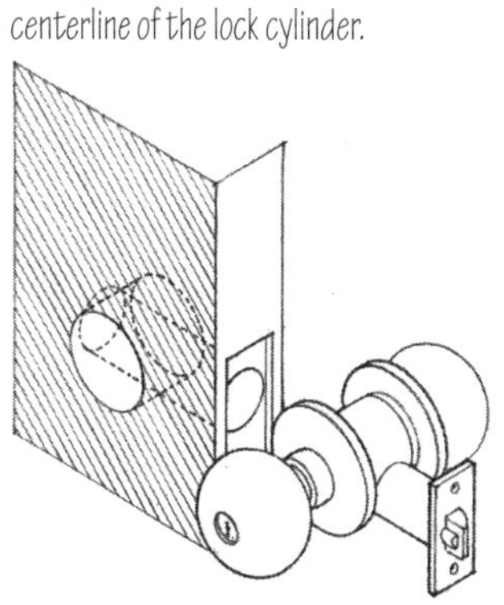

Cylinder Lock

- Cylinder lock is housed within two holes bored at right angles to each other, one through the lock stile of a door and the other in the door edge.
- Cylinder locks are relatively inexpensive and easy to install.
- Backset: 2-3/8" (60) for standard locksets; 2-3/4" (70) for heavy-duty locksets

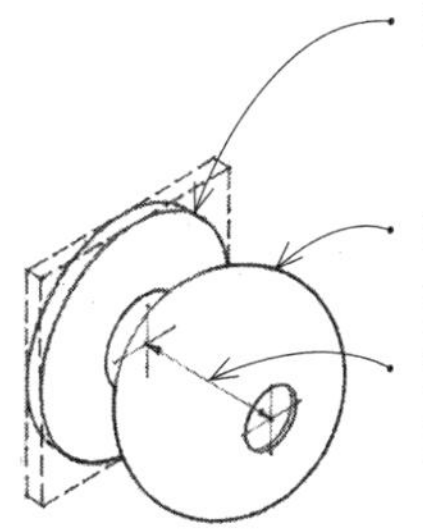

Door Knobs

- Rose refers to a round or square ornamental plate surrounding the shaft of a doorknob at the face of a door.
- Escutcheon is a protective or ornamental plate that may be substituted for a rose.

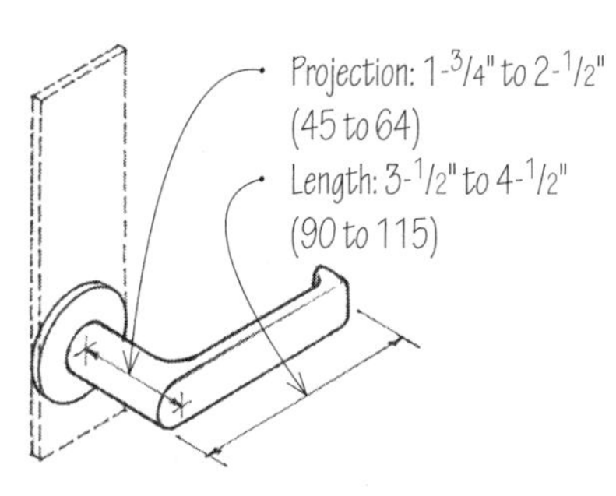

Lever Handles

- Lever-operated mechanisms, push-type mechanisms, and U-shaped handles are generally easier for people with disabilities to grasp.

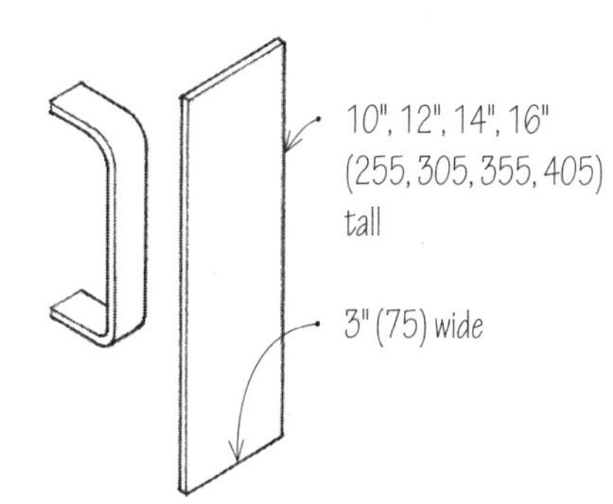

Pull Handles and Push Plates

ADA Accessibility Guidelines

- Door handles, pulls, latches, and locks should be easy to grasp with one hand without tight grasping, pinching, or twisting of the wrist.
- The force required for pushing open or pulling open a door should be no greater than 5.0 lbs. (22.2 N).

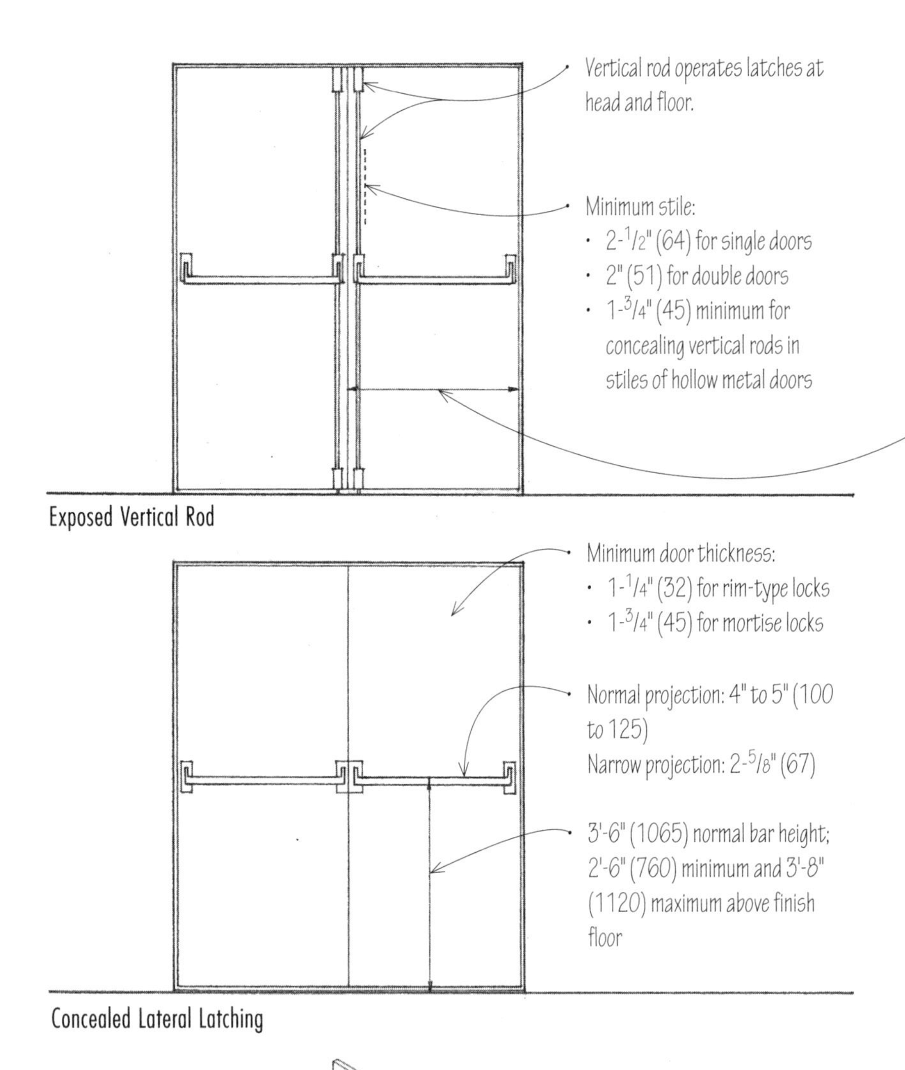

Exposed Vertical Rod

Concealed Lateral Latching

Panic Hardware

Panic hardware is a door-latching assembly that disengages when pressure is applied on a horizontal bar that spans the interior of an emergency exit door at waist height. The push bar should extend across at least one-half the width of the door leaf on which it is installed.

- Building codes require the use of panic hardware on emergency egress doors in certain building occupancies. Consult the applicable building code for details.
- The width, direction of swing, and location of required exit doors are also regulated by the building code according to the use and occupancy load of a building.

ADA Accessibility Guideline

- The force required for pushing open or pulling open a door should be no greater than 5.0 lbs. (22.2 N).

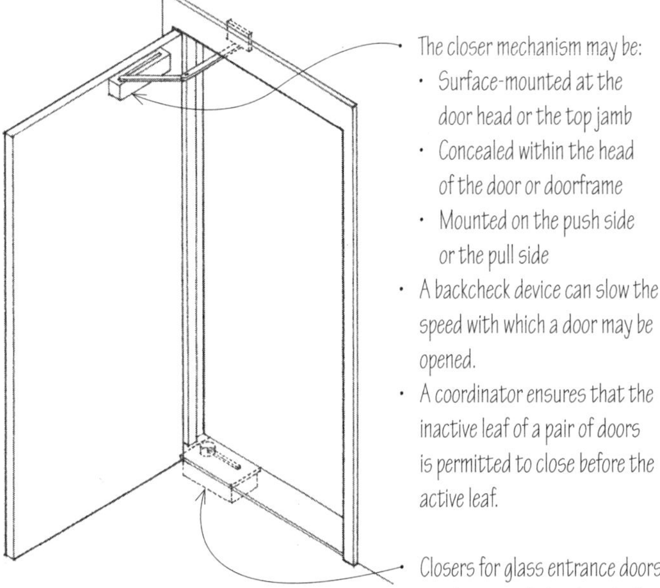

Door Closers

Door closers are hydraulic or pneumatic devices that automatically close doors quickly but quietly. They help reduce the shock a large, heavy, or heavily used door would otherwise transmit upon closing to its frame, hardware, and surrounding wall.

- Building codes require the use of self-latching, self-closing doors with UL-rated hardware to protect openings in fire walls and occupancy separations; see 2.07.

CSI MasterFormat 08 71 00: Door Hardware

Weatherstripping

Weatherstripping consists of metal, felt, vinyl, or foam rubber strips, placed between a door or window sash and its frame, to provide a seal against windblown rain and reduce the infiltration of air and dust.

- Weatherstripping may be fastened to the edge or face of a door, or to the doorframe and threshold.
- The weatherstripping material should be durable under extended use, noncorrosive, and replaceable.
- Basic types of weatherstripping include:
- Spring-tensioned strip of aluminum, bronze, or stainless or galvanized steel
- Vinyl or neoprene gaskets
- Foam plastic or rubber strips
- Woven pile strips

- Weatherstripping is often supplied and installed by the manufacturer of sliding glass doors, glass entrance doors, revolving doors, and overhead doors.
- Automatic door bottoms consist of a horizontal bar at the bottom of a door that drops automatically when the door is closed in order to seal the threshold and reduce noise transmission.

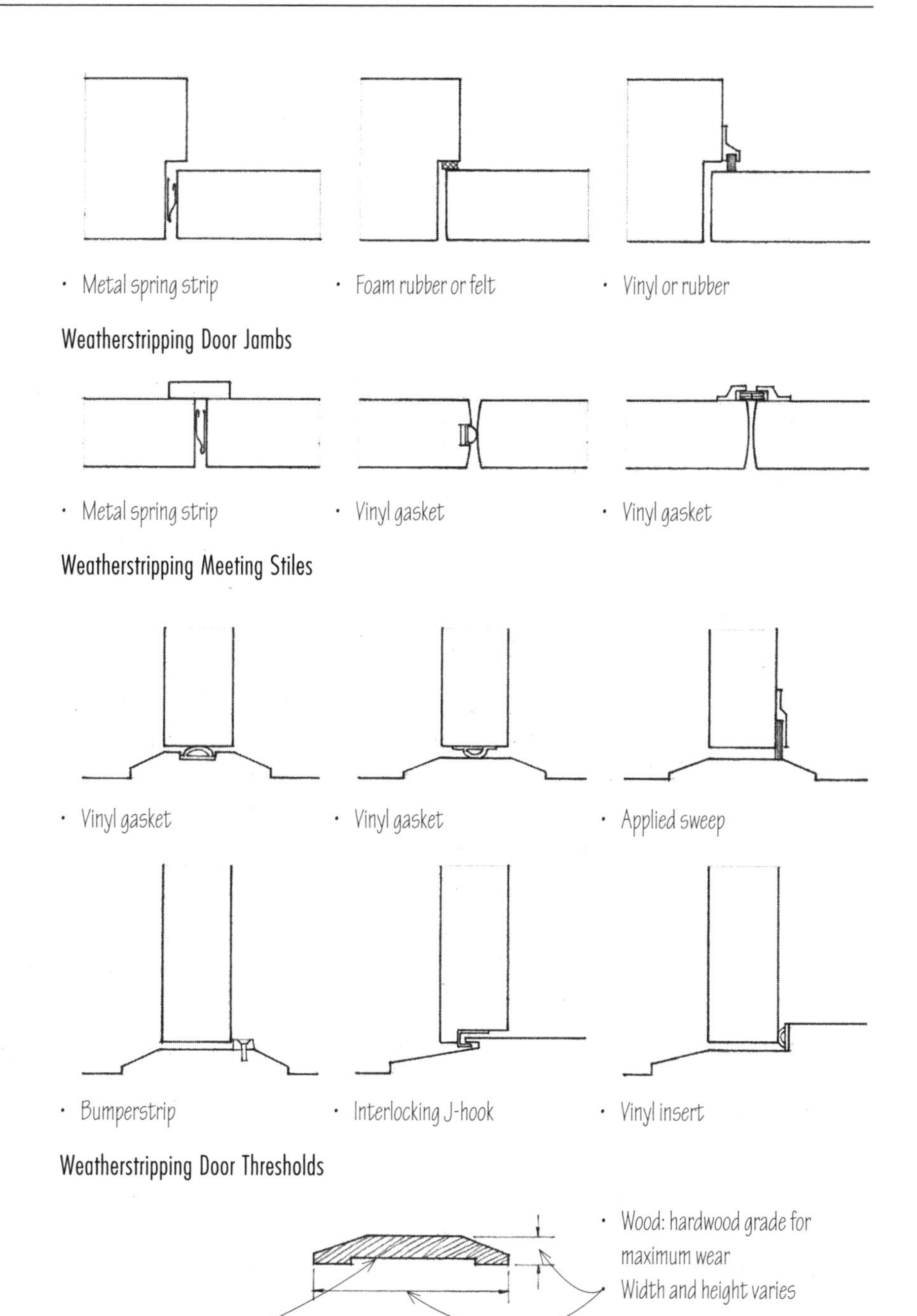

Weatherstripping Door Jambs

Weatherstripping Meeting Stiles

Weatherstripping Door Thresholds

Thresholds

Thresholds cover the joints between two flooring materials at doorways and serve as a weather barrier at exterior sills.

- Thresholds usually have recessed undersides to fit snugly against the flooring or sill.
- When installed at exterior sills, joint sealant is used for a tight seal.
- Metal thresholds may be cast or covered with abrasive material to provide a non-slip surface.

- Wood: hardwood grade for maximum wear
  Width and height varies
- Plain brass, bronze, or aluminum

2-1/4" to 6" (57 to 150)

3/16", 1/4", 1/2" (5, 6, 13)

- Fluted steel, aluminum, or bronze

3" to 7" (75 to 180)

5/16", 3/8", 1/2" (8, 10, 13)

ADA Accessibility Guideline

- Thresholds should be no higher than 1/2" (13) and be beveled with a slope not steeper than 1:2; thresholds for exterior residential sliding doors may be 3/4" (19) high.

Window Frame

- Metal window frames; see 8.24.
- Wood window frames; see 8.26.
- Insect screen may be on interior or exterior, depending on window operation

- Head is the uppermost member of a window frame.
- Jamb is either of the two side members of a window frame.

- Sill is the horizontal member beneath a door or window opening, having an upper surface sloped to shed rainwater.
- Subsill is an additional sill fitted to a window frame to cause rainwater to drip farther away from a wall surface.

- Exterior casing; not always used
- Drip cap or head casing
- Side casing

Sash and Glazing

- Sash refers to the fixed or movable framework of a window in which panes of glass are set. Its section profile varies with material, manufacturer, and type of operation.
- Pane is one of the divisions of a window, consisting of a single unit of glass set in a frame.
- Glazing refers to the panes or sheets of glass set in the sashes of a window. Single glazing offers little resistance to heat flow. For a reasonable thermal-resistance value (R-value), double glazing or a separate storm unit is required; using glass with a reflective coating or triple glazing is an option if a higher R-value is required. See 8.30.
- Just as important as the thermal-insulation rating of a window is its weathertightness. Operating sash should be weatherstripped against windblown rain and air infiltration. The joint between the window frame and the surrounding wall should be sealed and have a windbreak built into the detail.

Rough Opening

- Consult window manufacturer for required rough or masonry opening. Space is required at the top, sides, and bottom for leveling and shimming of the window unit.

Casing Trim

- Casing refers to the finishing trimwork around a window opening, consisting of head and jamb casings, window sills, and aprons; see 10.27.

Building Code Requirements

In selecting a window unit, review the building code requirements for:

- Natural light and ventilation of habitable spaces
- Thermal insulation value of the window assembly
- Structural resistance to wind loads
- Clear opening of any operable window that serves as an emergency exit for a residential sleeping space; such windows are typically required to be at least 5.7 sf (0.53 m^2) in area and have a minimum clear width of 20" (510), a minimum clear height of 24" (610), and a sill no higher than 44" (1120) above the floor.
- Safety glazing for a window that could be mistaken for an open doorway; any window that is more than 9 sf (0.84 m^2) and within 24" (610) of a doorway or less than 60" (1525) above the floor be safety glazed with tempered glass, laminated glass, or plastic.
- Type and size of glazing allowable in fire-rated walls and corridors.

ADA Accessibility Guidelines

- Windows that require operation by occupants in accessible spaces should have adequate clear floor space for maneuvering a wheelchair, be within reach, and be operable with one hand without requiring tight grasping, pinching, or twisting of the wrist.

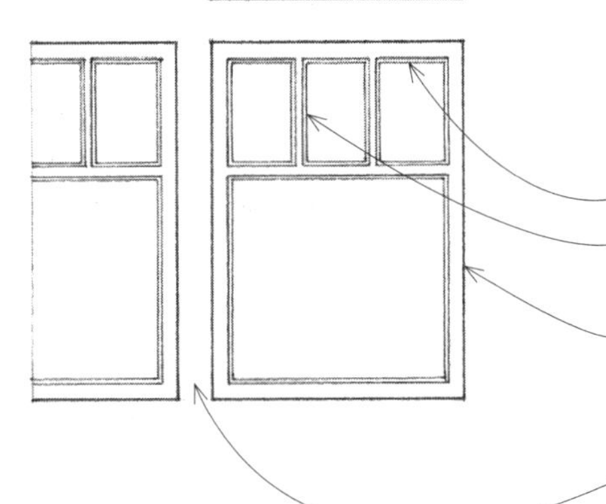

- Rails are the horizontal members framing a window sash.
- Top rail
- Muntins are the vertical members holding the edges of windowpanes within a sash.
- Stiles are the upright members framing a window sash or paneled door.
- Bottom rail
- Mullion is a vertical member separating a series of windows or doorways.

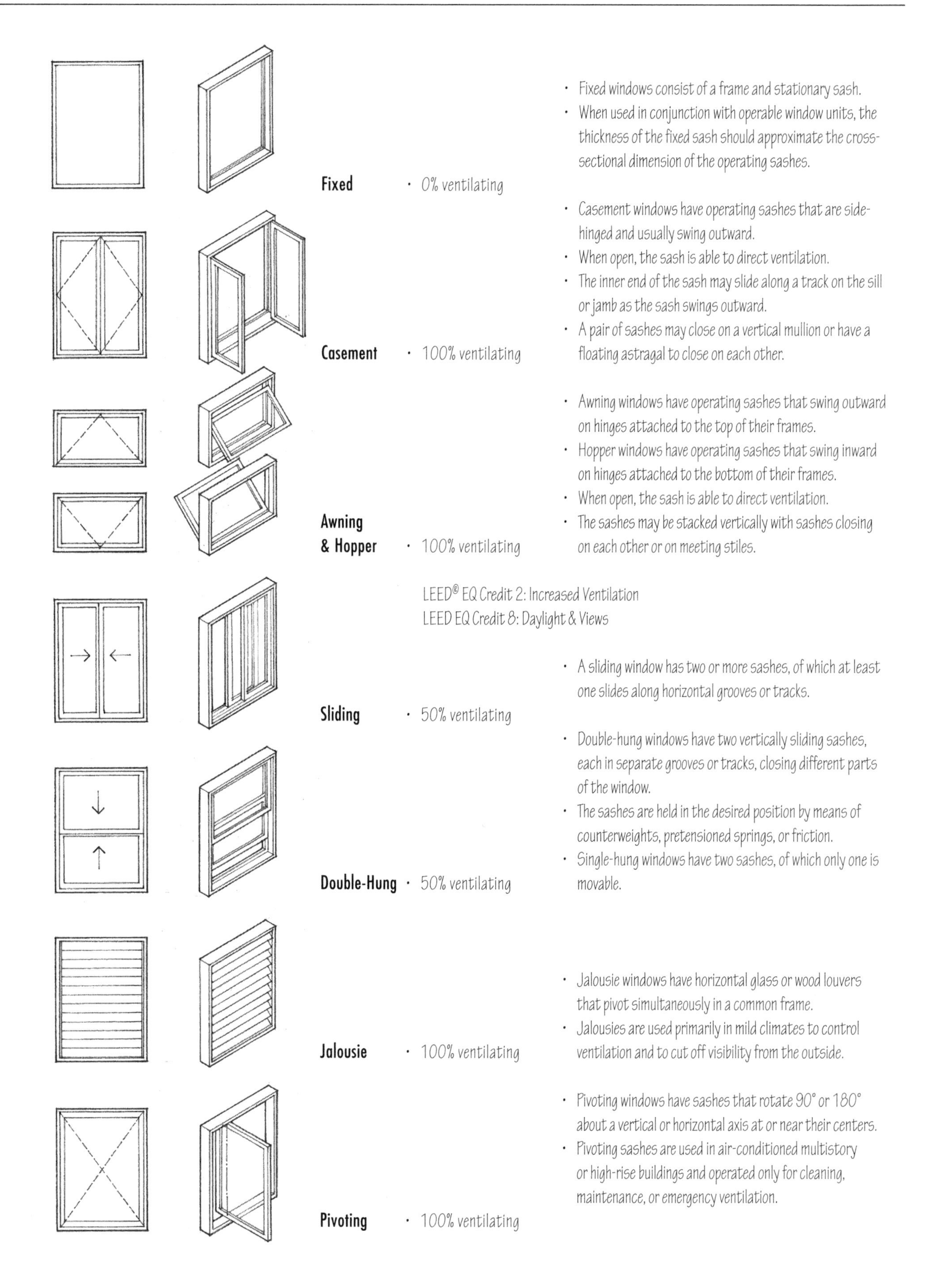

Fixed · 0% ventilating

- Fixed windows consist of a frame and stationary sash.
- When used in conjunction with operable window units, the thickness of the fixed sash should approximate the cross-sectional dimension of the operating sashes.

Casement · 100% ventilating

- Casement windows have operating sashes that are side-hinged and usually swing outward.
- When open, the sash is able to direct ventilation.
- The inner end of the sash may slide along a track on the sill or jamb as the sash swings outward.
- A pair of sashes may close on a vertical mullion or have a floating astragal to close on each other.

Awning & Hopper · 100% ventilating

- Awning windows have operating sashes that swing outward on hinges attached to the top of their frames.
- Hopper windows have operating sashes that swing inward on hinges attached to the bottom of their frames.
- When open, the sash is able to direct ventilation.
- The sashes may be stacked vertically with sashes closing on each other or on meeting stiles.

LEED® EQ Credit 2: Increased Ventilation
LEED EQ Credit 8: Daylight & Views

Sliding · 50% ventilating

- A sliding window has two or more sashes, of which at least one slides along horizontal grooves or tracks.

Double-Hung · 50% ventilating

- Double-hung windows have two vertically sliding sashes, each in separate grooves or tracks, closing different parts of the window.
- The sashes are held in the desired position by means of counterweights, pretensioned springs, or friction.
- Single-hung windows have two sashes, of which only one is movable.

Jalousie · 100% ventilating

- Jalousie windows have horizontal glass or wood louvers that pivot simultaneously in a common frame.
- Jalousies are used primarily in mild climates to control ventilation and to cut off visibility from the outside.

Pivoting · 100% ventilating

- Pivoting windows have sashes that rotate 90° or 180° about a vertical or horizontal axis at or near their centers.
- Pivoting sashes are used in air-conditioned multistory or high-rise buildings and operated only for cleaning, maintenance, or emergency ventilation.

- Aluminum window frames may have equal or unequal legs, depending on the nature of the wall construction.
- The fin created by unequal legs can serve as a windbreak for the joint between the window unit and the wall construction. The fin may also be used for securing the frame to the supporting structure.
- Sealant is required to weatherproof joints between the window frame and the wall construction.

- Head, jamb, and sill sections are usually similar in profile.

- Drips are required for horizontal members at the heads of ventilating sashes that are flush with the exterior face of the wall.
- Weatherstripping is set into integral grooves in the frame and sash sections.
- Thermal breaks
- Snap-in glazing bead
- See 8.28–8.29 for glazing systems.

- Because aluminum is susceptible to galvanic action, anchoring materials and flashing should be aluminum or a material compatible with aluminum, such as stainless steel or galvanized steel. Dissimilar metals, such as copper, should be insulated from direct contact with the aluminum by a waterproof, nonconductive material, such as neoprene or coated felt. For more information on galvanic action, see 12.09.
- Concealed aluminum in contact with concrete or masonry should also be protected by a coating of bituminous or aluminum paint or by a zinc chromate primer.

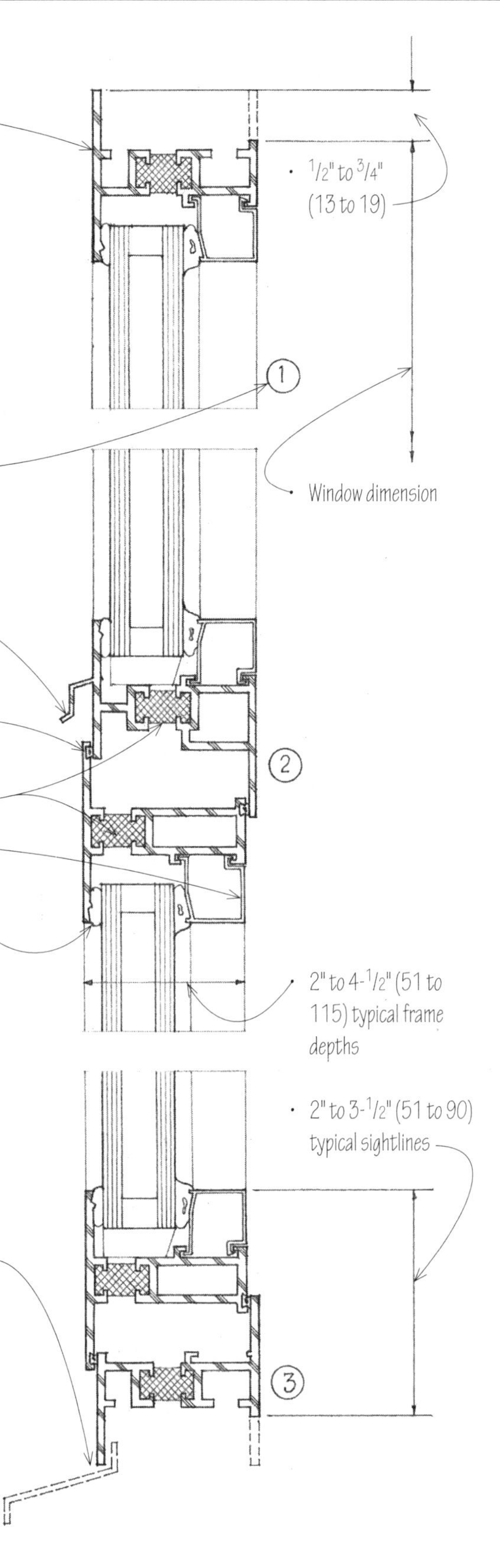

Metal windows may be fabricated and of aluminum, steel, or bronze. Shown on this and the following page are typical sections for aluminum windows and steel windows. Because window frame and sash sections vary greatly from one manufacturer to the next, refer to the manufacturer's literature for:

- Large-scale details of frame and sash profiles
- Alloy, weight, and thickness of sections
- Thermal performance of window assembly
- Resistance to corrosion, water pressure, air infiltration, and wind loading
- Glazing methods and options
- Finishes available
- Rough or masonry openings required; some manufacturers offer stock window sizes while others will fabricate custom sizes, shapes, and configurations.

Aluminum Windows

Aluminum window frames are relatively low in cost, lightweight, and corrosion resistant, but because they are such efficient conductors of heat, synthetic rubber or plastic thermal breaks are required to interrupt the flow of heat from the warm to the cool side of the frame. Aluminum frames may have anodized, baked enamel, or fluoropolymer resin finishes.

Consult the American Architectural Manufacturers Association (AAMA) for criteria developed for the performance of aluminum windows, including minimum provisions for frame strength and thickness, corrosion resistance, air infiltration, water resistance, and wind load capability.

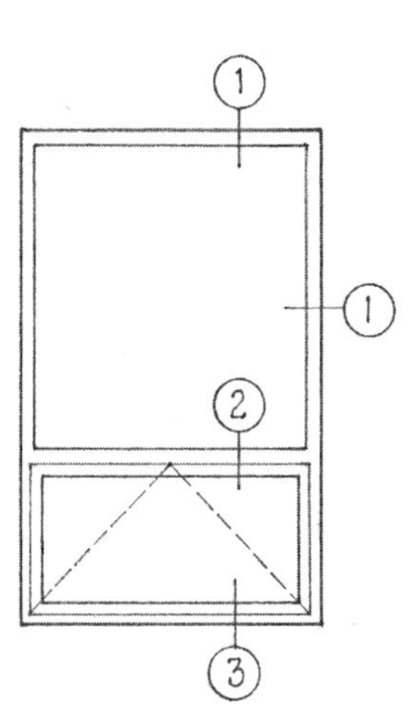

CSI MasterFormat 08 51 00: Metal Windows

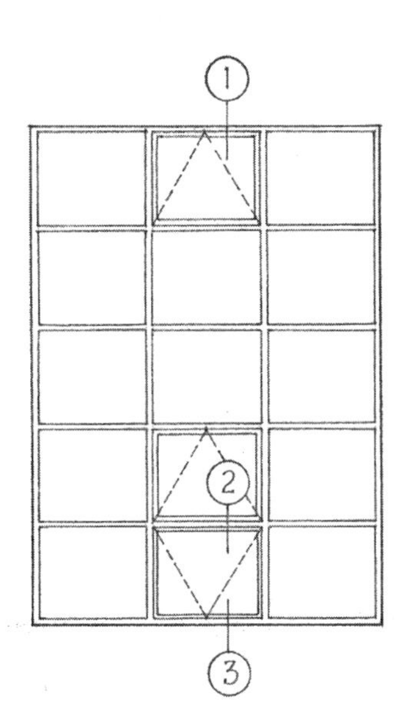

Steel Windows

Steel window frame and sash sections are manufactured from hot-rolled or cold-rolled steel. Because steel is stronger than aluminum, these sections are more rigid and thinner in profile than aluminum sections, offer narrower sightlines, and allow larger lights to be installed in a given rough or masonry opening. Steel also has a lower coefficient of heat transfer than aluminum and therefore steel window frames do not normally require thermal breaks.

The frame and sash sections are welded together and are usually galvanized or bonderized and primed for painting. Baked acrylic enamel, urethane, and polyvinyl chloride (PVC) finishes are also available.

Consult the Steel Window Institute (SWI) for the criteria and standards established for various weights of steel window frames and sashes.

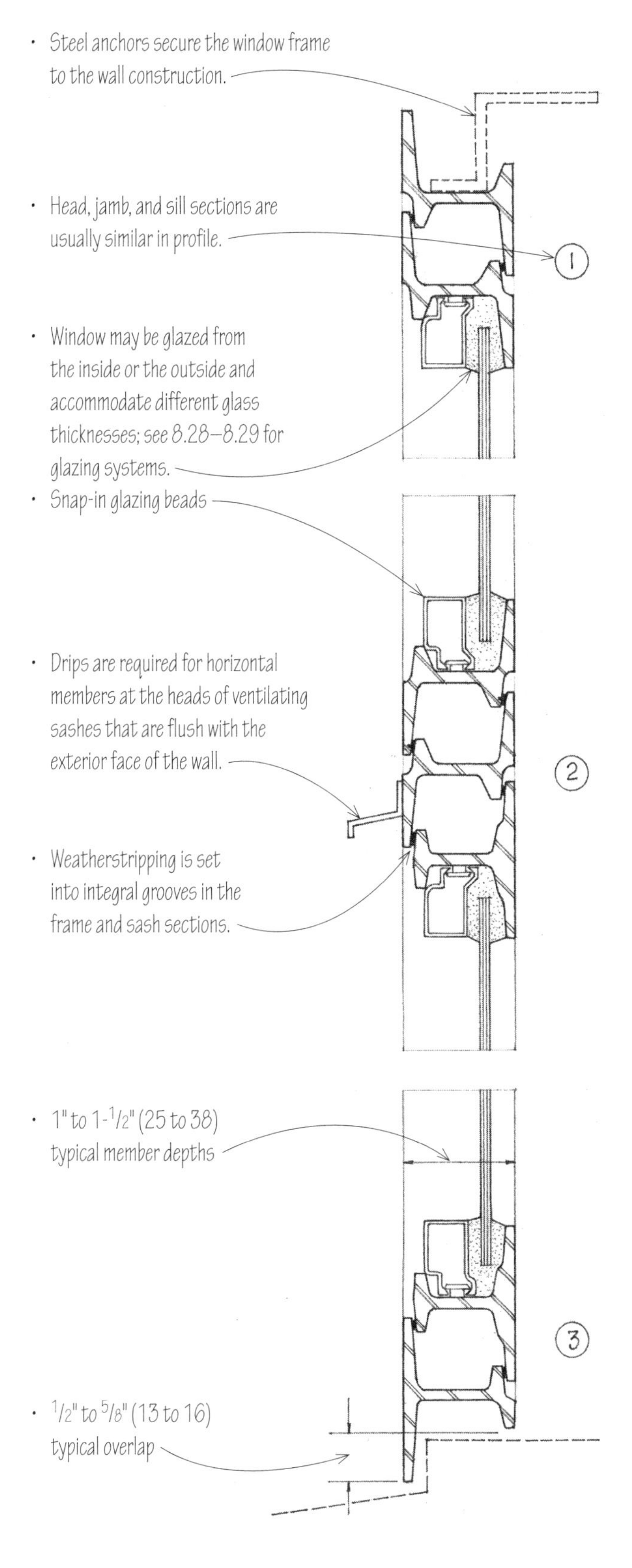

Wood frames are thicker than aluminum or steel frames, but they are also more effective as thermal insulators. The frames are usually of kiln-dried, clear, straight-grain wood, factory-treated with a water-repellant preservative. The wood may be stained, painted, or primed for painting on site. To minimize the need for maintenance, the majority of wood frames are now clad with vinyl or bonded to acrylic-coated aluminum sections that require no painting.

Most stock wood windows are manufactured according to standards established by the National Wood Window and Door Association (NWWDA) and adopted by the American National Standards Institute (ANSI). The exact profile and dimensions of the window frame and sash vary with the type of window operation and from manufacturer to manufacturer. Each manufacturer, however, usually has large-scale 1-1/2" or 3" = 1'-0" (1:10 or 1:5) details that can be used to work out specific window installations.

- Consult window manufacturer for stock window sizes and rough openings required. Some manufacturers will fabricate custom sizes, shapes, and configurations.

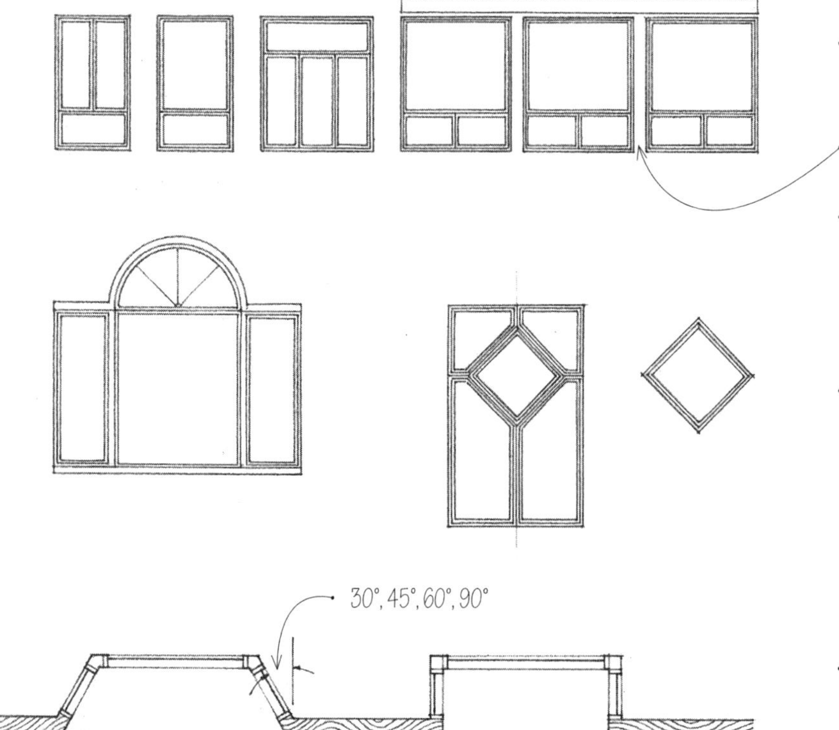

Window manufacturers offer various combinations of both fixed and venting units to cover large openings.

- Window units may be stacked vertically or be banked side by side.
- Structural supporting mullions may be used to reduce the span of the header or lintel.
- Reinforcement may be required when four windows meet @ a common corner
- Special shapes are available from many manufacturers.

- Angled or box bay windows

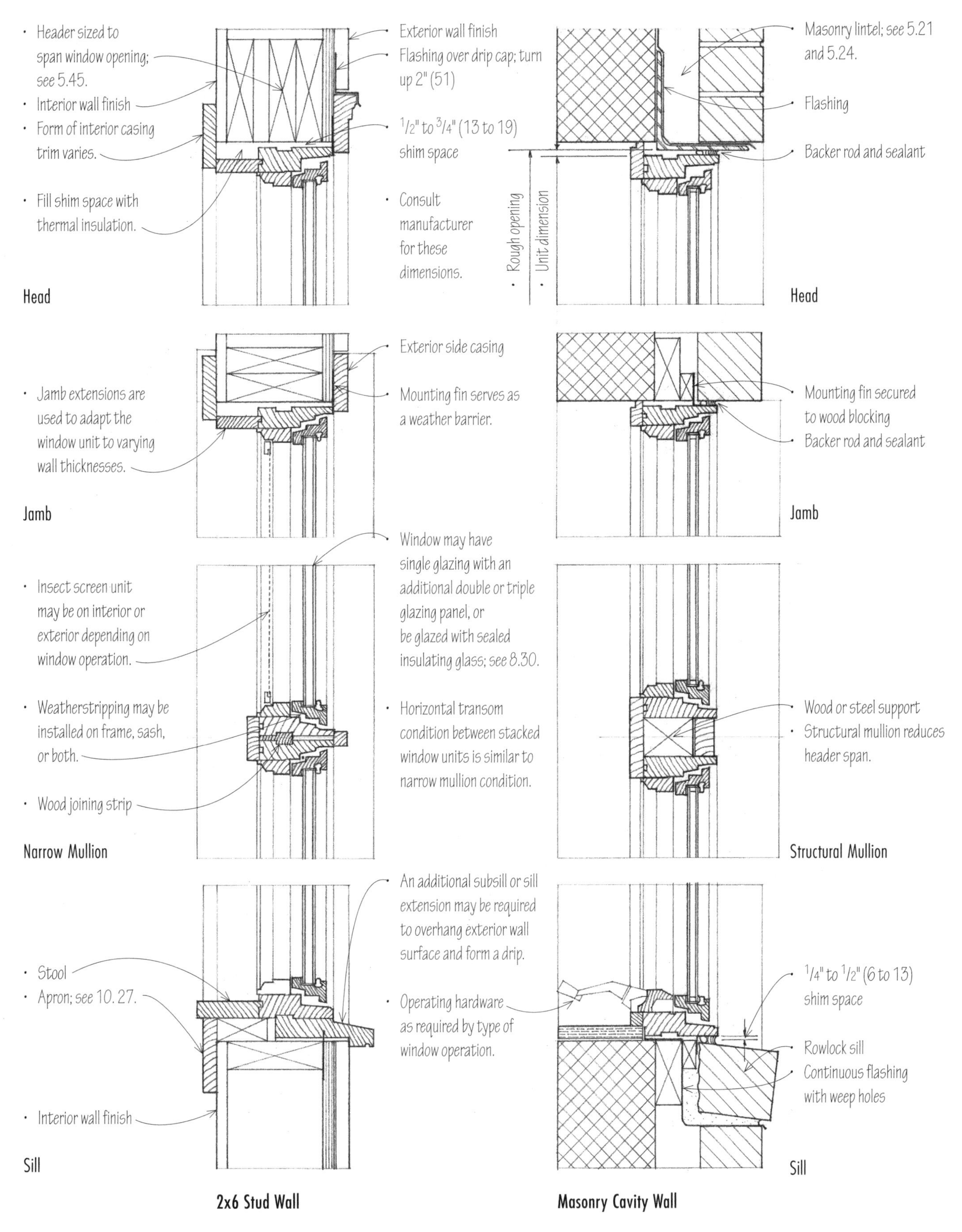

Header sized to span window opening; see 5.45.
Interior wall finish
Form of interior casing trim varies.
Fill shim space with thermal insulation.
Exterior wall finish
Flashing over drip cap; turn up 2" (51)
1/2" to 3/4" (13 to 19) shim space
Consult manufacturer for these dimensions.
Rough opening
Unit dimension
Masonry lintel; see 5.21 and 5.24.
Flashing
Backer rod and sealant
Head
Head
Jamb extensions are used to adapt the window unit to varying wall thicknesses.
Exterior side casing
Mounting fin serves as a weather barrier.
Mounting fin secured to wood blocking
Backer rod and sealant
Jamb
Jamb
Insect screen unit may be on interior or exterior depending on window operation.
Weatherstripping may be installed on frame, sash, or both.
Wood joining strip
Window may have single glazing with an additional double or triple glazing panel, or be glazed with sealed insulating glass; see 8.30.
Horizontal transom condition between stacked window units is similar to narrow mullion condition.
Wood or steel support
Structural mullion reduces header span.
Narrow Mullion
Structural Mullion
Stool
Apron; see 10.27.
Interior wall finish
An additional subsill or sill extension may be required to overhang exterior wall surface and form a drip.
Operating hardware as required by type of window operation.
1/4" to 1/2" (6 to 13) shim space
Rowlock sill
Continuous flashing with weep holes
Sill
Sill
2x6 Stud Wall
Masonry Cavity Wall

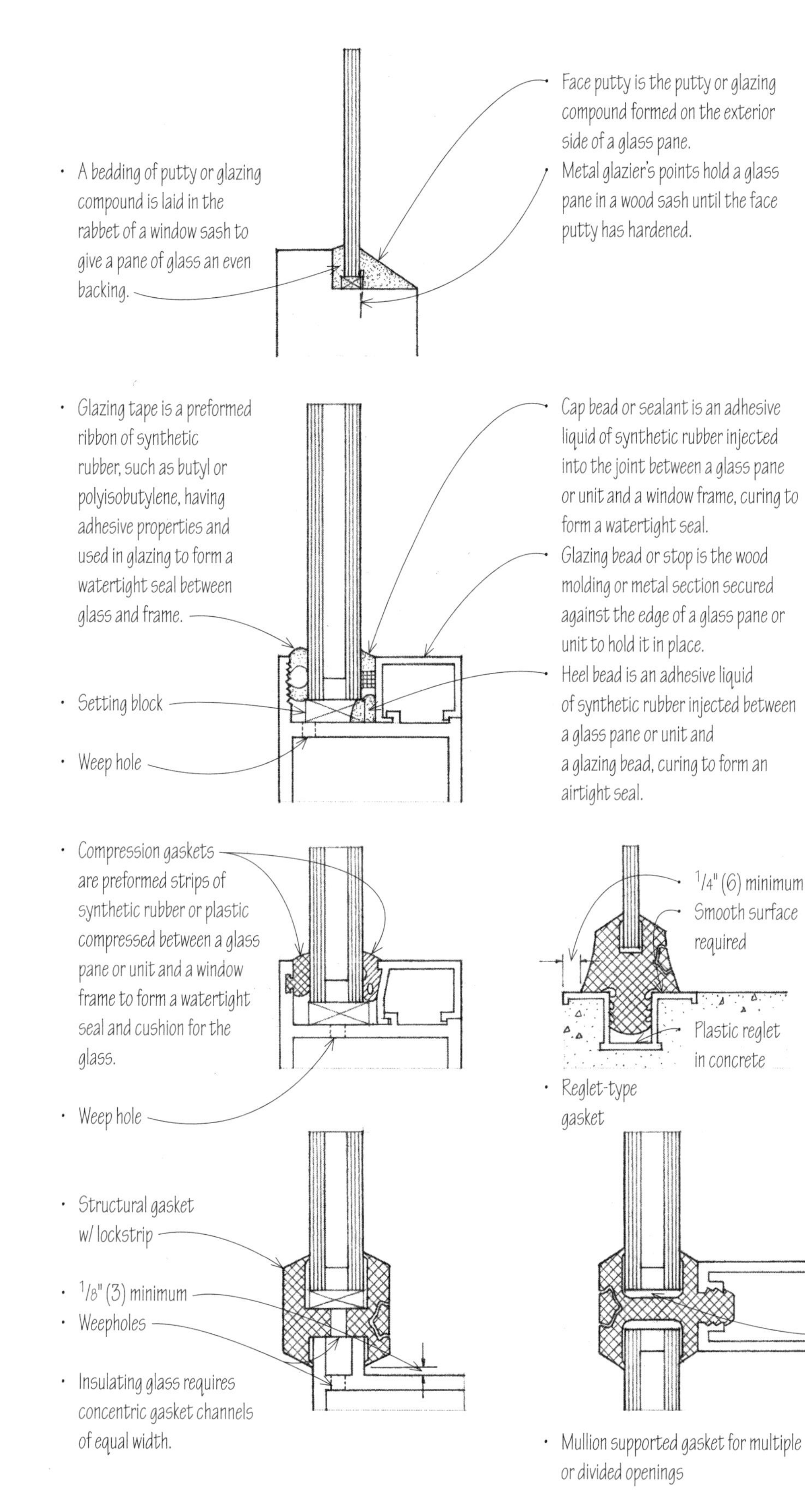

Face Glazing

Small glass panes may be set in a rabbeted frame, held in place with glazier's points, and sealed with a beveled bead of putty or glazing compound.

- Putty is a compound of whiting and linseed oil, of doughlike consistency when fresh, used in securing windowpanes or patching woodwork defects.
- Glazing compound is an adhesive compound used as putty, formulated so as not to become brittle with age.

Lights more than 6 sf (0.56 m^2) in area, must be wet- or dry-glazed.

Wet Glazing

Wet glazing is the setting of glass in a window frame with glazing tape or a liquid sealant.

Dry Glazing

Dry glazing is the setting of glass in a window frame with a compression gasket instead of glazing tape or a liquid sealant.

Structural Gaskets

Structural gaskets are preformed of synthetic rubber or other elastomeric material to secure a glass pane or unit in a window frame or opening. The gaskets are held in compression by forcing a keyed locking strip into a groove in the gasket. They require smooth contact surfaces and a frame or opening with exacting dimensional tolerances and true plane alignment. The glass must be supported on at least two sides by the frame or a supported gasket.

CSI MasterFormat 08 81 00: Glass Glazing

Both wet- and dry-glazing systems should allow the glass unit to float in its opening and be cushioned with a resilient glazing material. There should be no direct contact between the glass and the perimeter frame. The perimeter frame itself must support the glass against wind pressure or suction, and be strong enough that structural movements and thermal stresses are not transferred to the glass.

- Glass size is the size of a glass pane or unit required for glazing an opening, allowing for adequate edge clearances.
- United inches is the sum of one length and one width of a rectangular glass pane or unit, measured in inches.

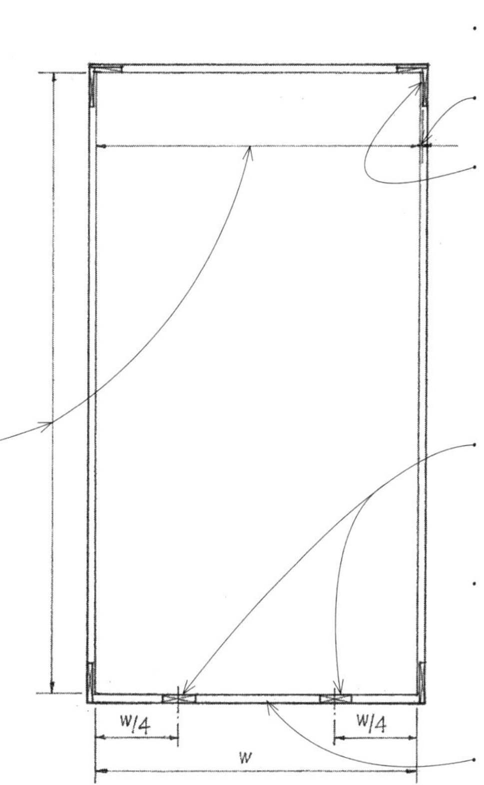

- Limit deflection to 1/175 of span.

1/8" (3) clearance

Edge blocks of synthetic rubber are placed between the side edges of a glass pane or unit and a frame to center it, maintain a uniform width of sealant, and limit lateral movement caused by building vibrations or thermal expansion or contraction; 4" (100) minimum length.

Setting blocks of lead or synthetic rubber are placed under the lower edge of a glass pane or unit to support it within a frame; two per panel at quarter points.

- Setting blocks should be as wide as glass thickness and 0.1" per square foot (25 mm per 0.09 m^2) of glass area in length; 4" (100) minimum.

Minimum of two 1/4" to 3/8" (6 to 10) Ø weep holes in glazing pocket

Glass Type		A	B	C
Sheet glass	SS	1/16" (2)	1/4" (6)	1/8" (3)
	DS	1/8" (3)	1/4" (6)	1/8" (3)
Plate glass	1/4" (6)	1/8" (3)	3/8" (10)	1/4" (6)
	3/8" (10)	3/16" (5)	7/16" (11)	5/16" (8)
	1/2" (13)	1/4" (6)	7/16" (11)	3/8" (10)
Insulating glass	1/2" (13)	1/8" (3)	1/2" (13)	1/8" (3)
	5/8" (16)	1/8" (3)	1/2" (13)	1/8" (3)
	3/4" (19)	3/16" (5)	1/2" (13)	1/4" (6)
	1" (25)	3/16" (5)	1/2" (13)	1/4" (6)

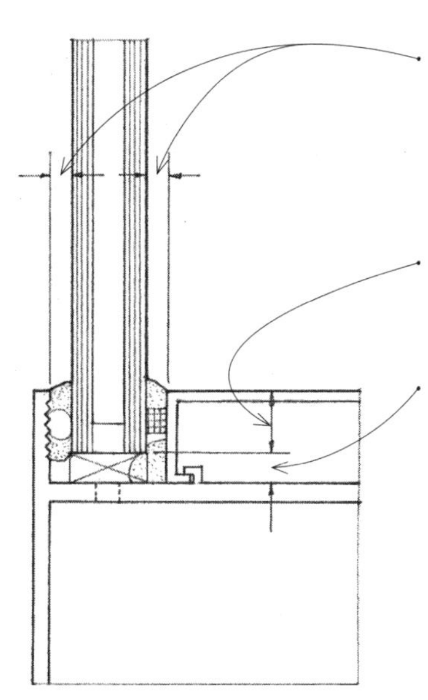

Face clearance (A) is the distance between the face of a glass pane or unit and the nearest face of its frame or stop, measured normal to the plane of the glass.

Bite (B) is the amount of overlap between the edge of a glass pane or unit and a window frame, stop, or lock-strip gasket.

Edge clearance (C) is the distance between the edge of a glass pane or unit and a window frame, measured in the plane of the glass.

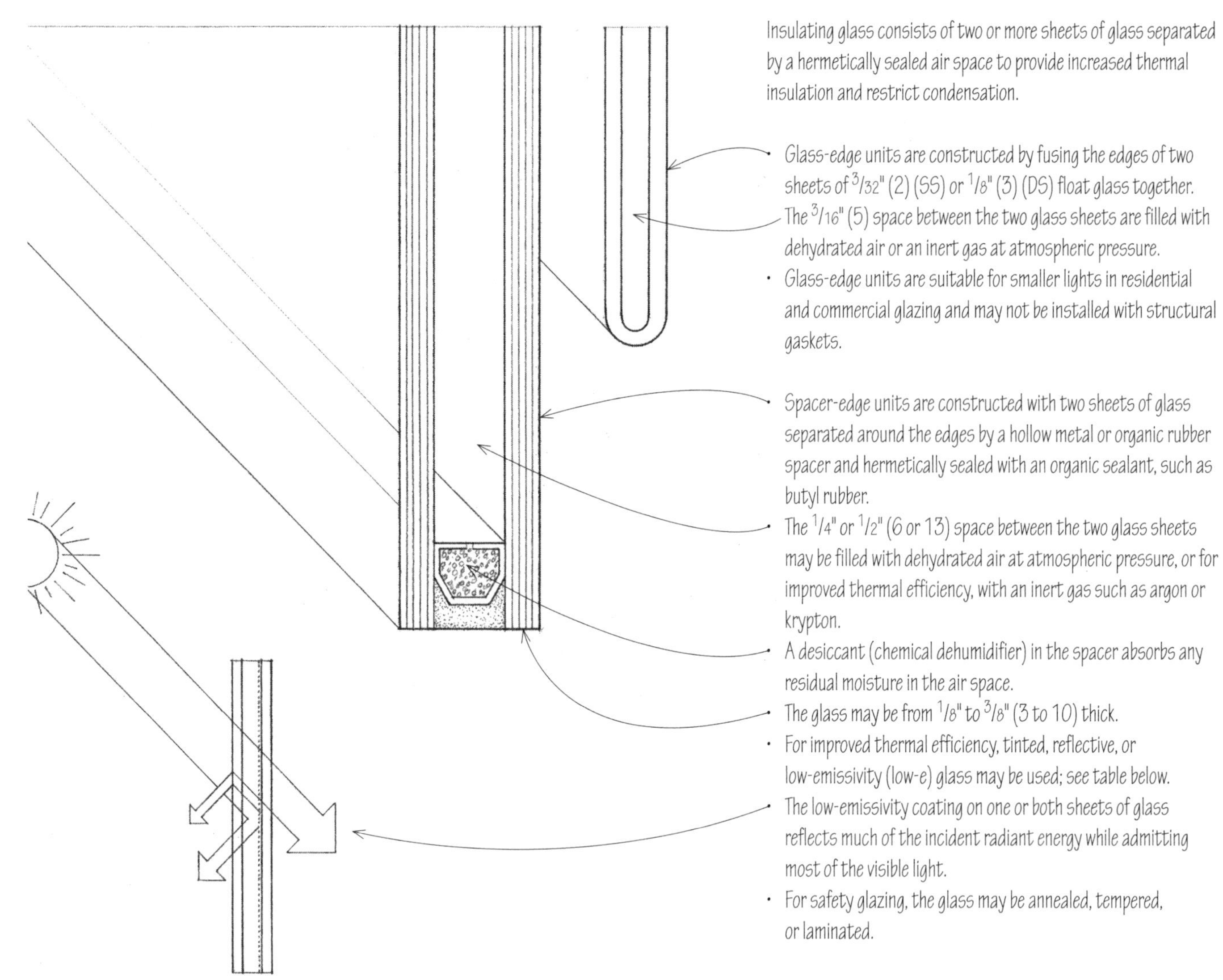

Insulating glass consists of two or more sheets of glass separated by a hermetically sealed air space to provide increased thermal insulation and restrict condensation.

- Glass-edge units are constructed by fusing the edges of two sheets of 3/32" (2) (SS) or 1/8" (3) (DS) float glass together. The 3/16" (5) space between the two glass sheets are filled with dehydrated air or an inert gas at atmospheric pressure.
- Glass-edge units are suitable for smaller lights in residential and commercial glazing and may not be installed with structural gaskets.

- Spacer-edge units are constructed with two sheets of glass separated around the edges by a hollow metal or organic rubber spacer and hermetically sealed with an organic sealant, such as butyl rubber.
- The 1/4" or 1/2" (6 or 13) space between the two glass sheets may be filled with dehydrated air at atmospheric pressure, or for improved thermal efficiency, with an inert gas such as argon or krypton.
- A desiccant (chemical dehumidifier) in the spacer absorbs any residual moisture in the air space.
- The glass may be from 1/8" to 3/8" (3 to 10) thick.
- For improved thermal efficiency, tinted, reflective, or low-emissivity (low-e) glass may be used; see table below.
- The low-emissivity coating on one or both sheets of glass reflects much of the incident radiant energy while admitting most of the visible light.
- For safety glazing, the glass may be annealed, tempered, or laminated.

- See 12.16 for other glass products.

Insulating Glass Type	Visible Light		Solar Radiation		U-value	
	% transmitted	% reflected	% transmitted	% reflected	Winter	Summer
clear + clear	78–82	14–15	60–76	11–15	0.42–0.61	169–192
clear + low-e	49–86	12–15	17–56	17–25	0.23–0.52	133–157
clear + tinted						
gray	13–56	5–13	22–56	7–9	0.49–0.60	74–152
bronze	19–62	8–13	26–57	8–9	0.49–0.60	76–152
blue	50–64	8–13	38–56	7–9	0.49–0.58	120–154
clear + coated						
silver	7–19	22–41	5–14	18–34	0.39–0.48	36–59
blue	12–27	16–32	12–18	15–20	0.42–0.46	58–73
copper	25	30–31	12	45	0.29–0.30	44

LEED EA Credit 1: Optimize Energy Performance

Glazed curtain walls are exterior nonloadbearing walls consisting of vision glass or opaque spandrel panels supported by metal framing. They may be categorized according to their method of assembly.

Stick Systems

The stick system consists of tubular metal mullions and rails assembled piece by piece on site to frame vision glass and spandrel units. It offers relatively low shipping and handling costs and can be adjusted more readily than other systems to on-site conditions.

Unit Systems

Unit systems consist of preassembled, framed wall units which may be preglazed or glazed after installation. Shipping bulk is greater than with the stick system, but less field labor and erection time is required.

Unit-and-Mullion Systems

In the unit-and-mullion system, one- or two-story-high mullions are installed before preassembled wall units are lowered into place behind the mullions. The panel units may be full-story height, preglazed or unglazed, or may be separate vision glass and spandrel units.

Column-Cover-and-Spandrel Systems

Column-cover-and-spandrel systems consist of vision-glass assemblies and spandrel units supported by spandrel beams between exterior columns clad with cover sections.

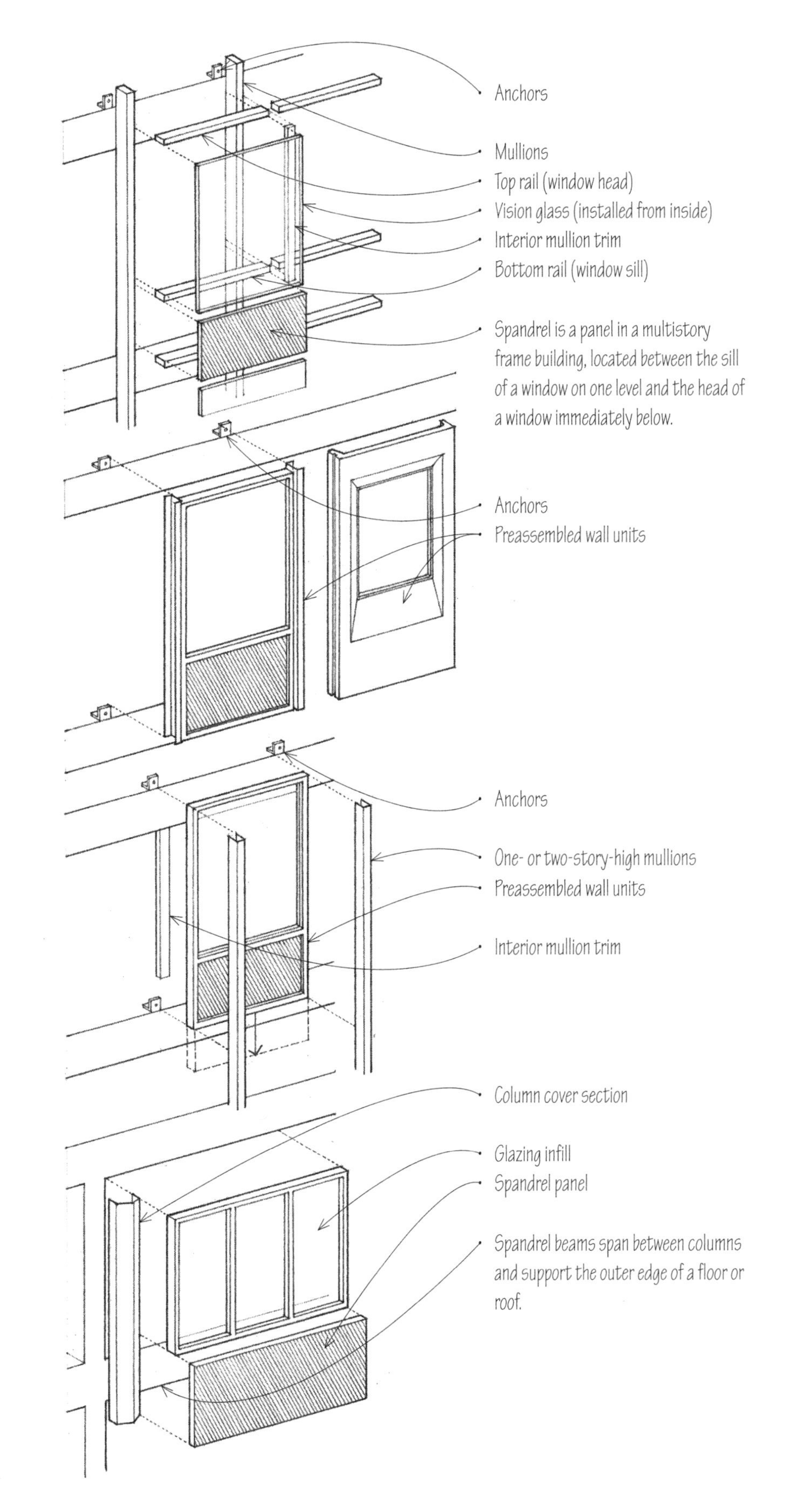

- See 7.24–7.26 for general conditions and requirements of curtain wall construction.

- Mullion sections are spliced with the lower mullion fixed to an internal spline and the upper mullion slipping down over the spline so that it is free to move.
- Angle anchors; see 7.26.
- All anchors and fasteners must be detailed to guard against galvanic action.
- Infill panel or spandrel glass, an opaque glass produced by fusing a ceramic frit to the interior surface of tempered or heat-strengthened glass
- A continuous firestop is secured between the wall and the edge of each floor slab or deck to prevent the spread of fire.

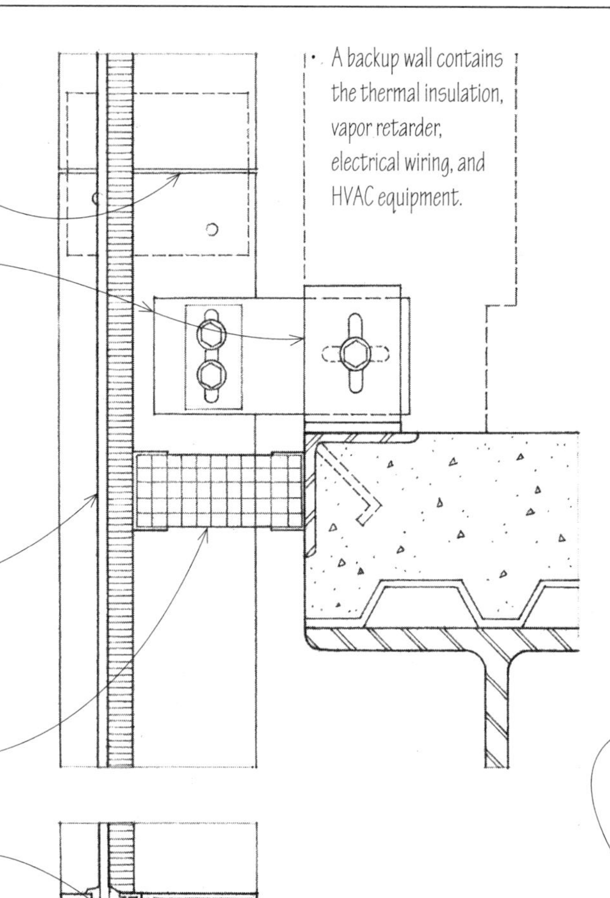

- A backup wall contains the thermal insulation, vapor retarder, electrical wiring, and HVAC equipment.

- See 7.25 for pressure-equalized design of curtain wall frames.
- Metal frames should have thermal breaks.
- Horizontal rails are provided w/ weep holes for drainage.

- Insulating glass
- Glass may be glazed from the outside using pressure bars or structural gaskets; see 8.29 and 8.31.

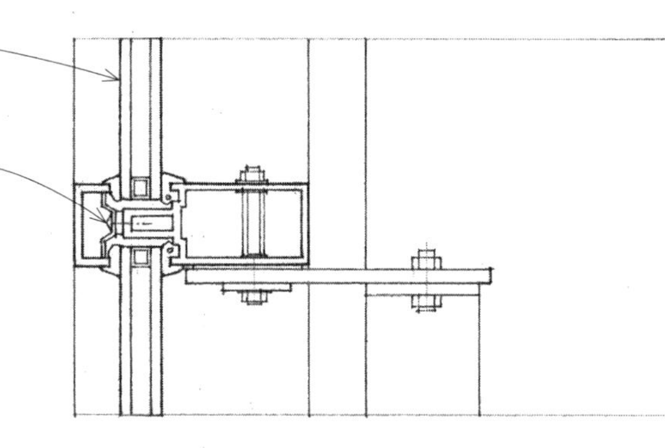

- For high-rise applications, interior glazing is more convenient and economical. It is accomplished by means of fixed exterior gaskets and interior wedge-shaped gaskets; snap-on covers conceal the inner frame and fasteners.
- Some curtain wall systems may be glazed from either the outside or the inside of the building.

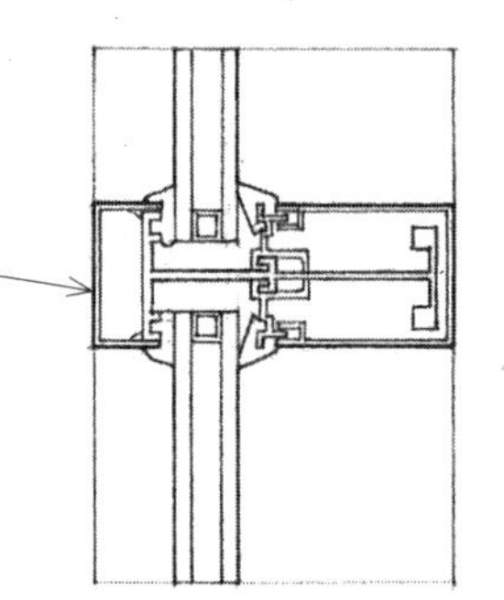

These details illustrate typical conditions of glazed curtain wall construction. When using standard fabricated wall systems, there is no need for extensive detailing except when components are modified. For more in-depth information, refer to the *Aluminum Curtain Wall Design Guide Manual*, published by the American Architectural Manufacturers Association (AAMA), the Flat Glass Marketing Association (FGMA), and standards developed by the American Society for Testing and Materials (ASTM). Things to note include:

- Overall wall pattern
- Type of glazing
- Type, size, and location of any operable window sash
- Type and finish of infill or spandrel panels
- Perimeter, corner, and anchorage conditions

- Snap-on covers may be used to conceal fasteners, provide uninterrupted profiles, and permit variations in metal finishes.

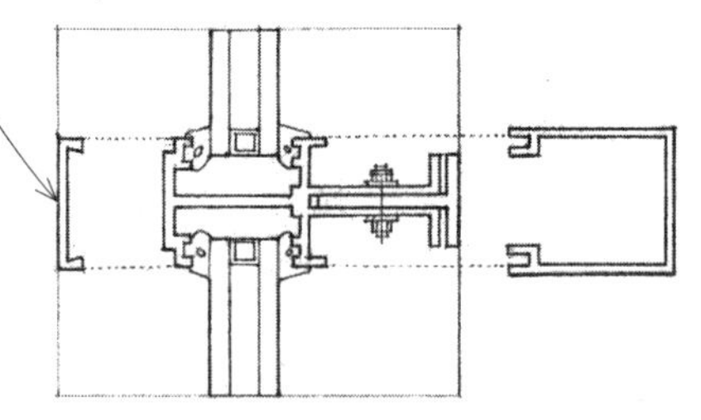

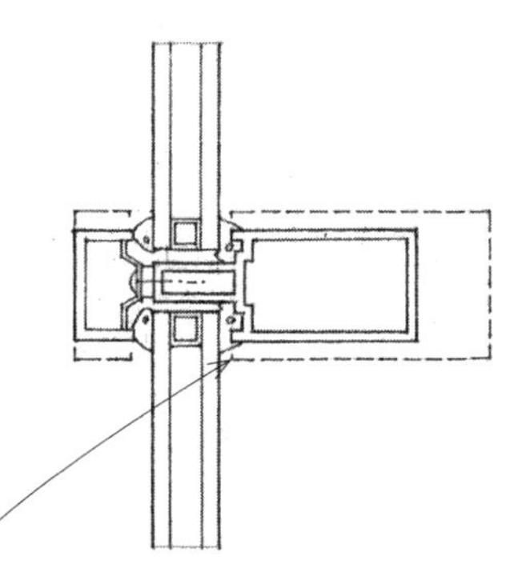

- The required size, strength, and stiffness of the curtain wall frame are determined by the loads the frame must carry—primarily lateral wind loads and relatively light gravity loads. Consult the manufacturer for the structural capacity of the curtain wall assembly, as well as its resistance to water and air infiltration.

A curtain wall system may utilize structural gaskets to glaze both fixed glass units and spandrel panels. The supporting frame members should be of the same thickness as the insulating glass unit to ensure balanced support.

When stacking insulating glass units vertically, the weight of the upper glass units can introduce stresses into the lower glass units. For this reason, the horizontal mullion rather than the gaskets should provide the necessary support for the glazing.

See 8.28 for more information on glazing with structural gaskets.

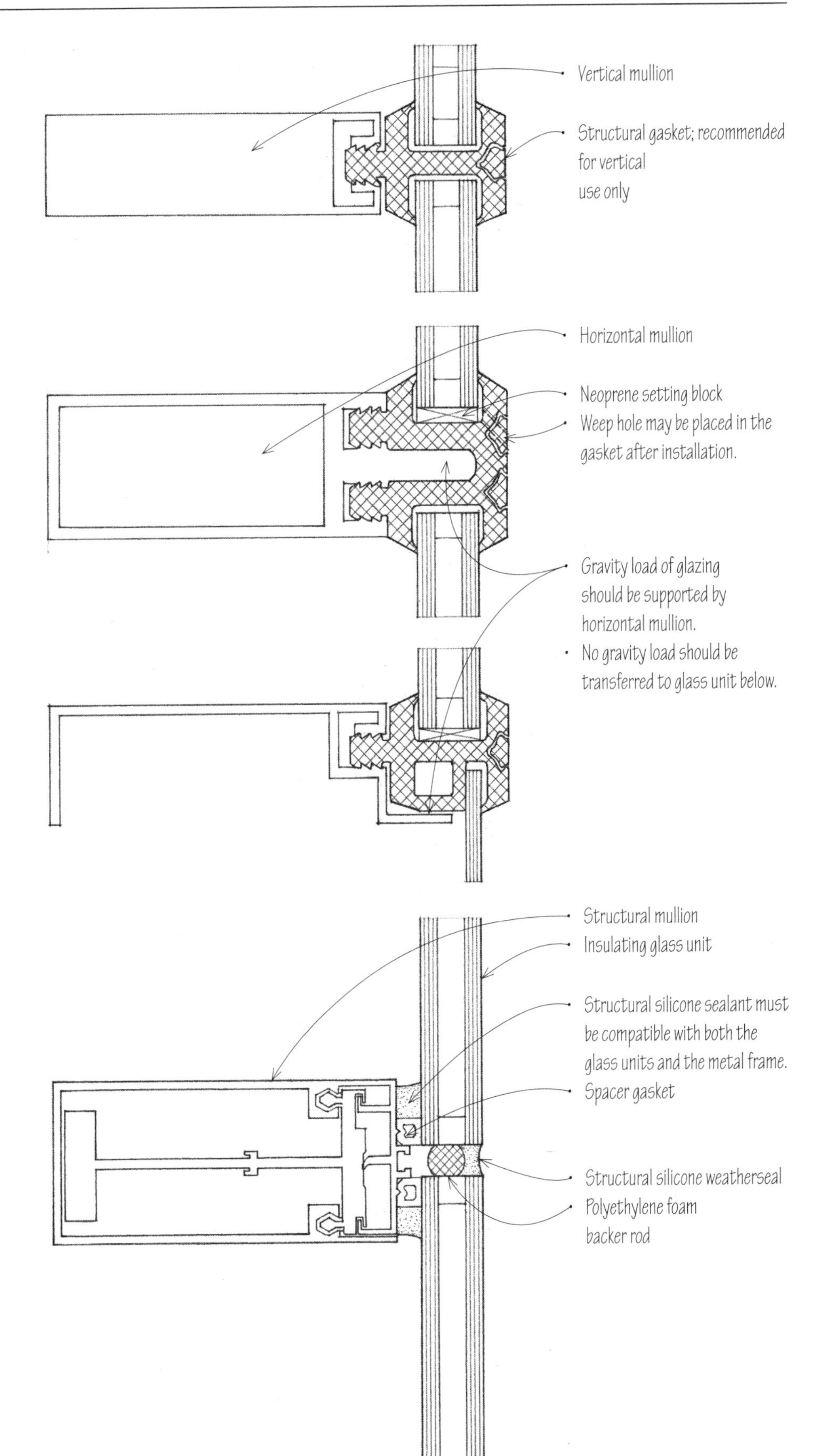

Flush Glazing

Flush glazing is a glazing system in which the metal framing members are set entirely behind the glass panes or units to form a flush exterior surface. The glass units adhere to the framing with a structural silicone sealant; the silicone sealant transfers wind and other loads from the glass to the metal curtain wall frame without mechanical fastenings. The design should allow for easy maintenance and replacement of broken glass units. Factory-glazing is preferred for better quality control. Consult manufacturer for details.

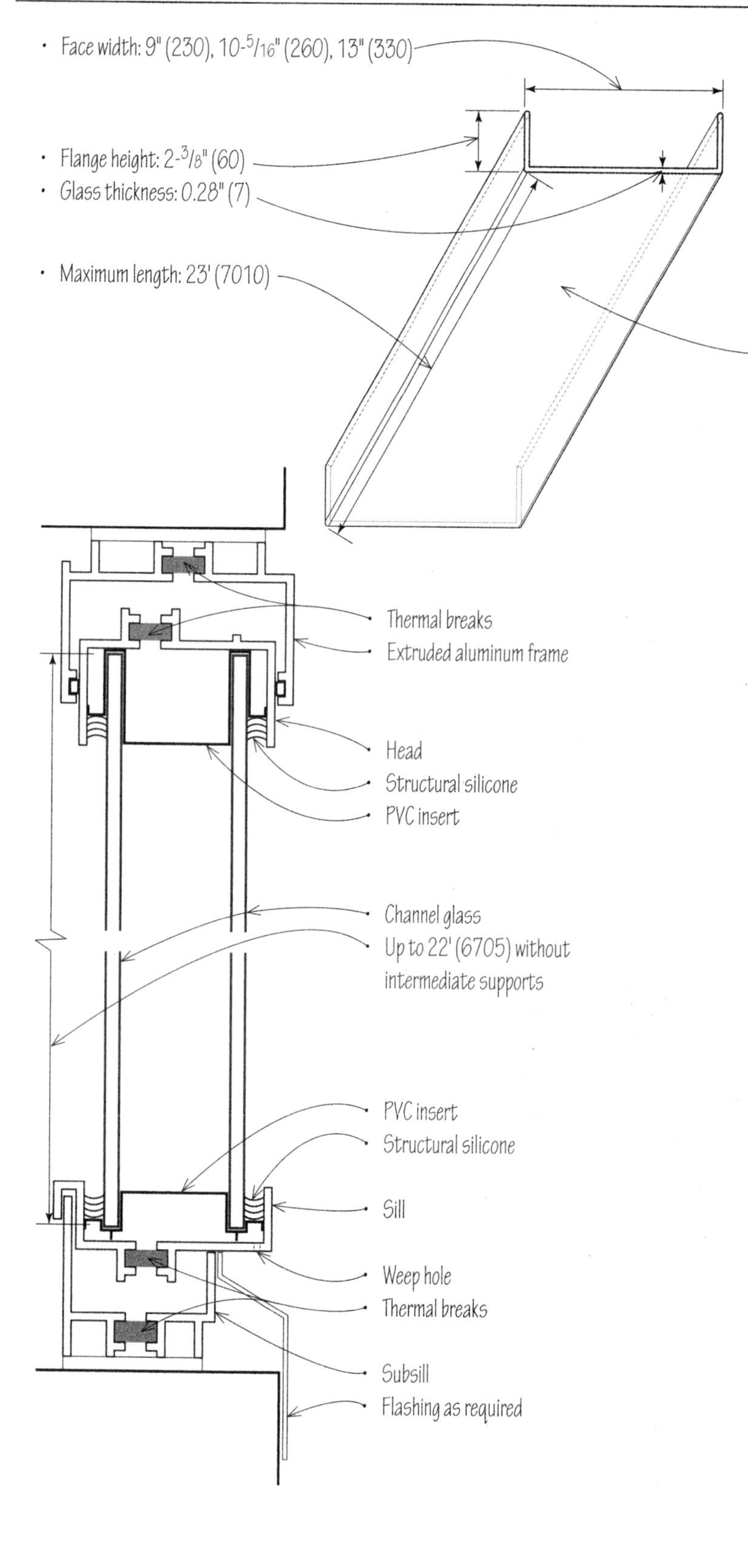

Channel Glass

Channel glass is cast by drawing molten glass over a series of steel rollers to form a continuous flat U-shape, which is cut to specified length after cooling. The translucent channel sections are available in widths from 9" to 19" (230 to 480) and in lengths up to 23 feet (7010). For exterior applications, the channel sections have 2-3/8" (60) flanges and come in three standard widths, 9" (230), 10-5/16" (260), and 13" (330).

Various surface textures offer a range of translucencies to obscure vision while allowing light to pass through. Channel glass may be annealed and tempered for increased compressive strength and use as safety glazing.

The thermal performance of channel glass can be enhanced with a low-e coating applied directly to the inside face of the glass. When greater performance is required a thermal insulating material can be inserted into the cavity of a double-glazed wall system, bringing down the U-value of the system to as low as 0.19.

Channel glass is suitable for interior and exterior applications. For exterior applications, channel glass systems can be constructed as double-glazed curtain walls and store fronts or as single-glazed rain screens and trombe walls.

The self-supporting sections may be installed vertically or horizontally within extruded aluminum perimeter frames. Vertical systems are site-built while horizontal systems are usually pre-assembled for better quality control and to shorten the project construction schedule.

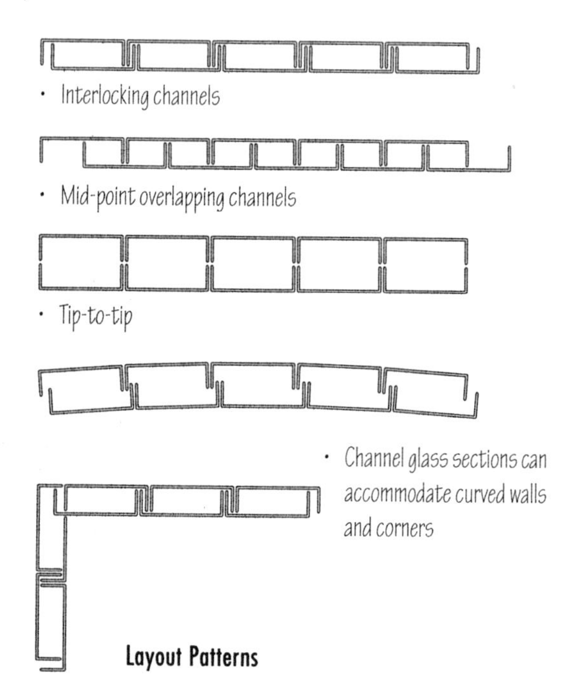

Layout Patterns

LEED EA Credit 1: Optimize Energy Performance
LEED IEQ Credits 6 & 7: Controllability of Systems & Thermal Comfort

CSI MasterFormat 08 45 11: Translucent Linear Channel Glazing System

Double-skin facades, also known as smart facades, refer to cladding systems designed to conserve and reduce the energy needed for heating, cooling, and lighting a building by integrating passive solar collection, solar shading, daylighting, thermal resistance, and natural ventilation into its assembly. The assembly usually consists of a double- or triple-glazed unit on the interior, an air space for harvesting heat and containing adjustable shading devices to control solar radiation and daylighting, and an exterior layer of safety or laminated glazing with operable panels and sometimes energy-generating photovoltaic technologies.

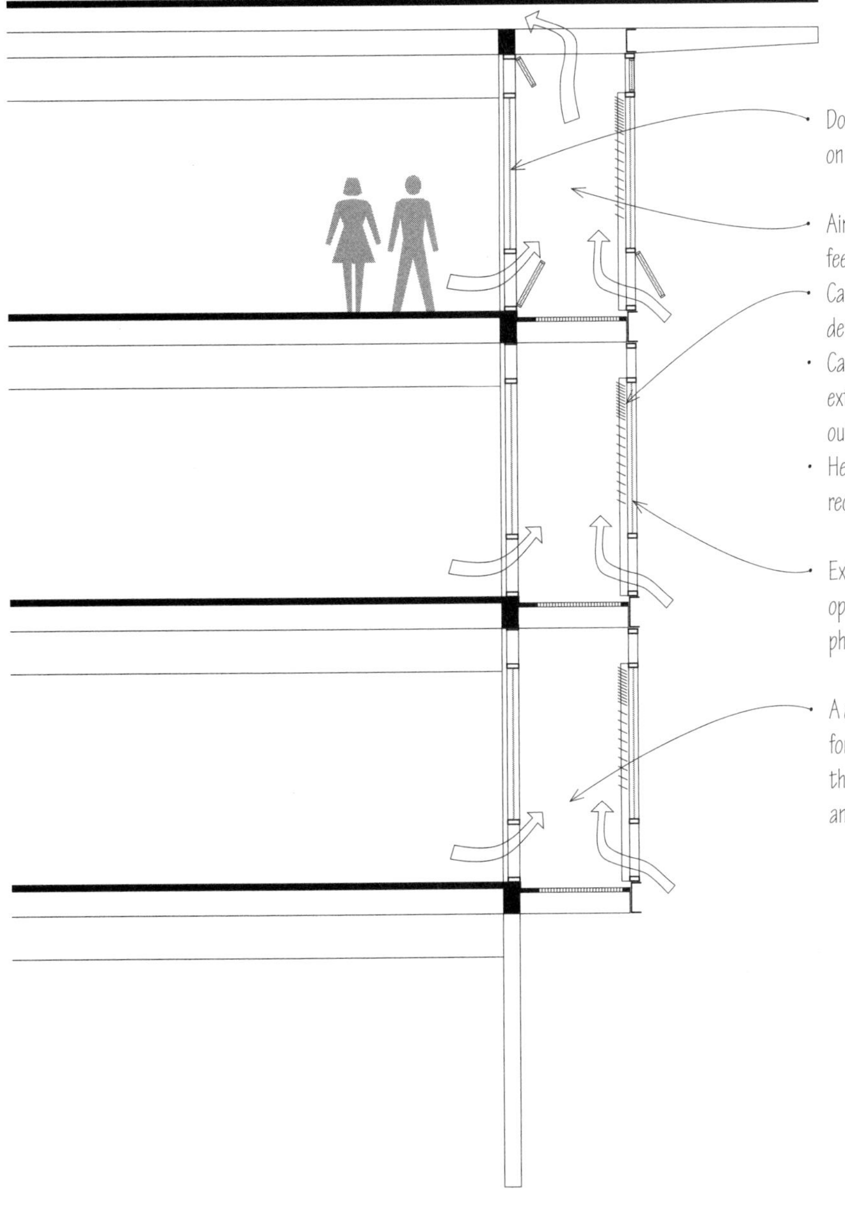

- Double- or triple-glazing constructed as a curtain wall on the interior side
- Air space can vary in width from a few inches to several feet for accessibility.
- Cavity contains adjustable blinds or other shading devices to control solar radiation and daylighting.
- Cavity serves as a thermal buffer between interior and exterior; it can also operate as a solar flue to draw air out of the adjacent spaces.
- Heat can be harvested through the use of an energy-recovery ventilating system.
- Exterior layer of safety or laminated glazing with operable panels and sometimes energy-generating photovoltaic technologies
- A double-skin facade can provide acoustic benefits for buildings while providing for natural ventilation through independently operable openings in both inner and outer layers.

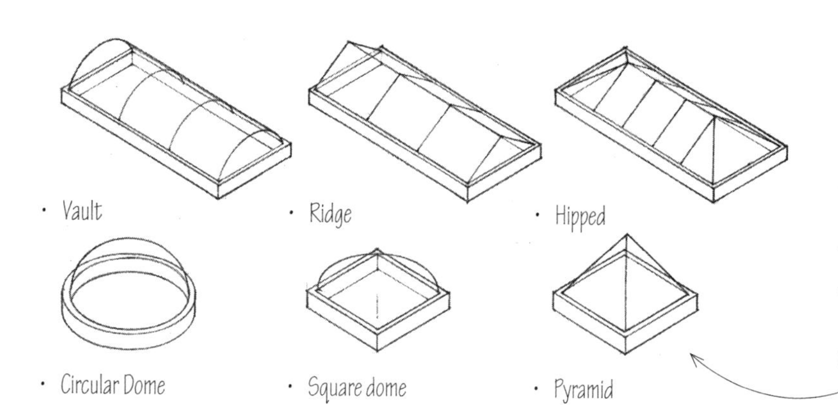

Skylight Forms

Glazed openings in a roof allow daylight to enter an interior space from above. This efficient and cost-effective source of lighting can be in place of or in addition to the normal daylighting from windows. Careful consideration, however, should be paid to the control of brightness and glare, which may require the use of louvers, shades, or reflector panels. Horizontal and south-facing skylights also increase solar heat gain in the winter, but in the summer, shading may again be required to prevent excessive heat gain.

Glazed openings may be constructed using the following elements:

- Skylights are metal-framed units preassembled with glass or plastic glazing and flashing. They are available in stock sizes and shapes but may also be custom-fabricated.
- Roof windows are stock wood windows designed for installation in a sloping roof. These windows either pivot or swing open for ventilation and cleaning. They are typically 2' to 4' (610 to 1220) wide and 3' to 6' high (915 to 1830) and available with shades, blinds, and electric operators.
- Sloped glazing systems are glazed curtain walls engineered to serve as pitched glass roofs.

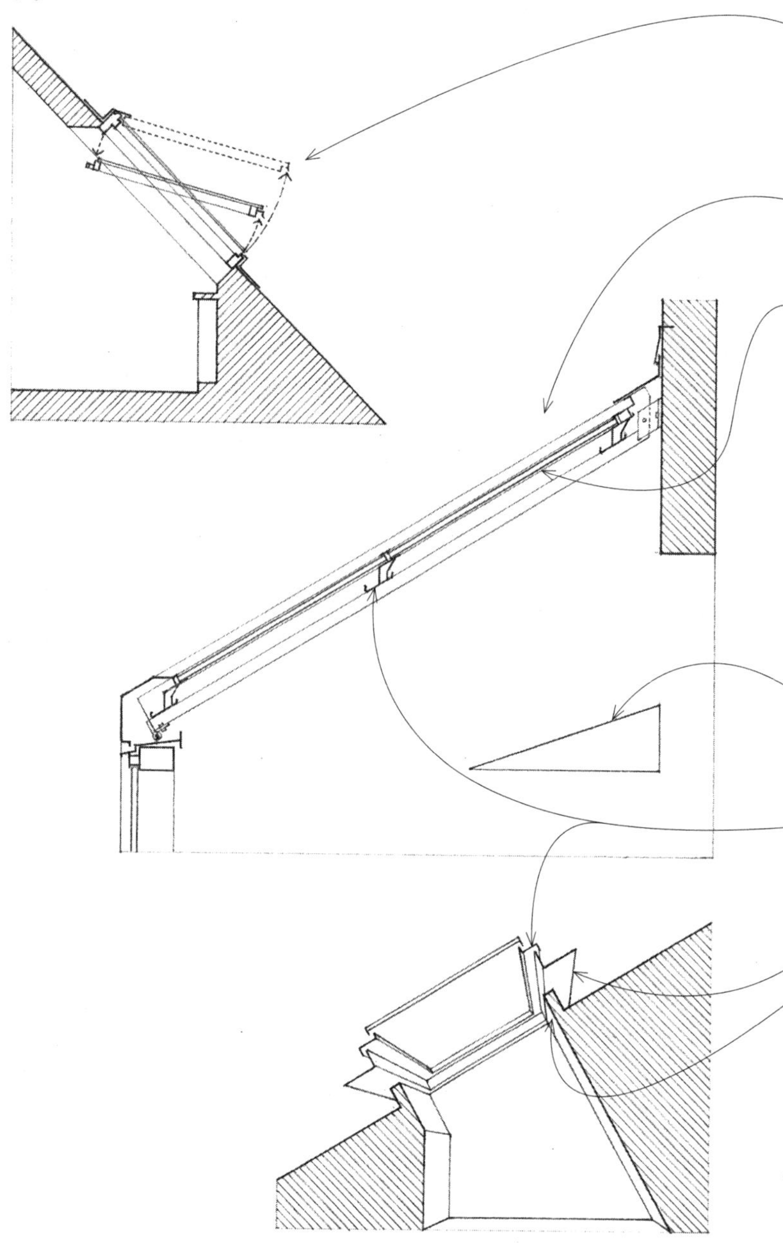

- Glazing may be of acrylic or polycarbonate plastic or of wired, laminated, heat-strengthened, or fully tempered glass. Building codes limit the maximum area of each glazed skylight panel.
- Double glazing is recommended for energy conservation and to reduce condensation.
- When wired glass, heat-strengthened glass, or fully tempered glass is used in a multiple-layer glazing system, the building code requires that wire screening be installed below the glazing to prevent the glass, if broken, from falling and injuring building occupants below; exceptions exist for individual dwelling units.

- The minimum slope for flat or corrugated plastic skylights is 4:12. Plastic domes should rise at least 10% of the span or at least 5" (125).

- The frames for skylights and sloped glazing systems should incorporate an internal guttering system to collect and drain infiltrating water and condensation through weep holes to the exterior.
- Roof flashing
- Skylights set at an angle of less than 45° require a curb at least 4" (100) high to elevate the skylight above the surrounding roof surface. This curb may be job-built or be an integral part of the skylight unit.

- Skylight units require a framed roof opening; both the supporting roof structure and the skylight units must be engineered to carry the anticipated roof loads.

CSI MasterFormat 08 61 00: Roof Windows
CSI MasterFormat 08 62 00: Unit Skylights
CSI MasterFormat 08 44 33: Sloped Glazing Assemblies

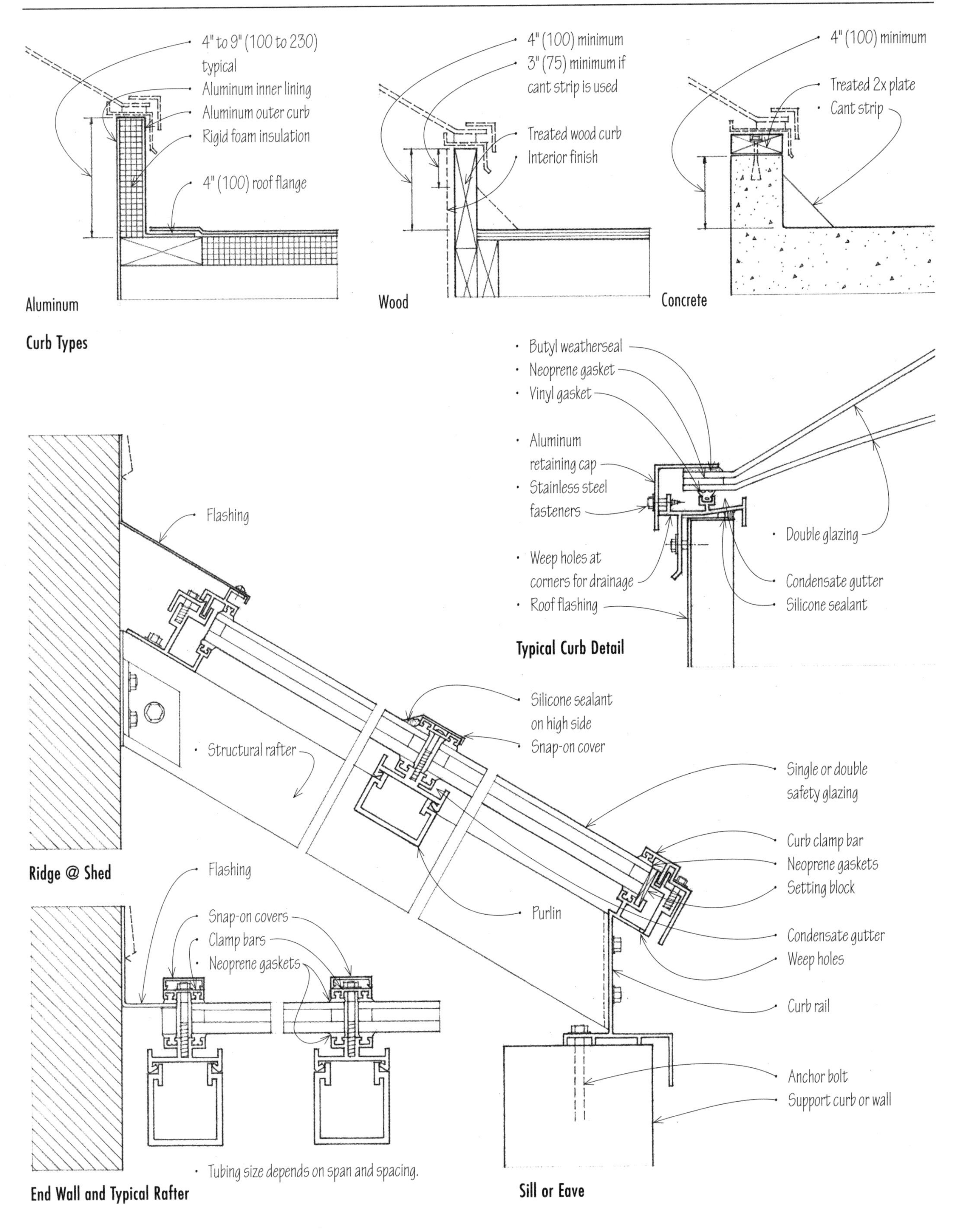
4" to 9" (100 to 230) typical
Aluminum inner lining
Aluminum outer curb
Rigid foam insulation
4" (100) roof flange
Aluminum
4" (100) minimum
3" (75) minimum if cant strip is used
Treated wood curb
Interior finish
Wood
4" (100) minimum
Treated 2x plate
Cant strip
Concrete
Curb Types
Butyl weatherseal
Neoprene gasket
Vinyl gasket
Aluminum retaining cap
Stainless steel fasteners
Weep holes at corners for drainage
Roof flashing
Double glazing
Condensate gutter
Silicone sealant
Typical Curb Detail
Flashing
Structural rafter
Silicone sealant on high side
Snap-on cover
Single or double safety glazing
Curb clamp bar
Neoprene gaskets
Setting block
Purlin
Condensate gutter
Weep holes
Curb rail
Ridge @ Shed
Flashing
Snap-on covers
Clamp bars
Neoprene gaskets
Anchor bolt
Support curb or wall
Tubing size depends on span and spacing.
End Wall and Typical Rafter
Sill or Eave

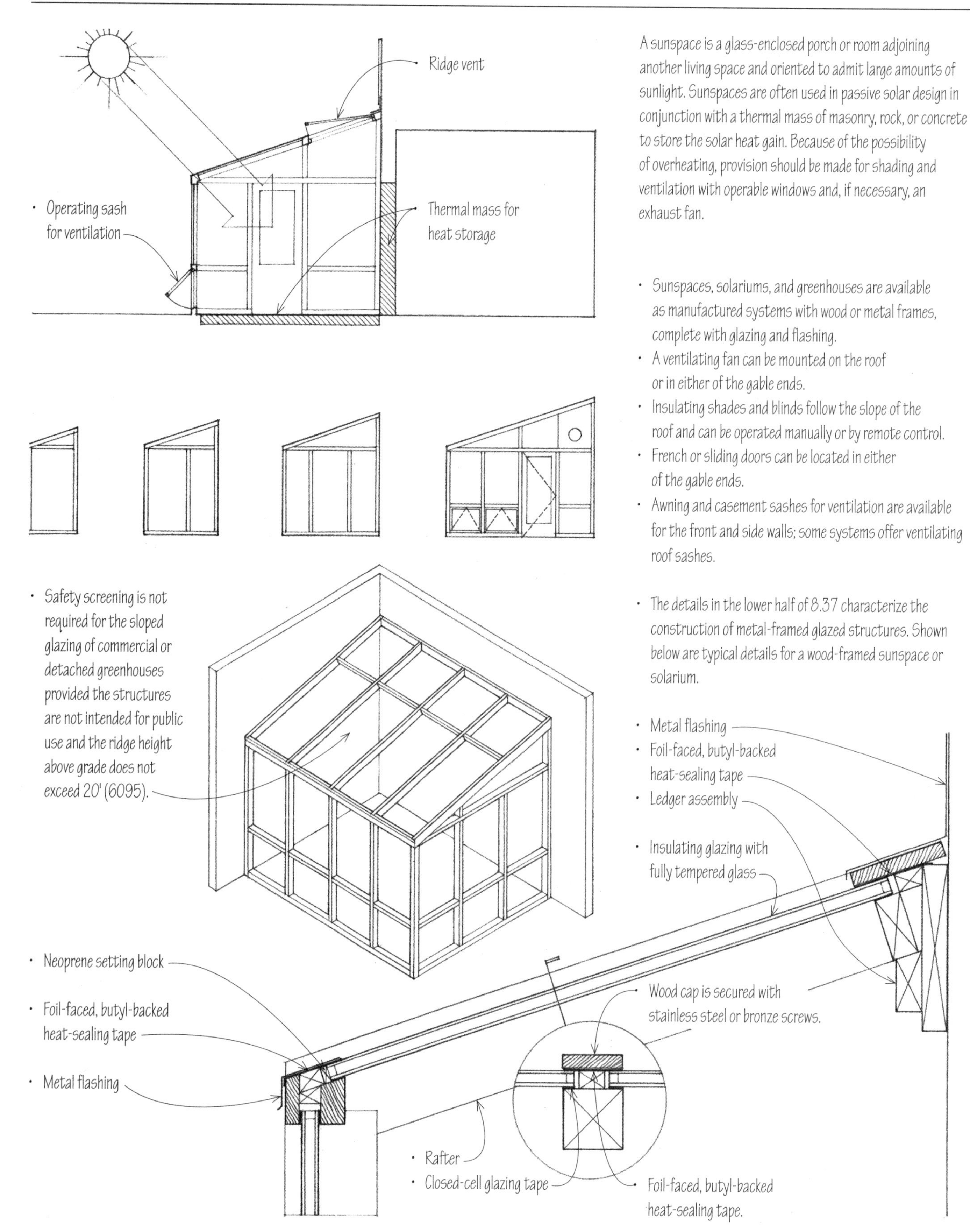

A sunspace is a glass-enclosed porch or room adjoining another living space and oriented to admit large amounts of sunlight. Sunspaces are often used in passive solar design in conjunction with a thermal mass of masonry, rock, or concrete to store the solar heat gain. Because of the possibility of overheating, provision should be made for shading and ventilation with operable windows and, if necessary, an exhaust fan.

- Sunspaces, solariums, and greenhouses are available as manufactured systems with wood or metal frames, complete with glazing and flashing.
- A ventilating fan can be mounted on the roof or in either of the gable ends.
- Insulating shades and blinds follow the slope of the roof and can be operated manually or by remote control.
- French or sliding doors can be located in either of the gable ends.
- Awning and casement sashes for ventilation are available for the front and side walls; some systems offer ventilating roof sashes.

- Safety screening is not required for the sloped glazing of commercial or detached greenhouses provided the structures are not intended for public use and the ridge height above grade does not exceed 20' (6095).

- The details in the lower half of 8.37 characterize the construction of metal-framed glazed structures. Shown below are typical details for a wood-framed sunspace or solarium.

9

SPECIAL CONSTRUCTION

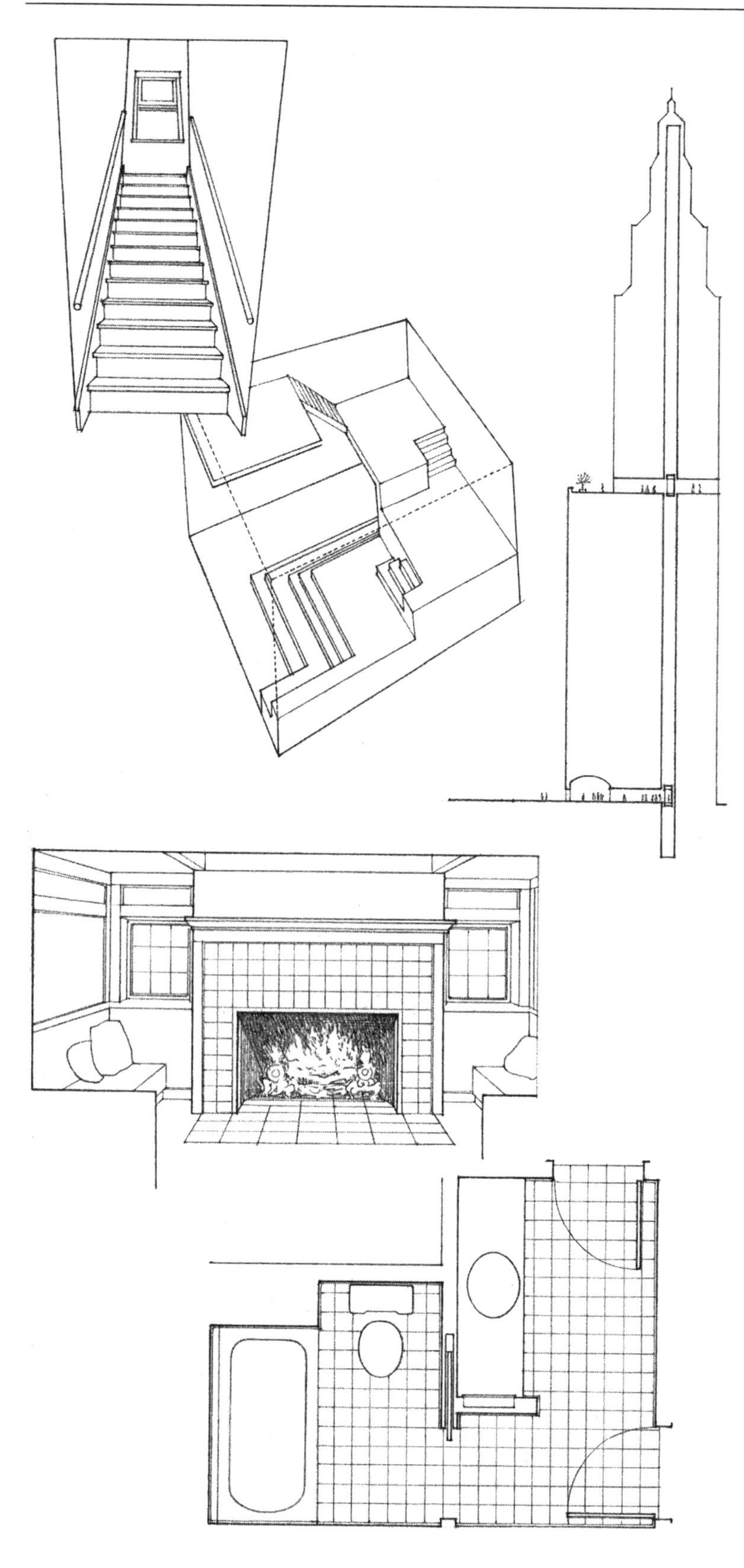

This chapter discusses those elements of a building that have unique characteristics and that therefore should be considered as separate entities. While not always affecting the exterior form of a building, they do influence the internal organization of spaces, the pattern of the structural system, and in some cases, the layout of heating, plumbing, and electrical systems.

Stairs provide means for moving from one level to another and are therefore important links in the overall circulation scheme of a building. Whether punctuating a two-story volume or rising through a narrow shaft, a stairway takes up a significant amount of space. The landings of a stairway should be logically integrated with the structural system to avoid overly complicated framing conditions. Safety and ease of travel are, in the end, the most important considerations in the design and placement of stairs.

Multistory buildings require elevators to move people, equipment, and freight from one floor to another. For accessibility to multistory public and commercial facilities by persons with disabilities, federal regulations mandate their installation. An alternative to elevators is the escalator, which can move a large number of people efficiently and comfortably between a limited number of floors.

Fireplaces and woodburning stoves are sources of heat and visual points of interest for any interior space. The placement and size of a fireplace or stove in a room should be related to the scale and use of the space. Both fireplaces and stoves must be located and constructed to draft properly. The damper and flue sizes should correspond to the size and proportions of the firebox and precautions should be taken against fire hazards and heat loss.

Kitchens and bathrooms are unique areas of a building that demand the careful integration of plumbing, electrical, and heating/ventilating systems with the functional and aesthetic requirements of the spaces. These areas also require special fixtures and equipment, as well as durability, ease of maintenance, and sanitary surfaces and finishes.

The dimensions of risers and treads in a stairway should be proportioned to accommodate our body movement. Their pitch, if steep, can make ascent physically tiring as well as psychologically forbidding, and can make descent precarious. If the pitch of a stairway is shallow, its treads should be deep enough to fit our stride.

Building codes regulate the minimum and maximum dimensions of risers and treads; see 9.04–9.05. For comfort, the riser and tread dimensions can be proportioned according to either of the following formulas:

- Tread (inches) + 2x riser (inches) = 24 to 25
- Riser (inches) x tread (inches) = 72 to 75

Exterior stairs are generally not as steep as interior stairs, especially where dangerous conditions such as snow and ice exist. The proportioning formula can therefore be adjusted to yield a sum of 26.

For safety, all risers in a flight of stairs should be the same rise and all treads should have the same run. Building codes limit the allowable variation in riser height or tread run to 3/8" (9.5 mm). Consult the building code to verify the dimensional guidelines outlined on this and the following page.

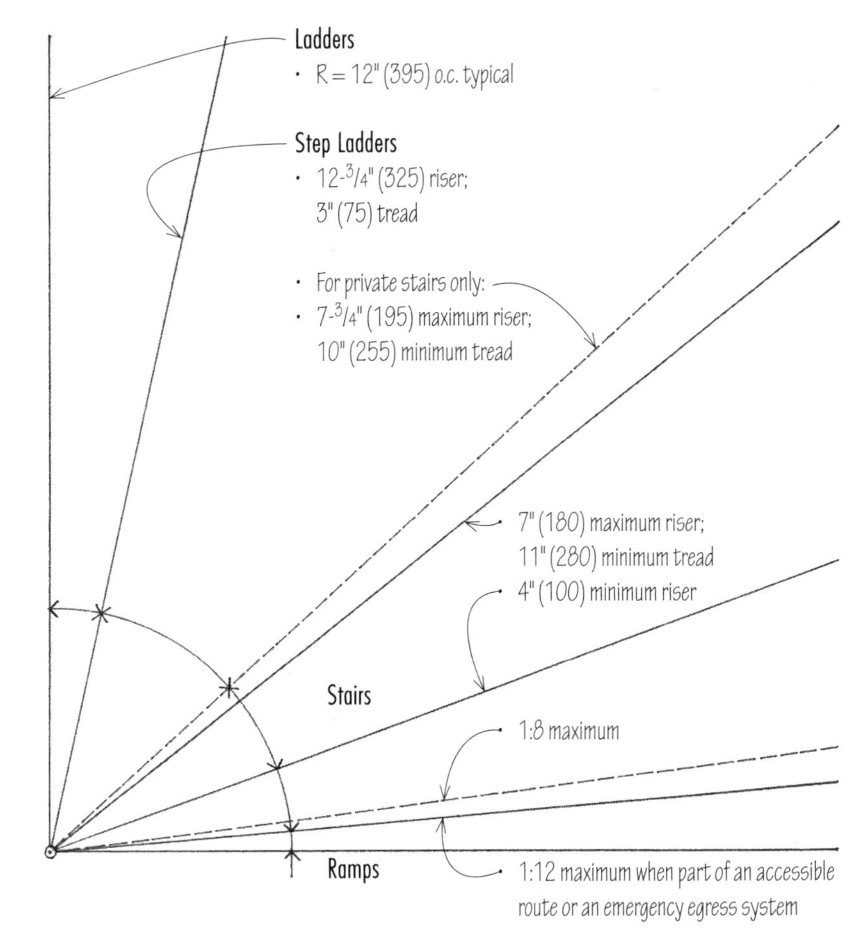

- The actual riser and tread dimensions for a set of stairs are determined by dividing the total rise or floor-to-floor height by the desired riser height. The result is rounded off to arrive at a whole number of risers. The total rise is then redivided by this whole number to arrive at the actual riser height.
- This riser height must be checked against the maximum riser height allowed by the building code. If necessary, the number of risers can be increased by one and the actual riser height recalculated.
- Once the actual riser height is fixed, the tread run can be determined by using the riser:tread proportioning formula.
- Since in any flight of stairs, there is always one less tread than the number of risers, the total number of treads and the total run can be easily determined.

Riser and Tread Dimensions

Riser inches (mm)	Tread inches (mm)	
5 (125)	15 (380)	
5-1/4 (135)	14-1/2 (370)	
5-1/2 (140)	14 (355)	
5-3/4 (145)	13-1/2 (340)	
6 (150)	13 (330)	
6-1/4 (160)	12-1/2 (320)	
6-1/2 (165)	12 (305)	
6-3/4 (170)	11-1/2 (290)	
7 (180)	11 (280)	Maximum riser height; minimum tread depth for accessible stairs and emergency egress
7-1/4 (185)	10-1/2 (265)	
7-1/2 (190)	10 (255)	

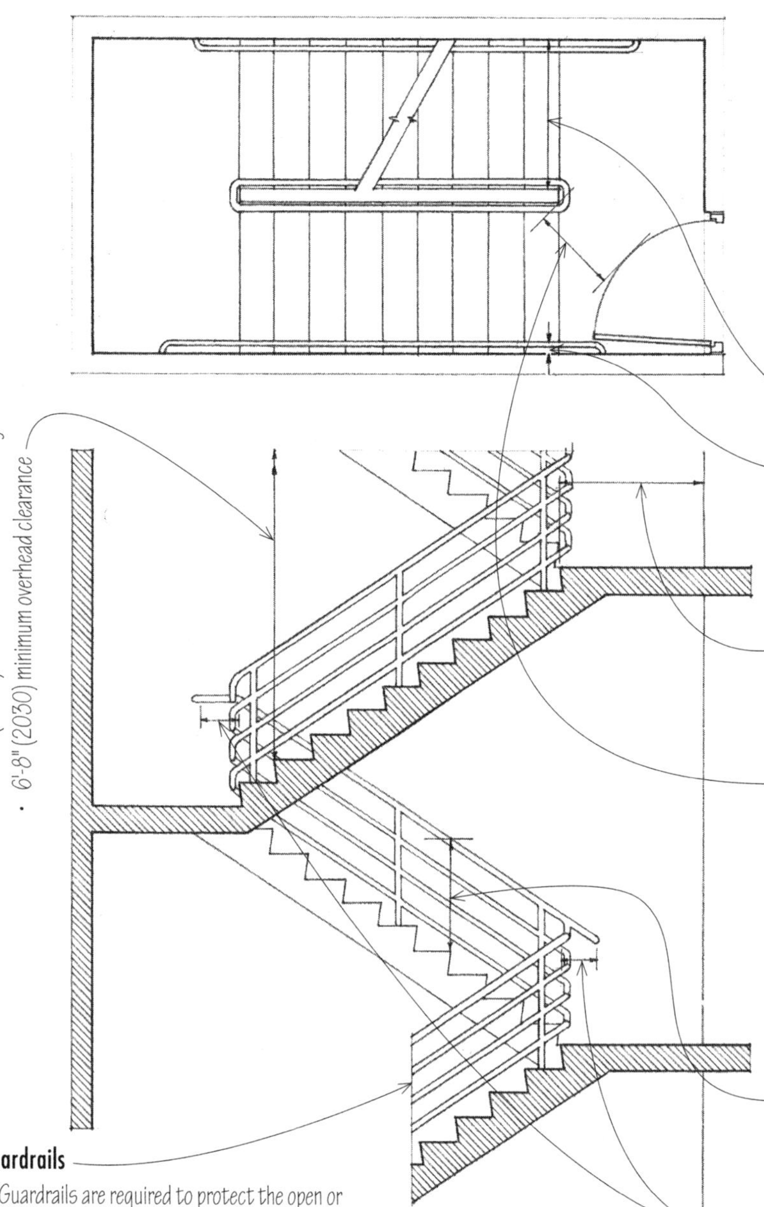

Stairway design is strictly regulated by the building code, especially when a stairway is an essential part of an emergency egress system. Because an accessible stairway should also serve as a means of egress during an emergency, the ADA accessibility requirements illustrated on the next page are similar to those of an emergency egress stairway.

Stairway Width

- The occupant load, which is based on the use group and the floor area served, determines the required width of an exit stairway. Consult the building code for details.
- 44" (1120) minimum width; 36" (915) minimum for stairways serving an occupant load of 49 or less.
- Handrails may project a maximum of 4-1/2" (115) into the required width; stringers and trim may project a maximum of 1-1/2" (38).

Landings

- Landings should be as least as wide as the stairway they serve and have a minimum length equal to the stair width, measured in the direction of travel. Landings serving straight-run stairs need not be longer than 48" (1220).
- Door should swing in the direction of egress. Door swing must not reduce the landing to less than one-half of its required width.
- When fully open, the door must not intrude into required width by more than 7" (180).

Handrails

- Handrails are required on both sides of the stair. The building code allows exceptions for stairs in individual dwelling units.
- 34" to 38" (865 to 965) height above the leading edge of the stair treads or nosings.
- Handrails should be continuous without interruption by a newel post or other obstruction.
- Handrails should extend at least 12" (305) horizontally beyond the top riser of a stair flight and extend at the slope of the stair run for a horizontal distance of at least one tread depth beyond the last riser nosing of the flight. The ends should return smoothly to a wall or walking surface, or continue to the handrail of an adjacent stair flight.
- See the next page for detailed handrail requirements.

Guardrails

- Guardrails are required to protect the open or glazed sides of stairways, ramps, porches, and unenclosed floor and roof openings.
- Guardrails should be at least 42" (1070) high; guardrails in dwellings may be 36" (915) high.
- Guardrails protecting the open or glazed side of a stairway may have the same height as the stair handrails.
- A 4" (100) sphere must not be able to pass through any opening in the railing from the floor up to 34" (865); from 34" to 42" (865 to 1070), the pattern may allow a sphere up to 8" (205) in diameter to pass.
- Guardrails should be able to withstand a concentrated load applied nonconcurrently to their top rails in both vertical and horizontal directions. Consult the building code for detailed requirements.

Treads, Risers, and Nosings

- A minimum of three risers per flight is recommended to prevent tripping and may be required by the building code.
- See the next page for detailed tread, riser, and nosing requirements.
- See 9.03 for tread and riser proportions.

ADA Accessibility Guidelines

Accessible stairs should also serve as a means of egress during an emergency, or lead to an accessible area of refuge where people who are unable to use stairs may remain temporarily in safety to await assistance during an emergency evacuation.

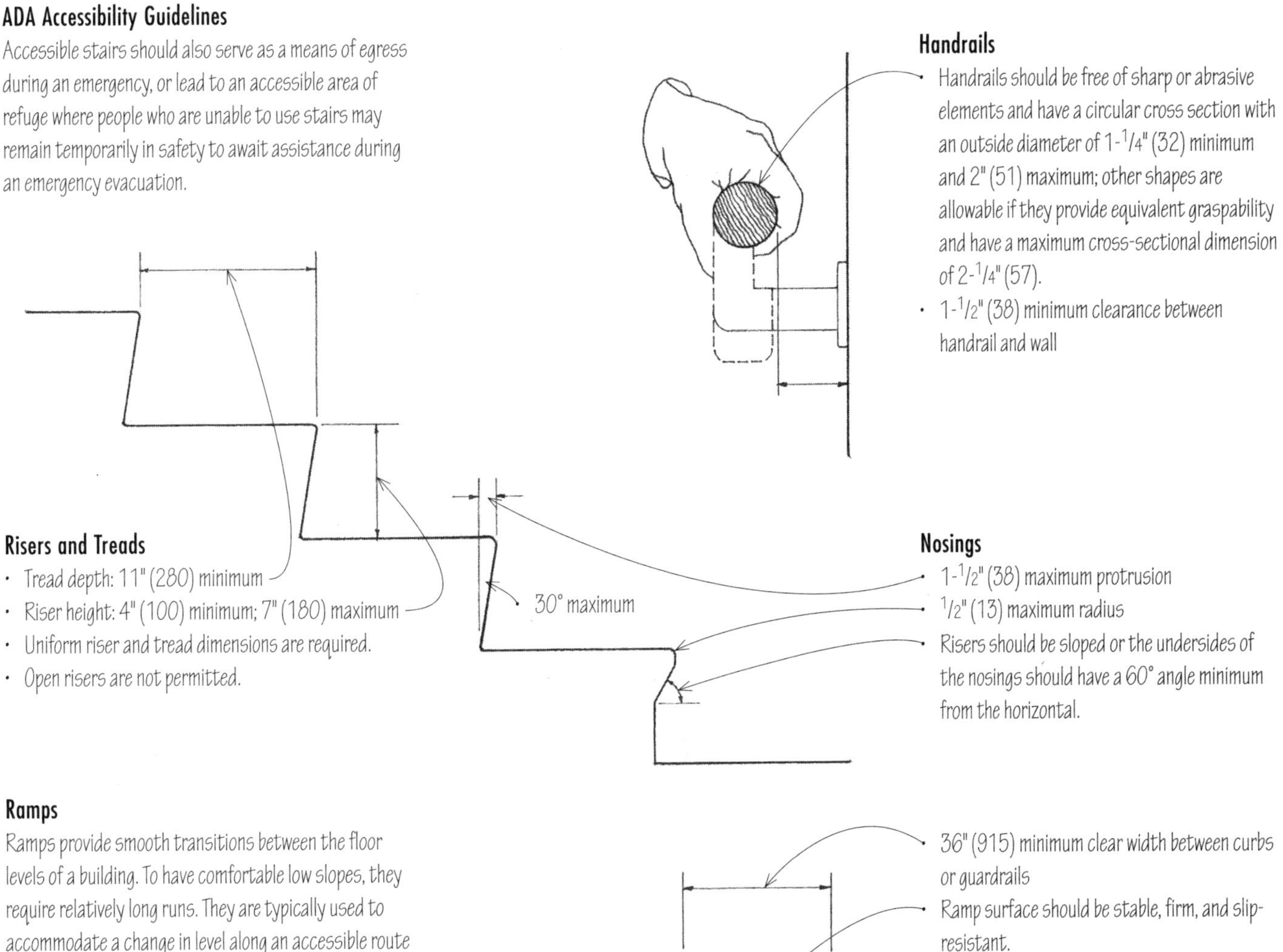

Handrails

- Handrails should be free of sharp or abrasive elements and have a circular cross section with an outside diameter of 1-1/4" (32) minimum and 2" (51) maximum; other shapes are allowable if they provide equivalent graspability and have a maximum cross-sectional dimension of 2-1/4" (57).
- 1-1/2" (38) minimum clearance between handrail and wall

Risers and Treads

- Tread depth: 11" (280) minimum
- Riser height: 4" (100) minimum; 7" (180) maximum
- Uniform riser and tread dimensions are required.
- Open risers are not permitted.

Nosings

- 1-1/2" (38) maximum protrusion
- 1/2" (13) maximum radius
- Risers should be sloped or the undersides of the nosings should have a 60° angle minimum from the horizontal.

Ramps

Ramps provide smooth transitions between the floor levels of a building. To have comfortable low slopes, they require relatively long runs. They are typically used to accommodate a change in level along an accessible route or to provide access for wheeled equipment. Short, straight ramps act as beams and may be constructed as wood, steel, or concrete floor systems. Long or curvilinear ramps are usually of steel or reinforced concrete.

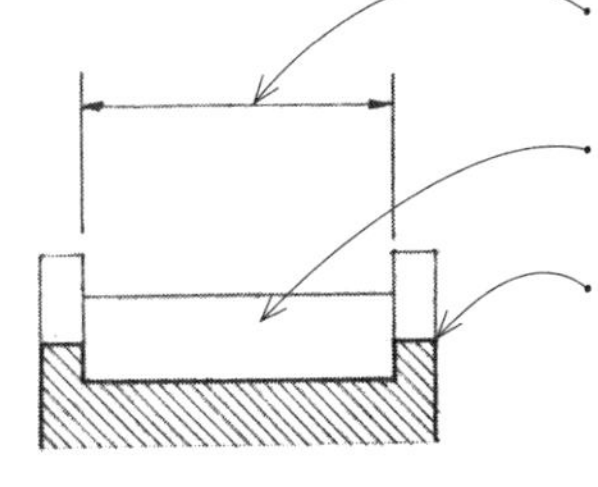

- 36" (915) minimum clear width between curbs or guardrails
- Ramp surface should be stable, firm, and slip-resistant.
- Curbs, guardrails, or walls are required to prevent people from slipping off of the ramp; 4" (100) minimum curb or barrier height.

- 1:12 maximum slope
- 30" (760) maximum rise between landings

Landings

- Ramps should have level landings at each end with a 60" (1525) minimum length.
- Landing should be as wide as the widest ramp leading to it.
- 60" x 60" (1525 x 1525) minimum landing where ramp changes direction

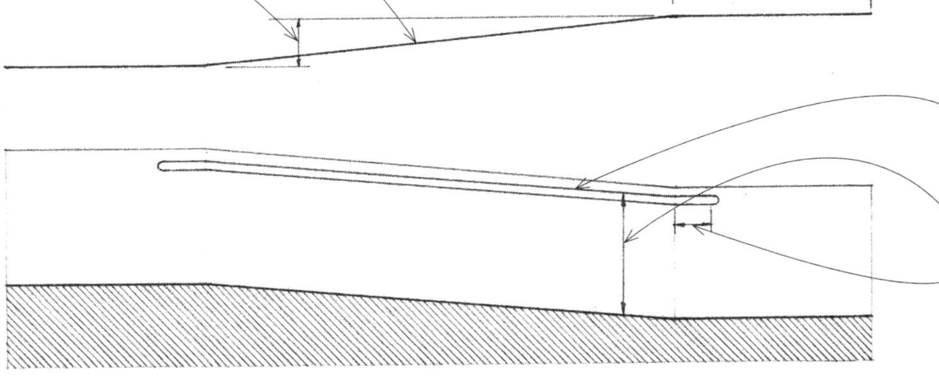

Handrails

- Ramps having a rise greater than 6" (150 or a run greater than 72" (1830) should have handrails along both sides.
- Handrail requirements are the same as for stairways.
- Extend handrails at least 12" (305) horizontally beyond the top and bottom of ramp runs.

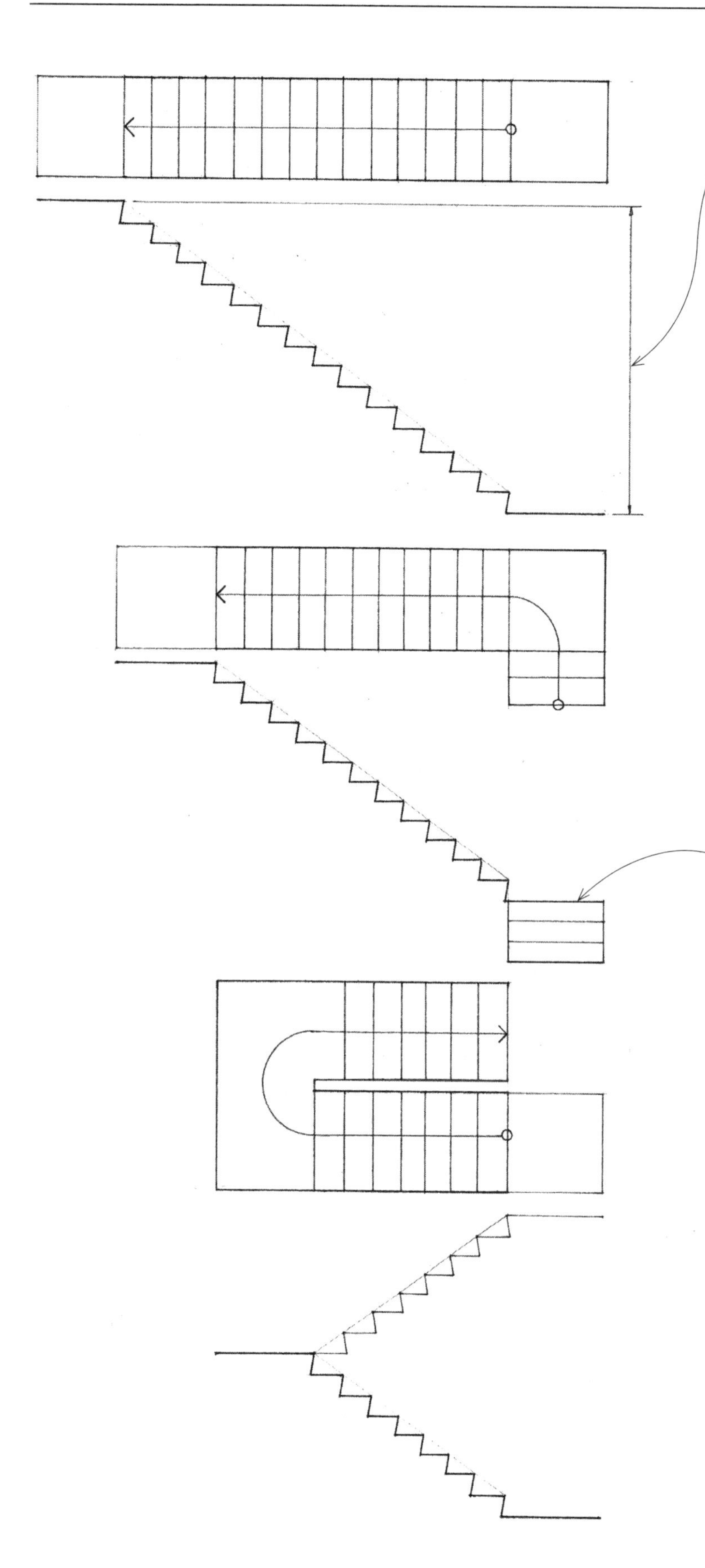

Straight-Run Stair

- A straight-run stair extends from one level to another without turns or winders.
- Building codes generally limit the vertical rise between landings to 12' (3660).

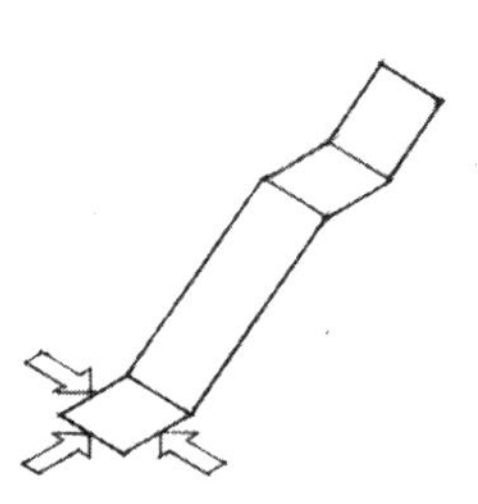

- A stairway may be approached or departed either axially or perpendicular to the stair run.

Quarter-Turn Stair

- A quarter-turn or L-shaped stair makes a right-angled turn in the path of travel.
- The two flights connected by an intervening landing may be equal or unequal, depending on the desired proportion of the stairway opening.

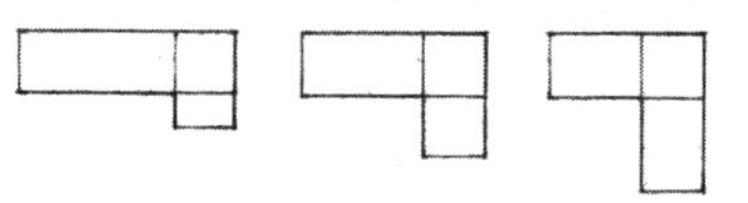

- Landings that are below normal eye level and provide a place to rest or pause are inviting.

Half-Turn Stair

- A half-turn stair turns 180° or through two right angles at an intervening landing.
- A half-return stair is more compact than a single straight-run stair.
- The two flights connected by the landing may be equal or unequal, depending on the desired proportion of the stairway opening.

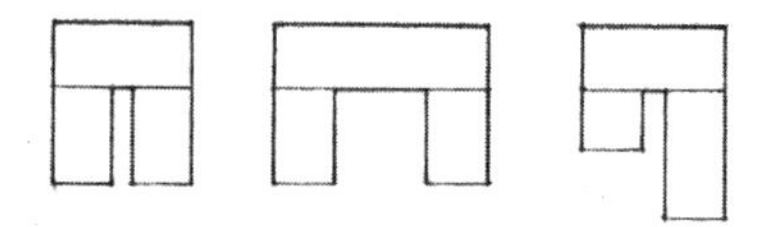

Winding Stair

- A winding stair is any stairway constructed with winders, as a circular or spiral stair. Quarter-turn and half-turn stairs may also use winders rather than a landing to conserve space when changing direction.
- Winders can be hazardous since they offer little foothold at their interior corners. Building codes generally restrict the use of winders to private stairs within individual dwelling units.

Circular Stair

- A circular stair, as its name implies, has a circular plan configuration. Even though a circular stair is constructed with winders, the building code may allow its use as part of the means of egress from a building if its inner radius is at least twice the actual width of the stairway.

Spiral Stair

- A spiral stair consists of wedge-shaped treads winding around and supported by a central post.
- Spiral stairs occupy a minimum amount of floor space, but building codes permit their use only as private stairs in individual dwelling units.
- See 9.12 for typical dimensions.

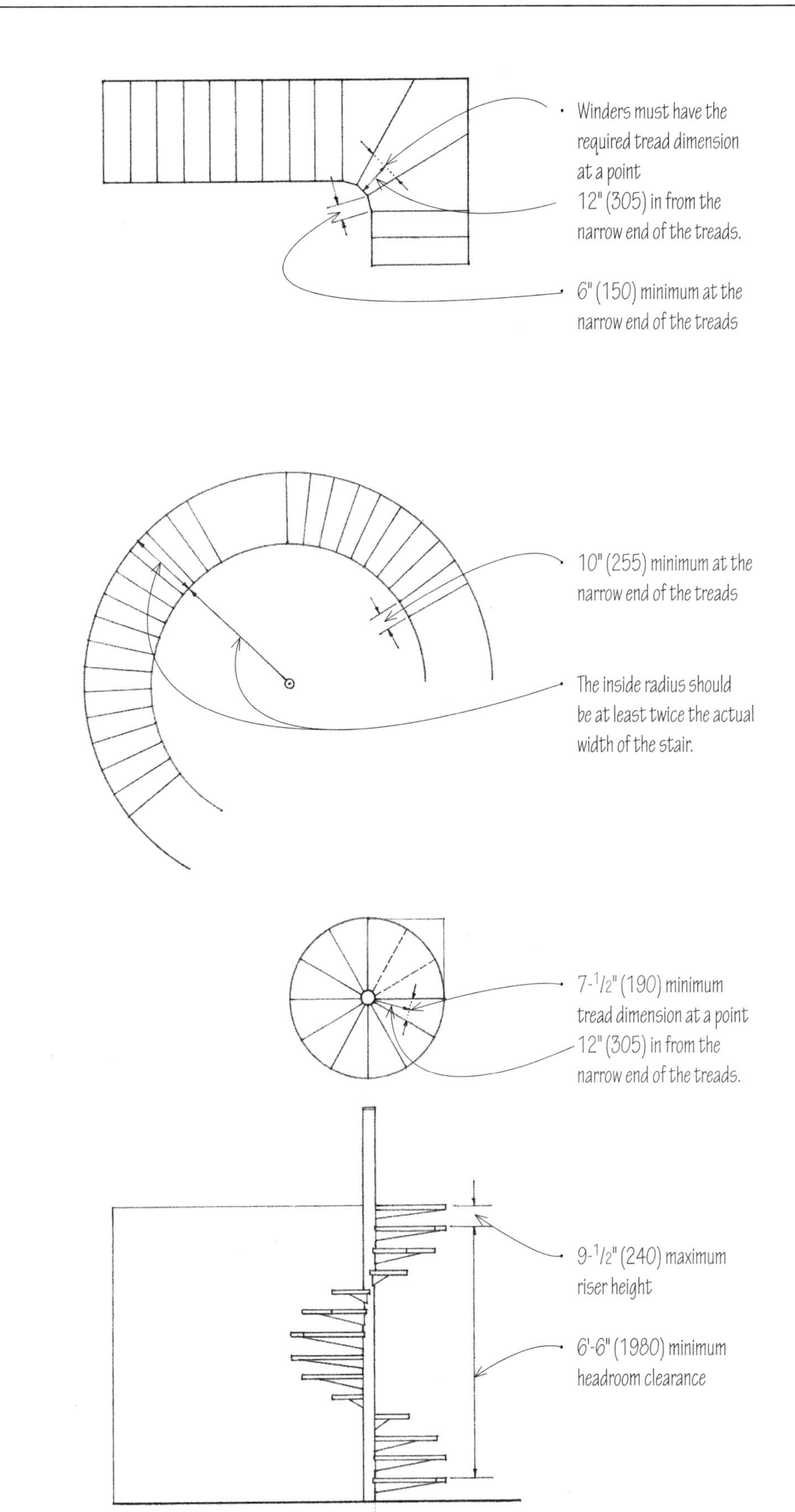

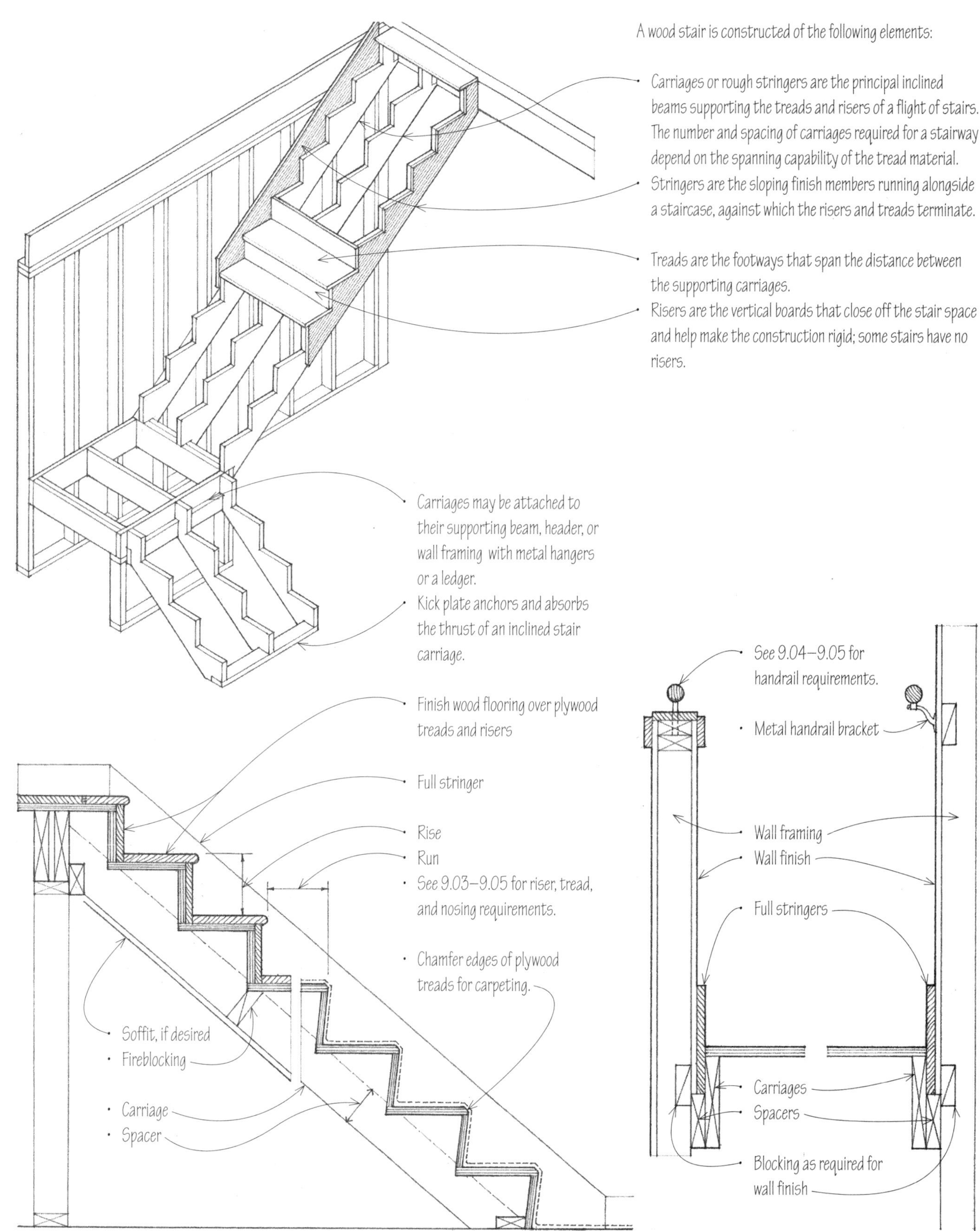

Closed-Riser Stair with Full Stringer

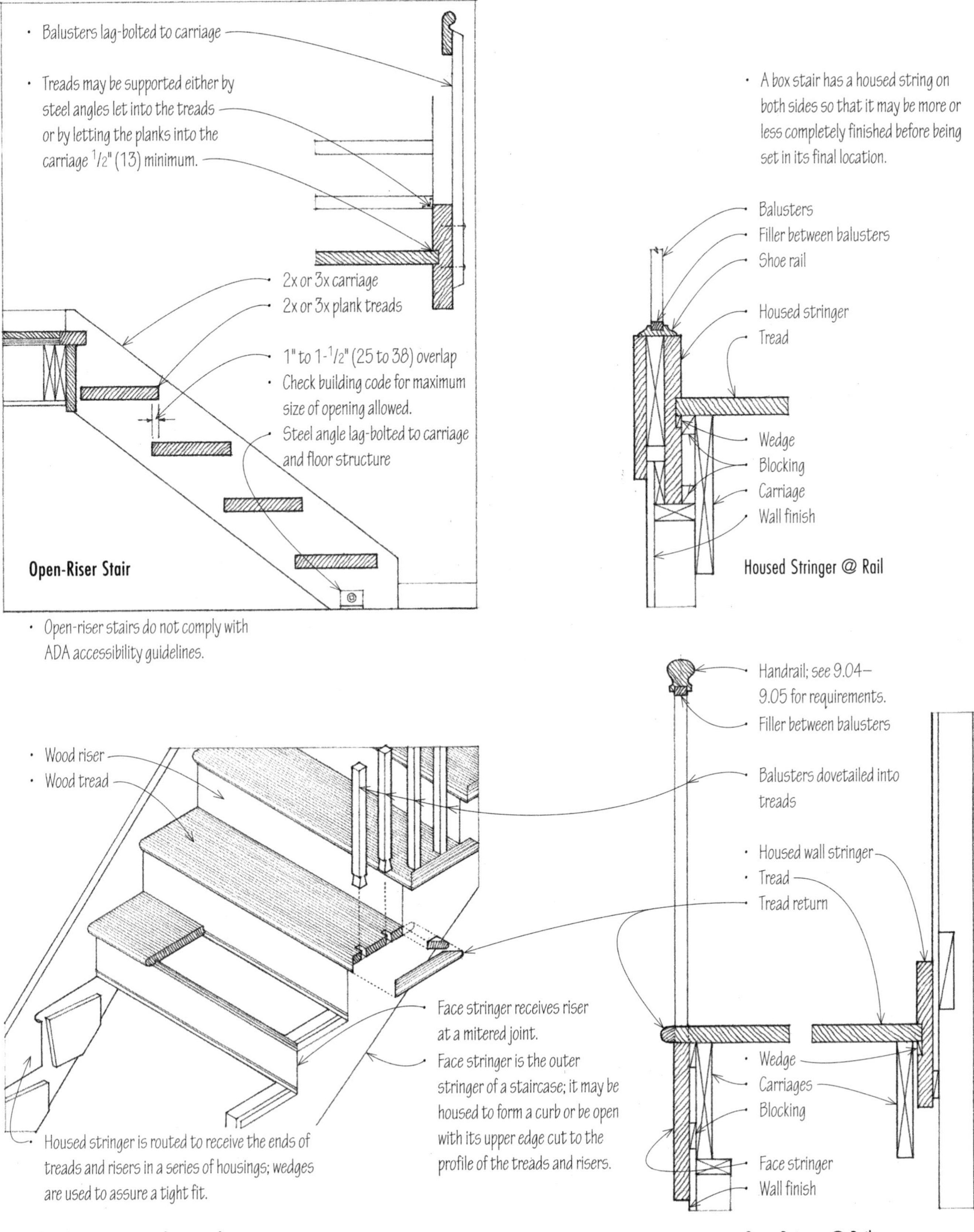
• Balusters lag-bolted to carriage
• Treads may be supported either by steel angles let into the treads or by letting the planks into the carriage 1/2" (13) minimum.
• 2x or 3x carriage
• 2x or 3x plank treads
• 1" to 1-1/2" (25 to 38) overlap
• Check building code for maximum size of opening allowed.
• Steel angle lag-bolted to carriage and floor structure
Open-Riser Stair
• Open-riser stairs do not comply with ADA accessibility guidelines.
• A box stair has a housed string on both sides so that it may be more or less completely finished before being set in its final location.
• Balusters
• Filler between balusters
• Shoe rail
• Housed stringer
• Tread
• Wedge
• Blocking
• Carriage
• Wall finish
Housed Stringer @ Rail
• Wood riser
• Wood tread
• Handrail; see 9.04–9.05 for requirements.
• Filler between balusters
• Balusters dovetailed into treads
• Housed wall stringer
• Tread
• Tread return
• Face stringer receives riser at a mitered joint.
• Face stringer is the outer stringer of a staircase; it may be housed to form a curb or be open with its upper edge cut to the profile of the treads and risers.
• Wedge
• Carriages
• Blocking
• Face stringer
• Wall finish
• Housed stringer is routed to receive the ends of treads and risers in a series of housings; wedges are used to assure a tight fit.
Closed-Riser Stair with Housed Stringer
Open Stringer @ Rail

A concrete stair is designed as an inclined, one-way reinforced slab with steps formed on its upper surface. If the stair is constructed after the floor beam or wall supports, it acts as a simple beam. If it is cast with the beam or slab supports, it is designed as a continuous beam. Concrete stairs require careful analysis of load, span, and support conditions; consult a structural engineer for final design requirements.

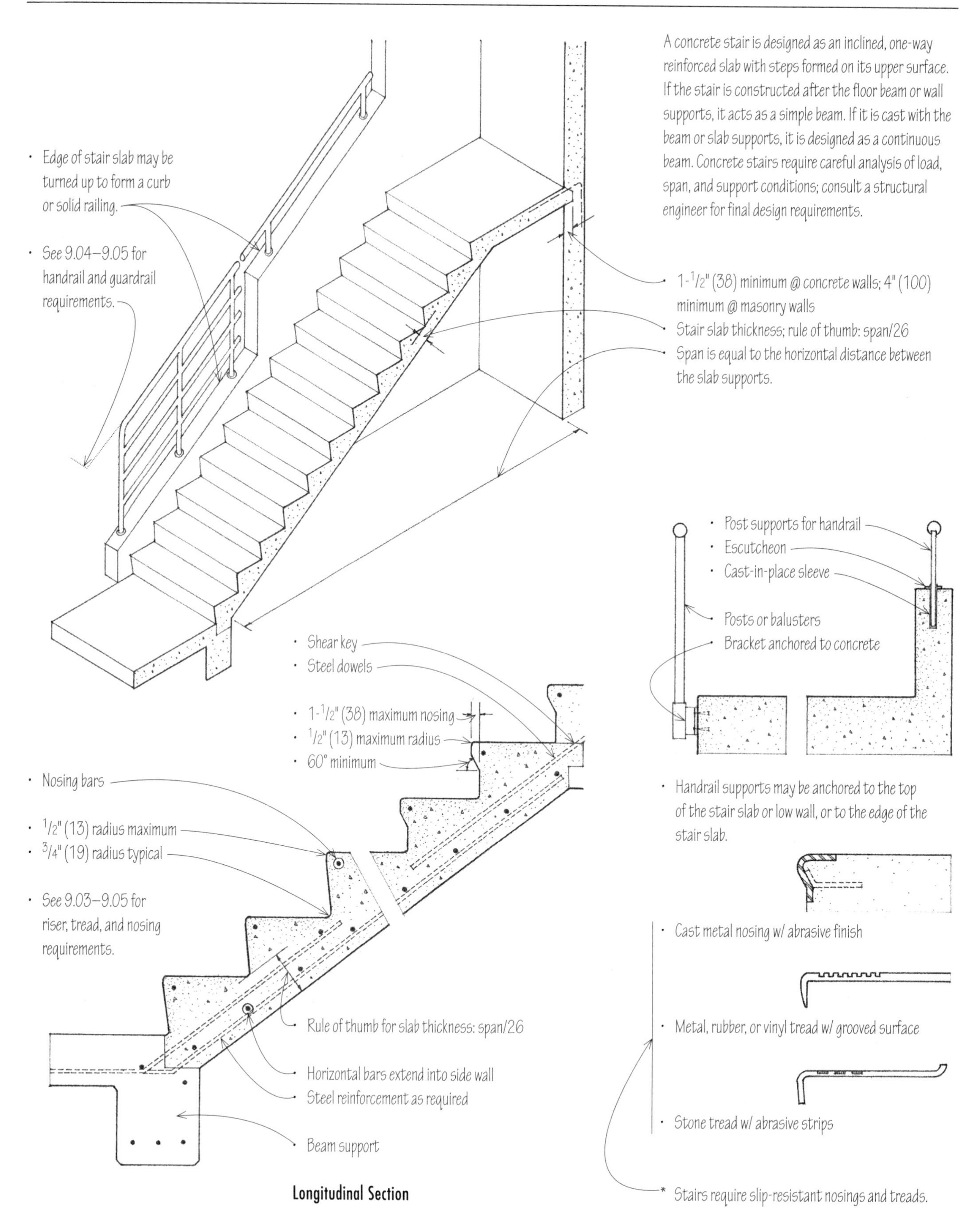

Longitudinal Section

CSI MasterFormat 03 30 00: Cast-in-place Concrete
CSI MasterFormat 03 11 23: Permanent Stair Forming

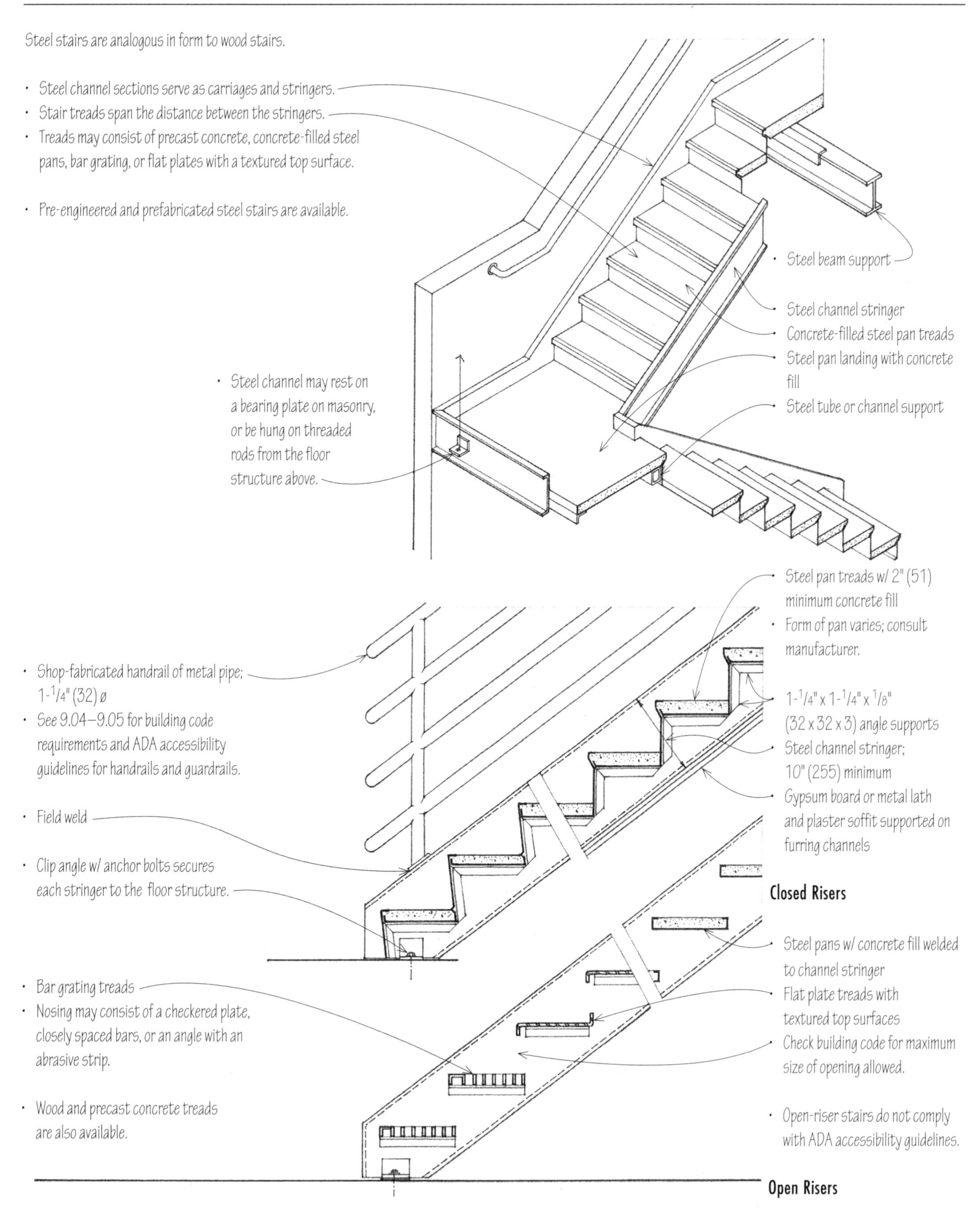

CSI MasterFormat 05 51 00: Metal Stairs

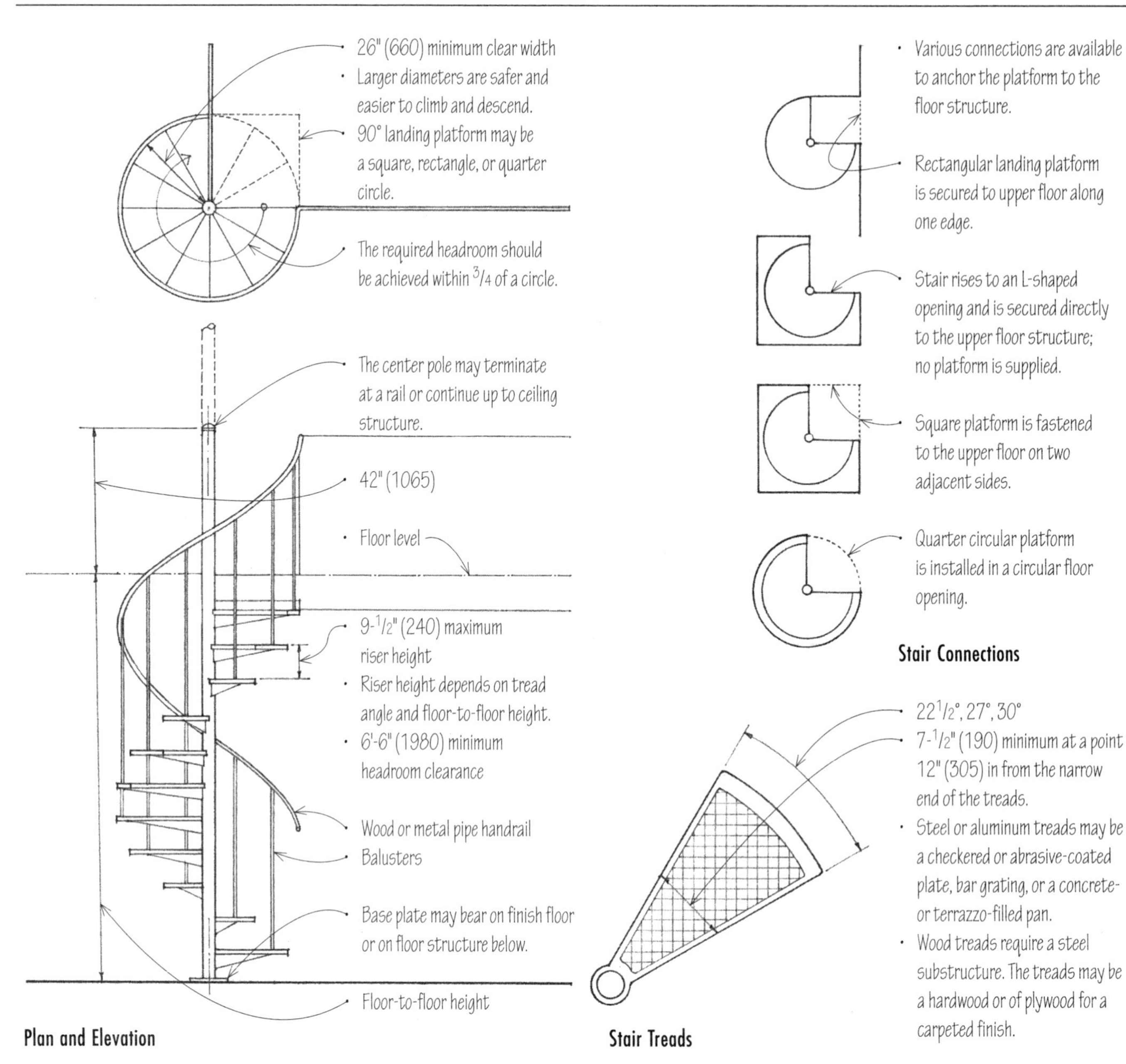

Plan and Elevation

Stair Connections

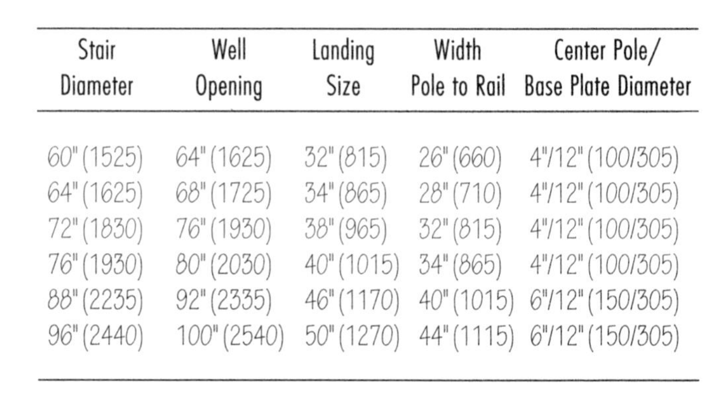

Stair Treads

Representative Sizes and Dimensions of Spiral Stairs*

Tread Angle	No. of Treads in 360°	Riser Height	Headroom
22 1/2°	16	7" (180)	7'-0" (2135)
27°	13	7-1/2" to 8" (190 to 205)	6'-9" (2055)
30°	12	8-1/2" to 9-1/2" (215 to 240)	6'-9" (2055)

*Consult manufacturer's literature to verify these dimensional guidelines.

Stair Diameter	Well Opening	Landing Size	Width Pole to Rail	Center Pole/ Base Plate Diameter
60" (1525)	64" (1625)	32" (815)	26" (660)	4"/12" (100/305)
64" (1625)	68" (1725)	34" (865)	28" (710)	4"/12" (100/305)
72" (1830)	76" (1930)	38" (965)	32" (815)	4"/12" (100/305)
76" (1930)	80" (2030)	40" (1015)	34" (865)	4"/12" (100/305)
88" (2235)	92" (2335)	46" (1170)	40" (1015)	6"/12" (150/305)
96" (2440)	100" (2540)	50" (1270)	44" (1115)	6"/12" (150/305)

CSI MasterFormat 05 71 13: Fabricated Metal Spiral Stairs

Ladders are used primarily in industrial construction and in utility and service areas. They may also be used in private residential construction where space is extremely tight and traffic is minimal.

The drawings on this page illustrate ladders built with metal components. The ladder forms may be translated into wood construction.

Safety considerations include:
- Proper riser height
- Adequate toe space
- Adequate support for strings and railings
- Slip-resistant treads

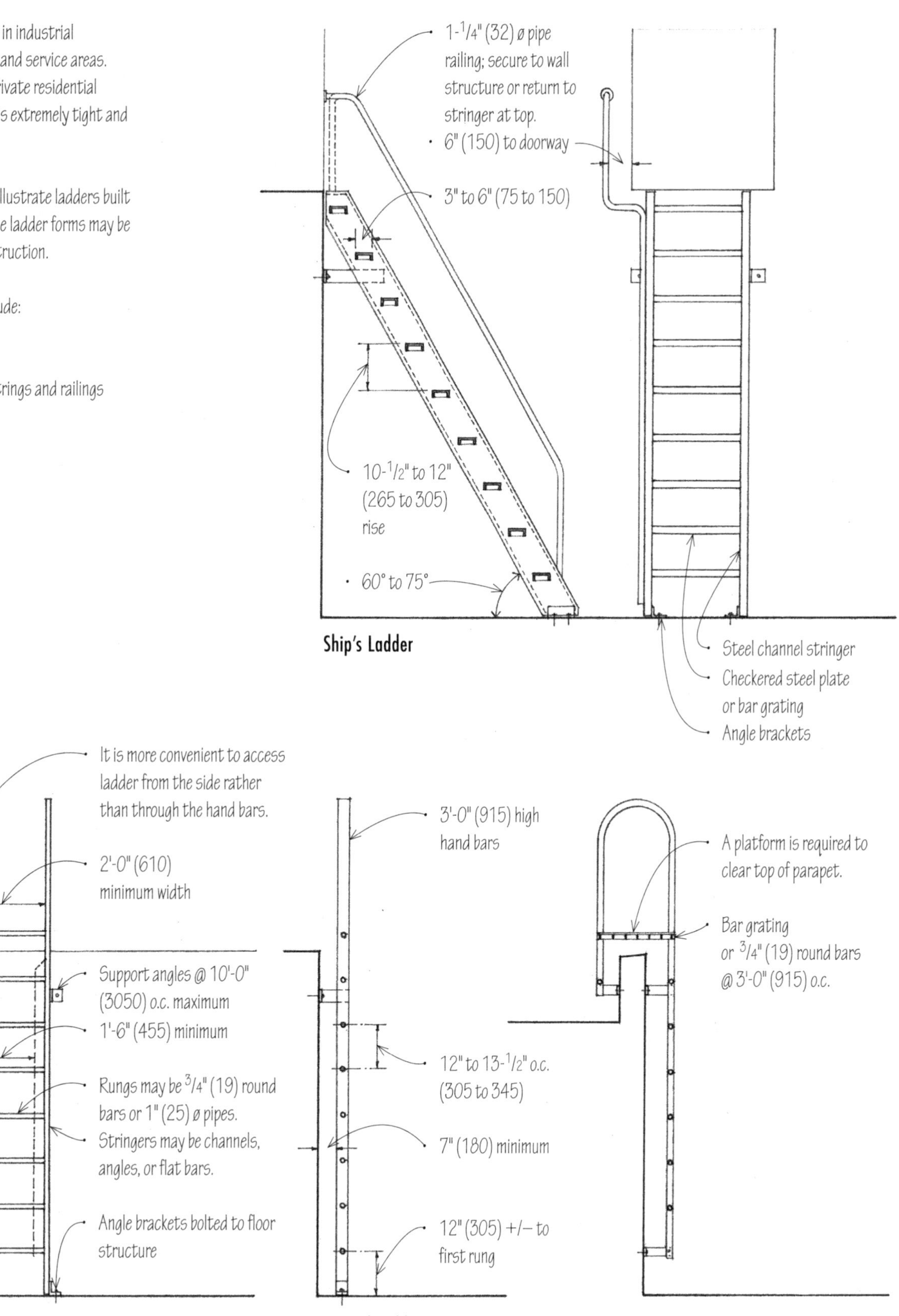

Elevators travel vertically to carry passengers, equipment, and freight from one level of a building to another. The two most common types are electric elevators and hydraulic elevators.

Electric Elevators

Electric elevators consist of a car that is mounted on guide rails, supported by hoisting cables, and driven by electric hoisting machinery in a penthouse. Geared traction elevators are capable of speeds up to 350 fpm (1.75 m/s) and are suitable for medium-rise buildings. Gearless traction elevators are available with speeds up to 1200 fpm (6 m/s) and typically serve high-rise buildings.

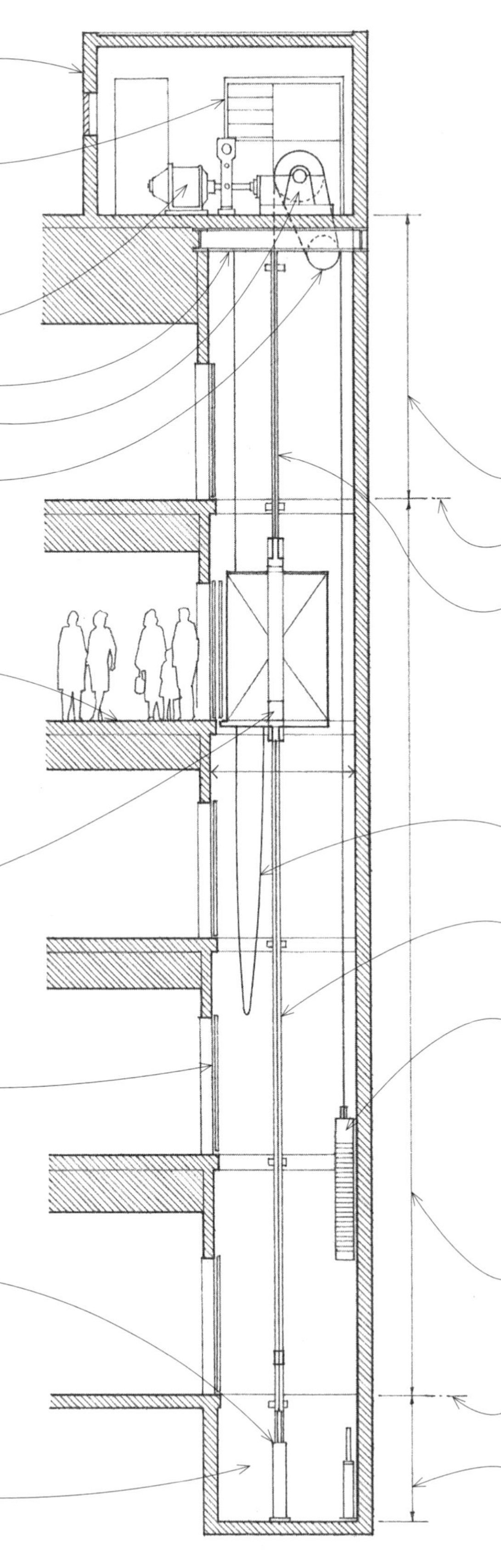

- A penthouse houses the hoisting machinery on the roof of a building.
- A control panel contains switches, buttons, and other equipment for regulating the hoisting machinery.
- The hoisting machinery for raising and lowering an elevator car consists of a motor-generator set, traction machine, speed governor, brake, driving sheave, and gears, if used.
- Heavy steel machine beams support the hoisting machinery for an elevator.
- Driving sheave is the hoisting pulley.
- Idle sheave tightens and guides the hoisting cables of the elevator system.
- Landing is the portion of a floor adjacent to an elevator hoistway, used for the receiving and discharge of passengers or freight.
- Elevator car safety is a mechanical device for slowing down and stopping an elevator car in the event of excessive speed or free fall, actuated by a governor and clamping the guide rails by a wedging action.
- Hoistway door between a hoistway and an elevator landing is normally closed except when an elevator car is stopped at the landing; 7'-0" and 8'-0" (2135 and 2440) heights are typical.
- Buffer is the piston or spring device that absorbs the impact of a descending elevator car or counterweight at the extreme lower limit of travel.
- Elevator pit is the portion of the shaft that extends from the level of the lowest landing to the floor of the hoistway.

- 16'-0" to 20'-0" (4875 to 6095)
- Top floor
- Hoisting cable is one of the wire cables or ropes used for raising and lowering an elevator car.
- Hoistway is the vertical enclosed space for the travel of one or more elevators.
- Traveling cable is one of the electric cables connecting an elevator car to a fixed electrical outlet in the hoistway.
- Guide rails are the vertical steel tracks controlling the travel of an elevator car or counterweight; they are secured to each floor with support brackets.
- Counterweights are rectangular cast-iron blocks mounted in a steel frame to counterbalance the load placed on the hoisting machine by an elevator car.
- A limit switch automatically cuts off current to an electric motor when an elevator car has passed a given point.
- Rise or travel is the vertical distance traversed by an elevator car from the lowest to the highest landings of the hoistway.
- Bottom floor
- 5'-0" to 11'-6" (1525 to 3505)

CSI MasterFormat 14 20 00: Elevators

Hydraulic Elevators

Hydraulic elevators consist of a car supported by a piston that is moved by or moves against a fluid under pressure. A penthouse is not required, but the hydraulic elevator's lower speed and piston length limit its use to buildings up to six stories in height.

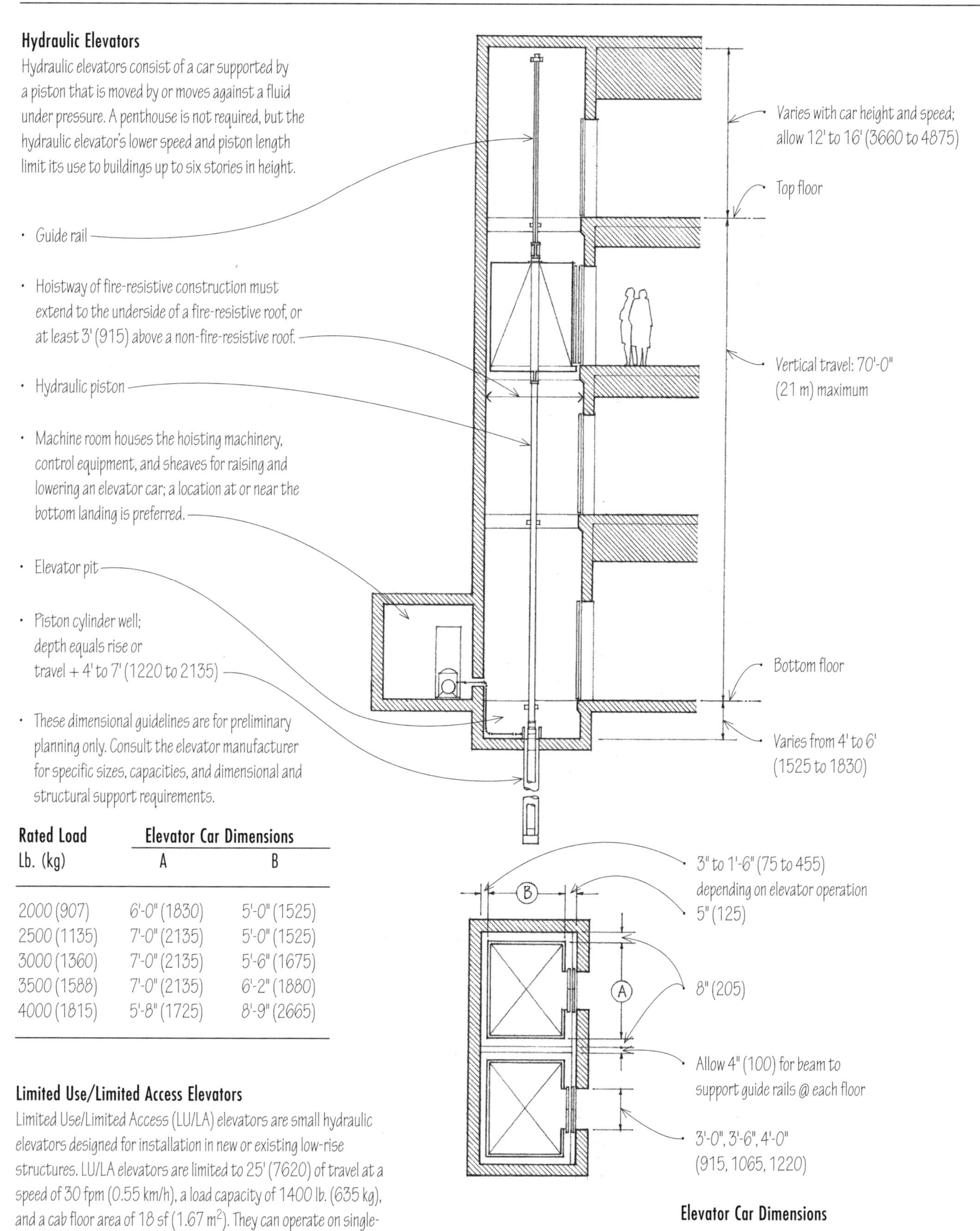

Rated Load	Elevator Car Dimensions	
Lb. (kg)	A	B
2000 (907)	6'-0" (1830)	5'-0" (1525)
2500 (1135)	7'-0" (2135)	5'-0" (1525)
3000 (1360)	7'-0" (2135)	5'-6" (1675)
3500 (1588)	7'-0" (2135)	6'-2" (1880)
4000 (1815)	5'-8" (1725)	8'-9" (2665)

Elevator Car Dimensions

Limited Use/Limited Access Elevators

Limited Use/Limited Access (LU/LA) elevators are small hydraulic elevators designed for installation in new or existing low-rise structures. LU/LA elevators are limited to 25' (7620) of travel at a speed of 30 fpm (0.55 km/h), a load capacity of 1400 lb. (635 kg), and a cab floor area of 18 sf (1.67 m^2). They can operate on single-phase power and require less pit depth and overhead space than do regular commercial elevators.

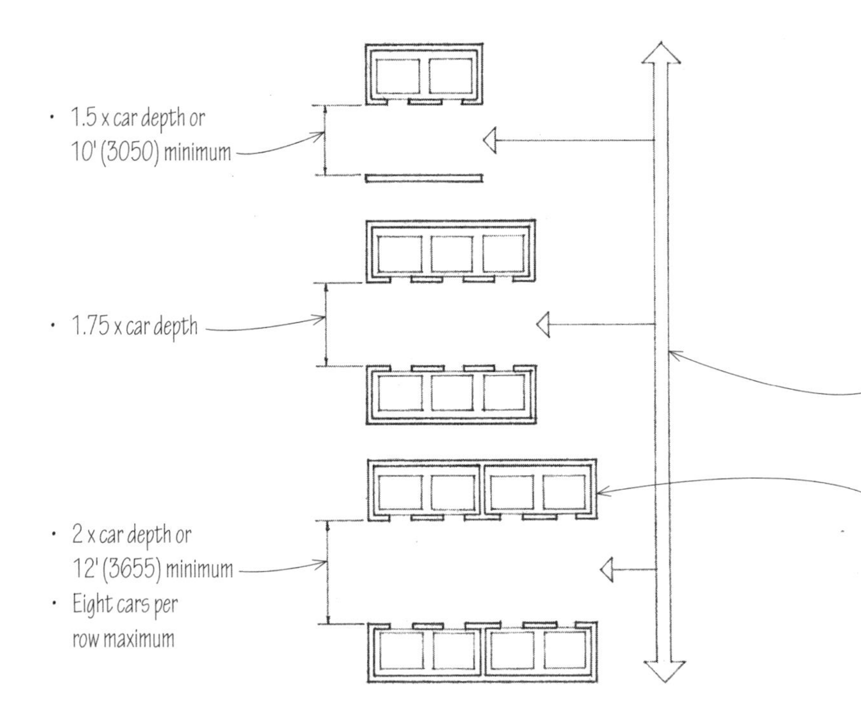

Elevator Layout

The type, size, number, speed, and arrangement of elevators are determined by:

- Type of occupancy
- Amount and tempo of traffic to be carried
- Total vertical distance of travel
- Round-trip time and speed desired

- Banks or rows of elevators in a high-rise building are controlled by a common operating system and respond to a single call button.
- Elevators should be centrally located near the main entrance to a building and be easily accessible on all floors, but also be placed off of the main circulation path.
- Two or more hoistways are required for four or more elevators.

- Consult elevator manufacturer for recommended type, size, layout, controls, and installation requirements and details.
- Consult the building code for structural requirements and shaftway requirements for fire separation, ventilation, and soundproofing.

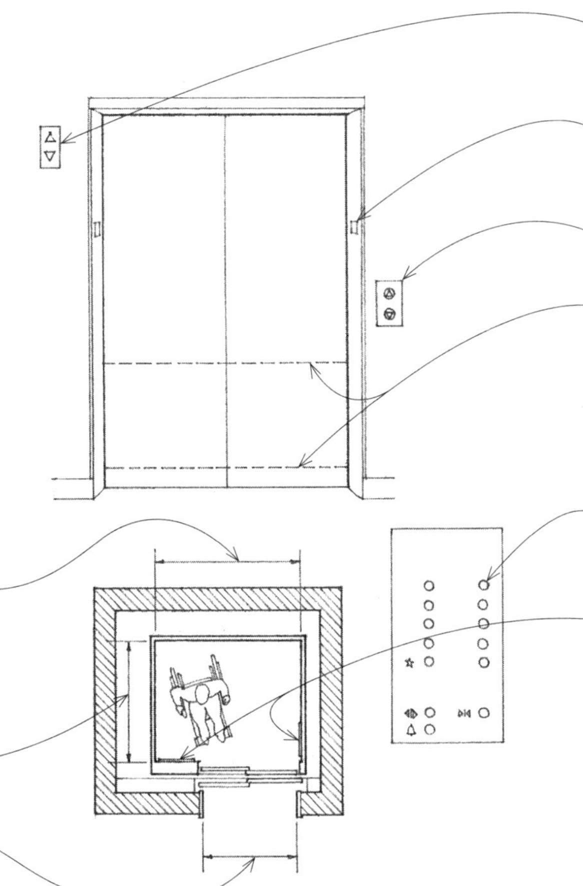

- 68" (1725) minimum width for cars with side opening doors; 80" (2030) minimum for cars with center opening doors
- 51" (1295) minimum clear car depth
- 36" (915) minimum clear doorway opening

ADA Accessibility Guidelines

- Visible and audible call signals or lanterns should be centered at least 72" (1830) above the floor at each hoistway entrance and be visible from the adjacent floor area.
- Raised characters and Braille floor designations should be provided on both jambs of elevator hoistway entrances and be centered at 60" (1525) above the floor.
- Call buttons for requesting an elevator should be centered 42" (1065) above the floor in each elevator lobby.
- Elevator doors should be provided with an automatic reopening device if the door becomes obstructed by an object or person.

- Elevator cars should be sized to allow wheelchair users to enter the car, maneuver within reach of controls, and exit from the car.
- Control buttons should be 3/4" (19) in the minimum dimension, be arranged with numbers in ascending order, with columns of numbers reading from left to right.
- Floor buttons should be located at least 35" (890) above the floor and be no higher than 48" (1220) for front approach and 54" (1370) for parallel approach.
- Raised and Braille designations should be placed immediately to the left of the button to which the designations apply.
- Audible and visible car position indicators should be provided in each elevator car.

Escalators are power-driven stairways consisting of steps attached to a continuously circulating belt. They can move a large number of people efficiently and comfortably between a limited number of floors; six floors are a practical limit. Because escalators move at a constant speed, there is practically no waiting period, but there should be adequate queuing space at each loading and discharge point. Escalators may not be used as required fire exits.

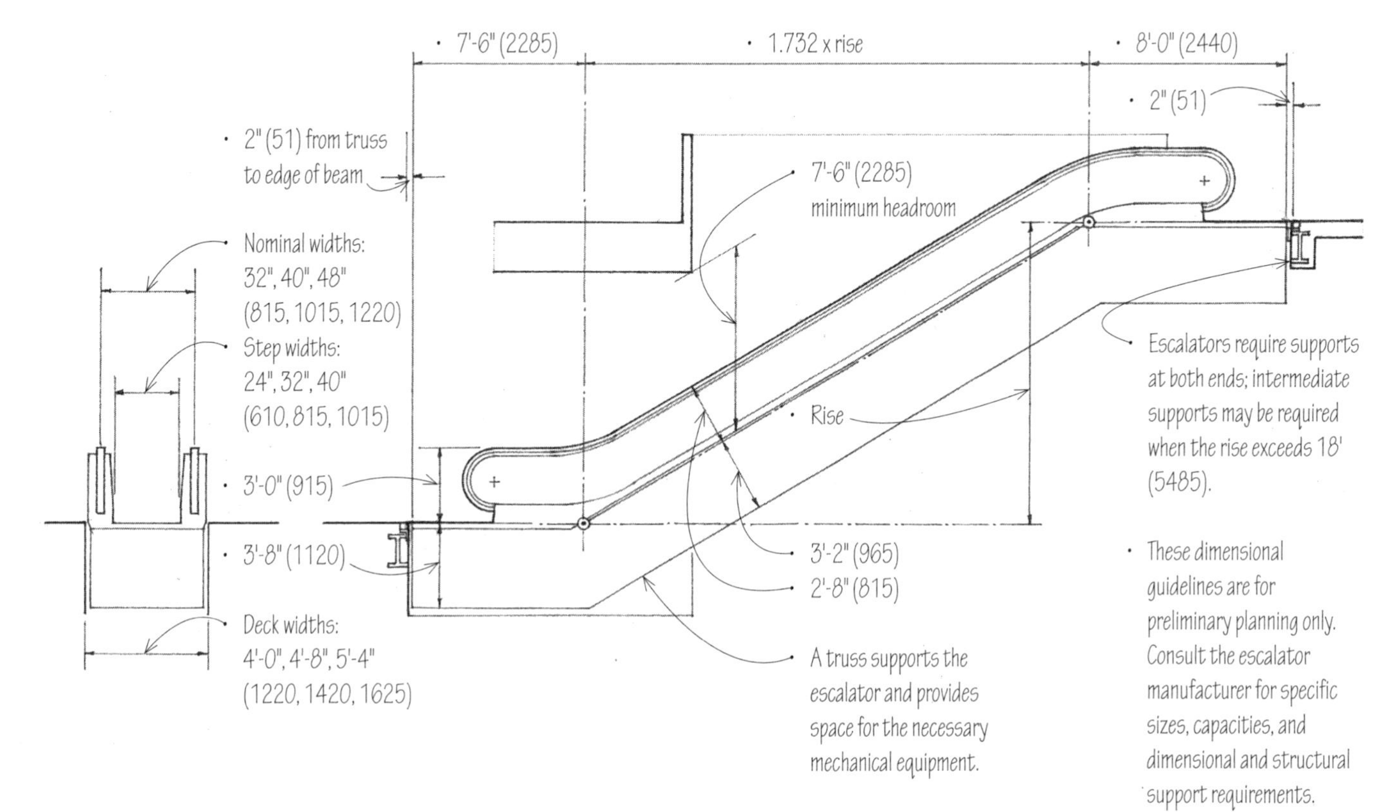

- Moving sidewalks are power-driven, continuously moving surfaces, similar to a conveyor belt, used for carrying pedestrians horizontally or along low inclines.

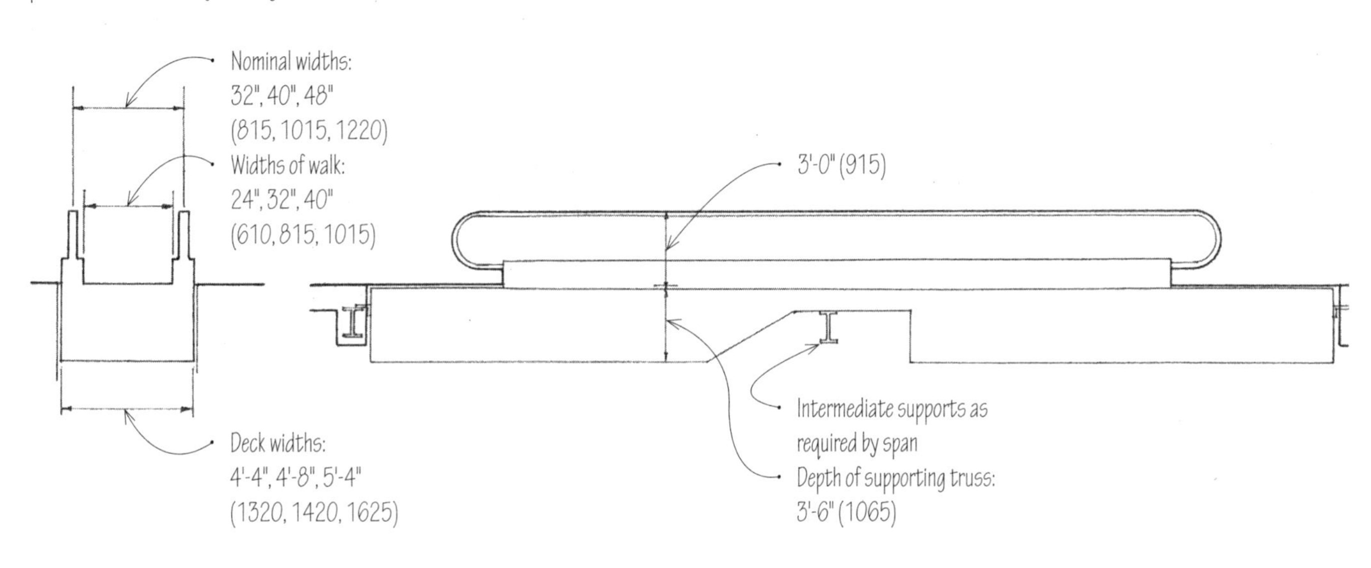

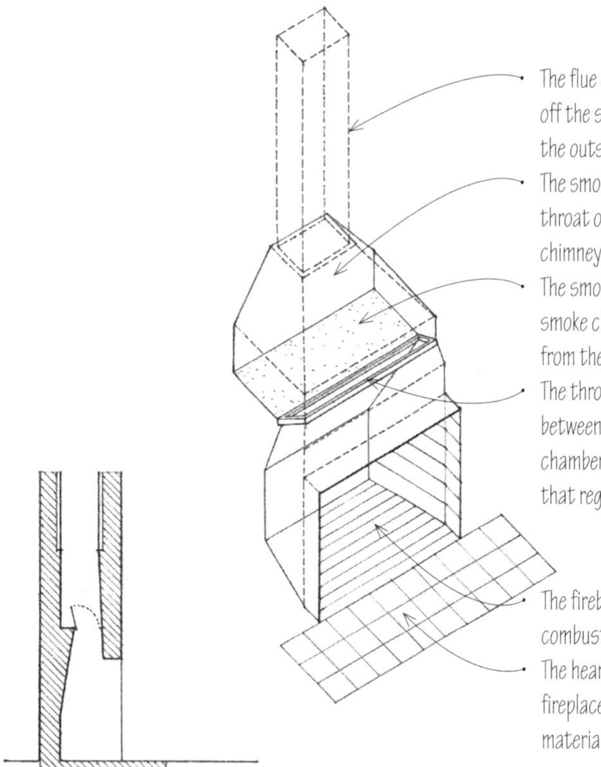

- The flue creates a draft and carries off the smoke and gases of a fire to the outside.
- The smoke chamber connects the throat of a fireplace to the flue of a chimney.
- The smoke shelf at the bottom of a smoke chamber deflects downdrafts from the chimney.
- The throat is the narrow opening between a firebox and the smoke chamber; it is fitted with a damper that regulates the draft in a fireplace.
- The firebox is the chamber where combustion takes place.
- The hearth extends the floor of a fireplace with a noncombustible material, such as brick, tile, or stone.

Open Front

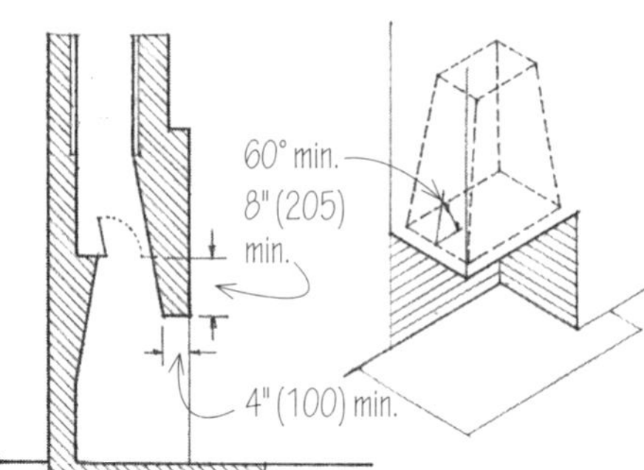

Open Front and Side

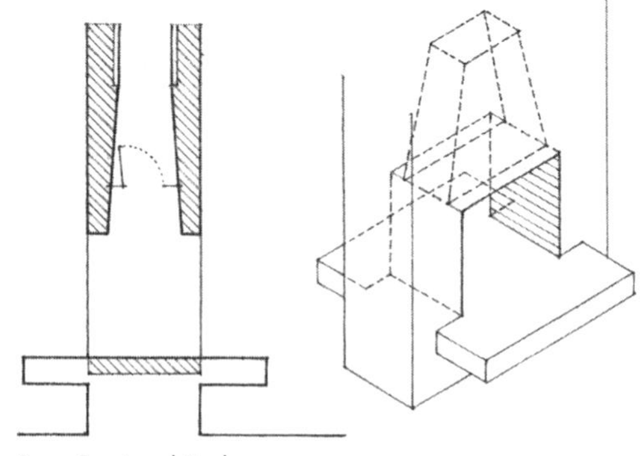

Open Front and Back

Types of Fireplaces

A fireplace is a framed opening in a chimney to hold an open fire. It must be designed and constructed to:

- Sustain the combustion of fuel;
- Draw properly to carry smoke and other combustive by-products to the outside;
- Radiate the maximum amount of heat comfortably into the room;
- Ensure proper distances from combustible materials.

Thus the dimensions and proportions of a fireplace and its flue and the arrangement of its components, are subject to the laws of nature and the requirements of the building and mechanical codes. The table below provides typical dimensions for three types of fireplaces. See facing page for key to letter heads (A) through (G).

Typical Fireplace Dimensions in Inches (mm)

Width (A)	Height (B)	Depth (C)	(D)	(E)	(F)	(G)	Flue Size
Open Front							
36 (915)	29 (735)	20 (510)	23 (560)	14 (355)	23 (560)	44 (1120)	12 x 12 (305 x 305)
42 (1065)	32 (815)	20 (510)	29 (735)	16 (405)	24 (610)	50 (1270)	16 x 16 (405 x 405)
48 (1220)	32 (815)	20 (510)	33 (840)	16 (405)	24 (610)	56 (1420)	16 x 16 (405 x 405)
54 (1370)	37 (940)	20 (510)	37 (940)	16 (405)	29 (735)	68 (1725)	16 x 16 (405 x 405)
60 (1525)	40 (1015)	22 (560)	42 (1065)	18 (455)	30 (760)	72 (1830)	16 x 20 (405 x 510)
72 (1830)	40 (1015)	22 (560)	54 (1370)	18 (455)	30 (760)	84 (2135)	20 x 20 (510 x 510)
Open Front and Side							
28 (710)	24 (610)	16 (405)					12 x 12 (305 x 305)
32 (815)	28 (710)	18 (455)					12 x 16 (305 x 405)
36 (915)	30 (760)	20 (510)					12 x 16 (305 x 405)
48 (1220)	32 (815)	22 (559)					16 x 16 (405 x 405)
Open Front and Back							
28 (710)	24 (610)	16 (405)					12 x 12 (305 x 305)
32 (815)	28 (710)	16 (405)					12 x 16 (305 x 405)
36 (915)	30 (760)	17 (430)					12 x 16 (305 x 405)
48 (1220)	32 (815)	19 (485)					16 x 16 (405 x 405)

Multifaced fireplaces are especially sensitive to drafts in a room; avoid placing their openings opposite an exterior door.

CSI MasterFormat 04 57 00: Masonry Fireplaces

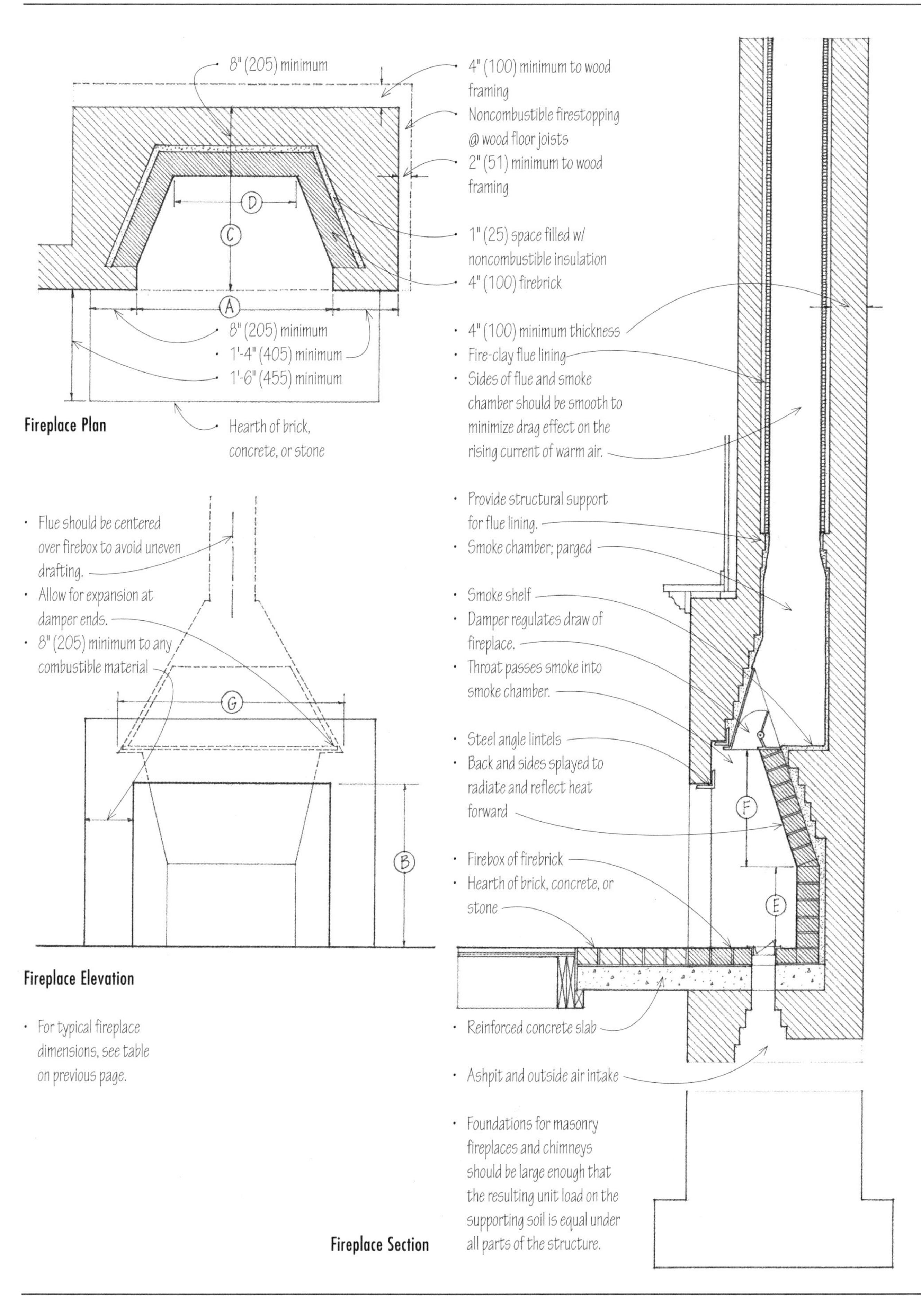
8" (205) minimum
4" (100) minimum to wood framing
Noncombustible firestopping @ wood floor joists
2" (51) minimum to wood framing
D
C
1" (25) space filled w/ noncombustible insulation
4" (100) firebrick
A
8" (205) minimum
1'-4" (405) minimum
1'-6" (455) minimum
Hearth of brick, concrete, or stone
Fireplace Plan
4" (100) minimum thickness
Fire-clay flue lining
Sides of flue and smoke chamber should be smooth to minimize drag effect on the rising current of warm air.
Provide structural support for flue lining.
Smoke chamber; parged
Smoke shelf
Damper regulates draw of fireplace.
Throat passes smoke into smoke chamber.
Steel angle lintels
Back and sides splayed to radiate and reflect heat forward
F
Firebox of firebrick
Hearth of brick, concrete, or stone
E
Reinforced concrete slab
Ashpit and outside air intake
Foundations for masonry fireplaces and chimneys should be large enough that the resulting unit load on the supporting soil is equal under all parts of the structure.
Fireplace Section
Flue should be centered over firebox to avoid uneven drafting.
Allow for expansion at damper ends.
8" (205) minimum to any combustible material
G
B
Fireplace Elevation
For typical fireplace dimensions, see table on previous page.

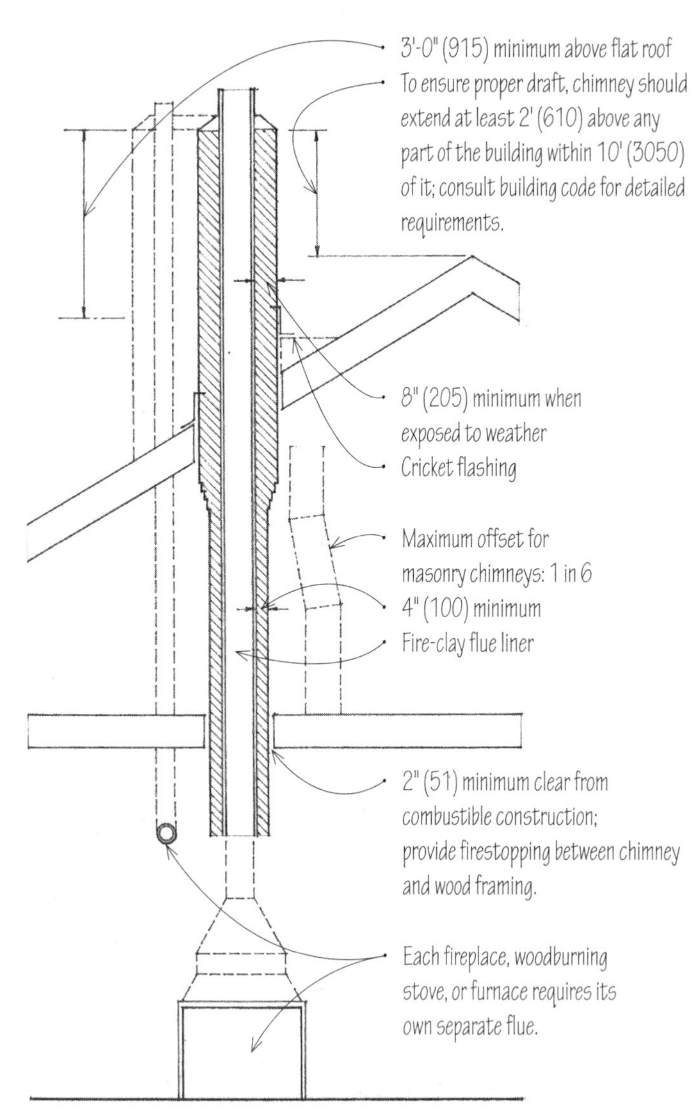

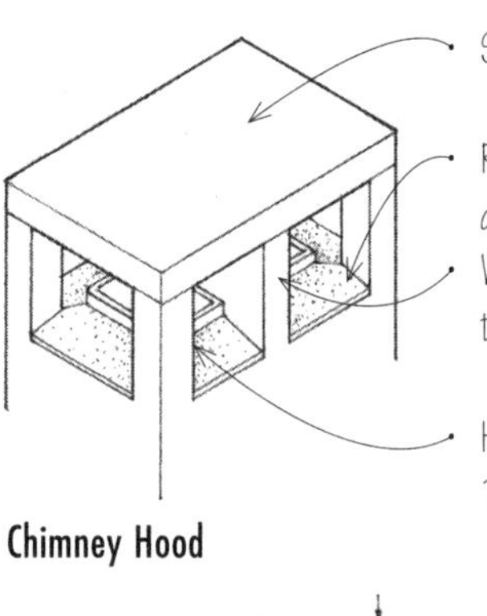

Chimney Hood

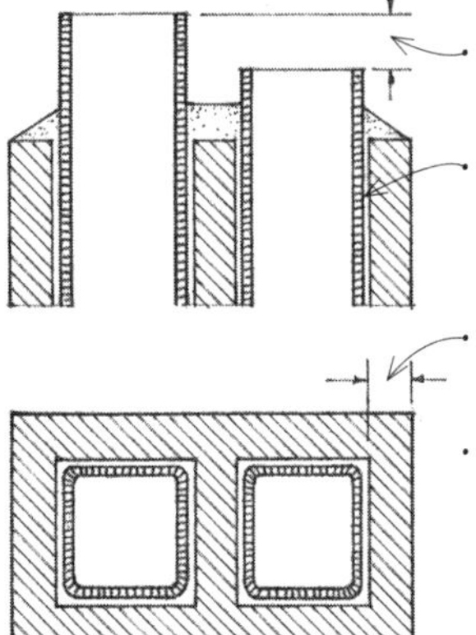

- Consult building or mechanical code for chimney requirements of high heat appliances such as incinerators.

- In certain seismic zones, masonry chimneys require reinforcement and anchorage to the structural frame of a building. Consult the building code for detail requirements.

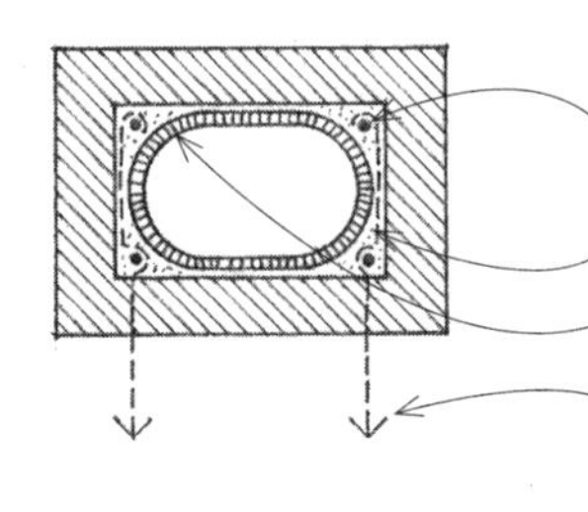

- Flue linings are smooth-surfaced units of heat-resistant fire clay or lightweight concrete.

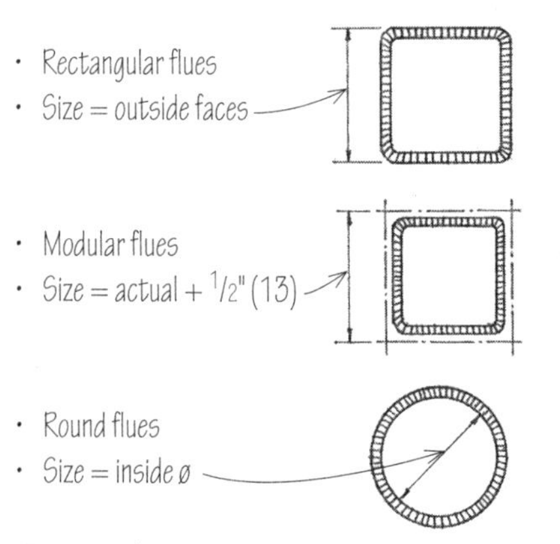

- Rectangular flues
- Size = outside faces

- Modular flues
- Size = actual + 1/2" (13)

- Round flues
- Size = inside ø

Minimum Flue Sizes

- Square or rectangular: 1/10th of fireplace opening
- Round: 1/12th of fireplace opening

Typical Flue Dimensions and Areas

Round		Rectangular		Modular	
Diameter in. (mm)	Area sq.in.*	Size in. (mm)	Area sq.in.*	Size in. (mm)	Area sq.in.*
8 (205)	47	8 1/2 x 8 1/2 (215 x 215)	51	8 x 12 (205 x 305)	57
10 (255)	74	8 1/2 x 13 (215 x 330)	79	12 x 12 (305 x 305)	87
12 (305)	108	13 x 13 (330 x 330)	125	12 x 16 (305 x 405)	120
15 (380)	171	13 x 18 (330 x 455)	168	16 x 16 (405 x 405)	162
18 (455)	240	18 x 18 (455 x 455)	232	16 x 20 (405 x 510)	208
20 (510)	298	20 x 20 (510 x 510)	279	20 x 20 (510 x 510)	262

* One square inch = 645.16 mm^2

CSI MasterFormat 04 51 00: Flue Liner Masonry

Prefabricated fireplaces and woodburning stoves should be certified by the Environmental Protection Agency (EPA) for burning efficiency and allowable particulate emissions.

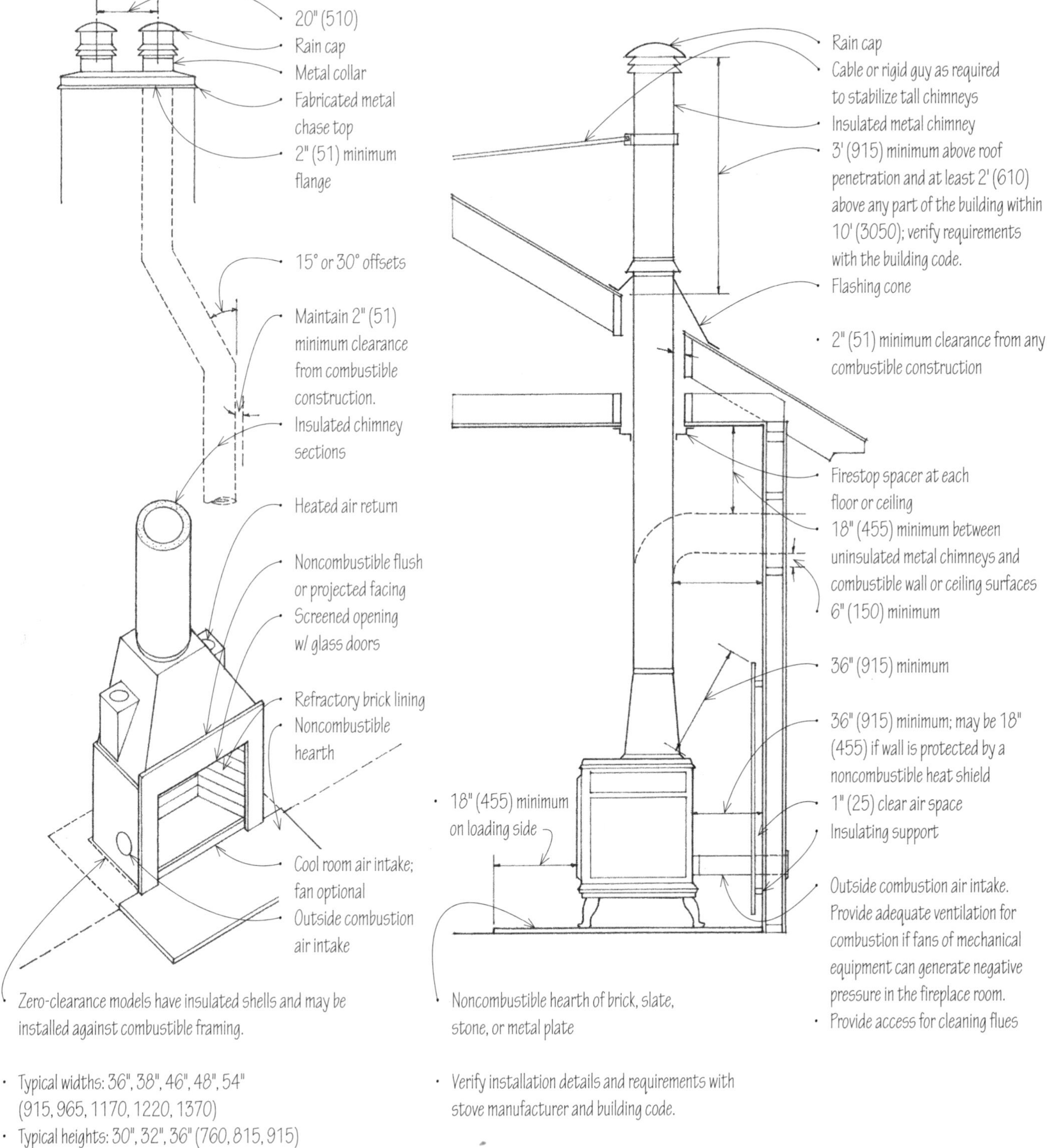

- Typical widths: 36", 38", 46", 48", 54" (915, 965, 1170, 1220, 1370)
- Typical heights: 30", 32", 36" (760, 815, 915)
- Typical depth: 24" (610)

Prefabricated Fireplaces

- Verify installation details and requirements with stove manufacturer and building code.

Woodburning Stoves

CSI MasterFormat 10 31 00: Manufactured Fireplaces
CSI MasterFormat 10 35 00: Stoves

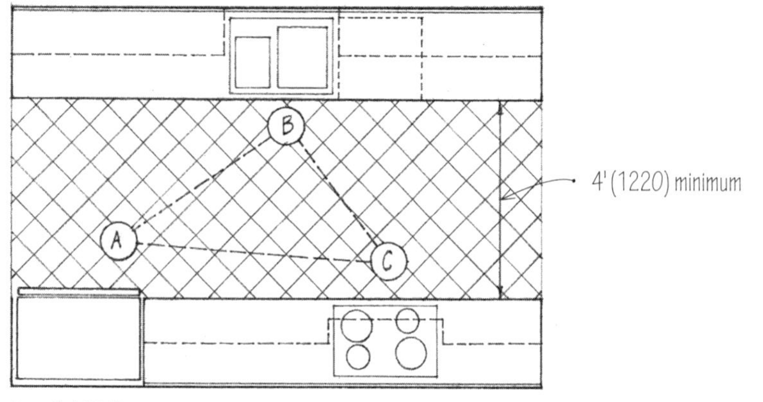

These plans illustrate the basic types of kitchen layouts. They can be readily adapted to various structural or spatial situations, but they are all based on a work triangle that connects the three major kitchen centers:

(A) Refrigerator center for receiving and preparing food
(B) Sink center for food preparation and clean up
(C) Range center for cooking and serving

The sum of the sides of the triangle should be not more than 22' (6705) nor less than 12' (3660).

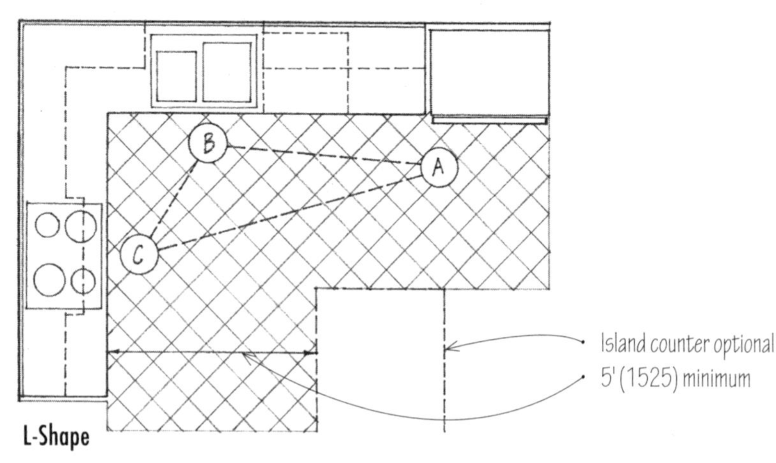

Additional factors to consider in laying out a kitchen space include:

- Amount of counter space and work surfaces required
- Type and quantity of under-counter and overhead storage required
- Requirements for natural light, views, and ventilation
- Type and degree of access desired
- Degree of enclosure envisioned for the space
- Integration of electrical, plumbing, and mechanical systems

Parallel Walls

L-Shape

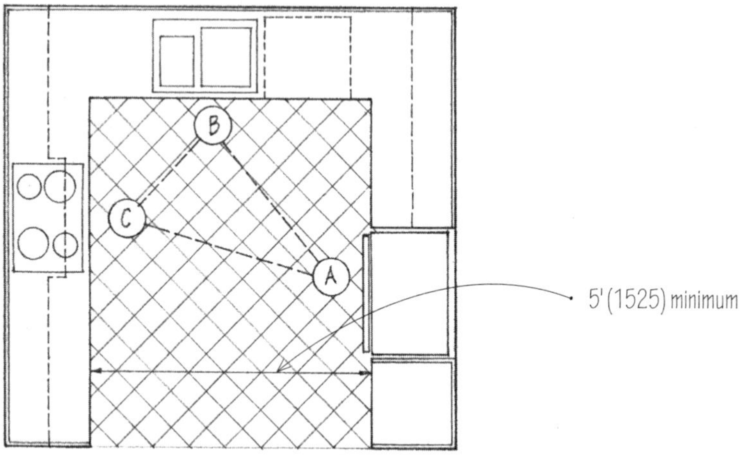

U-Shape

Single Wall

- Provide a 60" (1525) diameter space for turning a wheelchair in U-shaped kitchens.
- Provide at least one work surface 36" (915) wide, adjustable in height from 28" to 36" (710 and 915), or fixed at a height of 34" (865) max. above the floor.
- See also A.03 for general accessibility requirements.

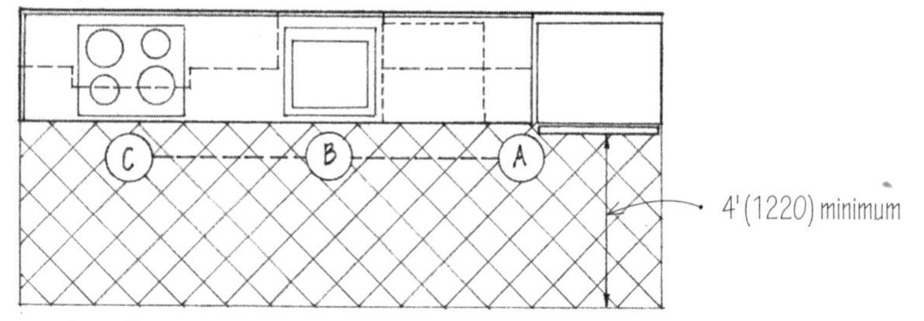

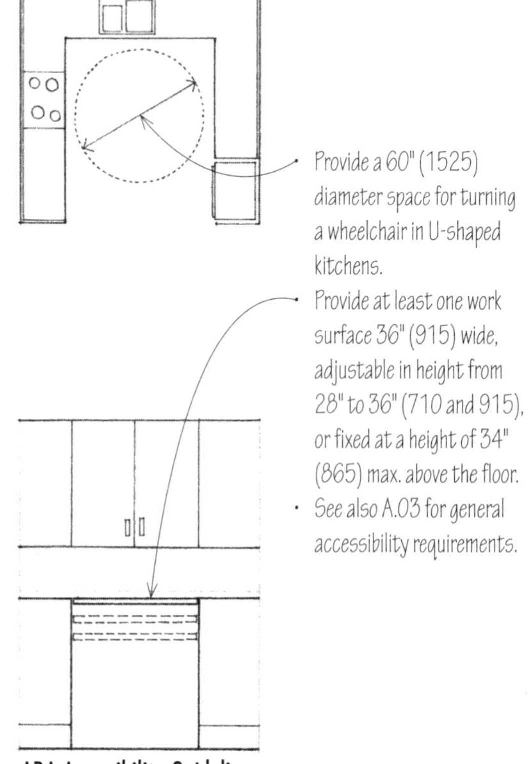

ADA Accessibility Guidelines

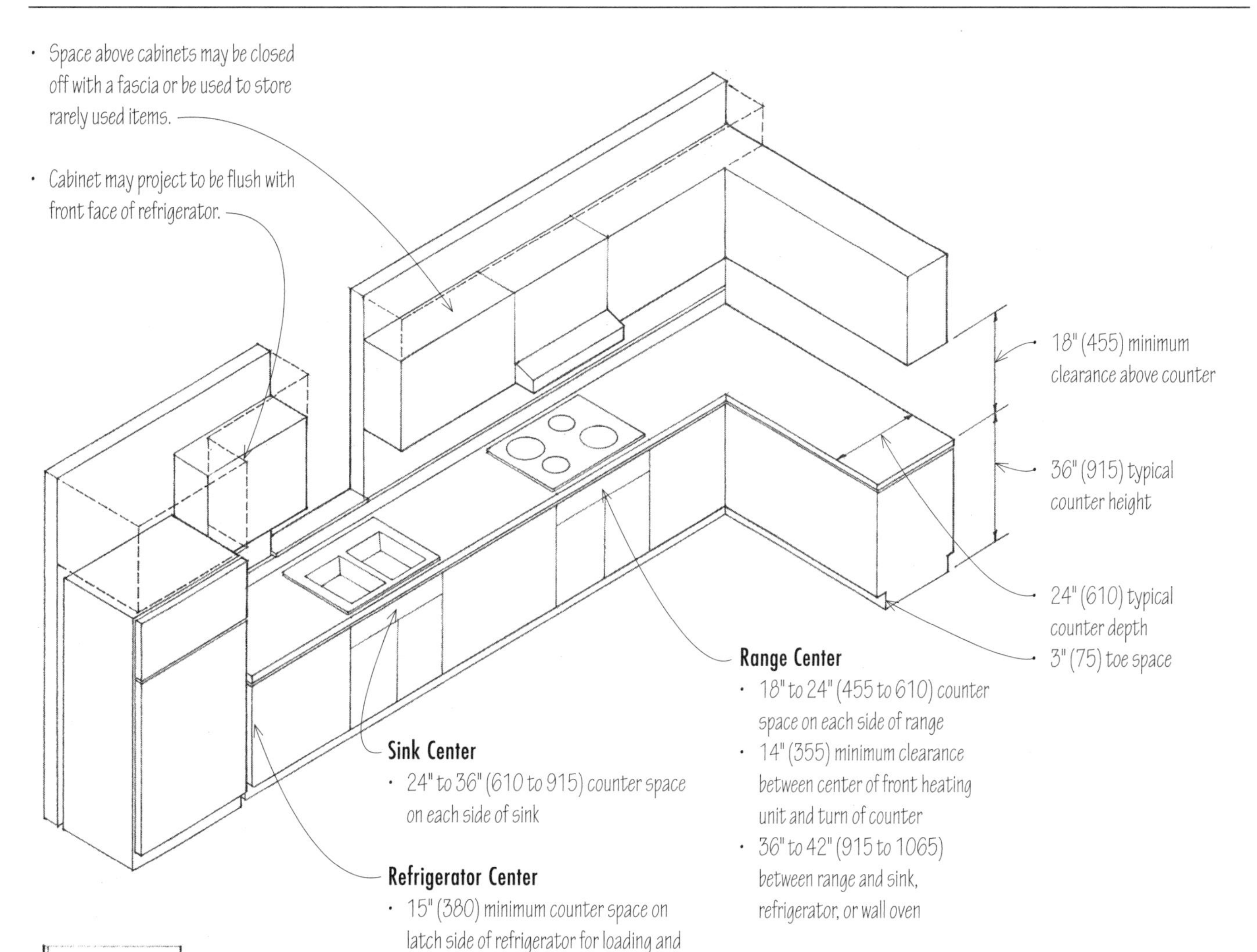

Range Center
- 18" to 24" (455 to 610) counter space on each side of range
- 14" (355) minimum clearance between center of front heating unit and turn of counter
- 36" to 42" (915 to 1065) between range and sink, refrigerator, or wall oven

Sink Center
- 24" to 36" (610 to 915) counter space on each side of sink

Refrigerator Center
- 15" (380) minimum counter space on latch side of refrigerator for loading and unloading
- 18" (455) minimum counter space between latch side of refrigerator and turn of counter

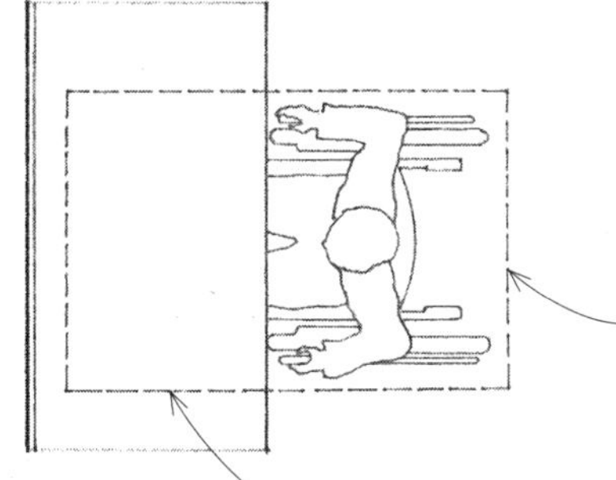

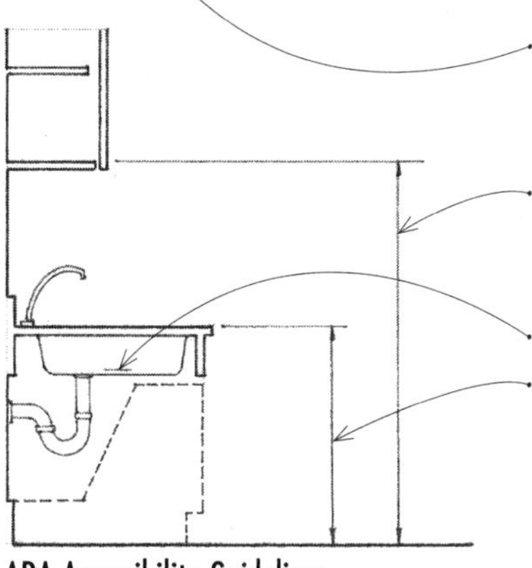

- 30" x 48" (760 x 1220) minimum clear floor space should be provided at sink, accessible work surface, and all appliances.
- Clear floor space may extend up to 19" (485) under the sink, accessible work surface, or appliance.
- At least one shelf of all cabinets mounted above work counters should be no more than 48" (1220) above the floor.
- 6-1/2" (165) maximum depth of bowl
- Rim of sink and surrounding counter should be adjustable in height from 28" to 36" (710 to 915), or be fixed at a height of 34" (865) maximum.

ADA Accessibility Guidelines

Appliances

Verify appliance dimensions when planning a kitchen layout. For preliminary planning purposes, the following range of widths can be used:
- Range: 33" to 40" (840 to 1015)
- Refrigerator: 32" to 36" (815 to 915)
- Dishwasher: 24" (610)
- Sink: 32" to 42" (815 to 1065)

- Countertop dimensions should be coordinated with standard cabinet sizes; see 9.24.

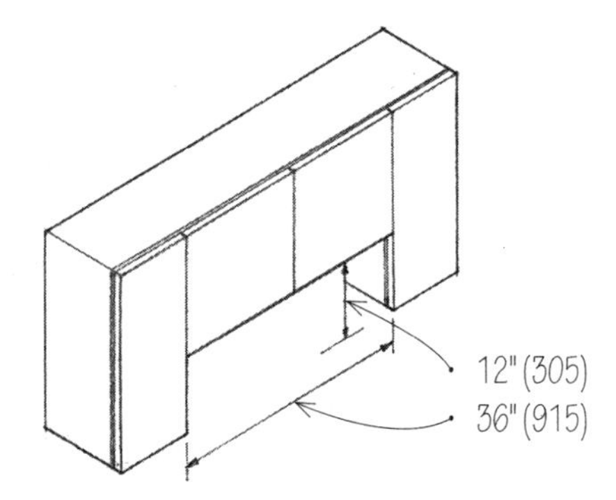

Combination Wall Unit
- For use over sinks and ranges
- 60" to 84" (1525 to 2135) long
- 30" (760) high

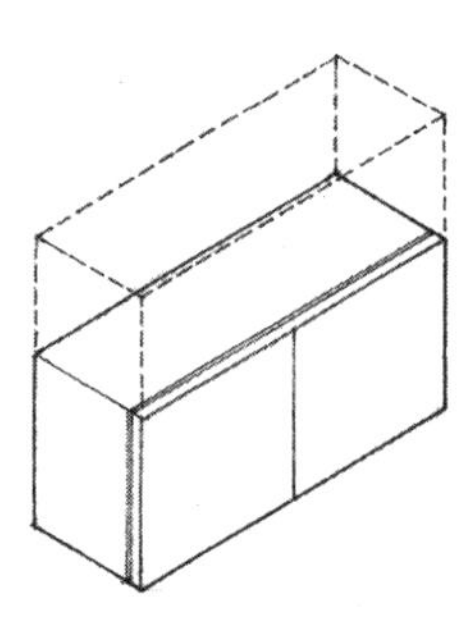

Basic Wall Unit
- 24" to 48" (610 to 1220) long in 3" (75) increments
- 12" to 33" (305 to 840) high

Kitchen cabinets may be constructed of wood or enameled steel. Wood cabinets usually have hardwood frames and plywood or particle board panels with plastic laminate, hardwood veneer, or lacquer finishes.

Stock kitchen cabinets are manufactured in 3" (75) modules and should conform to standards established by the National Kitchen Cabinet Association (NKCA). There are three basic types of units: base units, wall units, and special units. Consult manufacturer for available sizes, finishes, hardware, and accessories.

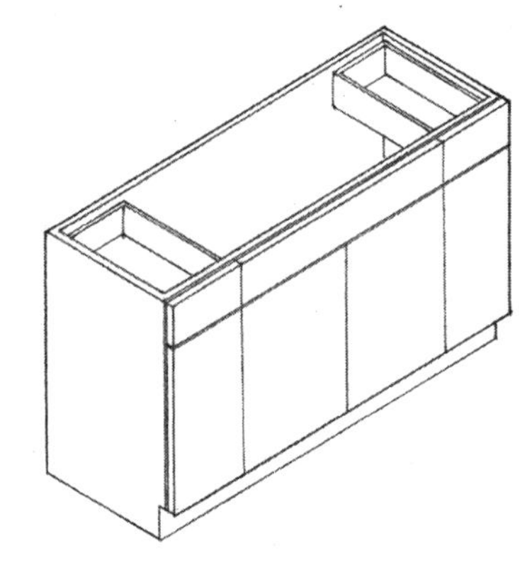

Sink Base Unit
- 54" to 84" (1370 to 2135) long in 3" (75) increments

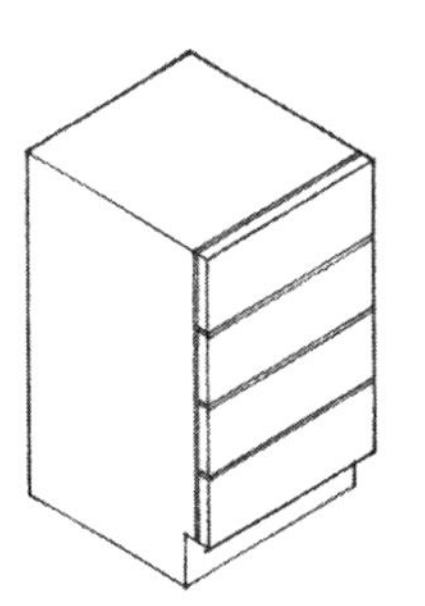

Drawer Unit
- 15" to 24" (380 to 610) wide

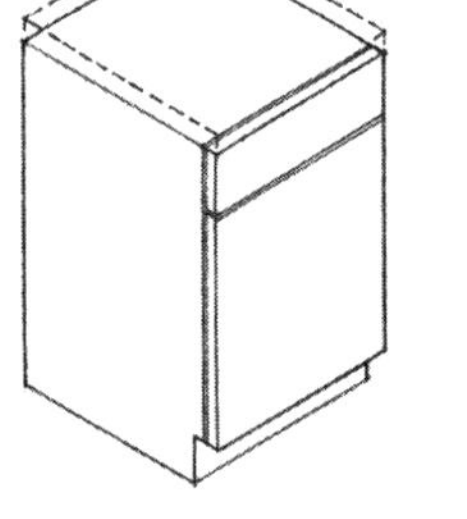

Basic Base Unit
- 12" to 24" (305 to 610) wide for one-door units
- 27" to 48" (685 to 1220) wide for two-door units
- 23" or 24" (585 or 610) deep

- 34-1/2" (875) height of base units allows for countertops up to 1-1/2" (38) thick.
- Base units for bathroom vanities are 30" (760) high and 21" (535) deep.
- Base units for buffets and desks are 28-1/2" (725) high.

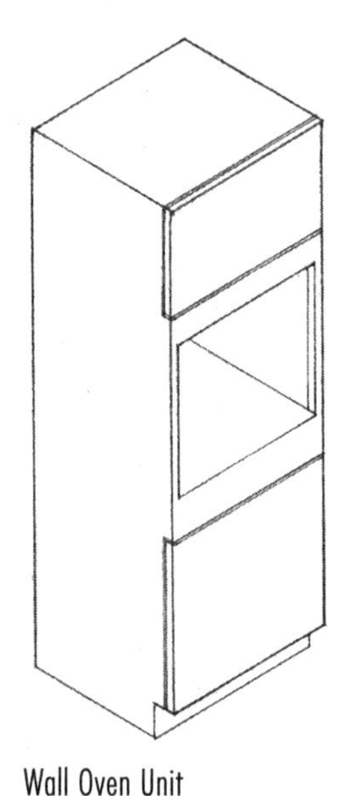

Wall Oven Unit
- 18" to 30" (455 to 760) wide
- 84" (2135) tall

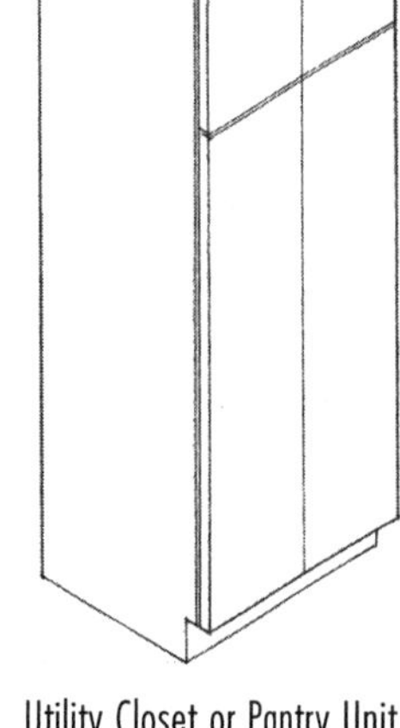

Utility Closet or Pantry Unit
- 12" and 24" (305 and 610) depths

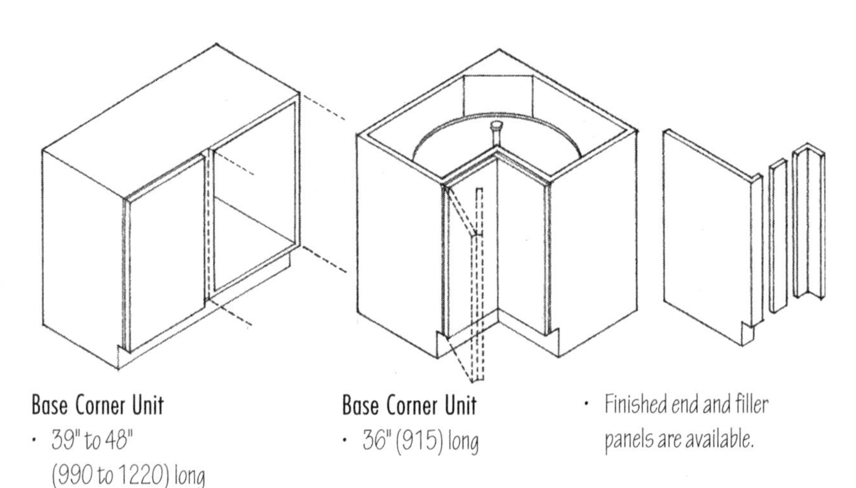

Base Corner Unit
- 39" to 48" (990 to 1220) long

Base Corner Unit
- 36" (915) long

- Finished end and filler panels are available.

CSI MasterFormat 12 35 30.13: Kitchen Casework

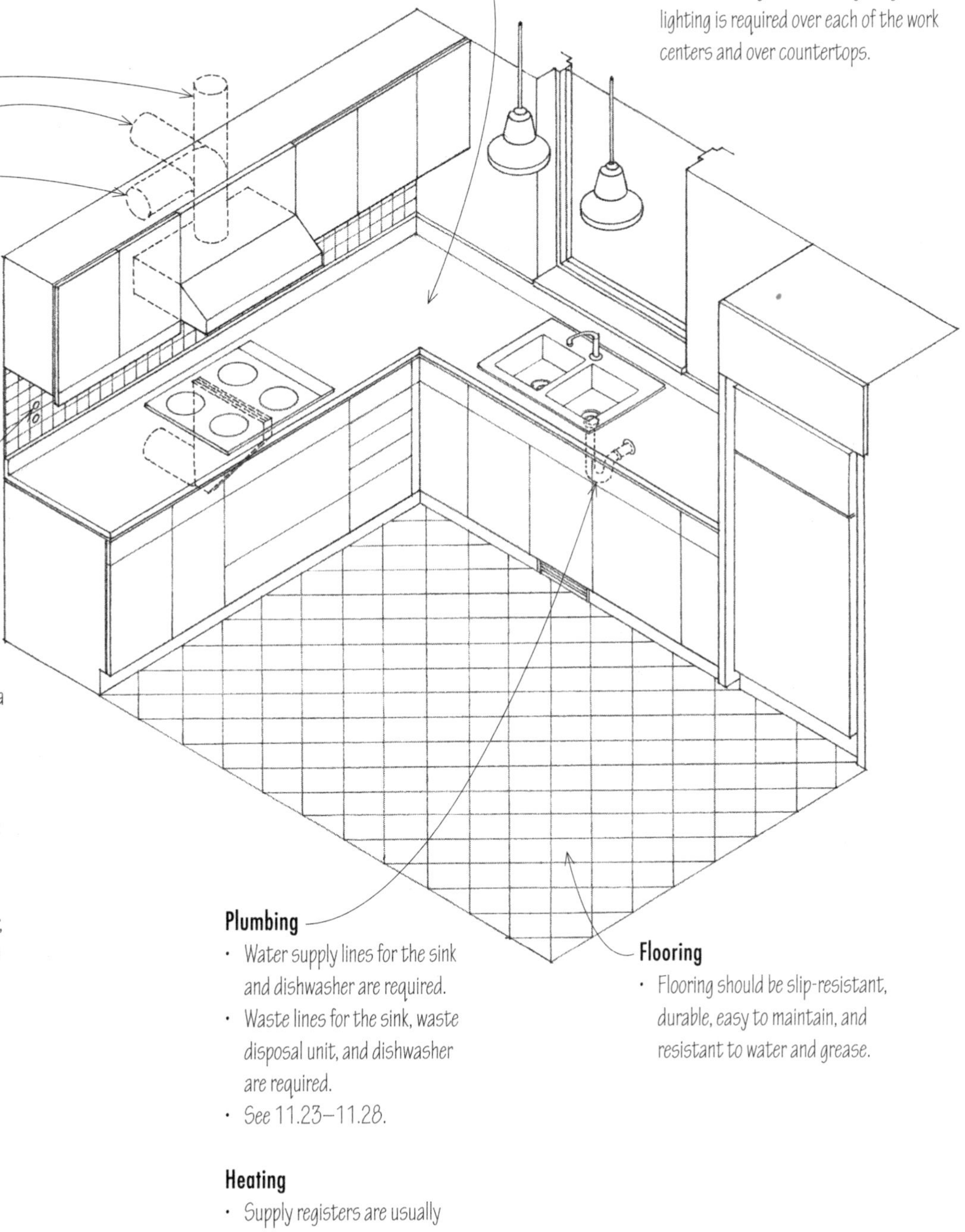

Ventilation
- Provide natural ventilation by means of openable exterior openings with an area not less than 1/20 of the floor area with a minimum of 5 s.f. (0.46 m^2).
- A mechanical ventilating system providing a minimum of two air changes per hour may be employed in lieu of natural ventilation.
- Range center may be ventilated by a hood with an exhaust fan:
  - Vertically through roof
  - Directly through exterior wall
  - Horizontally to outside through soffit above wall cabinets.
- Self-venting cooktops may exhaust directly to outside or, if in an interior location, through a duct in the floor system.

Electrical
- A minimum of two circuits for small appliances should be provided with outlets spaced 4' (1220) o.c. and about 6" (150) above the countertop. These circuits should be protected by a ground fault interrupter (GFI).
- Special, single-outlet circuits are required for permanently installed appliances such as electric ranges and ovens.
- Separate circuits are also required for appliances such as the refrigerator, dishwasher, garbage disposal unit, and microwave oven.

Gas
- Gas appliances require separate fuel supply lines.

Countertop Surfaces
- The countertop surface may be plastic laminate, butcher block, ceramic tile, marble or granite, synthetic stone, concrete, or stainless steel.
- Provide a heat-resistant surface next to the range.

Plumbing
- Water supply lines for the sink and dishwasher are required.
- Waste lines for the sink, waste disposal unit, and dishwasher are required.
- See 11.23–11.28.

Heating
- Supply registers are usually located under base cabinets.

Lighting
- Provide natural light by means of exterior glazed openings with an area not less than 1/10 of the floor area or a minimum of 10 s.f. (0.93 m^2).
- The building code typically allows residential kitchens to be illuminated solely with artificial lighting.
- In addition to general area lighting, task lighting is required over each of the work centers and over countertops.

Flooring
- Flooring should be slip-resistant, durable, easy to maintain, and resistant to water and grease.

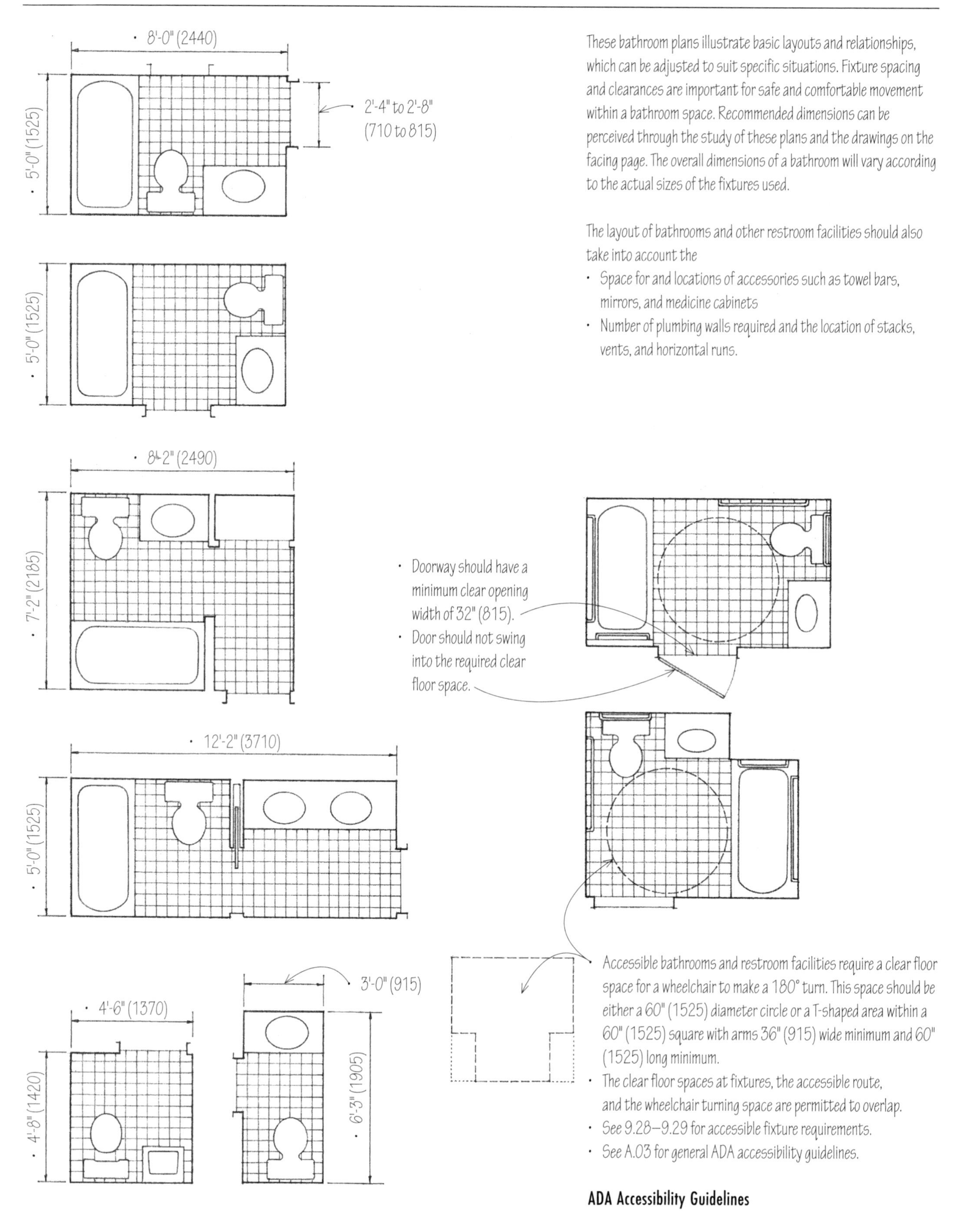

These bathroom plans illustrate basic layouts and relationships, which can be adjusted to suit specific situations. Fixture spacing and clearances are important for safe and comfortable movement within a bathroom space. Recommended dimensions can be perceived through the study of these plans and the drawings on the facing page. The overall dimensions of a bathroom will vary according to the actual sizes of the fixtures used.

The layout of bathrooms and other restroom facilities should also take into account the

- Space for and locations of accessories such as towel bars, mirrors, and medicine cabinets
- Number of plumbing walls required and the location of stacks, vents, and horizontal runs.

- Doorway should have a minimum clear opening width of 32" (815).
- Door should not swing into the required clear floor space.

- Accessible bathrooms and restroom facilities require a clear floor space for a wheelchair to make a 180° turn. This space should be either a 60" (1525) diameter circle or a T-shaped area within a 60" (1525) square with arms 36" (915) wide minimum and 60" (1525) long minimum.
- The clear floor spaces at fixtures, the accessible route, and the wheelchair turning space are permitted to overlap.
- See 9.28–9.29 for accessible fixture requirements.
- See A.03 for general ADA accessibility guidelines.

ADA Accessibility Guidelines

The range of fixture dimensions given below is for preliminary planning purposes only. Consult the fixture manufacturer for actual dimensions of specific models.

Plumbing fixtures may be made of the following materials.

- Water closets, urinals, and bidets: vitreous china
- Lavatories, bathtubs, and utility sinks: vitreous china, enameled cast iron, enameled steel
- Shower receptacles: terrazzo, enameled steel
- Shower enclosures: enameled steel, stainless steel, ceramic tile, fiberglass
- kitchen sinks: enameled cast iron, enameled steel, stainless steel

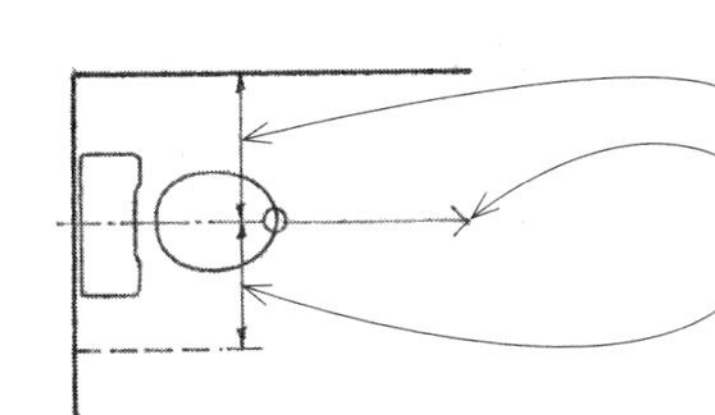

Water Closet
- 22" (559) to side wall; 15" (380) minimum
- 36" (915) to opposite wall; 18" (455) minimum
- 18" (455) to fixture; 12" (305) minimum
- Height of rim of water closet above floor: 14" to 15" (355 to 380)

Lavatory
- 22" (560) to side wall; 14" (355) minimum
- 30" (760) to opposite wall; 18" (455) minimum
- 6" (150) to fixture; 2" (51) minimum

Bathtub
- 34" (865) to opposite wall; 20" (508) minimum
- 8" (205) to fixture; 2" (51) minimum

Fixture Clearances

	Water Closet	Urinal	Bidet	Lavatory	Lavatory
Width	20" to 24" (510 to 610)	18" (455)	14" (355)	30" to 36" (760 to 915)	18" to 24" (455 to 610)
Depth	22" to 29" (560 to 735)	12" to 24" (305 to 610)	30" (760)	21" (535)	16" to 21" (405 to 535)
Height	20" to 28" (510 to 710)	24" (610) rim height	14" (355)	31" (785) above floor	31" (785) rim height

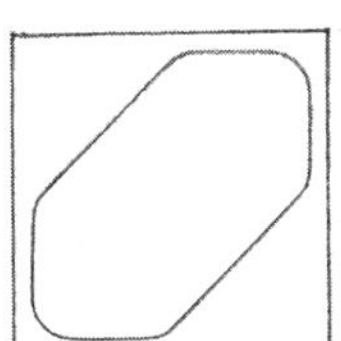
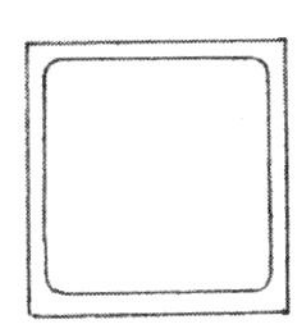
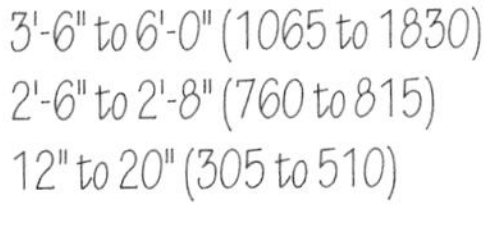

	Bathtub	Square Bathtub	Shower
Width	3'-6" to 6'-0" (1065 to 1830)	3'-8" to 4'-2" (1120 to 1270)	2'-6" to 3'-6" (760 to 1065)
Depth	2'-6" to 2'-8" (760 to 815)	3'-8" to 4'-2" (1120 to 1270)	2'-6" to 3'-6" (760 to 1065)
Height	12" to 20" (305 to 510)	12" to 16" (305 to 405)	6'-2" to 6'-8" (1880 to 2030)

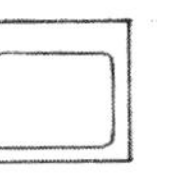
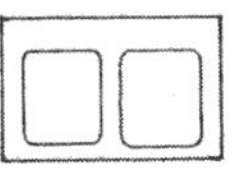
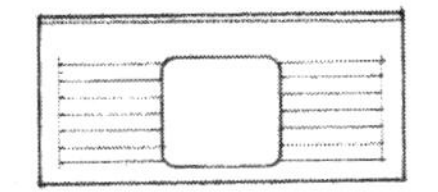

	Single Bowl Sink	Double Bowl Sink	Sink with Drainboards	Utility Sink
Width	12" to 33" (305 to 840)	28" to 46" (710 to 1170)	54" to 84" (1370 to 2135)	22" to 48" (560 to 1220)
Depth	13" to 21" (330 to 535)	16" to 21" (405 to 535)	21" to 25" (535 to 635)	18" to 22" (455 to 560)
Height	8" to 12" (205 to 305)	8" to 10" (205 to 255)	8" (205)	27" to 29" rim (685 to 735)

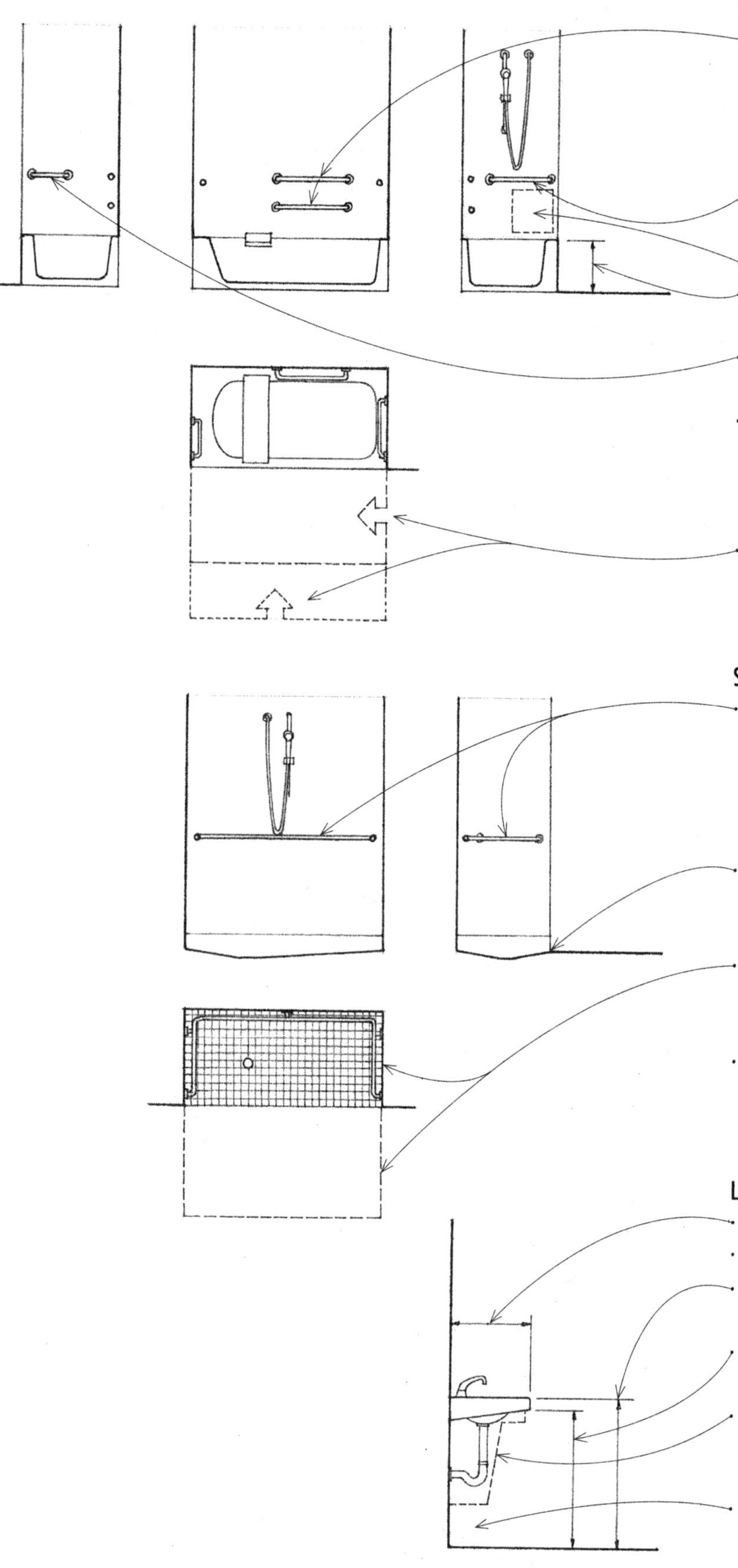

Bathtubs

- A grab bar 33" to 36" (840 to 915) above the floor and at least 24" (610) long should be installed on the back wall, 24" (610) maximum from the head end wall and 12" (305) maximum from the foot end wall. Another grab bar of equal length should be installed 9" (230) above the rim of the tub.
- A grab bar at least 24" (305) long should be installed on the foot end wall at the front edge of the tub.
- Control area
- Rim of tub should be 17" to 19" (430 to 480) above the floor.
- A grab bar at least 12" (305) long should be installed on the head end wall at the front edge of the tub.
- Diameter or width of grab bars should be 1-1/4" to 1-1/2" (32 to 38) with a 1-1/2" (38) space between the grab bar and the wall.
- 30" x 60" (760 x 1525) minimum clear floor space for a parallel approach to a bathtub and 48" x 60" (1220 x 1525) minimum for a forward approach

Shower Stalls

- Grab bars 33" to 36" (840 to 915) above the floor should be provided on three walls of roll-in type showers. For transfer type shower stalls, grab bar should extend across the control wall and back wall to a point 18" (455) from the control wall.
- 1/2" (13) maximum threshold beveled with a slope not steeper than 1:2
- 36" x 36" (915 x 915) minimum inside dimensions for transfer type shower stalls with 36" x 48" (915 x 1220) minimum clear floor space for access
- 30" x 60" (760 x 1525) minimum inside dimensions for roll-in type showers with 36" x 60" (915 x 1525) minimum clear floor space for access

Lavatories and Sinks

- 17" (430) minimum extension from wall
- 6-1/2" (165) maximum sink depth
- 34" (865) maximum rim height of lavatory or sink above the floor
- 29" (735) minimum clearance from the floor to the bottom of the front edge of the apron
- 8" (205) minimum depth for clear knee space at 27" (685) above the floor and 11" (280) minimum depth at 9" (230) above the floor
- 30" x 48" (760 x 1220) clear floor space should extend not more than 19" (485) under the sink or lavatory.

Water Closets

- Water closets should be mounted adjacent to a wall or partition. The distance from the centerline of the water closet to the wall or partition should be 18" (455).
- Top of toilet seat should be 17" to 19" (430 to 485) above the floor.
- 48" (1220) minimum clear floor space in front of water closet and 42" (1065) from the centerline of the water closet on the side not adjacent to a wall.

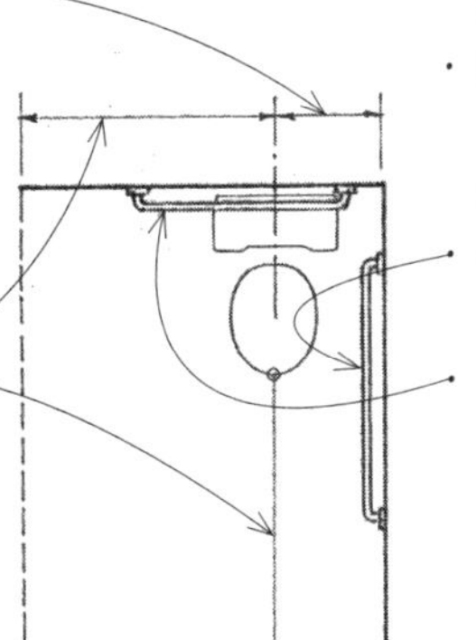

- Grab bars should be mounted in a horizontal position 33" to 36" (840 to 915) above the floor, on the rear wall and on the side wall closest to the water closet.
- Diameter or width of grab bars should be 1-1/4" to 1-1/2" (32 to 38) with a 1-1/2" (38) space between the grab bar and the wall.
- Grab bar on the side wall should be at least 42" (1065) long and 12" (305) from rear wall.
- Grab bar on the rear wall should be 24" (610) long minimum and centered on the water closet; where space permits, the bar should be 36" (915) long and extend on the transfer side of the water closet.

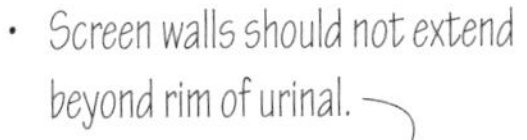

Toilet Stalls

- Wheelchair-accessible toilet stalls should be at least 60" (1525) wide and 56" (1420) deep minimum for wall-hung water closets and 59" (1500) deep minimum for floor-mounted water closets.
- Depth should be increased by 36" (915) if door swings into stall.
- Grab bars should be mounted in a horizontal position 33" to 36" (840 to 915) above the floor, on a rear wall and on the side wall closest to the water closet. See details above.
- Ambulatory-accessible stalls should be at least 36" (915) wide, 60" (1525) deep, and provided with grab bars on both sides of the stall.

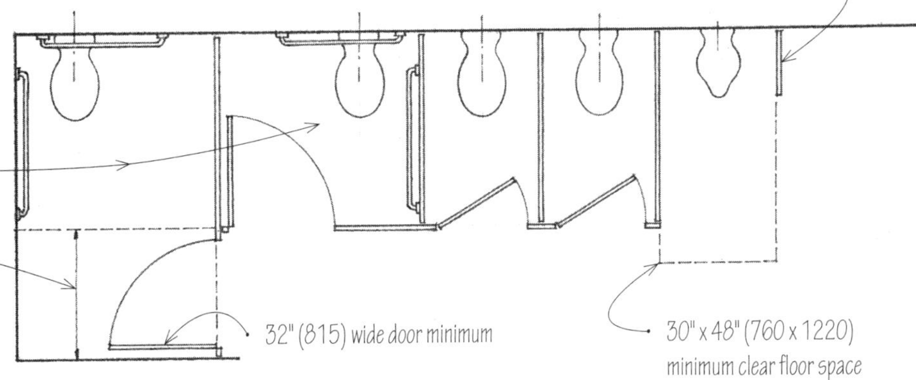

- 32" (815) wide door minimum
- 30" x 48" (760 x 1220) minimum clear floor space

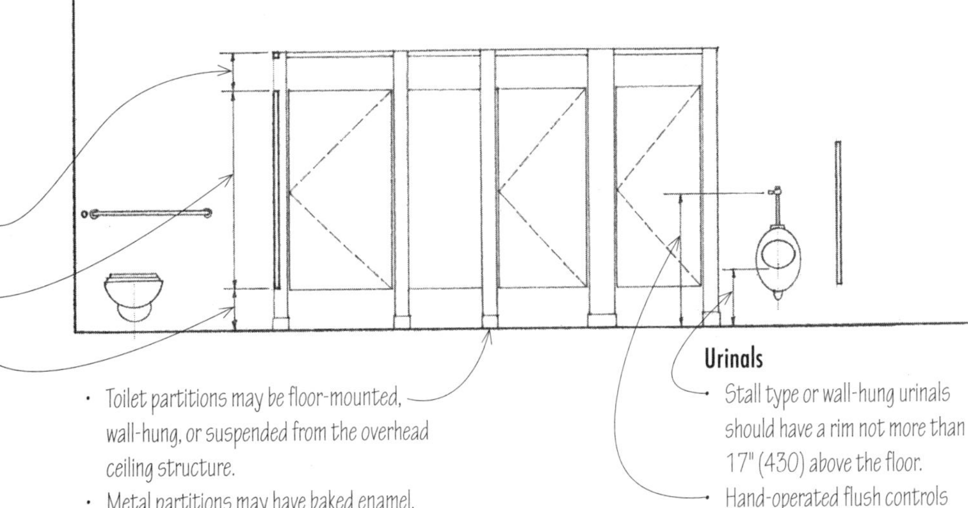

- 1'-0" (305)
- 4'-10" (1475)
- 1'-0" (305)

- Toilet partitions may be floor-mounted, wall-hung, or suspended from the overhead ceiling structure.
- Metal partitions may have baked enamel, porcelain enamel, or stainless steel finishes.
- Plastic laminate, tempered glass, and marble panels are also available.

Urinals

- Stall type or wall-hung urinals should have a rim not more than 17" (430) above the floor.
- Hand-operated flush controls should be mounted from 15" to 44" (380 to 1120) maximum above the floor.

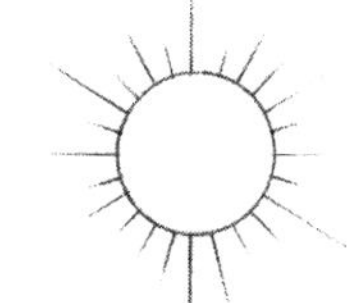

Lighting

- Natural lighting by means of exterior glazed openings is always desirable.
- A single overhead light fixture is usually not acceptable; auxiliary lighting is required over the tub or shower, over the lavatory and vanity counter, and over any compartmentalized toilet spaces.
- The light fixture over the tub or shower should be resistant to water vapor.

Ventilation

- Bathrooms require either natural or mechanical ventilation in order to remove stale air and supply fresh air.
- Provide natural ventilation by means of openable exterior openings with an area not less than 1/20 of the floor area or a minimum of 1-1/2 s.f. (0.14 m^2).
- A mechanical ventilating system may be employed in lieu of natural ventilation.
- The ventilating fan should be located close to the shower and high on an exterior wall opposite the bathroom door. It should be connected directly to the outside and be capable of providing five air changes per hour. The point of discharge should be at least 3' (915) away from any opening that allows outside air to enter the building.
- Residential exhaust fans are often combined with a light fixture, a fan-forced heater, or a radiant heat lamp.

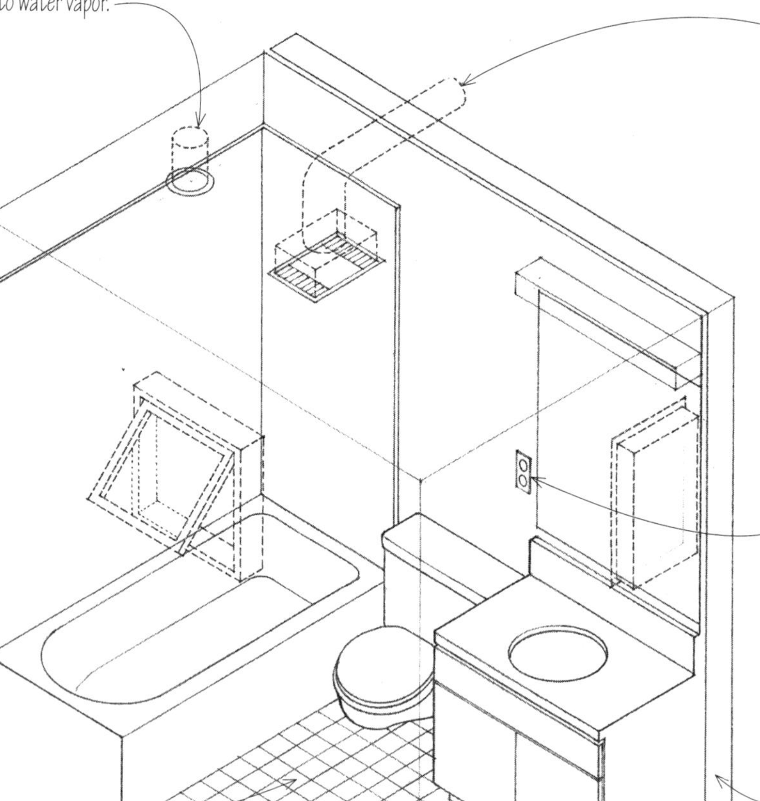

Electrical

- Electrical switches and convenience outlets should be located where they are needed but away from water or wet areas. They should not be accessible from a bathtub or shower.
- All convenience outlets should be protected by a ground fault interrupter (GFI); see 11.32.

Plumbing

- Plumbing walls should have sufficient depth to accommodate the required water supply and waste lines and vents.
- See 11.24–11.28.

- Space is required for accessories such as a medicine cabinet, mirror, towel bars, toilet paper holder, and soap dish.
- Storage space is required for towels, linen, and cleaning supplies.

Finishes

- Backing for bathtub or shower enclosure should be moisture resistant.
- All finishes should be durable, sanitary, and easy to clean, and flooring should have a nonslip surface.

Heating

- Heating may be supplied in the conventional manner through warm-air registers in the floor, hydronic or electric baseboard units, or electric resistance heaters in the wall.

CSI MasterFormat 12 35 30.23: Bathroom Casework

10

FINISH WORK

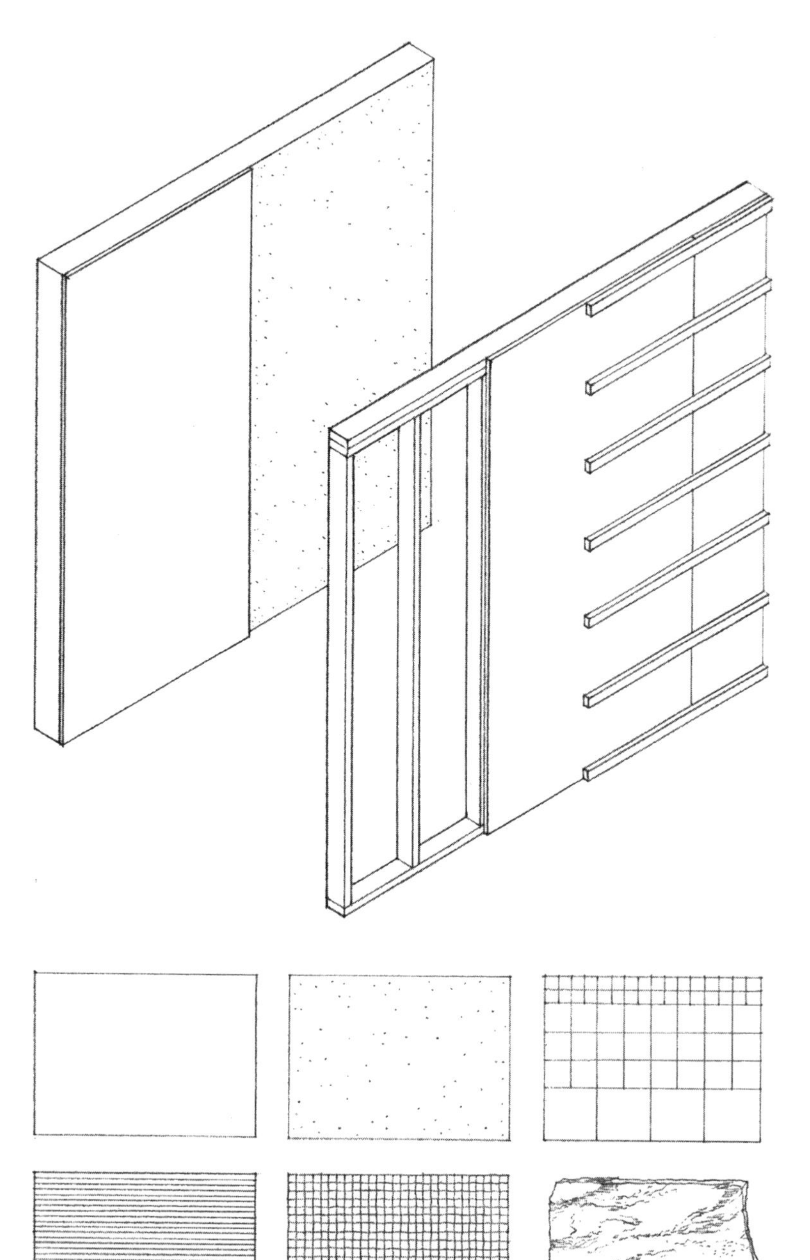

This chapter illustrates the major materials and methods used to finish the interior wall, ceiling, and floor surfaces of a building. Interior walls should be resistant to wear and be cleanable; floors should be durable, comfortable, and safe to walk on; ceilings should be relatively maintenance-free.

Because exterior wall surfaces, such as stucco and wood siding, must serve effectively as barriers against the penetration of water into the interior of a building, they are covered in Chapter 7 along with roof coverings.

Rigid finish materials capable of spanning short distances may be applied to a supporting grid of linear members. More flexible finish materials, on the other hand, require a solid, rigid backing. Additional technical factors to consider include the acoustic qualities, fire resistance, and thermal insulation value of a finish material.

Surface finishes have a critical influence on the aesthetic qualities of a space. In the selection and use of a finish material, we should carefully consider its color, texture, and pattern, and the way it meets and joins with other materials. If a finish material has modular characteristics, then its unit dimensions can be used to regulate the dimensions of a wall, floor, or ceiling surface.

LEED EQ Credit 4: Low-Emitting Materials

Plaster refers to any of various mixtures applied in a pasty form to the surfaces of walls or ceilings in a plastic state and allowed to harden and dry. The most common type of plaster used in construction is gypsum plaster, which is made by mixing calcined gypsum with water, fine sand or lightweight aggregate, and various additives to control its setting and working qualities. Gypsum plaster is a durable, relatively lightweight, and fire-resistant material that can be used on any wall or ceiling surface that is not subject to moist or wet conditions. Portland cement plaster, also known as stucco, is used on exterior walls and in areas subject to wet or moist conditions; see 7.36.

- Plaster is applied in layers, the number of which depends on the type and strength of base used.

Two-Coat Plaster

- Plaster is applied in two coats, a basecoat followed by a finish coat.

Three-Coat Plaster

- Plaster is applied in three successive coats, a scratch coat followed by a brown coat and a finish coat.

- Finish coat is the final coat of plaster, serving either as a finished surface or as a base for decoration.
- Brown coat is a roughly finished, leveling coat of plaster, either the second coat in three-coat plaster or the base coat in two-coat plaster applied over gypsum lath or masonry.
- Basecoat refers to any plaster coat applied before the finish coat.
- Scratch coat is the first coat in three-coat plaster, which must adhere firmly to the lath and be raked to provide a better bond for the second or brown coat.

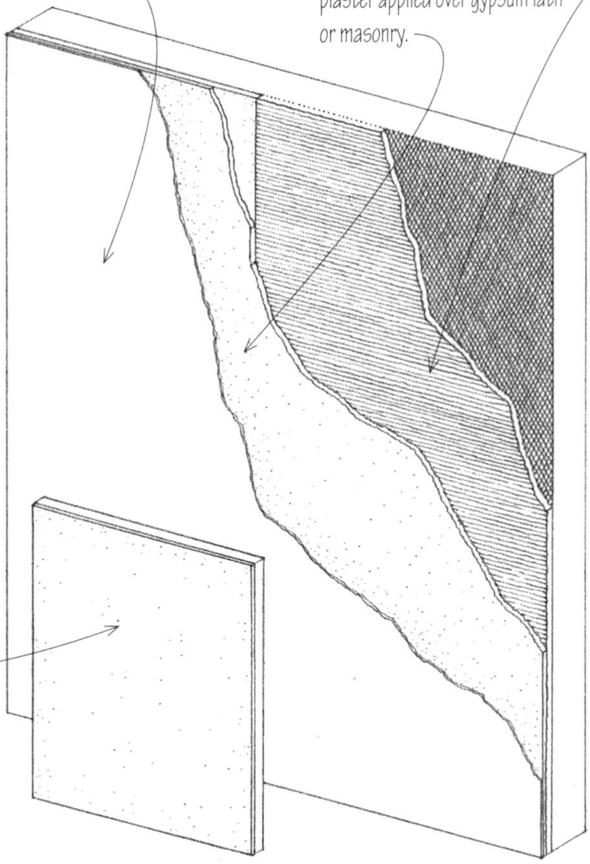

- Hard finish refers to a finish coat of lime putty and Keene's cement or gauging plaster, troweled to a smooth, dense finish.
- Keene's cement is a trademark for a brand of white anhydrous gypsum plaster that produces an exceptionally strong, dense, crack-resistant finish.
- Gauging plaster is a specially ground gypsum plaster for mixing with lime putty, formulated to control the setting time and counteract shrinkage in a finish coat of plaster.
- White coat refers to a finish coat of lime putty and white gauging plaster, troweled to a smooth, dense finish.

- Veneer or thin-coat plaster is a ready-mixed gypsum plaster applied as a very thin, one- or two-coat finish over a veneer base.

- Acoustical plaster is a low-density plaster containing vermiculite or other porous material to enhance its ability to absorb sound.
- Molding plaster, consisting of very finely ground gypsum and hydrated lime, is used for ornamental plasterwork.

- The final appearance of a plaster surface depends on both its texture and its finish. It may be troweled to produce a smooth, nonporous finish, floated to a sandy, lightly textured finish, or sprayed on for a rougher finish. The finish may be painted; smooth finishes will accept textile or paper wall coverings.

- Wood-fibered plaster is mill-mixed gypsum basecoat plaster containing coarse cellulose fibers for greater bulk, strength, and fire resistance, used neat or mixed with sand to obtain a basecoat of superior hardness.
- Neat plaster is a gypsum basecoat plaster having no admixture except hair or other fiber, used for on-the-job mixing with aggregates.
- Ready-mixed plaster is a mill-prepared plaster mix of calcined gypsum and an aggregate, such as perlite or vermiculite. It requires only the addition of water.
- The addition of perlite or vermiculite reduces the weight and increases the thermal and fire resistance of the plaster.

CSI MasterFormat™ 09 21 00: Plaster and Gypsum Board Assemblies
CSI MasterFormat 09 23 00: Gypsum Plastering

Metal Lath Type	Weight psf*	Support Spacing in In. (mm) o.c.	
		Vertical	Horizontal
Diamond mesh	0.27	16 (405)	12 (305)
Diamond mesh	0.38	16 (405)	16 (405)
1/8" flat rib lath	0.31	16 (405)	12 (305)
1/8" flat rib lath	0.38	24 (610)	19 (485)
3/8" rib lath	0.38	24 (610)	24 (610)
Welded or woven wire	0.19	16 (405)	16 (405)
Wire fabric w/ paper backing	0.19	16 (405)	16 (405)

*1 psf = 47.88 Pa

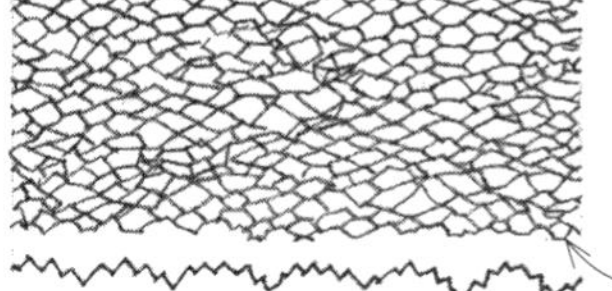

Metal Lath

Metal lath is a plaster base fabricated of expanded metal or of wire fabric, galvanized or coated with a rust-inhibiting paint for corrosion resistance.

- The weight and strength of the metal lath used is related to the spacing and rigidity of its supports.
- Expanded metal lath is fabricated by slitting and expanding a sheet of steel alloy to form a stiff network with diamond-shaped openings.
- Rib lath is an expanded-metal lath having V-shaped ribs to provide greater stiffness and permit wider spacing of the supporting framing members.
- Self-centering lath is a rib lath used over steel joists as formwork for concrete slabs, or as lathing in solid plaster partitions.
- Self-furring lath is expanded-metal, welded-wire, or woven-wire lath that is dimpled to space itself from the supporting surface, creating a space for the keying of plaster or stucco.
- Paper-backed lath is expanded-metal or wire lath having a backing of perforated or building paper, used as a base for ceramic tile and exterior stucco walls.

- Corner beads reinforce external angles of plasterwork and gypsum board surfaces.
- 1-1/4" to 3-3/8" (32 to 86) expanded flanges
- 1/8" (3) radius
- 3/4" (19) radius bullnose
- Flexible corner beads may be bent for curved edges.

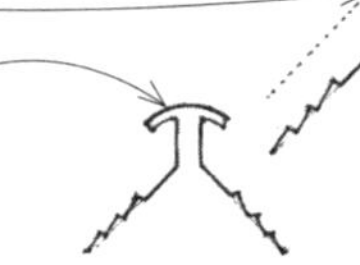

- Casing beads reinforce the edges of plasterwork and gypsum board surfaces.
- 3-1/8" (79) expanded flange
- Square end
- 1/2", 5/8", 3/4", 7/8" (13, 16, 19, 22) depths
- Square end w/ 1/4" (6) 45° break

- A variety of moldings create reveals at the corners and edges of plasterwork.
- F-reveal
- Corner mold
- 3/4" (19)

- Base screeds separate a plastered surface from another material.
- 1/2", 3/4", 7/8" (13, 19, 22) depths

- Gypsum plaster expands slightly as it hardens, requiring expansion joints to control cracking.
- 1/2", 3/4", 7/8" (13, 19, 22) depths

Gypsum Lath

Gypsum lath is a panel having an air-entrained core of hardened gypsum plaster faced with fibrous, absorbent paper to which plaster adheres.

- Perforated gypsum lath is punched with 3/4" (19) ø holes @ 4" (100) o.c. to provide a mechanical key for the plaster.
- Insulating gypsum lath has an aluminum foil backing that serves as a vapor retarder and reflective thermal insulator.
- Type X lath has glass fibers and other additives for greater fire resistance.
- Veneer base is a gypsum lath having a special paper facing for receiving veneer plaster.

- 3/8" or 1/2" (10 or 13) thick
- 16" (405) wide x 48" (1220) long
- 24" (610) widths and lengths up to 12'-0" (3660) are available.

Trim Accessories

Various accessories made of galvanized steel or zinc alloy are used to protect and reinforce the edges and corners of plaster surfaces. These trim accessories also serve as grounds that help the plasterer level the finish coat and bring it up to the proper thickness. For this reason, all grounds should be securely fastened to their supports and installed straight, level, and plumb. Wood grounds may be used where a nailable base is required for the addition of wood trim.

CSI MasterFormat 09 22 36: Lath

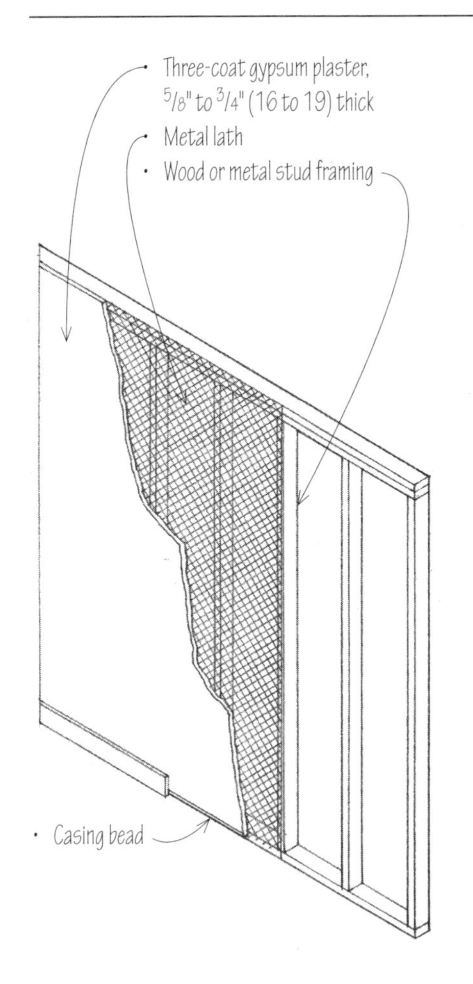

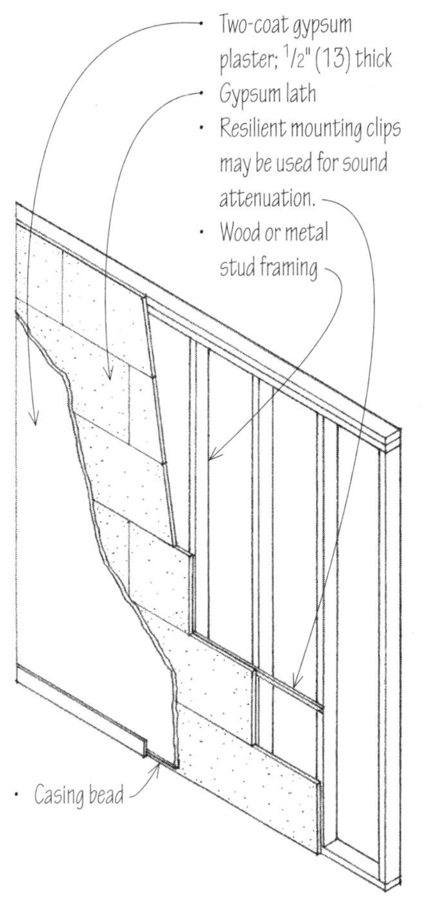

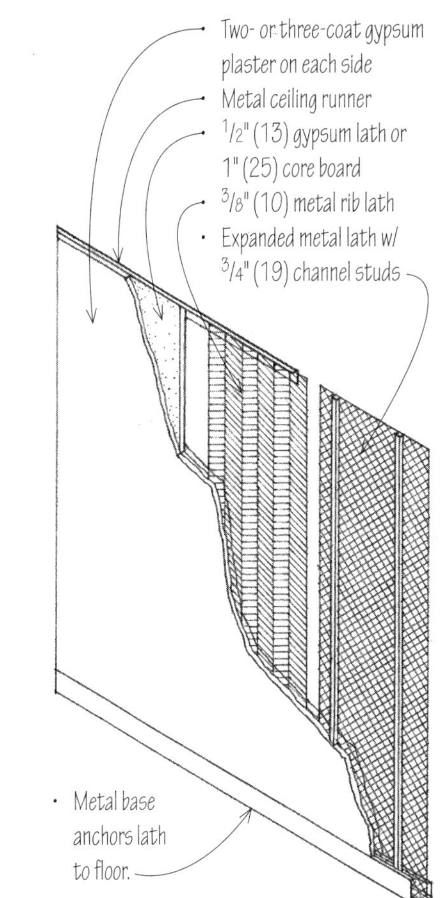

Plaster over Metal Lath

- Three-coat plaster is applied over metal lath.
- Wood or metal studs are spaced 16" or 24" (405 or 610) o.c., depending on the weight of metal lath used; see table on 10.04.
  The frame should be sturdy, rigid, plane, and level; deflection should be limited to 1/360th of the support spacing.
- The long dimension or ribs of the lath are laid across the supports.

Plaster over Gypsum Lath

- Two-coat plaster is normally used over gypsum lath. Veneer plaster can also be applied as a 1/16" to 1/8" (2 to 3) thick one-coat finish over a special gypsum board base.
- Supports may be spaced 16" (405) o.c. for 3/8" (10) lath, and 24" (610) o.c. for 1/2" (13) lath.
- The long dimension of the lath is laid across the supports; ends of lath should bear on a support or be supported by sheet metal clips.

Solid Plaster Partition

- 2" (51) total thickness of partition conserves floor space.
- Three-coat plaster is applied to both sides of metal or gypsum lath.
- Proprietary ceiling runners and metal base anchors are required to stabilize the partition.

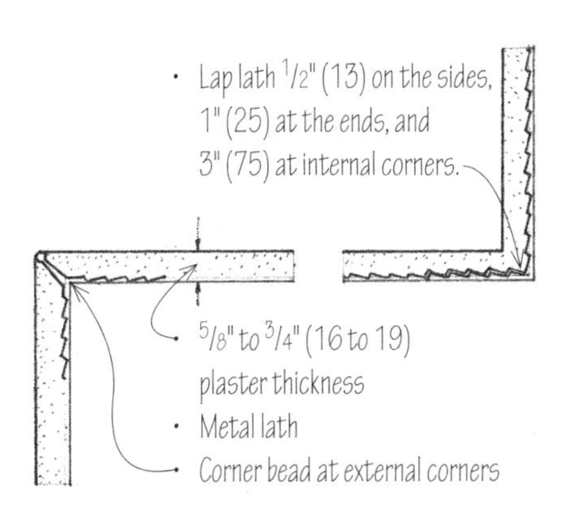

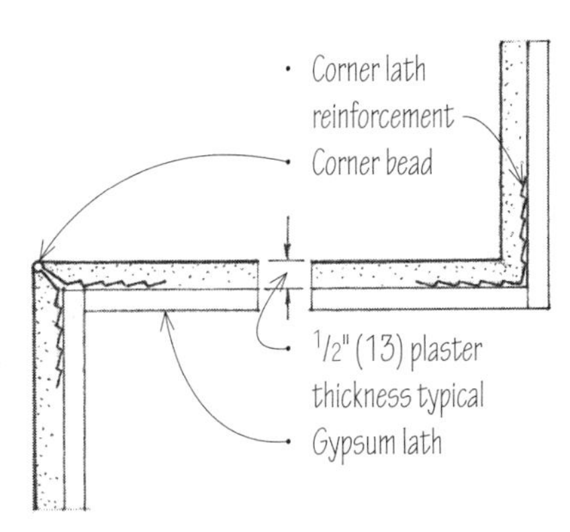

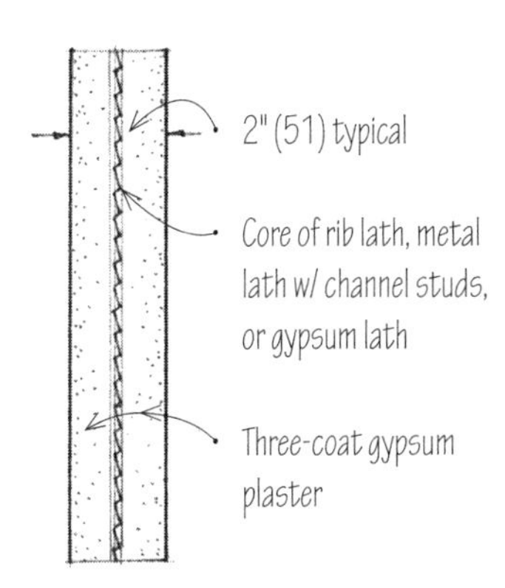

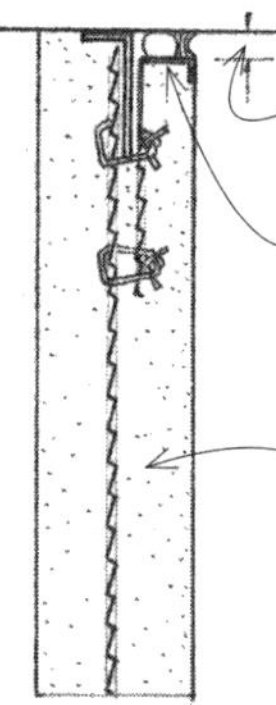

Ceiling Detail

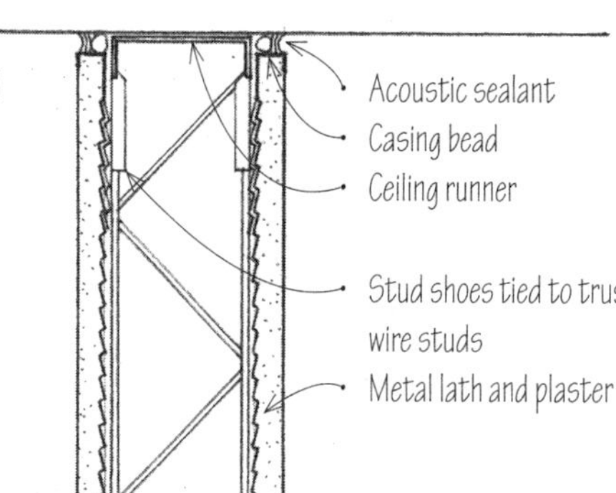

Ceiling Detail

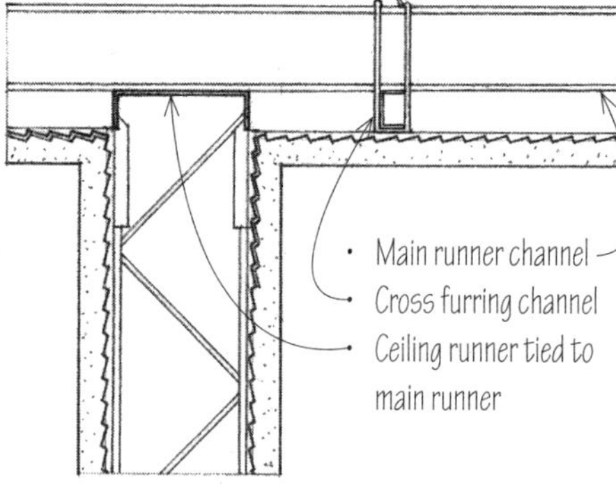

Ceiling Detail

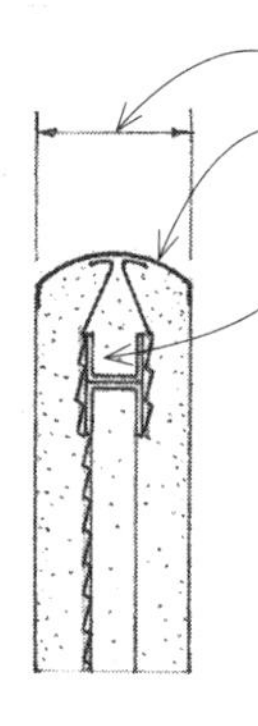

Partition End Detail

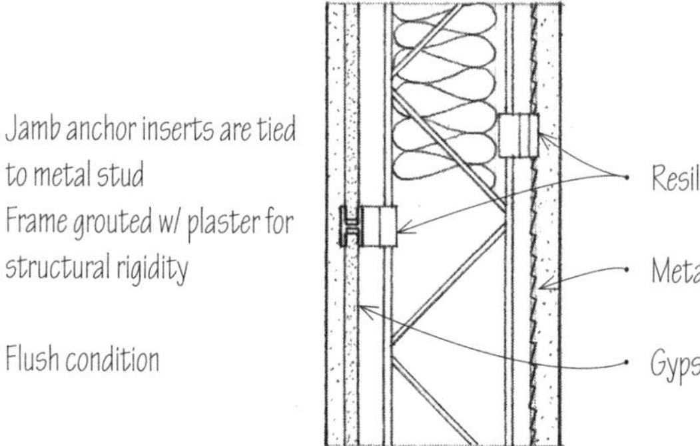

Metal Door Frame

Acoustic Wall Treatment

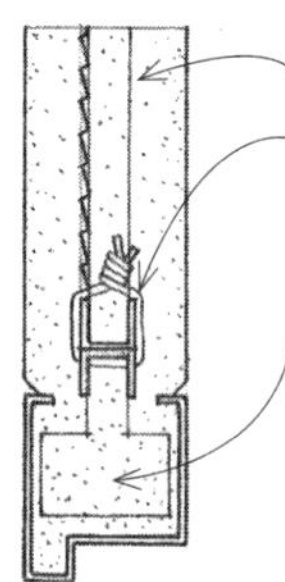

Metal Door Frame

Similar to above

Backbend extends 1/8" (3) into face of plaster

Metal Door Frame

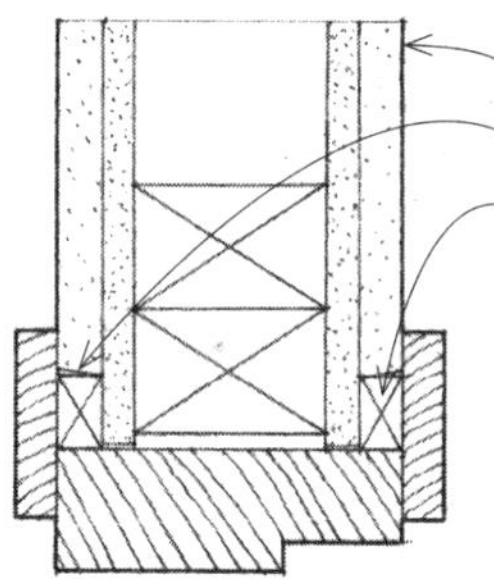

Wood Door Frame

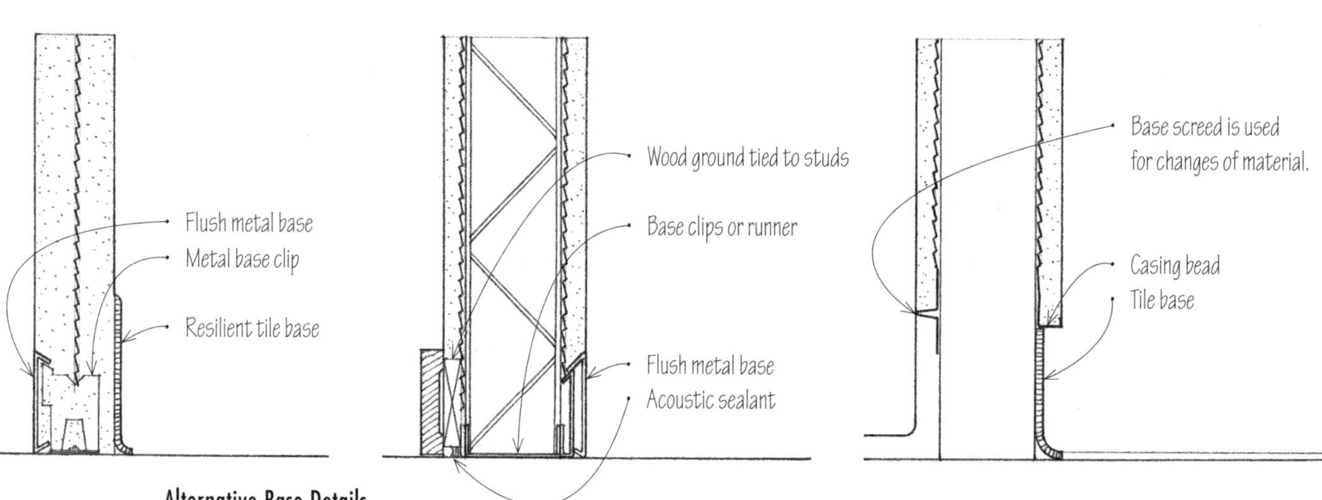

Alternative Base Details

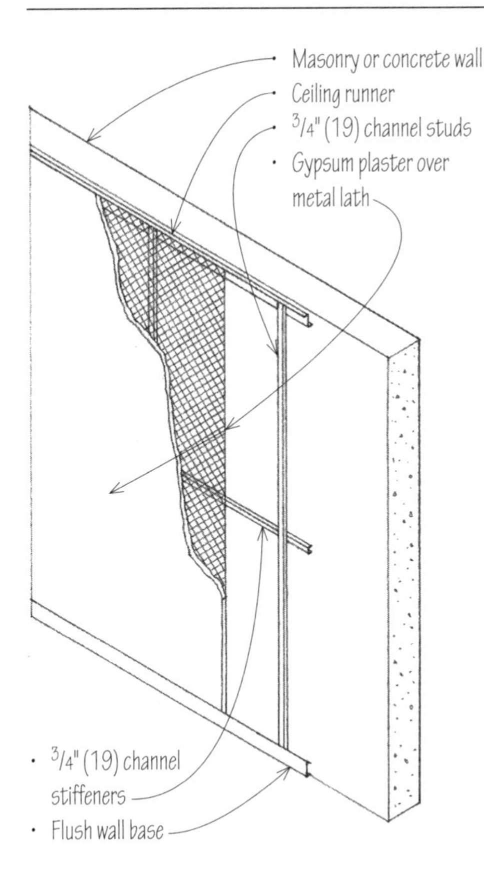

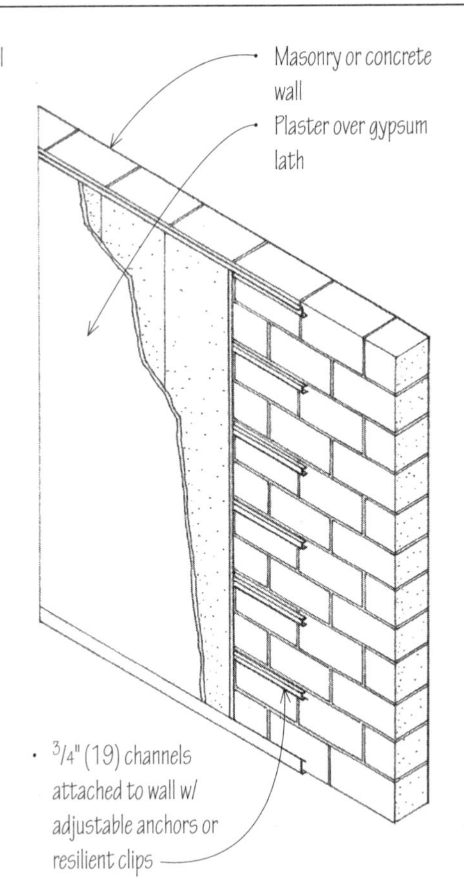

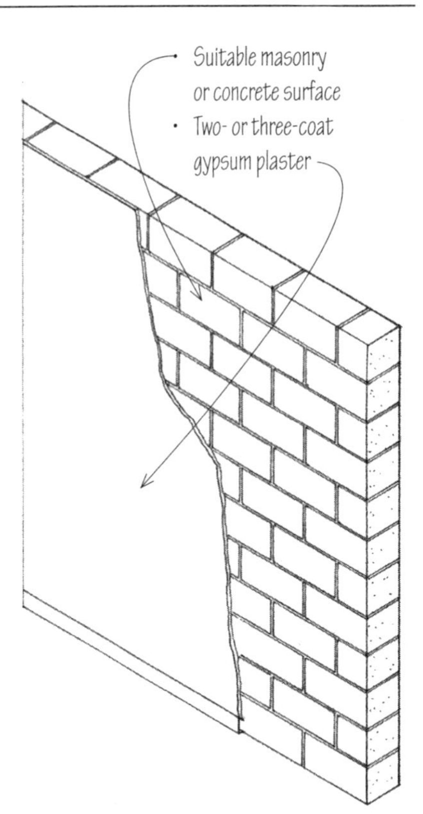

Plaster over Furring

Plaster should be applied over lath and furring when:

- The masonry surface is not suitable for direct application.
- The possibility exists that moisture or condensation might penetrate the wall.
- Additional air space or space for insulating material is required.
- A resilient wall surface is desired for acoustical treatment of the space.

- Wood or metal furring may be applied vertically or horizontally.
- Plaster requires either metal or gypsum lath over the furring; the application and support spacing are similar to the examples shown on 10.06.
- Wall anchors are available that adjust to various furring depths.

Direct Application

- Two-coat plaster, 5/8" (16) thick, is normally applied directly over masonry.
- Plaster may be applied directly to brick, clay tile, or concrete masonry if the surface is sufficiently rough and porous to allow for a good bond.
- A bonding agent is required when applying plaster directly to dense, nonporous surfaces such as concrete.

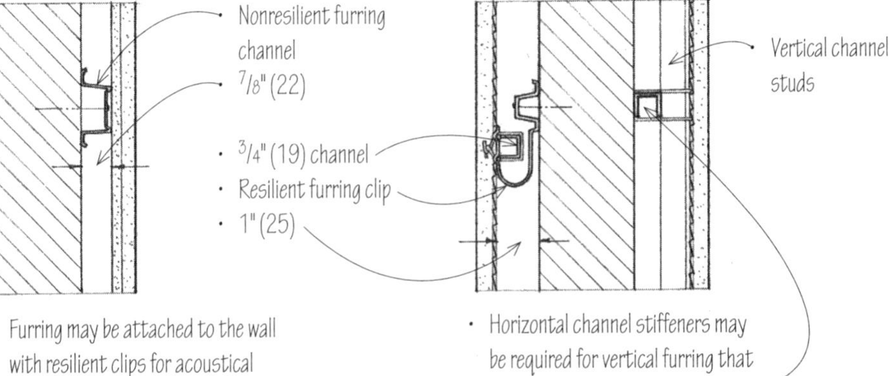

- Furring may be attached to the wall with resilient clips for acoustical treatment and independent movement between the plaster and masonry.

- Horizontal channel stiffeners may be required for vertical furring that is installed away from the wall.

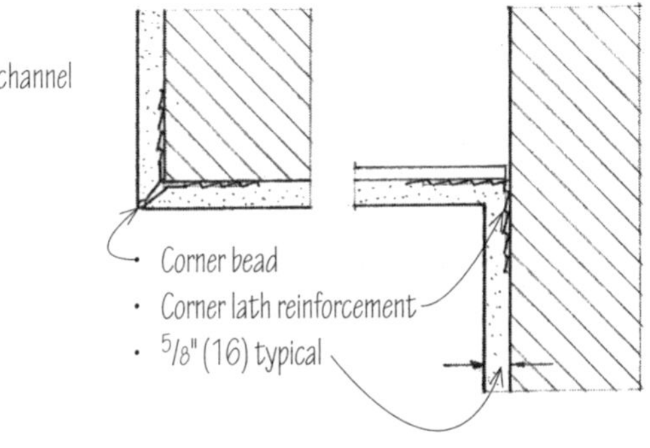

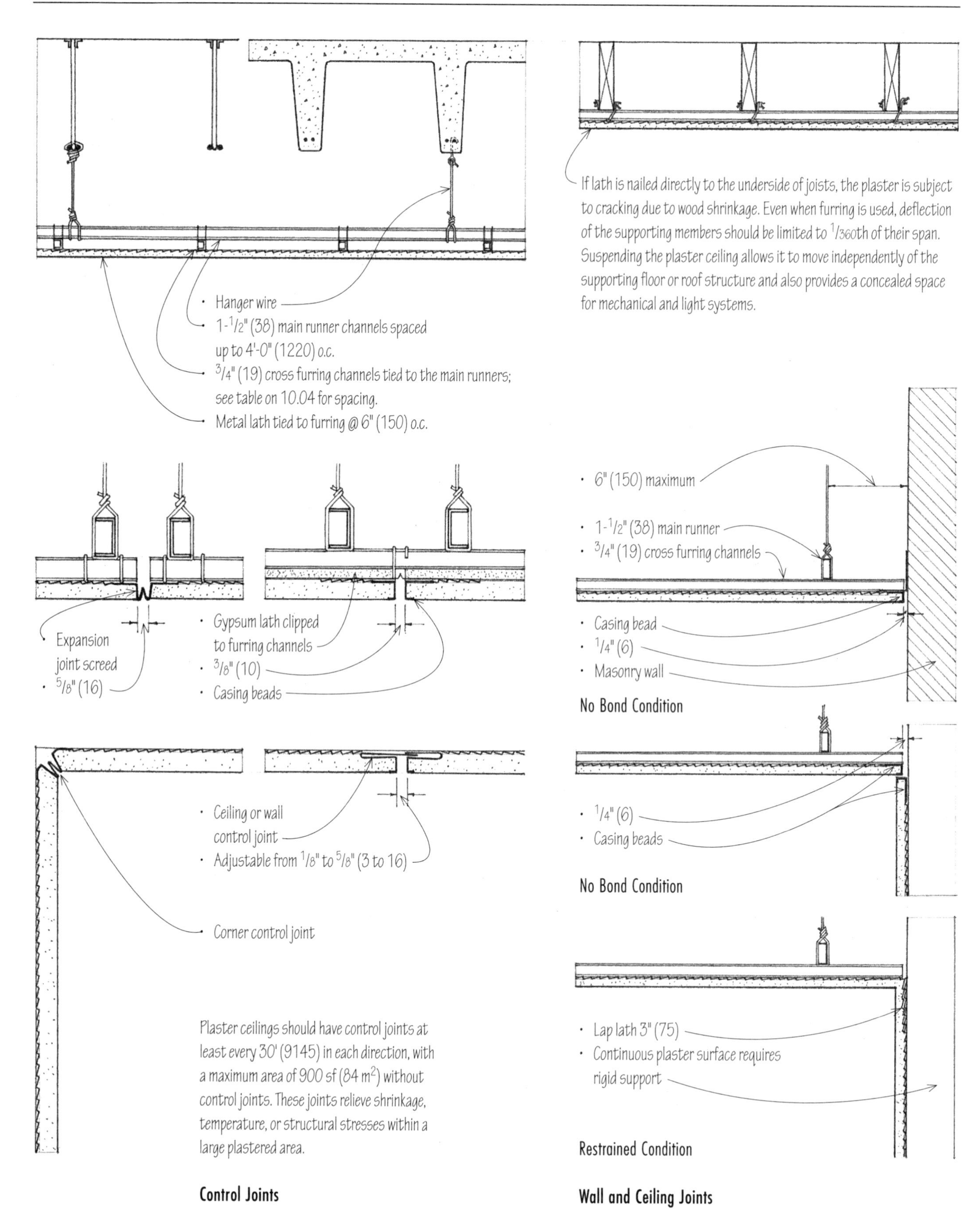

Control Joints

Wall and Ceiling Joints

Gypsum board is a sheet material used for covering walls or as lath. It consists of a gypsum core surfaced and edged to satisfy specific performance, location, application, and appearance requirements. It has good fire resistance and dimensional stability. In addition, its relatively large sheet size makes it an economical material to install. Gypsum wallboard is often referred to as drywall because of its low moisture content, and little or no water is used in its application to interior walls or ceilings. Sheetrock is a trademark for a brand of gypsum board.

Gypsum board may have different edge conditions. Base or intermediate boards in multilayer construction may have square or tongue-and-groove edges. Prefinished boards may have square or beveled edges. Most commonly, however, gypsum board has a tapered edge. The tapered edge allows the joints to be taped and filled to produce strong, invisible seams. Gypsum board thus can form smooth surfaces that are monolithic in appearance and that can be finished by painting or applying a paper, vinyl, or fabric wall covering.

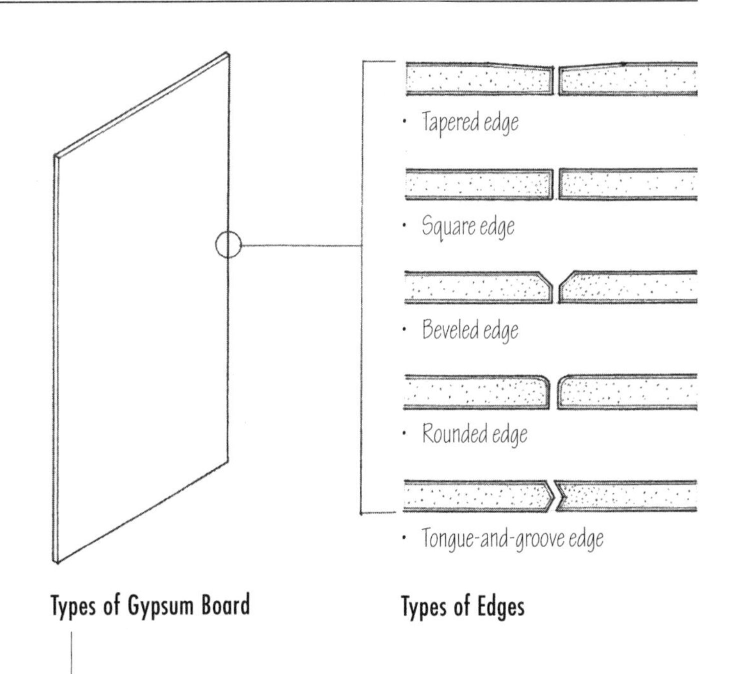

Types of Gypsum Board

Types of Edges

Regular Wallboard
- Tapered edge
- 4' (1220) wide, 8' to 16' (2440 to 4875) long
- 1/4" (6) board is used as the base layer in sound-control walls; 3/8" (10) board is used in multilayer construction, and for remodeling projects; 1/2" and 5/8" (13 and 16) boards are for single-layer construction.

Coreboard
- Square or tongue-and-groove edge
- 1" (25) thick
- 2' (610) wide, 4' to 16' (1220 to 4875) long
- Used to line elevator shafts, stairways, and mechanical chases, and as a base in solid gypsum partitions

Foil-Backed Board
- Square or tapered edge
- 3/8", 1/2", 5/8" (10, 13, 16) thick
- 4' (1220) wide, 8' to 16' (2440 to 4875) long
- Aluminum-foil backing serves as a vapor retarder and as a reflective thermal insulator when the foil faces a 3/4" (19) minimum dead air space.

Water-Resistant Board
- Tapered edge
- 1/2", 5/8" (13, 16) thick
- 4' (1220) wide, 8' to 12' (2440 to 3660) long
- Used as a base for ceramic or other nonabsorbent tile in high-moisture areas

Type-X Board
- Tapered or rounded edge
- 1/2", 5/8" (13, 16) thick
- 4' (1220) wide, 8' to 16' (2440 to 4875) long
- Core has glass fibers and other additives to increase its fire-resistance; available with foil backing.

Prefinished Board
- Square edge
- 5/16" (8) thick
- 4' (1220) wide, 8' (2440) long
- Vinyl or printed paper surface in various colors, patterns, and textures

Backing Board
- Square or tongue-and-groove edge
- 3/8", 1/2", 5/8" (10, 13, 16) thick
- 4' (1220) wide, 8' (2440) long
- Used as the base layer in a multilayer assembly for increased rigidity, sound insulation, and fire resistance; available with regular or Type-X cores, or with foil backing

Sheathing Board
- Square or tongue-and-groove edge
- 1/2", 5/8" (13, 16) thick
- 2' or 4' (610 or 1220) wide, 8' to 10' (2438 to 3048) long
- Has a fire-resistant core and is faced with a water-repellent paper for use as exterior sheathing; available with regular or Type X-cores

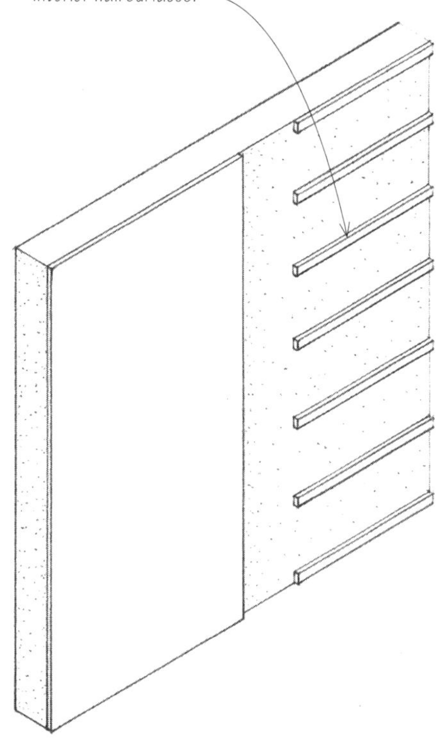

• Vertical application: board length parallel to framing
• Horizontal application: board length perpendicular to framing

• 1x2 minimum wood furring; use 2x2s or metal channels for greater stiffness.

• Support spacing:
16" (405) maximum for 3/8" (10) gypsum board;
24" (610) maximum for 1/2" (13) gypsum board

Masonry or Concrete Base

Gypsum board may be applied to above-grade masonry or concrete walls whose surfaces are dry, smooth, even, and free of oil or other parting materials.

Stud Wall Base

Gypsum board may be fastened directly to wood or metal stud framing that is structurally sound and rigid enough to prevent buckling or cracking of the gypsum board. The face of the frame should form a flat and even plane.

Horizontal application is preferred for greater stiffness if it results in fewer joints. Butt-end joints, which should be kept to a minimum, must fall over a support.

Wood or metal furring is required when:

- The frame or masonry base is not sufficiently flat and even.
- The framing supports are spaced too far apart.
- Additional space for thermal or acoustic insulation is desired.
- The use of resilient furring channels is needed to improve the acoustic performance of the assembly.

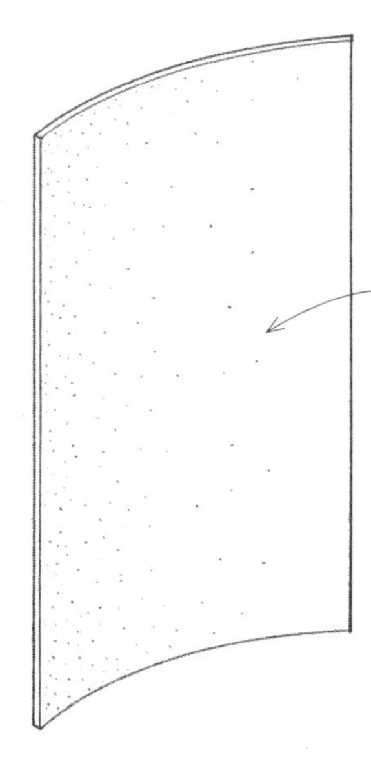

• Gypsum board can be bent and attached to a curving line of studs. The maximum bending radius is as follows:

Board Thickness	Lengthwise	By Width
1/4" (6)	5'-0" (1525)	15'-0" (4570)
3/8" (10)	7'-6" (2285)	25'-0" (7620)
1/2" (13)	20'-0" (6095)	

CSI MasterFormat 09 21 16: Gypsum Board Assemblies

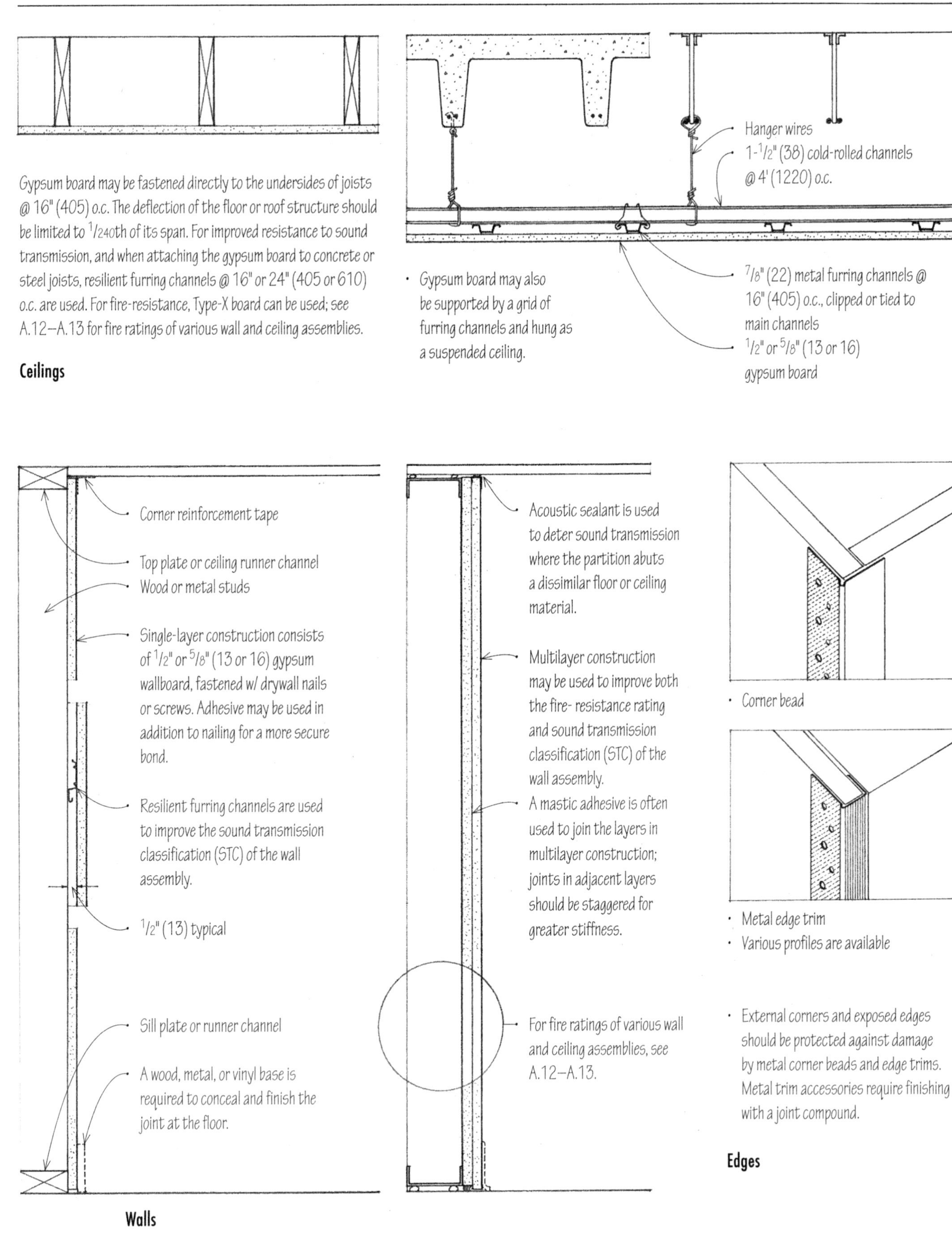

Gypsum board may be fastened directly to the undersides of joists @ 16" (405) o.c. The deflection of the floor or roof structure should be limited to 1/240th of its span. For improved resistance to sound transmission, and when attaching the gypsum board to concrete or steel joists, resilient furring channels @ 16" or 24" (405 or 610) o.c. are used. For fire-resistance, Type-X board can be used; see A.12–A.13 for fire ratings of various wall and ceiling assemblies.

- Gypsum board may also be supported by a grid of furring channels and hung as a suspended ceiling.

Ceilings

Walls

- External corners and exposed edges should be protected against damage by metal corner beads and edge trims. Metal trim accessories require finishing with a joint compound.

Edges

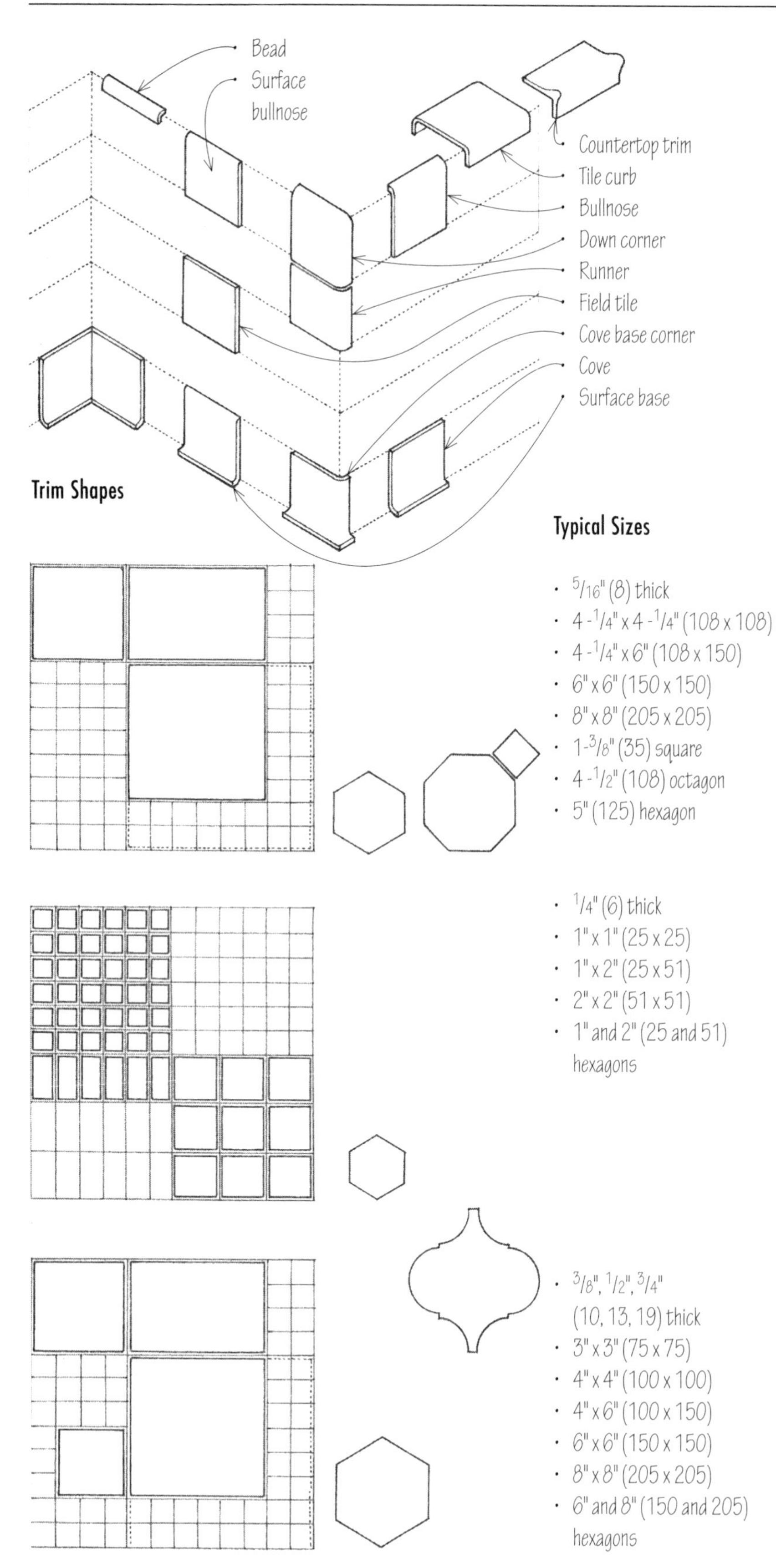

Ceramic tiles are relatively small, modular surfacing units made of clay or other ceramic material. The tiles are fired in a kiln at very high temperatures. The result is a durable, tough, dense material that is water-resistant, difficult to stain, and easy to clean; its colors generally do not fade.

Ceramic tile is available glazed or unglazed. Glazed tile has a face of ceramic material fused into the body of the tile, and may have glossy, matte, or crystalline finishes in a wide range of colors. Unglazed tiles are hard and dense, and derive their color from the body of the clay material. These colors tend to be more muted than those of glazed tiles.

Typical Sizes

Types of Ceramic Tile

Glazed Wall Tile

- 5/16" (8) thick
- 4-1/4" x 4-1/4" (108 x 108)
- 4-1/4" x 6" (108 x 150)
- 6" x 6" (150 x 150)
- 8" x 8" (205 x 205)
- 1-3/8" (35) square
- 4-1/2" (108) octagon
- 5" (125) hexagon

Glazed wall tile has a nonvitreous body and a bright, matte, or crystalline glaze, used for surfacing interior walls and light-duty floors. Exterior tiles are weatherproof and frostproof, and can be used for both exterior and interior walls.

Ceramic Mosaic Tile

- 1/4" (6) thick
- 1" x 1" (25 x 25)
- 1" x 2" (25 x 51)
- 2" x 2" (51 x 51)
- 1" and 2" (25 and 51) hexagons

Ceramic mosaic tile has a porcelain or natural clay body, glazed for surfacing walls or unglazed for use on both floors and walls. Porcelain tiles have bright colors, while natural clay tiles have more muted colors. To facilitate the handling and speed of installation, small tiles are usually faced with paper or backed with mesh to form 1' x 1' (305 x 305) or 1' x 2' (305 x 610) sections with the proper tile spacing.

Quarry and Paver Tiles

- 3/8", 1/2", 3/4" (10, 13, 19) thick
- 3" x 3" (75 x 75)
- 4" x 4" (100 x 100)
- 4" x 6" (100 x 150)
- 6" x 6" (150 x 150)
- 8" x 8" (205 x 205)
- 6" and 8" (150 and 205) hexagons

Quarry tile is unglazed floor tile of natural clay or porcelain. The tiles are impervious to dirt, moisture, and stains, and resistant to freezing and abrasion. Pavers are similar in composition to ceramic mosaic tiles but thicker and larger. They are weatherproof and can be used on floors subjected to heavy-duty loads.

- Consult tile manufacturer for exact sizes, shapes, colors, and glazes.

CSI MasterFormat 09 30 13: Ceramic Tiling

Thinset Process

In the thinset process, ceramic tile is bonded to a continuous, stable backing with a thin coat of dry-set mortar, latex-portland cement mortar, epoxy mortar, or an organic adhesive.

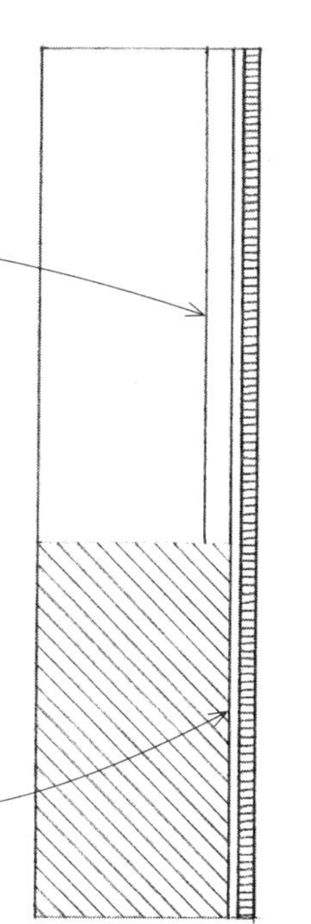

- Thinset installations require a solid, dimensionally stable backing of gypsum plaster, gypsum board, or plywood.
- In wet areas around bathtubs and showers, use 1/2" (13) thick glass-fiber-reinforced concrete backerboard and set the tile with latex-portland cement or dry-set mortar.
- Masonry surfaces should be clean, sound, and free of efflorescence. When dry-set or latex-portland cement mortar is used to set the tile, the surface should be roughened to ensure a good bond.

Thickset Process

In the thickset process, ceramic tile is applied over a bed of portland cement mortar. This relatively thick bed allows for accurate slopes and true planes in the finished work. The mortar bed is also not affected by prolonged contact with water.

- Suitable backings for cement mortar bed installations include brick or concrete block masonry, monolithic concrete, plywood, gypsum plaster, and gypsum board. Open stud framing and furring can also be used with metal lath.
- The setting bed, which is a field mix of portland cement, sand, water, and sometimes hydrated lime, is 3/4" to 1" (19 to 25) thick on walls.
- Tiles may be laid with a 1/16" (2) bond coat of neat portland cement or dry-set mortar while the mortar bed is still plastic, or set with a 1/8" to 1/4" (3 to 6) coat of latex-portland cement after the mortar bed is fully cured.

- Setting bed is 1-1/4" to 2" (32 to 51) thick on floors.

- Suitable floors for thinset installations include concrete slabs and double wood floors.
- Concrete slabs should be smooth, level, and properly reinforced and cured; a leveling topping can be used if required.

- Double wood floors consist of a 5/8" (16) minimum plywood subfloor and an underlayment of 1/2" or 5/8" (13 or 16) exterior grade plywood. A 1/4" (6) space should be left between the underlayment and vertical surfaces. When using epoxy mortar, provide 1/4" (6) gaps between underlayment panels, and fill with epoxy.
- Maximum deflection of the floor under full load should be limited to 1/360th of the span.

- Suitable floors for cement mortar bed installations include properly reinforced and cured concrete slabs and structurally sound plywood subflooring.
- Maximum deflection of the floor under full load should be limited to 1/360th of the span.

- A cleavage membrane isolates the mortar bed from damaged or unstable backings and allows some independent movement of the supporting construction to occur.
- The mortar bed should be reinforced with metal lath or mesh whenever it is backed by a membrane.

CSI MasterFormat 09 31 00: Thinset Tiling
CSI MasterFormat 09 32 00: Mortar-Bed Tiling

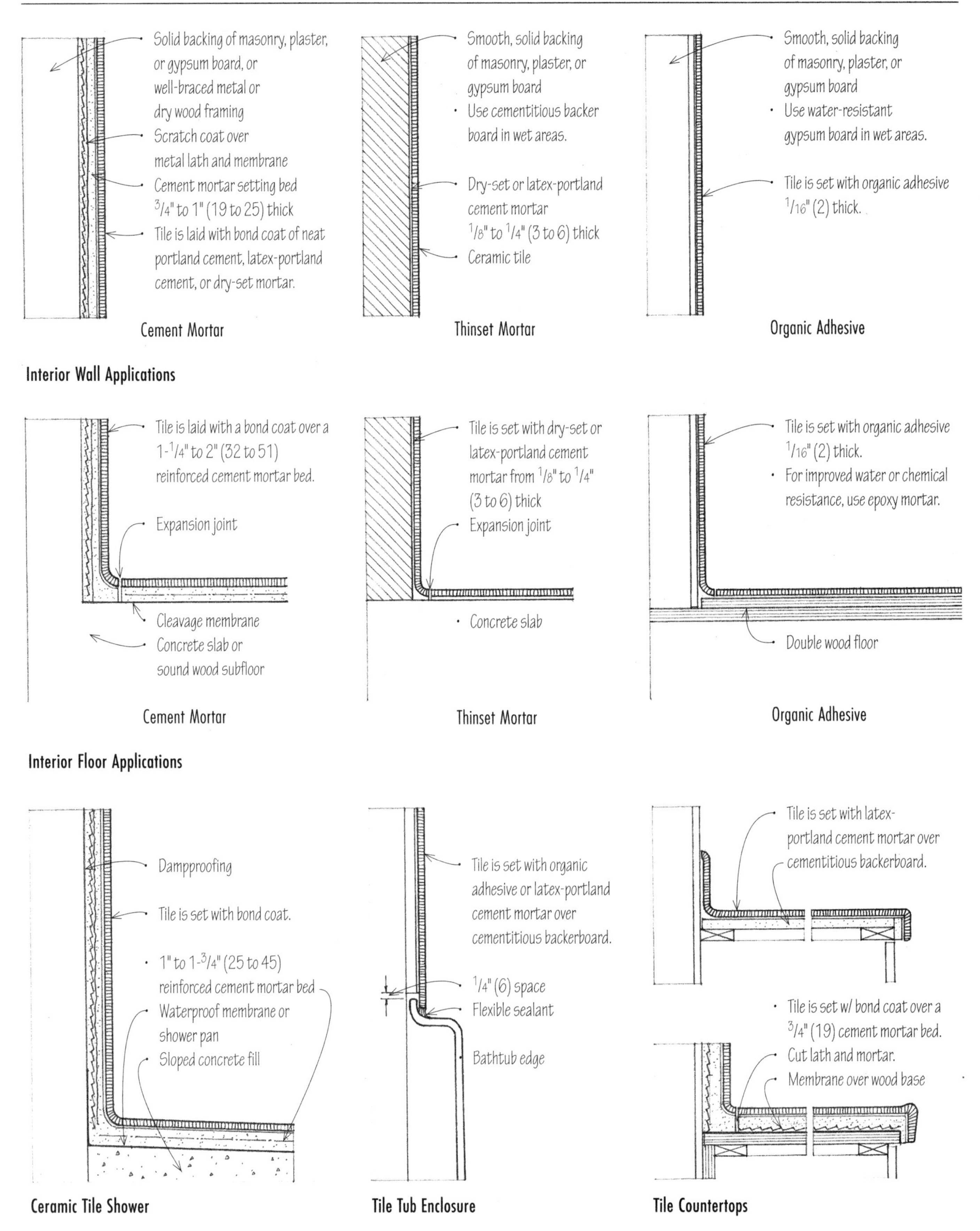

Solid backing of masonry, plaster, or gypsum board, or well-braced metal or dry wood framing
Scratch coat over metal lath and membrane
Cement mortar setting bed 3/4" to 1" (19 to 25) thick
Tile is laid with bond coat of neat portland cement, latex-portland cement, or dry-set mortar.
Cement Mortar
Smooth, solid backing of masonry, plaster, or gypsum board
• Use cementitious backer board in wet areas.
Dry-set or latex-portland cement mortar 1/8" to 1/4" (3 to 6) thick
Ceramic tile
Thinset Mortar
Smooth, solid backing of masonry, plaster, or gypsum board
• Use water-resistant gypsum board in wet areas.
Tile is set with organic adhesive 1/16" (2) thick.
Organic Adhesive
Interior Wall Applications
Tile is laid with a bond coat over a 1-1/4" to 2" (32 to 51) reinforced cement mortar bed.
Expansion joint
Cleavage membrane
Concrete slab or sound wood subfloor
Cement Mortar
Tile is set with dry-set or latex-portland cement mortar from 1/8" to 1/4" (3 to 6) thick
Expansion joint
• Concrete slab
Thinset Mortar
Tile is set with organic adhesive 1/16" (2) thick.
• For improved water or chemical resistance, use epoxy mortar.
Double wood floor
Organic Adhesive
Interior Floor Applications
Dampproofing
Tile is set with bond coat.
• 1" to 1-3/4" (25 to 45) reinforced cement mortar bed
Waterproof membrane or shower pan
Sloped concrete fill
Ceramic Tile Shower
Tile is set with organic adhesive or latex-portland cement mortar over cementitious backerboard.
1/4" (6) space
Flexible sealant
Bathtub edge
Tile Tub Enclosure
Tile is set with latex-portland cement mortar over cementitious backerboard.
• Tile is set w/ bond coat over a 3/4" (19) cement mortar bed.
Cut lath and mortar.
Membrane over wood base
Tile Countertops

Terrazzo is a mosaic floor or paving composed of marble or other stone chips, set in a cementitious or resinous matrix and ground and polished when dry. It provides a dense, extremely durable, smooth flooring surface whose mottled coloring is controlled by the size and colors of the aggregate and the color of the binder.

Terrazzo Finishes

- Standard terrazzo is a ground and polished terrazzo finish consisting mainly of relatively small stone chips.

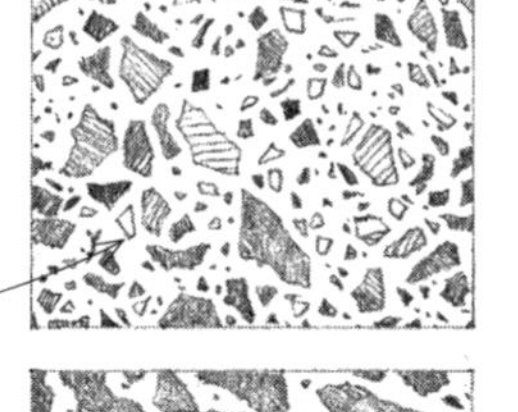

- Venetian terrazzo is a ground and polished terrazzo finish consisting mainly of large stone chips, with smaller chips filling the spaces between.

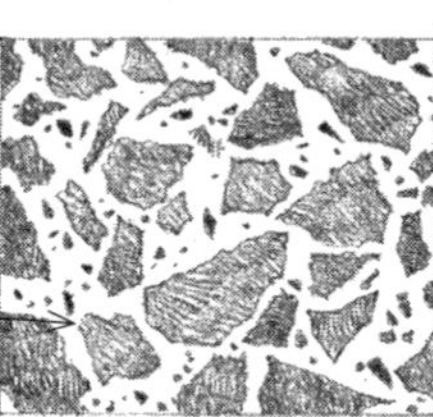

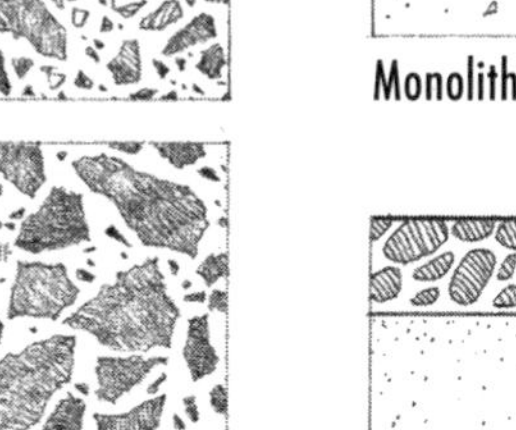

Metal or plastic-tipped divider strips are used:

- To localize shrinkage cracking;
- To serve as construction joints;
- To separate the different colors of a floor pattern;
- To act as decorative elements.

- Expansion joints are required over isolation or control joints in the subfloor. They consist of a pair of divider strips separated by a resilient material such as neoprene.

- ¹/₄" to ¹/₂" (6 to 13) resinous topping
- Divider strip at all control joints
- Wood, metal, or concrete subfloor

Thinset Terrazzo

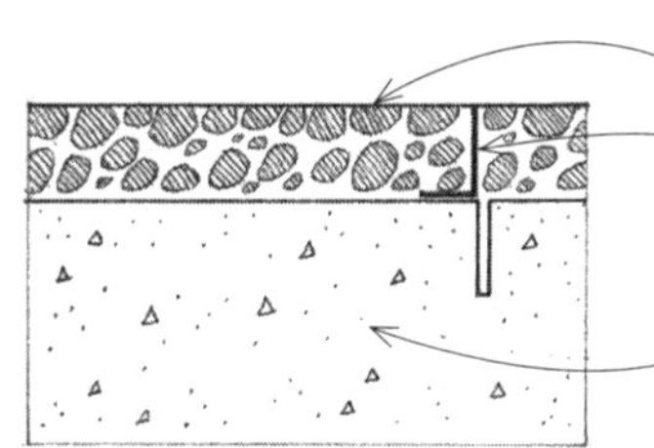

Monolithic Terrazzo

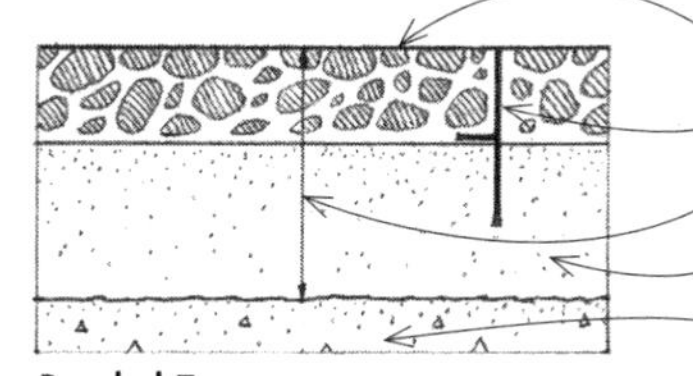

Bonded Terrazzo

- ⁵/₈" (16) or thicker portland cement topping
- Divider strips as per monolithic terrazzo
- Saw-cut control joint
- Smooth-finished slab w/ a chemical bonding agent if the concrete surface is too smooth for a mechanical bond

Chemically Bonded Terrazzo

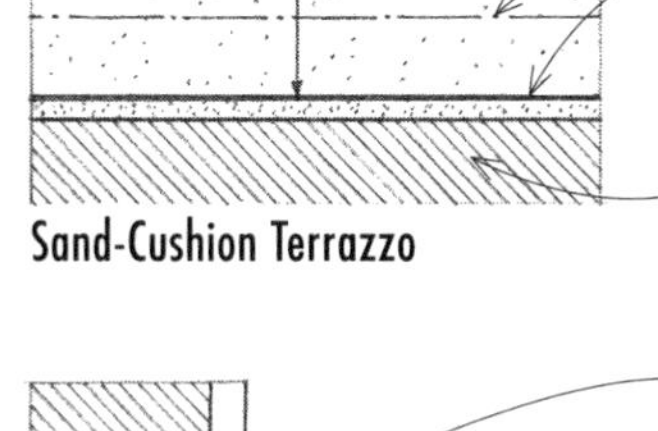

Sand-Cushion Terrazzo

- ⁵/₈" (16) terrazzo topping
- Wall stringer
- 2" (51) minimum
- 1-¹/₂" (38)
- ³/₄" (19)
- ⁷/₈" (22) underbed
- Concrete sub-stair

Terrazzo Stair

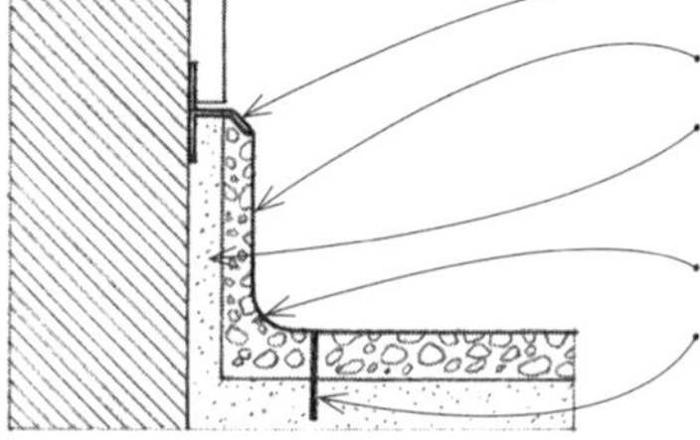

Terrazzo Base

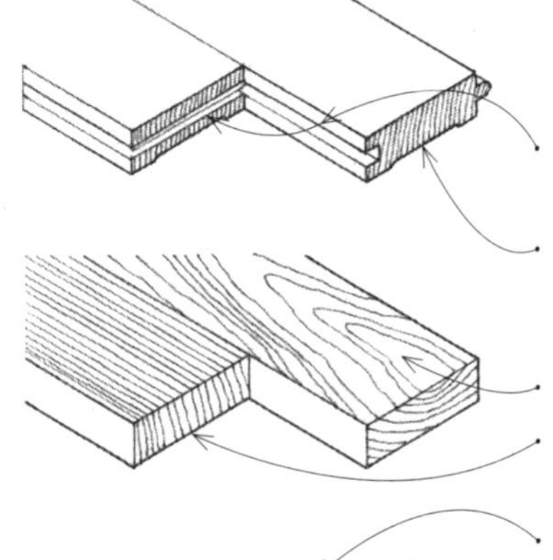

Face widths:
1-1/2", 2", 2-1/4", 3-1/4" (38, 51, 57, 85)
Thicknesses:
Light duty: 3/8", 1/2", 5/8" (10, 13, 16)
Normal service: 25/32" (20)
Heavy duty: 33/32", 41/32", 53/32" (25, 33, 41)

Wood flooring combines durability and wear resistance with comfort and warmth. Durable, hard, close-grained species of both hardwood and softwood are used for flooring. Common species of hardwood flooring species include oak, maple, birch, pecan, and cherry. Common species of softwood flooring include southern pine, Douglas fir, and hemlock. Whenever possible, woods used for flooring should be from certified sustainable sources. While technically not a wood, bamboo is a relatively fast-growing grass product that qualifies as a renewable resource. (LEED® MR Credit 6: Rapidly Renewable Materials)

The various species of wood flooring are appearance-graded, but not according to the same standards. The best grades—clear or select—typically minimize or exclude defects such as knots, streaks, checks, and torn grain. Consult the flooring manufacturer for precise standards and specifications.

Solid wood flooring is available in strips and planks.

Strip Flooring

Strip flooring is composed of long wood strips 3-1/4" (85) or less in face width.

Plank Flooring

Plank flooring refers to flooring boards that are wider than 3-1/4" (85). End- and side-matched boards are blind-nailed. The boards may also be face-nailed or screwed and then plugged. Some new plank flooring systems can be laid with mastic or adhesive. To minimize the effect of variations in humidity on the wide planks, 3-ply laminated planks are available.

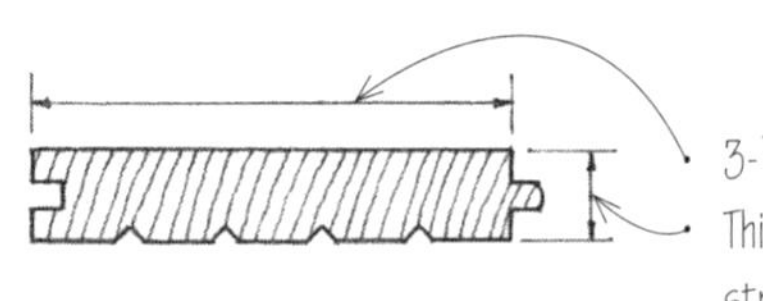

Wood flooring is most often finished with clear polyurethane, varnish, or a penetrating sealer. Finishes can range from high gloss to satin. Ideally, the finish should enhance the durability of the wood and its resistance to water, dirt, and staining, without concealing the wood's natural beauty. Stains are used to add color to the natural color of the wood without obscuring the wood grain. Wood flooring can also be waxed, painted, or stenciled, but painted surfaces require more maintenance.

Engineered Flooring

Engineered flooring is impregnated with acrylic or sealed with urethane or vinyl. Laminated flooring assembles high-pressure laminates, including wood veneers, into durable, acrylic-urethane sealed panels. Bamboo is also laminated under high pressure, milled into planks, immersed in polyurethane, and coated with acrylic polyurethane.

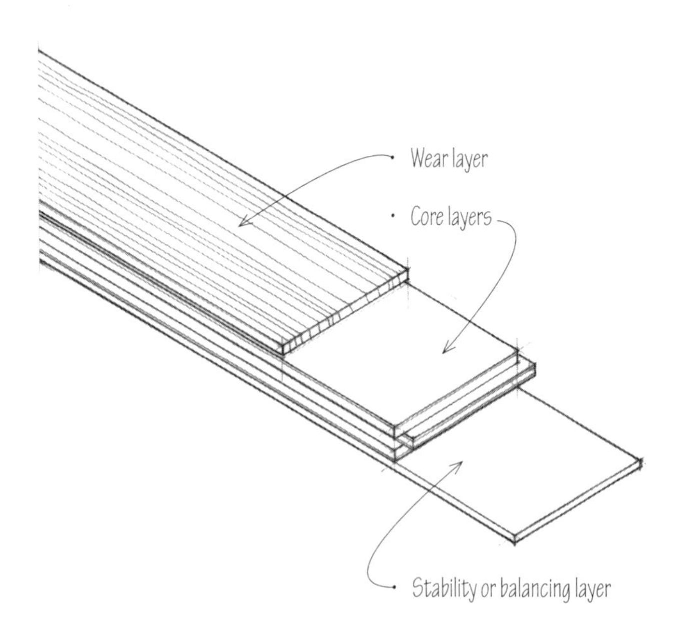

CSI MasterFormat 09 62 23: Bamboo Flooring
CSI MasterFormat 09 64 00: Wood Flooring

Wood strip and plank flooring requires a wood subfloor or a base of spaced wood sleepers. Plywood or panel subfloors, integral parts of a wood joist floor system, may be laid over other floor systems as well to receive the wood flooring. Treated wood sleepers are usually required over concrete slabs to receive a wood subfloor or the finish wood flooring. This is especially important to protect the flooring from dampness when it is installed on concrete slabs on or below grade.

Wood block flooring requires a clean, dry, smooth, flat surface such as a plywood subfloor or underlayment. While block tiles can be applied to the surface of a dry concrete slab, it is best, especially in basements, to lay the flooring over a plywood subfloor and a vapor barrier set on treated wood sleepers.

Wood flooring will shrink and swell as its moisture content changes with variations in atmospheric humidity. It should not be installed until the building is enclosed, permanent lighting and the heating plant are installed, and all building materials are dry. The wood flooring should be stored for several days in the space where it will be installed to allow the flooring to become acclimated to the interior conditions. As the flooring is installed, space should be provided along the perimeter for ventilation and expansion of the flooring.

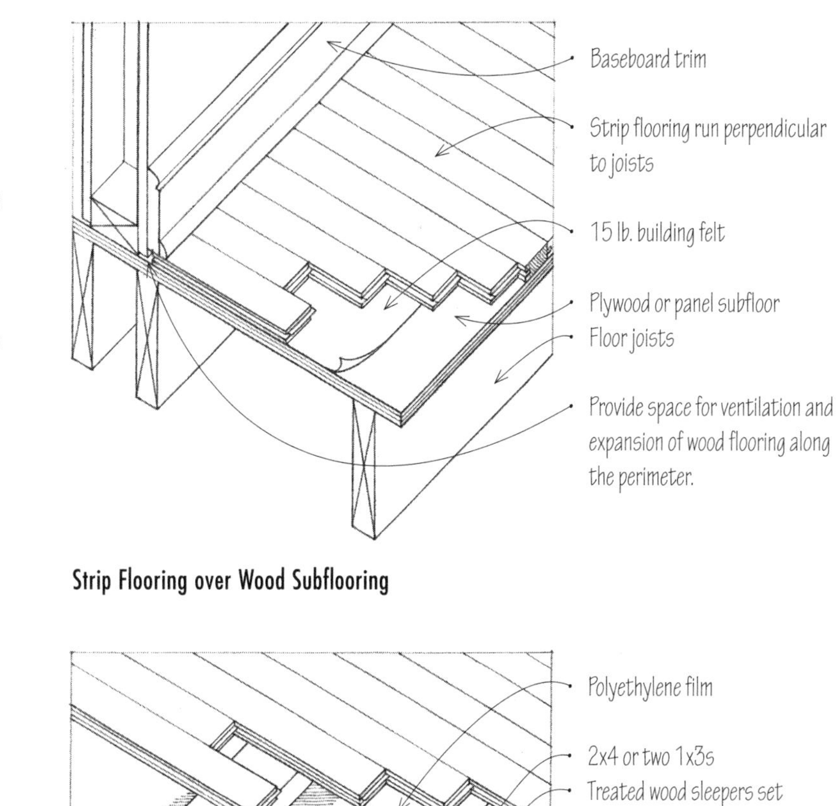

Strip Flooring over Wood Subflooring

Strip Flooring over Concrete Slab

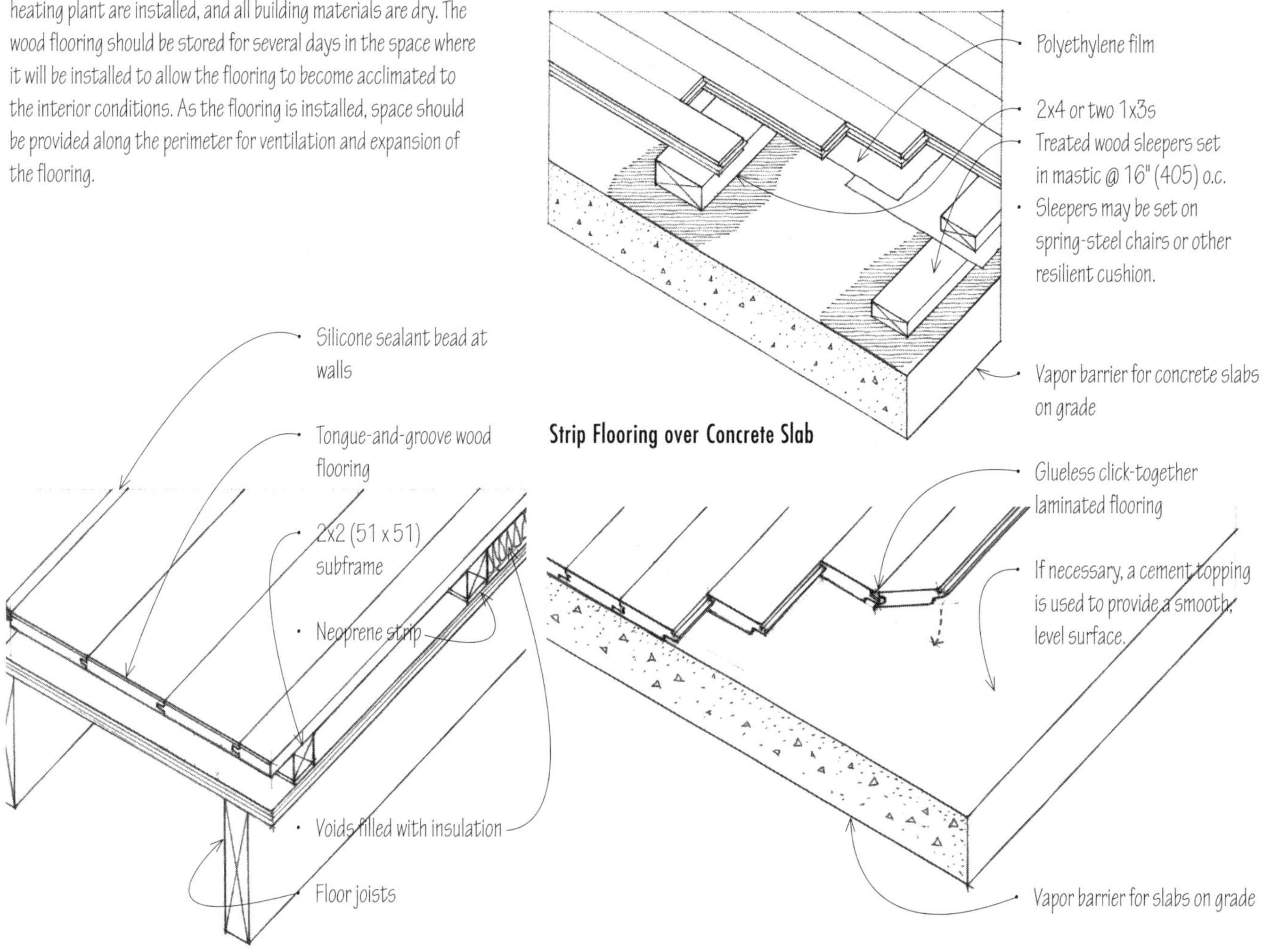

Floating Wood Flooring Installation

Glueless Laminated Wood Flooring Installation

Stone flooring may consist of sandstone, limestone, polished marble or granite, or split-face slate. Consideration should be given to the color and texture of the stone finish, its abrasion- and slip-resistance, as well as the additional dead load the stone will impose on the floor structure. Stone flooring 1/2" (13) thick weighs about 7.5 psf (359 Pa).

The tiles or slabs may be laid in regular or irregular patterns over a portland cement mortar bed in a manner similar to the installation of ceramic tile flooring. Thin stone flooring may also be applied using the thinset process. See 10.13.

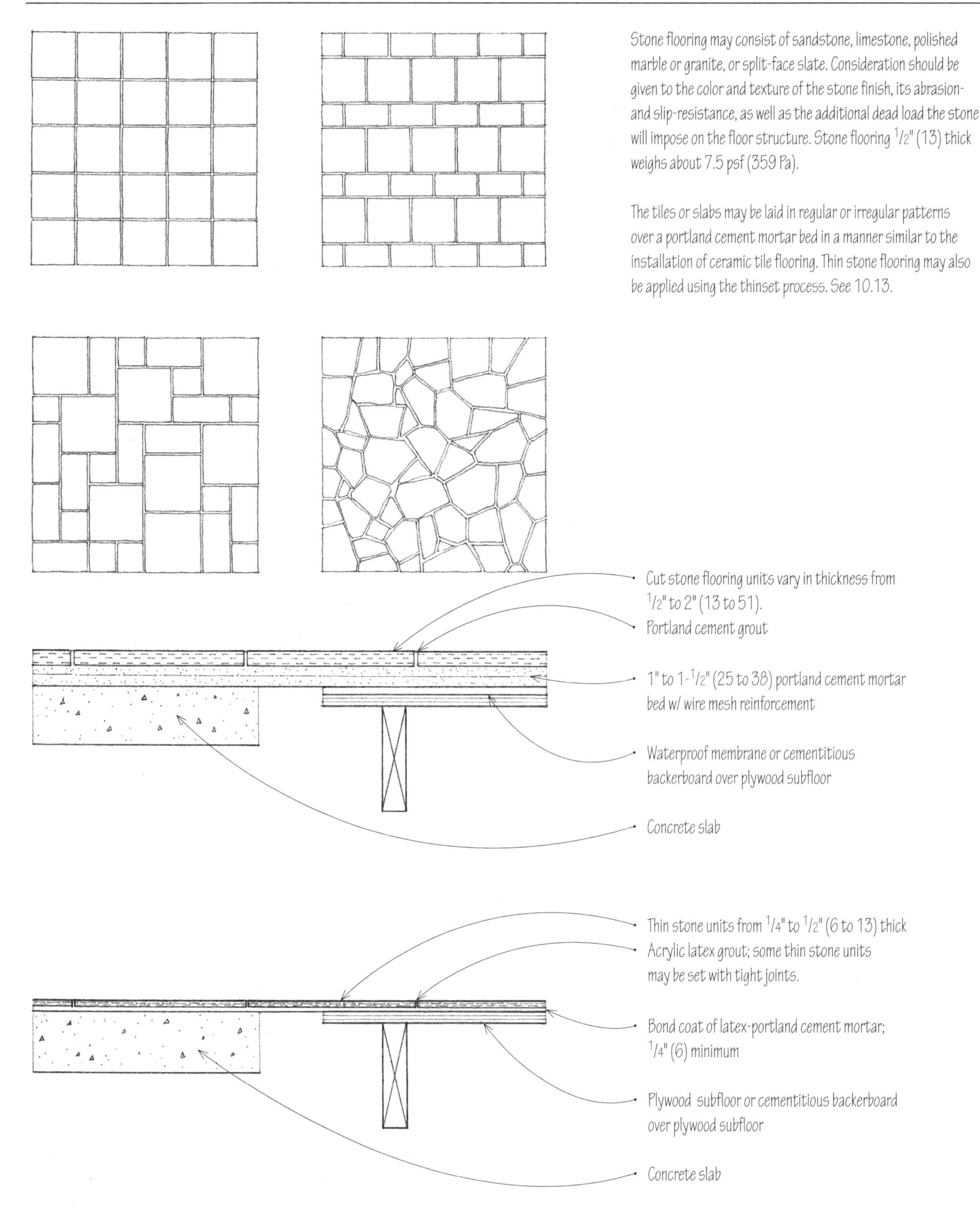

CSI MasterFormat 09 63 40: Stone Flooring

Resilient flooring materials provide an economical, relatively dense, nonabsorbent flooring surface that is durable and easy to maintain. Their degree of resilience enables them to resist permanent indentation and contributes to their quietness and comfort underfoot. How comfortable a resilient floor covering is, however, depends not only on its resilience but also on its backing and the hardness of the supporting substrate.

None of the resilient flooring types is superior in all respects. Listed below are the types that perform well in specific areas.

- Resilience and quietness: cork tile, rubber tile, homogeneous vinyl tile
- Resistance to indentation: homogeneous vinyl tile, vinyl sheet, cork tile w/ vinyl coating
- Stain resistance: rubber tile, homogeneous vinyl tile, vinyl composition tile, linoleum
- Alkali resistance: cork tile w/ vinyl coating, vinyl sheet, homogeneous vinyl tile, rubber tile
- Grease resistance: vinyl sheet, homogeneous vinyl tile, cork tile w/ vinyl coating, linoleum
- Durability: homogeneous vinyl tile, vinyl sheet, vinyl composition tile, rubber tile
- Ease of maintenance: vinyl sheet, homogeneous vinyl tile, vinyl composition tile, cork tile w/ vinyl coating

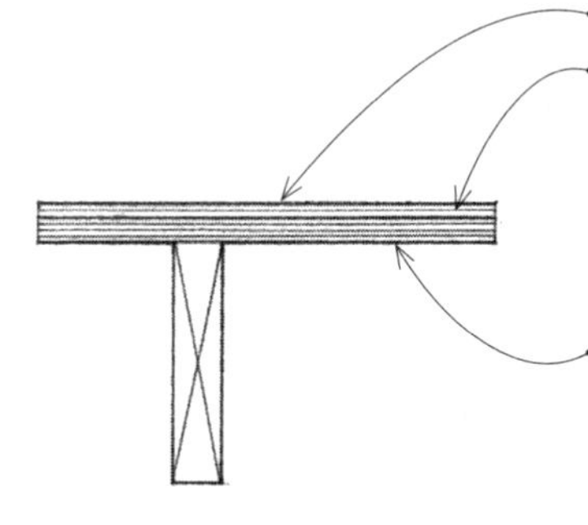

- Surface must be smooth, firm, clean, and dry.
- Double layer wood floor consists of hardboard underlayment at least 1/4" (6) thick or sanded plywood underlayment at least 3/8" (10) thick, laid with the face grain perpendicular to floor joists or to flooring boards.
- Single layer wood floor consists of combination subfloor/underlayment panels at least 5/8" (16) thick, laid with the face grain perpendicular to floor joists or to flooring boards; see 4.32.

Wood Subfloors

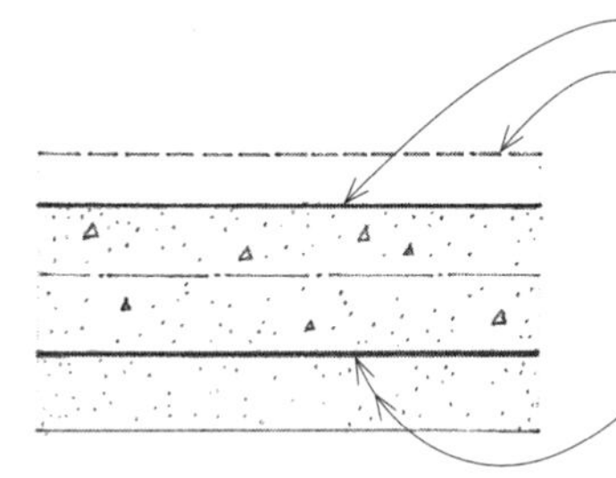

- Surface must be smooth, dense, clean, and dry.
- Provide a 2" to 3" (51 to 75) reinforced concrete topping over precast slabs; over lightweight concrete slabs, provide a 1" (25) concrete topping.
- Provide a moisture barrier and a gravel base under concrete slabs on grade.
- For concrete slabs below grade, provide a waterproofing membrane and a 2" (51) mudslab.

Concrete Subfloors

LEED EQ Credit 4.1: Low-Emitting Materials, Adhesives & Sealants

Flooring Type	Components	Thickness	Sizes	Permissible Location
Vinyl sheet	vinyl resins w/ fiber back	.065" to .160" (2 to 4)	6' to 15' (1830 to 4570) wide	B O S
Homogeneous vinyl tile	vinyl resins	1/16" to 1/8" (2 to 3)	9" x 9" (230 x 230) 12" x 12" (305 x 305)	B O S
Vinyl composition tile	vinyl resins w/ fillers	.050" to .095" (1 to 2)	9" x 9" (230 x 230) 12" x 12" (305 x 305)	B O S
Cork tile	raw cork and resins	1/8" to 1/4" (3 to 6)	6" x 6" (150 x 150) 9" x 9" (230 x 230)	S
Cork tile w/ vinyl coating	raw cork, vinyl resins	1/8", 3/16" (3, 5)	9" x 9" (230 x 230) 12" x 12" (305 x 305)	S
Rubber tile	rubber compound	3/32" to 3/16" (2 to 5)	9" x 9" (230 x 230) 12" x 12" (305 x 305)	B O S
Linoleum sheet	linseed oil, cork, rosin	1/8" (3)	6' (1830) wide	S
Linoleum tile	linseed oil, cork, rosin	1/8" (3)	9" x 9" (230 x 230) 12" x 12" (305 x 305)	S

S: Suspended
O: On grade
B: Below grade

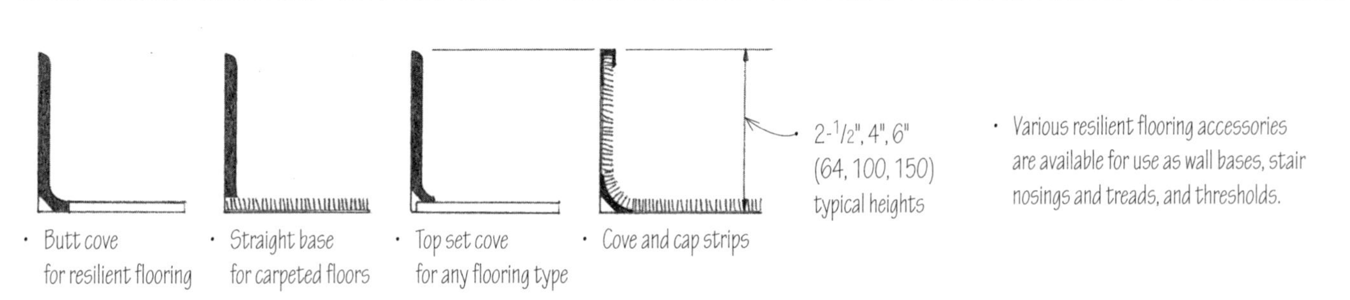

- Butt cove for resilient flooring
- Straight base for carpeted floors
- Top set cove for any flooring type
- Cove and cap strips

- Various resilient flooring accessories are available for use as wall bases, stair nosings and treads, and thresholds.

CSI MasterFormat 09 65 00: Resilient Flooring

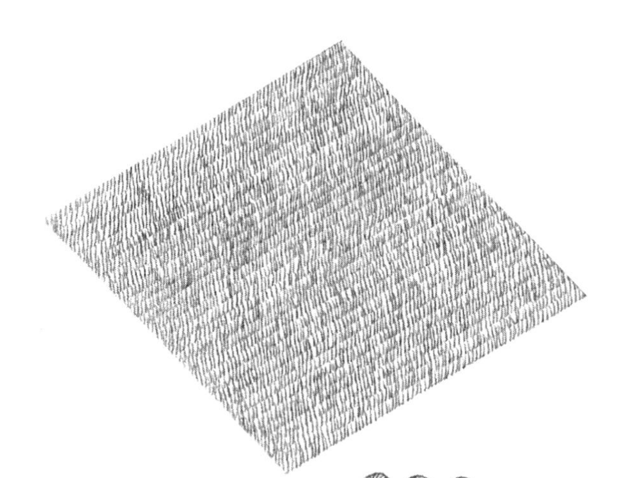

Carpeting provides floors with both visual and textural softness, resilience, and warmth in a wide range of colors and patterns. These qualities, in turn, enable carpeting to absorb sound, reduce impact noise, and provide a comfortable and safe surface to walk on. As a group, carpeting is also fairly easy to maintain.

Carpeting is normally installed wall to wall, covering the entire floor of a room. It can be laid directly over a subfloor and underlayment pad, obviating the need for a finish floor. It can also be laid over an existing floor.

Carpet Fibers

- Nylon: predominant face fiber; excellent strength and wearability; soil- and mildew-resistant; anti-static properties achieved through the use of conductive filaments
- Polypropylene (Olefin): good resistance to abrasion, soil, and mildew; used extensively in outdoor carpeting
- PET polyester: durable form of polyester made from recycled plastic containers; resists soiling, abrasion, stains, and fading.
- Wool: excellent resilience and warmth; good soil, flame, and solvent resistance; cleanable
- Cotton: not as durable as other face fibers, but softness and colorability used to advantage in flat-woven rugs

- Plastic fibers are a source of gases harmful to the respiratory system; some also yield toxic fumes when burned. Select carpets, carpet adhesives, and carpet pads that have passed the Carpet and Rug Institute's tests for indoor air quality. The Carpet and Rug Institute also recommends that ventilating fans be operated at full capacity and, if possible, doors and windows be opened for maximum ventilation during and for 48 to 72 hours after installation.

- LEED EQ Credit 4.3: Low-Emitting Materials, Carpet Systems

Carpet Construction

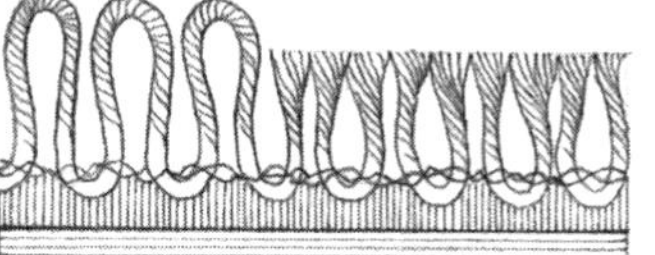

- Tufted carpet is made by mechanically stitching pile yarn through a primary fabric backing and bonding the yarn with latex to a secondary backing. The majority of carpet produced today is tufted.

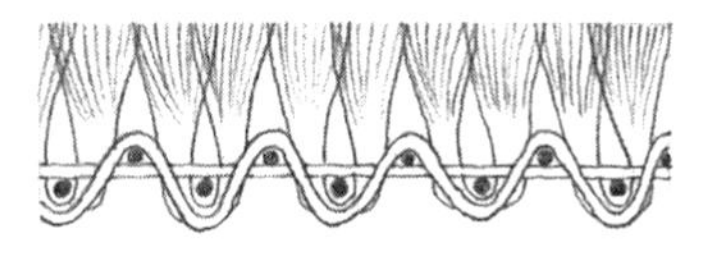

- Woven carpet is made by simultaneously interweaving the backing and pile yarns on a loom. Woven carpet is longer-wearing and more stable than tufted carpet, but it is more expensive to produce.

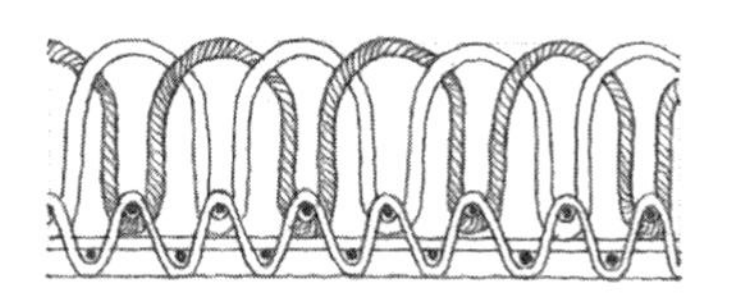

- Knitted carpet is made by looping the backing, stitching, and pile yarns with three sets of needles.

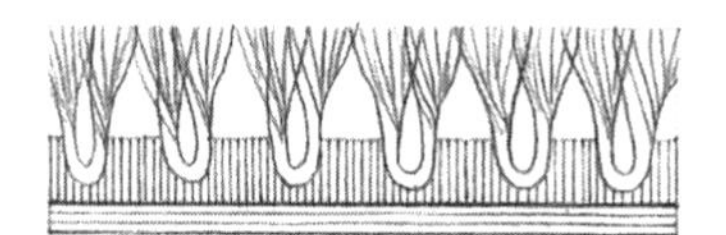

- Fusion-bonded carpet is made by heat-fusing face yarns to a vinyl backing supported by other materials.

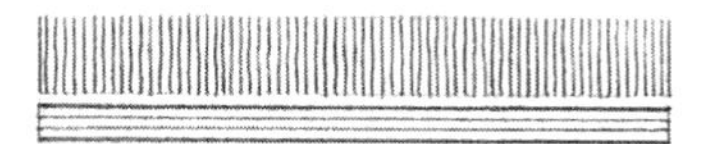

- Flocked carpet is made by propelling short strands of pile fiber electrostatically against an adhesive-coated backing.

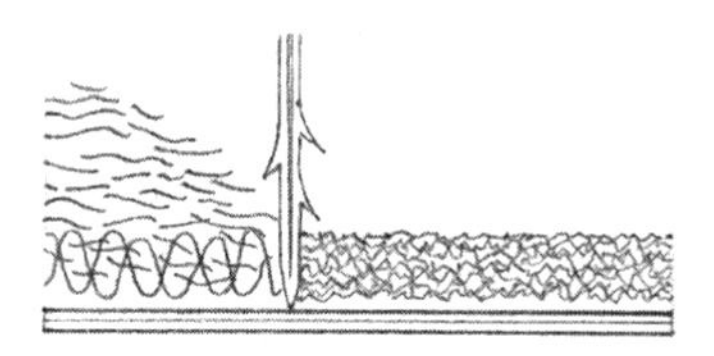

- Needle-punched carpet is made by punching carpet fibers back and forth through a woven polypropylene sheet with barbed needles to form a felted fiber mat.

- Pile refers to the upright tufts of yarn forming the surface of a carpet.
- Pile weight is the average weight of pile yarn in a carpet, stated in ounces per square yard.
- Pile density is the weight of pile yarn per unit volume of carpet, stated in ounces per cubic yard.
- Pitch is the crosswise number of tuft-forming pile yarns in a 27" (685) width of woven carpet.
- Gauge is the spacing of tufts across the width of a tufted or knitted carpet, expressed in fractions of an inch.

Carpet pad is a pad of cellular rubber, felted animal hair, or jute, over which carpet is installed to increase resilience and comfort, improve durability of the carpet, and reduce impact sound transmission.

Backing is the foundation material securing the pile yarns of a carpet and providing it with stiffness, strength, and dimensional stability.

Carpet Terminology

Carpet Textures

After color, texture is the prime visual characteristic of a carpet. The various carpet textures available are a result of the pile construction, pile height, and the manner in which the carpet is cut. There are three major groups of carpet textures—cut pile, loop pile, and a combination of cut and loop pile.

- Cut pile is created by cutting each loop of pile yarn, producing a range of textures from informal shags to short, dense velvets.

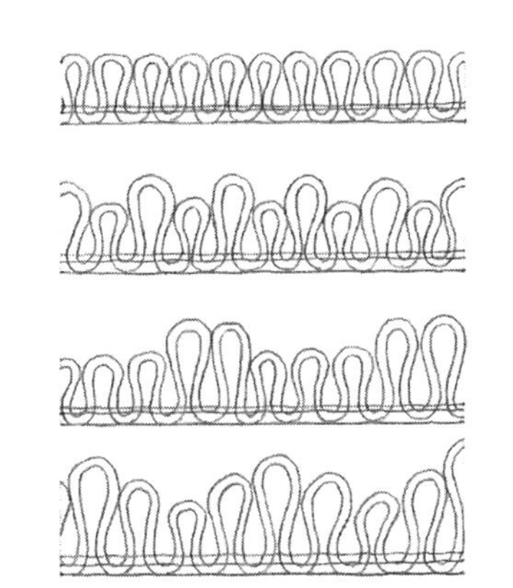

- Plush: smooth cut pile; cut yarn ends blend; called velvet plush when dense pile is cut closely
- Saxony plush: texture between plush and shag; thicker yarn
- Twist or frieze; heavier, rougher texture than plush; twist set into yarn
- Shag: heavily textured surface created by long, twisted yarns

- Loop pile is created by weaving, tufting, or knitting the pile yarn into loops. Loop pile is tougher and more easily maintained than cut pile but is less versatile in color and pattern.

- Level loop: looped tufts are at the same height; very sturdy; little textural variation
- Ribbed loop: creates directional, ribbed, or corrugated texture
- Hi-lo loop: adds another dimension to the loop texture
- Multilevel loop: capable of producing sculptured patterns

- Combination loop and pile adds a degree of warmth to all-loop pile. It can be produced in tufted and woven constructions.

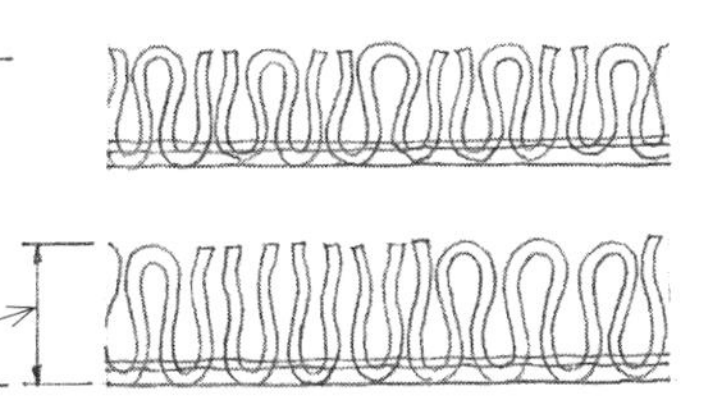

- Cut and loop: cut and uncut loops alternate in a uniform fashion; adds a degree of softness and warmth to loop texture; symmetrical geometric figures may be created by cut rows.

ADA Accessibility Guidelines

- Securely attach carpet to a firm underlayment.
- Carpet should have a level cut pile, level loop, textured loop, or cut-and-loop texture, with a maximum pile height of 1/2" (13).
- Fasten and trim all exposed edges to the floor surface.
- Bevel edge trim to a slope of 1:20 if more than 1/4" (6) high.

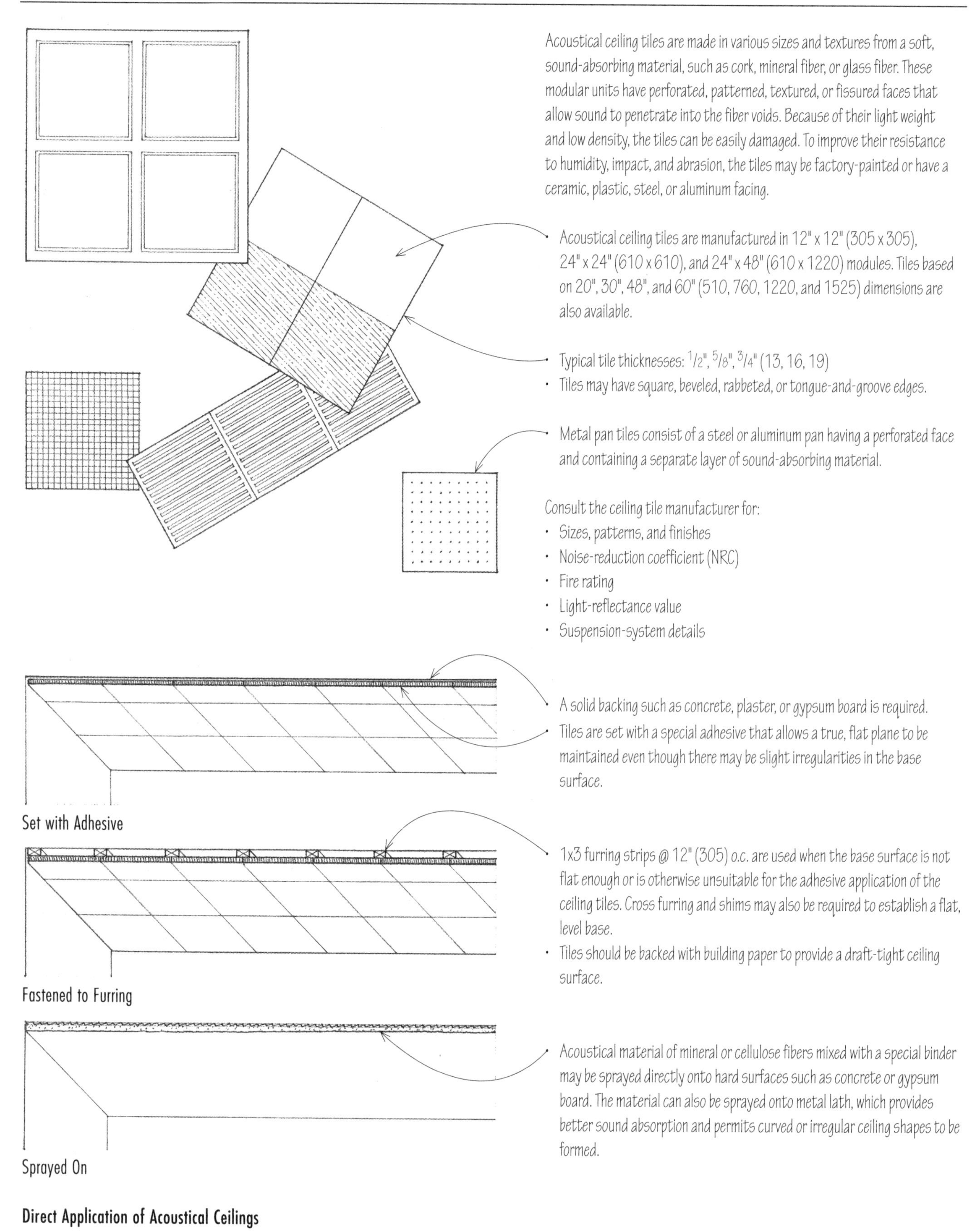

Acoustical ceiling tiles are made in various sizes and textures from a soft, sound-absorbing material, such as cork, mineral fiber, or glass fiber. These modular units have perforated, patterned, textured, or fissured faces that allow sound to penetrate into the fiber voids. Because of their light weight and low density, the tiles can be easily damaged. To improve their resistance to humidity, impact, and abrasion, the tiles may be factory-painted or have a ceramic, plastic, steel, or aluminum facing.

- Acoustical ceiling tiles are manufactured in 12" x 12" (305 x 305), 24" x 24" (610 x 610), and 24" x 48" (610 x 1220) modules. Tiles based on 20", 30", 48", and 60" (510, 760, 1220, and 1525) dimensions are also available.

- Typical tile thicknesses: 1/2", 5/8", 3/4" (13, 16, 19)
- Tiles may have square, beveled, rabbeted, or tongue-and-groove edges.

- Metal pan tiles consist of a steel or aluminum pan having a perforated face and containing a separate layer of sound-absorbing material.

Consult the ceiling tile manufacturer for:
- Sizes, patterns, and finishes
- Noise-reduction coefficient (NRC)
- Fire rating
- Light-reflectance value
- Suspension-system details

- A solid backing such as concrete, plaster, or gypsum board is required.
- Tiles are set with a special adhesive that allows a true, flat plane to be maintained even though there may be slight irregularities in the base surface.

Set with Adhesive

- 1x3 furring strips @ 12" (305) o.c. are used when the base surface is not flat enough or is otherwise unsuitable for the adhesive application of the ceiling tiles. Cross furring and shims may also be required to establish a flat, level base.
- Tiles should be backed with building paper to provide a draft-tight ceiling surface.

Fastened to Furring

- Acoustical material of mineral or cellulose fibers mixed with a special binder may be sprayed directly onto hard surfaces such as concrete or gypsum board. The material can also be sprayed onto metal lath, which provides better sound absorption and permits curved or irregular ceiling shapes to be formed.

Sprayed On

Direct Application of Acoustical Ceilings

CSI MasterFormat 09 51 00: Acoustical Ceilings

Acoustical ceiling tiles can be suspended from an overhead floor or roof structure to provide a concealed space for mechanical ductwork, electrical conduit, and plumbing lines. Light fixtures, sprinkler heads, fire detection devices, and sound systems can be recessed into the ceiling plane. The ceiling membrane can be fire-rated and provide fire protection for the supporting floor and roof structure. Thus, the ceiling system is able to integrate the functions of lighting, air distribution, acoustical control, and fire protection.

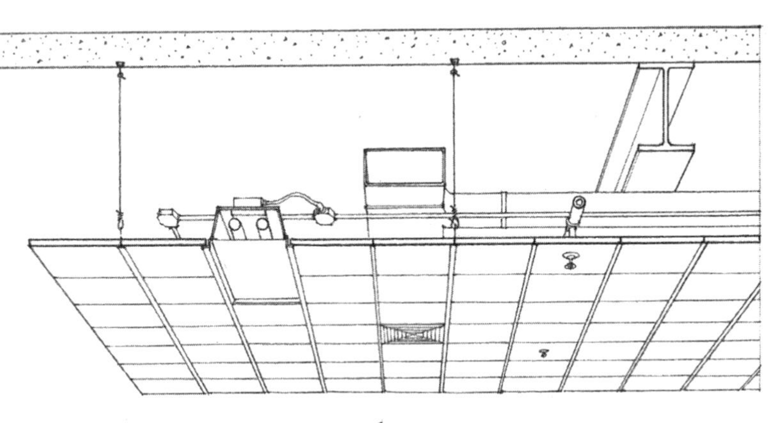

Although the suspension systems of each manufacturer may vary in their details, they all consist of a grid of main channels or runners, cross tees, and splines. This grid, suspended from the overhead floor or roof structure, may be exposed, recessed, or fully concealed. In most suspension systems, the acoustical tiles are removable for replacement or for access into the ceiling space.

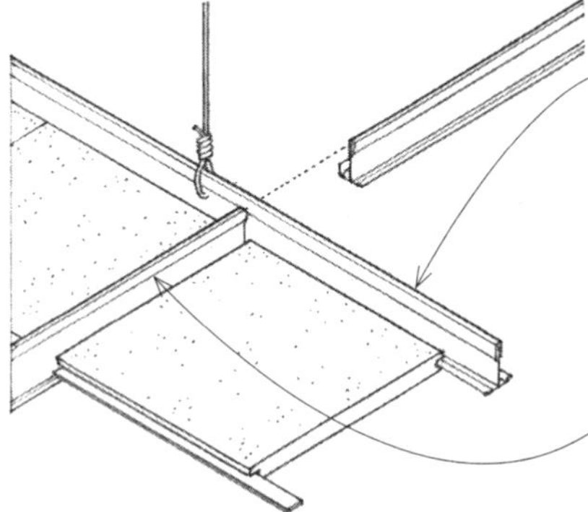

- Main runners are the principal supporting members of a suspended ceiling system, usually consisting of sheet-metal tees or channels suspended by hanger wires from the overhead structure.
- Cross tees are the secondary supporting members, usually consisting of sheet-metal tees carried by the main runners.

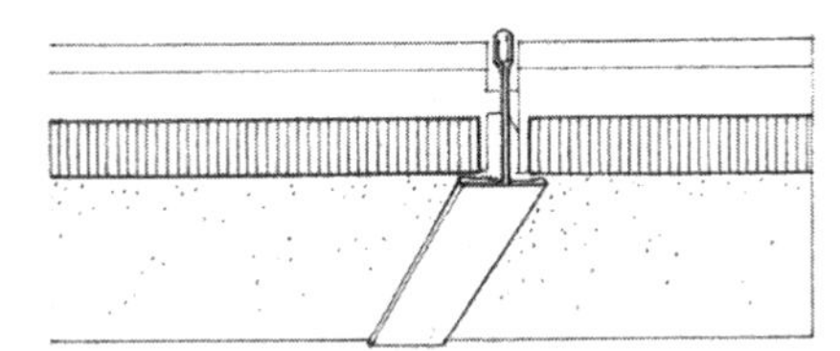

- Exposed grid suspension systems support the acoustical tiles with inverted tees.

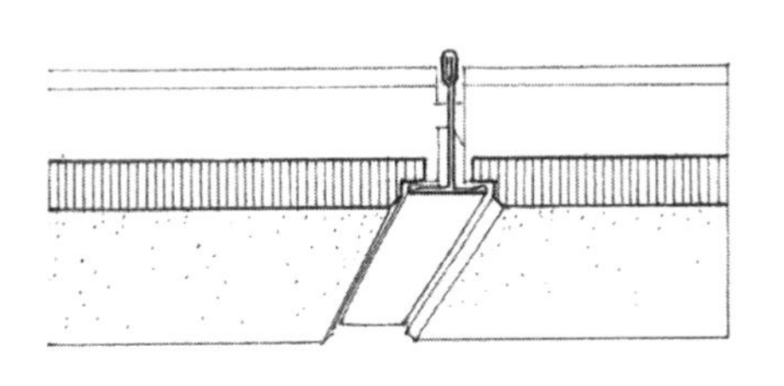

- Recessed grid suspension systems support acoustical tiles within rabbeted joints.

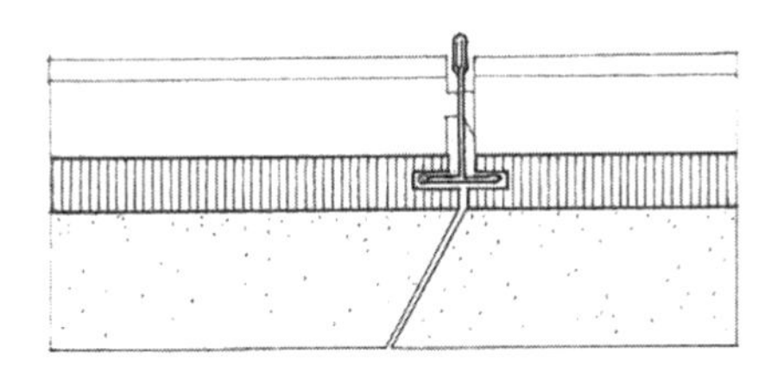

- Concealed grid suspension systems are hidden within kerfs cut into the edges of the acoustical tiles.

Ceiling canopies and clouds are made of fabric, acoustical tile, metal, translucent plastic, or other materials. Suspended from pendants or wires, they allow ceilings to be dropped over small areas, even below other ceiling finish materials, and usually allow access to equipment above.

Integrated ceiling systems incorporate acoustical, lighting, and air-handling components into a unified whole. The suspension systems, which typically form a 60" x 60" (1525 x 1525) grid, may support either flat or coffered acoustical panels. Air-handling components may be integral parts of modular luminaires and disperse conditioned air along the edges of the fixtures, or be integrated into the suspension system and diffuse conditioned air through long, narrow slots between the ceiling panels.

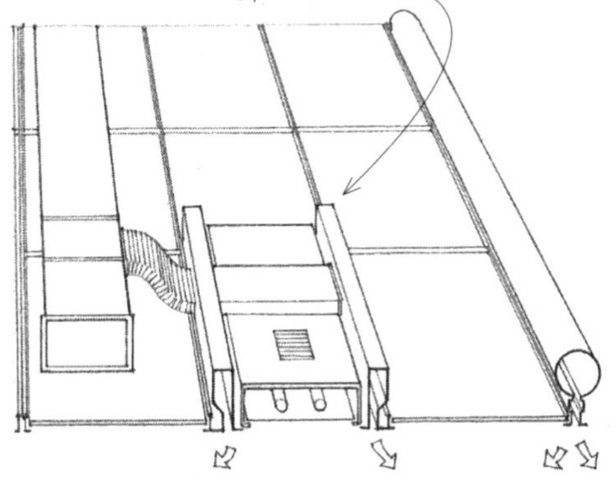

Linear metal ceilings consist of narrow anodized aluminum, painted steel, or stainless steel strips. The slots between the spaced strips may be open or closed. Open slots permit sound to be absorbed by a backing of batt insulation in the ceiling space. Linear metal ceiling systems usually incorporate modular lighting and air-handling components.

CSI MasterFormat 09 53 00: Acoustical Ceiling Suspension Assemblies
CSI MasterFormat 09 54 00: Specialty Ceilings

The strength and rigidity of ordinary wood framing are more important than its appearance because it is normally covered with a finish surface. In finish trim, cabinetry, and furniture work, however, the appearance of a wood joint becomes just as important as its strength. Small-scale work requires more sophisticated and refined joints that present a clean appearance.

Wood joints can express the manner in which the members are connected, or they can be relatively inconspicuous. In either case, they must remain tight. If they open due to the shrinkage or structural movement of the wood, they become both weaker and more noticeable.

In designing and constructing a wood joint, it is important to understand the basic nature of the compressive, tensile, shear forces acting on the joint, and to comprehend their relationship to the direction of the wood grain. See 12.11.

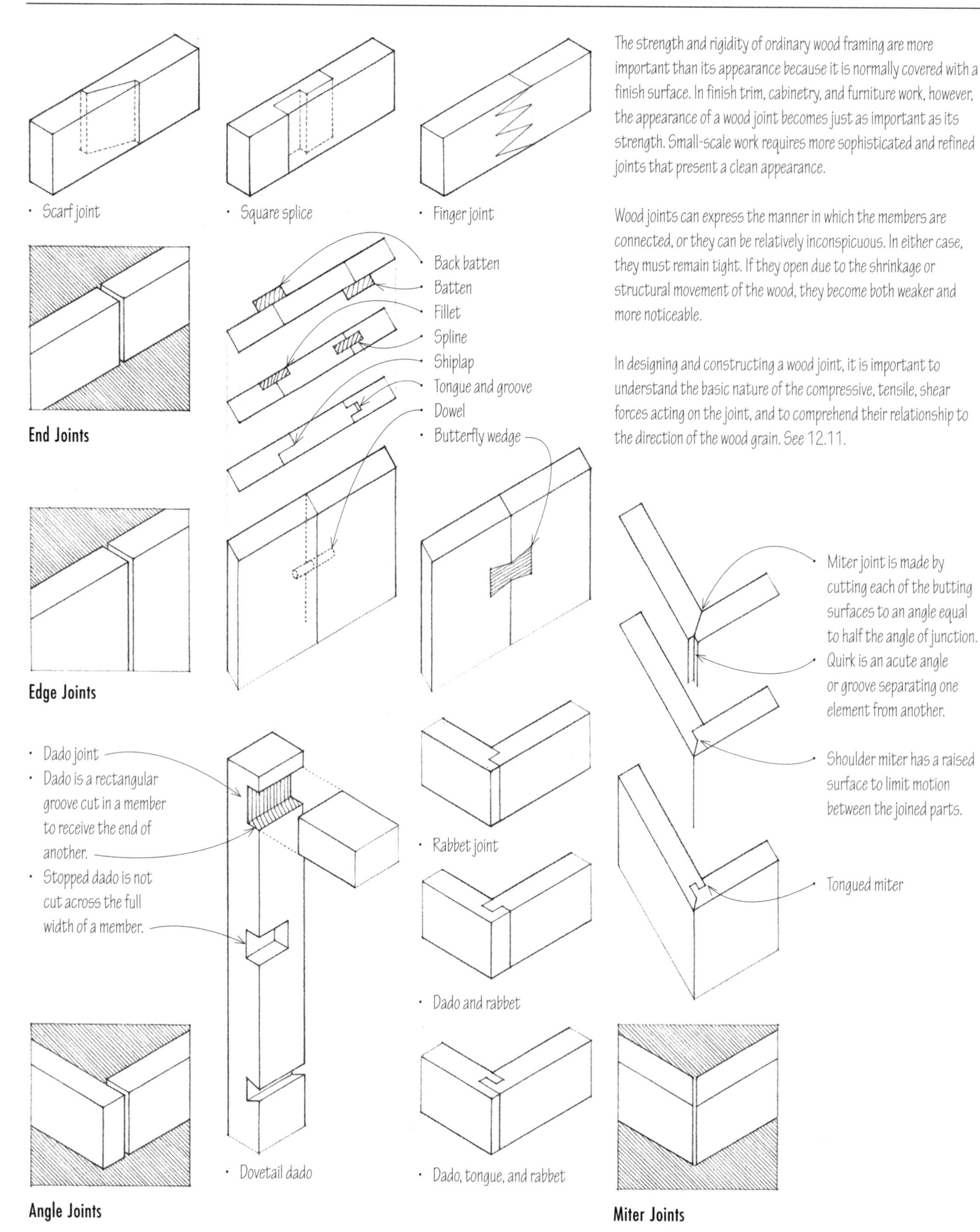

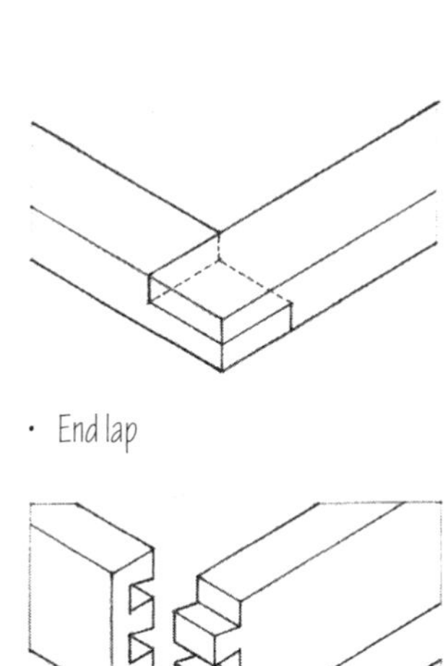

• End lap

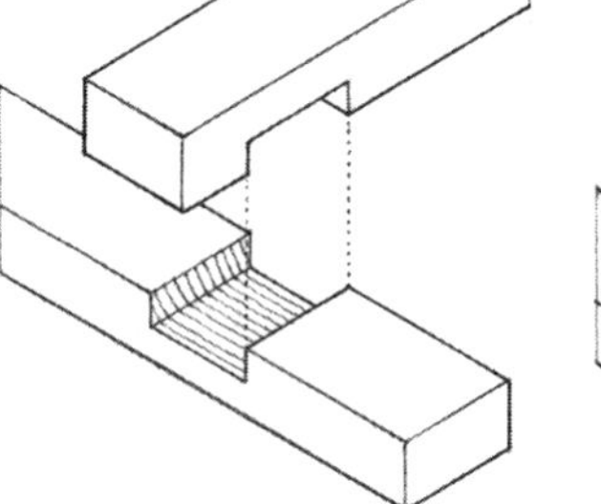

• Cross lap

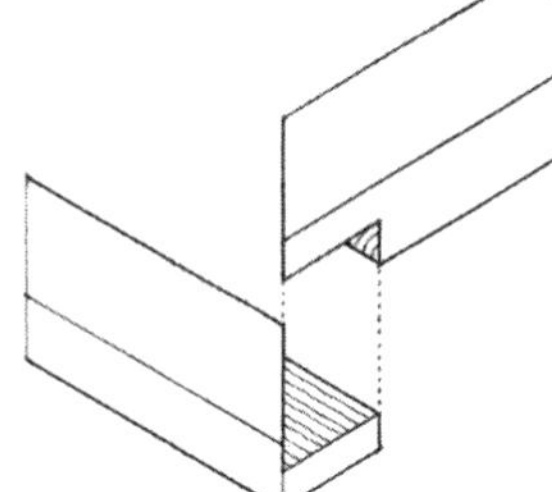

• Mitered halving

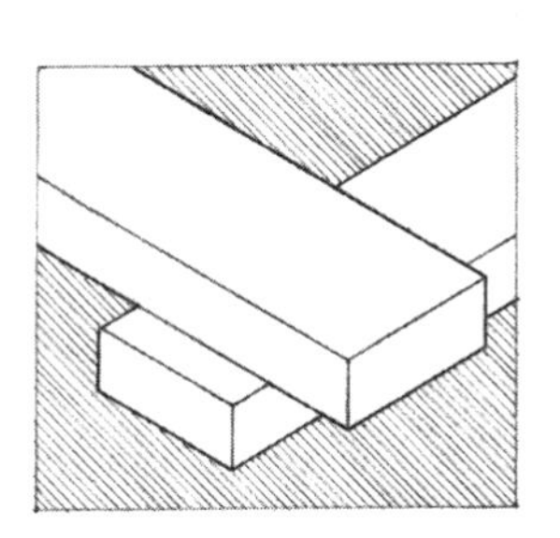

Lap Joints

• Lap or half-blind dovetail

• Secret dovetail or blind miter

• Halving refers to cutting away half of each member at the place of joining so that a flush surface results.

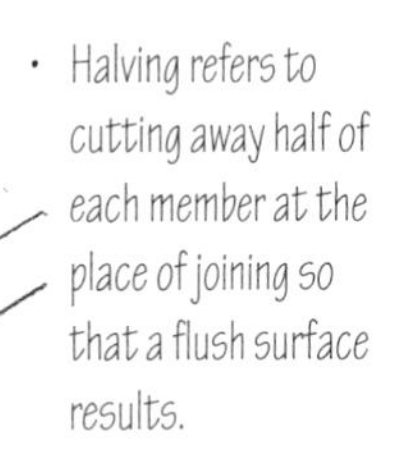

• Dovetail halving

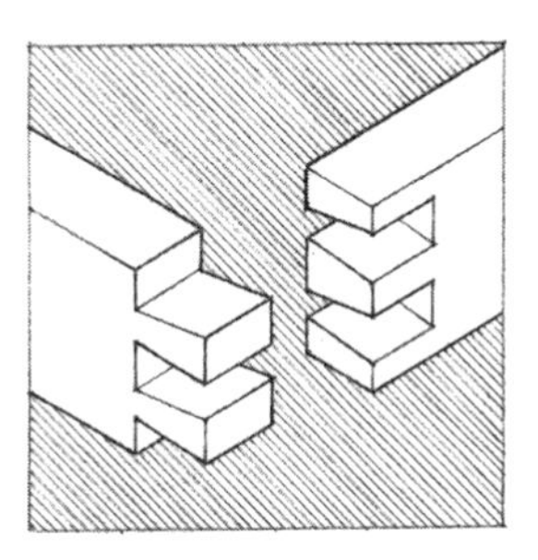

Dovetail Joints

• Blind mortise and stub tenon

• Foxtail wedge spreads and secures a stub tenon when driven into a blind mortise.

• Pinned joint

• Haunched tenon is strengthened by a beveled shoulder at a blind joint.

• Through tenon

• Keyed joint

• Open or slip mortise

Mortise-and-Tenon Joints

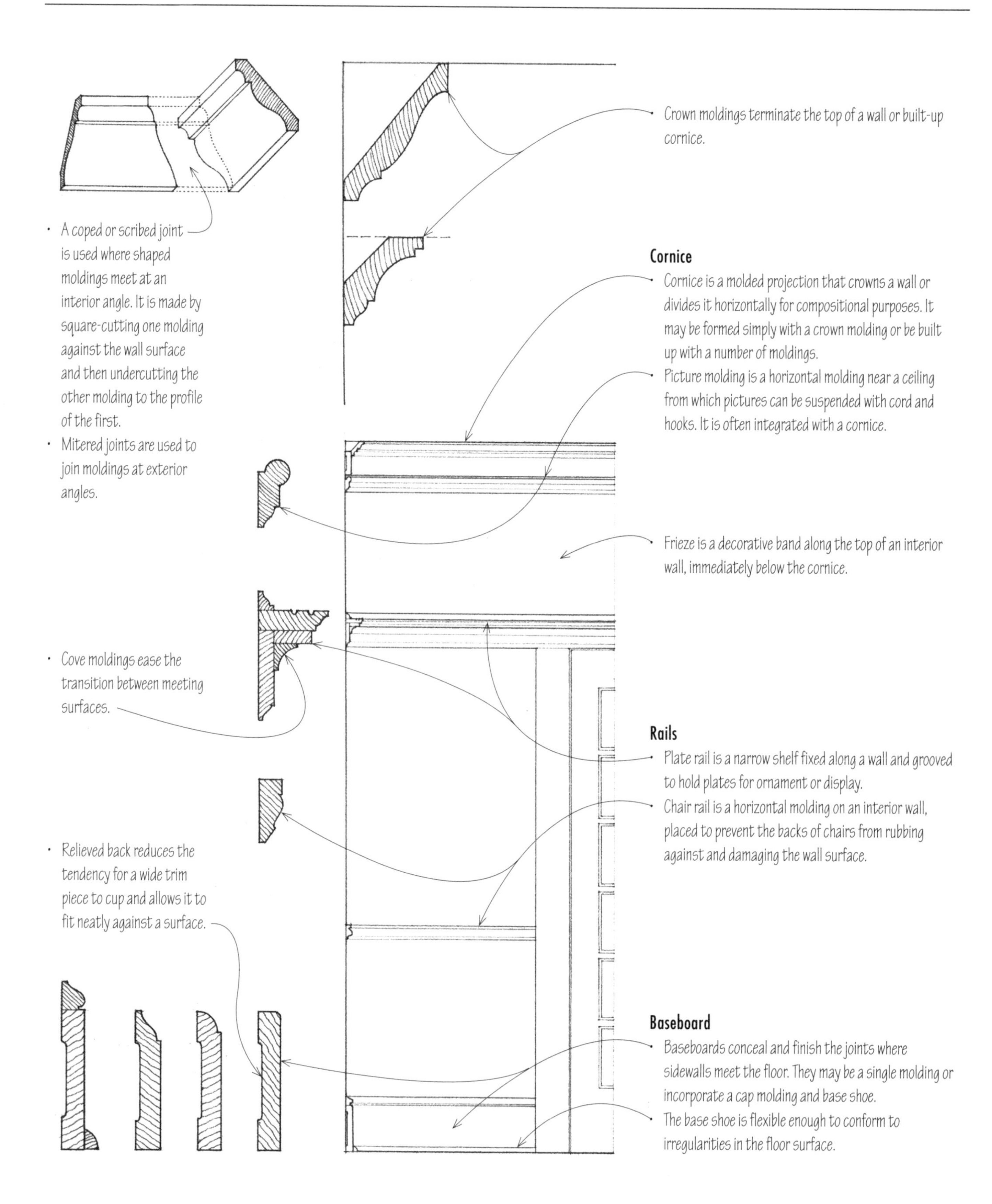

• A coped or scribed joint is used where shaped moldings meet at an interior angle. It is made by square-cutting one molding against the wall surface and then undercutting the other molding to the profile of the first.
• Mitered joints are used to join moldings at exterior angles.
• Crown moldings terminate the top of a wall or built-up cornice.
Cornice
• Cornice is a molded projection that crowns a wall or divides it horizontally for compositional purposes. It may be formed simply with a crown molding or be built up with a number of moldings.
• Picture molding is a horizontal molding near a ceiling from which pictures can be suspended with cord and hooks. It is often integrated with a cornice.
• Frieze is a decorative band along the top of an interior wall, immediately below the cornice.
• Cove moldings ease the transition between meeting surfaces.
Rails
• Plate rail is a narrow shelf fixed along a wall and grooved to hold plates for ornament or display.
• Chair rail is a horizontal molding on an interior wall, placed to prevent the backs of chairs from rubbing against and damaging the wall surface.
• Relieved back reduces the tendency for a wide trim piece to cup and allows it to fit neatly against a surface.
Baseboard
• Baseboards conceal and finish the joints where sidewalls meet the floor. They may be a single molding or incorporate a cap molding and base shoe.
• The base shoe is flexible enough to conform to irregularities in the floor surface.

For use as trim, a variety of stock wood moldings are available at millwork shops. They vary in section, length, and species of wood. They can be used singly or be combined to form more complex sections. In addition to these stock sections, wood moldings can be milled to custom specifications.

The type of wood used for trim depends on the type of finish to be applied to the woodwork. For painted finishes, the wood should be close-grained, smooth, and free of pitch streaks or other imperfections. If the woodwork is to receive a transparent or natural finish, the wood should have a uniform color, an attractive figure, and a degree of hardness.

Interior trim is normally applied after the finish walls, ceiling, and flooring are in place. Although decorative in nature, interior trim also serves to conceal, finish, and perfect the joints between interior materials.

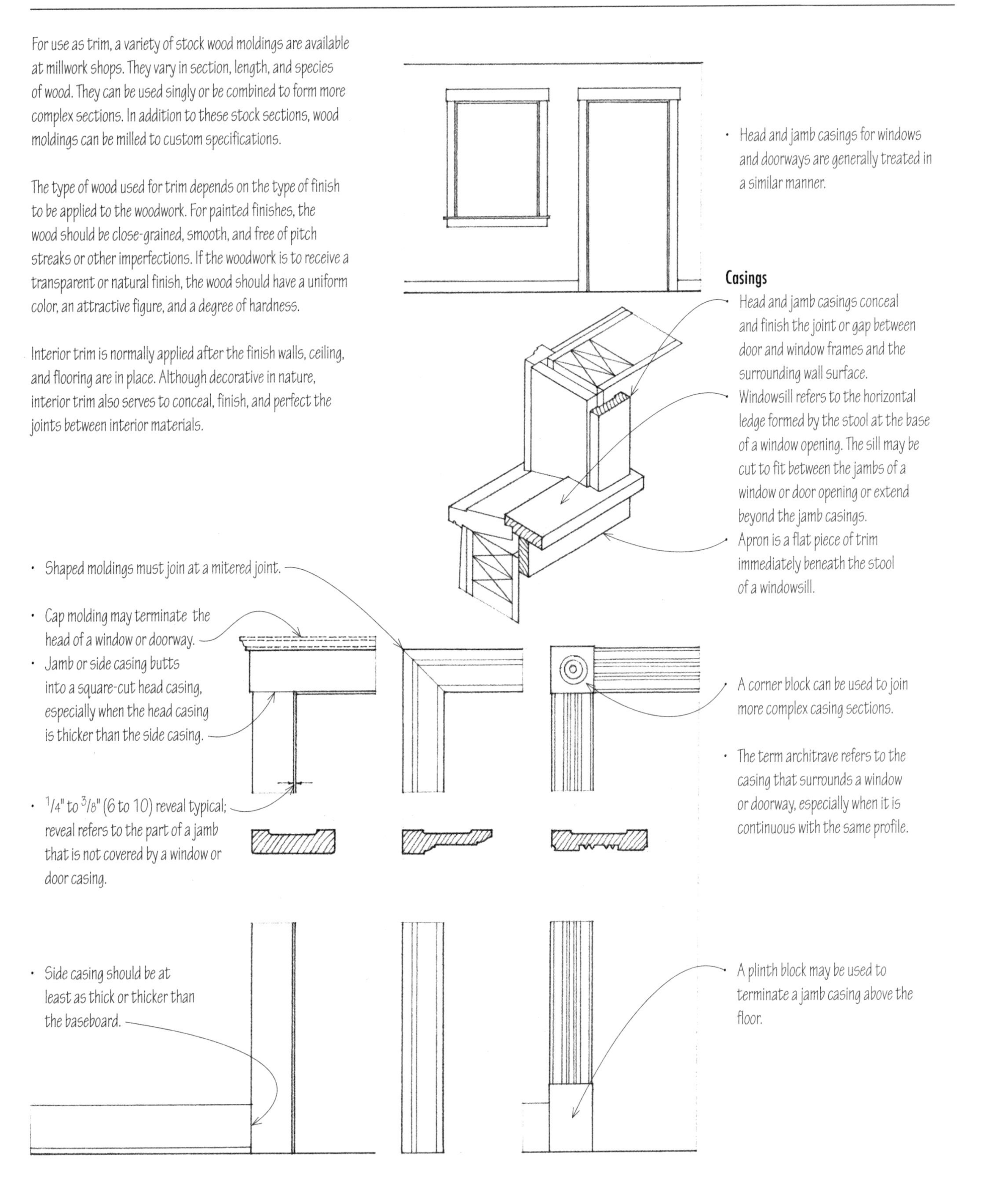

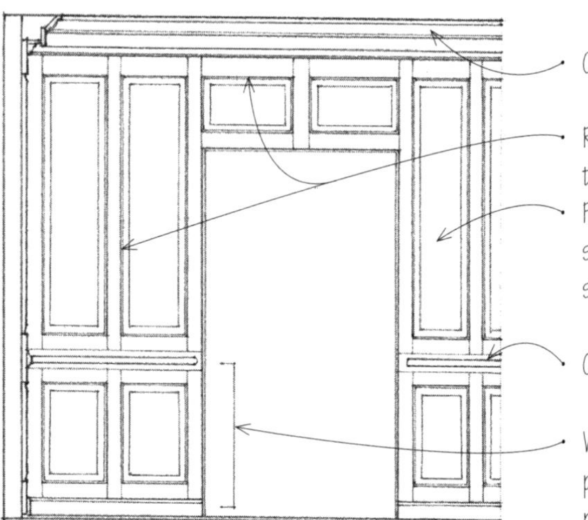

Interior wood paneling may consist of veneer-faced panels applied directly to wood or metal framing or furring. Furring is required over masonry or concrete walls. Furring may also be used over frame walls when improved thermal insulation properties, greater acoustical isolation, or additional wall depths are desired. The panels are normally fastened with nails or screws although adhesives can be used for greater rigidity. The final appearance of the paneled wall will depend on the treatment of the joints and the grain or figure of the wood panels.

Solid wood planks may also be used for interior paneling. The planks may have square cut, tongue-and-groove, or shiplap edges. The resulting wall pattern and texture will depend on the plank width, orientation, spacing, and joint details.

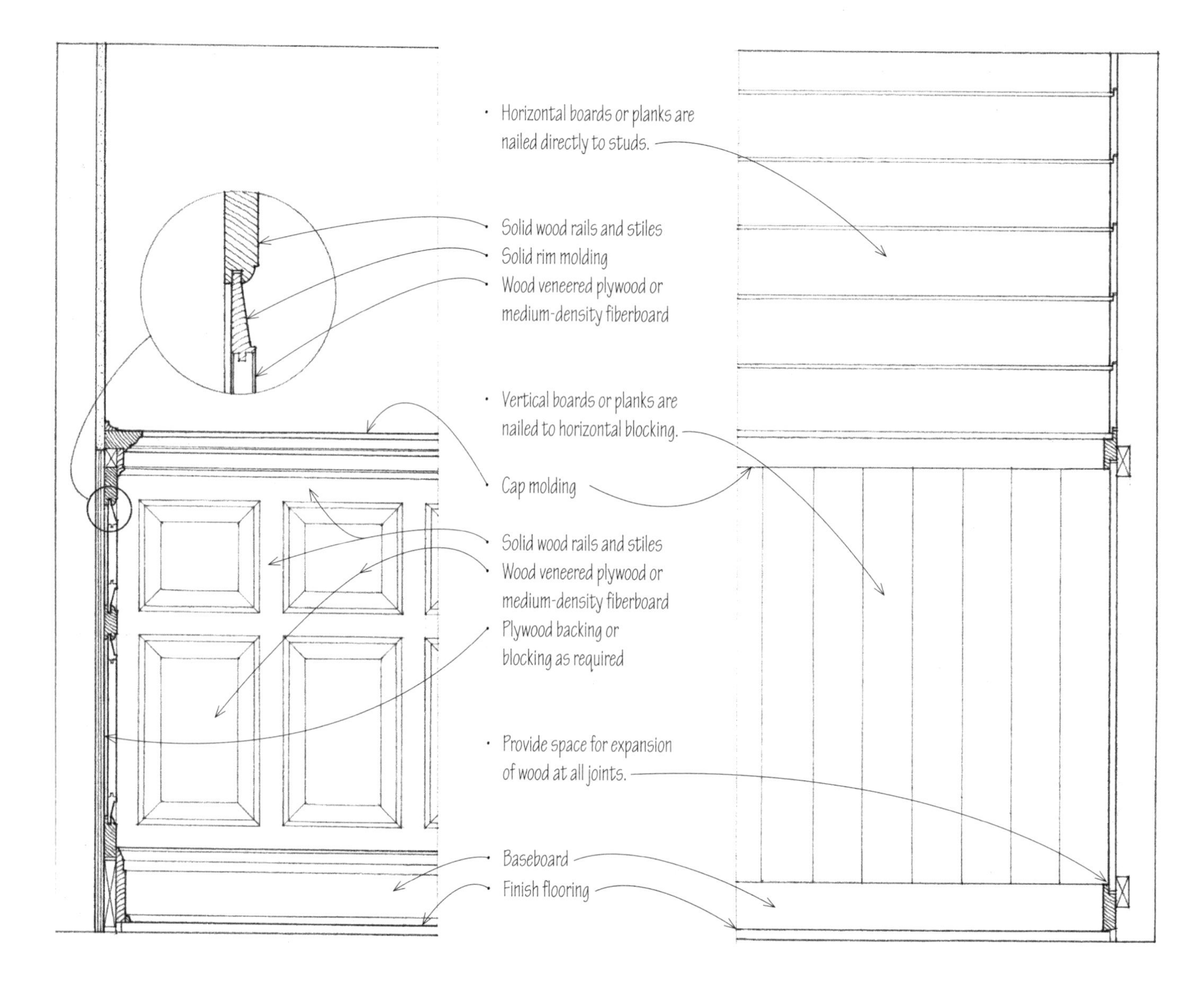

CSI MasterFormat 06 42 00: Wood Paneling

Decorative plywood panels are available with hardwood or softwood face veneers for use as wall paneling, cabinetry, and furniture work. The panels are typically 4' x 8' (1220 x 2440) and available in 1/4", 3/8", 1/2", and 3/4" (6, 10, 13, and 19) thicknesses.

Matching Patterns

The appearance of naturally finished plywood paneling depends on the species of wood used for the face veneer and the way in which the sheets of veneer are arranged so as to emphasize the color and figure of the wood.

- Book matching arranges veneers from the same flitch alternately face up and face down to produce symmetrical mirror images about the joints between adjacent sheets.
- Herringbone matching is book matching in which the figures in adjacent sheets slope in opposite directions.
- Slip matching arranges adjacent sheets of veneer from the same flitch side by side without turning so as to repeat the figure.
- Diamond matching arranges four diagonally cut sheets of a veneer to form a diamond pattern about a center.
- Random matching arranges veneers to intentionally create a casual, unmatched appearance.

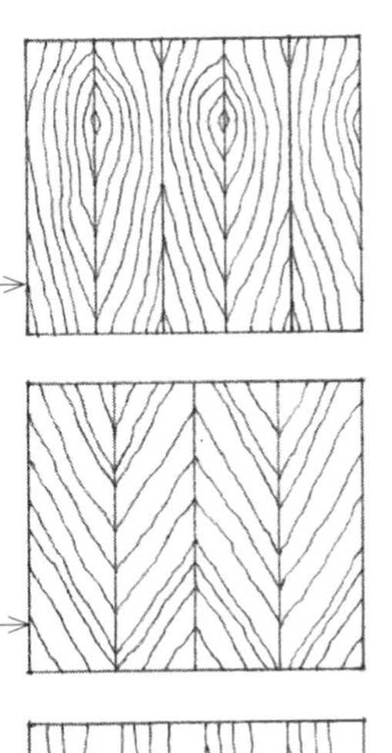

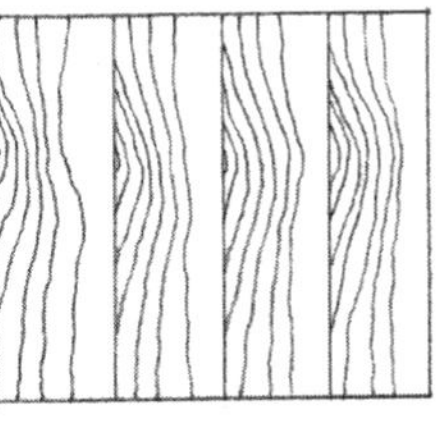

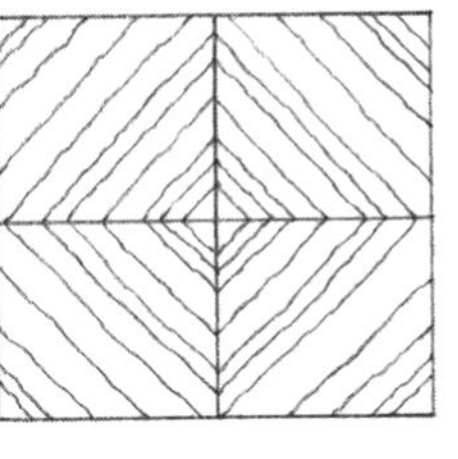

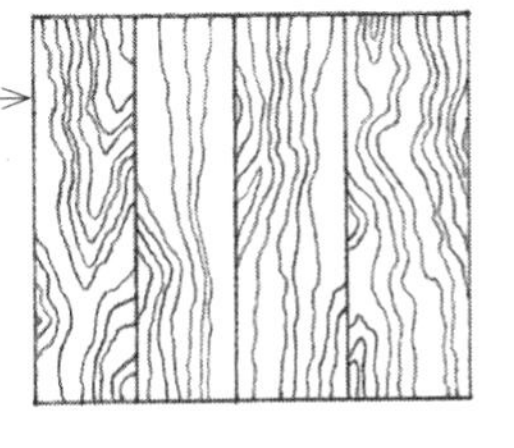

Softwood Veneer Grades

N – select, smooth surface for natural finishes
A – smooth face suitable for painting
B – solid surface utility panel

Hardwood Veneer Grades

- Premium grade permits only a few small burls, knots, and inconspicuous patches.
- Good grade is similar to premium grade except that matching of veneer faces is not required and no sharp contrasts in color are allowed.
- Sound grade is a smooth veneer free of open defects but containing streaks, discoloration, patches, and small, sound, and tight knots.
- Utility grade permits discoloration, streaks, patches, tight knots, small knotholes, and splits.
- Backing grade permits larger defects not affecting the strength or durability of the panel.

Grain Figures

Figure refers to the natural pattern on a sawn wood surface produced by the intersection of annual rings, knots, burls, rays, and other growth characteristics. Different figures may be produced by varying the way in which a wood veneer is cut from a log.

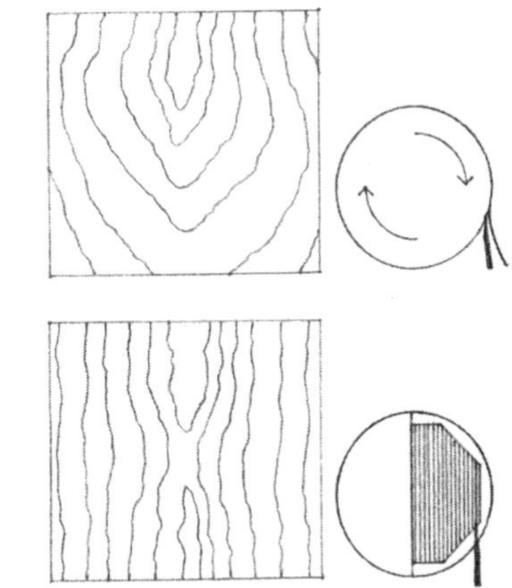

- Rotary cutting against the edge of a knife in a lathe produces a continuous veneer with a variegated ripple figure.
- Flat or plain slicing of a half-log parallel to a line through its center produces a variegated wavy figure.

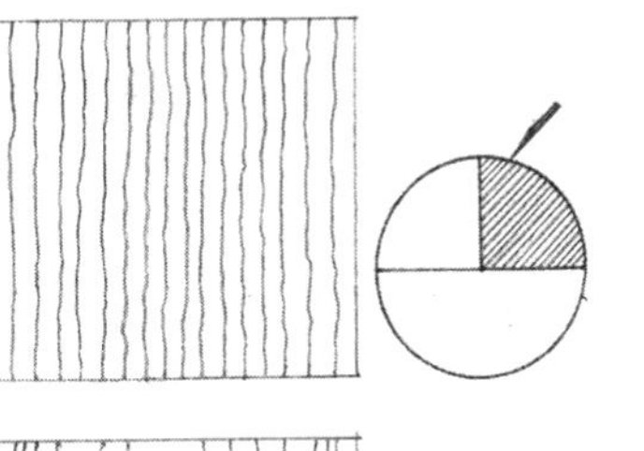

- Quarter slicing of a log perpendicular to the annual rings produces a series of straight or varied stripes in the veneer.

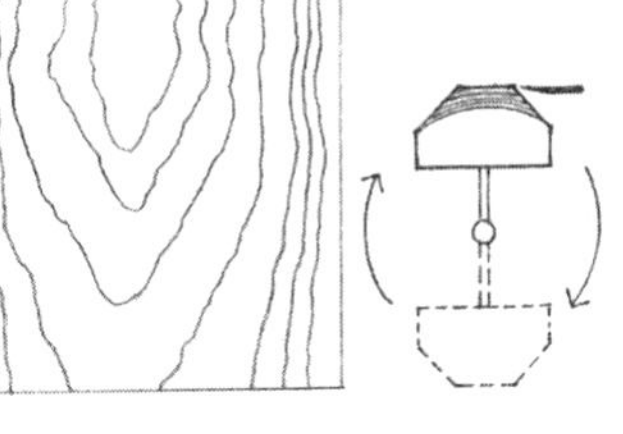

- Half-round slicing of a flitch mounted off-center in the lathe, slightly across the annual rings, produces characteristics of both rotary cutting and flat slicing.

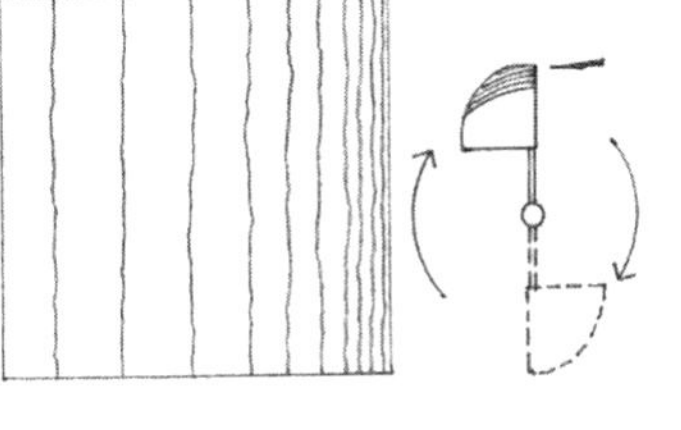

- Rift cutting is the slicing of oak and similar species perpendicular to the conspicuous, radiating rays so as to minimize their appearance.

LEED EQ Credit 4.4: Low-Emitting Materials, Composite Wood & Agrifiber Products

CSI MasterFormat 06 42 16: Wood-Veneer Paneling

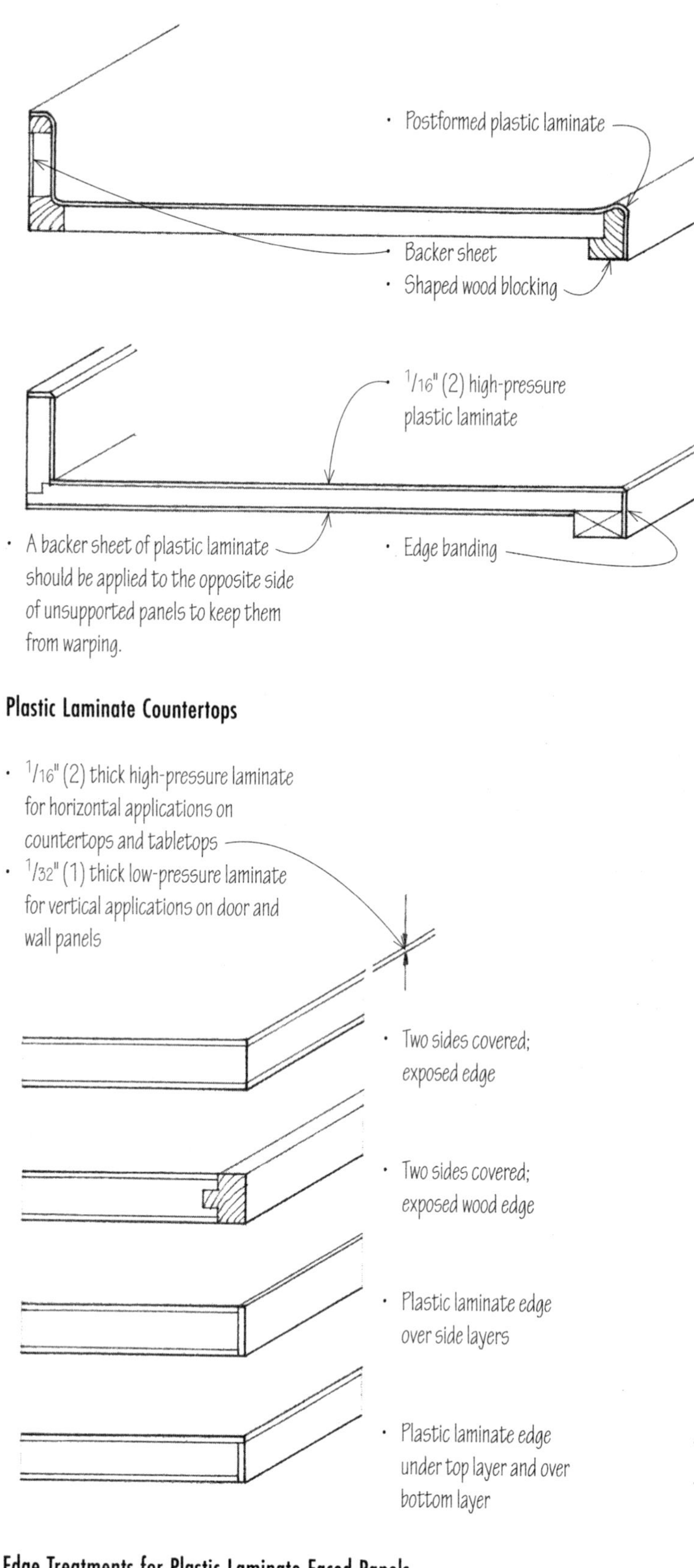

Plastic laminate is a hard surfacing material consisting of superposed layers of kraft paper, foil, printed paper, wood veneer, or fabric impregnated with melamine and phenolic resins, fused together under heat and pressure. Plastic laminates provide a durable, heat- and water-resistant surface covering for countertops, furniture, doors, and wall panels. They may be applied to smooth plywood, hardboard, particleboard, and other common core materials. They may be bonded with contact adhesive in the field or with thermosetting adhesive, under pressure, in the shop.

- High-pressure laminate is molded and cured in the range of pressures from 1200 to 2000 psi (84 to 140 kg/m^2), and used for surfacing countertops and tabletops.
- Low-pressure laminate is molded and cured with a maximum pressure of 400 psi (28 kg/m^2), and used in vertical and low-wear applications.
- Formica is a trademark for a brand of plastic laminate.
- Plastic laminate surfaces with tight rolls and bends should be postformed during manufacture and bonded with thermosetting adhesive. Postformed plastic laminate 1/20" (1.2 mm) thick may be bent to a radius as small as 3/4" (19). Plastic laminate edge banding may be bent to a radius of 3" (75) or smaller if heated.
- A wide range of colors and patterns is available in glossy, satin, low-glare, or textured finishes.

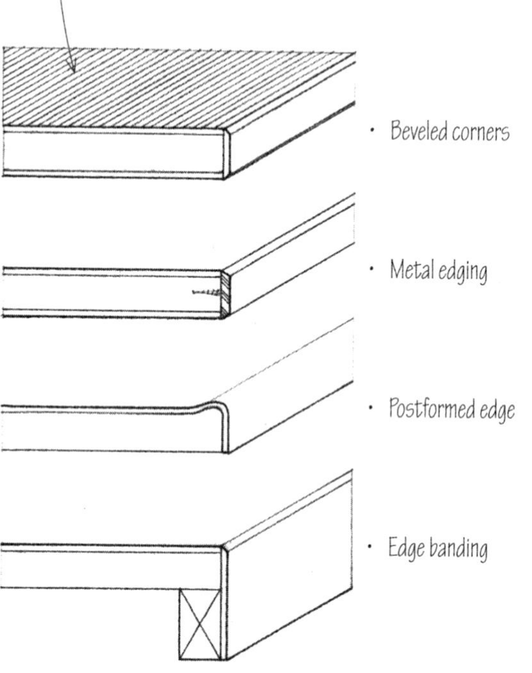

Edge Treatments for Plastic-Laminate-Faced Panels

CSI MasterFormat 06 41 16: Plastic-Laminate-Clad Architectural Cabinets
CSI MasterFormat 09 62 19: Laminate Flooring
CSI MasterFormat 12 36 23.13: Plastic-Laminate-Clad Countertops

EQ Credit 4.1: Low-Emitting Materials, Adhesives & Sealants

11

MECHANICAL & ELECTRICAL SYSTEMS

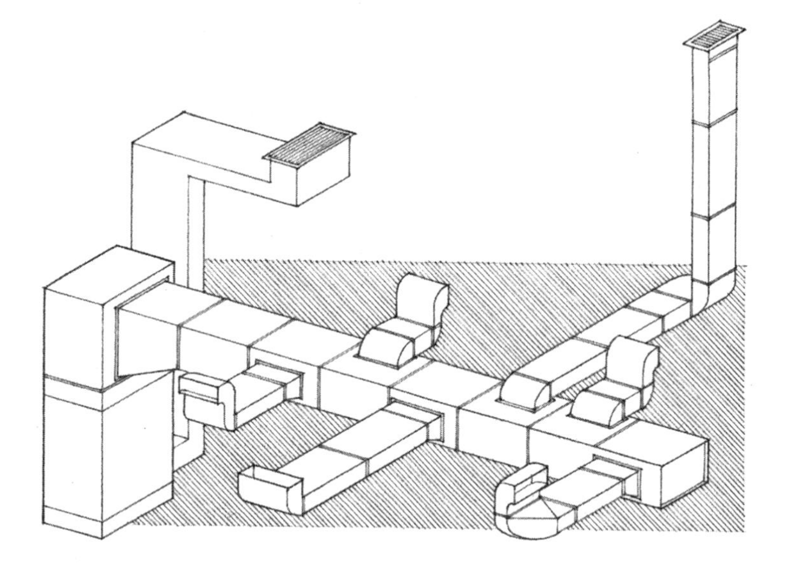

This chapter discusses the mechanical and electrical systems that are required to maintain the necessary conditions of environmental comfort, health, and safety for the occupants of a building. The intent is not to provide a complete design manual but to outline those factors that should be considered for the successful operation of these systems and their integration with other building systems.

Heating, ventilating, and air-conditioning systems condition the interior spaces of a building for the environmental comfort of the occupants. A potable water supply is essential for human consumption and sanitation. The efficient disposal of fluid waste and organic matter is necessary in order to maintain sanitary conditions within a building and in the surrounding area. Electrical systems furnish light and heat for a building's occupants, and power to run its machines.

These systems require a significant amount of space. Because much of the hardware for these systems is normally hidden from view—within concealed construction spaces or special rooms—the layout of these systems should be carefully integrated with each other as well as with the structural and enclosure systems of the building.

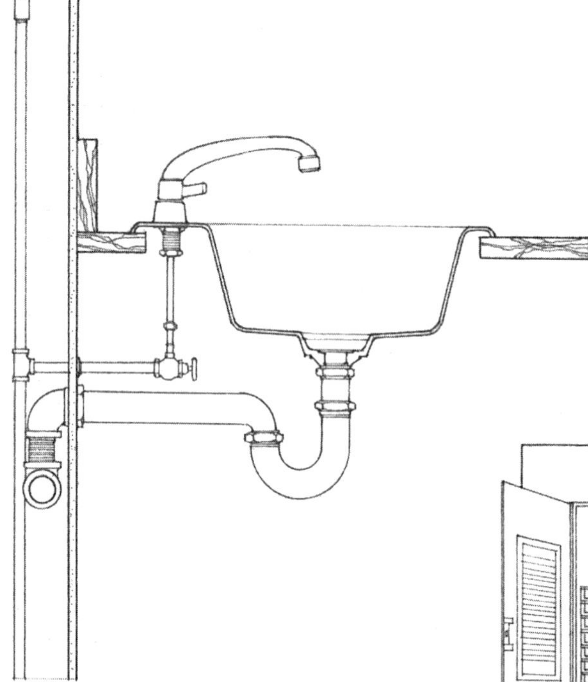

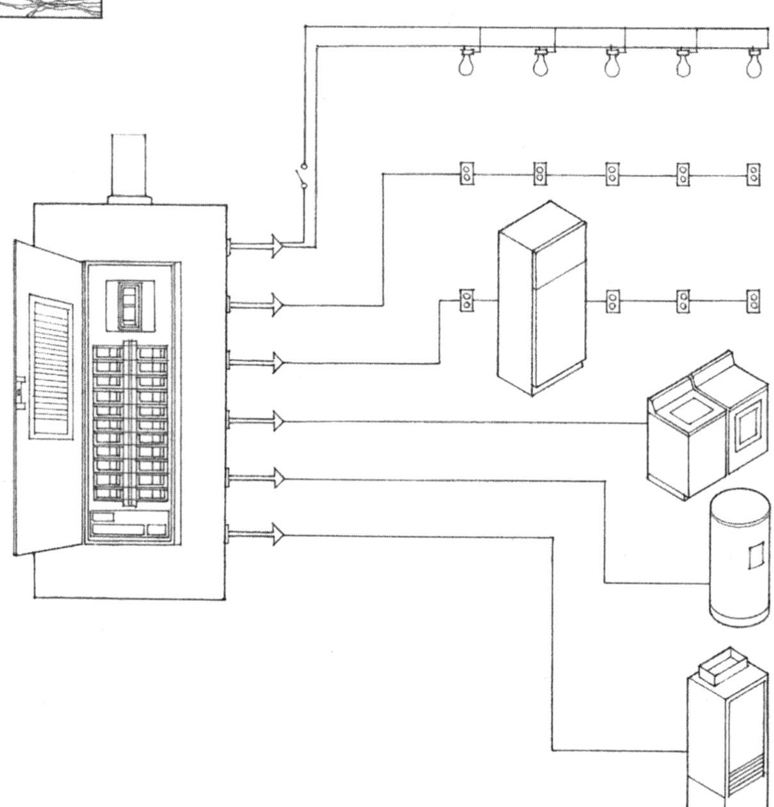

At rest, the human body produces about 400 Btu/h (117 W) of energy. Moderate activities like walking can raise this amount to 750 Btu/h (220 W), while strenuous activities can cause the body to generate up to 1200 Btu/h (351 W). Thermal comfort is achieved when the human body is able to dissipate the heat and moisture it produces by metabolic action in order to maintain a stable, normal body temperature. In other words, thermal equilibrium must exist between the body and its environment.

The human body loses or transfers heat to the surrounding air and surfaces in the following ways.

Conduction

- Conduction is the transfer of heat from the warmer to the cooler particles of a medium or of two bodies in direct contact, occurring without perceptible displacement of the particles themselves.
- Conduction accounts for a very small portion of the total heat loss from the body.

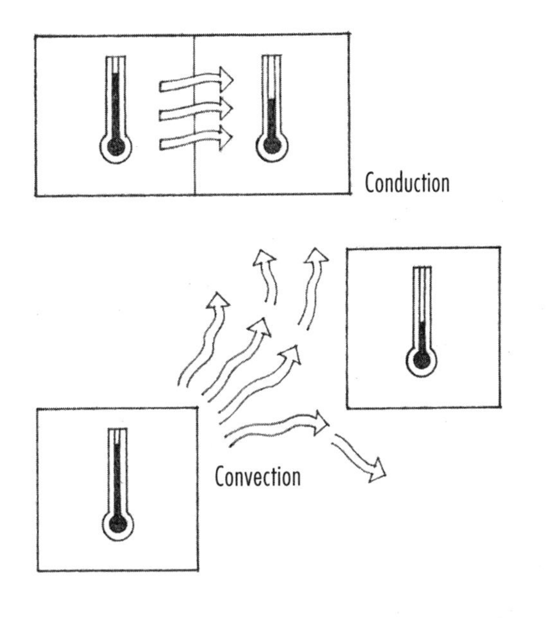

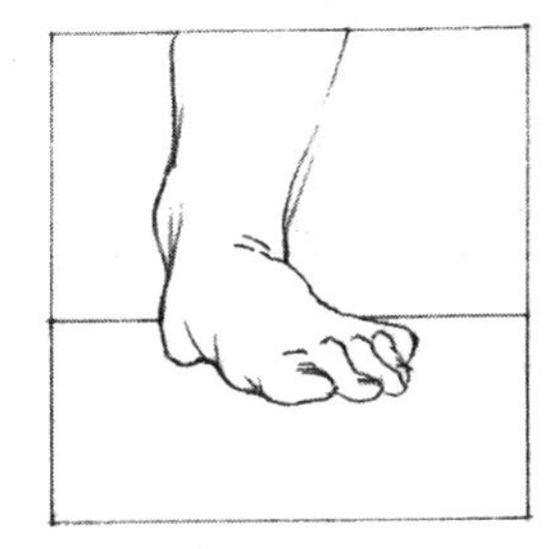

Convection

- Convection is the transfer of heat by the circulatory motion of the heated parts of a liquid or gas owing to a variation in density and the action of gravity. In other words, the body gives off heat to the surrounding cooler air.
- A large differential between air and skin temperature and increased air motion induce more heat transmission by convection.

Radiation

- Radiation is the process by which heat energy in the form of electromagnetic waves is emitted by a warm body, transmitted through an intervening space, and absorbed by a cooler body. No air motion is required for the transfer of heat.
- Light colors reflect while dark colors absorb heat; poor reflectors make good radiators.
- Radiant heat cannot travel around corners and is not affected by air motion.

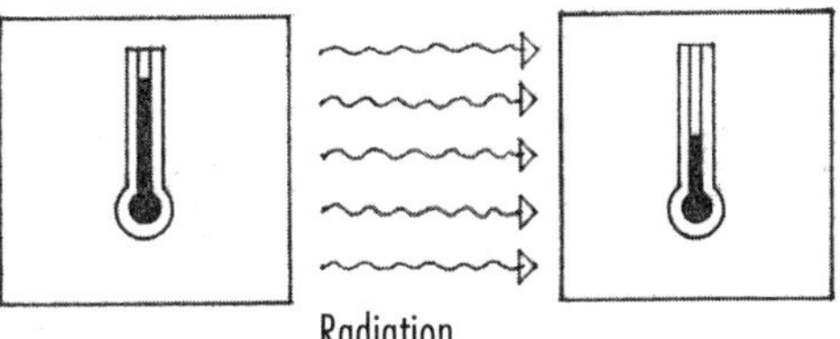

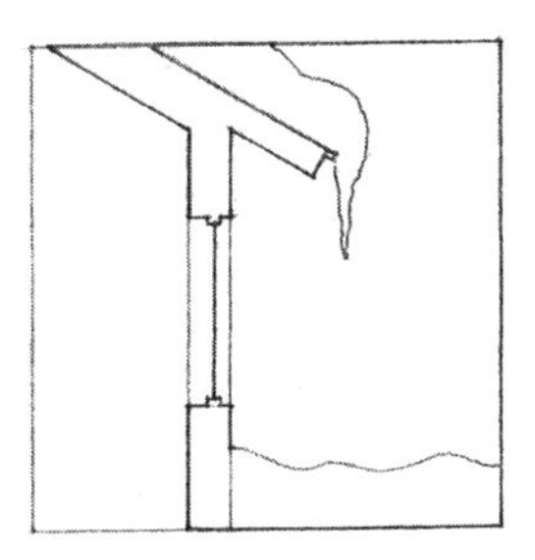

Evaporation

- Heat is required for the evaporative process of converting body moisture into a vapor.
- Heat loss by evaporation increases with air motion.
- Evaporative cooling is especially beneficial when high air temperatures, humidity, and activity levels exist.

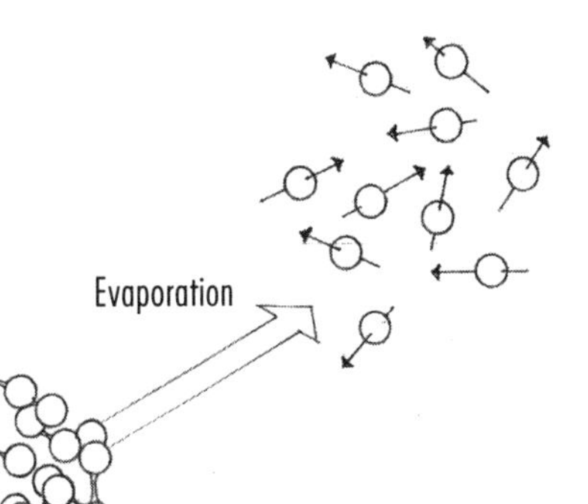

Factors affecting human comfort include air temperature, relative humidity, mean radiant temperature, air motion, air purity, sound, vibration, and light. Of these, the first four are of primary importance in determining thermal comfort. Certain ranges of air temperature, relative humidity, mean radiant temperature, and air motion have been judged to be comfortable by a majority of Americans and Canadians tested. These comfort zones are described by the following graphs of the interaction between the four primary thermal comfort factors. Note that a specific level of comfort for any individual is a subjective judgment of these thermal comfort factors and will vary with prevailing and seasonal variations in climate as well as the age, health, clothing, and activity of the individual.

LEED® EQ Credit 7: Thermal Comfort

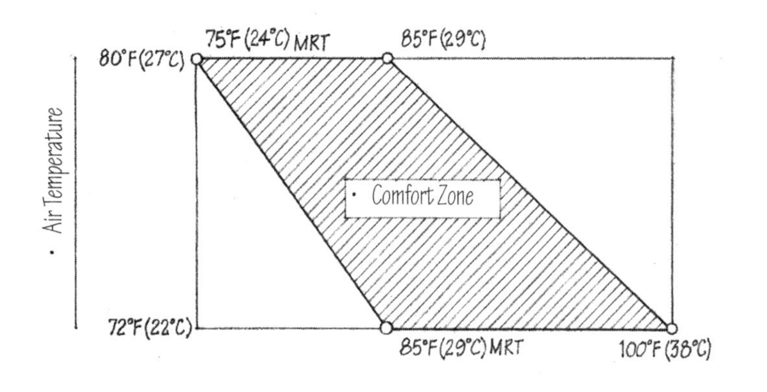

Air Temperature and Mean Radiant Temperature

- Mean radiant temperature (MRT) is important to thermal comfort because the human body receives radiant heat from or loses heat by radiation to the surrounding surfaces if their MRT is significantly higher or lower than the air temperature. See diagram on following page.
- The higher the MRT of the surrounding surfaces, the cooler the air temperature should be.
- MRT has about 40% more effect on comfort than air temperature.
- In cold weather, the MRT of the interior surfaces of exterior walls should not be more than 5°F (2.8°C) below the indoor air temperature.

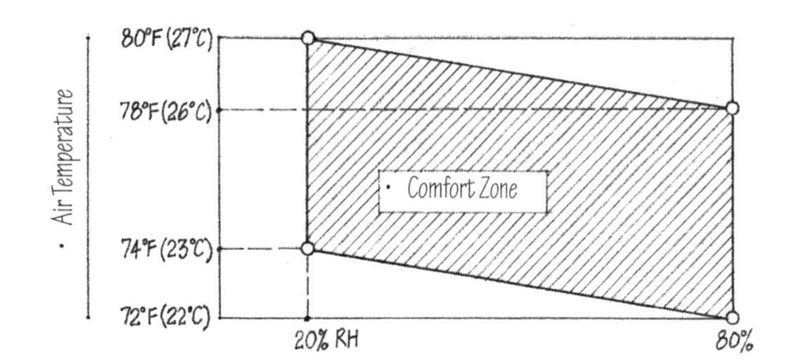

Air Temperature and Relative Humidity

- Relative humidity (RH) is the ratio of the amount of water vapor actually present in the air to the maximum amount that the air could hold at the same temperature, expressed as a percentage.
- The higher the relative humidity of a space, the lower the air temperature should be.
- Relative humidity is more critical at high temperatures than within the normal temperature range.
- Low humidity (<20%) can have undesirable effects such as the buildup of static electricity and the drying out of wood; high humidity can cause condensation problems.

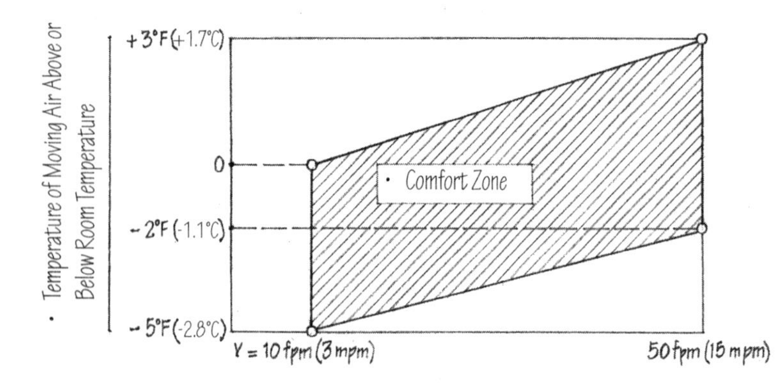

Air Temperature and Air Motion

- Air motion (V) increases heat loss by convection and evaporation.
- The cooler the moving air stream is, relative to the room air temperature, the less velocity it should have.
- Air velocity should range between 10 and 50 feet per minute (fpm) [3 and 15 meters per minute (mpm)]; higher velocities can cause drafty conditions.
- Air motion is especially helpful for cooling evaporation in hot, humid weather.

A psychrometer is an instrument for measuring atmospheric humidity, consisting of two thermometers, the bulb of one being dry and the bulb of the other being kept moist and ventilated so that the cooling that results from evaporation makes it register a lower temperature than the dry one, with the difference between the readings being a measure of atmospheric humidity. Psychrometric charts relate the wet-bulb and dry-bulb readings from a psychrometer to relative humidity, absolute humidity, and dew point. Mechanical engineers use psychrometric charts to determine the amount of heat that must be added or removed by an HVAC system to achieve an acceptable level of thermal comfort in a space.

- Effective temperature represents the combined effect of ambient temperature, relative humidity, and air motion on the sensation of warmth or cold felt by the human body, equivalent to the dry-bulb temperature of still air at 50% relative humidity that induces an identical sensation.

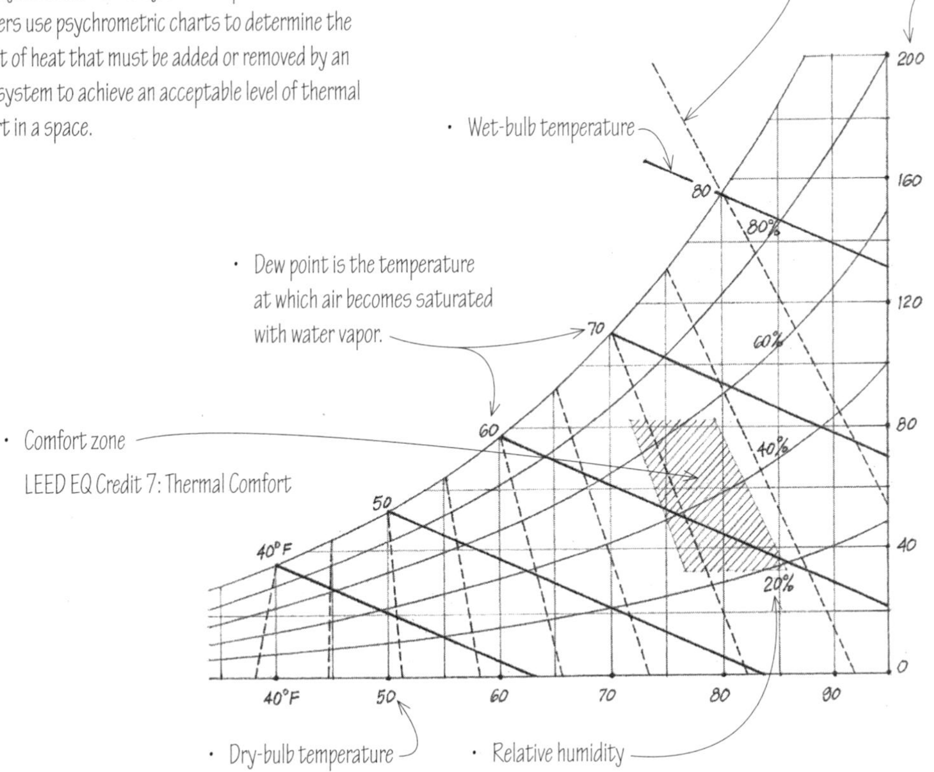

- Humidity ratio in grains of water vapor per pound of dry air (1 lb. = 7000 grains).
- Wet-bulb temperature
- Dew point is the temperature at which air becomes saturated with water vapor.
- Comfort zone

LEED EQ Credit 7: Thermal Comfort

- Dry-bulb temperature
- Relative humidity
- Enthalpy is a measure of the total heat contained in a substance, equal to the internal energy of the substance plus the product of its volume and pressure. The enthalpy of air is equal to the sensible heat of the air and the water vapor present in the air plus the latent heat of the water vapor, expressed in Btu per pound (2.326 Btu/lb. = 1 kJ/kg) of dry air.

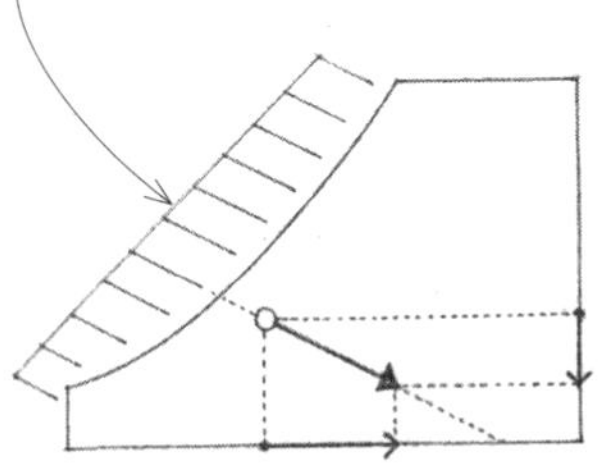

- Adiabatic heating is a rise in temperature occurring without the addition or removal of heat, as when excess water vapor in the air condenses and the latent heat of vaporization of the water vapor is converted to sensible heat in the air.

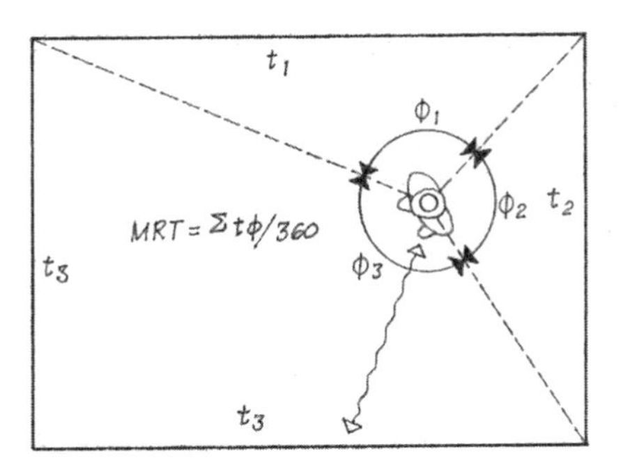

- Mean radiant temperature (MRT) is the sum of the temperatures of the surrounding walls, floor, and ceiling of a room, weighted according to the solid angle subtended by each at the point of measurement.

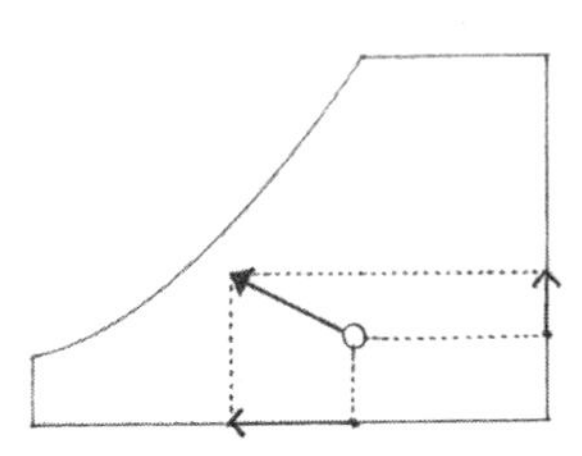

- Evaporative cooling is a drop in temperature occurring without the addition or removal of heat, as when moisture evaporates and the sensible heat of the liquid is converted to latent heat in the vapor.

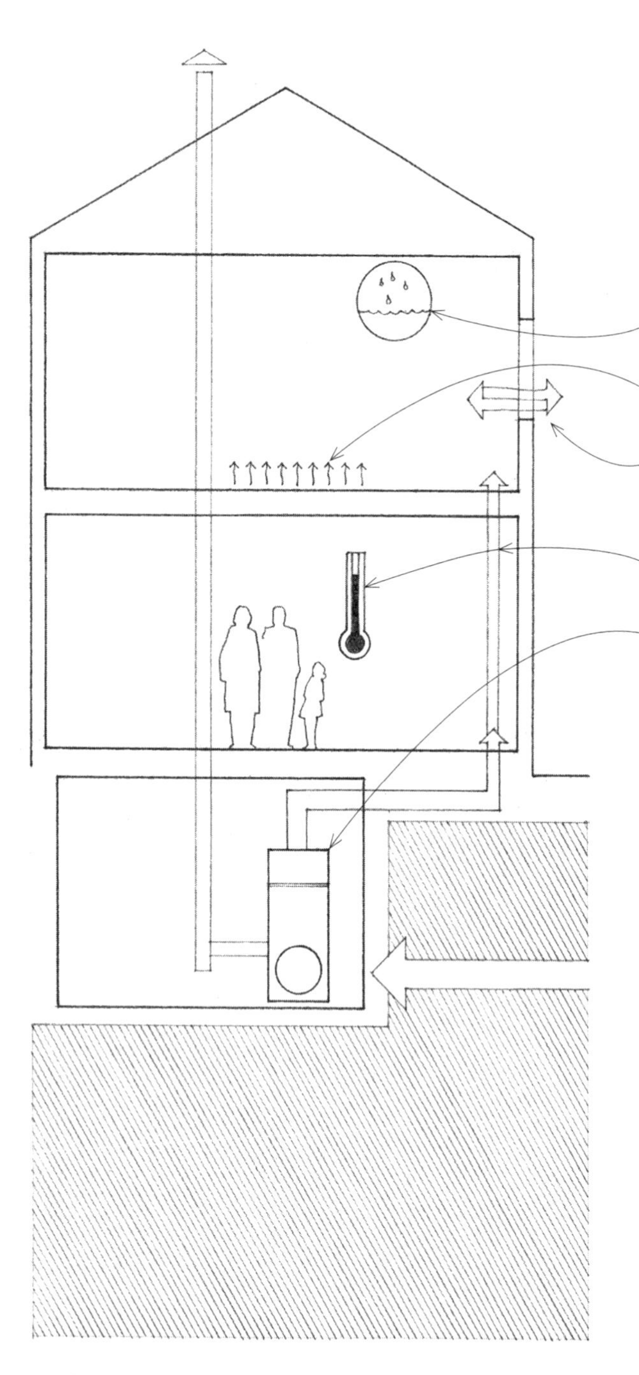

LEED EA Credit 1: Optimize Energy Performance

The siting, orientation, and construction assemblies of a building should minimize heat loss to the outside in cold weather and minimize heat gain in hot weather. Any excessive heat loss or heat gain must be balanced by passive energy systems or by mechanical heating and cooling systems in order to maintain conditions of thermal comfort for the occupants of a building. While heating and cooling to control the air temperature of a space is perhaps the most basic and necessary function of a mechanical system, attention should be paid to the other three factors that affect human comfort—relative humidity, mean radiant temperature, and air motion.

- Relative humidity can be controlled by introducing water vapor through humidifying devices, or removing it by ventilation.
- The mean radiant temperature of room surfaces can be raised by using radiant heating panels or lowered by radiant cooling.
- Air motion can be controlled by natural or mechanical ventilation.

Heating and Cooling

- Air temperature is controlled by the supply of a fluid medium—warm or cool air, or hot or chilled water—to a space.
- Furnaces heat air; boilers heat water or produce steam; electric heaters employ resistance to convert electric energy into heat. See 11.16 for cooling systems.
- The size of heating and cooling equipment required for a building is determined by the heating and cooling loads anticipated; see 11.09.

The traditional fossil fuels—gas, oil, and coal—continue to be the most commonly used to produce the energy for heating and cooling buildings. Natural gas burns cleanly and does not require storage or delivery except through a pipeline. Propane gas is also a clean-burning fuel that is slightly more expensive than natural gas. Oil is an efficient fuel choice, but it requires delivery by trucks to storage tanks located in or near the point of utilization. Coal is rarely used for heating in new residential construction; its use fluctuates in commercial and industrial construction.

Electricity is a clean energy source requiring no combustion or fuel storage at the site. It is also a compact system, being distributed through small wires and using relatively small and quiet equipment. However, the cost to electrically heat or cool a building can be prohibitive and most electric power must be generated by using other sources of energy—nuclear fission or the burning of fossil fuels—to drive turbines. Nuclear energy, despite continuing concerns with the safety of its installations and the disposal of nuclear waste material, may still become an important source of power. A small percentage of turbines are driven by flowing water (hydropower), wind, and the gases produced by burning natural gas, oil, and coal.

Of increasing concern are the uncertain cost and availability of conventional energy sources, the impact of energy extraction and production on environmental resources, and the burning of greenhouse-gas-emitting fossil fuels (see 1.06). Because more than 40% of all energy and more than 65% of all electricity in the United States are consumed in buildings, the design professions, construction industry, and governmental agencies are exploring strategies for reducing the energy consumption of buildings and evaluating alternative, renewable sources of energy: solar, wind, biomass, hydrogen, hydropower, ocean, and geothermal.

Solar Energy

Solar energy can be used directly for passive heating, daylighting, hot water heating, and generating electricity with photovoltaic (solar cell) systems. The conversion efficiency is low with present technology but some systems may be able to produce enough electricity to operate off-grid or to sell the extra electricity to the public utility. Businesses and industry can employ larger-scale applications of solar technology for preheating ventilation air, solar process heating, and solar cooling. Utilities and power plants are also taking advantage of the sun's energy in concentrating solar power systems to produce electricity on a larger scale. These large-scale systems require sizable installations as well as a means of storing the electricity when the sun is not available to produce it.

Wind Power

Wind power is the process by which a turbine converts the kinetic energy of wind flow into mechanical power that a generator can use to produce electricity. The technology consists of blades, sails, or hollow drums that catch the flow of winds and rotate, causing a shaft connected to a generator to turn. Small wind turbines can be used to pump water and power homes and telecommunication dishes; some can be connected to a utility power grid or be combined with a photovoltaic (solar cell) system. For utility-scale sources of wind energy, a large number of wind turbines are usually built close together to form a wind plant. Like solar power, wind power is dependent upon location and weather and can be intermittent; the electricity generated when the wind is blowing cannot be stored without batteries. The best sites for wind farms are often remote and distant from where the electricity is needed. Additional concerns include the aesthetics of wind turbines, noise, and the potential for birdkill.

Biomass Energy

Biomass, the organic matter that makes up plants, can be used to produce electricity, transportation fuels, and chemicals that would otherwise be made from fossil fuels. Properly harvested wood is one example of a natural and sustainable biomass, but its burning can create air pollution and harm indoor air quality. Wood-burning appliances should meet Environmental Protection Agency (EPA) regulations for emissions. Wood pellets made from wood by-products burn cleanly and should be considered as an alternative. Other viable sources of biomass include food crops, such as corn for ethanol and soybeans for biodiesel, grassy and woody plants, residues from forestry or agriculture, and the organic component of municipal and industrial wastes.

Some consider biomass to be a carbon-neutral fuel because its burning does not release more carbon dioxide than that captured in its own growth and released by its natural biodegradation. The conversion process of biomass into fuel, however, can be energy negative if more energy is required for the conversion process than is obtained from the product itself. Using grain such as corn also precludes it from being used as food for humans or livestock.

Hydrogen

Hydrogen is the most abundant element on earth and can be found in many organic compounds as well as water. While it does not occur naturally as a gas, once separated from another element, hydrogen can be burned as a fuel or used by fuel cells to electrochemically combine with oxygen to produce electricity and heat, emitting only water vapor in the process. Because hydrogen has very high energy for its weight, but very low energy for its volume, new technology is needed to more efficiently store and transport it.

H_2

LEED EA Credit 2: On-Site Renewable Energy
LEED EA Credit 6: Green Power

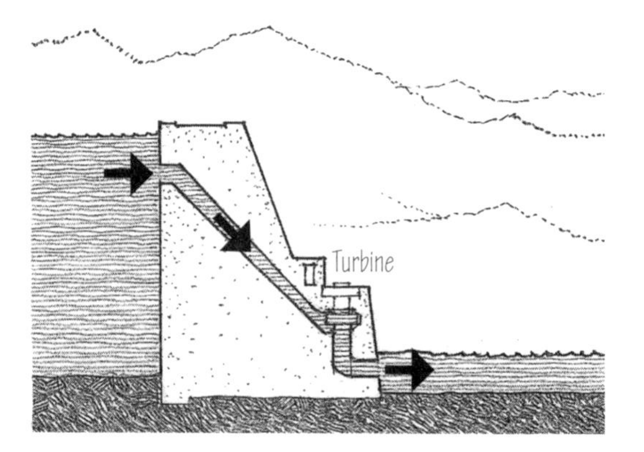

Hydropower

Hydroelectric power, or hydropower, is created and controlled by the damming of rivers. As the water stored behind a dam is released at high pressure, its kinetic energy is transformed into mechanical energy and used by turbine blades to generate electricity. Because the water cycle is an endless, constantly recharging system, hydropower is considered a clean, renewable energy source, but hydropower plants can be impacted by drought. Benefits of hydropower include flood control and the recreational opportunities afforded by the reservoirs created by dams. Disadvantages include enormous installation costs, loss of farmland, disruption of fish migration, and uncertain effects on riparian habitats and historical sites.

Ocean Energy

Covering more than 70% of the earth's surface, the ocean can produce thermal energy from the sun's heat and mechanical energy from its tides and waves. Ocean Thermal Energy Conversion (OTEC) is a process for generating electricity from the heat energy stored in the earth's oceans. The process works best in tropical coastal areas, where the surface of the ocean is warm and the depths are cold enough to create a modest temperature differential. OTEC utilizes this temperature differential to run a heat engine—pumping warm surface-seawater through a heat exchanger where a low-boiling-point fluid, such as ammonia, is vaporized, with the vapor expanding to rotate a turbine connected to a generator. Cold deep-seawater—pumped through a second heat exchanger—condenses the vapor back into a liquid, which is then recycled through the system. Because its conversion efficiency is very low, an OTEC plant would have to be vast and move an enormous amount of water while anchored in the deep open ocean subject to storms and corrosion.

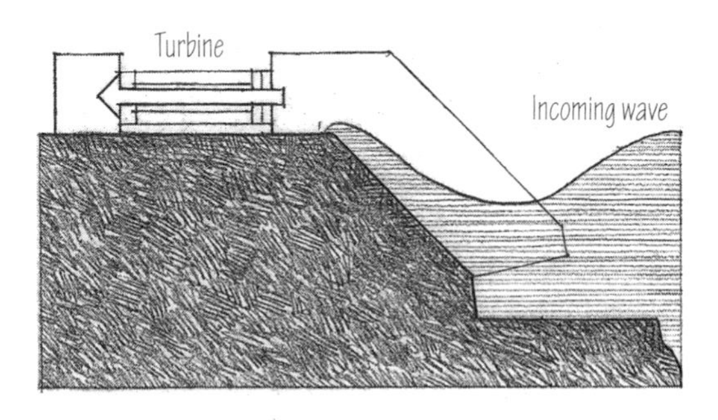

Similar to more conventional hydroelectric dams, the tidal process uses the natural motion of the tides to fill reservoirs, which are then discharged through electricity-producing turbines. Because seawater has a much higher density than air, ocean currents carry significantly more energy than wind currents. Utilizing tidal power requires a high tide and special coastline conditions present in both the northeastern and northwestern coasts of the United States. Damming estuaries would have considerable environmental impact, affecting both sea life migration and fisheries.

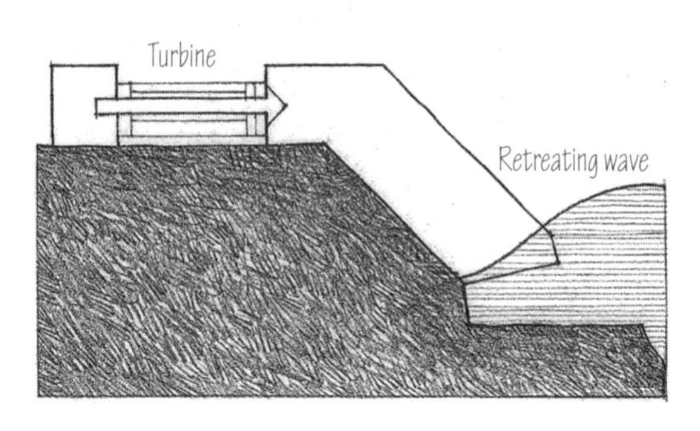

Wave energy can be converted into electricity through both offshore and onshore systems. Offshore systems are situated in deep water and use either the bobbing motion of the waves to power a pump or the funneling of waves through internal turbines on floating platforms to create electricity. Onshore wave power systems are built along shorelines to extract the energy in breaking waves by using the alternating compression and depressurization of an enclosed air column to drive turbines. The potential energy of waves can be effectively harvested in only certain areas of the world, such as the northeastern and northwestern coasts of the United States. Careful site selection is the key to keeping the environmental impacts of wave power systems to a minimum, preserving scenic shorefronts, and avoiding altering flow patterns of sediment on the ocean floor.

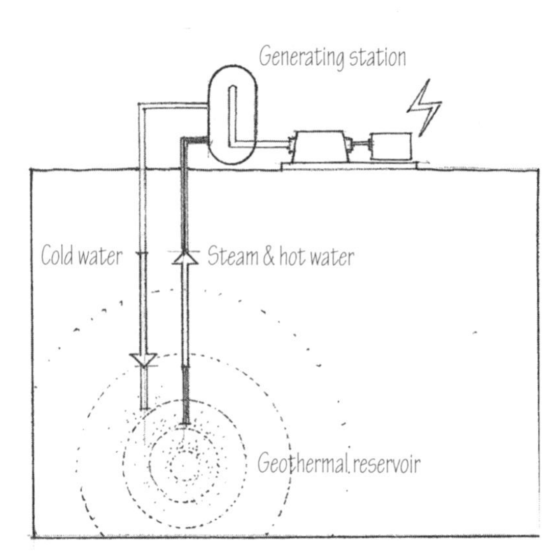

Geothermal Energy

Geothermal energy—the earth's internal heat—can yield warmth and power for a variety of uses without burning fuels, damming rivers, or harvesting forests. The shallow ground near the earth's surface maintains a relatively constant temperature of 50°–60°F (10°–16°C), heat that can be used to provide direct heating and cooling in homes and other buildings. Steam, heat or hot water from deeper geothermal reservoirs can provide the force that spins turbine generators to produce electricity. The used geothermal water is then returned down an injection well into the reservoir to be reheated, to maintain pressure, and to sustain the reservoir.

Calculating heat loss in cold weather and heat gain in hot weather is necessary to size the heating and cooling equipment required for a building. It takes into account the differential between desired indoor air temperature and outdoor design temperature, the daily temperature range, the solar orientation and thermal resistance of wall, window, and roof assemblies, and the use and occupancy of inhabited spaces. The more heat loss and heat gain can be reduced by the siting, layout, and orientation of a building, the less energy will be consumed by smaller heating and cooling equipment. Other energy-conscious design strategies include using thermal insulation and thermal mass to effectively control the transmission of heat through building assemblies; making wise choices in selecting energy-efficient HVAC systems, water heaters, appliances, and lighting systems; and employing "smart" systems to control thermal conditions and lighting.

LEED EA Credit 1: Optimize Energy Performance

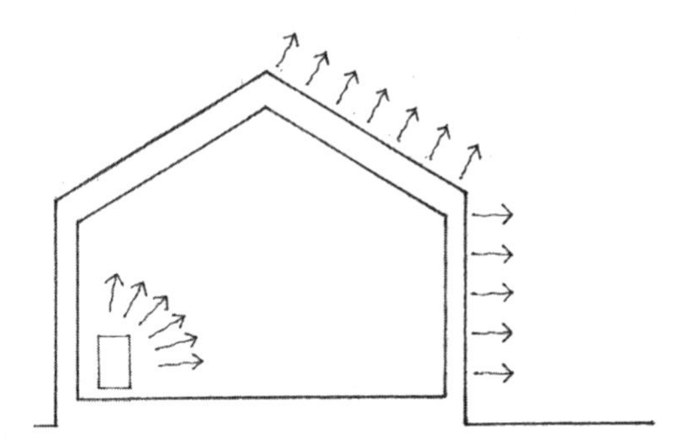

Heating Load

- Heating load is the hourly rate of net heat loss in an enclosed space, expressed in Btu per hour and used as the basis for selecting a heating unit or system.
- British thermal unit (Btu) is the quantity of heat required to raise the temperature of 1 lb. (0.4 kg) of water 1°F.
- Degree-day is a unit that represents 1 degree of departure in the mean daily outdoor temperature from a given standard temperature. It is used to compute heating and cooling loads, size HVAC systems, and calculate yearly fuel consumption.
- Heating degree-day is one degree-day below the standard temperature of 65°F (18°C), used in estimating fuel or power consumption by a heating system.

Heat Loss

The primary sources of heat loss in cold weather are:

- Convection, radiation, and conduction of heat through exterior wall, window, and roof assemblies to the outside, and through floors over unheated spaces
- Infiltration of air through cracks in exterior construction, especially around windows and doorways

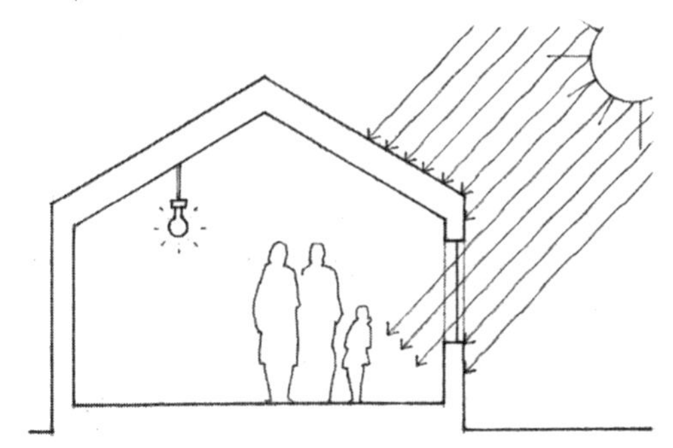

Cooling Load

- Cooling load is the hourly rate of heat gain in an enclosed space, expressed in Btu per hour and used as the basis for selecting an air-conditioning unit or cooling system.
- Cooling degree-day is one degree-day above the standard temperature of 75°F (24°C), used in estimating energy requirements for air-conditioning and refrigeration.
- Ton of refrigeration is the cooling effect obtained when 1 ton of ice at 32°F (0°C) melts to water at the same temperature in 24 hours, equivalent to 12,000 Btu/hr (3.5 kW).
- Energy efficiency rating is an index of the efficiency of a refrigerating unit, expressing the Btu removed per watt of electrical energy input.

- For more detailed information on the calculation of heating and cooling loads, refer to the handbook published by the American Society of Heating, Refrigerating and Air-Conditioning Engineers (ASHRAE).

Heat Gain

Sources of heat gain in warm or hot weather include:

- Convection, radiation, and conduction through exterior wall, window, and roof assemblies when outdoor temperatures are high; varies with the time of day, the solar orientation of the assemblies, and the effect of thermal lag
- Solar radiation on glazing; varies with solar orientation and the effectiveness of any shading devices used
- Building occupants and their activities
- Lighting and other heat-producing equipment
- Ventilation of spaces that may be required to remove odors and pollutants
- Latent heat, requiring energy to condense the moisture in warm air so that the relative humidity in a space will not be excessive

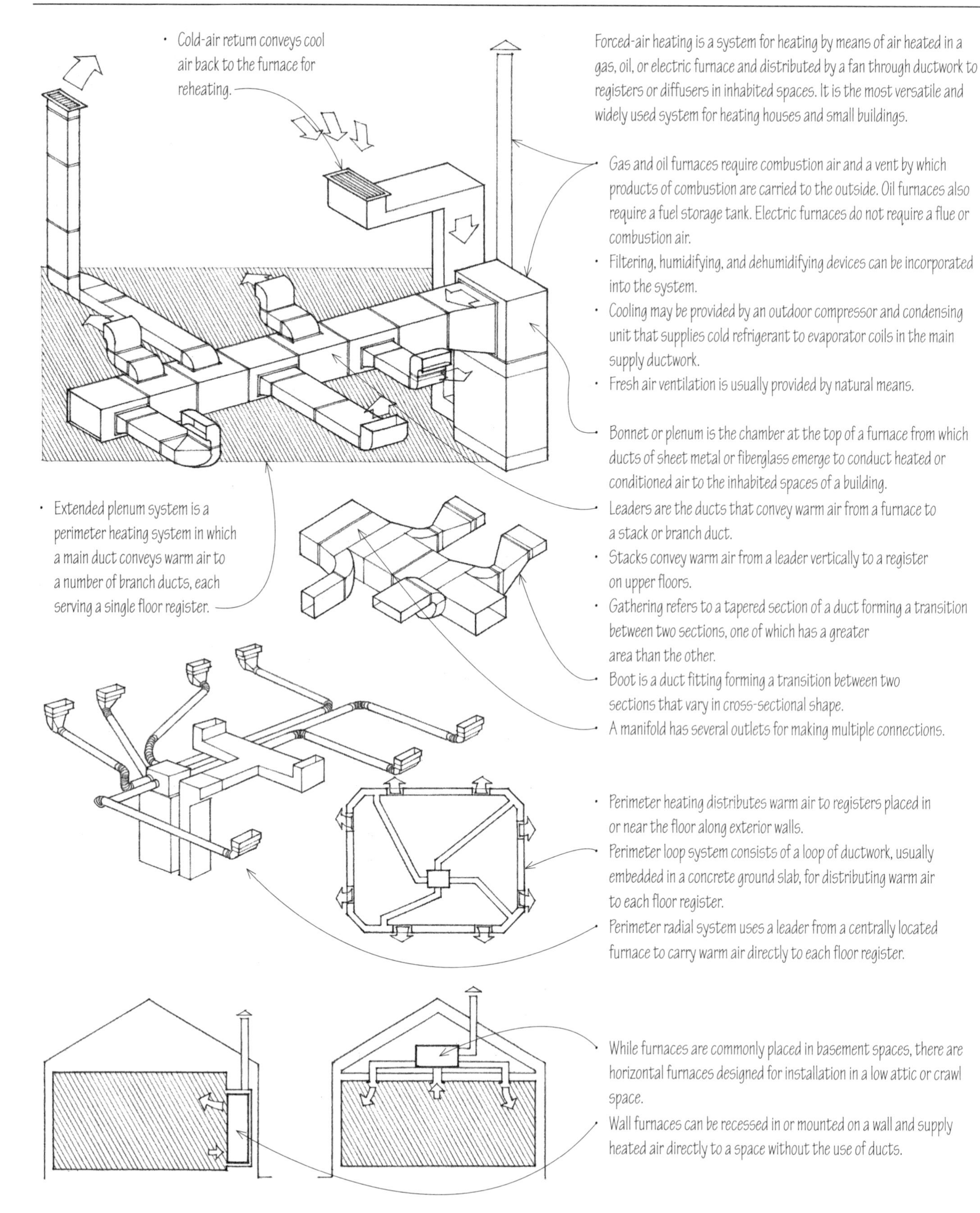

Forced-air heating is a system for heating by means of air heated in a gas, oil, or electric furnace and distributed by a fan through ductwork to registers or diffusers in inhabited spaces. It is the most versatile and widely used system for heating houses and small buildings.

- Cold-air return conveys cool air back to the furnace for reheating.

- Gas and oil furnaces require combustion air and a vent by which products of combustion are carried to the outside. Oil furnaces also require a fuel storage tank. Electric furnaces do not require a flue or combustion air.
- Filtering, humidifying, and dehumidifying devices can be incorporated into the system.
- Cooling may be provided by an outdoor compressor and condensing unit that supplies cold refrigerant to evaporator coils in the main supply ductwork.
- Fresh air ventilation is usually provided by natural means.

- Bonnet or plenum is the chamber at the top of a furnace from which ducts of sheet metal or fiberglass emerge to conduct heated or conditioned air to the inhabited spaces of a building.
- Leaders are the ducts that convey warm air from a furnace to a stack or branch duct.
- Stacks convey warm air from a leader vertically to a register on upper floors.
- Gathering refers to a tapered section of a duct forming a transition between two sections, one of which has a greater area than the other.
- Boot is a duct fitting forming a transition between two sections that vary in cross-sectional shape.
- A manifold has several outlets for making multiple connections.

- Extended plenum system is a perimeter heating system in which a main duct conveys warm air to a number of branch ducts, each serving a single floor register.

- Perimeter heating distributes warm air to registers placed in or near the floor along exterior walls.
- Perimeter loop system consists of a loop of ductwork, usually embedded in a concrete ground slab, for distributing warm air to each floor register.
- Perimeter radial system uses a leader from a centrally located furnace to carry warm air directly to each floor register.

- While furnaces are commonly placed in basement spaces, there are horizontal furnaces designed for installation in a low attic or crawl space.
- Wall furnaces can be recessed in or mounted on a wall and supply heated air directly to a space without the use of ducts.

CSI MasterFormat™ 23 30 00: HVAC Air Distribution
CSI MasterFormat 23 50 00: Central Heating Equipment
CSI MasterFormat 23 55 00: Fuel-Fired Heaters

Hot-water or hydronic heating is a system for heating a building by means of water heated in a boiler and circulated by a pump through pipes to radiators or convectors. Steam heating is similar in principle, using steam generated in a boiler and circulating it through piping to radiators. In large cities and building complexes, hot water or steam generated at a central boiler plant may be available via underground pipelines. This availability would eliminate the need for an on-site boiler.

- Boiler is a closed vessel or arrangement of vessels and tubes in which water is heated or steam is generated. The heat may be supplied by the combustion of gas or oil, or by electric-resistance coils. Safety relief valves on boilers open when actuated by a vapor pressure above a predetermined level, allowing the vapor to escape until its pressure is reduced to a safe or acceptable level.

- One-pipe system is a hot-water heating system in which a single pipe supplies hot water from a boiler to each radiator or convector in sequence.

- Two-pipe system is a hot-water heating system in which one pipe supplies hot water from a boiler to the radiators or convectors and a second pipe returns the water to the boiler.
- Direct return is a two-pipe hot-water system in which the return pipe from each radiator or convector takes the shortest route back to the boiler.
- Reverse return is a two-pipe hot-water system in which the lengths of the supply and return pipes for each radiator or convector are nearly equal.
- Dry return is a return pipe in a steam-heating system that carries both air and water of condensation.

- Radiators consist of a series or coil of pipes through which hot water or steam passes. The heated pipes warm a space primarily by radiation. Convectors, on the other hand, are heating units in which air heated by contact with a radiator or a fin-tube circulates by convection.
- Fin-tube convectors are baseboard convectors having horizontal tubes with closely spaced vertical fins to maximize heat transfer to the surrounding air. Cool room air is drawn in from below by convection, heated by contact with the fins, and discharged out the top.

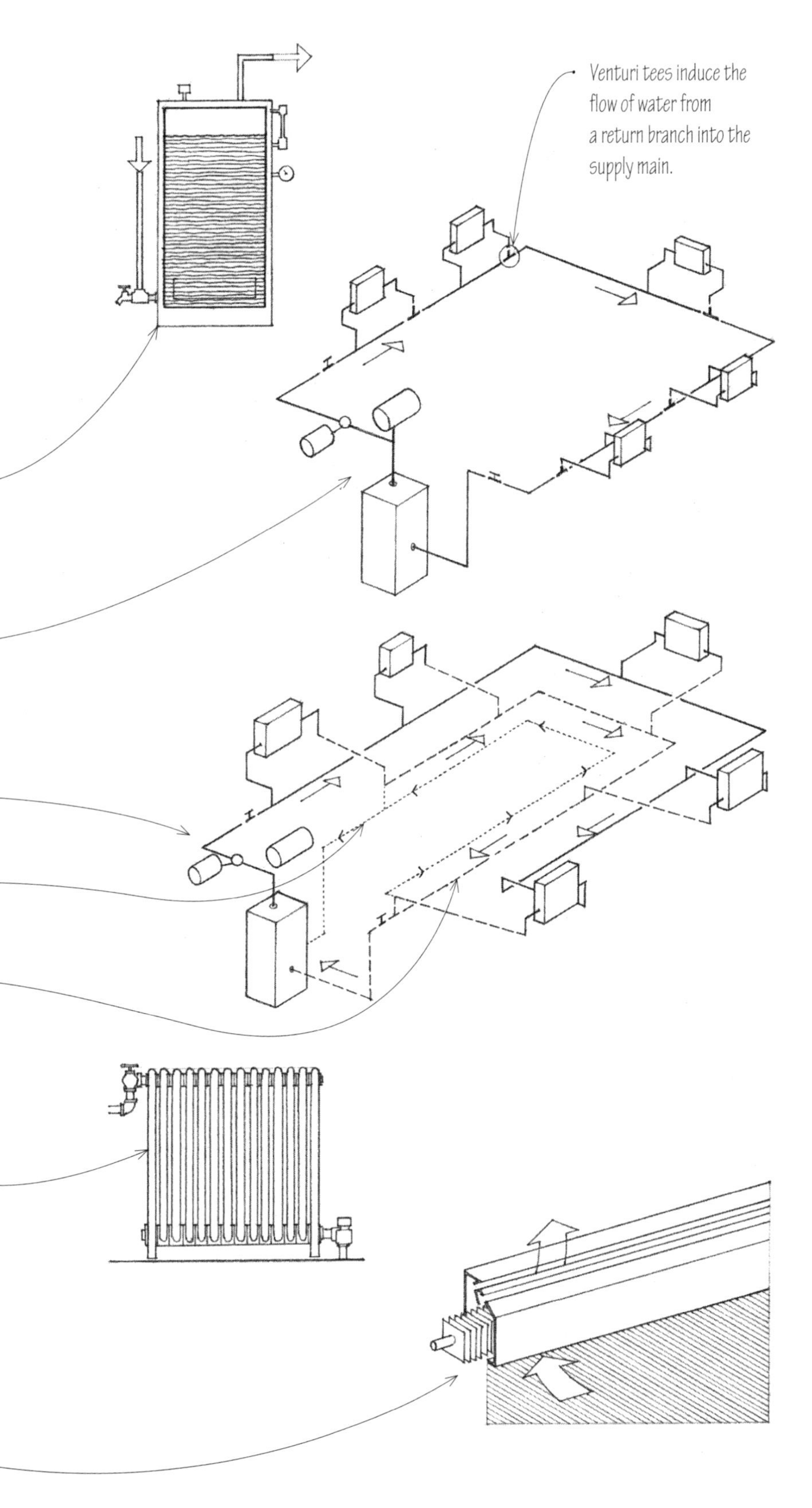

Comparative Heat Values of Energy Sources

Fuel	Heat Value
Anthracite Coal	14,600 Btu/lb
Oil	139,000 Btu/gal
Natural Gas	1,052 Btu/cf
Electricity	1 watt = 3.41 Btu/hr

Electric heating is more accurately described as electric-resistance heating. Resistance is the property of a conductor by virtue of which the passage of current is opposed, causing electric energy to be converted into heat. Electric-resistance heating elements may be exposed to the air stream in a furnace or ductwork in a forced-air heating system or provide the heat for a boiler in a hydronic heating system. More direct means of heating with electric energy involves housing the resistance wires or coils in space-heating units. While compact and versatile, these electric-resistance heaters have no provision for controlling humidity and air quality.

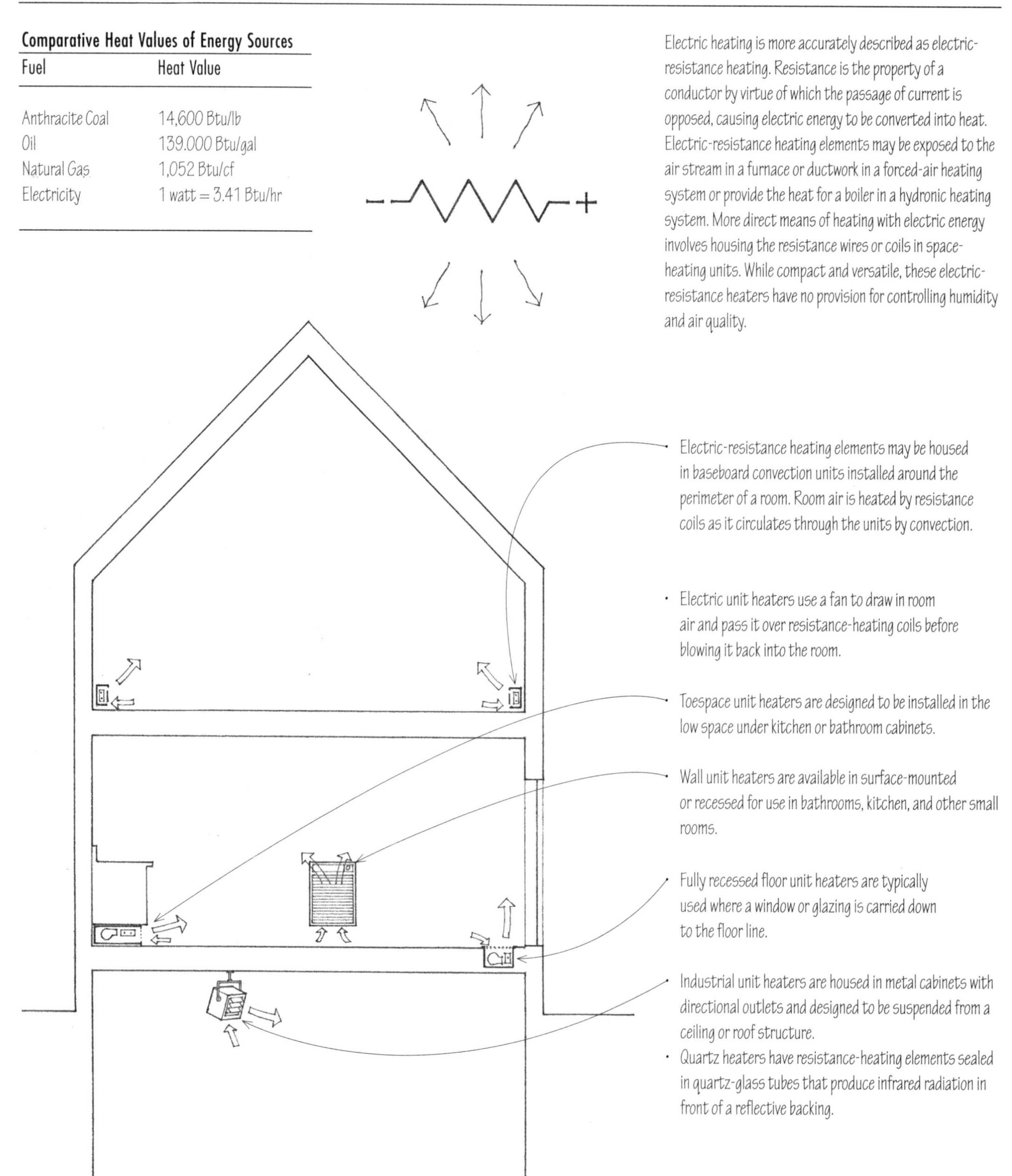

- Electric-resistance heating elements may be housed in baseboard convection units installed around the perimeter of a room. Room air is heated by resistance coils as it circulates through the units by convection.
- Electric unit heaters use a fan to draw in room air and pass it over resistance-heating coils before blowing it back into the room.
- Toespace unit heaters are designed to be installed in the low space under kitchen or bathroom cabinets.
- Wall unit heaters are available in surface-mounted or recessed for use in bathrooms, kitchen, and other small rooms.
- Fully recessed floor unit heaters are typically used where a window or glazing is carried down to the floor line.
- Industrial unit heaters are housed in metal cabinets with directional outlets and designed to be suspended from a ceiling or roof structure.
- Quartz heaters have resistance-heating elements sealed in quartz-glass tubes that produce infrared radiation in front of a reflective backing.

CSI MasterFormat 23 83 00: Radiant Heating Units
CSI MasterFormat 23 83 33: Electric Radiant Heaters

Radiant heating systems use heated ceilings, floors, and sometimes walls, as radiating surfaces. The heat source may be pipes or tubing carrying hot water or electric-resistance heating cables embedded within the ceiling, floor, or wall construction. The radiant heat is absorbed by surfaces and objects in the room, reradiates from the warmed surfaces, and raises the mean radiant temperature (MRT) as well as the ambient temperature in the space.

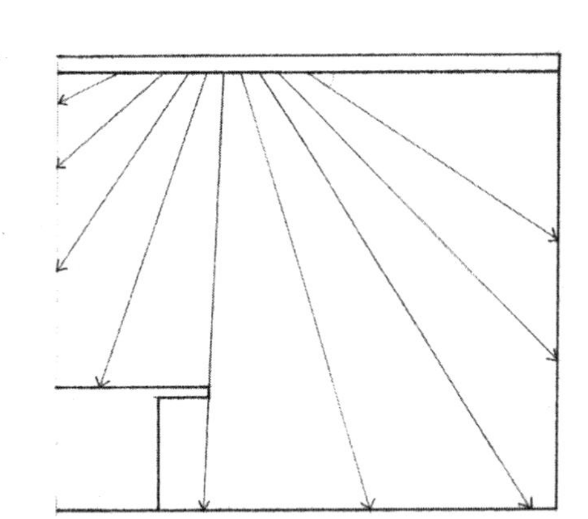

Radiant heat:

- Travels in a direct path;
- Cannot travel around corners and may therefore be obstructed by physical elements within the space such as furniture;
- Cannot counteract cold downdrafts along exterior glass areas;
- Is not affected by air motion.

Floor installations are effective in warming concrete slabs. In general, however, ceiling installations are preferred because ceiling constructions have less thermal capacity and can respond faster. Ceiling panels can also be heated to a higher surface temperature than floor slabs. In both electric and hot-water radiant systems, the installations are completely concealed except for thermostats or balancing valves.

Because radiant panel heating systems cannot respond quickly to changing temperature demands, they may be supplemented with perimeter convector units. For complete air-conditioning, separate ventilating, humidity control, and cooling systems are required.

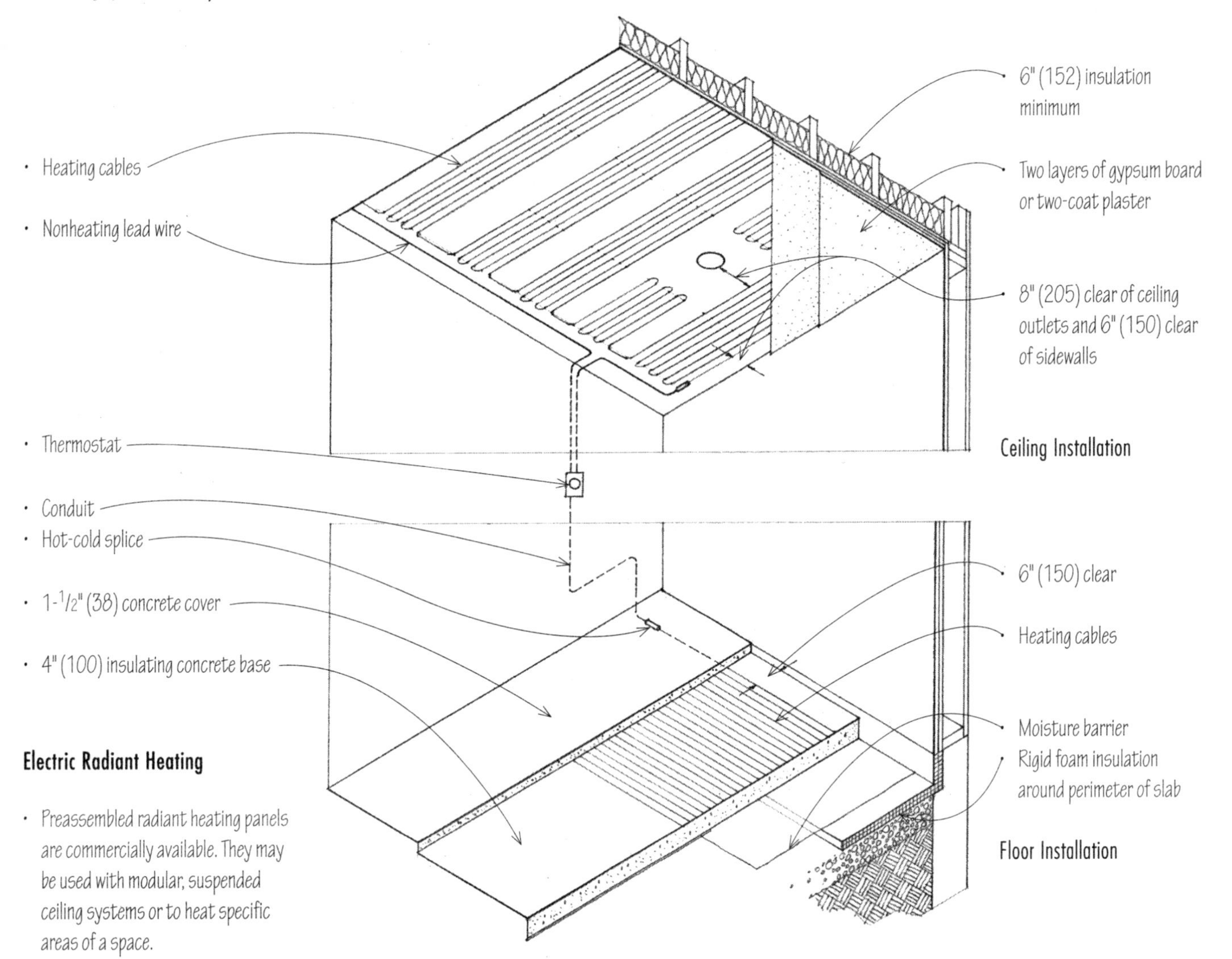

Electric Radiant Heating

- Preassembled radiant heating panels are commercially available. They may be used with modular, suspended ceiling systems or to heat specific areas of a space.

CSI MasterFormat 23 83 13: Radiant-Heating Electric Cables
CSI MasterFormat 23 83 23: Radiant-Heating Electric Panels

Liquid radiant heating systems circulate warm water through metal or plastic pipes either encased in a concrete slab that serves as a thermal mass or secured to the underside of subflooring with conductive heat plates. The supply water may be heated in a boiler, heat pump, solar collector, or geothermal system. In response to the thermostat setting, a control valve adjusts the supply water temperature by mixing it with the circulating water from the pipe loops.

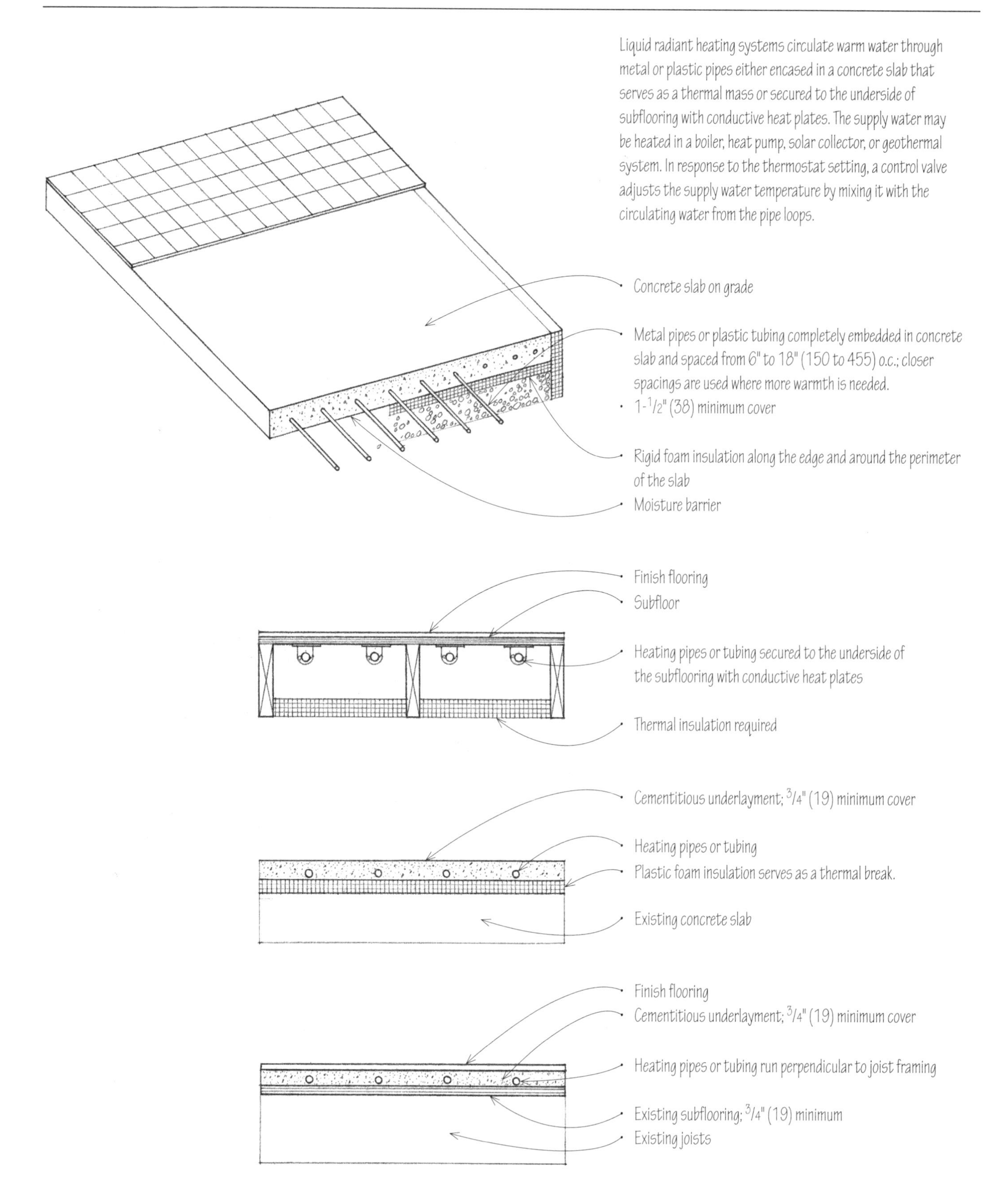

CSI MasterFormat 23 83 16: Radiant-Heating Hydronic Piping

Active solar energy systems absorb, transfer, and store energy from solar radiation for building heating and cooling. They normally consist of the following components:

- Solar collector panels
- Circulation and distribution system for the heat transfer medium
- Heat exchanger and storage facility

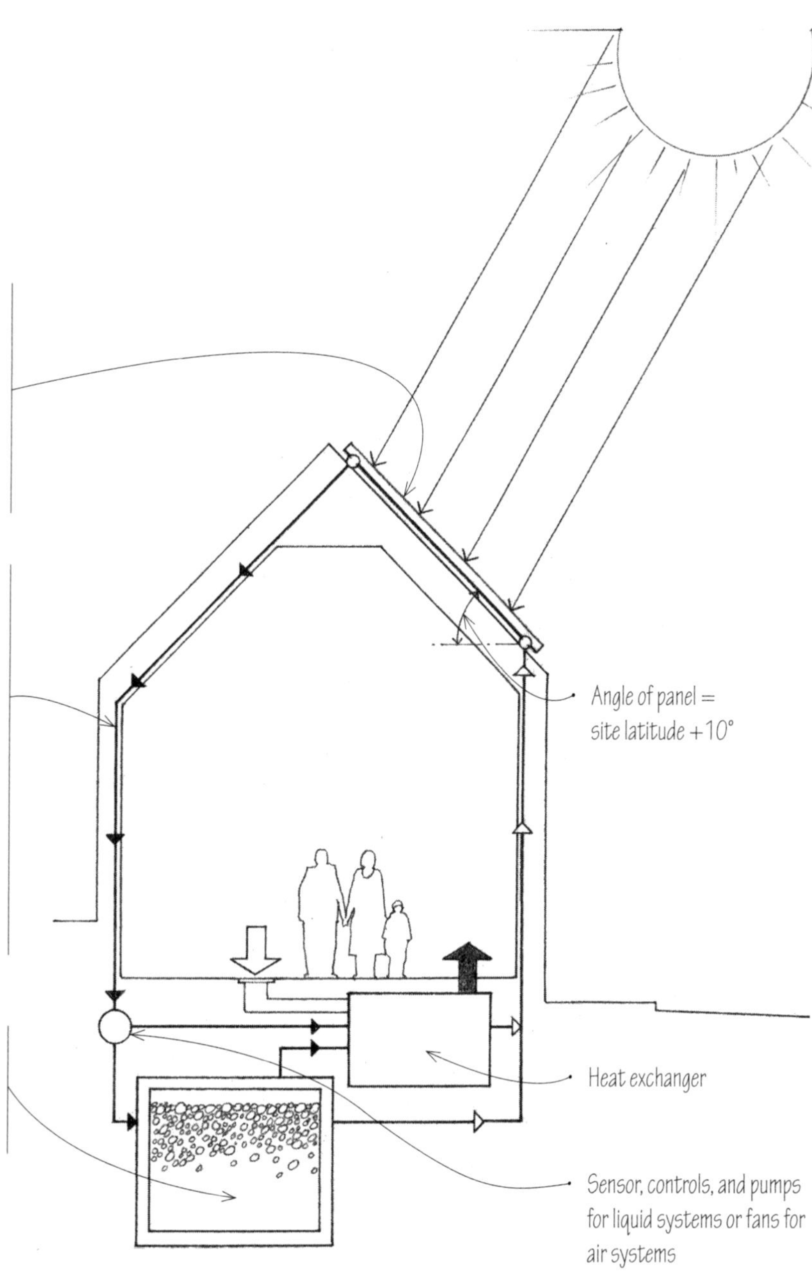

Solar Collector Panels

- The solar collector panels should be oriented within 20° of true south and not be shaded by nearby structures, terrain, or trees. The required collector surface area depends on the heat exchange efficiency of the collector and heat transfer medium, and the heating and cooling load. Current recommendations range from 1/3 to 1/2 of the net floor area of the building.

Heat Transfer Medium

- The heat transfer medium may be air, water, or other liquid. It carries the collected heat energy from the solar panels to the heat exchange equipment or to a storage utility for later use.
- Liquid systems use pipes for circulation and distribution. An antifreeze solution provides freeze protection; a corrosion-retarding additive is required for aluminum pipes.
- The ductwork for air systems requires more installation space. Larger collector surfaces are also required since the heat transfer coefficient for air is less than that of liquids. The construction of the collector panels, however, is simpler and not subject to problems of freezing, leakage, and corrosion.

Storage Facility

- An insulated storage facility holds heat for use at night or on overcast days. It may be in the form of a tank filled with water or other liquid medium, or a bin of rocks or phase-change salts for air systems.

- The heat distributing components of the solar energy system are similar to those of conventional systems.
- Heat may be delivered by an all-air or an air-water system.
- For cooling, a heat pump or absorption cooling unit is required.
- A backup heating system is recommended.

- For an active solar energy system to be efficient, the building itself must be thermally efficient and well insulated. Its siting, orientation, and window openings should take advantage of the seasonal solar radiation.

- See 1.16–1.17 for passive solar design.
- See 11.32 for photovoltaic technology.

Compressive Refrigeration

Compressive refrigeration is a process in which cooling is effected by the vaporization and expansion of a liquid refrigerant.

Heat Pumps

Heat pumps are electrically powered heating and cooling units. For cooling, the normal compressive refrigeration cycle is used to absorb and transfer excess heat to the outdoors. For heating, heat energy is drawn from the outdoor air by reversing the cooling cycle and switching the heat exchange functions of the condenser and evaporator.

Heat pumps are most efficient in moderate climates where heating and cooling loads are almost equal. In freezing temperatures, a heat pump requires an electric resistance heater to keep the outdoor coils from freezing.

Absorption Refrigeration

Absorption refrigeration uses an absorber and a generator instead of a compressor to transfer heat and produce cooling.

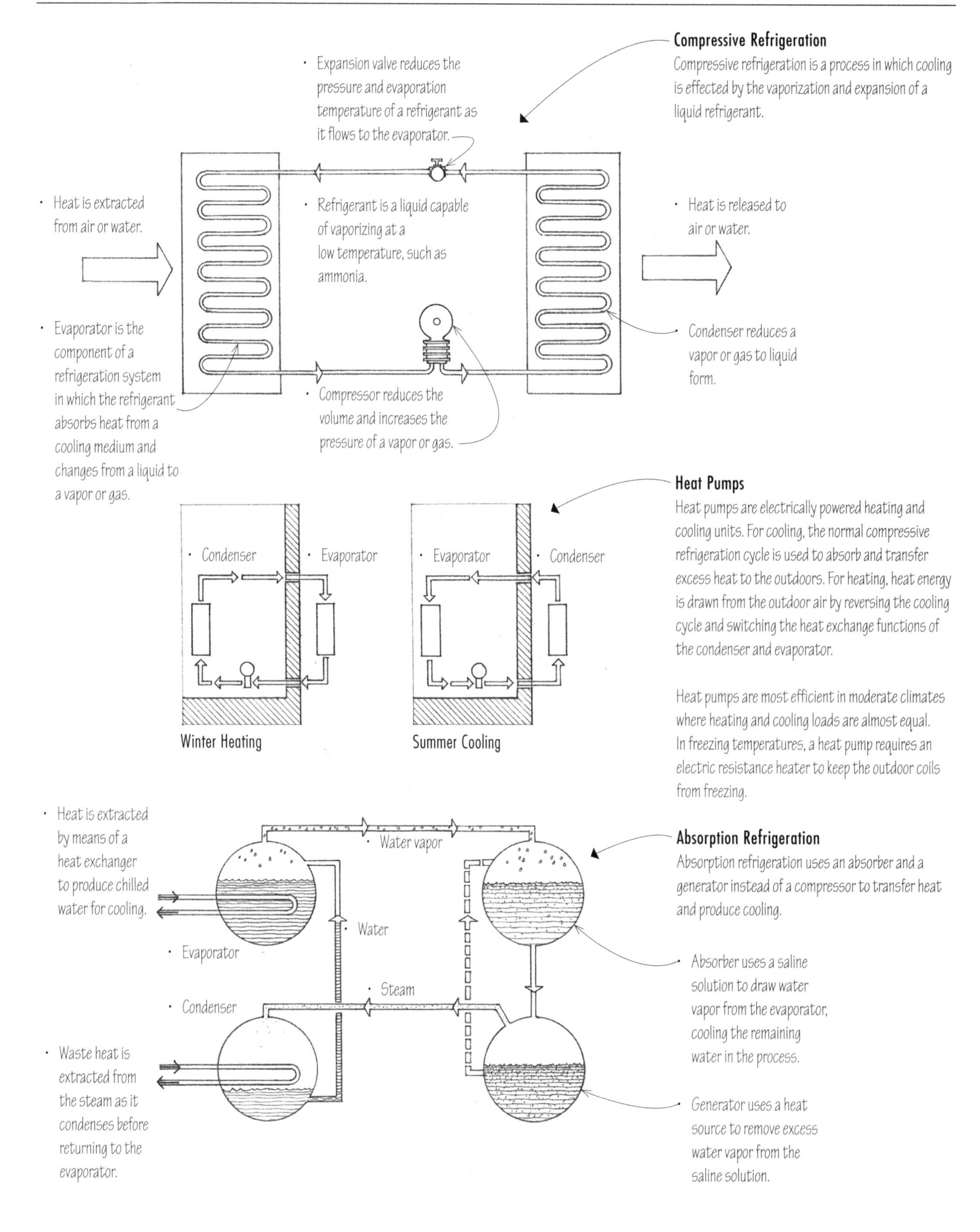

CSI MasterFormat 23 60 00: Central Cooling Equipment

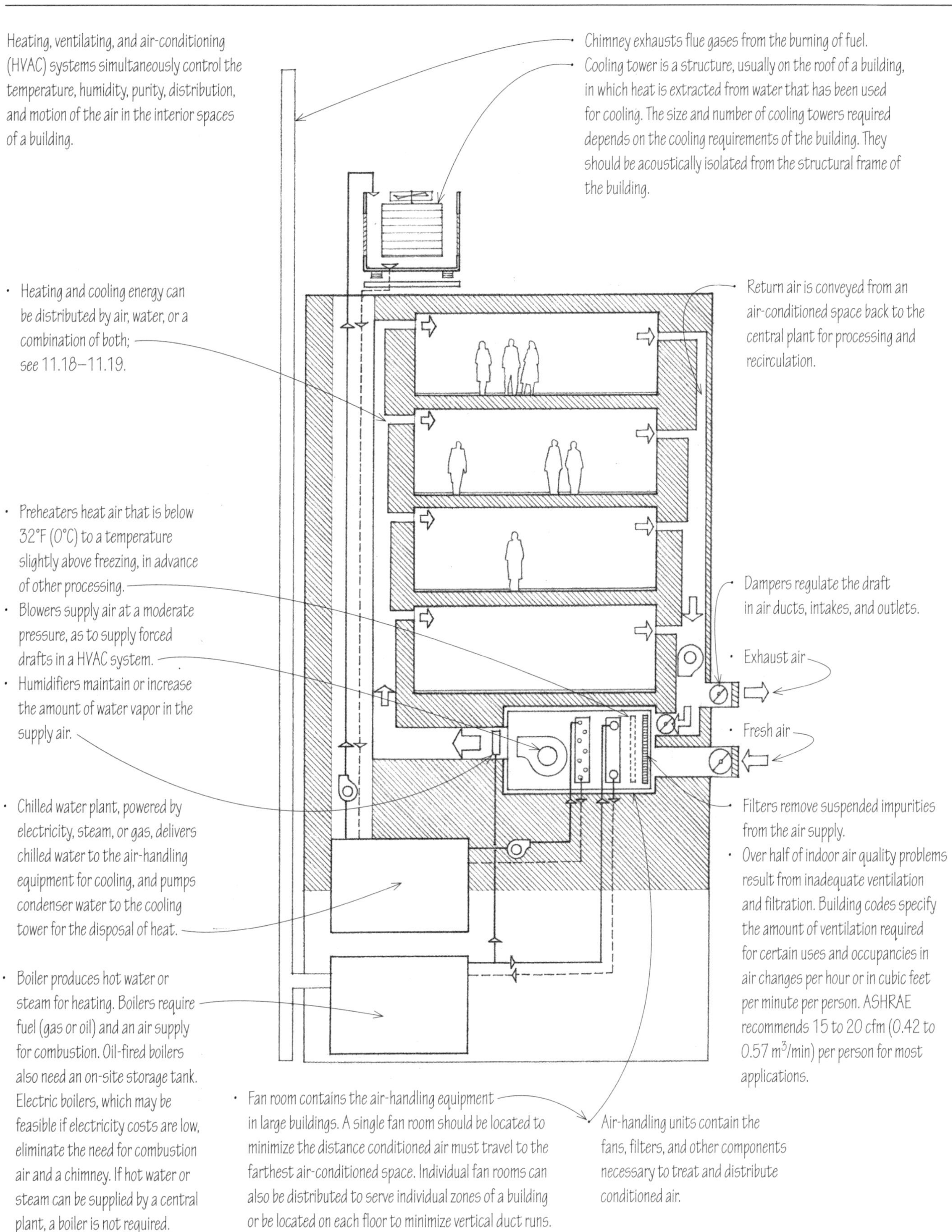
Heating, ventilating, and air-conditioning (HVAC) systems simultaneously control the temperature, humidity, purity, distribution, and motion of the air in the interior spaces of a building.
Chimney exhausts flue gases from the burning of fuel.
Cooling tower is a structure, usually on the roof of a building, in which heat is extracted from water that has been used for cooling. The size and number of cooling towers required depends on the cooling requirements of the building. They should be acoustically isolated from the structural frame of the building.
Heating and cooling energy can be distributed by air, water, or a combination of both; see 11.18–11.19.
Return air is conveyed from an air-conditioned space back to the central plant for processing and recirculation.
Preheaters heat air that is below 32°F (0°C) to a temperature slightly above freezing, in advance of other processing.
Blowers supply air at a moderate pressure, as to supply forced drafts in a HVAC system.
Humidifiers maintain or increase the amount of water vapor in the supply air.
Dampers regulate the draft in air ducts, intakes, and outlets.
Exhaust air
Fresh air
Chilled water plant, powered by electricity, steam, or gas, delivers chilled water to the air-handling equipment for cooling, and pumps condenser water to the cooling tower for the disposal of heat.
Filters remove suspended impurities from the air supply.
Over half of indoor air quality problems result from inadequate ventilation and filtration. Building codes specify the amount of ventilation required for certain uses and occupancies in air changes per hour or in cubic feet per minute per person. ASHRAE recommends 15 to 20 cfm (0.42 to 0.57 m³/min) per person for most applications.
Boiler produces hot water or steam for heating. Boilers require fuel (gas or oil) and an air supply for combustion. Oil-fired boilers also need an on-site storage tank. Electric boilers, which may be feasible if electricity costs are low, eliminate the need for combustion air and a chimney. If hot water or steam can be supplied by a central plant, a boiler is not required.
Fan room contains the air-handling equipment in large buildings. A single fan room should be located to minimize the distance conditioned air must travel to the farthest air-conditioned space. Individual fan rooms can also be distributed to serve individual zones of a building or be located on each floor to minimize vertical duct runs.
Air-handling units contain the fans, filters, and other components necessary to treat and distribute conditioned air.

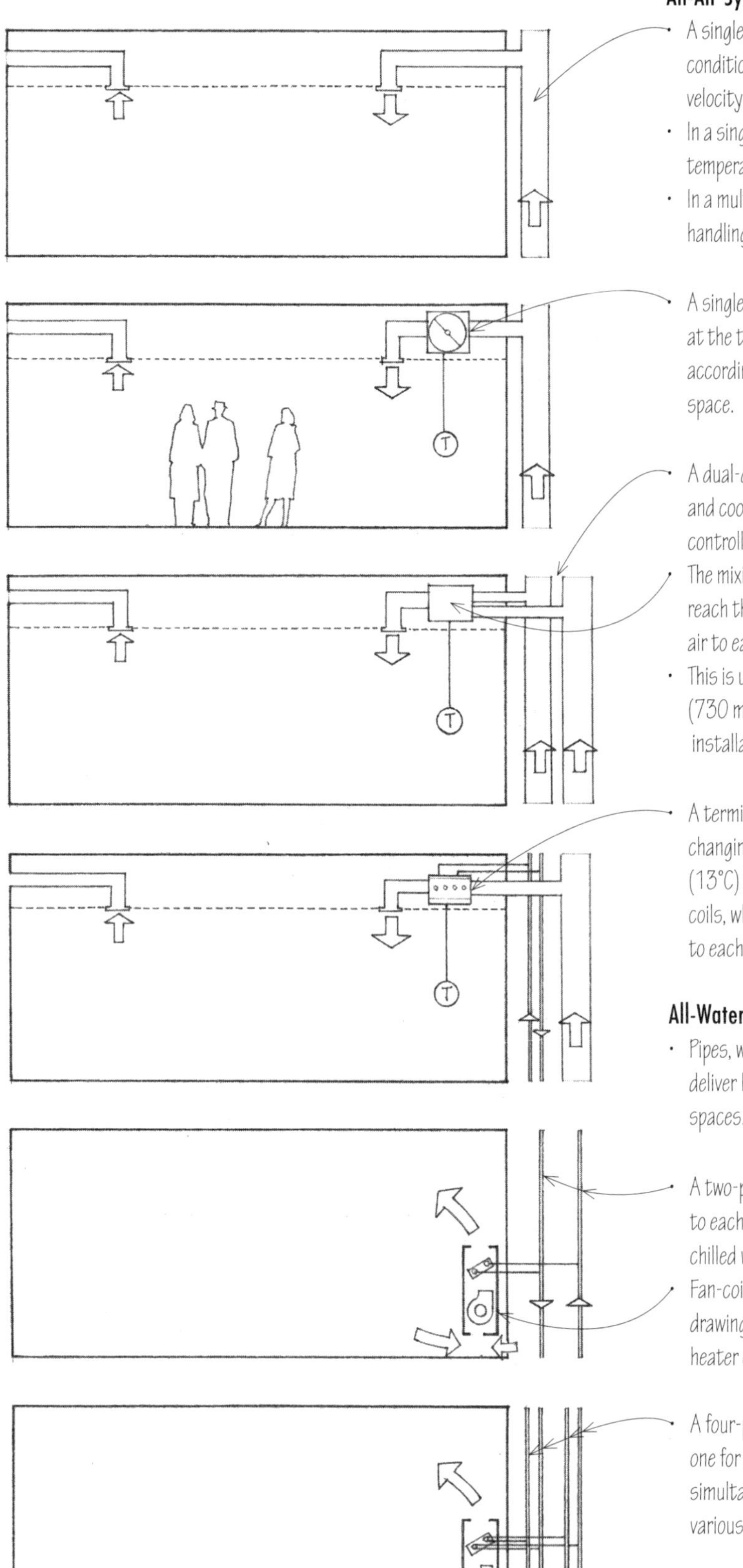

All-Air Systems

- A single-duct, constant-air-volume (CAV) system delivers conditioned air at a constant temperature through a low-velocity duct system to the served spaces.
- In a single-zone system, a master thermostat regulates the temperature for the entire building.
- In a multizone system, separate ducts from a central air-handling unit serve each of a number zones.

- A single-duct, variable-air-volume (VAV) system uses dampers at the terminal outlets to control the flow of conditioned air according to the temperature requirements of each zone or space.

- A dual-duct system uses separate ducts to deliver warm air and cool air to mixing boxes, which contain thermostatically controlled dampers.
- The mixing boxes proportion and blend the warm and cold air to reach the desired temperature before distributing the blended air to each zone or space.
- This is usually a high-velocity system [2400 fpm (730 m/min) or higher] to reduce duct sizes and installation space.

- A terminal reheat system offers more flexibility in meeting changing space requirements. It supplies air at about 55°F (13°C) to terminals equipped with electric or hot-water reheat coils, which regulate the temperature of the air being furnished to each individually controlled zone or space.

All-Water Systems

- Pipes, which require less installation space than air ducts, deliver hot or chilled water to fan-coil units in the served spaces.

- A two-pipe system uses one pipe to supply hot or chilled water to each fan-coil unit and another to return it to the boiler or chilled water plant.
- Fan-coil units contain an air filter and a centrifugal fan for drawing in a mixture of room air and outside air over coils of heater or chilled water and then blowing it back into the space.

- A four-pipe system uses two separate piping circuits—one for hot water and one for chilled water—to provide simultaneous heating and cooling as needed to the various zones of a building.

- Ventilation is provided through wall openings, by infiltration, or by a separate duct system.

Air-Water Systems

- Air-water systems use high-velocity ducts to supply conditioned primary air from a central plant to each zone or space, where it mixes with room air and is further heated or cooled in induction units.
- The primary air draws in room air through a filter and the mixture passes over coils that are heated or chilled by secondary water piped from a boiler or chilled-water plant.
- Local thermostats control water flow over the coils to regulate air temperature.

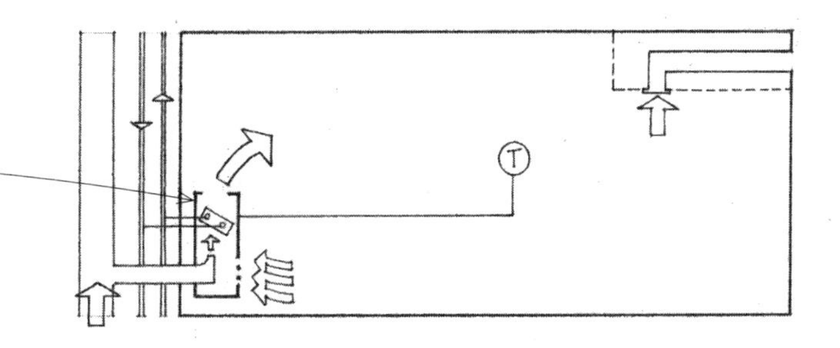

Packaged Systems

- Packaged systems are self-contained, weatherproof units incorporating a fan, filters, compressor, condenser, and evaporator coils for cooling. For heating, the unit may operate as a heat pump or contain auxiliary heating elements. Packaged systems are powered by electricity or by a combination of electricity and gas.

- Packaged systems may be mounted as a single piece of equipment on the roof or on a concrete pad alongside an exterior wall of a building.

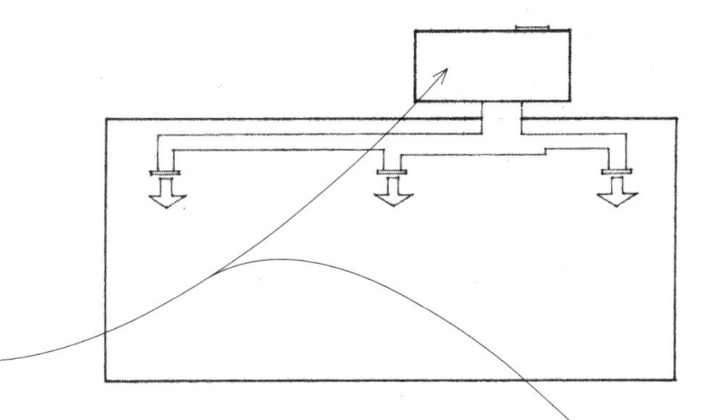

- Rooftop packaged units may be placed at intervals to serve long buildings.
- Packaged systems with vertical shafts that connect to horizontal branch ducts can serve buildings up to four or five stories in height.

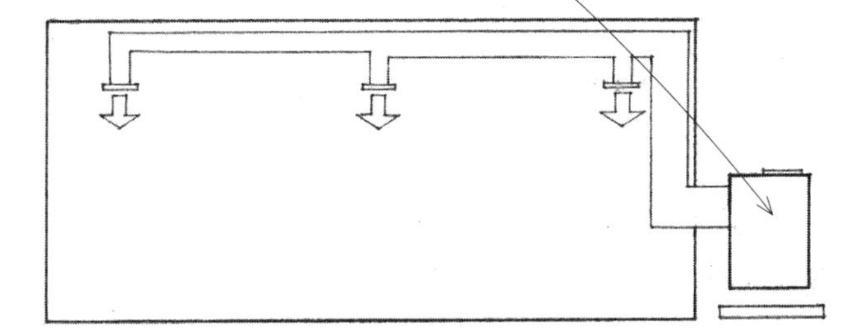

- Split-packaged systems consist of an outdoor unit incorporating the compressor and condenser and an indoor unit that contains the cooling and heating coils and the circulating fan; insulated refrigerant tubing and control wiring connect the two parts.

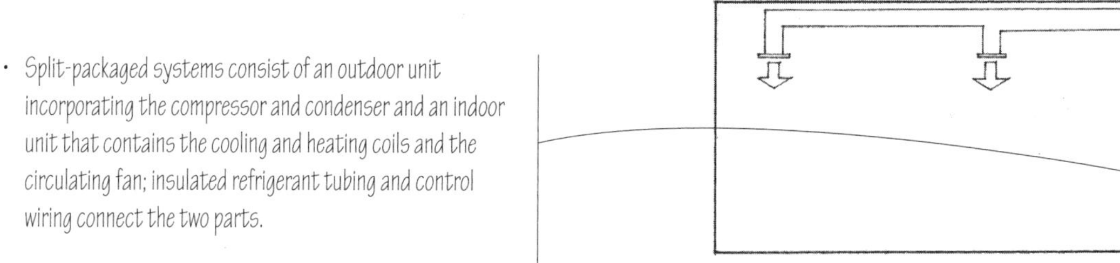

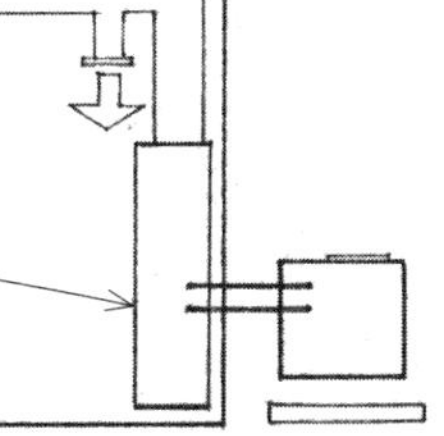

- Small terminal units may be mounted directly below a window or in openings cut into the exterior wall of each served space. Window-mounted units are typically used for retrofitting existing buildings.

Factors to consider in the selection, design, and installation of a heating, ventilating, and air-conditioning system include:

- Performance, efficiency, and both the initial and life costs of the system
- Fuel, power, air, and water required and the means for their delivery and storage; some equipment may require direct access to the outdoors.
- Flexibility of the system to service different zones of a building, which may have different demands because of use or site orientation. Decentralized or local systems are economical to install, require short distribution runs, and allow each space or zone to have individual temperature control, while central systems are generally more energy-efficient, easier to service, and offer better control of air quality.
- Type and layout of the distribution system used for the heating and cooling media. To minimize friction loss, ductwork and piping should have short, direct runs with a minimum of turns and offsets.
- Space requirements for the mechanical equipment and the distribution system. The heating, ventilating, and air-conditioning equipment of a building can often occupy 10% to 15% of the area of a building; some pieces of equipment also require space or a domain for access, service, and maintenance. Air duct systems require more space than either pipes carrying hot or chilled water or wiring for electric resistance heating. Ductwork should therefore be carefully laid out to be integrated with the structure and spaces of a building, as well as with its plumbing and electrical systems.
- Access required for service and maintenance
- Construction requirements for the enclosure of the mechanical plant, fire resistance, and noise and vibration control
- Structural requirements imposed by the weight of the equipment
- Degree of visibility, whether concealed within the construction or exposed to view. If ductwork is to be left exposed, the layout should have a visually coherent order and be coordinated with the physical elements of the space (e.g., structural elements, lighting fixtures, surface patterns).

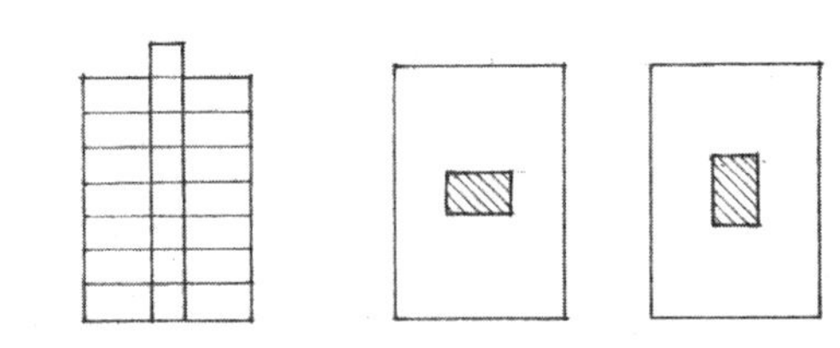

- A single core is often used in high-rise office buildings to leave a maximum amount of unobstructed rentable area.
- Central locations are ideal for short runs and efficient distribution patterns.

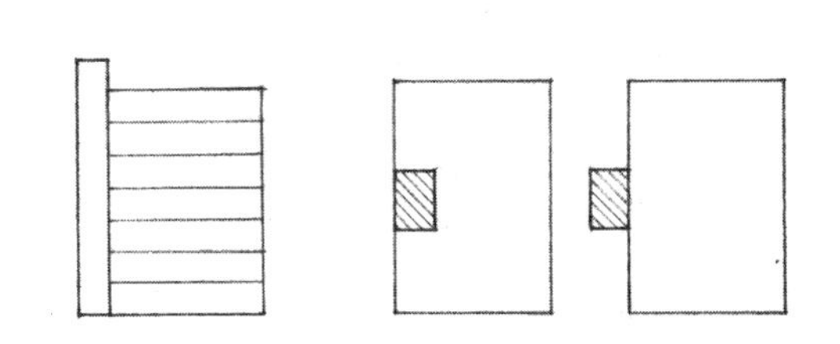

- Placing the core along an edge leaves an unobstructed floor space but occupies a portion of the daylit perimeter.

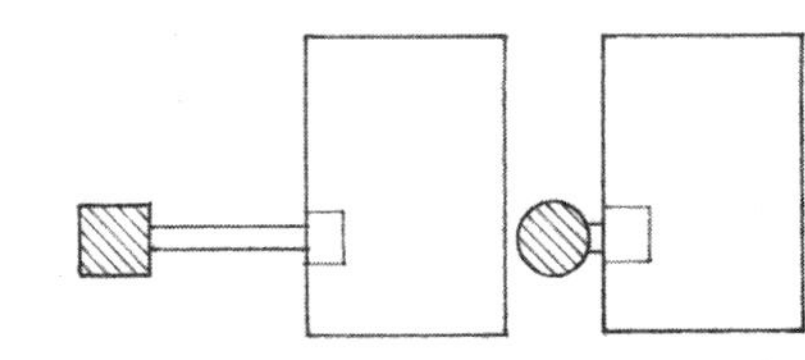

- Detached cores leave a maximum amount of floor space but require long service runs and cannot serve as lateral bracing.

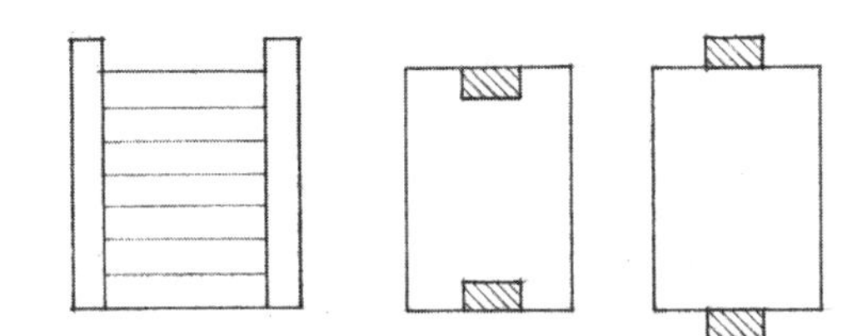

- Two cores may be symmetrically placed to reduce service runs and to serve effectively as lateral bracing, but the remaining floor area loses some flexibility in layout and use.

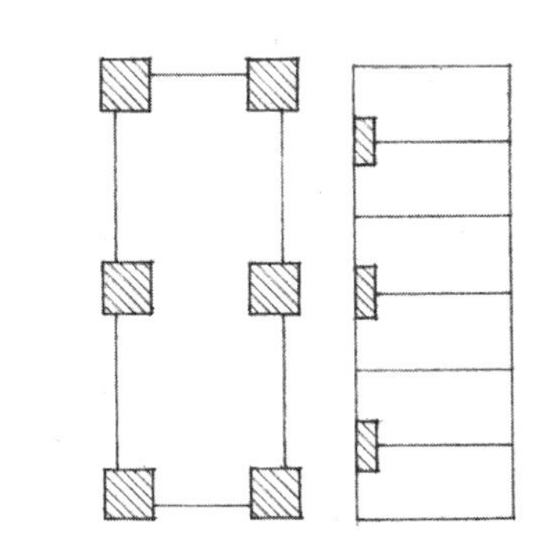

- Multiple cores are often used in broad, low-rise buildings in order to avoid long horizontal runs.
- The cores may be dispersed to better serve spaces or zones that have different demands and load requirements.
- In apartment buildings and other structures housing repetitive units, the cores may be situated between the units or along interior corridors.

- The service core or cores of a building house the vertical distribution of mechanical and electrical services, elevator shafts, and exit stairways. These cores must be coordinated with the structural layout of columns, bearing walls, and shear walls or lateral bracing as well as with the desired patterns of space, use, and activity. Shown above are some basic ways in which we can lay out the service cores of a building.

Air for heating, cooling, and ventilating is supplied through registers and diffusers. They should be evaluated in terms of their air-flow capacity and velocity, pressure drop, noise factor, and appearance.

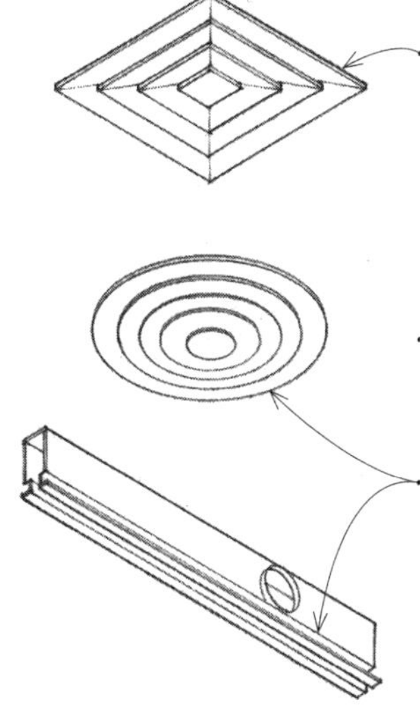

- Diffusers have slats at different angles for deflecting warm or conditioned air from an outlet in various directions.
- Ceiling diffusers discharge low-velocity air in a spreading pattern.
- Diffusers may be round, square, or linear, or be in the form of perforated ceiling tiles.

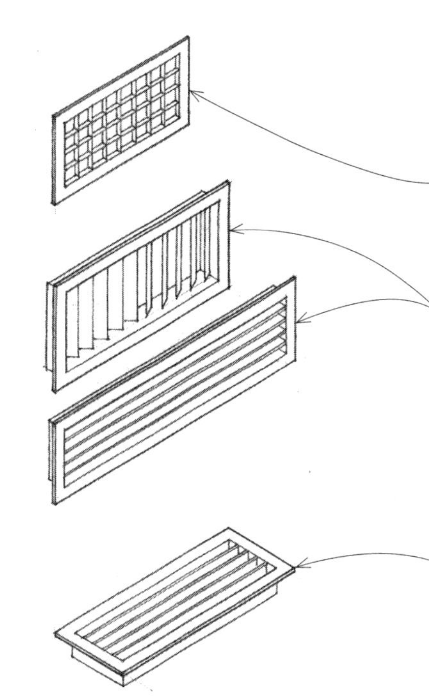

- Grills are simply gratings or perforated screens for covering and protecting an opening.
- Registers control the flow of warm or conditioned air from an outlet, composed of a grill with a number of parallel blades that may be adjusted so as to overlap and close the opening.
- Floor registers are used to control heat loss and condensation along exterior windows and walls.

Air-supply outlets should be located to distribute warm or cool air to the occupied areas of a space comfortably, without noticeable drafts, and without stratification. The throw distance and spread or diffusion pattern of the supply outlet should be carefully considered along with any obstructions that might interfere with the air distribution.

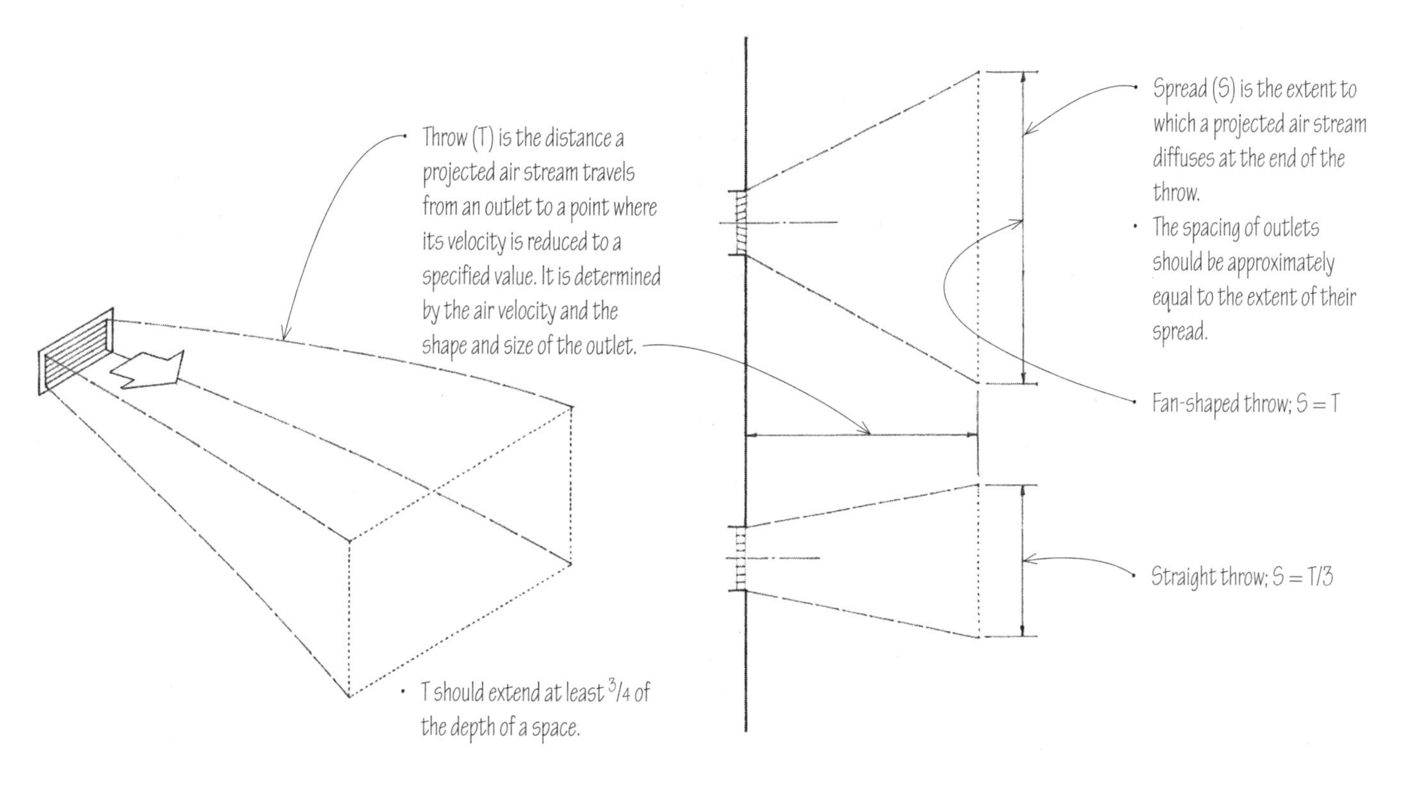

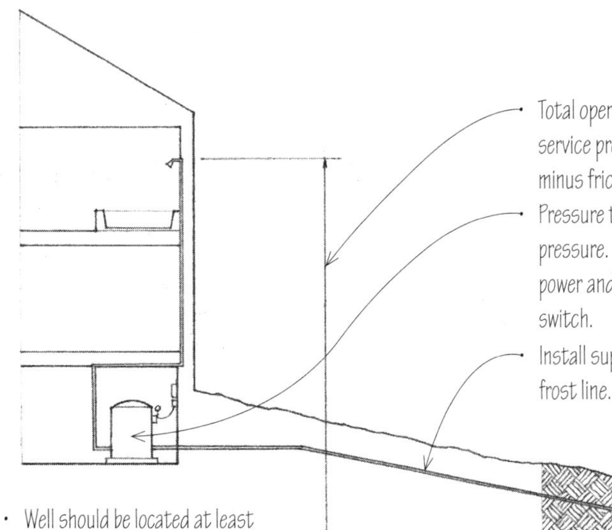

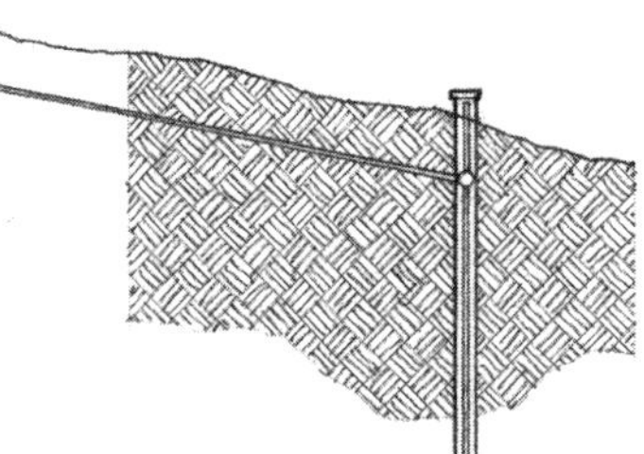

- Total operating head = service pressure minus friction head loss
- Pressure tank maintains service pressure. It requires electrical power and a fused disconnect switch.
- Install supply pipe below frost line.

- Well should be located at least 100' (30 m) away from building sewers, septic tanks, and sewage disposal fields, and should be accessible to permit the removal of the well casings or pump for maintenance or repair.
- Check applicable codes that govern well locations and installations.

Private Well

Water is used in a building in the following ways:

- Water is consumed by drinking, cooking, and washing.
- HVAC systems circulate water for heating and cooling, and maintaining a desirable level of humidity.
- Fire-protection systems store water for extinguishing fires.

Water must be supplied to a building in the correct quantity, and at the proper flow rate, pressure, and temperature, to satisfy the above requirements. For human consumption, water must be potable—free of harmful bacteria—and palatable. To avoid the clogging or corrosion of pipes and equipment, water may have to be treated for hardness or excessive acidity.

If water is supplied by a municipal or public system, there can be no direct control over the quantity or quality of water supplied until it reaches the building site. If a public water system is not available, then either drilled or bored wells or rainwater storage tanks are required.

Well water, if the source is deep enough, is usually pure, cool, and free of discoloration and taste or odor problems. A sample should be checked for bacteria and chemical content by the local health department before a well is put into operation.

- Water main is the conduit through which a public or community water system conveys water to all service connections.
- Service pipe connects a building to a water main, usually installed by or under the jurisdiction of a public utility.
- Building shutoff valve
- Corporation stop is a valve controlling the flow of water from a main to a service pipe.
- Curb box provides access to a water meter that measures and records the quantity of water that passes through a service pipe, and the control valve for shutting off the water supply to a building in case of an emergency.

Public Water Supply

CSI MasterFormat 33 10 00: Water Utilities
CSI MasterFormat 33 21 00: Water Supply Wells

Water supply systems operate under pressure. The service pressure of a water supply system must be great enough to absorb pressure losses due to vertical travel and friction as the water flows through pipes and fittings, and still satisfy the pressure requirement of each plumbing fixture. Public water systems usually supply water at about 50 psi (345 kPa). This pressure is the approximate upper limit for most private well systems.

If water is supplied at 50 psi (345 kPa), upfeed distribution is feasible for low-rise buildings up to six stories in height. For taller buildings, or where the water service pressure is insufficient to maintain adequate fixture service, water is pumped up to an elevated or rooftop storage tank for gravity downfeed. Part of this water is often used as a reserve for fire-protection systems.

There must be sufficient pressure at each fixture to ensure its satisfactory operation. Fixture pressure requirements vary from 5 to 30 psi (35 to 207 kPa). Too much pressure is as undesirable as insufficient pressure. Water supply pipes are therefore sized to use up the differential between the service pressure, allowing for the pressure loss due to vertical lift or hydraulic friction, and the pressure requirement for each fixture. If the supply pressure is too high, pressure reducers or regulators may be installed on plumbing fixtures.

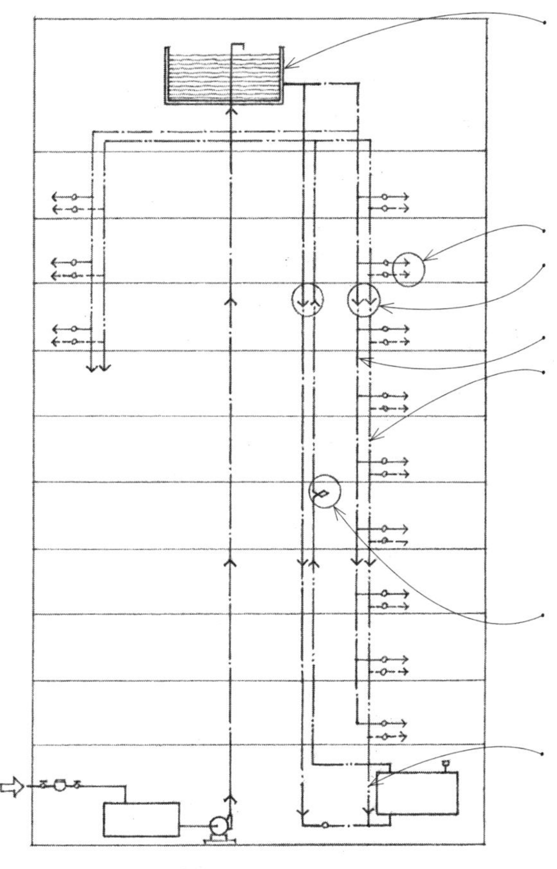

- Gravity downfeed system sets a water source at a height sufficient to maintain adequate supply pressure throughout the water distribution system.
- Branch lines
- Risers
- Cold-water supply
- Hot-water supply; hot water circulates by virtue of its natural rising action. In long, low-rise buildings, pumps may be required for hot-water circulation and distribution.
- Expansion bends permit thermal expansion to occur in long runs of hot-water piping.
- Hot-water return line to heater or storage tank in two-pipe systems

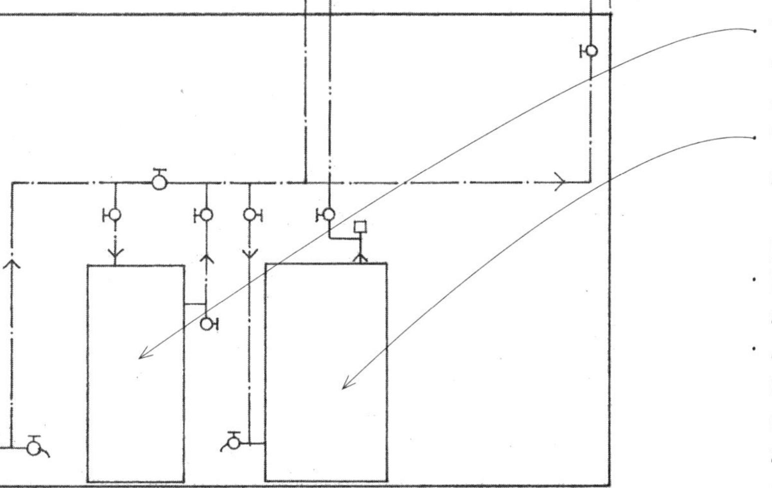

- Exterior hose bibbs should be frostproof in cold climates.
- Water softener removes calcium and magnesium salts from hard water by ion exchange; hard water can clog pipes, corrode boilers, and inhibit the sudsing action of soap.
- Water heaters are electric or gas appliances (CSI 22 33 00 and 22 34 00) for heating water to a temperature between 120°F and 140°F (49°C and 60°C) and storing it for use. Safety pressure-relief valves are required for all water heaters.
- Hot-water storage tanks may be required for large installations and widespread fixture groupings.
- An alternative to standard water heaters is an on-demand water-heating system that heats water at the time and point of use. These systems are energy-efficient and require no space for a storage tank, but they do need an exhaust vent for the natural gas heater.
- A third alternative is a solar water-heating system able to satisfy the typical hot-water needs of a household in sunny climates. In temperate climates, solar water-heating systems can effectively serve as a preheating system backed up by a standard water-heating system.

- Upfeed system distributes water from a water main or an enclosed storage tank under pressure from compressed air.

CSI MasterFormat 22 00 00: Plumbing
CSI MasterFormat 22 11 00: Facility Water Distribution

- The pressure loss due to hydraulic friction depends on the diameter of the supply pipe, the distance of water flow, and the number of valves, tees, and elbow fittings through which the water passes. Runs should be short, straight, and as direct as possible.

Maximum pressure required at any fixture [5 to 30 psi (35 to 207 kPa)]
\+ Pressure loss through water meter
\+ Pressure loss due to static head or vertical lift
\+ Pressure loss by hydraulic friction in pipe runs and fittings
= Water service pressure

Individual Demand

Fixture Type	Load in Fixture Units	Minimum Supply Pipe in. (mm)
Bathtub	2 to 4	1/2 (13)
Shower head	2 to 4	1/2 (13)
Lavatory	1 to 2	3/8 or 1/2 (10 or 13)
Water closet, tank type	3 to 5	3/8 (10)
Water closet, flush valve	6 to 10	1 (25)
Urinal	5 to 10	1/2 or 3/4 (13 or 19)
Kitchen sink	2 to 4	1/2 or 3/4 (13 or 19)
Clothes washer	2 to 4	1/2 (13)
Service sink	3	1/2 (13)
Hose Bibb	2 to 4	1/2 (13)

Estimating Total Demand

Total Load Fixture Units	Total Demand gpm (m^3/m)
10	8 (0.03)
20	14 (0.06)
40	25 (0.10)
60	32 (0.13)
80	38 (0.15)
100	44 (0.18)
120	48 (0.19)
140	53 (0.21)
160	55 (0.22)
200	65 (0.26)

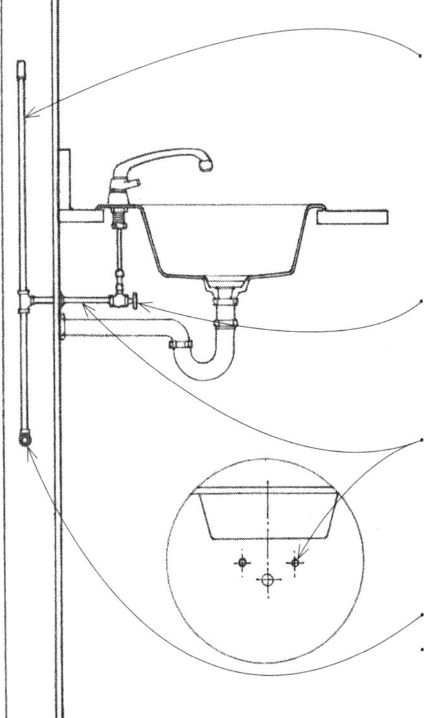

- Water hammer is the concussion and banging noise that can occur when a volume of water moving in a pipe is shut off abruptly. Air chambers are installed at fixture branches to prevent water hammer. The trapped air elastically compresses and expands to equalize the pressure and flow of water in the system.
- Fixture shutoff valve controls the flow of water at each fixture; additional valves can be installed to isolate one or more fixtures from the water-supply system for repair and maintenance.
- Fixture runout; the rough-in dimensions for each plumbing fixture should be verified with the fixture manufacturer so that the fixture supplies can be accurately installed during the proper phase of construction.
- Branch supply line
- If a water supply pipe must be located in an exterior wall, it should be placed on the warm side of the wall insulation.

Water supply lines may be of copper, galvanized steel, or plastic. Copper piping is commonly used for water supply lines because of its corrosion resistance, strength, low friction loss, and small outside diameter. Plastic pipes are lightweight, easily joined, produce low friction, and do not corrode, but not all types are suitable for carrying potable water. Polybutylene (PB), polyethylene (PE), polyvinyl chloride (PVC), and chlorinated polyvinyl chloride (CPVC) pipes may be used for cold-water supply lines; only PB and CPVC are suitable for hot-water lines.

Engel-method crosslinked polyethylene (PEX-A) tubing is suitable for both cold- and hot-water lines. It is flexible, immune to corrosion and mineral buildup, retains more heat in hot-water lines, resists condensation in cold-water lines, and dampens water noise.

Water pipes are sized according to the number and types of plumbing fixtures served and pressure losses due to hydraulic friction and static head. Each type of fixture is assigned a number of fixture units. Based on the total number of fixture units for a building, an equivalent demand in gallons per minute (gpm) is estimated. Because it is assumed that not all fixtures will be used at the same time, the total demand is not directly proportional to the total load in fixture units.

The water supply system can usually be accommodated within floor and wall construction spaces without too much difficulty. It should be coordinated with the building structure and other systems, such as the parallel but bulkier sanitary drainage system.

Water supply pipes should be supported at every story vertically and every 6' to 10' (1830 to 3050) horizontally. Adjustable hangers can be used to ensure proper pitch along horizontal runs for drainage.

Cold-water pipes should be insulated to prevent heat flow into the water from the warmer surrounding air. Hot-water pipes should be insulated against heat loss and should be no closer than 6" (150) to parallel cold-water pipes.

In very cold climates, water pipes in exterior walls and unheated buildings can freeze and rupture. Provision should be made for their drainage to a low point in the system where a drainage faucet is located.

Fire-alarm systems are installed in a building to automatically sound an alarm when actuated by a fire-detection system. The fire-detection system may consist of heat sensors such as thermostats, or smoke detectors that are actuated by-products of combustion. Most jurisdictions require the installation and hard-wiring of smoke detectors in residential occupancies and hotel/motel units. Refer to the National Fire Protection Association's (NFPA's) *Life Safety Code* for recommendations concerning the type and placement of heat and smoke detectors.

In large commercial and institutional buildings where public safety is an issue, building codes often require a fire sprinkler system; some codes allow an increase in floor area if an approved sprinkler system is installed. Some jurisdictions require the installation of fire sprinkler systems in multifamily housing as well.

Fire sprinkler systems consist of pipes that are located in or below ceilings, connected to a suitable water supply, and supplied with valves or sprinkler heads made to open automatically at a certain temperature. The two major types of sprinkler systems are wet-pipe systems and dry-pipe systems.

- Wet-pipe systems contain water at sufficient pressure to provide an immediate, continuous discharge through sprinkler heads that open automatically in the event of a fire.
- Dry-pipe systems contain pressurized air that is released when a sprinkler head opens in the event of fire, allowing water to flow through the piping and out the opened nozzle. Dry-pipe systems are used where the piping is subject to freezing.
- Preaction systems are dry-pipe sprinkler systems through which water flow is controlled by a valve operated by fire-detection devices more sensitive than those in the sprinkler heads. Preaction systems are used when an accidental discharge would damage valuable materials.
- Deluge systems have sprinkler heads open at all times, through which water flow is controlled by a valve operated by a heat-, smoke-, or flame-sensing device.

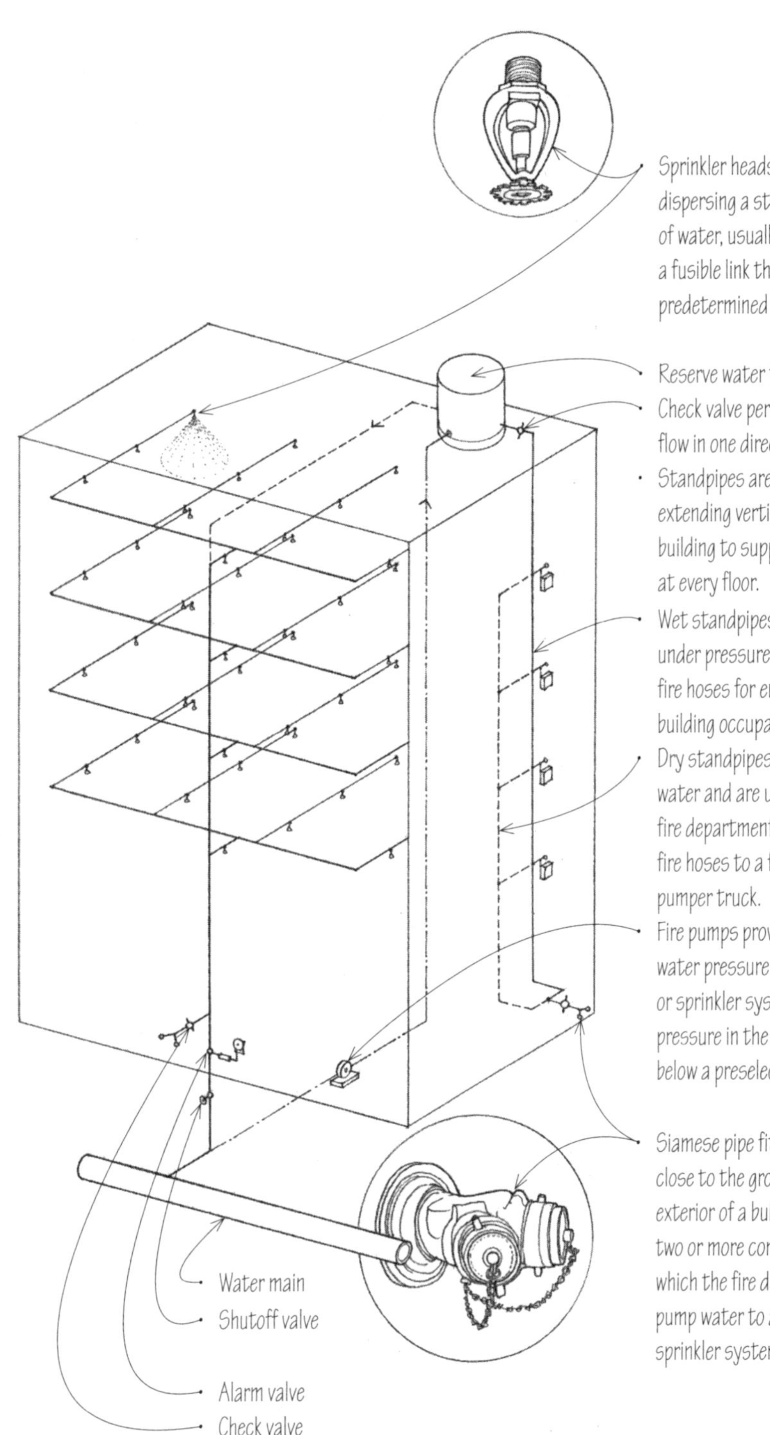

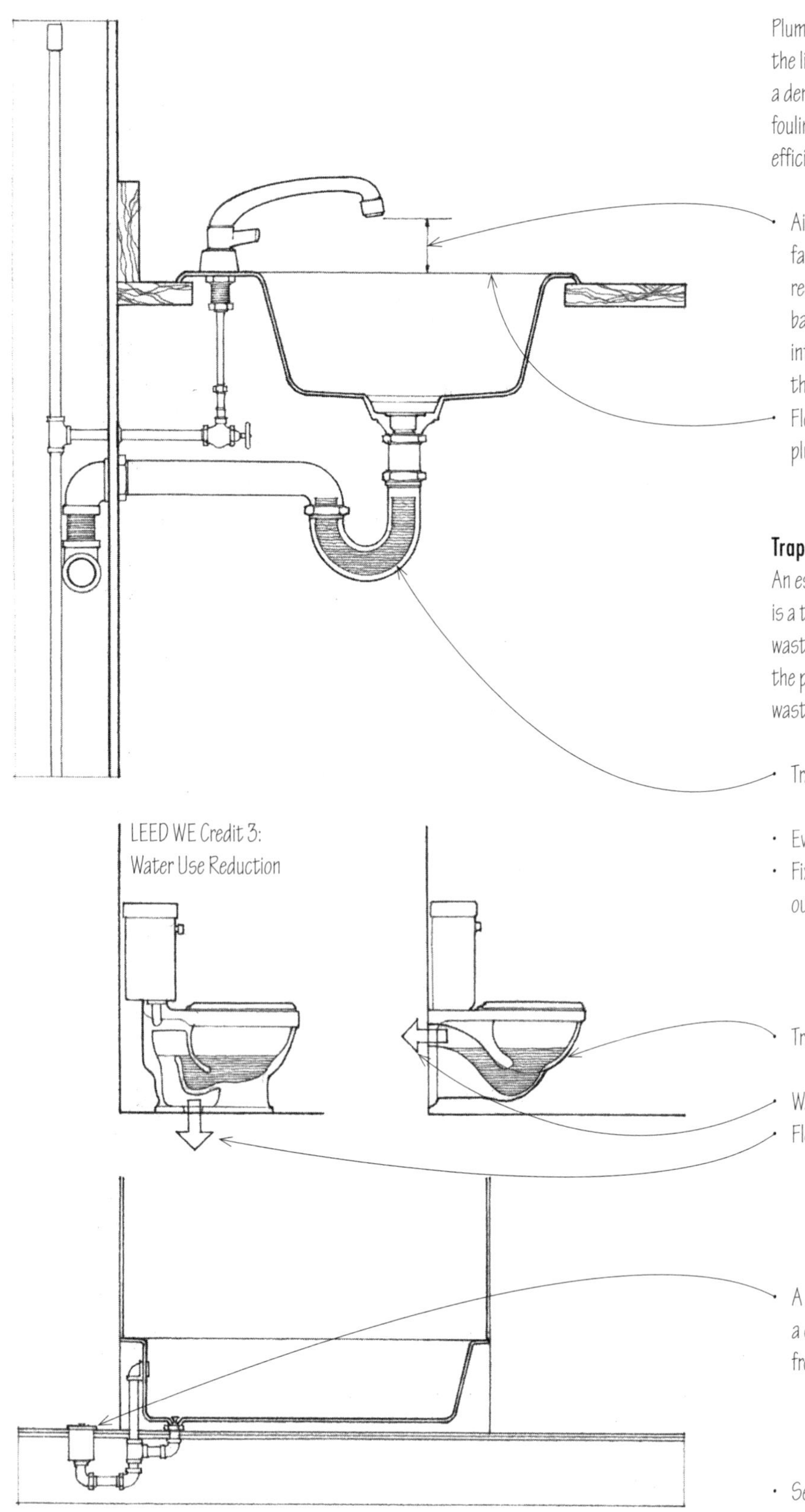

Plumbing fixtures receive water from a supply system and discharge the liquid waste into a sanitary drainage system. They should be of a dense, smooth, nonabsorbent material, and be free of concealed fouling surfaces. Some building codes mandate the use of water-efficient fixtures and valves in order to conserve water resources.

- Air gap is the clear vertical distance between the spout of a faucet or other outlet of a supply pipe and the flood level of a receptacle. Air gaps are required to prevent back-siphonage or backblow of used or contaminated water from a plumbing fixture into a pipe supplying potable water due to negative pressure in the pipe.
- Flood level is the level at which water would overflow the rim of a plumbing fixture.

Traps

An essential feature of the sanitary drains from plumbing fixtures is a trap, a U-shaped or S-shaped section of drainpipe in which wastewater remains. This wastewater forms a seal that prevents the passage of sewer gas without affecting the normal flow of wastewater or sewage through it.

- Trap with water seal

- Every plumbing fixture requires a trap.
- Fixtures should have sufficient water flow to periodically clean out their traps and prevent sediment from collecting.

- Traps are cast into water closets.

- Wall outlet for wall-hung water closets
- Floor outlet for other types of water closets

- A drum trap is a cylindrical trap closed on the bottom and having a cover plate for access, commonly installed on the drain line from a bathtub.

- See 9.27 for typical sizes of plumbing fixtures.

The water supply system terminates at each plumbing fixture. After water has been drawn and used, it enters the sanitary drainage system. The primary objective of this drainage system is to dispose of fluid waste and organic matter as quickly as possible.

Since a sanitary drainage system relies on gravity for its discharge, its pipes are much larger than the water supply lines, which are under pressure. Drainage lines are sized according to their location in the system and the total number and types of fixtures served. Always consult the plumbing code for allowable pipe materials, pipe sizing, and restrictions on the length and slope of horizontal runs and on the types and number of turns allowed.

Drainage lines may be of cast iron or plastic. Cast iron, the traditional material for drainage piping, may have hubless or bell-and-spigot joints and fittings. The two types of plastic pipe that are suitable for drainage lines are polyvinyl chloride (PVC) and acrylonitrile-butadiene-styrene (ABS). Some building codes also permit the use of galvanized wrought iron or steel.

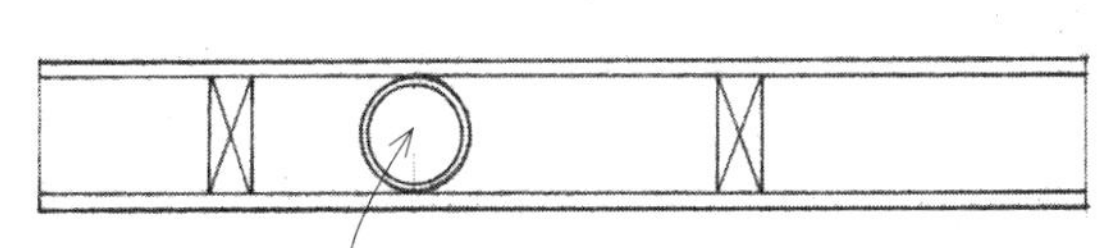

4" (100) stud wall
- 3" (75) ø hubless cast iron or plastic pipe
- 2" (51) ø bell-and-spigot cast iron pipe

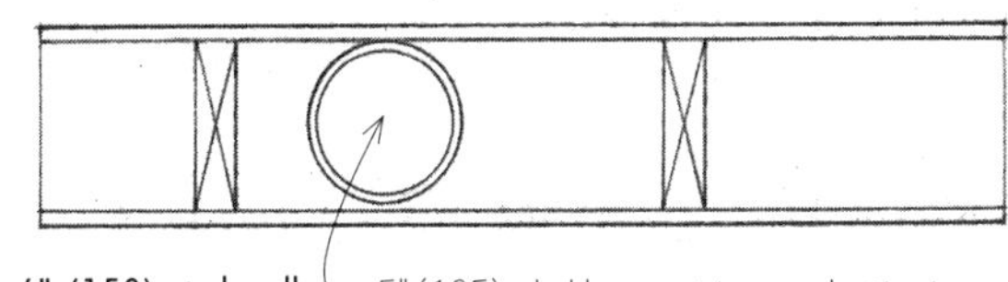

6" (150) stud wall
- 5" (125) ø hubless cast iron or plastic pipe
- 3" (75) ø bell-and-spigot cast iron pipe

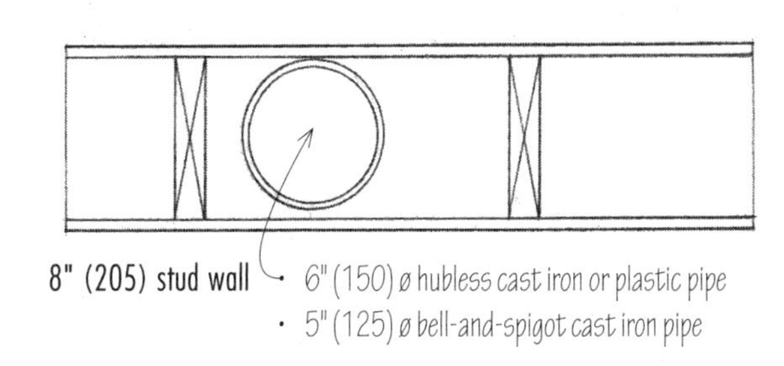

8" (205) stud wall
- 6" (150) ø hubless cast iron or plastic pipe
- 5" (125) ø bell-and-spigot cast iron pipe

Maximum Pipe Sizes

- The plumbing or wet wall behind fixtures should be deep enough to accommodate branch lines, fixture runouts, and air chambers.

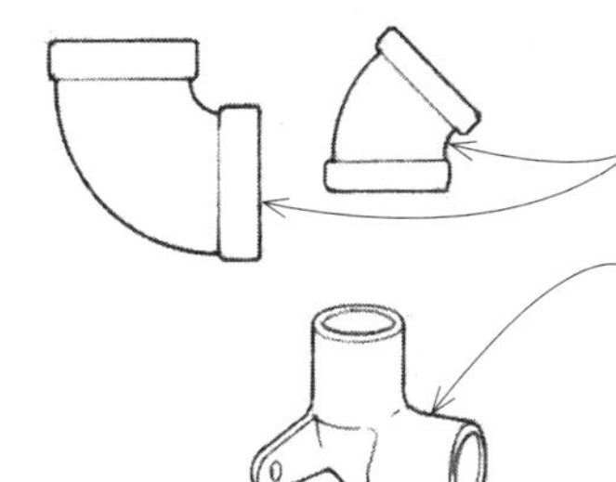

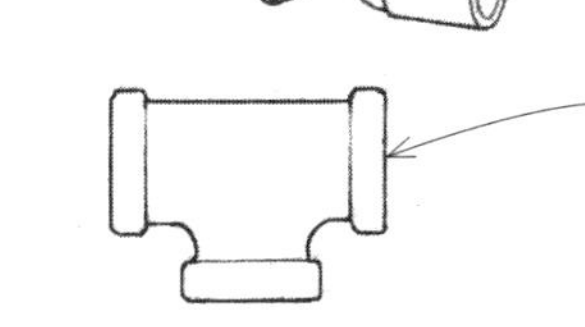

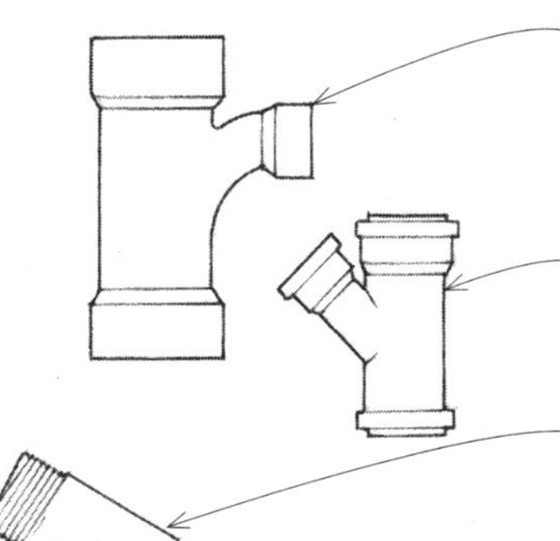

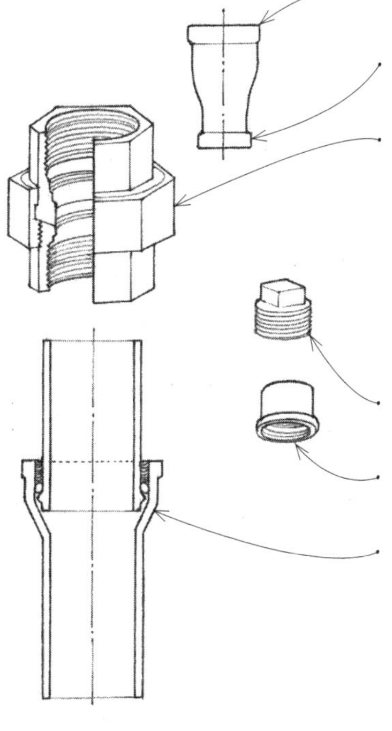

- Elbows have an angled bend, usually 45° or 90°.
- Drop elbow has lugs for attachment to a wall or joist.
- Closet bend is a 90° soil fitting installed directly beneath a water closet.
- Tees are T-shaped pipe fittings for making three-way joints.
- Drop tee has lugs for attachment to a wall or joist.
- Sanitary tees have a slight curve in the 90° transition to channel the flow from a branch pipe in the direction of the main.
- Wyes are Y-shaped pipe fittings for joining a branch pipe with a main, usually at a 45° angle.
- Nipples are short lengths of pipe with threads on each end, used for joining couplings or other pipe fittings.
- Couplings are short lengths of pipe having each end threaded on the inside, used for joining two pipes of the same diameter.
- Increaser is a coupling increasing in diameter at one end.
- Reducer is a coupling decreasing in diameter at one end.
- Unions connect two pipes neither of which can be turned, consisting of two internally threaded end pieces that are tightened around the pipe ends to be joined, and an externally threaded center piece, which draws the two end pieces together as it is rotated.
- Plugs are externally threaded fittings for closing the end of a pipe.
- Caps are internally threaded fittings for enclosing the end of a pipe.
- Bell-and-spigot pipe joints are made by fitting the end (spigot) of one pipe into the enlarged end (bell) of another pipe and sealing with a caulking compound or a compressible ring.

Pipe Fittings

CSI MasterFormat 33 30 00: Sanitary Sewerage Utilities

The layout of the sanitary drainage system should be as direct and straightforward as possible to prevent the deposit of solids and clogging. Cleanouts should be located to allow pipes to be easily cleaned if they do clog.

- Branch drain connects one or more fixtures to a soil or waste stack.
- Horizontal drain lines should slope 1/8" per foot (1:100) for pipes up to 3" (75) ø, and 1/4" per foot (1:50) for pipes larger than 3" (75) ø.
- Fixture drain extends from the trap of a plumbing fixture to a junction with a waste or soil stack.
- Soil stack carries the discharge from water closets or urinals to the building drain or building sewer.
- Waste stack carries the discharge from plumbing fixtures other than water closets or urinals.
- Minimize bends in all stacks.
- Branch interval refers to a length of soil or waste stack corresponding to a story height but never less than 8' (2440), within which the horizontal branch drains from one floor are connected.
- Fresh-air inlet admits fresh air into the drainage system of a building, connected to the building drain at or before the building trap.
- Building sewer connects a building drain to a public sewer or private treatment facility.

- Stack vent is an extension of a soil or waste stack above the highest horizontal drain connected to the stack; extend 12" (305) above roof surface and keep away from vertical surfaces, operable skylights, and roof windows.

Vents

The vent system permits septic gases to escape to the outside and supplies a flow of fresh air into the drainage system to protect trap seals from siphonage and back pressure.

- Relief vent provides circulation of air between a drainage and a venting system by connecting a vent stack to a horizontal drain between the first fixture and the soil or waste stack.
- Loop vent is a circuit vent that loops back and connects with a stack vent instead of a vent stack.
- Common vent serves two fixture drains connected at the same level.
- Vent stack is a vertical vent installed primarily to provide circulation of air to or from any part of a drainage system.
- Branch vent connects one or more individual vents with a vent stack or stack vent.
- Continuous vent is formed by a continuation of the drain line to which it connects.
- Back vent is installed on the sewer side of a trap.
- Circuit vent serves two or more traps and extends from in front of the last fixture connection of a horizontal branch to the vent stack.
- Wet vent is an oversized pipe functioning both as a soil or waste pipe and a vent.
- Cleanouts

- Sump pump removes the accumulations of liquid from a sump pit. It is required for fixtures located below the street sewer.

- Sanitary sewers convey only the sewage from plumbing fixtures and exclude storm water; storm sewers convey rainfall drained from roofs and paved surfaces; combined sewers carry both sewage and storm water.

- Building trap is installed in the building drain to prevent the passage of sewer gases from the building sewer to the drainage system of a building. Not all plumbing codes require a building trap.

- Building drain is the lowest part of a drainage system that receives the discharge from soil and waste stacks inside the walls of a building and conveys it by gravity to the building sewer.
- Building storm drain conveys only rainwater or similar discharge to a building storm sewer, which in turn leads to a public storm sewer, combined sewer, or other point of disposal.

CSI MasterFormat 33 31 00: Sanitary Utility Sewerage Piping

Sanitary sewers usually convey sewage from plumbing fixtures to a public facility for treatment and disposal. When this is not possible, a private sewage disposal system is required. Its type and size depend on the number of fixtures served and the permeability of the soil as determined by a percolation test. Sewage disposal systems are designed by sanitary engineers and must be approved and inspected by the health department before being put into use. Consult the building and health codes for specific regulations and requirements.

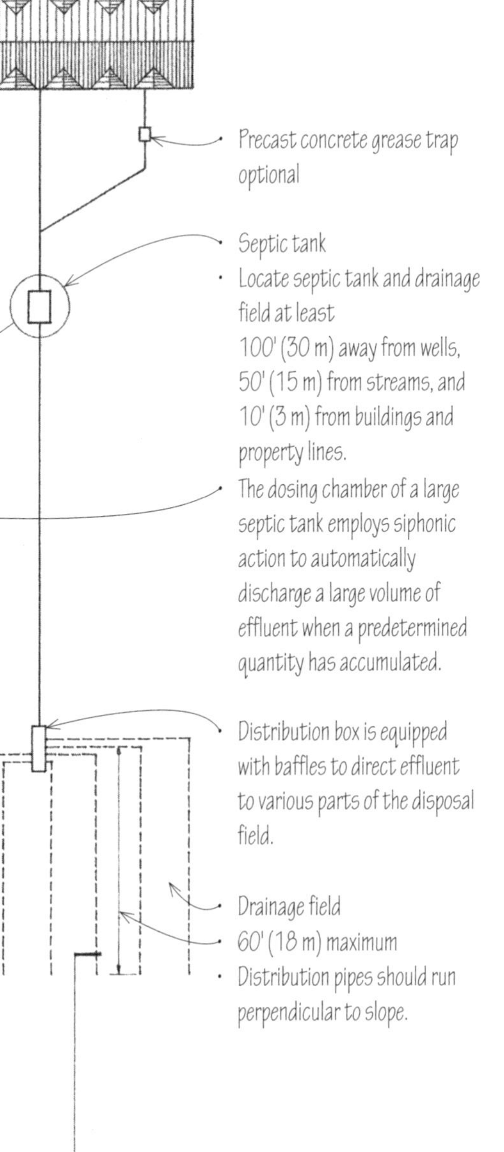

A septic tank is a covered watertight tank for receiving the discharge from a building sewer, separating out the solid organic matter, which is decomposed and purified by anaerobic bacteria, and allowing the clarified liquid to discharge for final disposal.

The liquid effluent, which is about 70% purified, may flow into one of the following systems:

- A drainfield is an open area containing an arrangement of absorption trenches through which effluent from a septic tank may seep or leach into the surrounding soil.
- A seepage pit lined with a perforated masonry or concrete wall is sometimes used as a substitute for a drainfield when the soil is absorbent and the highest level of water table is at least 2' (610) below the bottom of the pit.
- A subsurface sand filter consists of distribution pipes surrounded by graded gravel, an intermediate layer of clean, coarse sand, and a system of underdrains to carry off the filtered effluent. Sand filters are used only where other systems are not feasible.

- Absorption trenches are 18" to 30" (455 to 760) wide and 30" (760) deep; they contain coarse aggregate and a perforated distribution pipe through which the effluent from a septic tank is allowed to seep into the soil.
- 2'-0" (610) minimum to water table

- Graywater refers to the wastewater from sinks, baths, showers, and dishwashers, which can be treated and recycled for such uses as toilet flushing and irrigation. To date, few communities have adopted code provisions allowing the reuse of graywater. Graywater systems should be used in conjunction with other water-conservation strategies, such as specifying water-efficient fixtures and capturing rainwater and surface runoff in cisterns and reservoirs for use in landscaping.

LEED WE Credit 2: Innovative Wastewater Technologies

CSI MasterFormat 33 36 00: Utility Septic Tanks

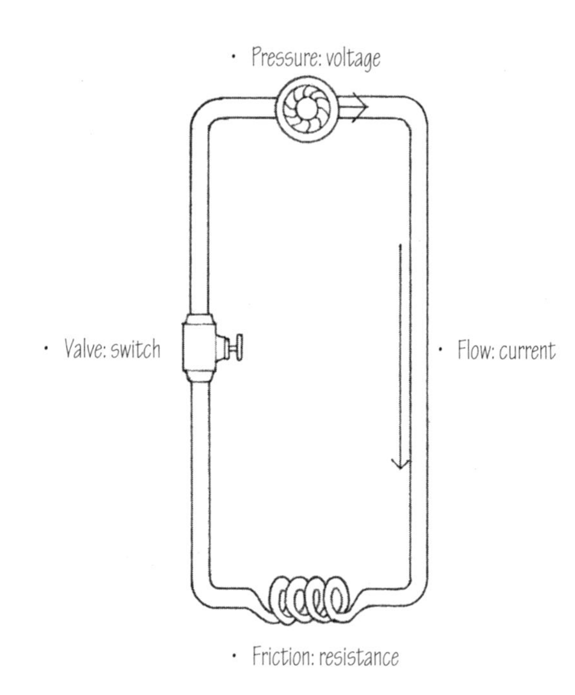

Hydraulic Analogy to Electric Circuit

The electrical system of a building supplies power for lighting, heating, and the operation of electrical equipment and appliances. This system must be installed according to the building and electrical codes in order to operate safely, reliably, and effectively. All electrical equipment should meet Underwriters' Laboratories (UL) standards. Consult the National Electrical Code for specific requirements in the design and installation of any electrical system.

Electrical energy flows through a conductor because of a difference in electrical charge between two points in a circuit.

- Volt (V) is the SI unit of electromotive force, defined as the difference of electric potential between two points of a conductor carrying a constant current of one ampere, when the power dissipated between the points is equal to one watt.
- Ampere (A) is the basic SI unit of electric current, equivalent to a flow of one coulomb per second or to the steady current produced by one volt applied across a resistance of one ohm.
- Watt (W) is the SI unit of power, equal to one joule per second or to the power represented by a current of one ampere flowing across a potential difference of one volt.
- Ohm is the SI unit of electrical resistance, equal to the resistance of a conductor in which a potential difference of one volt produces a current of one ampere. Symbol: Ω

Power is usually supplied to a building by the electric utility company. The schematic diagram below illustrates several voltage systems that may be furnished by the public utility according to the load requirements of a building. A large installation may use its own transformer to step down from a more economical, higher supply voltage to the service voltage. Generator sets may be required to supply emergency electrical power for exit lighting, alarm systems, elevators, telephone systems, fire pumps, and medical equipment in hospitals.

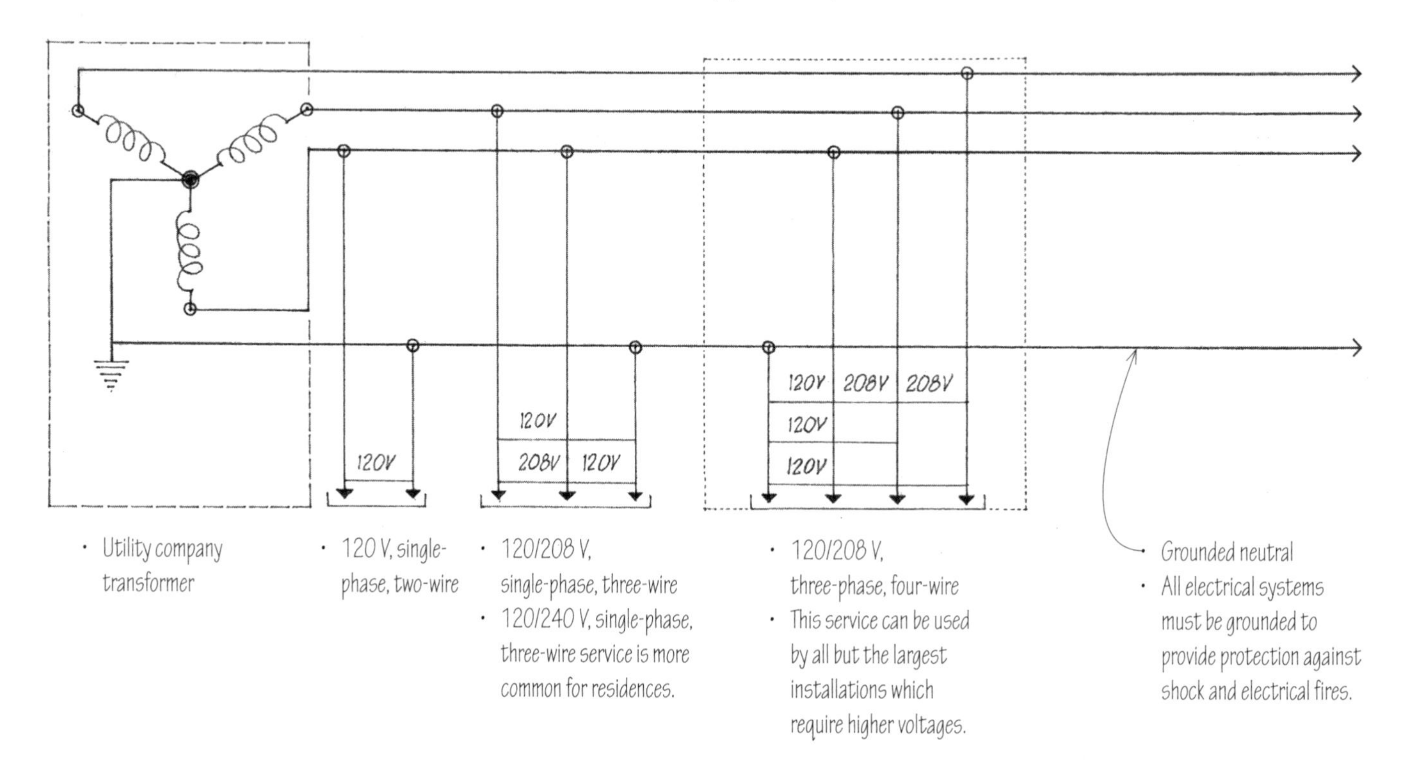

- Utility company transformer
- 120 V, single-phase, two-wire
- 120/208 V, single-phase, three-wire
- 120/240 V, single-phase, three-wire service is more common for residences.
- 120/208 V, three-phase, four-wire
- This service can be used by all but the largest installations which require higher voltages.
- Grounded neutral
- All electrical systems must be grounded to provide protection against shock and electrical fires.

The public utility company should be notified of the estimated total electrical load requirements for a building during the planning phase to confirm service availability and to coordinate the location of the service connection and meter.

The service connection may be overhead or underground. Overhead service is less expensive, easily accessible for maintenance, and can carry high voltages over long runs. Underground service is more expensive but is used in high load-density situations such as urban areas. The service cables are run in pipe conduit or raceways for protection and to allow for future replacement. Direct burial cable may be used for residential service connections.

- A transformer is used by medium-sized and large buildings to step down from a high-supply voltage to the service voltage. To reduce costs, maintenance, and noise and heat problems, a transformer may be placed on an outdoor pad. If located within a building, oil-filled transformers require a well-ventilated, fire-rated vault with two exits and located on an exterior wall adjacent to the switchgear room.
- Dry-type transformers used in small- and medium-sized buildings may be placed together with a disconnect switch and switchgear in a unit substation.

- The service switch is the main disconnect for the entire electrical system of a building, except for any emergency power systems.
- The service equipment includes a main disconnect switch and secondary switches, fuses, and circuit breakers for controlling and protecting the electric power supply to a building. It is located in a switchgear room near the entrance of the service conductors.
- The main switchboard is a panel on which are mounted switches, overcurrent devices, metering instruments, and busbars for controlling, distributing, and protecting a number of electric circuits. It should be located as close as possible to the service connection to minimize voltage drop and for wiring economy.

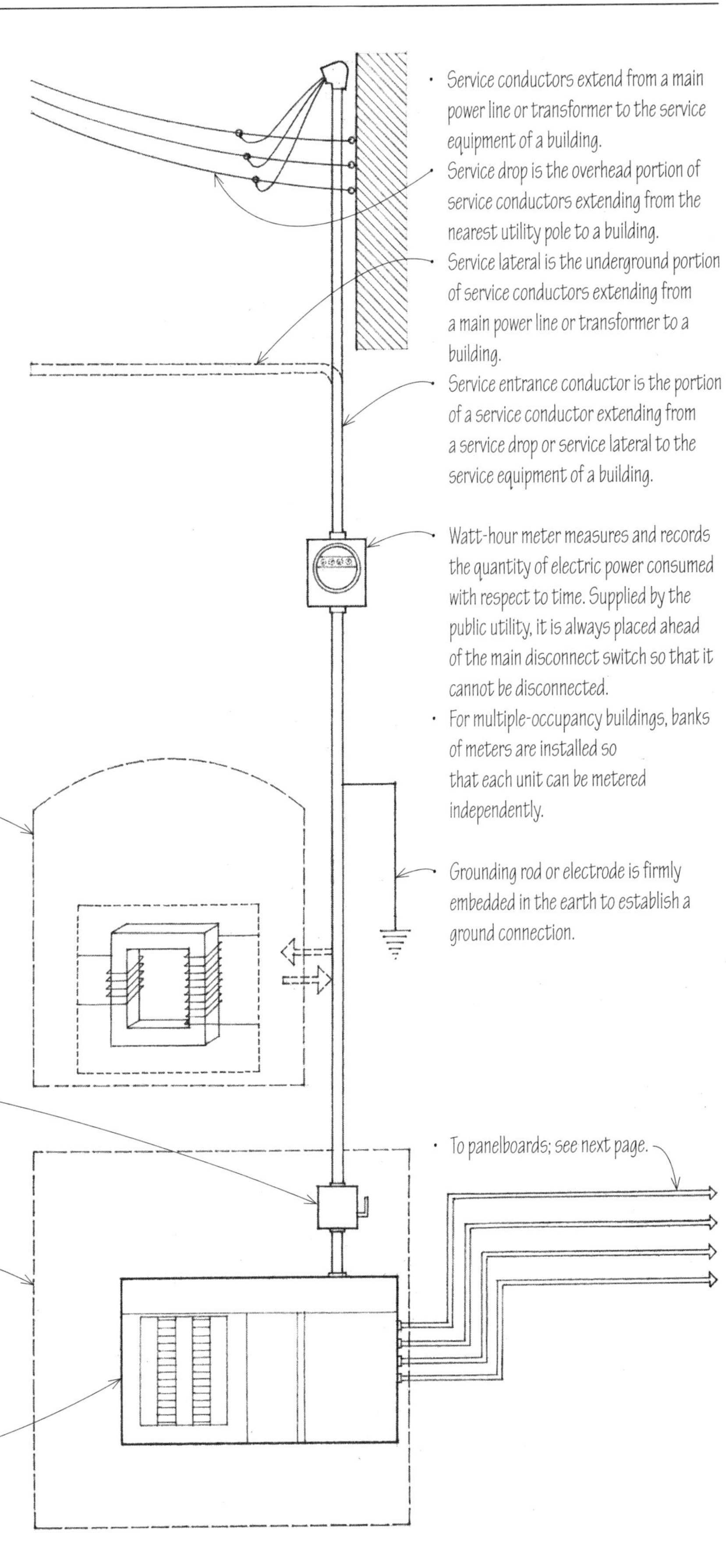

CSI MasterFormat 26 05 26: Grounding and Bonding for Electrical Systems

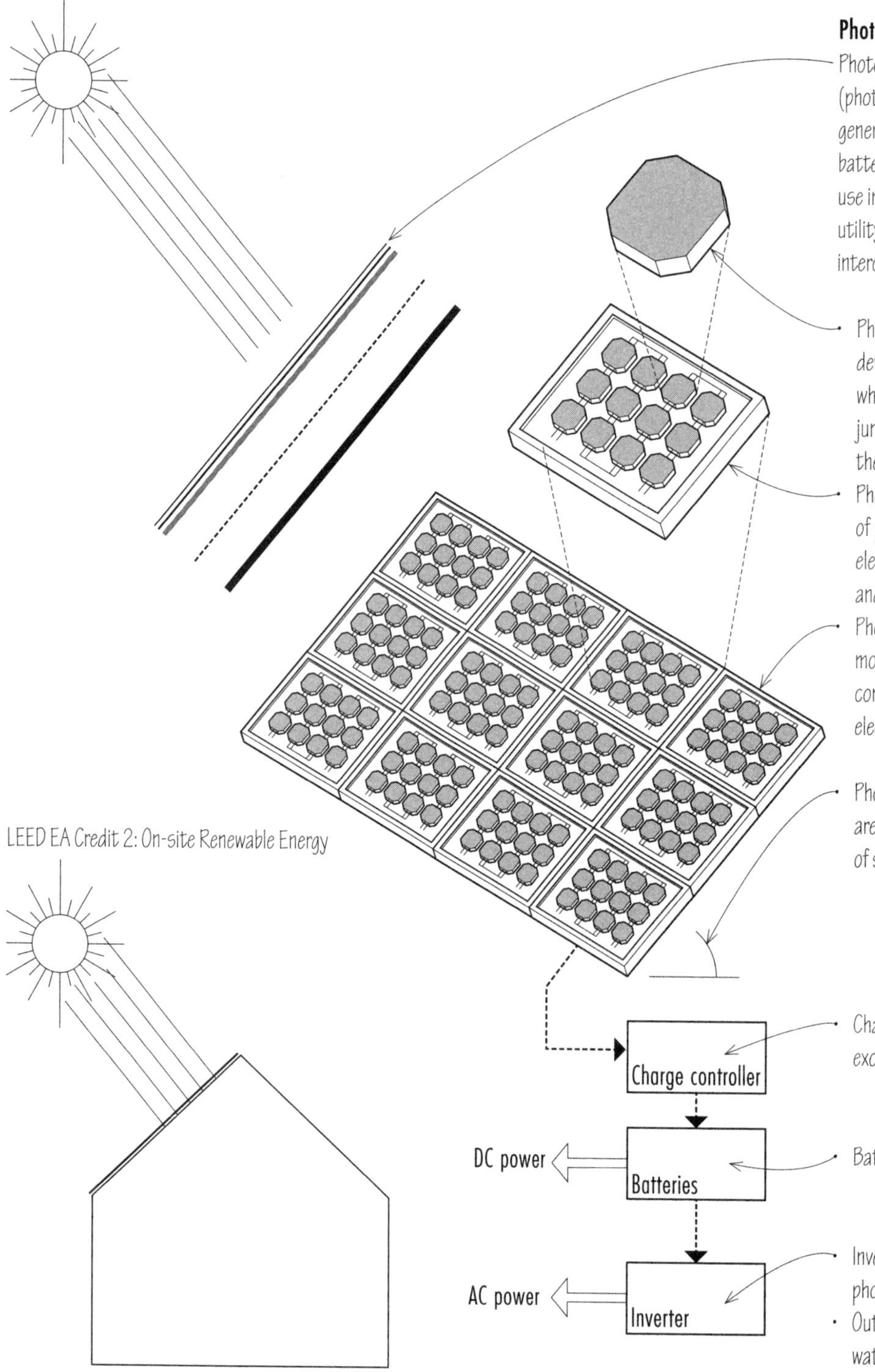

Photovoltaic Technology

Photovoltaic (PV) technology converts solar radiant energy (photons) directly into electricity (voltage). The electricity generated is direct current (DC), which is either stored in a battery system or converted to alternating current (AC) for use in commercial and residential buildings. For large electric utility or industrial applications, hundreds of solar arrays are interconnected to form a large utility-scale PV system.

- Photovoltaic cells, also called a solar cells, are solid-state devices that convert solar energy into electrical energy when the incidence of light or other radiant energy upon the junction of two types of semiconducting materials induces the generation of an electromotive force.
- Photovoltaic module or solar panel consists of a number of photovoltaic cells housed in a protective structure and electrically connected in series to obtain a certain voltage and in parallel to provide the desired amount of current.
- Photovoltaic array consists of multiple photovoltaic modules typically mounted on rooftops and electrically connected to generate and supply the required amount of electricity in commercial and residential buildings.

- Photovoltaic modules are inclined at an angle as close to the area's latitude as possible to absorb the maximum amount of solar energy year-round.

- Charge controller prevents batteries from overcharging or excessive discharging.

- Batteries store energy and provide direct current.

- Inverter converts the direct current supplied by the photovoltaic modules into alternating current.
- Output of a photovoltaic array is typically measured in watts or kilowatts.

Building-Integrated Photovoltaics

Second-generation thin-film solar cells are made from amorphous silicon or nonsilicon materials such as cadmium telluride. Because of their flexibility, thin-film solar cells can be incorporated into the roof, walls, or windows of a building as either a principal or ancillary source of electrical power, often replacing conventional building materials. They can be can be integrated with flexible roofing membranes, shaped and used as roofing shingles or tiles, serve as components of a curtain-wall system, or used for the glazing of skylights.

- Net metering is a policy of some public utilities that promotes investment in renewable energy–generating technologies by allowing customers to offset their consumption over a billing period when they generate electricity in excess of their demand.

CSI MasterFormat 48 14 13: Solar Energy Collectors

Once the electrical power requirements for the various areas of a building are determined, wiring circuits must be laid out to distribute the power to the points of utilization.

- Branch circuits are the portions of an electrical system extending from the final overcurrent device protecting a circuit to the outlets served by the circuit. Each branch circuit is sized according to the amount of load it must carry. About 20% of its capacity is reserved for flexibility, expansion, and safety. To avoid an excessive drop in voltage, a branch circuit should not exceed 100' (30 m) in length.

- General purpose circuits supply current to a number of outlets for lighting and appliances.
- Receptacles in wet locations, such as in bathrooms, should be protected by a ground fault interrupter (GFI). A GFI is a circuit breaker that senses currents caused by ground faults and instantaneously shuts off power before damage or injury can occur. This protection may be provided by a GFI receptacle or by a GFI breaker at the service panel.

- Appliance circuits supply current to one or more outlets specifically intended for appliances.
- Individual circuits supply current only to a single piece of electrical equipment.

Panelboards control, distribute, and protect a number of similar branch circuits in an electrical system. In large buildings, they are located in electrical closets close to the load ends of circuits. In residences and small installations, the panelboard is combined with the switchboard to form a service panel.

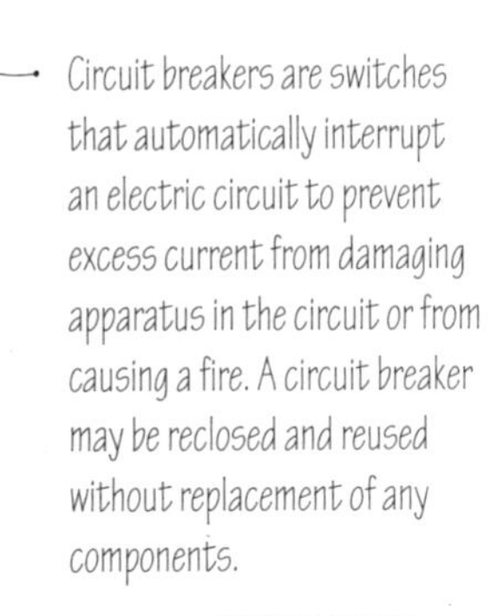

Circuit breakers are switches that automatically interrupt an electric circuit to prevent excess current from damaging apparatus in the circuit or from causing a fire. A circuit breaker may be reclosed and reused without replacement of any components.

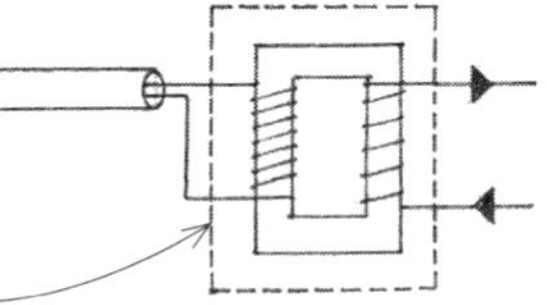

- Low-voltage circuits carry alternating current below 50 V, supplied by a step-down transformer from the normal line voltage. These circuits are used in residential systems to control doorbells, intercoms, heating and cooling systems, and remote lighting fixtures. Low-voltage wiring does not require a protective raceway.
- Low-voltage switching is used when a central control point is desired from which all switching may take place. The low-voltage switches control relays that do the actual switching at the service outlets.

- Load requirements for lighting fixtures and electrically powered appliances and equipment are specified by their manufacturer. The design load for a general purpose circuit, however, depends on the number of receptacles served by the circuit and how they are used. Consult the National Electrical Code for requirements.

- Separate wiring circuits are required for the sound and signal equipment of telephone, cable, intercom, and security or fire alarm systems.
- Telephone systems should have their outlets located and wired during construction. Large installations also require a service connection, terminal enclosures, riser spaces, etc., similar to electrical systems. Large systems are usually designed, furnished, and installed by a telecommunications company.
- Cable television systems may receive their signals from an outdoor antenna or satellite dish, a cable company, or a closed circuit system. If several outlets are required, a 120 V outlet is supplied to serve an amplifier. Coaxial cables in a nonmetallic conductor raceway transmit the amplified signal to the various outlets.

CSI MasterFormat 26 10 00: Medium-Voltage Electrical Distribution
CSI MasterFormat 26 18 00: Medium-Voltage Circuit Protection Devices

Metals, offering little resistance to the flow of electric current, make good conductors. Copper is most often used. The various forms of conductors—wire, cable, and busbars—are sized according to their safe current-carrying capacity and the maximum operating temperature of their insulation. They are identified according to:

- Voltage class
- Number and size of conductors
- Type of insulation

A conductor is covered with insulation to prevent its contact with other conductors or metal, and to protect it against heat, moisture, and corrosion. Materials with a high resistance to the flow of electric current, such as rubber, plastics, porcelain, and glass are commonly used to insulate electrical wiring and connections.

Conduit provides support for wires and cables and protects them against physical damage and corrosion. Metal conduit also provides a continuous grounded enclosure for the wiring. For fireproof construction, rigid metal conduit, electrical metallic tubing, or flexible metal conduit can be used. For frame construction, armored or nonmetallic sheathed cable is used. Plastic tubing and conduits are most commonly used for underground wiring.

Being relatively small, conduit can be easily accommodated in most construction systems. Conduit should be adequately supported and laid out as directly as possible. Codes generally restrict the radius and number of bends a run of conduit may have between junction or outlet boxes. Coordination with a building's mechanical and plumbing systems is required to avoid conflicting paths.

Electrical conductors are often run within the raceways of cellular steel decking to allow for the flexible placement of power, signal, and telephone outlets in office buildings. Flat conductor cable systems are also available for installation directly under carpet tiles.

For exposed installations, special conduit, raceways, troughs, and fittings are available. As with exposed mechanical systems, the layout should be visually coordinated with the physical elements of the space.

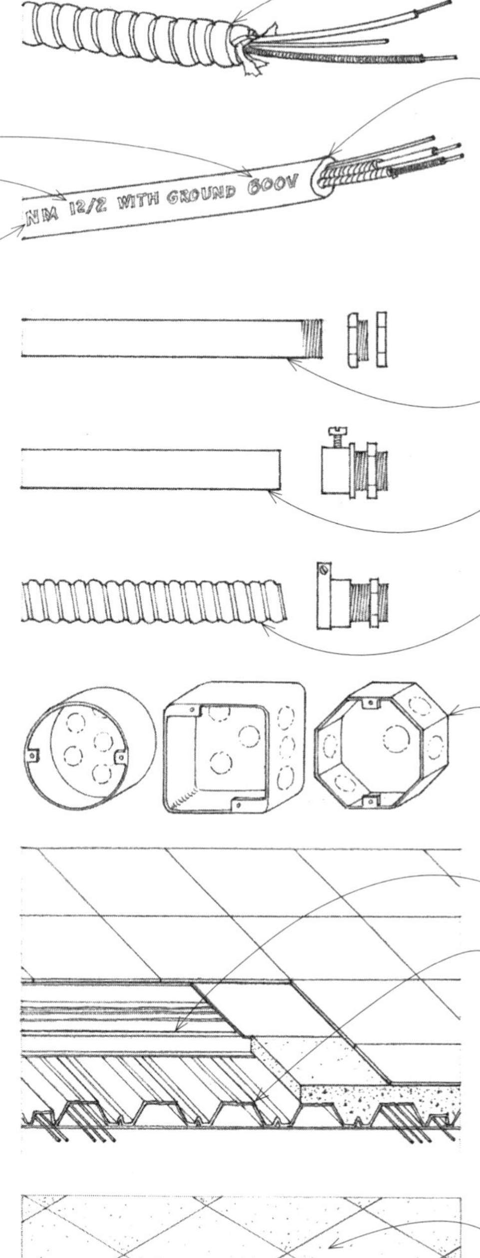

- Armored cable, also called BX cable, consists of two or more insulated conductors protected by a flexible, helically wound metal wrapping.
- Nonmetallic sheathed cable, also called Romex cable, consists of two or more insulated conductors enclosed in a nonmetallic, moisture-resistant, flame-retardant sheath.
- Mineral-insulated cable consists of a tubular copper sheath containing one or more conductors embedded in a highly compressed, insulating refractory mineral.
- Rigid metal conduit is heavy-walled, steel tubing joined by screwing directly into a threaded hub with locknuts and bushings.
- Electrical metallic tubing is thin-walled, steel tubing joined by compression or setscrew couplings.
- Flexible metal conduit is a helically wound metal conduit, used for connections to motors or other vibrating equipment.
- Junction boxes are enclosures for housing and protecting electric wires or cables that are joined together in connecting or branching electric circuits.

- Trench header perpendicular to raceways
- Floor outlets are located on a preset module.
- Cellular steel floor decking

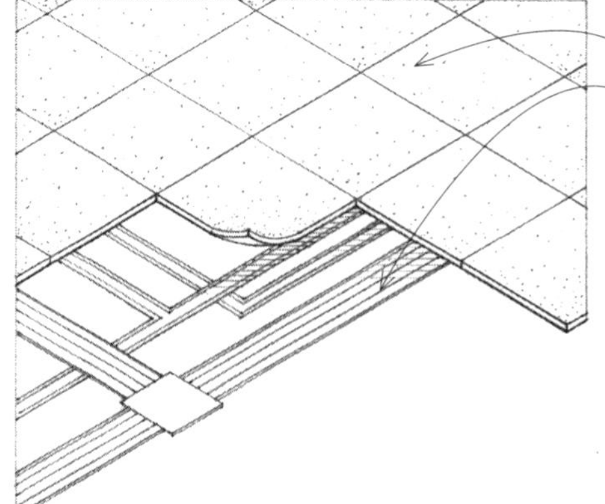

- Carpet squares
- 1, 2, or 3 circuit flat conductor cables with low-profile outlets

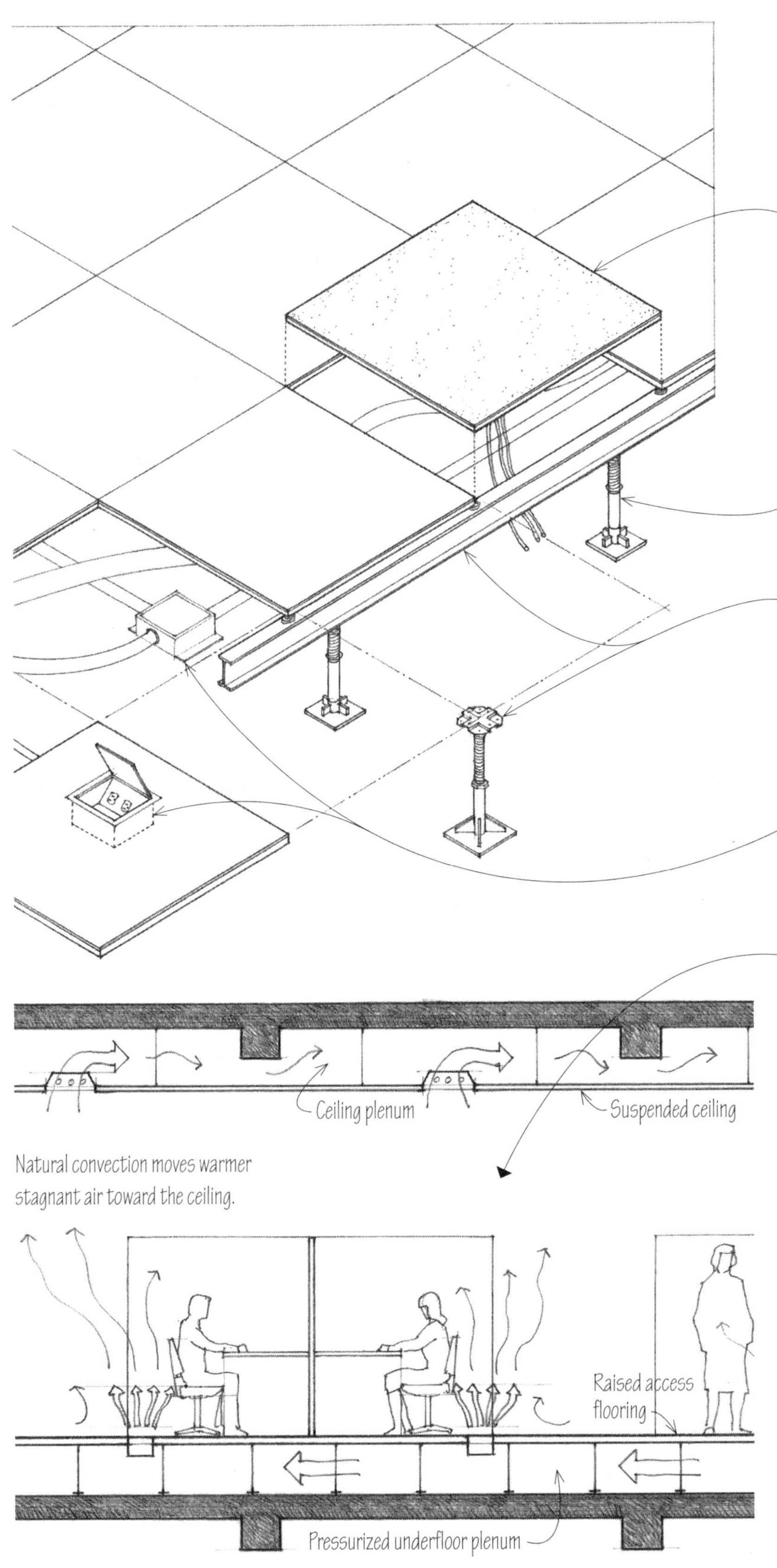

Access flooring systems are typically used in office spaces, hospitals, laboratories, computer rooms, and television and communication centers to provide accessibility and flexibility in the placement of desks, workstations, and equipment. Equipment can be moved and reconnected fairly easily with modular wiring systems.

- Access flooring systems consist essentially of removable and interchangeable floor panels supported on adjustable pedestals to allow free access to the space beneath.
  The floor panels are typically 24" or 600 mm square and constructed of steel, aluminum, a wood core encased in steel or aluminum, or lightweight reinforced concrete. The panels may be finished with carpet tile, vinyl tile, or high-pressure laminate; fire-rated and electrostatic-discharge-control coverings are also available.
- The pedestals are adjustable to provide finished floor heights from 12" to 30" (305 to 455); a minimum finished floor height as low as 8" (203) is also available.
- Systems using stringers have greater lateral stability than stringerless systems; seismic pedestals are available to meet building code requirements for lateral stability.
- Design loads range from 250 to 625 psf (1220 to 3050 kg/m^2), but are available up to 1125 psf (5490 kg/m^2) to accommodate heavier loadings.

- The underfloor space is used for the installation of electrical conduit, junction boxes, and the cables for computer, security, and communication systems.

- The space can also be used as a plenum to distribute the supply air of the HVAC system, allowing the ceiling plenum to be used only for return air. Separating cool supply air from warmer return air in this manner can reduce energy consumption. Lowering the overall height of service plenums also reduces the floor-to-floor height of new construction.

- Consult manufacturer for installation details and available accessories, such as ramps and steps.

CSI MasterFormat 09 69 00: Access Flooring

Lighting fixtures, wall switches, and convenience outlets are the most visible parts of an electrical system. Switches and receptacle outlets should be located for convenient access, and coordinated with visible surface patterns. Wall plates for these devices may be of metal, plastic, or glass, and are available in a variety of colors and finishes.

The design load for a general purpose circuit depends on the number of outlets served by the circuit and how they are used. Consult the National Electrical Code for calculating the required number and spacing of convenience outlets.

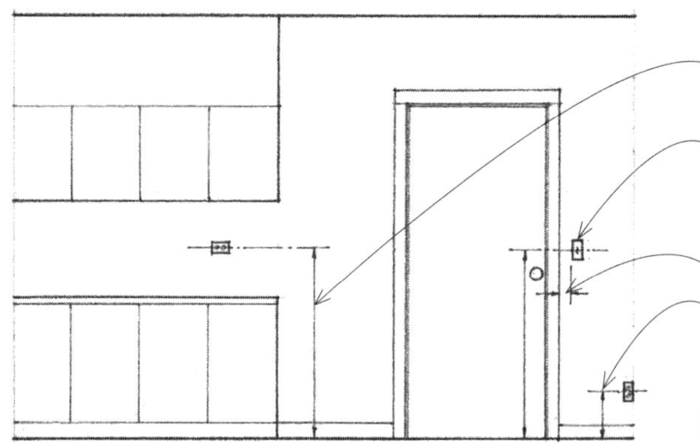

- Outlet above counter: 4'-0" (1220); 3'-6" (1220) for accessibility
- Switch on latch side of door: 4'-0" (1220) maximum for accessibility
- 2-1/2" (64) minimum
- Outlet above floor: 12" (305); 18" (455) for accessibility

Heights of Switches and Outlets

Residences
- One outlet every 12' (3660) along walls in a living space
- One outlet every 4' (1220) along countertops in kitchens
- One GFI-protected outlet in bathrooms

Offices
- One outlet every 10' (3050) along walls, or
- One outlet every 40 sf (3.7 m^2) of floor area for the first 400 sf (37 m^2) and one outlet for every 100 sf (9.3 m^2) thereafter

Number of Convenience Outlets

Switches

- Toggle switch has a lever or knob that moves through a small arc and causes the contacts to open or close an electric circuit.
- Three-way switch is a single-pole, double-throw switch used in conjunction with another to control lights from two locations.
- Four-way switch is used in conjunction with two three-way switches to control lights from three locations.
- Dimmer is a rheostat or similar device for regulating the intensity of an electric light without appreciably affecting spatial distribution.

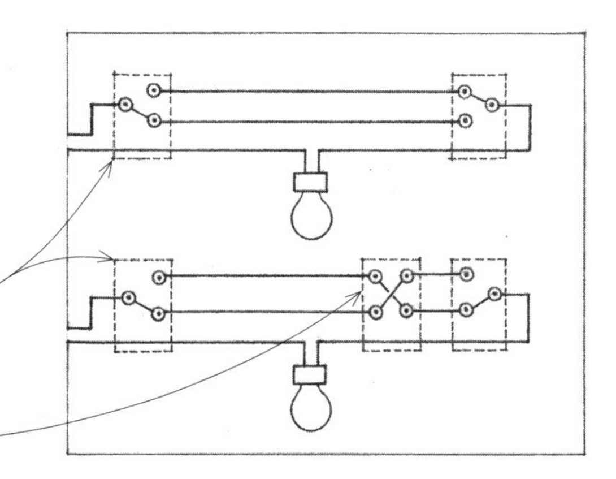

Receptacles

- Duplex receptacles, also called convenience outlets, are usually mounted on a wall and house one or more receptacles for portable lamps or appliances.
- Split-wired receptacles contain one outlet that is always energized and a second outlet that is controlled by a wall switch.
- Special receptacles designed to serve a specific type of appliance will be polarized and have a specific configuration so that only attachments from that appliance will fit the receptacle.
- Outdoor receptacles should have a water-resistant cover.
- In all wet locations, receptacles should be protected by a ground fault interrupter (GFI).

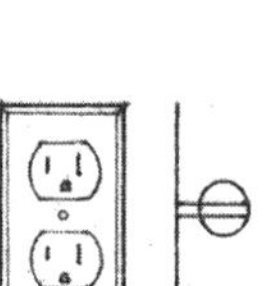

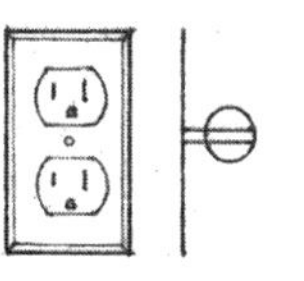

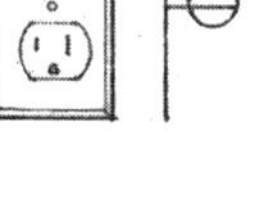

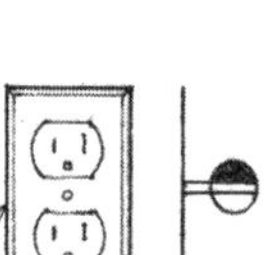

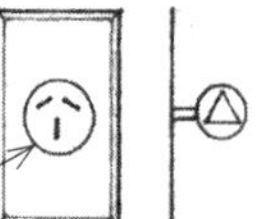

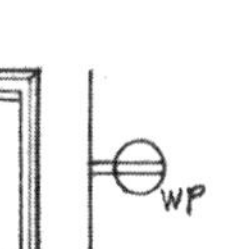

- Panelboard, recessed
- Panelboard, surface
- Power panel
- Lighting panel
- Transformer
- Generator
- Motor
- Disconnect switch
- Single-pole switch
- Three-way switch
- Switched receptacle
- Dimmer switch
- Duplex outlet
- Floor duplex outlet
- Telephone outlet
- Fluorescent fixture
- Ceiling incandescent
- Wall incandescent
- Track light
- Recessed light
- Exit light outlet
- Special purpose outlet
- Television outlet
- Chime
- Pushbutton
- Fan receptacle
- Junction box
- Underfloor junction box
- Thermostat
- Computer data outlet

Typical Electrical Plan Symbols

Light is electromagnetic radiation that the unaided human eye can perceive, having a wavelength in the range from about 370 to 800 nm and propagating at a speed of 186,281 miles per second (299,972 km/sec). It radiates equally in all directions and spreads over a larger area as it emanates from its source. As it spreads, it also diminishes in intensity.

- Luminous intensity is the luminous flux emitted per unit solid angle by a light source, expressed in candelas.
- Candela is the basic SI unit of luminous intensity, equal to the luminous intensity of a source that emits monochromatic radiation of frequency 540 x 1012 hertz and that has a radiant intensity of 1/683 watt per steradian.
- A steradian is a solid angle at the center of a sphere subtending an area on the surface equal to the square of the radius of the sphere.

- Luminous flux is the rate of flow of visible light per unit time, expressed in lumens.
- Lumen is the SI unit of luminous flux, equal to the light emitted in a solid angle of one steradian by a uniform point source having an intensity of one candela.

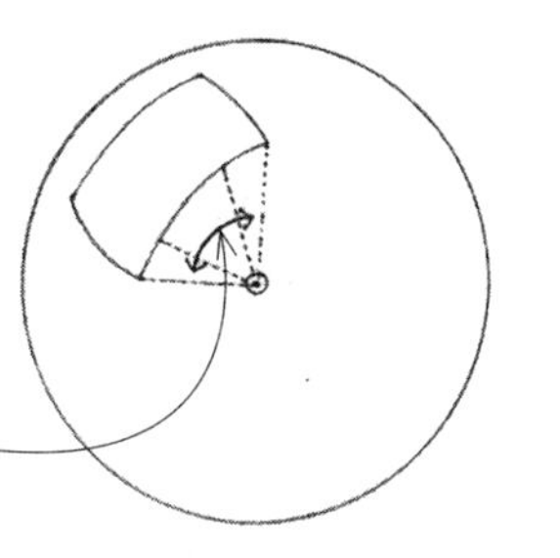

- Inverse square law states that the illumination produced on a surface by a point source varies inversely as the square of the distance of the surface from the source.
- Cosine law, also called Lambert's law, states that the illumination produced on a surface by a point source is proportional to the cosine of the angle of incidence.

- Illumination is the intensity of light falling at any given place on a lighted surface, equal to the luminous flux incident per unit area and expressed in lumens per unit of area.
- Lux is the SI unit of illumination, equal to one lumen per square meter.
- Foot-candle is a unit of illumination on a surface that is everywhere one foot from a uniform point source of one candela and equal to one lumen incident per square foot.

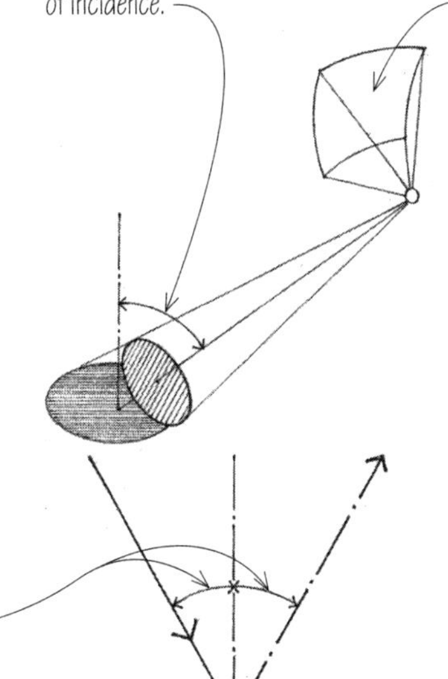

- Law of reflection is the principle that when light is reflected from a smooth surface, the angle of incidence is equal to the angle of reflection, and the incident ray, the reflected ray, and the normal to the surface all lie in the same plane.

- Angle of refraction is the angle that a refracted ray makes with a normal to the interface between two media at the point of incidence.

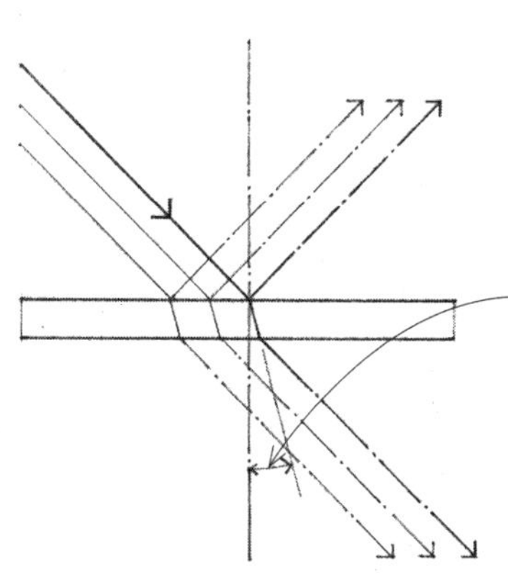

- Reflectance is the ratio of the radiation reflected by a surface to the total incident on the surface.
- Absorptance is the ratio of the radiation absorbed by a surface to the total incident on the surface.
- Transmittance is the ratio of the radiation transmitted through and emerging from a body to the total incident on it, equivalent to one minus the absorptance.

Light reveals to our eyes the shape, texture, and color of objects in space. An object in its path will reflect, absorb, or allow the light striking its surface to pass through. Luminance is the quantitative measure of brightness of a light source or an illuminated surface, equal to the luminous intensity per unit projected area of the source or surface viewed from a given direction.

- Lambert is a unit of luminance or brightness equal to 0.32 candela per square centimeter.
- Foot-lambert is a unit of luminance or brightness equal to 0.32 candela per square foot.
- Brightness is affected by both color and texture. Shiny, light-colored surfaces reflect more light than dark, matte or rough-textured surfaces, even though both surfaces may receive the same amount of illumination.

Brightness is the sensation by which an observer is able to distinguish between differences in luminance. Visual acuity increases with object brightness. Of equal importance is the ratio of the luminance of an object being viewed and that of its background. To discern shape and form, some degree of contrast or brightness ratio is required. Contrast is especially critical for visual tasks that require discrimination of shape and contour. For seeing tasks requiring discrimination of texture and detail, less contrast is desirable since our eyes adjust automatically to the average brightness of a scene. When the contrast or brightness ratio is too high, glare can result.

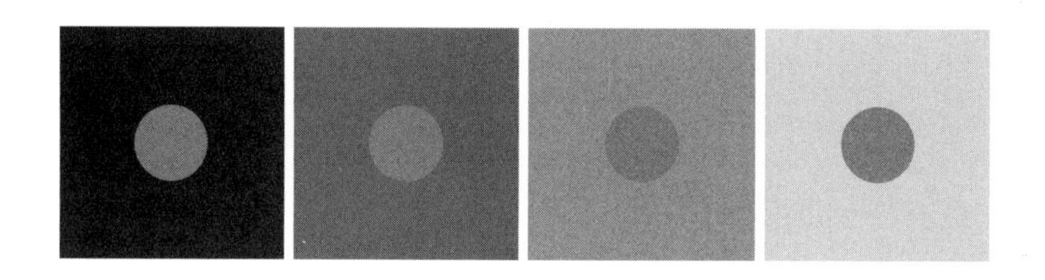

Glare is the sensation produced by any brightness within the visual field that is sufficiently greater than the luminance to which the eyes are adapted to cause annoyance, discomfort, or loss of visibility. There are two types of glare: direct and reflected.

- Direct glare results from a high brightness ratio or an insufficiently shielded light source in the visual field.
- Strategies to control or minimize glare include using shielded luminaires to cut off direct view of lamps and using luminaires with diffusers or lenses that lower their brightness levels.

- Reflected or indirect glare results from the specular reflection of a light source within the visual field.
- A specific type of reflected glare is veiling reflectance, which occurs on a task surface and reduces the contrast necessary for seeing details.
- To prevent veiling reflectance, locate the light source in such a way that incident light rays are reflected away from the observer.

- Bulb is the glass housing of an incandescent lamp, filled with an inert gas mixture, usually of argon and nitrogen, to retard evaporation of the filament. Its shape is designated by a letter, followed by a number that indicates the lamp diameter in eighths of an inch.

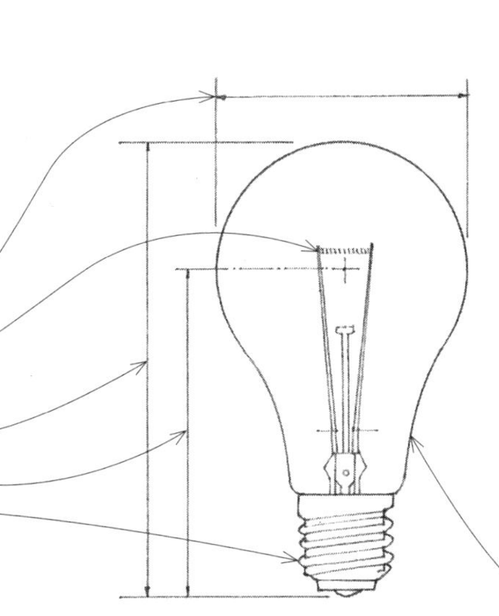

- Filament
- Maximum overall length
- Light center length
- Lamp base

Artificial light is natural light that is produced by manufactured elements. The quantity and quality of light produced differ according to the type of lamp used. The light is further modified by the housing that holds and energizes the lamp. There are three major types of artificial light sources—incandescent, fluorescent, and high-intensity discharge (HID) lamps. For accurate, current data on lamp sizes, wattages, lumen output, and average life, consult the lamp manufacturer.

Incandescent Lamps

Incandescent lamps contain a filament that gives off light when heated to incandescence by the passage of an electric current. They provide point sources of light, have low efficacy, render color well, and are easy to dim with rheostats.

- Efficacy is a measure of the effectiveness with which a lamp converts electric power into luminous flux, equal to the ratio of flux emitted to power input and expressed in lumens per watt.
- Rated life is the average life in hours of a given type of lamp, based on laboratory tests of a representative group under controlled conditions.
- Extended-service lamps are designed for reduced energy consumption and a life longer than the conventional value for its general class.
- Three-way lamp is an incandescent lamp having two filaments so that it can be switched to three successive degrees of illumination.

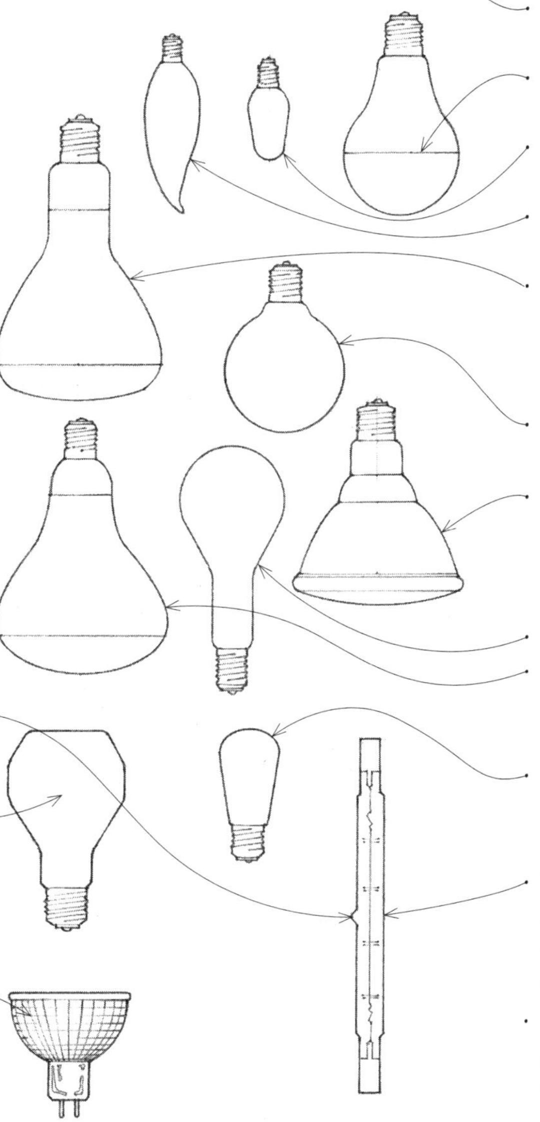

- A bulb: standard rounded shape for general-service incandescent lamps
- A/SB bulb: A bulb having a hemispherical, reflective silver bowl opposite the lamp base to decrease glare
- C bulb: cone-shaped bulb for low-wattage, decorative incandescent lamps
- CA bulb: candle-flame shaped bulb for low-wattage, decorative incandescent lamps
- ER bulb: ellipsoidal reflector bulb for incandescent lamps, having a precisely formed internal reflector that collects light and redirects it into a dispersed pattern at some distance in front of the light source
- G bulb: globe-shaped bulb for incandescent lamps, having a low brightness for exposed use
- PAR bulb: parabolic aluminized reflector bulb for incandescent and HID lamps, having a precisely formed internal reflector and a lensed front to provide the desired beam spread
- PS bulb: pear-shaped bulb for large incandescent lamps
- R bulb: reflector bulb for incandescent and HID lamps, having an internal reflective coating and either a clear or frosted glass front to provide the desired beam spread
- S bulb: straight-sided bulb for low-wattage, decorative incandescent lamps

- T bulb: a tubular, quartz bulb for tungsten-halogen lamps
- TB bulb: a quartz bulb for tungsten-halogen lamps similar in shape to the A bulb but having an angular profile
- MR bulb: a multifaceted reflector bulb for tungsten-halogen lamps, having highly polished reflectors arranged in discrete segments to provide the desired beam spread

- Tungsten-halogen lamps have a tungsten filament and a quartz bulb containing a small amount of a halogen that vaporizes on heating and redeposits any evaporated tungsten particles back onto the filament.
- IR lamp is a tungsten-halogen lamp having an infrared dichroic coating for reflecting infrared energy back to the filament, raising lamp efficiency, and reducing radiant heat in the emitted light beam.

CSI MasterFormat 26 51 13: Interior Lighting Fixtures, Lamps, and Ballasts

Discharge lamps produce light by the discharge of electricity between electrodes in a gas-filled glass enclosure. The two major types of discharge lamps are fluorescent lamps and a variety of high-intensity discharge lamps.

Fluorescent Lamps

Fluorescent lamps are tubular low-discharge lamps in which light is produced by the fluorescence of phosphors coating the inside of the tube. Because fluorescent lamps contain mercury, they require special handling for recycling. The amount of mercury used continues to be reduced, and T5 lamps now have low mercury content.

Fluorescent lamps are more efficient and have a longer life (6,000–24,000+ hours) than incandescent lamps. They produce little heat and are available in a variety of types and wattages. Common lengths range from a 6" (150) 4-watt T5 to an 8' (2440) 125-watt T12. Fluorescent lamps require a ballast to regulate electric current through the lamp. Some lamps have pin bases, while others have screw-in bases.

- Ballasts maintain the current through a fluorescent or high-intensity discharge lamp at the desired constant value.
- Preheat lamps require a separate starter to preheat the cathodes before opening the circuit to the starting voltage.
- Rapid-start lamps are designed to operate with a ballast having a low-voltage winding for continuous heating of the cathodes, which allows the lamps to be started more rapidly than a preheat lamp.
- Instant-start lamps are designed to operate with a ballast having a high-voltage transformer to initiate the arc directly without any preheating of the cathodes.
- High-output lamps are rapid-start fluorescent lamps designed to operate on a current of 800 milliamperes, resulting in a corresponding increase in luminous flux per unit length of lamp.
- Very-high-output lamps are designed to operate on a current of 1500 milliamperes, providing a corresponding increase in luminous flux per unit length of lamp.

Compact fluorescent lamps are any of various small, improved-efficiency fluorescent lamps having a single, double, or U-shaped tube, and often an adapter for fitting an incandescent lampholder.

- Available from 5 to 80 watts.
- High efficacy (typically 60 to 72 lumens per watt).
- Good color rendering.
- Very long lives (6000 to 15,000 hours).
- Tubular or spiral types.
- Many are available with built-in ballast and screw bases for direct replacement of incandescent lamps.

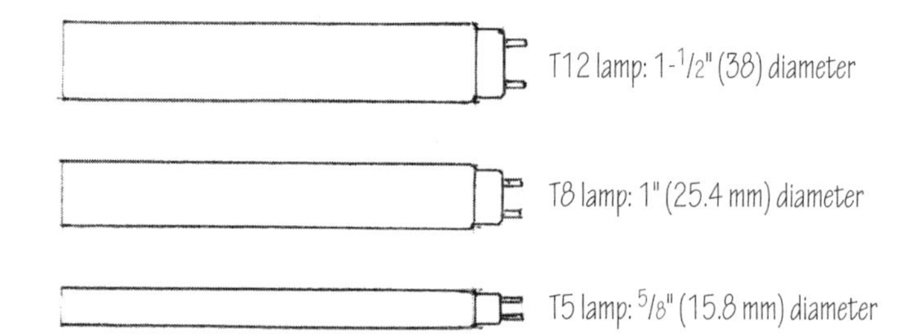

- T bulb: tubular bulb for fluorescent and HID lamps
- The standard T12 lamp is now being replaced by smaller and more efficient T8 and T5 lamps.

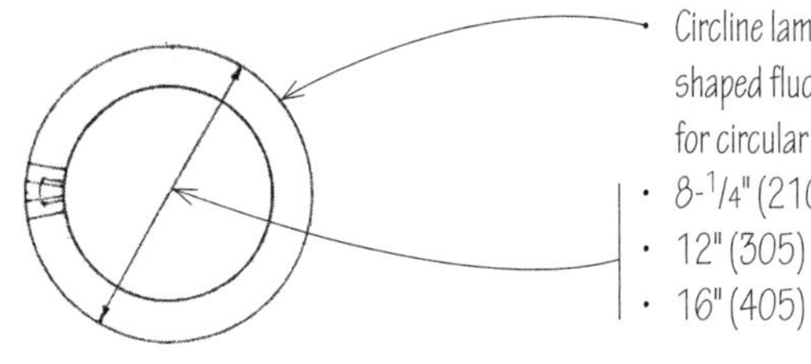

Circline lamp: doughnut-shaped fluorescent lamp for circular luminaires

- 8-1/4" (210) 22W
- 12" (305) 32W
- 16" (405) 40W

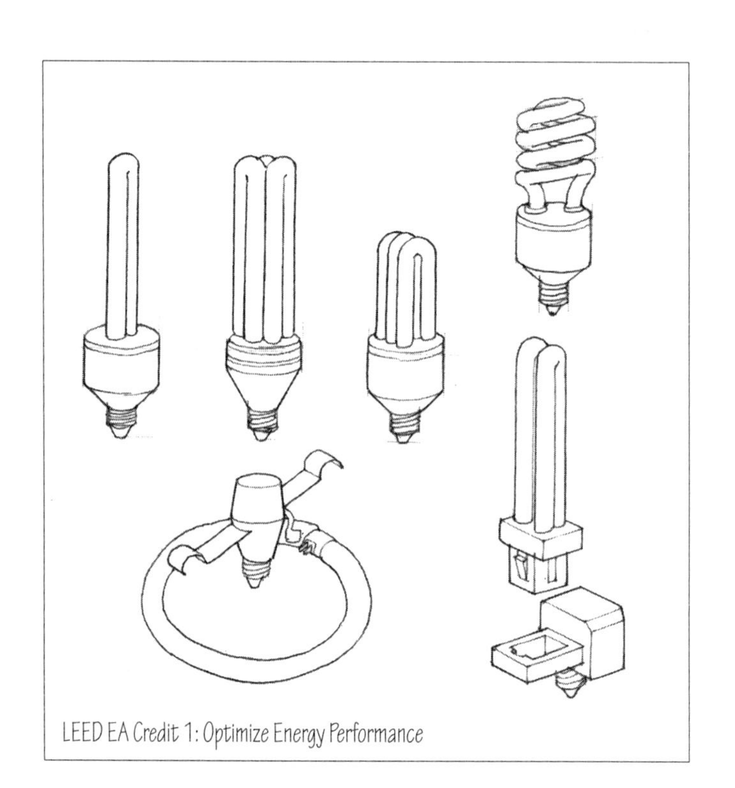

LEED EA Credit 1: Optimize Energy Performance

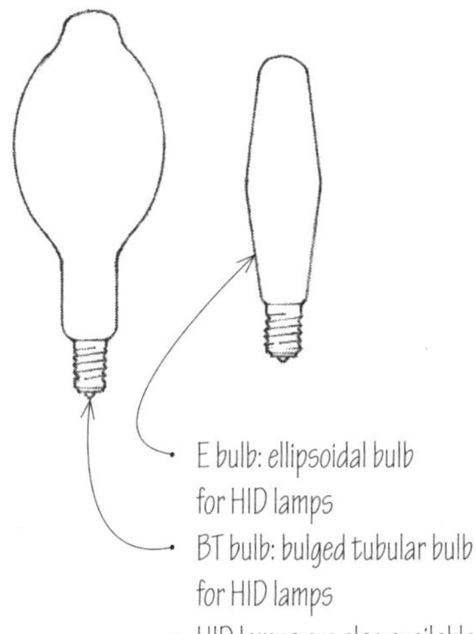

- HID lamps are also available in B and T shapes.

High-Intensity Discharge Lamps

High-intensity discharge (HID) lamps are discharge lamps in which a significant amount of light is produced by the discharge of electricity through a metallic vapor in a sealed glass enclosure. HID lamps combine the form of an incandescent lamp with the efficacy of a fluorescent.

- Mercury lamps produce light by means of an electric discharge in mercury vapor.
- Metal-halide lamps are similar in construction to a mercury lamp, but have an arc tube to which various metal halides are added to produce more light and improve color rendering.
- High-pressure sodium (HPS) lamps produce a broad spectrum of golden-white light by means of an electric discharge in sodium vapor.

Color Rendering Index (CRI) of Various Light Sources

CRI	Light Source
100	Noon sunlight; average daylight
93	500-watt incandescent
89	Cool-white deluxe fluorescent
78	Warm-white deluxe fluorescent
62	Cool-white fluorescent
52	Warm-white fluorescent

Correlated Color Temperature (CCT)

CCT in °Kelvins	Light Source
2700	Incandescent
3000	Halogen
2700–6500	Fluorescent
3000–4000	Metal halide
2800–6000	LED
5500–7500	Daylight

Light and Color

The spectral distribution of artificial light varies with the type of lamp. For example, an incandescent bulb produces a yellow-white light while a cool-white fluorescent produces a blue-white light. The spectral distribution of a light source is important because if certain wavelengths of color are missing, then those colors cannot be reflected and will appear to be missing in any surface illuminated by that light.

- Color rendering index is a measure of the ability of an electric lamp to render color accurately when compared with a reference light source of similar color temperature. A tungsten lamp operating at a color temperature of 3200°K, noon sunlight having a color temperature of 4800°K, and average daylight having a color temperature of 7000°K all have an index of 100 and are considered to render color perfectly.
- Spectral distribution curves plot the radiant energy in each wavelength of a particular light source.

100 Noon sunlight
500-watt incandescent
Cool-white deluxe fluorescent
Warm-white deluxe fluorescent
Cool-white fluorescent
Warm-white fluorescent
Relative power in watts
400 500 600 700
Wavelength in nanometers

Fiber Optics

The optical glass or plastic fibers in fiber-optic lighting transmit light from one end to the other by reflecting light rays back and forth inside their cores in a zigzag pattern. Each small-diameter fiber is protected by a transparent sheath and combined with others into flexible bundles.

A typical fiber-optic lighting system includes:

- A light projector, which may have a color wheel
- A tungsten-halogen or metal halide light source
- An optical-fiber harness
- Bundles of optical fibers and their fittings

Light-Emitting Diodes

Light-emitting diodes (LEDs) radiate very little heat and are highly energy-efficient. LEDs have an extremely long life, typically about ten years. High-powered white-light LEDs are used for illumination. They are insensitive to vibration and temperature, are shock-resistant, and contain no mercury. The tiny 1/8" (3.2) lamps can be combined into larger groups to mix colors and increase illumination. LEDs operate on DC voltage, which is transformed into AC voltage within the fixture.

LEDs are used for both residential and commercial lighting. They can be designed to focus light, and are widely used for task lights. LED downlights, step lighting, and exit signs are also available.

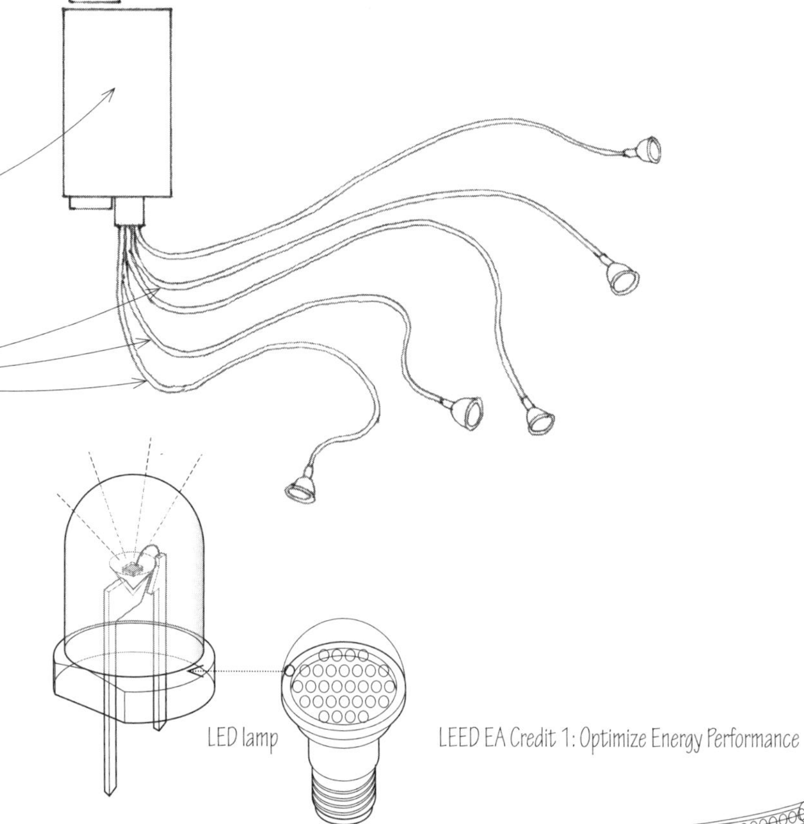

LED lamp

LEED EA Credit 1: Optimize Energy Performance

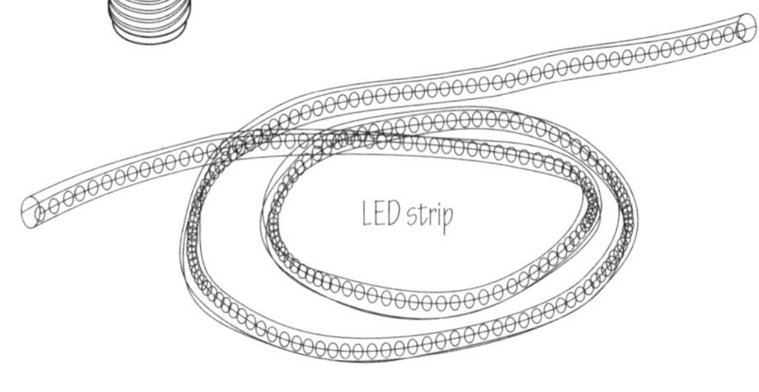

LED strip

LED tube

LED step lighting

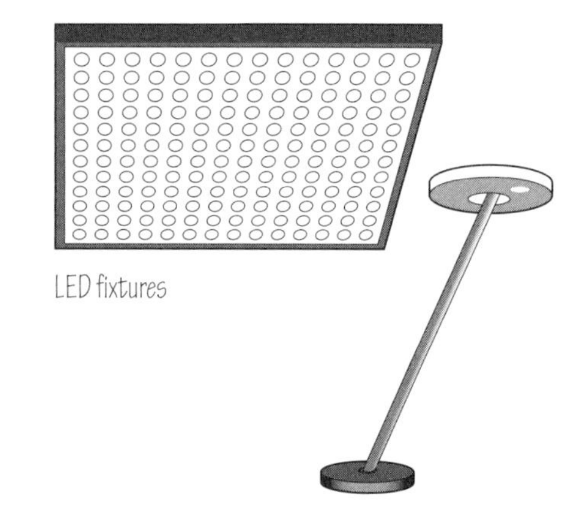

LED fixtures

A luminaire, also commonly called a lighting fixture, consists of one or more electric lamps with all of the necessary parts and wiring for positioning and protecting the lamps, connecting the lamps to a power supply, and distributing the light.

- Reflectors control the distribution of light emitted by a lamp.
- Parabolic reflectors spread, focus, or collimate (make parallel) the rays from a light source, depending on the location of the source.
- Elliptical reflectors focus the rays from a light source.

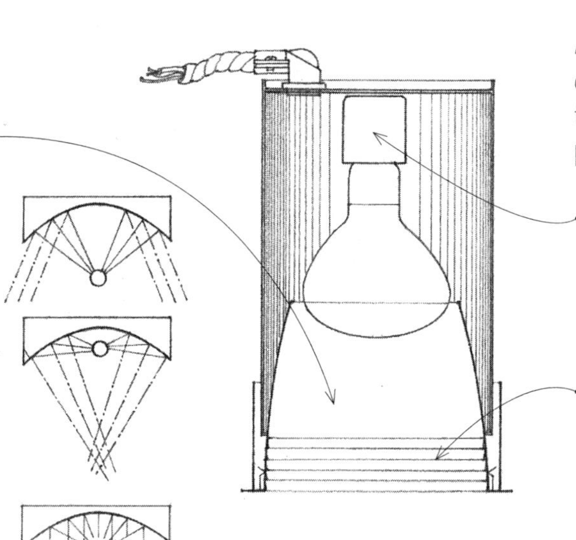

- Lamp holder mechanically supports and makes the electrical contact with a lamp.
- Ridged baffles are a series of circular ridges for reducing the brightness of a light source at an aperture.
- Lens of glass or plastic have two opposite surfaces either or both of which are curved. They are used in luminaires to focus, disperse, or collimate the emitted light.
- Fresnel lenses have concentric, prismatic grooves to concentrate light from a small source.
- Prismatic lenses have a multifaceted surface with parallel prisms to redirect the rays from a light source.
- Baffles are louvered devices for shielding a light source from view at certain angles. They may consist of a series of parallel fins or form an egg-crate pattern.
- Shielding angle is the angle between a horizontal line through the light center and the line of sight at which the lamp first becomes visible.
- Cutoff angle is the angle between a vertical axis and the line of sight at which the lamp first becomes visible.

- In order to evaluate problems with direct glare, the visual comfort probability factor was developed. It rates the likelihood that a lighting system will not cause direct glare, expressed as the percentage of people who may be expected to experience visual comfort when seated in the least-favorable visual position.

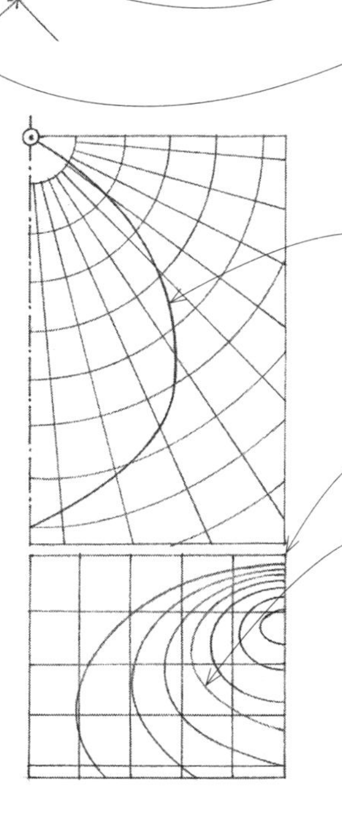

- Candlepower distribution curve is a polar plot of the luminous intensity emitted by a lamp, luminaire, or window in a given direction from the center of the light source, measured in a single plane for a symmetrical light source, and in a parallel, perpendicular, and sometimes a 45° plane for an asymmetrical source.
- Isochart plots the pattern of illumination produced on a surface by a lamp or luminaire.
- Isolux line is a line through all points on a surface where the level of illumination is the same; called isofoot-candle line if illumination is expressed in foot-candles.
- Luminaire efficiency is the ratio of luminous flux emitted by a luminaire to the total flux emitted by the lamps in the luminaire.

The primary purpose of a lighting system is to provide sufficient illumination for the performance of visual tasks. Recommended levels of illumination for certain tasks specify only the quantity of light to be supplied. How this amount of light is supplied affects how a space is revealed or how an object is seen.

Recommended Illumination Levels

Task Difficulty	Foot-candles	Lux
Casual (dining)	20	215
Ordinary (reading)	50	538
Moderate (drafting)	100	1076
Difficult (sewing)	200	2152
Severe (surgery)	>400	>4034

Luminaires may be categorized according to the percentage of light emitted above and below a horizontal plane. The actual light distribution of a specific luminaire is determined by the type of lamp, lens, and reflector housing used. Consult the luminaire manufacturer for candlepower distribution curves.

Diffused light emanates from broad or multiple light sources and reflecting surfaces. The flat, fairly uniform illumination minimizes contrast and shadows and can make the reading of textures difficult.

Directional light, on the other hand, enhances our perception of shape, form, and texture by producing shadows and brightness variations on the surfaces of the objects illuminated.

While diffused lighting is useful for general vision, it can be monotonous. Some directional lighting can help relieve this dullness by providing visual accents, introducing variations in luminance, and brightening task surfaces. A mix of both diffused and directional lighting is often desirable and beneficial, especially when a variety of tasks are to be performed.

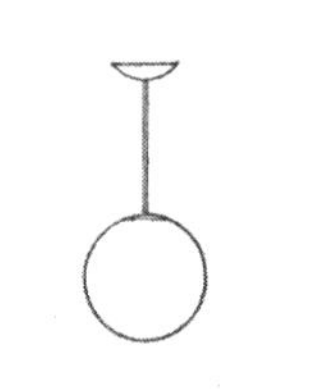

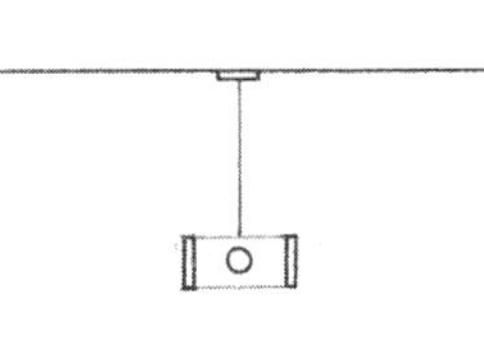

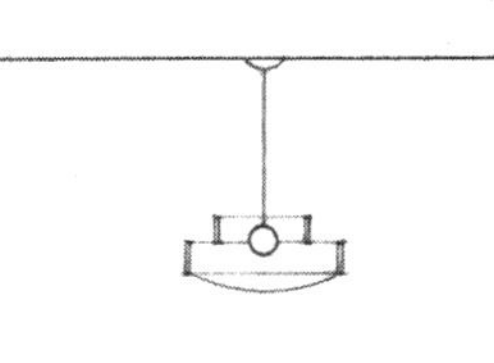

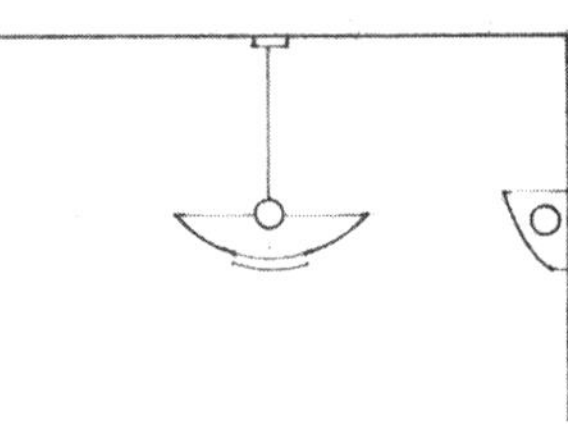

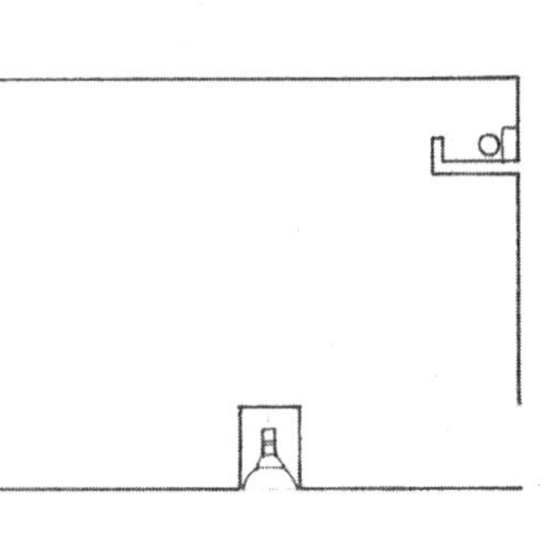

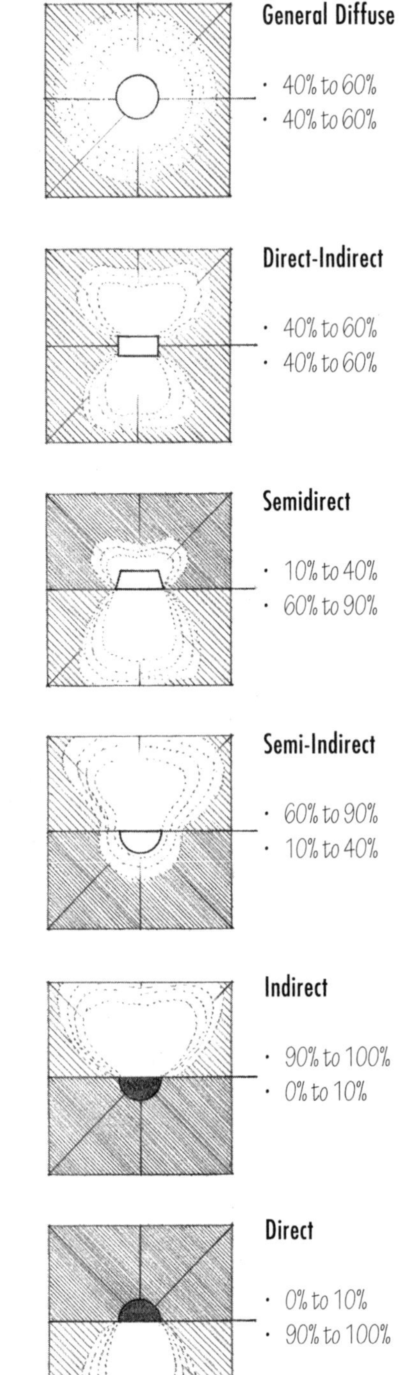

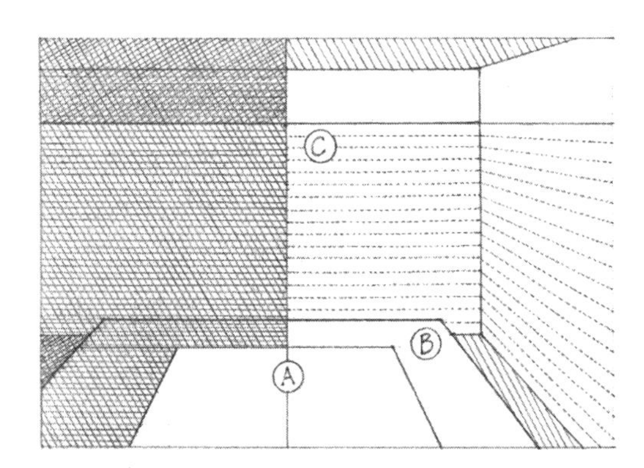

- 3:1 is the maximum recommended brightness ratio between the visual task area (A) and its immediate background (B).
- The surrounding area (C) should range from 1/5 to five times the brightness of the visual task area (A).

Brightness Ratios

Daylight Harvesting

Daylight harvesting is a method of lighting control that reduces energy consumption by using photosensors to detect daylighting levels and automatically adjusting the output level of electric lighting to create the desired or recommended level of illumination for a space. If the daylighting from windows is sufficient to meet users' needs, the lighting control system can automatically turn off all or a portion of the electric lighting or dim the lighting, and immediately reactivate the lighting if the daylighting falls below a preset level. Daylight harvesting controls can be integrated with occupancy sensors for automated on/off control to further increase energy savings as well as with manual override controls to allow for adjustment of lighting levels by occupants. Some control systems can also adjust the color balance of the light by varying the intensity of individual LED lamps of different colors installed in the overhead fixtures.

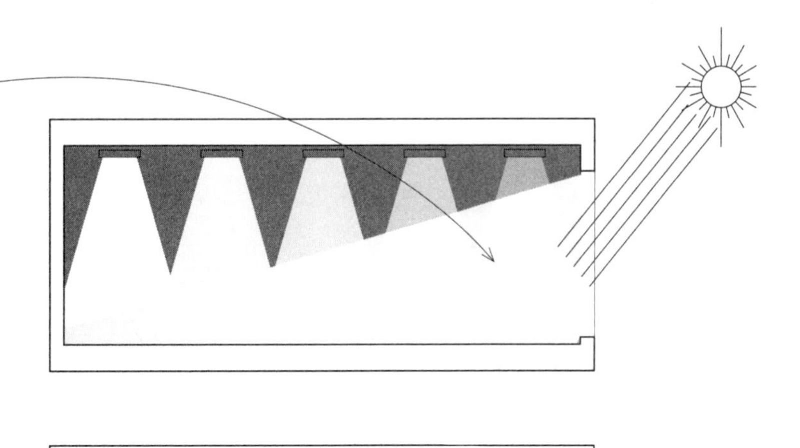

Bi-Level Switching

Bi-level switching is a lighting control system that provides two levels of lighting power in a space, not including off. The switching system may control alternate ballasts or lamps in a luminaire, alternate luminaires, or alternate lighting circuits independently by such means as: photosensors that detect the light level from available daylighting; occupancy sensors that detect user presence; time-based control panels; or manual switches controllable by occupants or the facility operator. Many energy codes in the United States require light-level reduction controls, such as bi-level switching, in enclosed spaces of certain occupancies.

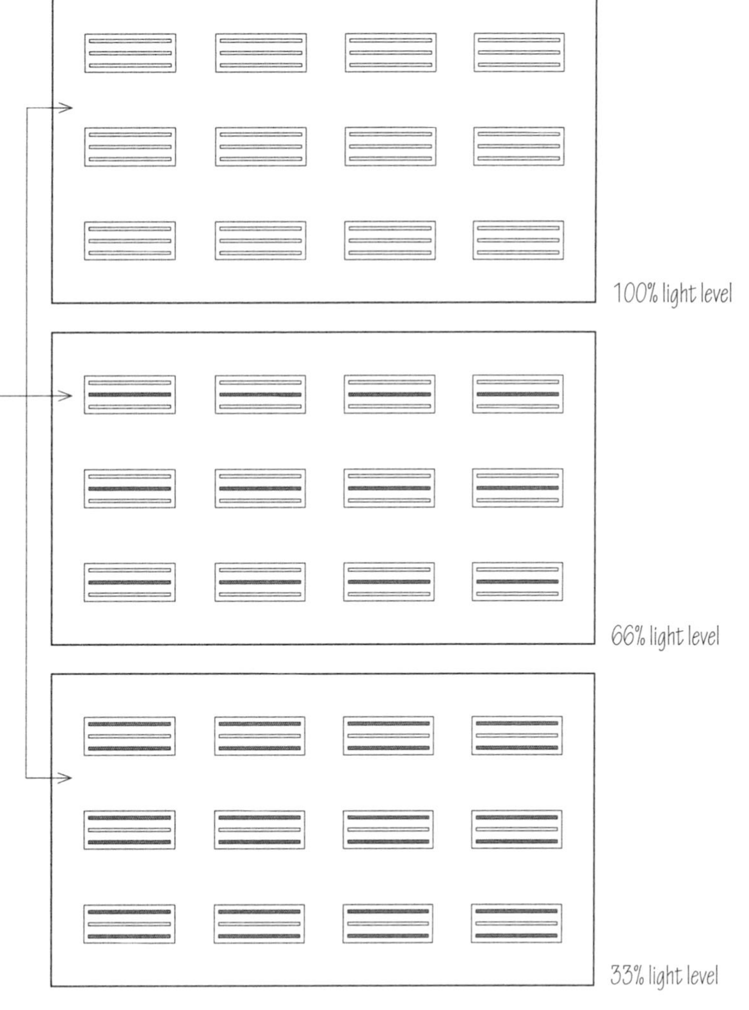

Multi-Level Switching

A form of bi-level switching in which multiple lamps in a single light fixture can be switched on and off independently of each other, allowing for one or two steps between full output and zero illumination while maintaining the required uniform distribution of light suitable for work. For example, a series of three-lamp fixtures with split-ballast wiring can provide four light levels: 100% (all lamps lit), 66% (2 lamps in each fixture lit), 33% (1 lamp in each fixture lit), and 0% (all lamps extinguished). Multi-level switching provides greater flexibility and lessens the abrupt changes in light level of bi-level switching.

Continuous Dimming

Continuous dimming is a method of lighting control that maintains the desired or recommended level of illumination for a space by modulating the output from electric lamps and fixtures in proportion to the amount of available daylight detected by light-level sensors. Continuous dimming systems minimize the abrupt changes in light level created by bi-level and multi-level switching systems.

Occupancy Control

Occupancy controls are automatic lighting control systems that use motion or occupancy sensors to turn lights on when human activity is detected and turn lights off when a space is vacated. Occupancy sensors can replace wall-mounted light switches or can be mounted remotely, retaining the normal switching for use as override switches, which allows the lighting to be kept off even when the space is occupied.

LEED IEQ Credit 6.1: Controllability of Systems, Lighting

- Beam spread is the angle of a light beam that intersects the candlepower distribution curve at points where the luminous intensity equals a stated percent of a maximum reference intensity.

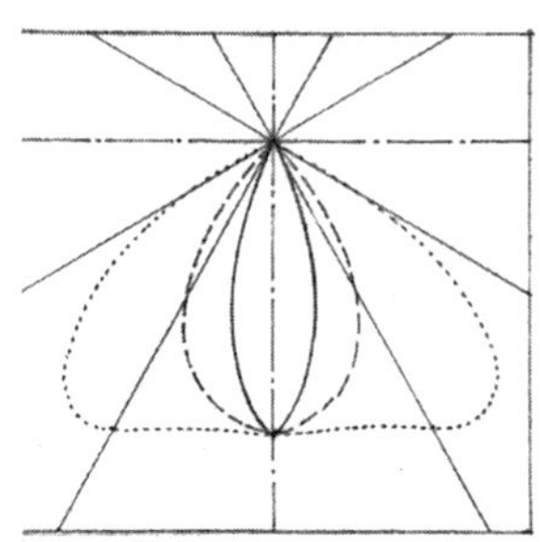

Spacing criteria is a formula for determining how far apart luminaires may be installed for uniform lighting of a surface or area, based on mounting height.

- Spacing criteria (SC) = Spacing (S)/Mounting Height (MH)

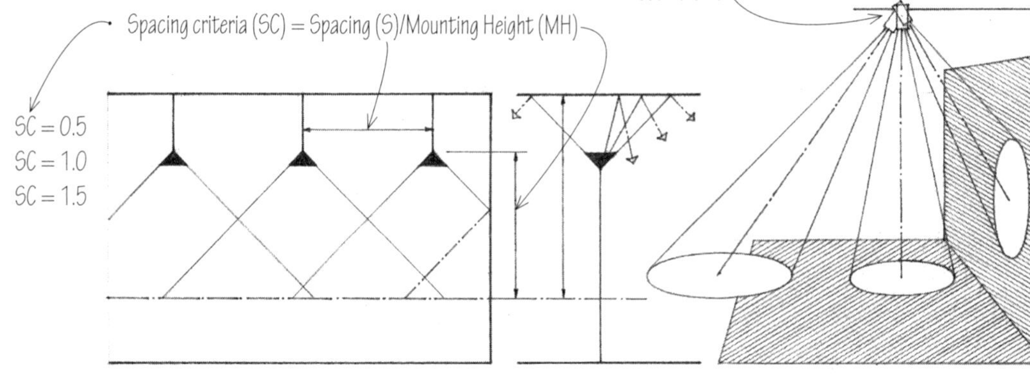

- S/MH ratios are calculated and supplied by the luminaire manufacturer.

The point method is a procedure for calculating the illumination produced on a surface by a point source from any angle, based on the inverse square and cosine laws.

- Ceiling cavity is formed by a ceiling, a plane of suspended luminaires, and wall surfaces between these two planes.
- Room cavity is formed by a plane of luminaires, the work plane, and the wall surfaces between these two planes.
- Floor cavity is formed by the work plane, the floor, and the wall surfaces between these two planes.

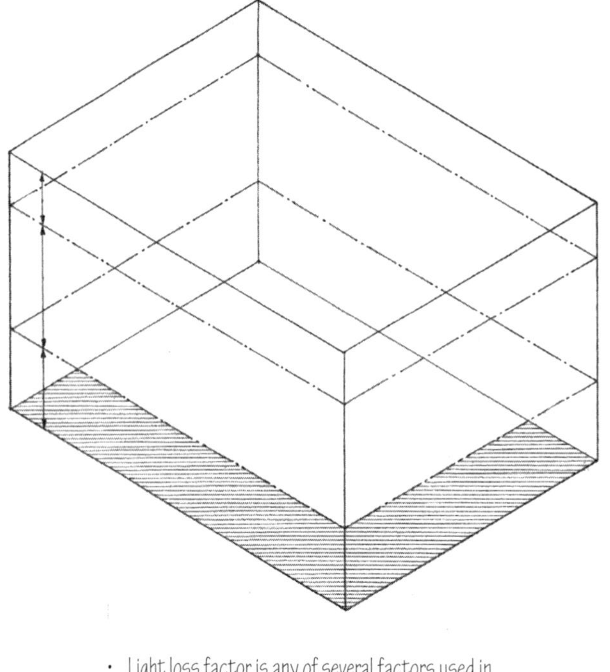

- Room cavity ratio is a single number derived from the dimensions of a room cavity for use in determining the coefficient of utilization.
- Coefficient of utilization (CU) is the ratio of the luminous flux reaching a specified work plane to the total lumen output of a luminaire, taking into account the proportions of a room and the reflectances of its surfaces.

- Light loss factor is any of several factors used in calculating the effective illumination provided by a lighting system after a given period of time and under given conditions.
- Recoverable light loss factors (RLLF) may be recovered by relamping or maintenance.

The lumen method, also called the zonal cavity method, is a procedure for determining the number and types of lamps, luminaires, or windows required to provide a uniform level of illumination on a work plane, taking into account both direct and reflected luminous flux.

- Work plane is the horizontal plane at which work is done and on which illumination is specified and measured, usually assumed to be 30" (762) above the floor.

- Lamp lumen depreciation represents the decrease in luminous output of a lamp during its operating life, expressed as a percentage of initial lamp lumens.
- Luminaire dirt depreciation represents the decrease in luminous output of a luminaire resulting from the accumulation of dirt on its surfaces, expressed as a percentage of the illumination from the luminaire when new or clean.
- Room surface dirt depreciation represents the decrease in reflected light resulting from the accumulation of dirt on a room's surfaces, expressed as a percentage of the light reflected from the surfaces when clean.
- Nonrecoverable light loss factor (NRLLF) is any of several permanent light loss factors that take into account the effects of temperature, voltage drops or surges, ballast variations, and partition heights.

- $$\text{Average maintained illuminance} = \frac{\text{Initial lamp lumens}^* \times \text{CU} \times \text{RLLF} \times \text{NRLLF}}{\text{Work area}}$$

* Initial lamp lumens = lumens per lamp x lamps per luminaire

LEED EQ Credit 6.1: Controllability of Lighting Systems

12

NOTES ON MATERIALS

This chapter describes the major types of building materials, their physical properties, and their uses in building construction. The criteria for selecting and using a building material include those listed below.

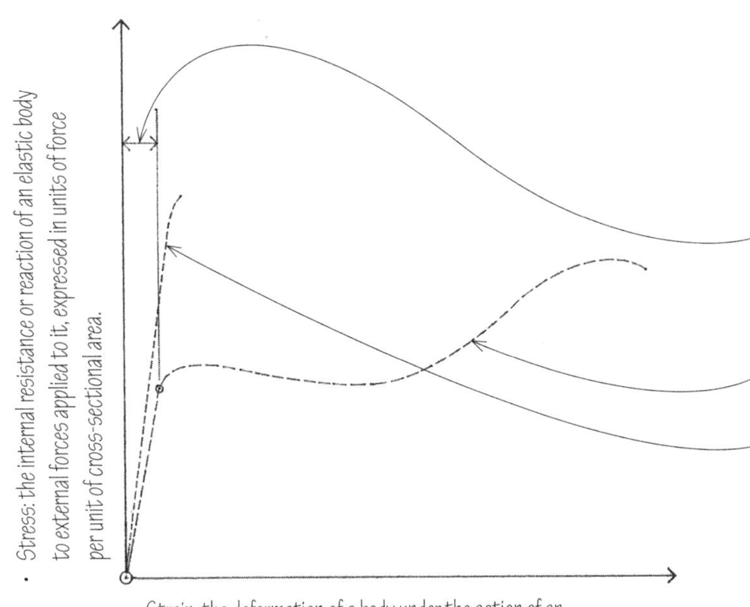

- Strain: the deformation of a body under the action of an applied force, equal to the ratio of the change in size or shape to the original size or shape of a stressed element

- Each material has distinct properties of strength, elasticity, and stiffness. The most effective structural materials are those that combine elasticity with stiffness.
- Elasticity is the ability of a material to deform under stress—bend, stretch, or compress—and return to its original shape when the applied stress is removed. Every material has its elastic limit beyond which it will permanently deform or break.
- Materials that undergo plastic deformation before actually breaking are termed ductile.
- Brittle materials, on the other hand, have low elastic limits and rupture under loads with little visible deformation. Because brittle materials have less reserve strength than ductile materials, they are not as suitable for structural purposes.
- Stiffness is a measure of the extent to which an elastic body resists deformation. The stiffness of a solid body is dependent on its structural shape as well as the elasticity of its material and is an important factor when considering the relationship between span and deflection under loading.

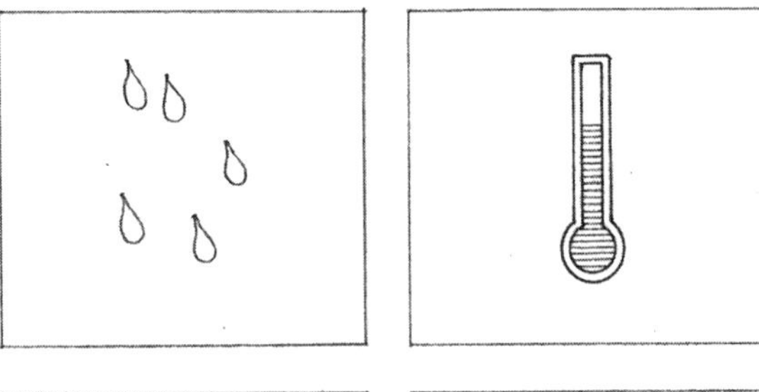

- The dimensional stability of a material as it responds to changes in temperature and moisture content affects the manner in which it is detailed and constructed to join with other materials.
- The resistance of a material to water and water vapor is an important consideration when it is exposed to weather or used in moist environments.
- The thermal conductivity or resistance of a material must be assessed when it is used in constructing the exterior envelope of a building.

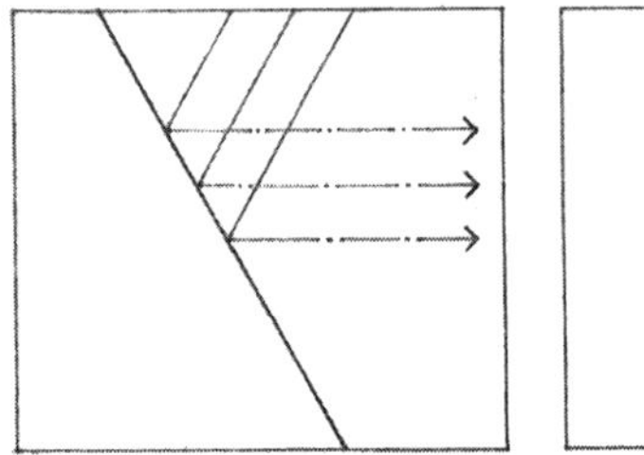

- A material's transmission, reflection, or absorption of visible light and radiant heat should be evaluated when the material is used to finish the surfaces of a room.
- The density or hardness of a material determines its resistance to wear and abrasion, its durability in use, and the costs required to maintain it.

- The ability of a material to resist combustion, withstand exposure to fire, and not produce smoke and toxic gases, must be evaluated before using it as a structural member or an interior finish.

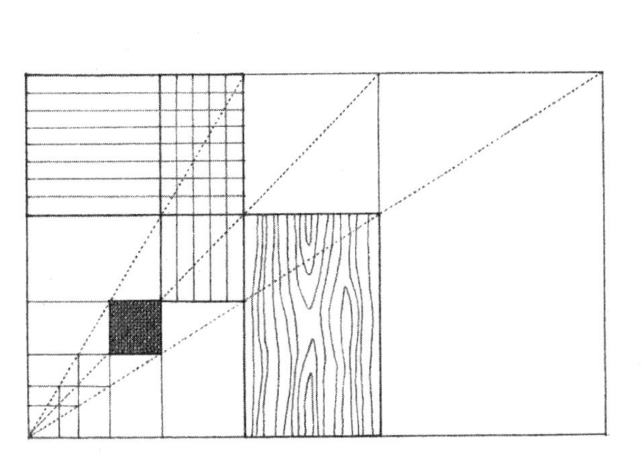

- The color, texture, and scale of a material are obvious considerations in evaluating how it fits within the overall design scheme.
- Many building materials are manufactured in standard shapes and sizes. These stock dimensions, however, may vary slightly from one manufacturer to the next. They should be verified in the planning and design phases of a building so that unnecessary cutting or wasting of material can be minimized during construction.

The evaluation of building materials should extend beyond their functional, economic, and aesthetic aspects and include assessing the environmental consequences associated with their selection and use. This examination, called a life-cycle assessment, encompasses the extraction and processing of raw materials, the manufacturing, packaging, and transportation of the finished product to the point of use, maintaining the material in use, the possible recycling and reuse of the material, and its final disposal. This assessment process consists of three components: inputs, life-cycle inventory, and outputs.

- Embodied energy includes all of the energy expended during the life cycle of a material.
- Refer to the *Environmental Resource Guide*, a project of the American Institute of Architects, for more information.

Embodied Energy in Building Materials

Material	Energy Content Btu/lb.*
Sand & gravel	18
Wood	185
Lightweight concrete	940
Gypsum board	1830
Brickwork	2200
Cement	4100
Glass	11,100
Plastic	18,500
Steel	19,200
Lead	25,900
Copper	29,600
Aluminum	103,500

*1 Btu/lb = 2.326 kJ/kg

Inputs

- Raw materials
- Energy
- Water

Acquisition of Raw Materials

- What impact does the extraction, mining, or harvesting process have on health and the environment?
- Is the material renewable or nonrenewable?
- Nonrenewable resources include metals and other minerals.
- Renewable resources, such as timber, vary in their rate of renewal; their rate of harvest should not exceed their rate of growth.

Processing, Manufacturing, and Packaging

- How much energy and water is required to process, manufacture, and package the material or product?

Transportation and Distribution

- Is the material or product available regionally or locally, or does it have to be shipped a long distance?

Construction, Use, and Maintenance

- Does the material perform its intended function efficiently and effectively?
- How does the material affect the indoor air quality and energy consumption of a building?
- How durable is the material or product and how much maintenance is required for its upkeep?
- What is the material's useful life?

Disposal, Recycling, and Reuse

- Usable products

+

- How much waste and how many toxic by-products result from the manufacture and use of the material or product?

Outputs

- Waterborne effluents
- Atmospheric emissions
- Solid wastes
- Other environmental releases

Life-Cycle Inventory

Evaluating the choice of a building material is a complex matter that cannot be reduced to a simple formula yielding a precise and valid answer with certainty. For example, using less of a material with a high energy content may be more effective in conserving energy and resources than using more of a lower-energy material. Using a higher-energy material that will last longer and require less maintenance, or one that can be recycled and reused, may be more compelling than using a lower-energy material.

Reduce, reuse, and recycle best summarize the kinds of strategies that are effective in achieving the goal of sustainability.

- Reduce building size through more efficient layout and use of spaces.
- Reduce construction waste. LEED® MR Credit 2: Construction Waste Management
- Specify products that use raw materials more efficiently. LEED MR Credit 5: Regional Materials
- Substitute plentiful resources for scarce resources. LEED MR Credit 6: Rapidly Renewable Materials
- Reuse building materials from demolished buildings. LEED MR Credit 3: Materials Reuse
- Rehabilitate existing buildings for new uses. LEED MR Credit 1: Building Reuse
- Recycle new products from old. LEED MR Credit 3: Materials Reuse

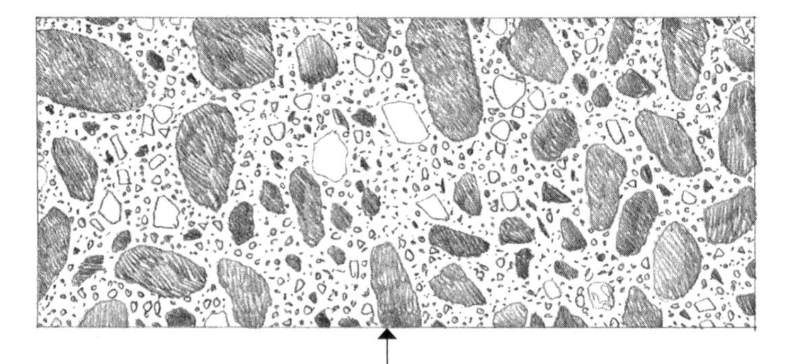

Concrete is made by mixing cement and various mineral aggregates with sufficient water to cause the cement to set and bind the entire mass. While concrete is inherently strong in compression, steel reinforcement is required to handle tensile and shear stresses. It is capable of being formed into almost any shape with a variety of surface finishes and textures. In addition, concrete structures are relatively low in cost and inherently fire-resistant. Concrete's liabilities include its weight—150 pcf (2400 kg/m^3) for normal reinforced concrete—and the forming or molding process that is required before it can be placed to set and cure.

Cement

- Portland cement is a hydraulic cement made by burning a mixture of clay and limestone in a rotary kiln and pulverizing the resulting clinker into a very fine powder.

- Type I normal portland cement is used for general construction, having none of the distinguishing qualities of the other types.
- Type II moderate portland cement is used in general construction where resistance to moderate sulfate action is required or where heat buildup can be damaging, as in the construction of large piers and heavy retaining walls.
- Type III high-early-strength portland cement cures faster and gains strength earlier than normal portland cement; it is used when the early removal of formwork is desired, or in cold-weather construction to reduce the time required for protection from low temperatures.
- Type IV low-heat portland cement generates less heat of hydration than normal portland cement; it is used in the construction of massive concrete structures, as in gravity dams, where a large buildup in heat can be damaging.
- Type V sulfate-resisting portland cement is used where resistance to severe sulfate action is required.
- Air-entraining portland cement is a Type I, Type II, or Type III portland cement to which a small quantity of an air-entraining agent has been interground during manufacture; it is designated by the suffix A.

Water

- The water used in a concrete mix must be free of organic material, clay, and salts; a general criterion is that the water should be fit for drinking.
- Cement paste is a mixture of cement and water for coating, setting, and binding the aggregate particles together in a concrete mix.

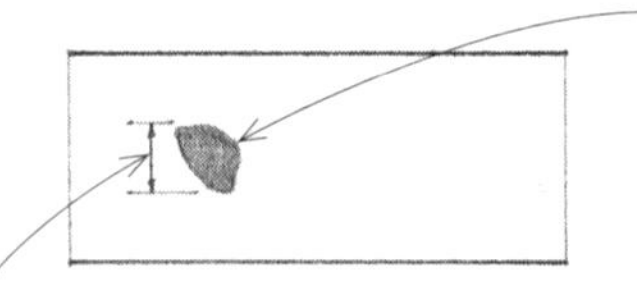

- 1/3 the depth of a slab, 1/5 the thickness of a wall, or 3/4 of the clear space between reinforcing bars or between the bars and the formwork

Lightweight Concrete

- Structural lightweight concrete, made with expanded shale or slate aggregate, has a unit weight from 85 to 115 pcf (1362 to 1840 kg/m^3) and compressive strength comparable to that of normal concrete.
- Insulating concrete, made with perlite aggregate or a foaming agent, has a unit weight of less than 60 pcf (960 kg/ m^3) and low thermal conductivity.

Aggregate

- Aggregate refers to any of various inert mineral materials, as sand and gravel, added to a cement paste to make concrete. Because aggregate represents from 60% to 80% of the concrete volume, its properties are important to the strength, weight, and fire-resistance of the hardened concrete. Aggregate should be hard, dimensionally stable, and free of clay, silt, and organic matter that can prevent the cement matrix from binding the particles together.
- Fine aggregate consists of sand having a particle size smaller than 1/4" (6).
- Coarse aggregate consists of crushed stone, gravel, or blast-furnace slag having a particle size larger than 1/4" (6).
- The maximum size of coarse aggregate in reinforced concrete is limited by the size of the section and the spacing of the reinforcing bars.

Admixtures

Admixtures may be added to a concrete mix to alter its properties or those of the hardened product.

- Air-entraining agents disperse microscopic, spherical air bubbles in a concrete mix to increase workability, improve resistance of the cured product to the cracking induced by free-thaw cycles or the scaling caused by deicing chemicals, and in larger amounts, to produce lightweight, insulating concrete.
- Accelerators hasten the setting and strength development of a concrete mix, while retarders slow the setting of a concrete mix in order to allow more time for placing and working the mix.
- Surface-active agents, or surfactants, reduce the surface tension of the mixing water in a concrete mix, thereby facilitating the wetting and penetrating action of the water or aiding in the emulsifying and dispersion of other additives in the mix.
- Water-reducing agents, or superplasticizers, reduce the amount of mixing water required for the desired workability of a concrete or mortar mix. Lowering the water-cement ratio in this manner generally results in increased strength.
- Coloring agents are pigments or dyes added to a concrete mix to alter or control its color.

Water-Cement Ratio

Water-cement ratio is the ratio of mixing water to cement in a unit volume of a concrete mix, expressed by weight as a decimal fraction or as gallons of water per sack of cement. The water-cement ratio controls the strength, durability, and watertightness of hardened concrete. According to Abrams' law, formulated by D. A. Abrams in 1919 from experiments at the Lewis Institute in Chicago, the compressive strength of concrete is inversely proportional to the ratio of water to cement. If too much water is used, the concrete mix will be weak and porous after curing. If too little water is used, the mix will be dense but difficult to place and work. For most applications, the water-cement ratio should range from 0.45 to 0.60.

Concrete is normally specified according to the compressive strength it will develop within 28 days after placement (7 days for high-early-strength concrete).

- Slump test is a method for determining the consistency and workability of freshly mixed concrete by measuring the slump of a test specimen, expressed as the vertical settling, in inches, of a specimen after it has been placed in a slump cone, tamped in a prescribed manner, and the cone is lifted.
- Compression test for determining the compressive strength of a concrete batch uses a hydraulic press to measure the maximum load a test cylinder 6" (150) ø and 12" (305) high can support in axial compression before fracturing.

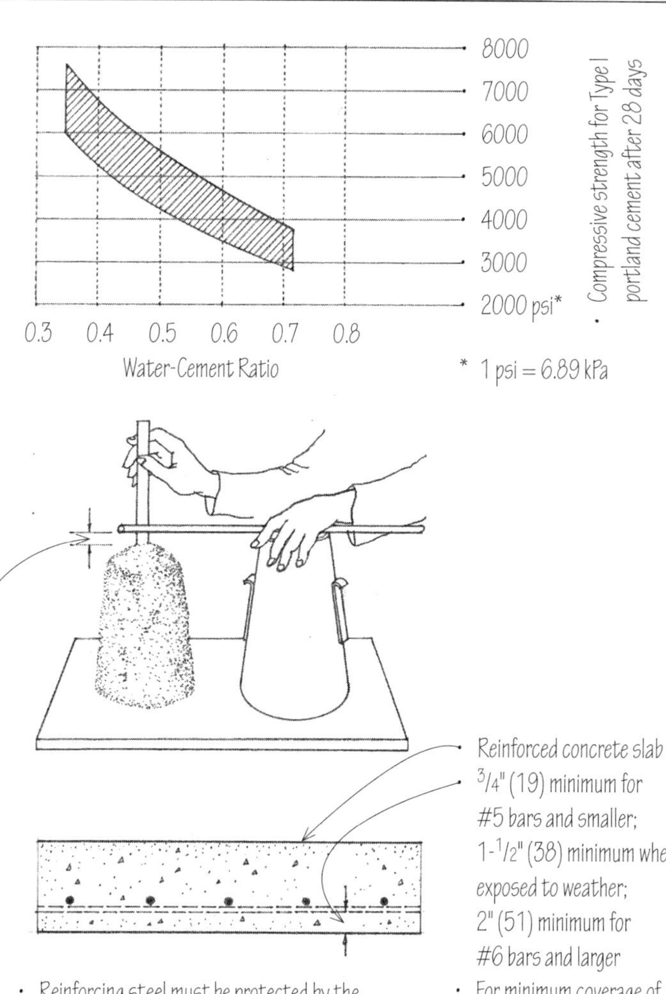

- Reinforcing steel must be protected by the surrounding concrete against corrosion and fire. Minimum requirements for cover and spacing are specified by the American Concrete Institute (ACI) *Building Code Requirements for Reinforced Concrete* according to the concrete's exposure, and the size of the coarse aggregate and steel used.
- For minimum coverage of steel reinforcement in other concrete members, see 3.08 for spread footings, 4.04 for concrete beams, 5.04 for concrete columns, and 5.06 for concrete walls.

Steel Reinforcement

Because concrete is relatively weak in tension, reinforcement consisting of steel bars, strands, or wires is required to absorb tensile, shearing, and sometimes the compressive stresses in a concrete member or structure. Steel reinforcement is also required to tie vertical and horizontal elements, reinforce the edges around openings, minimize shrinkage cracking, and control thermal expansion and contraction. All reinforcement should be designed by a qualified structural engineer.

- Reinforcing bars are steel sections hot-rolled with ribs or other deformations for better mechanical bonding to concrete. The bar number refers to its diameter in eighths of an inch—for example, a #5 bar is 5/8" (16) in diameter.
- Welded wire fabric consists of a grid of steel wires or bars welded together at all points of intersection. The fabric is typically used to provide temperature reinforcement for slabs but the heavier gauges can also be used to reinforce concrete walls. The fabric is designated by the size of the grid in inches followed by a number indicating the wire gauge or cross-sectional area; see 3.18 for typical sizes.

ASTM Standard Reinforcing Bars

Bar Size	Nominal Dimensions		
	Diameter inches (mm)	Cross-Sectional Area sq. in. (mm²)	Weight plf (N/m)
#3	0.375 (10)	0.11 (71)	0.38 (5.5)
#4	0.50 (13)	0.20 (129)	0.67 (9.7)
#5	0.625 (16)	0.31 (200)	1.04 (15.2)
#6	0.75 (19)	0.44 (284)	1.50 (21.9)
#7	0.875 (22)	0.60 (387)	2.04 (29.8)
#8	1.00 (25)	0.79 (510)	2.67 (39.0)
#9	1.125 (29)	1.00 (645)	3.40 (49.6)
#10	1.25 (32)	1.27 (819)	4.30 (62.8)

- Common brick, also called building brick, is made for general building purposes and not specially treated for color and texture.
- Face brick is made of special clays for facing a wall, often treated to produce the desired color and surface texture.

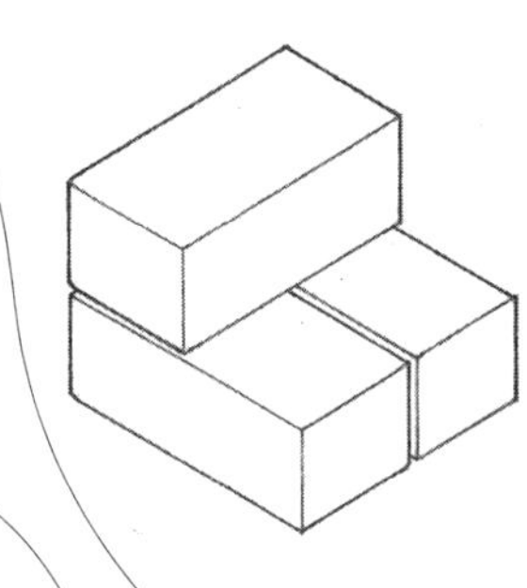

Masonry refers to building with units of various natural or manufactured products, such as brick, stone, or concrete block, usually with the use of mortar as a bonding agent. The modular aspect (i.e., uniform sizes and proportional relationships) of unit masonry distinguishes it from most of the other building materials discussed in this chapter. Because unit masonry is structurally most effective in compression, the masonry units should be laid up in such a way that the entire masonry mass acts as an entity.

Brick Types

- Brick type designates the permissible variation in size, color, chippage, and distortion allowed in a face brick unit.
- FBX is face brick suitable for use where a minimum variation in size, narrow color range, and high degree of mechanical perfection are required.
- FBS is face brick suitable for use where a wider color range and greater variation in size are permitted than for type FBX.
- FBA is face brick suitable for use where particular effects are desired resulting from nonuniformity in size, color, and texture of the individual units.

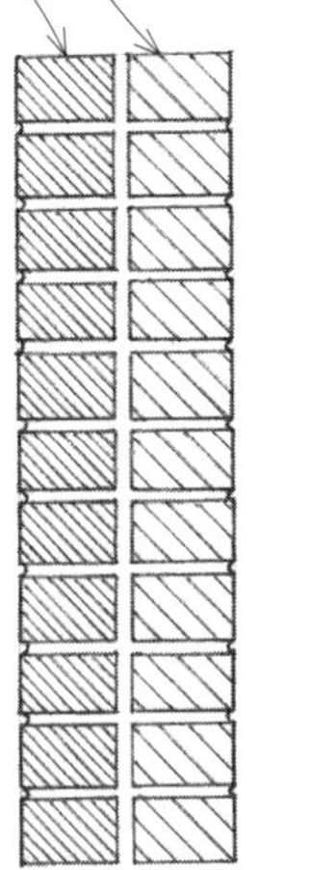

Brick

Brick is a masonry unit of clay, formed into a rectangular prism while plastic and hardened by firing in a kiln or drying in the sun.

- Soft-mud process refers to forming brick by molding relatively wet clay having a moisture content of 20% to 30%.
- Sandstruck brick is formed in the soft-mud process with a mold lined with sand to prevent sticking, producing a matte-textured surface.
- Waterstruck brick is formed in the soft-mud process with a mold lubricated with water to prevent sticking, producing a smooth, dense surface.
- Stiff-mud process refers to forming brick and structural tile by extruding stiff but plastic clay having a moisture content of 12% to 15% through a die and cutting the extrusion to length with wires before firing.
- Dry-press process refers to forming brick by molding relatively dry clay having a moisture content of 5% to 7% under high pressure, resulting in sharp-edged, smooth-surfaced bricks.

- Efflorescence is a white, powdery deposit that forms on an exposed masonry or concrete surface, caused by the leaching and crystallization of soluble salts from within the material. Reducing moisture absorption is the best assurance against efflorescence.

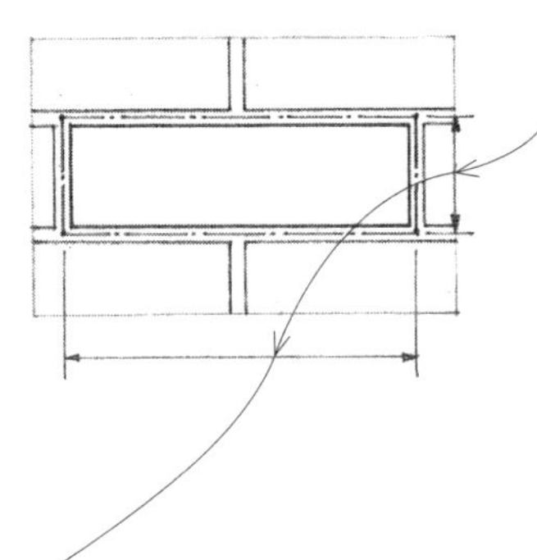

- The actual dimensions of brick units vary due to shrinkage during the manufacturing process. The nominal dimensions given in the table include the thickness of the mortar joints, which vary from 1/4" to 1/2" (6 to 13).

Brick Unit	Nominal Dimensions thickness x height x length inches	mm	Modular Coursing inches	mm
Modular	4 x 2-2/3 x 8	100 x 68 x 205	3C = 8	205
Norman	4 x 2-2/3 x 12	100 x 68 x 305	3C = 8	205
Engineer	4 x 3-1/5 x 8	100 x 81 x 205	5C = 16	405
Norwegian	4 x 3-1/5 x 12	100 x 81 x 305	5C = 16	405
Roman	4 x 2 x 12	100 x 51 x 305	2C = 4	100
Utility	4 x 4 x 12	100 x 100 x 305	1C = 4	100

- See 5.26 for modular brick coursing and 5.27 for masonry bonding patterns.

Brick Grades

- Brick grade designates the durability of a brick unit when exposed to weathering. The United States is divided into three weathering regions—severe, moderate, and negligible—according to annual winter rainfall and the annual number of freezing-cycle days. Brick is graded for use in each region according to compressive strength, maximum water absorption, and maximum saturation coefficient.
- SW is brick suitable for exposure to severe weathering, as when in contact with the ground or used on surfaces likely to be permeated with water in subfreezing temperatures; minimum compressive strength of 2500 psi (17,235 kPa).
- MW is brick suitable for exposure to moderate weathering, as when used above grade on surfaces unlikely to be permeated with water in subfreezing temperatures; minimum compressive strength of 2200 psi (15,167 kPa).
- NW is brick suitable for exposure to negligible weathering, as when used as a backup or in interior masonry; minimum compressive strength of 1250 psi (8,618 kPa).
- The allowable compressive stresses in masonry walls are much less than the values given here because the quality of the masonry units, mortar, and workmanship may vary. See table on 5.15 for these values.

Concrete Masonry

Concrete masonry units (CMU) are precast of portland cement, fine aggregate, and water, molded into various shapes to satisfy various construction conditions. The availability of these types varies with locality and manufacturer.

- Concrete block, often incorrectly referred to as cement block, is a hollow concrete masonry unit having a compressive strength from 600 to 1500 psi (4137 to 10,342 kPa).
- Normal-weight block is made from concrete weighing more than 125 pcf (2000 kg/m^3).
- Medium-weight block is made from concrete weighing from 105 to 125 pcf (1680 to 2000 kg/m^3).
- Lightweight block is made from concrete weighing 105 pcf (1680 kg/m^3) or less.

CMU Grades

- Grade N is a loadbearing concrete masonry unit suitable for use both above and below grade in walls exposed to moisture or weather; grade N units have a compressive strength from 800 to 1500 psi (5516 to 10,342 kPa).
- Grade S is a loadbearing concrete masonry unit limited to use above grade, in exterior walls with weather-protective coatings, or in walls not exposed to moisture or weather; grade S units have a compressive strength from 600 to 1000 psi (4137 to 6895 kPa).

CMU Types

- Type I is a concrete masonry unit manufactured to a specified limit of moisture content in order to minimize the drying shrinkage that can cause cracking.
- Type II is a concrete masonry unit not manufactured to a specified limit moisture content.

- Concrete brick is a solid rectangular concrete masonry unit usually identical in size to a modular clay brick but also available in 12" (305) lengths; concrete brick units have a compressive strength from 2000 to 3000 psi (13,790 to 20,685 kPa).

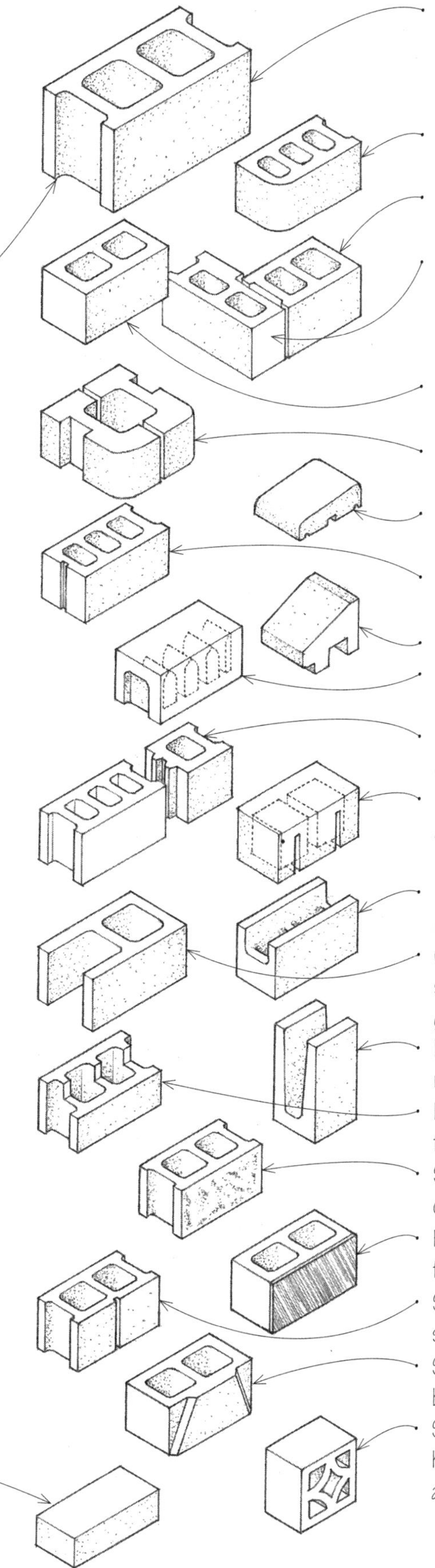

- Stretcher blocks have two or three cores and nominal dimensions of 8" x 8" x 16" (205 x 205 x 405); 4", 6", 10" and 12" (100, 150, 255 and 305) wide units are also available.
- Bullnose blocks have one or more rounded exterior corners.
- Corner blocks have a solid end face for use in constructing the end or corner of a wall.
- Corner-return blocks are used at the corners of 6", 10", and 12" (150, 255, and 305) walls to maintain horizontal coursing with the appearance of full- and half-length units.
- Double-corner blocks have solid faces at both ends and are used in constructing a masonry pier.
- Pilaster blocks are used in constructing a plain or reinforced masonry pilaster.
- Coping blocks are used in constructing the top or finishing course of a masonry wall.
- Sash or jamb blocks have an end slot or rabbet to receive the jamb of a door or window frame.
- Sill blocks have a wash to shed rainwater from a sill.
- Cap blocks have a solid top for use as a bearing surface in the finishing course of a foundation wall.
- Control-joint blocks are used in constructing a vertical control joint.
- Sound-absorbing masonry units have a solid top and a slotted face shell, and sometimes a fibrous filler, for increased sound absorption.
- Bond-beam blocks have a depressed section in which reinforcing steel can be placed for embedment in grout.
- Open-end blocks have one end open in which vertical steel reinforcement can be placed for embedment in grout.
- Lintel blocks have a U-shaped section in which reinforcing steel can be placed for embedment in grout.
- Header blocks have a portion of one face shell removed to receive headers in a bonded masonry wall.
- Split-face blocks are split lengthwise by a machine after curing to produce a rough, fractured face texture.
- Faced blocks have a special ceramic, glazed, or polished face.
- Scored blocks have one or more vertical grooves that simulate raked joints.
- Shadow blocks have a face shell with a pattern of beveled recesses.
- Screen blocks, used especially in tropical architecture, have a decorative pattern of transverse openings for admitting air and excluding sunlight.

- W shape (wide flange)
- S shape (American standard)
- C shape (American standard channel)

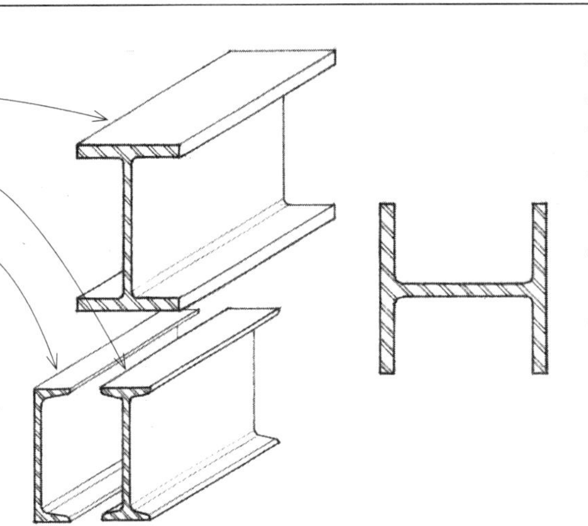

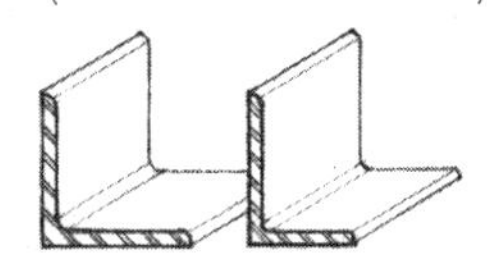

- L shapes (equal and unequal leg angles)

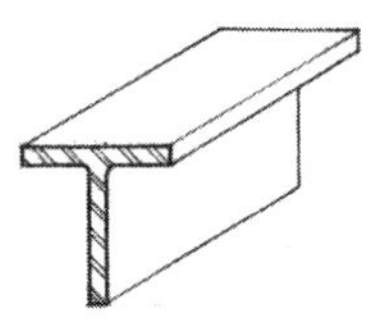

- WT shape (structural tee cut from W shape)

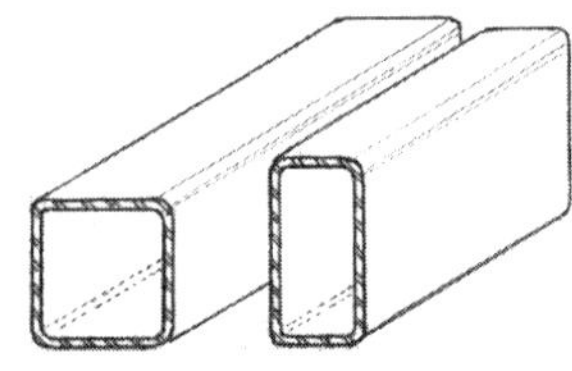

- Structural tubing (square or rectangular)

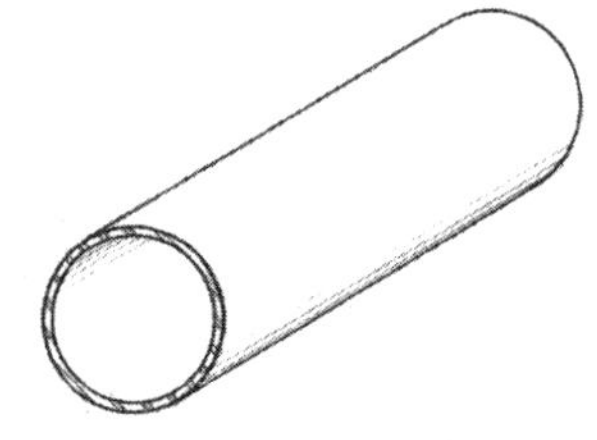

- Structural tubing (circular pipe)

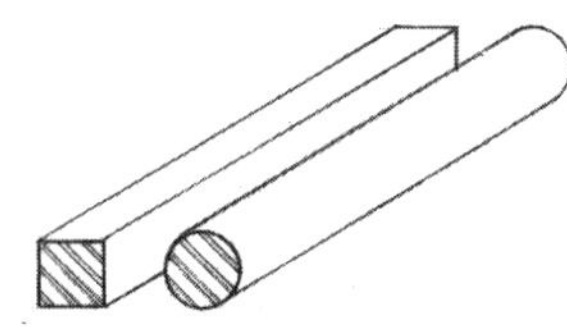

- Bars (square, round, and flat)

Steel Shapes
- Refer to the American Institute of Steel Construction (AISC) *Manual of Steel Construction* for complete listing of sizes and weights.

Steel refers to any of various iron-based alloys having a carbon content less than that of cast iron and more than that of wrought iron, and having qualities of strength, hardness, and elasticity varying according to composition and heat treatment. Steel is used for light and heavy structural framing, as well as a wide range of building products such as windows, doors, hardware, and fastenings. As a structural material, steel combines high strength and stiffness with elasticity. Measured in terms of weight to volume, it is probably the strongest low-cost material available. Although classified as an incombustible material, steel becomes ductile and loses its strength when subject to temperatures over 1000°F (538°C). When used in buildings requiring fire-resistive construction, structural steel must be coated, covered, or enclosed with fire-resistant materials; see A.12. Because it is normally subject to corrosion, steel must be painted, galvanized, or chemically treated for protection against oxidation.

- Mild or soft steel is a low-carbon steel containing from 0.15% to 0.25% carbon.
- Medium steel is a carbon steel containing from 0.25% to 0.45% carbon; most structural steel is medium-carbon steel; ASTM A36 is the most common strength grade with a yield point of 36,000 psi (248,220 kPa).
- Hard steel is a high-carbon steel containing from 0.45% to 0.85% carbon.
- Spring steel is a high-carbon steel containing 0.85% to 1.8% carbon.

- Carbon steel is unalloyed steel in which the residual elements, such as carbon, manganese, phosphorus, sulfur, and silicon, are controlled. Any increase in carbon content increases the strength and hardness of the steel but reduces its ductility and weldability.

- Stainless steel contains a minimum of 12% chromium, sometimes with nickel, manganese, or molybdenum as additional alloying elements, so as to be highly resistant to corrosion.
- High-strength low-alloy steel is a low-carbon steel containing less than 2% alloys in a chemical composition specifically developed for increased strength, ductility, and resistance to corrosion; ASTM A572 is the most common strength grade with a yield point of 50,000 psi (344,750 kPa).
- Weathering steel is a high-strength, low-alloy steel that forms an oxide coating when exposed to rain or moisture in the atmosphere; this coating adheres firmly to the base metal and protects it from further corrosion. Structures using weathering steel should be detailed to prevent the small amounts of oxide carried off by rainwater from staining adjoining materials.
- Tungsten steel is an alloy steel containing 10% to 20% tungsten for increased hardness and heat retention at high temperatures.

- Alloy steel refers to a carbon steel to which various elements, such as chromium, cobalt, copper, manganese, molybdenum, nickel, tungsten, or vanadium, have been added in a sufficient amount to obtain particular physical or chemical properties.

Other ferrous metals used in building construction include:
- Cast iron, a hard, brittle, nonmalleable iron-based alloy containing 2.0% to 4.5% carbon and 0.5% to 3% silicon, cast in a sand mold and machined to make many building products, such as piping, grating, and ornamental work
- Malleable cast iron, which has been annealed by transforming the carbon content into graphite or removing it completely
- Wrought iron, a tough, malleable, relatively soft iron that is readily forged and welded, having a fibrous structure containing approximately 0.2% carbon and a small amount of uniformly distributed slag
- Galvanized iron, which is coated with zinc to prevent rust

CSI MasterFormat 05 12 00: Structural Steel Framing

Nonferrous metals contain no iron. Aluminum, copper, and lead are nonferrous metals commonly used in building construction.

Aluminum is a ductile, malleable, silver-white metallic element that is used in forming many hard, light alloys. Its natural resistance to corrosion is due to the transparent film of oxide that forms on its surface; this oxide coating can be thickened to increase corrosion resistance by an electrical and chemical process known as anodizing. During the anodizing process, the naturally light, reflective surface of aluminum can be dyed a number of warm, bright colors. Care must be taken to insulate aluminum from contact with other metals to prevent galvanic action. It should also be isolated from alkaline materials such as wet concrete, mortar, and plaster.

Aluminum is widely used in extruded and sheet forms for secondary building elements such as windows, doors, roofing, flashing, trim, and hardware. For use in structural framing, high-strength aluminum alloys are available in shapes similar to those of structural steel. Aluminum sections may be welded, bonded with adhesives, or mechanically fastened.

Copper is a ductile, malleable metallic element that is widely used for electrical wiring, water piping, and in the manufacture of alloys, as bronze and brass. Its color and resistance to corrosion also make it an excellent roofing and flashing material. However, copper will corrode aluminum, steel, stainless steel, and zinc. It should be fastened, attached, or supported only with copper or carefully selected brass fittings. Contact with red cedar in the presence of moisture will cause premature deterioration of the copper.

Brass refers to any of various alloys consisting essentially of copper and zinc, used for windows, railings, trim, and finish hardware. Alloys that are brass by definition may have names that include the word bronze, as architectural bronze.

Lead is a heavy, soft, malleable, bluish-gray metallic element used for flashing, sound isolation, and radiation shielding. Although lead is the heaviest of the common metals, its pliability makes it desirable for application over uneven surfaces. Lead dust and vapors are toxic.

Galvanic Action

Galvanic action can occur between two dissimilar metals when enough moisture is present for electric current to flow. This electric current will tend to corrode one metal while plating the other. The severity of the galvanic action depends on how far apart the two metals are on the galvanic series table.

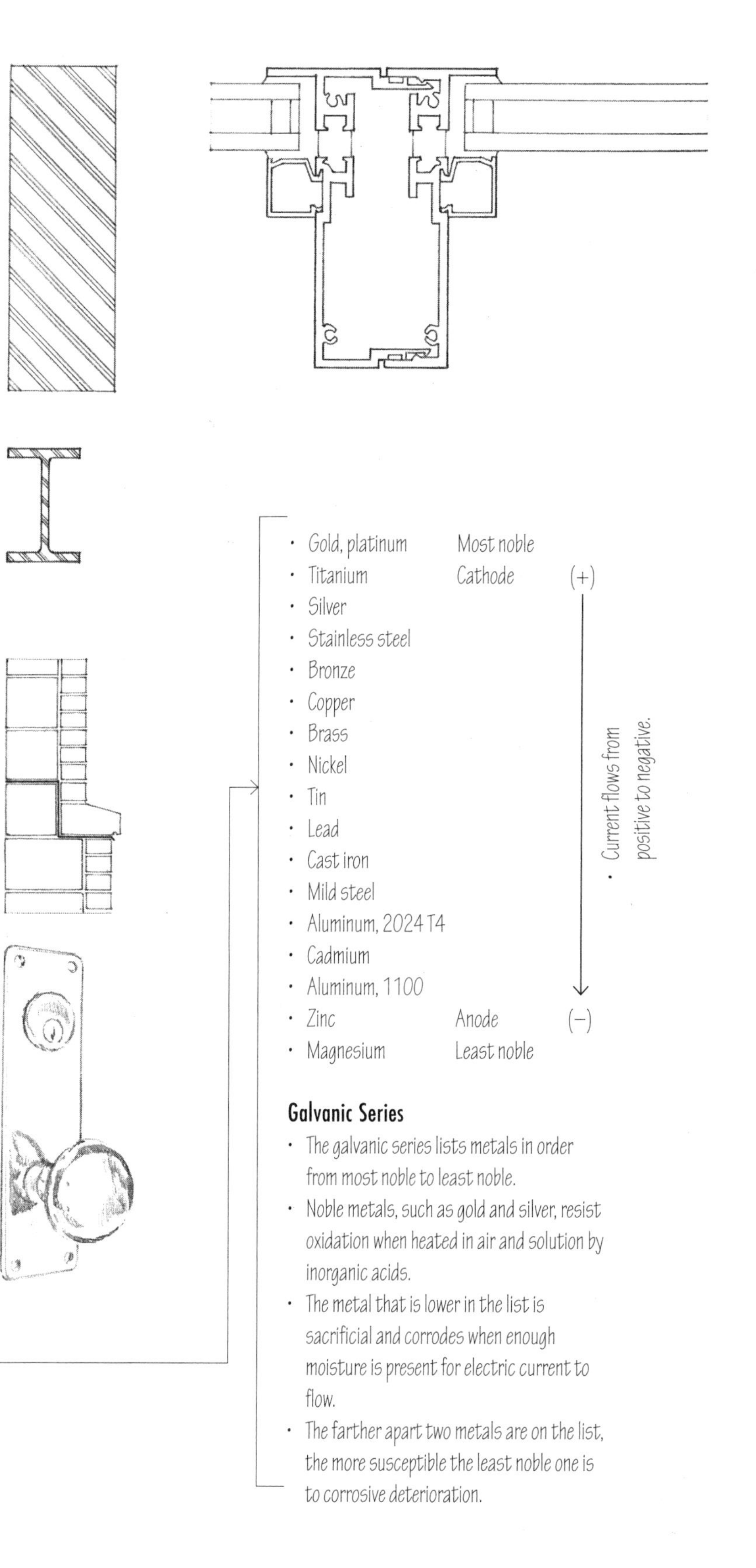

Galvanic Series

- The galvanic series lists metals in order from most noble to least noble.
- Noble metals, such as gold and silver, resist oxidation when heated in air and solution by inorganic acids.
- The metal that is lower in the list is sacrificial and corrodes when enough moisture is present for electric current to flow.
- The farther apart two metals are on the list, the more susceptible the least noble one is to corrosive deterioration.

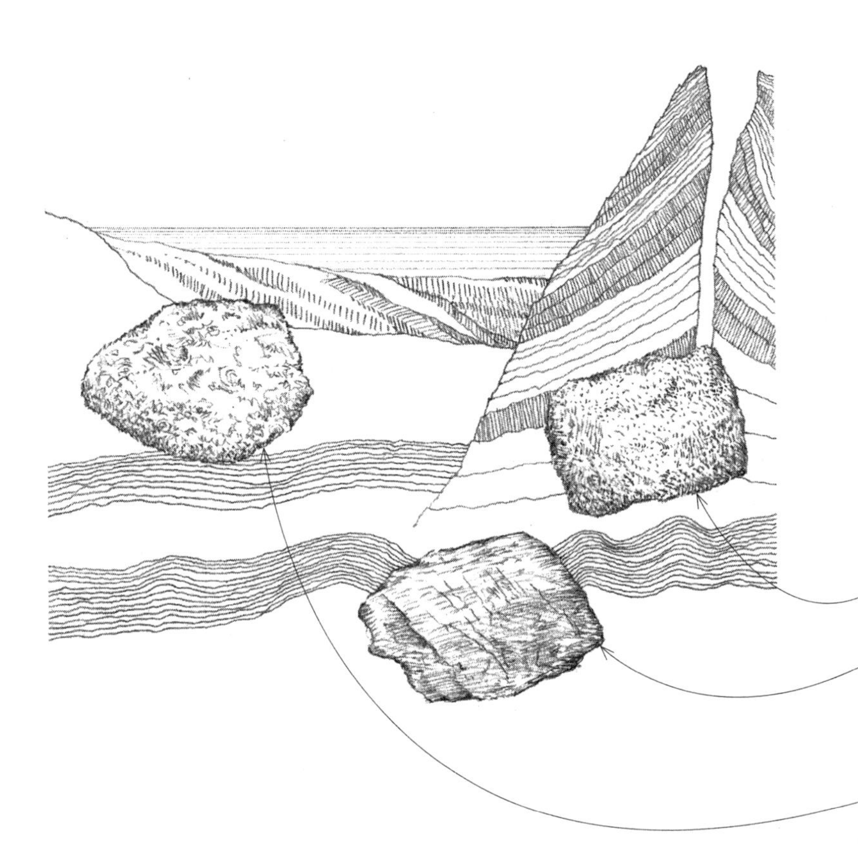

Stone is an aggregate or combination of minerals, each of which is composed of inorganic chemical substances. To qualify as a construction material, stone should have the following qualities:

- Strength: Most types of stone have more than adequate compressive strength. The shear strength of stone, however, is usually about 1/10 of its compressive strength.
- Hardness: Hardness is important when stone is used for flooring, paving, and stair treads.
- Durability: Resistance to the weathering effects of rain, wind, heat, and frost action is necessary for exterior stonework.
- Workability: A stone's hardness and grain texture must allow it to be quarried, cut, and shaped.
- Density: A stone's porosity affects its ability to withstand frost action and staining.
- Appearance: Appearance factors include color, grain, and texture.

Stone may be classified according to geological origin into the following types:

- Igneous rock, such as granite, obsidian, and malachite, is formed by the crystallization of molten magma.
- Metamorphic rock, such as marble and slate, has undergone a change in structure, texture, or composition due to natural agencies, such as heat and pressure, especially when the rock becomes harder and more crystalline.
- Sedimentary rock, such as limestone, sandstone, and shale, is formed by the deposition of sediment by glacial action.

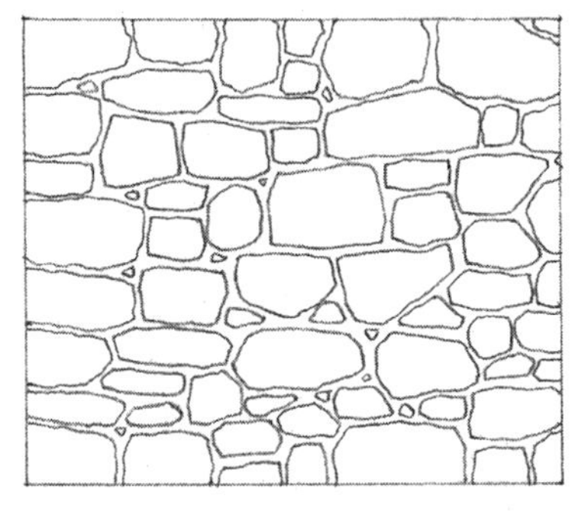

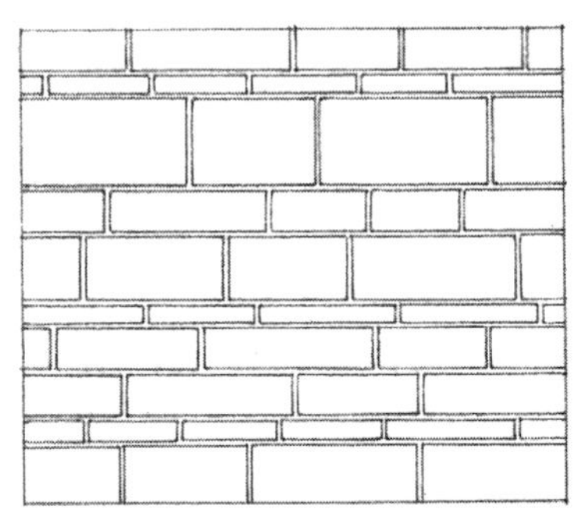

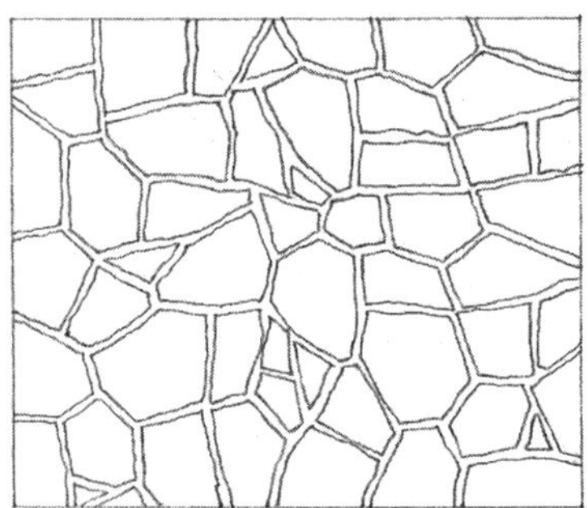

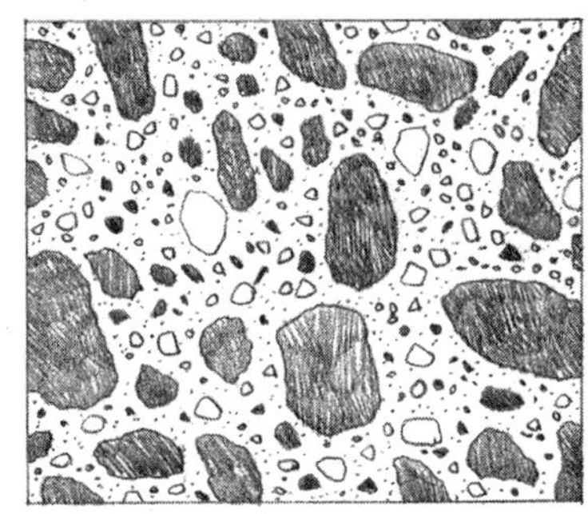

As a loadbearing wall material, stone is similar to modular unit masonry. Although stone masonry is not necessarily uniform in size, it is laid up with mortar and used in compression. Almost all stone is adversely affected by sudden changes in temperature and should not be used where a high degree of fire resistance is required.

Stone is used in construction in the following forms:

- Rubble consists of rough fragments of broken stone that have at least one good face for exposure in a wall.
- Dimension stone is quarried and squared stone 2' (610) or more in length and width and of specified thickness, used commonly for wall panels, cornices, copings, lintels, and flooring.
- Flagstone refers to flat stone slabs used for flooring and horizontal surfacing.
- Crushed stone is used as aggregate in concrete products.

- See 5.33 for types of stone masonry.

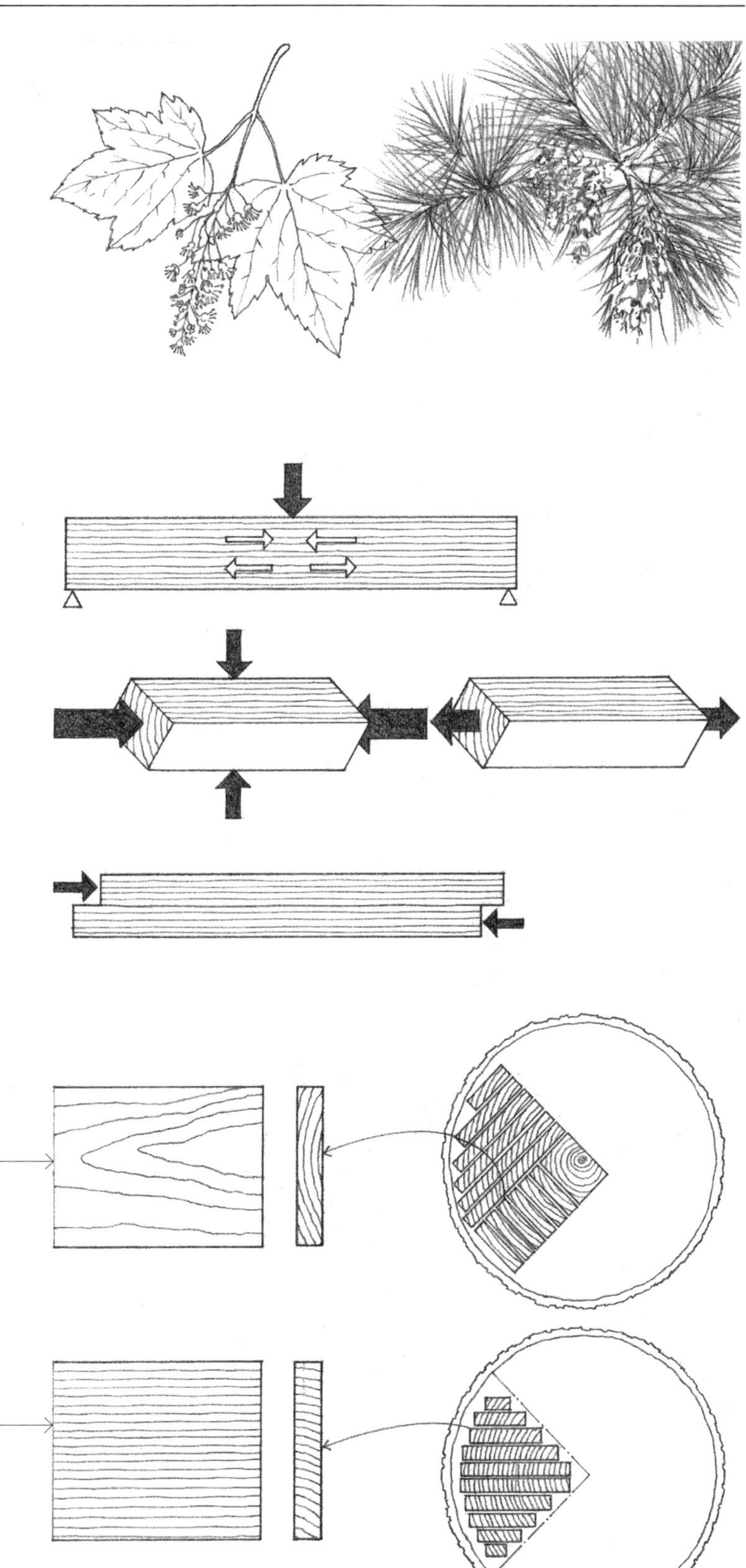

As a construction material, wood is strong, durable, light in weight, and easy to work. In addition, it offers natural beauty and warmth to sight and touch. Although it has become necessary to employ conservation measures to ensure a continued supply, wood is still used on construction in many and varied forms.

There are two major classes of wood—softwood and hardwood. These terms are not descriptive of the actual hardness, softness, or strength of a wood. Softwood is the wood from any of various predominantly evergreen, cone-bearing trees, such as pine, fir, hemlock, and spruce, used for general construction. Hardwood is the wood from a broad-leaved flowering tree, such as cherry, maple, or oak, typically used for flooring, paneling, furniture, and interior trim.

The manner in which a tree grows affects its strength, its susceptibility to expansion and contraction, and its effectiveness as an insulator. Tree growth also affects how pieces of sawn wood may be joined to form the structure and enclosure of a building.

Grain direction is the major determining factor in the use of wood as a structural material. Tensile and compressive forces are best handled by wood in a direction parallel to the grain. Typically, a given piece of wood will withstand 1/3 more force in compression than in tension parallel to its grain. The allowable compressive force perpendicular to its grain is only about 1/5 to 1/2 of the allowable compressive force parallel to the grain. Tensile forces perpendicular to the grain will cause the wood to split. The shear strength of wood is greater across its grain than parallel to the grain. It is therefore more susceptible to horizontal shear than to vertical shear.

The manner in which lumber is cut from a log affects its strength as well as its appearance. Plainsawing a squared log into boards with evenly spaced parallel cuts results in flat grain lumber that:

- May have a variety of noticeable grain patterns;
- Tends to twist and cup, and wears unevenly;
- Tends to have raised grain;
- Shrinks and swells less in thickness, more in width.

Quartersawing logs approximately at right angles to the annual rings results in edge or vertical grain lumber that:

- Has more even grain patterns;
- Wears more evenly with less raised grain and warping;
- Shrinks and swells less in width, more in thickness;
- Is less affected by surface checks;
- Results in more waste in cutting and is more expensive.

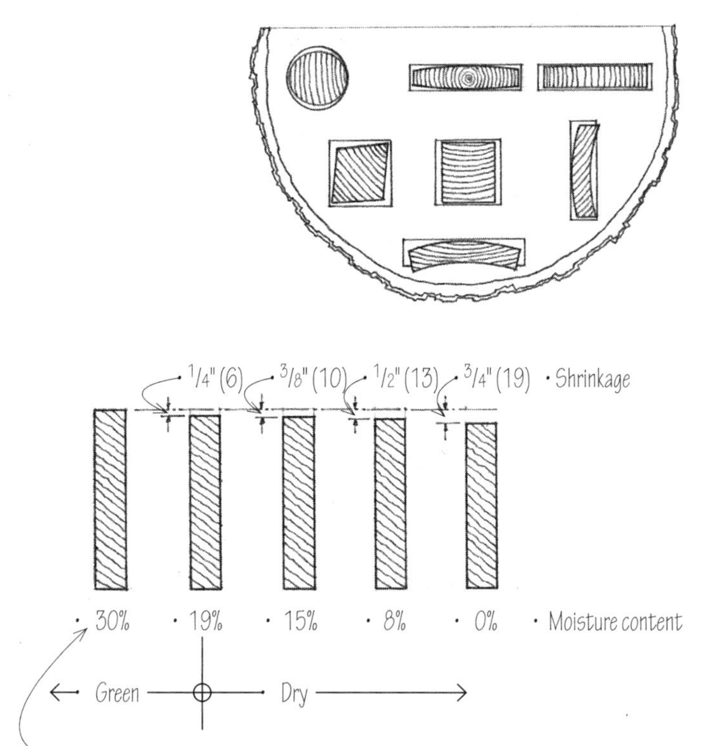

• Fiber saturation point is the stage at which the cell walls are fully saturated but the cell cavities are void of water, ranging from a moisture content of 25% to 32% for commonly used species. Further drying results in shrinkage and generally greater strength, stiffness, and density of the wood.

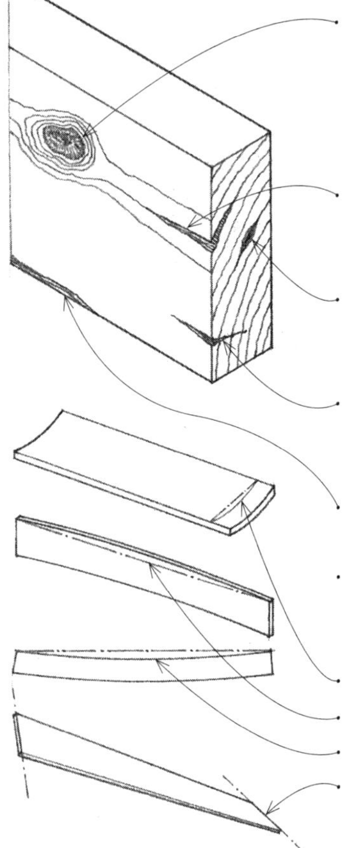

• Knots are hard nodes of wood that occur where branches join the trunk of a tree, appearing as circular, cross-grained masses in a piece of sawn lumber. In the structural grading of a wood piece, knots are restricted by size and location.
• Shakes are separations along the grain of a wood piece, usually between the annual rings, caused by stresses on a tree while standing or during felling.
• Pitch pockets are well-defined openings between the annual rings of a softwood, containing or having once contained solid or liquid pitch.
• Checks are lengthwise separations of wood across the annual rings, caused by uneven or rapid shrinkage during the seasoning process.
• Wane is the presence of bark or absence of wood at a corner or along an edge of a piece.
• Warping is usually caused by uneven drying during the seasoning process or by a change in moisture content.
• Cup is a curvature across the face of a wood piece.
• Bow is a curvature along the length of a wood piece.
• Crook is a curvature along the edge of a wood piece.
• Twist results from the turning of the edges of a wood piece in opposite directions.

To increase its strength, stability, and resistance to fungi, decay, and insects, wood is seasoned—dried to reduce its moisture content—by air-drying or kiln-drying under controlled conditions of heat, air circulation, and humidity. It is impossible to completely seal a piece of wood to prevent changes in its moisture content. Below a moisture content of about 30%, wood expands as it absorbs moisture and shrinks as it loses moisture. This possibility of shrinkage and swelling must always be taken into account when detailing and constructing wood joints, both in small- and large-scale work.

Shrinkage tangential to the wood grain is usually twice as much as radial shrinkage. Vertical grain lumber shrinks uniformly while plainsawn cuts near a log's perimeter will cup away from the center. Because the thermal expansion of wood is generally much less than volume changes due to changes in moisture content, moisture content is therefore the controlling factor.

Wood is decay-resistant when its moisture content is under 20%. If installed and maintained below this moisture content level, wood will usually not rot. Species that are naturally resistant to decay-causing fungi include redwood, cedar, bald cypress, black locust, and black walnut. Insect-resistant species include redwood, eastern red cedar, and bald cypress.

Preservative treatments are available to further protect wood from decay and insect attack. Of these, pressure treatment is the most effective, especially when the wood is in contact with the ground. There are three types of preservatives.
- Water-borne preservatives leave the wood clean, odorless, and readily paintable; preservatives do not leach out when exposed to weather.
  - AWPB (American Wood Preservers Bureau)
  - LP-2 (LP-22 for ground contact)
- Oil-borne preservatives may color the wood, but treated wood is paintable; pentachlorophenol is highly toxic.
  - AWPB LP-3 (LP-33 for ground contact)
- Creosote treatment leaves wood with colored, oily surfaces; odor remains for a long period; used especially in marine and saltwater installations.
  - AWPB LP-5 (LP-55 for ground contact)

Defects affect the grading, appearance, and use of wood members. They may also affect a wood's strength, depending on their number, size, and location. Defects include the natural characteristics of wood, such as knots, shakes, and pitch pockets, as well as the effects of manufacturing, such as checks and warping.

Because of the diversity of its applications and its use for remanufacturing purposes, hardwood is graded according to the amount of clear, usable lumber in a piece that may be cut into smaller pieces of a certain grade and size. Softwood is classified in the following manner.

- Yard lumber: softwood lumber intended for general building purposes, including boards, dimension lumber, and timbers
  - Boards: less than 2" (51) thick and 2" (51) or more wide, graded for appearance rather than strength and used as siding, subflooring, and interior trim
  - Dimension lumber: from 2" to 4" (51 to 100) thick and 2" (51) or more wide, graded for strength rather than appearance, and used for general construction
    - Joists and planks: from 2" to 4" (51 to 100) thick and more than 4" (100) wide, graded primarily with respect to bending strength when loaded either on the narrow face as a joist or on the wide face as a plank
    - Light framing: 2" to 4" (51 to 100) thick and 2" to 4" (51 to 100) wide, intended for use where high strength values are not required
    - Decking: 2" to 4" (51 to 100) thick and 4" (100) or more wide, graded primarily with respect to bending strength when loaded on the wide face
  - Timbers: 5" (125) or more in the least dimension, graded for strength and serviceability, and often stocked in green, undressed condition
    - Beams and stringers: at least 5" (125) thick and a width more than 2" (51) greater than the thickness, graded primarily with respect to bending strength when loaded on the narrow face
    - Posts and timbers: 5" x 5" (125 x 125) or larger and a width not more than 2" (51) greater than the thickness, graded primarily for use as columns carrying an axial load
- Structural lumber: dimension lumber and timbers graded either by visual inspection or mechanically on the basis of strength and intended use
- Factory and shop lumber: sawn or selected primarily for further manufacture into doors, windows, and millwork, and graded according to the amount of usable wood that will produce cuttings of a specified size and quality

Lumber is specified by species and grade. Each piece of lumber is graded for structural strength and appearance. Structural lumber may be graded visually by trained inspectors according to quality-reducing characteristics that affect strength, appearance, or utility, or by a machine that flexes a test specimen, measures its resistance to bending, calculates its modulus of elasticity, and electronically computes the appropriate stress grade, taking into account such factors as the effects of knots, slope of grain, density, and moisture content.

- Each piece of lumber has a grademark indicating the assigned stress grade, mill of origin, moisture content at time of manufacture, species or species group, and the grading authority.
- Stress grade is a grade of structural lumber for which a set of base values and corresponding modulus of elasticity is established for a species or group of species by a grading agency.

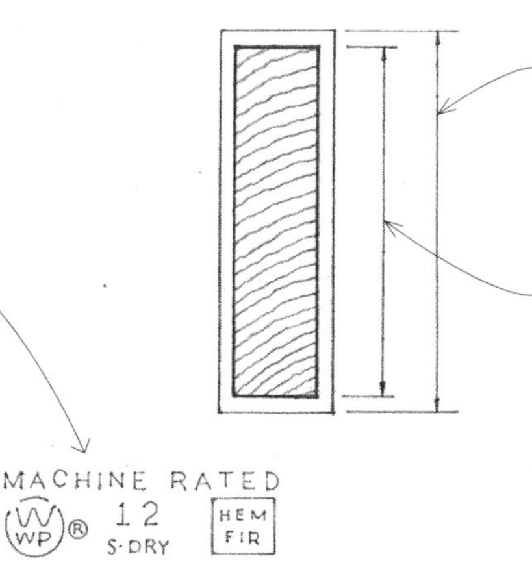

- Design value: any of the allowable unit stresses for a species and grade of structural lumber obtained by modifying the base value by factors related to size and conditions of use

=

- Base value: any of the allowable unit stresses for bending, compression perpendicular and parallel to grain, tension parallel to grain, horizontal shear, and corresponding modulus of elasticity, established by a grading agency for various species and grades of structural lumber

x

- Base values must be adjusted first for size and then for conditions of use. Size-adjusted values are increased for repetitive members and members subject to short-term loading, and decreased for members exceeding a moisture content of 19% in use.

- Lumber is measured in board feet; 1 board foot is equal to the volume of a piece whose nominal dimensions are 12" (305) square and 1" (25) thick.
- Nominal dimensions are the dimensions of a piece of lumber before drying and surfacing, used for convenience in defining size and computing quantity. Nominal dimensions are always written without inch marks (").
- Dressed sizes are the actual dimensions of a piece of lumber after seasoning and surfacing, from 3/8" to 3/4" (10 to 19) less than the nominal dimensions.
- For dressed sizes:
  - Subtract 1/4" (6) from nominal dimensions up to 2" (51);
  - Subtract 1/2" (13) from nominal dimensions of 2" to 6" (51 to 150);
  - Subtract 3/4" (19) from nominal dimensions greater than 6" (150).
- Lumber is generally available in lengths from 6' to 24' (1830 to 7315), in multiples of 2' (610).

Wood panel products are less susceptible to shrinking or swelling, require less labor to install, and make more efficient use of wood resources than solid wood products. The following are the major types of wood panel products.

- Plywood is made by bonding veneers together under heat and pressure, usually with the grain of adjacent plies at right angles to each other and symmetrical about the center ply.
- Gradestamp is a trademark of the American Plywood Association (APA), stamped on the back of a structural wood panel product to identify the panel grade, thickness, span rating, exposure durability classification, mill number, and National Research Board (NRB) report number.

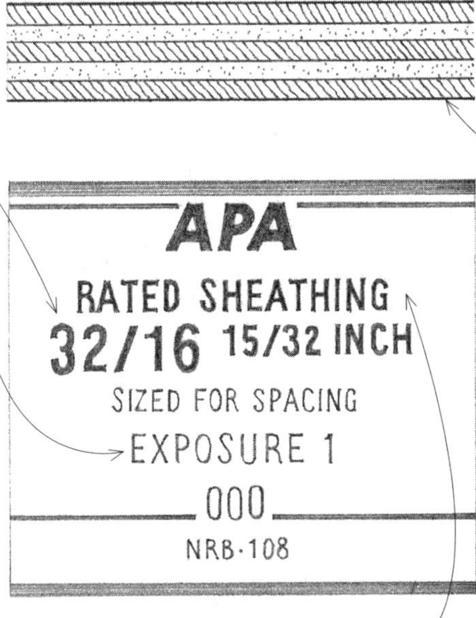

- Span rating specifies the maximum recommended center-to-center spacing in inches of the supports for a structural wood panel spanning with its long dimension across three or more supports.
- Exposure durability is a classification of a wood panel product according to its ability to withstand exposure to weather or moisture without weakening or warping.
- Exterior: structural wood panels manufactured with a waterproof glueline for use as siding or other continuously exposed applications
- Exposure 1: structural wood panels manufactured with an exterior glueline for use in protected construction subject to repeated wetting
- Exposure 2: structural wood panels manufactured with an intermediate glueline for use in fully protected construction subject to a minimum of wetting
- Panel grade identifies the intended use or the veneer grade of a wood panel product.

- Veneer grades define the appearance of a veneer in terms of growth characteristics and the number and size of repairs that may be made during manufacture.
- N-grade: a smooth softwood veneer of all heartwood or all sapwood, free from open defects with only a few well-matched repairs
- A-grade: a smooth, paintable softwood veneer with a limited number of neatly made repairs parallel to the grain
- B-grade: softwood veneer having a solid surface with circular repair plugs, tight knots, and minor splits permitted
- C-grade: softwood veneer having tight knots and knotholes of limited size, synthetic or wood repairs, and discoloration and sanding defects that do not impair the strength of the panel
- C-plugged grade: an improved C-grade softwood veneer having smaller knots and knotholes, some broken grain, and synthetic repairs
- D-grade: a softwood veneer having large knots and knotholes, pitch pockets, and tapering splits

- Engineered grades have relatively high shear strength for loads perpendicular to the panel face and are used for sheathing, subflooring, or in the fabrication of box beams and stressed-skin panels.

- High-density overlay (HDO) is an exterior wood panel having a resin-fiber overlay on both sides providing a smooth, hard, abrasion-resistant surface, used for concrete forms, cabinets, and countertops.
- Medium-density overlay (MDO) is an exterior wood panel having a phenolic or melamine resin overlay on one or both sides, providing a smooth base for painting.
- Specialty panel refers to any of various wood panel products, as grooved or rough-sawn plywood, intended for use as siding or paneling.

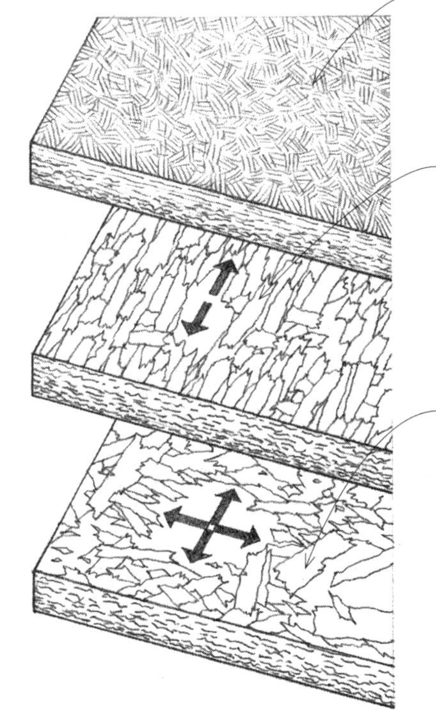

- Particleboard is a nonveneered wood panel product made by bonding small wood particles under heat and pressure, commonly used as a core material for decorative panels and cabinetwork, and as underlayment for floors.
- Oriented strandboard (OSB) is a nonveneered wood panel product commonly used for sheathing and as subflooring, made by bonding layers of long, thin wood strands under heat and pressure using a waterproof adhesive. The surface strands are aligned parallel to the long axis of the panel, making the panel stronger along its length.
- Waferboard is a nonveneered panel product composed of large, thin wood flakes bonded under heat and pressure with a waterproof adhesive. The planes of the wafers are generally oriented parallel to the plane of the panel but their grain directions are random, making the panel approximately equal in strength and stiffness in all directions in the plane of the panel.

Plastics are any of the numerous synthetic or natural organic materials that are mostly thermoplastic or thermosetting polymers of high molecular weight and that can be molded, extruded, or drawn into objects, films, or filaments. As a class, plastics are tough, resilient, lightweight, and resistant to corrosion and moisture. Many plastics also emit gases harmful to the respiratory system and release toxic fumes when burned.

While there are many types of plastics with a wide range of characteristics, they can be divided into two basic categories:

- Thermosetting plastics go through a pliable stage, but once they are set or cured, they become permanently rigid and cannot be softened again by reheating.
- Thermoplastics are capable of softening or fusing when heated without a change in any inherent properties, and of hardening again when cooled.

In the table below are listed the plastics that are commonly used in construction and their primary uses.

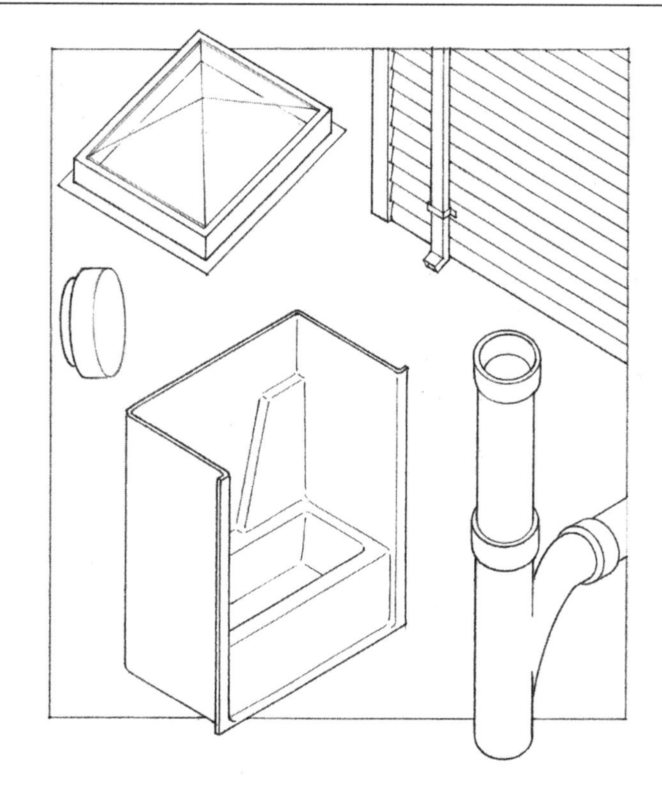

Thermosetting Plastics	Uses
Epoxies (EP)	Adhesives and surface coatings
Melamines (MF)	High-pressure laminates, molded products, adhesives, coatings
Phenolics (PF)	Electrical parts, laminates, foam insulation, adhesives, coatings
Polyesters	Fiberglass-reinforced plastics, skylights, plumbing fixtures, films
Polyurethanes (UP)	Foam insulation, sealants, adhesives, coatings
Silicones (SI)	Waterproofing, lubricants, adhesives, synthetic rubber

Thermoplastics	Uses
Acrylonitrile-butadiene-styrene (ABS)	Pipe and pipe fittings, door hardware
Acrylics (Polymethylmethacrylate – PMMA)	Glazing, adhesives, caulking, latex paints
Cellulosics (Cellulose acetate-butyrate – CAB)	Pipe and pipe fittings, adhesives
Nylons (Polyamides – PA)	Synthetic fibers and filaments, hardware
Polycarbonates (PC)	Safety glazing, lighting fixtures, hardware
Polyethylene (PE)	Dampproofing, vapor retarder, electrical insulation
Polypropylene (PP)	Pipe fittings, electrical insulation, carpeting fibers
Polystyrene (PS)	Lighting fixtures, foam insulation
Vinyls (Polyvinyl chloride – PVC)	Flooring, siding, gutters, window frames, insulation, piping

CSI MasterFormat 06 50 00: Structural Plastics
CSI MasterFormat 06 60 00: Plastic Fabrications

Glass is a hard, brittle, chemically inert substance produced by fusing silica together with a flux and a stabilizer into a mass that cools to a rigid condition without crystallization. It is used in building construction in various forms. Foamed or cellular glass is used as rigid, vaporproof thermal insulation. Glass fibers are used in textiles and for material reinforcement. In spun form, glass fibers form glass wool, which is used for acoustical and thermal insulation. Glass block is used to control light transmission, glare, and solar radiation. Glass, however, is used most commonly to glaze the window, sash, and skylight openings of buildings.

The three major types of flat glass are the following:

- Sheet glass is fabricated by drawing the molten glass from a furnace (drawn glass), or by forming a cylinder, dividing it lengthwise, and flattening it (cylinder glass). The fire-polished surfaces are not perfectly parallel, resulting in some distortion of vision. To minimize this distortion, glass should be glazed with the wave distortion running horizontally.
- Plate glass is formed by rolling molten glass into a plate that is subsequently ground and polished after cooling. Plate glass provides virtually clear, undistorted vision.
- Float glass is manufactured by pouring molten glass onto a surface of molten tin and allowing it to cool slowly. The resulting flat, parallel surfaces minimize distortion and eliminate the need for grinding and polishing. Float glass is the successor to plate glass and accounts for the majority of flat-glass production.

Other types of glass include the following:

- Annealed glass is cooled slowly to relieve internal stresses.
- Heat-strengthened glass is annealed glass that is partially tempered by a process of reheating and sudden cooling. Heat-strengthened glass has about twice the strength of annealed glass of the same thickness.
- Tempered glass is annealed glass that is reheated to just below the softening point and then rapidly cooled to induce compressive stresses in the surfaces and edges of the glass and tensile stresses in the interior. Tempered glass has three to five times the resistance of annealed glass to impact and thermal stresses but cannot be altered after fabrication. When fractured, it breaks into relatively harmless pebble-sized particles.
- Laminated or safety glass consists of two or more plies of flat glass bonded under heat and pressure to interlayers of polyvinyl butyral resin that retains the fragments if the glass is broken. Security glass is laminated glass that has exceptional tensile and impact strength.
- Wired glass is flat or patterned glass having a square or diamond wire mesh embedded within it to prevent shattering in the event of breakage or excessive heat. Wired glass is considered a safety glazing material and may be used to glaze fire doors and windows.
- Patterned glass has a linear or geometric surface pattern formed in the rolling process to obscure vision or to diffuse light.
- Obscure glass has one or both sides acid-etched or sand-blasted to obscure vision. Either process weakens the glass and makes it difficult to clean.
- Spandrel glass is an opaque glass for concealing the structural elements in curtain wall construction, produced by fusing a ceramic frit to the interior surface of tempered or heat-strengthened glass.

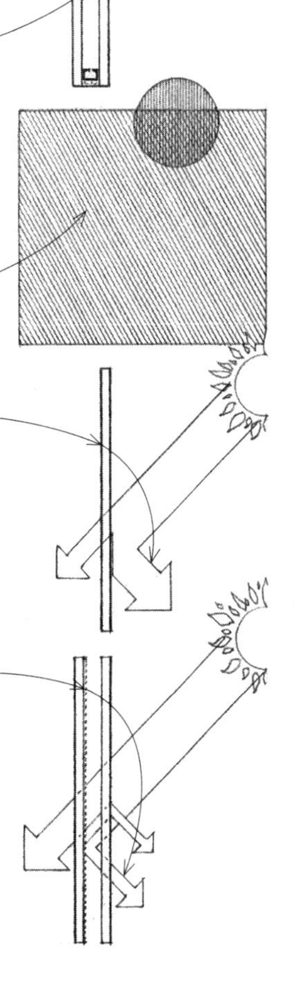

- Insulating glass is a glass unit consisting of two or more sheets of glass separated by a hermetically sealed air space to provide thermal insulation and restrict condensation; glass edge units have a 3/16" (5) air space; metal edge units have a 1/4" or 1/2" (6 or 13) air space.
- Tinted or heat-absorbing glass has a chemical admixture to absorb a portion of the radiant heat and visible light that strike it. Iron oxide gives the glass a pale blue-green tint; cobalt oxide and nickel impart a grayish tint; selenium infuses a bronze tint.
- Reflective glass has a thin, translucent metallic coating to reflect a portion of the light and radiant heat that strike it. The coating may be applied to one surface of single glazing, in between the plies of laminated glass, or to the exterior or interior surfaces of insulating glass.
- Low-emissivity (low-e) glass transmits visible light while selectively reflecting the longer wavelengths of radiant heat, produced by depositing a low-e coating either on the glass itself or over a transparent plastic film suspended in the sealed air space of insulating glass.

Glass Product	Type	Nominal Thickness inch	(mm)	Maximum Area inches	(mm)	Weight psf*	Notes
Sheet Glass	AA, A, B	SS 3/32	(2.4)	60 x 60	(1525 x 1525)	1.22	• Verify maximum sizes with glass manufacturer.
		DS 1/8	(3.2)	60 x 80	(1525 x 2030)	1.63	• Any glass 1/8" (3.2) or thicker can be tempered, except for patterned or wired glass; tempered glass can also be incorporated into insulating or laminated glass units.
Float or Plate	Mirror	1/4	(6.4)	75 sf	(7 m^2)	3.28	• Reflective coatings may be applied to float, plate, tempered, laminated, or insulating glass.
	Glazing	1/8	(3.2)	74 x 120	(1880 x 3050)	1.64	
		1/4	(6.4)	128 x 204	(3250 x 5180)	3.28	
Heavy Float or Plate	Glazing	5/16	(7.9)	124 x 200	(3150 x 5080)	4.10	
		3/8	(9.5)	124 x 200	(3150 x 5080)	4.92	
		1/2	(12.7)	120 x 200	(3050 x 5080)	6.54	
		5/8	(15.9)	120 x 200	(3050 x 5080)	8.17	
		3/4	(19.1)	115 x 200	(2920 x 5080)	9.18	
		7/8	(22.2)	115 x 200	(2920 x 5080)	11.45	
Patterned Glass	Various patterns	1/8	(3.2)	60 x 132	(1525 x 3355)	1.60	
		7/32	(5.6)	60 x 132	(1525 x 3355)	2.40	
Wired Glass	Polished-mesh	1/4	(6.4)	60 x 144	(1525 x 3660)	3.50	
	Patterned-mesh	1/4	(6.4)	60 x 144	(1525 x 3660)	3.50	
	Parallel wires	7/32	(5.6)	54 x 120	(1370 x 3050)	2.82	
		1/4	(6.4)	60 x 144	(1525 x 3660)	3.50	
		3/8	(9.5)	60 x 144	(1525 x 3660)	4.45	
Laminated Glass	(2) 1/8" float	1/4	(6.4)	72 x 120	(1830 x 3050)	3.30	
	Heavy float	3/8	(9.5)	72 x 120	(1830 x 3050)	4.80	
		1/2	(12.7)	72 x 120	(1830 x 3050)	6.35	
		5/8	(15.9)	72 x 120	(1830 x 3050)	8.00	
Tinted Glass	Bronze	1/8	(3.2)	35 sf	(3 m^2)	1.64	• Solar energy transmission reduced 35% to 75%
		3/16	(4.8)	120 x 144	(3050 x 3660)	2.45	• Visible light transmission reduced 32% to 72%
		1/4	(6.4)	128 x 204	(3250 x 5180)	3.27	
		3/8	(9.5)	124 x 200	(3150 x 5080)	4.90	
		1/2	(12.7)	120 x 200	(3050 x 5080)	6.54	
	Gray	1/8	(3.2)	35 sf	(3 m^2)	1.64	
		3/16	(4.8)	120 x 144	(3050 x 3660)	2.45	
		1/4	(6.4)	128 x 204	(3250 x 5180)	3.27	
		3/8	(9.5)	124 x 200	(3150 x 5080)	4.90	
		1/2	(12.7)	120 x 200	(3050 x 5080)	6.54	
Insulating Glass	Glass edge units						
(2) 3/32" sheets	3/16" air space	3/8	(9.5)	10 sf	(0.9 m^2)	2.40	• R-value = 1.61
(2) 1/8" sheets	3/16" air space	7/16	(11.1)	24 sf	(2.2 m^2)	3.20	• R-value = 1.61
	Metal edge units						
(2) 1/8"	1/4" air space	1/2	(12.7)	22 sf	(2.0 m^2)	3.27	• R-value = 1.72
Sheet, plate, or float	1/2" air space	3/4	(19.1)	22 sf	(2.0 m^2)	3.27	• R-value = 2.04
(2) 3/16"	1/4" air space	5/8	(15.9)	34 sf	(3.2 m^2)	4.90	• R-values for units w/ 1/2" (12.7) air space and low-e coating:
Plate or float	1/2" air space	7/8	(22.2)	42 sf	(3.8 m^2)	4.90	e = 0.20, R = 3.13
(2) 1/4"	1/4" air space	3/4	(19.1)	50 sf	(4.6 m^2)	6.54	e = 0.40, R = 2.63
Plate or float	1/2" air space	1	(25.4)	70 sf	(6.5 m^2)	6.54	e = 0.60, R = 2.33

*1 psf = 47.88 Pa

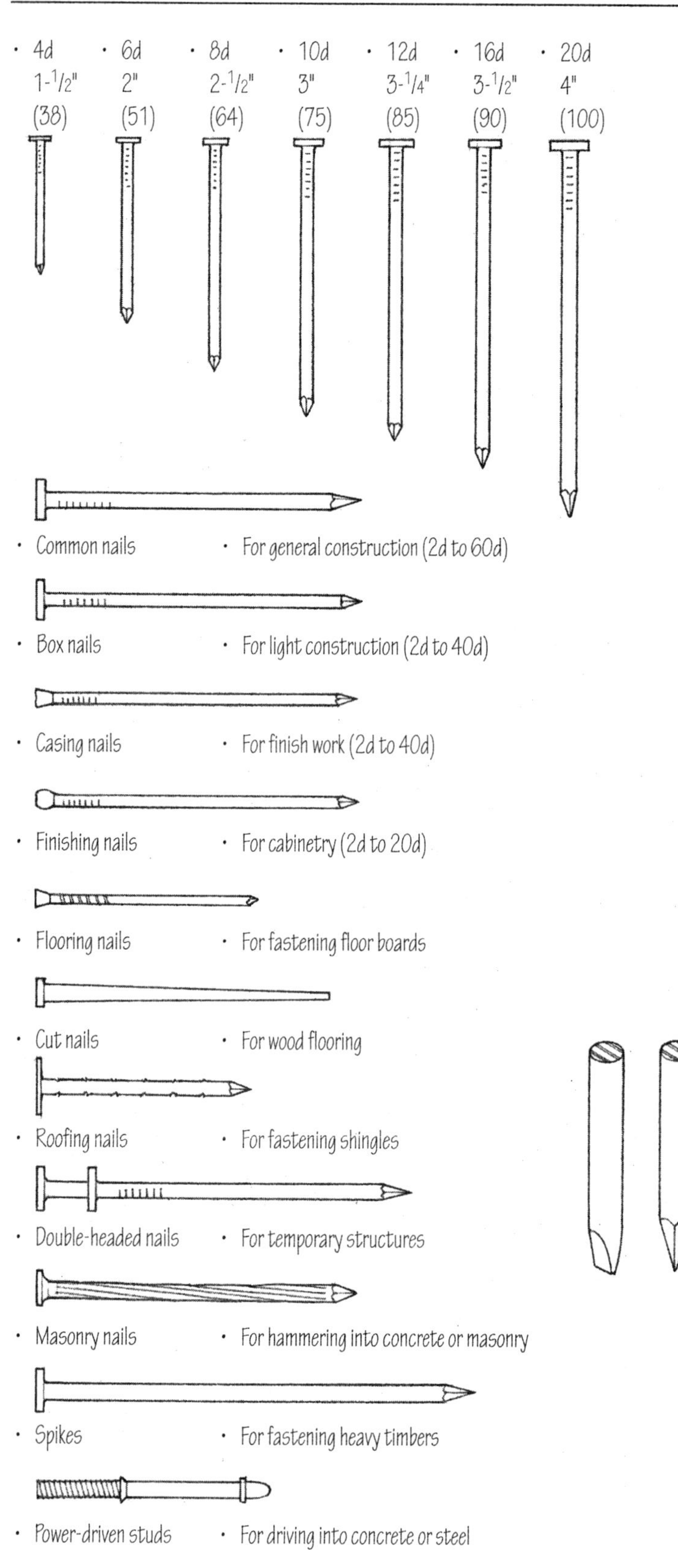

Nails are straight, slender pieces of metal having one end pointed and the other enlarged and flattened for hammering into wood or other building materials as a fastener.

Material

- Nails are usually of mild steel, but may also be of aluminum, copper, brass, zinc, or stainless steel.
- Tempered, high-carbon steel nails are used for greater strength in masonry applications.
- The type of metal used should be compatible with the materials being secured to avoid loss of holding power and prevent staining of the materials.

Length and Diameter of the Shank

- Nail lengths are designated by the term penny (d).
- Nails range in length from 2d, about 1" (25) long, to 60d, about 6" (150) long.
- Nail length should be about 3 x thickness of the material being secured.
- Large diameter nails are used for heavy work while lighter nails are used for finish work; thinner nails are used for hardwood rather than for softwood.

Form of the Shank

- For greater gripping strength, nail shafts may be serrated, barbed, threaded, fluted, or twisted.
- Nail shafts may be cement-coated for greater resistance to withdrawal, or be zinc-coated for corrosion resistance.

Nail Heads

- Flat heads provide the largest amount of contact area and are used when exposure of the heads is acceptable.
- The heads of finish nails are only slightly larger than the shaft and may be tapered or cupped.
- Double-headed nails are used for easy removal in temporary construction and concrete formwork.

Nail Points

- Most nails have diamond-shaped points.
- Sharp-pointed nails have greater holding strength but may tend to split some woods; blunt points should be used for easily split woods.

Power-Driven Fasteners

- Pneumatic nailers and staplers, driven by a compressor, are capable of fastening materials to wood, steel, or concrete.
- Powder-driven fasteners use gunpowder charges to drive a variety of studs into concrete or steel.

CSI MasterFormat 06 06 00: Schedules for Wood, Plastics, and Composites

Screws

Screws are metal fasteners having tapered, helically threaded shanks and slotted heads, designed to be driven into wood or the like by turning, as with a screwdriver. Because of their threaded shafts, screws have greater holding power than nails, and are more easily removable. The more threads they have per inch, the greater their gripping strength. Screws are classified by use, type of head, material, length, and diameter.

- Material: steel, brass, aluminum, bronze, stainless steel
- Lengths: 1/2" to 6" (13 to 150)
- Diameters: up to 24 gauge

The length of a wood screw should be about 1/8" (3) less than the combined thickness of the boards being joined, with 1/2 to 2/3 of the screw's length penetrating the base material. Fine-threaded screws are generally used for hardwoods while coarse-threaded ones are used for softwoods.

Holes for screws should be predrilled and be equal to the base diameter of the threads. Some screws, such as self-tapping and drywall screws, are designed to tap corresponding female threads as they are driven.

Bolts

Bolts are threaded metal pins or rods, usually having a head at one end, designed to be inserted through holes in assembled parts and secured by a mating nut. Carriage bolts are used where the head may be inaccessible during tightening. Lag bolts or screws are used in areas inaccessible to the placement of a nut or where an exceptionally long bolt would be needed to penetrate a joint fully.

- Lengths: 3/4" to 30" (75 to 760)
- Diameters: 1/4" to 1-1/4" (6 to 32)

- Washers are perforated disks of metal, rubber, or plastic, used under the head of a nut or bolt or at a joint to distribute pressure, prevent leakage, relieve friction, or insulate incompatible materials.
- Lock washers are specially constructed to prevent a nut from shaking loose.
- Load-indicating washers have small projections that are progressively flattened as a bolt is tightened, the gap between the head or nut and the washer indicating the tension in the bolt.

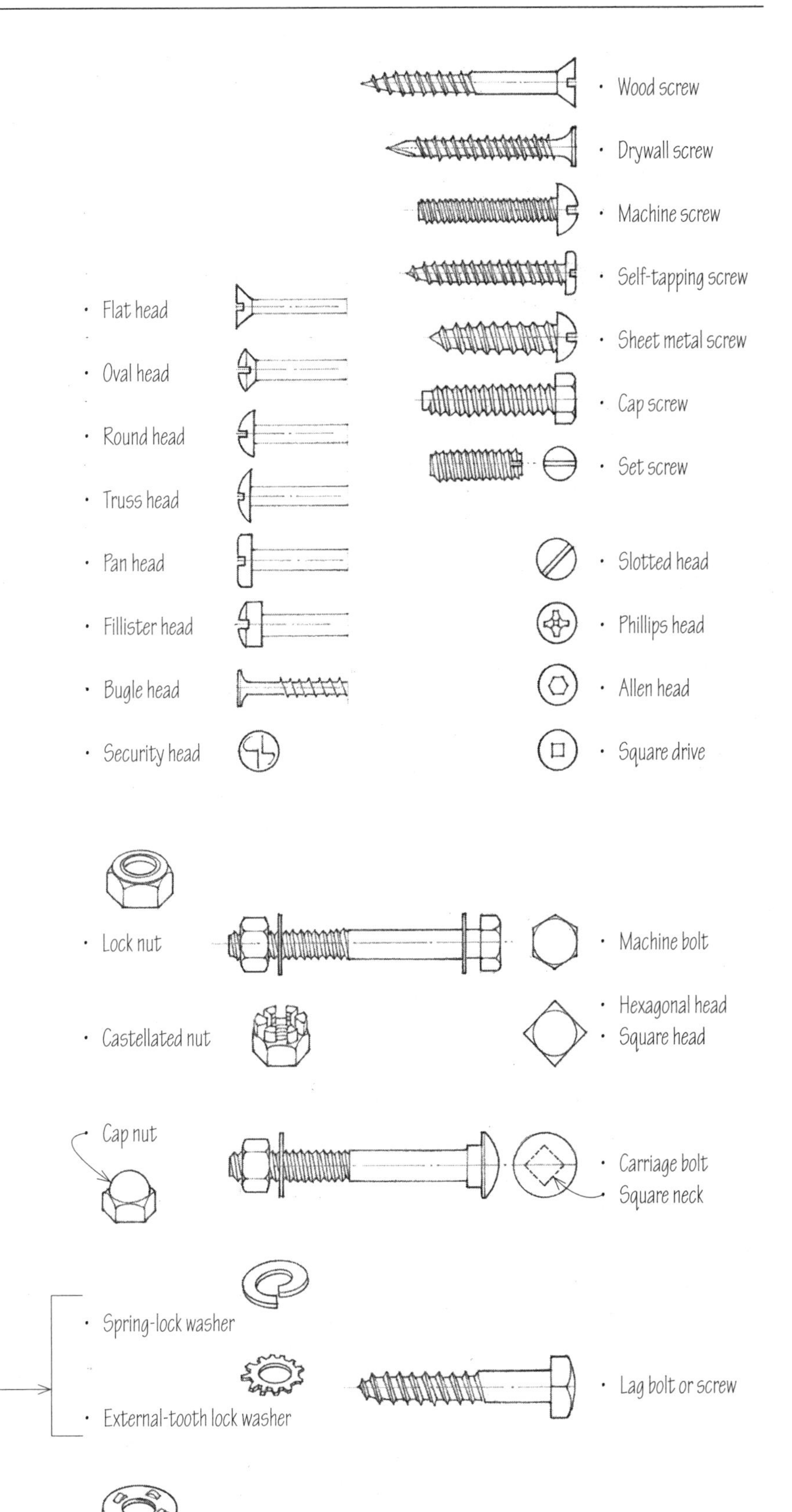

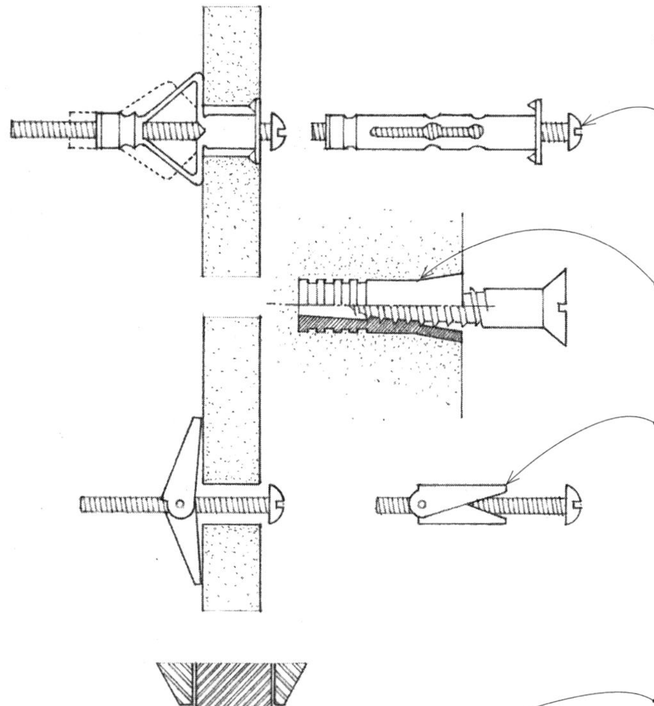

- Expansion bolts are anchor bolts having a split casing that expands mechanically to engage the sides of a hole drilled in masonry or concrete.
- Molly is a trademark for a brand of expansion bolt having a split, sleevelike sheath threaded so that turning the bolt draws the ends of the sheath together and spreads the sides to engage a hole drilled in masonry or the inner surface of a hollow wall.

- Expansion shields are lead or plastic sleeves inserted into a predrilled hole and expanded by driving a bolt or screw into it.

- Toggle bolts are used to fasten materials to plaster, gypsum board, and other thin wall materials. They have two hinged wings that close against a spring when passing through a predrilled hole and open as they emerge to engage the inner surface of a hollow wall.

- Rivets are metal pins that are used for permanently joining two or more structural steel members by passing a headed shank through a hole in each piece and hammering down the plain end to form a second head. Their use has been largely superseded by the less labor-intensive techniques of bolting or welding.
- Explosive rivets, used when a joint is accessible from one side only, have an explosive-filled shank that is detonated by striking the head with a hammer to expand the shank on the far side of the hole.

Common types of adhesives:

- Animal or fish glues are primarily for indoor use where temperature and humidity do not vary greatly; they may be weakened by exposure to heat or moisture.
- White or polyvinyl glue sets quickly, does not stain, and is slightly resilient.
- Epoxy resins are extremely strong and may be used to secure both porous and nonporous materials; they may dissolve some plastics. Unlike other adhesives, epoxy glues will set at low temperatures and under wet conditions.
- Resorcin resins are strong, waterproof, and durable for outdoor use, but they are flammable and their dark color may show through paint.
- Contact cement forms a bond on contact and therefore does not require clamping. It is generally used to secure large sheet materials such as plastic laminate.

Adhesives

Adhesives are used to secure the surfaces of two materials together. Numerous types of adhesives are available, many of them being tailor-made for use with specific materials and under specified conditions. They may be supplied in the form of a solid, liquid, powder, or film; some require a catalyst to activate their adhesive properties. Always follow the manufacturer's recommendations in the use of an adhesive. Important considerations in the selection of an adhesive include:

- Strength: Adhesives are usually strongest in resisting tensile and shear stresses and weakest in resisting cleavage or splitting stresses.
- Curing or setting time: This ranges from immediate bonding to curing times of up to several days.
- Setting temperature range: Some adhesives will set at room temperature while others require baking at elevated temperatures.
- Method of bonding: Some adhesives bond on contact while others require clamping or higher pressures.
- Characteristics: Adhesives vary in their resistance to water, heat, sunlight, and chemicals, as well as their aging properties.

The purpose of a coating is to protect, preserve, or visually enhance the surface to which it is applied. The principal types of coating are paints, stains, and varnishes.

Paints

Paint is a mixture of a solid pigment suspended in a liquid vehicle and applied as a thin, usually opaque coating to a surface for protection and decoration.

- Pigment: a finely ground, insoluble substance suspended in a liquid vehicle to impart color and opacity to the coating;

+

- Vehicle: a liquid in which pigment is dispersed before being applied to a surface in order to control consistency, adhesion, gloss, and durability.
  - Binder is the nonvolatile part of a paint vehicle that bonds particles of pigment into a cohesive film during the drying process.
  - Solvent or thinner is the volatile part of a paint vehicle that ensures the desired consistency for application by brush, roller, or spray.

- Primers are basecoats applied to a surface to improve the adhesion of subsequent coats of paint or varnish.
- Sealers are basecoats applied to a surface to reduce the absorption of subsequent coats of paint or varnish, or to prevent bleeding through the finish coat.

- Oil paints use a drying oil that oxidizes and hardens to form a tough elastic film when exposed in a thin layer to air.
- Alkyd paints have as a binder an alkyd resin, such as a chemically modified soy or linseed oil.
- Latex paints have as a binder an acrylic resin that coalesces as water evaporates from the emulsion.

- Epoxy paints have an epoxy resin as a binder for increased resistance to abrasion, corrosion, and chemicals.
- Rust-inhibiting paints and primers are specially formulated with anticorrosive pigments to prevent or reduce the corrosion of metal surfaces.
- Fire-retardant paints are specially formulated with silicone, polyvinyl chloride, or other substance to reduce the flame-spread of a combustible material.
- Intumescent coatings, when exposed to the heat of a fire, swell to form a thick insulating layer of inert foam that retards flame spread and combustion.
- Heat-resistant paints are specially formulated with silicone resins to withstand high temperatures.

Stains

Stain is a solution of dye or suspension of pigment in a vehicle, applied to penetrate and color a wood surface without obscuring the grain.

- Penetrating stains permeate a wood surface, leaving a very thin film on the surface.
- Water stain is a penetrating stain made by dissolving dye in a water vehicle.
- Spirit stain is a penetrating stain made by dissolving dye in an alcohol or spirit vehicle.

- Pigmented or opaque stain is an oil stain containing pigments capable of obscuring the grain and texture of a wood surface.
- Oil stain is made by dissolving dye or suspending pigment in a drying oil or oil varnish vehicle.

Varnishes

Varnish is a liquid preparation consisting of a resin dissolved in an oil (oil varnish) or in alcohol (spirit varnish), that when spread and allowed to dry forms a hard, lustrous, usually transparent coating.

- Spar or marine varnish is a durable, weather-resistant varnish made from durable resins and linseed or tung oil.
- Polyurethane varnish is an exceptionally hard, abrasion-resistant, and chemical-resistant varnish made from a plastic resin of the same name.
- Lacquer refers to any of various clear or colored synthetic coatings consisting of nitrocellulose or other cellulose derivative dissolved in a solvent that dries by evaporation to form a high-gloss film.
- Shellac is a spirit varnish made by dissolving purified lac flakes in denatured alcohol.

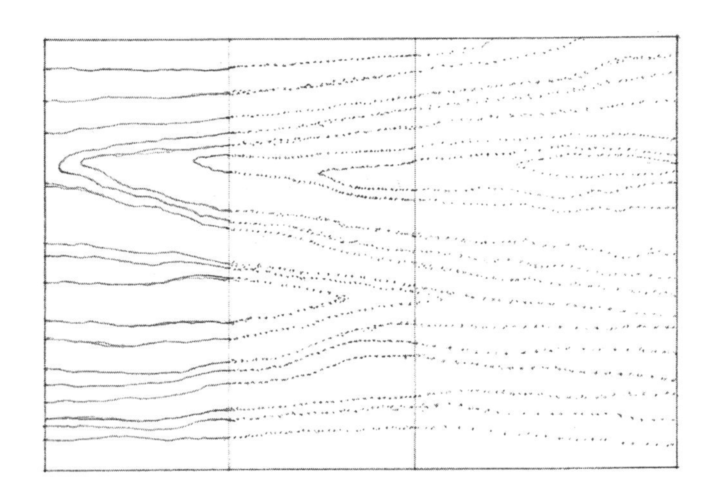

CSI MasterFormat 09 90 00: Painting and Coating

All materials to receive paint or other coating must be properly prepared and primed to ensure adhesion of the coating to their surfaces and to maximize the life of the coating. In general, surfaces should be dry and free of contaminants, such as dirt, grease, moisture, and mildew. The following are recommendations for various materials:

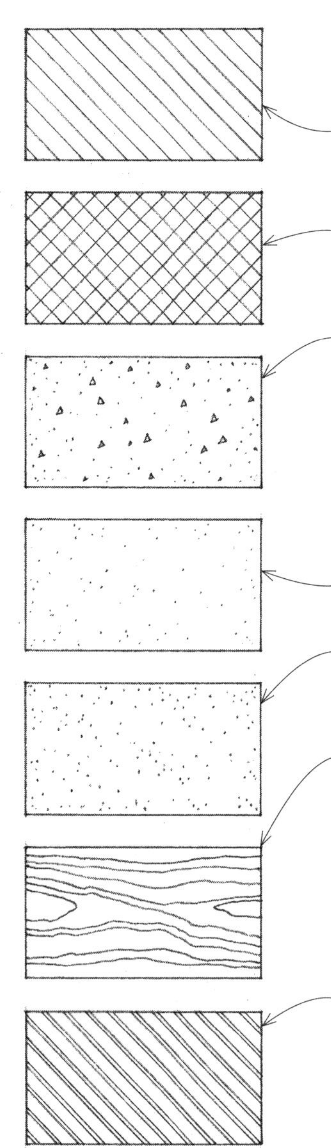

- Brick surface should have dirt, loose mortar, efflorescence, and other foreign matter removed by wire brushing, air pressure, or steam cleaning. Seal with a latex primer-sealer or a clear silicone water-repellant.
- Concrete masonry should be thoroughly dry and free of dirt and loose or excess mortar. Porous surfaces may require a block filler or cement grout primer if the acoustical value of a rough surface is not important.
- Concrete surface should be well-cured and free of dirt, form oils, and curing compounds. Porous surfaces may require a block filler or cement grout primer. Prime grouted surfaces with a latex, alkyd, or oil primer-sealer. Concrete surfaces may also be sealed with a clear silicone water-repellant.
- Concrete floors should be free of dirt, wax, grease, and oils, and should be etched with a muriatic acid solution to improve adhesion of the coating. Prime with an alkali-resistant coating.
- Gypsum board surfaces should be clean and dry. Use a latex primer-sealer to avoid raising the fibers of the paper surface.
- Plaster and stucco surfaces should be allowed to dry thoroughly and be completely cured. Prime with a latex, alkyd, or oil primer-sealer. Fresh plaster should be primed with an alkali-resistant coating.
- Wood should be clean, dry, well-seasoned lumber. Knots and pitch stains should be sanded and sealed before priming. Surfaces to be painted should be primed or sealed to stabilize the moisture content of the wood and prevent the absorption of succeeding coats; stains and some paints may be self-priming. All nail holes, cracks, and other small holes should be filled after the prime coat.
- Old paint surfaces should be clean, dry, and roughened by sanding or washing with a detergent solution.
- Ferrous metal surfaces should be free of rust, metal burrs, and foreign matter. Clean with solvents or by wire brushing, sandblasting, flame cleaning, or pickling with acids. Prime with a rust-inhibitive primer.
- Galvanized iron should have all grease, residue, and corrosion removed with a solvent or chemical wash. Prime with a zinc oxide or portland cement paint. If weathered, galvanized iron should be treated as a ferrous metal.

In addition to the surface preparation and priming required, other considerations in the selection of a coating include:

- Compatibility of the coating with the surface to which it is applied
- The method of application and drying time required
- Conditions of use and the required resistance to water, heat, sunlight, temperature variation, mildew, chemicals, and physical abrasion
- The possible emission of harmful volatile organic compounds

A

APPENDIX

Our body dimensions, and the way we move through and perceive space, are prime determinants of the scale, proportions, and spatial layout of a building. It should be noted that there is a difference between the structural dimensions of our bodies and those dimensional requirements that result from how we reach for something on a shelf, sit down at a table, walk down a stairway, or interact with other people. These functional dimensions will vary according to the nature of our activity and the social situation.

Caution should always be exercised when using a set of dimensional tables or illustrations such as these. These are based on average measurements, which may have to be adjusted to satisfy specific user needs. Variations from the norm will always exist due to the differences between men and women, among various age and racial groups, and from one individual to the next.

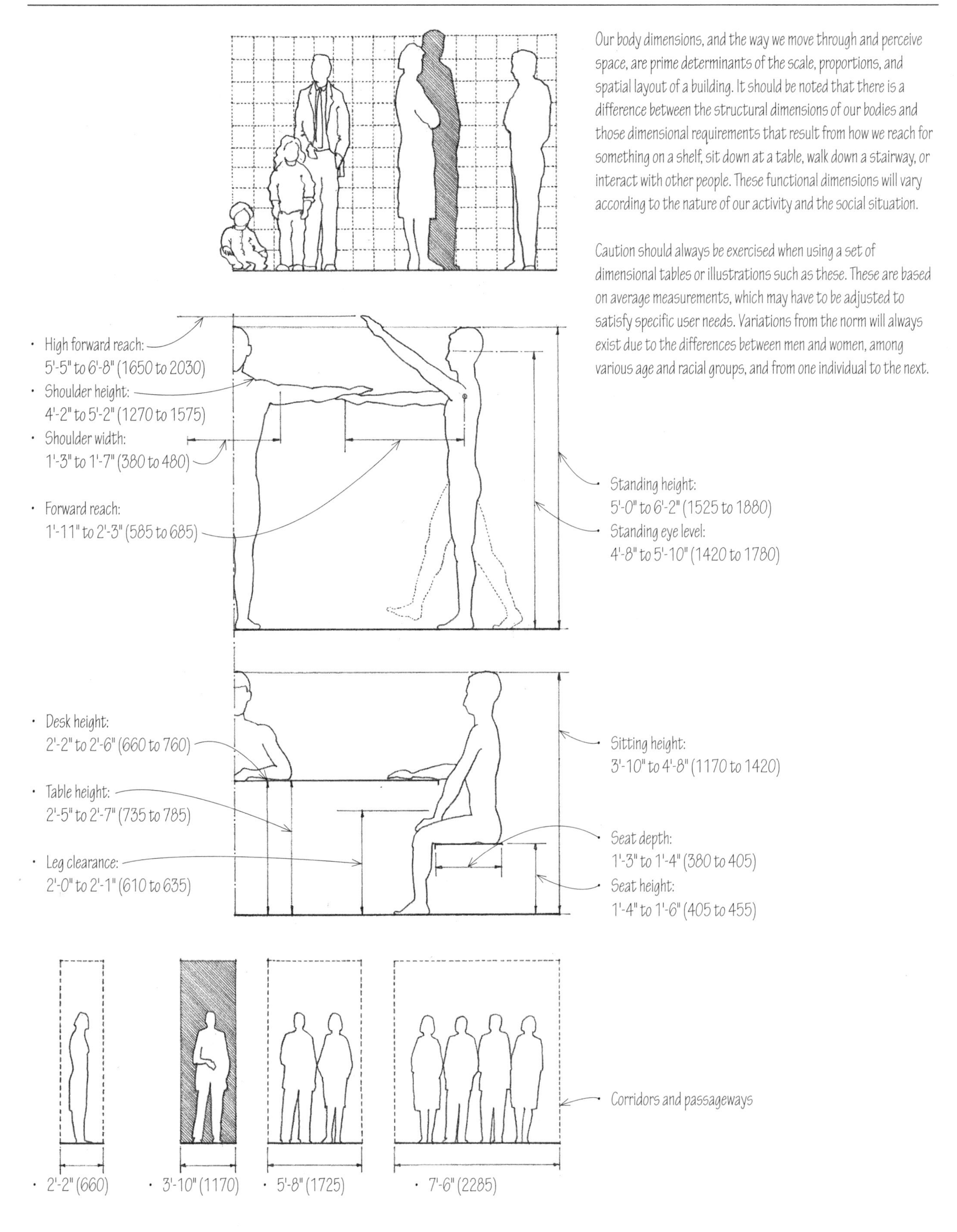

The *Americans with Disabilities Act* (ADA) of 1990 is federal civil-rights legislation requiring that buildings be accessible to persons with physical disabilities and certain defined mental disabilities. The *ADA Accessibility Guidelines* (*2010 ADA Standards for Accessible Design*) are maintained by the U.S. Access Board, an independent federal agency, and the regulations are administered by the U.S. Department of Justice. Federal facilities must comply with standards issued under the *Architectural Barriers Act* (ABA). In its last update, the Access Board harmonized the ADA guidelines with its guidelines for facilities covered by the ABA and published them jointly as the *ADA-ABA Accessibility Guidelines*. Related legislation includes the Federal Fair Housing Act (FFHA) of 1988, which contains Department of Housing and Urban Development (HUD) regulations requiring residential complexes of four or more dwelling units to be adaptable for use by persons with disabilities.

Facilities should be accessible to those confined to a wheelchair and the ambulatory.

- Accessible routes consist of walking surfaces with a maximum slope of 1:20, marked crossings at vehicular roadways, clear floor space at accessible elements, access aisles, ramps, curb ramps, and elevators.
- Floor surfaces should be firm, stable, and slip-resistant.
- Avoid changes in level and the use of stairs.
- Use ramps only where necessary.

Facilities should be identifiable to the blind.

- Use raised lettering, audible warning signals, and textured surfaces to indicate stairs or hazardous openings.

Facilities should be usable.

- Circulations spaces should be adequate for comfortable movement.
- All public facilities should have fixtures designed for use by persons with disabilities.

For ADA Accessibility Guidelines for other building elements or components, see the following:

- Vehicular parking: 1.29
- Doors: 8.03
- Door hardware: 8.17, 8.19, 8.20
- Thresholds: 8.21
- Windows: 8.22
- Stairs and ramps: 9.05, 9.09
- Elevators: 9.16
- Kitchens: 9.22–9.23
- Toilet and bathing facilities: 9.26
- Carpeting: 10.21

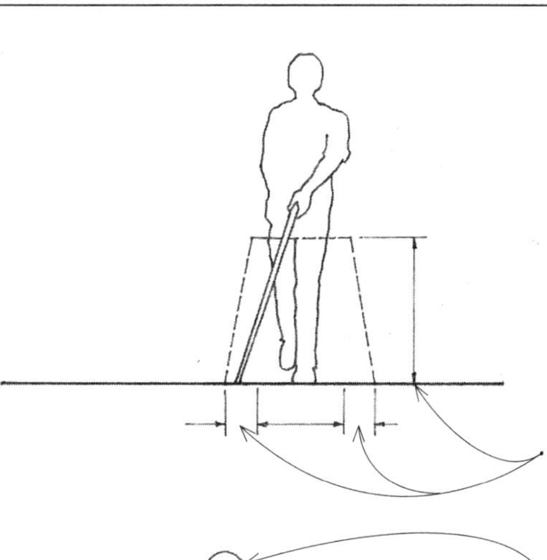

- Cane range: 6" (150) minimum to either side and 27" (685) high
- Up to a 1/4" (6) change in level may be vertical.
- Changes in level from 1/4" to 1/2" (6 to 13) should be beveled with a slope not steeper than 1:2.
- Changes in level greater than 1/2" (13) must be ramped.

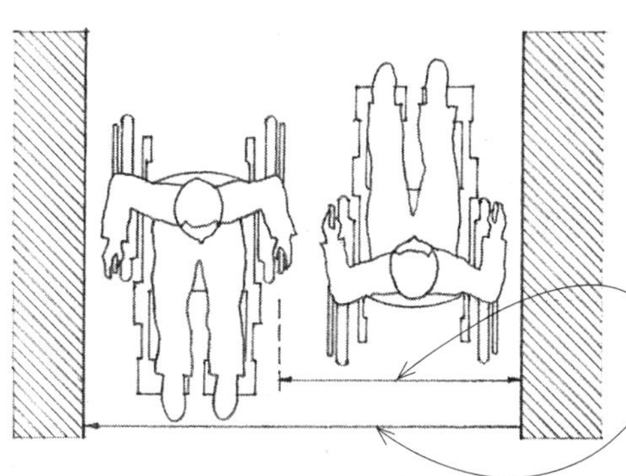

- 36" (915) minimum clear width for passage
- 60" (1525) minimum clear width for two wheelchairs to pass

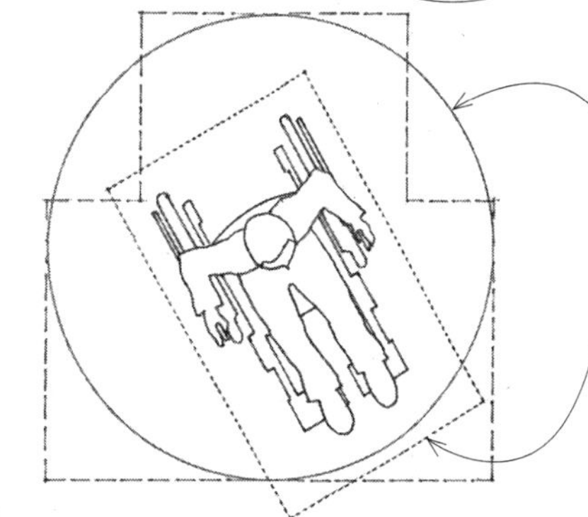

- 60" (1525) minimum clear ø, or a T-shaped space with arms at least 36" (915) wide and 60" (1525) long, to allow a wheelchair to turn
- 30" x 48" (760 x 1220) minimum clear floor space required for either forward or parallel approach to an object

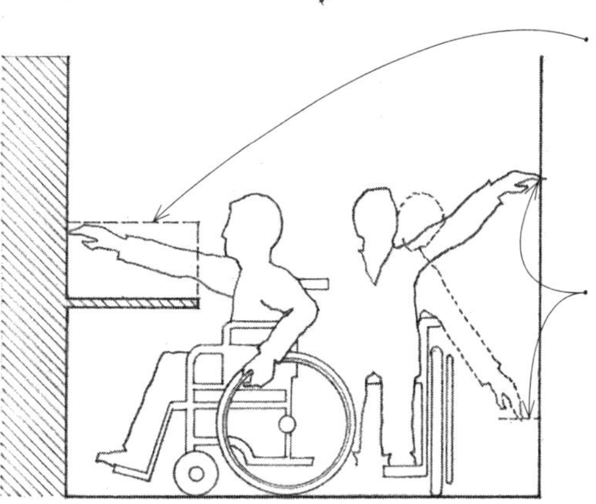

- Maximum reach height of 48" (1220) for reach depths up to 20" (510); maximum reach height of 44" (1120) for reach depths from 20" to 25" (510 to 635)
- 54" (1370) maximum and 15" (380) minimum side reach above the floor

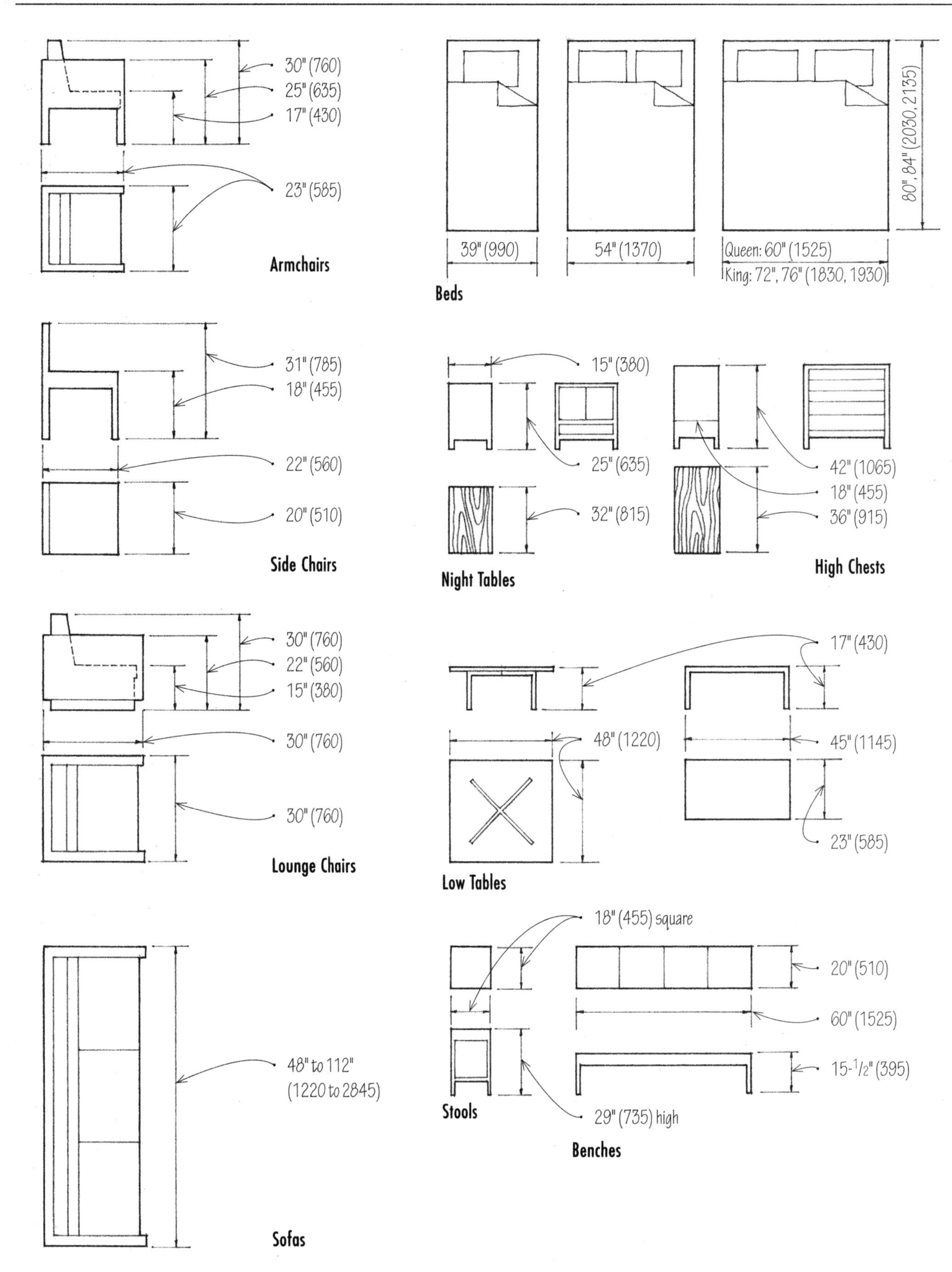
30" (760)
25" (635)
17" (430)
23" (585)
Armchairs
39" (990)
54" (1370)
Queen: 60" (1525)
King: 72", 76" (1830, 1930)
80", 84" (2030, 2135)
Beds
31" (785)
18" (455)
22" (560)
20" (510)
Side Chairs
15" (380)
25" (635)
32" (815)
Night Tables
42" (1065)
18" (455)
36" (915)
High Chests
30" (760)
22" (560)
15" (380)
30" (760)
30" (760)
Lounge Chairs
17" (430)
48" (1220)
45" (1145)
23" (585)
Low Tables
48" to 112"
(1220 to 2845)
Sofas
18" (455) square
20" (510)
60" (1525)
15-1/2" (395)
29" (735) high
Stools
Benches

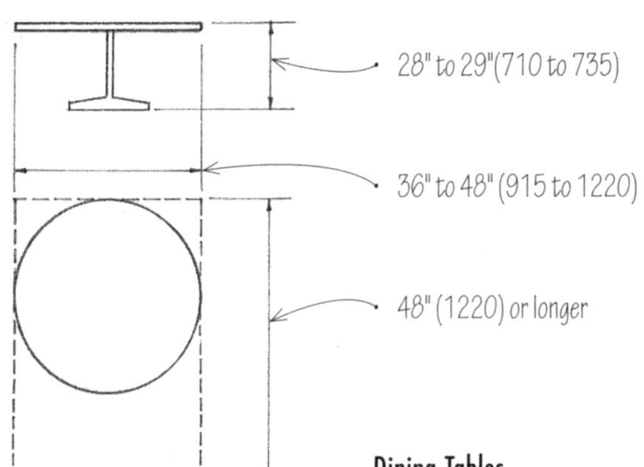

Dining Tables

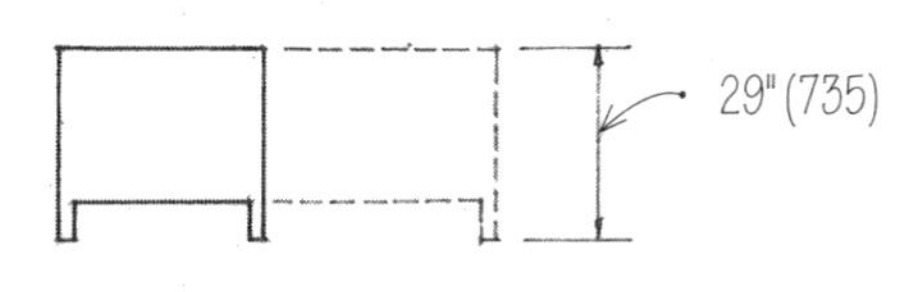

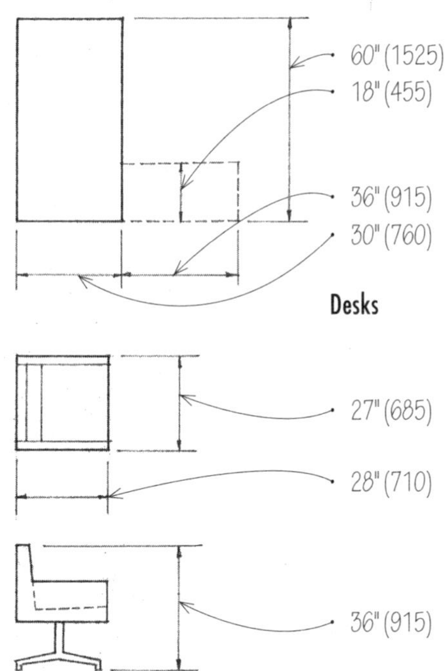

Desks

Swivel Chairs

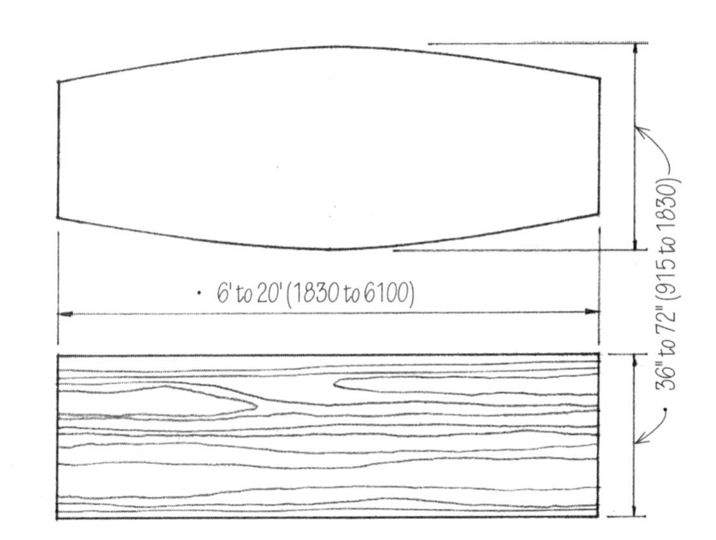

Conference Tables

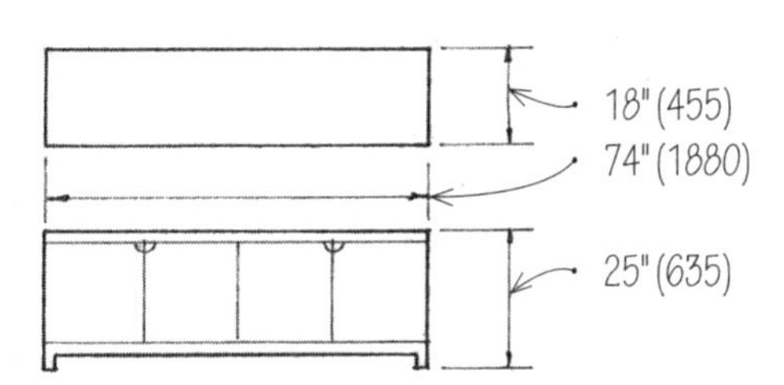

Credenzas

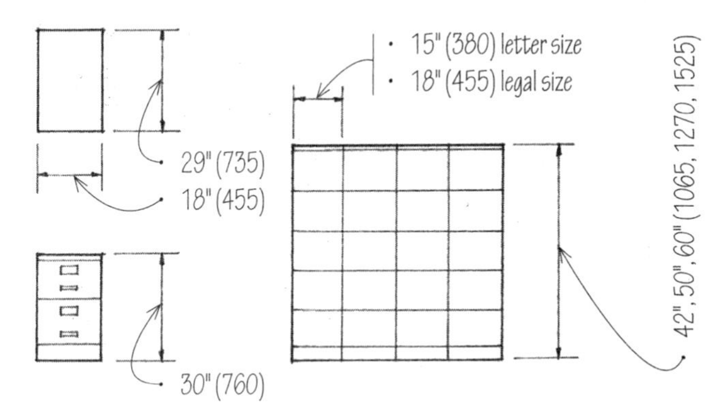

File Cabinets

- All dimensions are typical. Verify with furniture manufacturer.
- Furniture may serve as space-defining elements, define circulation paths, or be built-in or set as objects in space.
- Selection factors include function, comfort, scale, color, and style.

Minimum Uniformly Distributed Loads (psf*)

*1 psf = 47.88 Pa

Assembly Facilities

- Theaters with fixed seating........................60
- Auditoriums and gyms w/ movable seats......... 100
- Corridors and lobbies........................... 100
- Stages.. 150

Libraries

- Reading rooms....................................60
- Book stacks.................................... 150

Offices

- Office spaces....................................80
- Lobbies.. 100

Residential Facilities

- Private dwellings.................................40
- Apartment units, hotel rooms.....................40
- Public rooms................................... 100
- Corridors...60

Schools

- Classrooms......................................40
- Corridors...................................... 100

Sidewalks and Vehicular Drives.................... 250

Stairs, Fire Escapes, Exitways...................... 100

Storage Warehouses

- Light... 125
- Heavy... 250

Manufacturing Facilities........................... 125

Stores

- Retail: first floor............................... 100
- Upper floors.....................................75

Roof Loads

Minimum, not including wind or seismic loads.........20

Roof gardens..................................... 100

- In the design of a building, the assumed live loads should be the maximum expected to be produced by the intended use or activity. In some instances, such as with parking garages, concentrated loads will take precedence.
- Always verify live load requirements with the building code.

Average Weights of Materials (pcf*)

*1 pcf = 16 kg/m^3

Soil, Sand and Gravel

- Cinder...45
- Clay, damp..................................... 110
- Clay, dry...63
- Earth, dry and loose..............................76
- Earth, moist and packed..........................96
- Sand and gravel, dry and loose................... 105
- Sand and gravel, wet........................... 120

Wood

- Cedar..22
- Douglas fir.......................................32
- Hemlock...29
- Maple...42
- Oak, red...41
- Oak, white.......................................46
- Pine, southern...................................29
- Redwood..26
- Spruce..27

Metals

- Aluminum...................................... 165
- Brass, red..................................... 546
- Bronze, statuary................................ 509
- Copper... 556
- Iron, cast...................................... 450
- Iron, wrought.................................. 485
- Lead... 710
- Nickel.. 565
- Stainless steel................................. 510
- Steel, rolled.................................... 490
- Tin... 459
- Zinc.. 440

Concrete

- Stone, plain.................................... 144
- Stone, reinforced............................... 150
- Cinder... 100
- Lightweight, expanded shale..................... 105
- Lightweight, perlite..........................35–50

Stone

- Granite.. 175
- Limestone...................................... 165
- Marble... 165
- Sandstone...................................... 147
- Slate... 175

Water

- Maximum density @ 4°C..........................62
- Ice..56
- Snow...8

Average Weights of Materials (psf*)
*1 psf = 47.88 Pa

Walls and Partitions

- Brick, per 4" (100) of thickness 35
- Concrete masonry units
  w/ stone or gravel aggregate
  4" (100) 34
  6" (150) 50
  8" (205) 58
  12" (305) 90
  w/ lightweight aggregate
  4" (100) 22
  6" (150) 31
  8" (205) 38
  12" (305) 55
- Glass block, 4" (100) 18
- Gypsum board, 1/2" (51) 2
- Metal lath 0.5
- Metal studs with gypsum board 6
- Plaster, 1" (25)
  Cement plaster 10
  Gypsum plaster 5
- Plywood, 1/2" (13) 1.5
- Stone
  Granite, 4" (100) 59
  Limestone, 6" (150) 55
  Marble, 1" (25) 13
  Sandstone, 4" (100) 49
  Slate, 1" (25) 14
- Tile, ceramic 2.5
  Glazed wall tile 3
- Tile, structural clay
  4" (100) 18
  6" (150) 28
  8" (205) 34
- Wood studs, 2x4, with
  gypsum board on both sides 8

Insulation

- Batt or blanket, per inch 0.3
- Fiberboard 2
- Foamed board, per inch 0.2
- Loose 0.5
- Poured in place 2
- Rigid 0.8

Glass

- See 12.17

Average Weights of Materials (psf*)
*1 psf = 47.88 Pa

Floor and Roof Construction

- Concrete, reinforced, per inch (25)
  Stone 12.5
  Perlite 6–10
  Concrete, plain, per inch
  Stone 12
  Lightweight 3–9
- Concrete, precast
  6" (150) hollow core, stone 40
  6" (150) hollow core, lightweight 30
  2" (51) cinder concrete plank 15
  2" (51) gypsum plank 12
- Steel deck 2–4

Roofing

- Built-up, 5-ply felt and gravel 6
- Copper or tin 2
- Corrugated iron 2
- Corrugated fiberglass 0.5
- Monel metal 1.5
- Shingles
  Composition 3
  Slate 10
  Wood 2
- Tile
  Concrete 16
  Clay 14

Ceilings

- Acoustical tile, 3/4" (19) 1
- Acoustic plaster on gypsum lath 10
- Channel suspended system 1

Floor Finishes

- Cement finish, 1" (25) 12
- Marble 30
- Terrazzo, 1" (25) 13
- Wood
  Hardwood 25/32" (20) 4
  Softwood 3/4" (19) 4.5
  Wood block 3" (75) 15
- Vinyl tile 1.33

Factor	Multiples	Prefixes	Symbols
Thousand million	10^9	Giga	G
One million	10^6	Mega	M
One thousand	10^3	Kilo	k
One hundred	10^2	Hecto	h
Ten	10	Deca	da
One-tenth	10^{-1}	Deci	d
One-hundredth	10^{-2}	Centi	c
One-thousandth	10^{-3}	Milli	m
One-millionth	10^{-6}	Micro	u

The International System of Units (SI), more commonly known as the metric system, is an internationally accepted system of coherent physical units, using the meter, gram, second, ampere, kelvin, and candela as the basic units of the fundamental quantities of length, mass, time, electric current, temperature, and luminous intensity. The metric system is universally used in science and mandatory for use in a large number of countries.

- The meter is the basic unit of length in the metric system, equivalent to 39.37 inches. It was originally defined as one ten-millionth of the distance from the equator to the pole measured on the meridian, later as the distance between two lines on a platinum-iridium bar preserved at the International Bureau of Weights and Measures near Paris, and now as 1/299,972,458 of the distance light travels in a vacuum in one second.
- A centimeter is equal to 1/100th of a meter or 0.3937 inch. The centimeter is not recommended for use in construction.
- A millimeter is equal to 1/1000th of a meter or 0.03937 inch.
- A foot, divided into 12 inches, is equal to 304.8 millimeters.

Measurement	Imperial Unit	Metric Unit	Symbol	Conversion Factor
Length	mile	kilometer	km	1 mile = 1.609 km
	yard	meter	m	1 yard = 0.9144 m = 914.4 mm
	foot	meter	m	1 foot = 0.3408 m = 304.8 mm
		millimeter	mm	1 foot = 304.8 mm
	inch	millimeter	mm	1 inch = 25.4 mm
Area	square mile	sq kilometer	km^2	1 sq mile = 2.590 km^2
		hectare	ha	1 sq mile = 259.0 ha (1 ha = 10,000 m^2)
	acre	hectare	ha	1 acre = 0.4047 ha
		square meter	m^2	1 acre = 4046.9 m^2
	square yard	square meter	m^2	1 sq yard = 0.8361 m^2
	square foot	square meter	m^2	1 sq foot = 0.0929 m^2
		sq centimeter	cm^2	1 sq foot = 929.03 cm^2
	square inch	sq centimeter	cm^2	1 sq inch = 6.452 cm^2
Volume	cubic yard	cubic meter	m^3	1 cu yard = 0.7646 m^3
	cubic foot	cubic meter	m^3	1 cu foot = 0.02832 m^3
		liter	liter	1 cu foot = 28.32 liters (1000 liters = 1 m^3)
		cubic decimeter	dm^3	1 cu foot = 28.32 dm^3 (1 liter = 1 dm^3)
	cubic inch	cubic millimeter	mm^3	1 cu inch = 16390 mm^3
		cubic centimeter	cm^3	1 cu inch = 16.39 cm^3
		milliliter	ml	1 cu inch = 16.39 ml
		liter	liter	1 cu inch = 0.01639 liter

Measurement	Imperial Unit	Metric Unit	Symbol	Conversion Factor
Mass	ton	kilogram	kg	1 ton = 1016.05 kg
	kip (1000 lb)	metric ton (1000 kg)	kg	1 kip = 453.59 kg
	pound	kilogram	kg	1 lb = 0.4536 kg
	ounce	gram	g	1 oz = 28.35 g
per length	pound/lf	kilogram/meter	kg/m	1 plf = 1.488 kg/m
per area	pound/sf	kilogram/meter^2	kg/m^2	1 psf = 4.882 kg/m^2
Mass density	pound/cu ft	kilogram/meter^3	kg/m^3	1 pcf = 16018 kg/m^3
Capacity	quart	liter	liter	1 qt = 1.137 liter
	pint	liter	liter	1 pt = 0.568 liter
	fluid ounce	cubic centimeter	cm^3	1 fl oz = 28.413 cm^3
Force	pound	Newton	N	1 lb = 4.488 N
				1 N = kg m/s^2
per length	pound/lf	Newton/meter	N/m	1 plf = 14.594 N/m
Pressure	pound/sf	Pascal	Pa	1 psf = 47.88 Pa
				1 Pa = N/m^2
	pound/sq in	kiloPascal	kPa	1 psi = 6.894 kPa
Moment	foot-pound	Newton-meter	Nm	1 ft-lb = 1.356 Nm
Mass	pound-feet	kilogram-meter	kg m	1 lb-ft = 0.138 kg m
Inertia	pound-feet^2	kilogram-meter^2	kg m^2	1 lb-ft^2 = 0.042 kg m^3
Velocity	miles/hour	kilometer/hour	km/h	1 mph = 1.609 km/h
	feet/minute	meter/minute	m/min	1 fpm = 0.3408 m/min
	feet/second	meter/second	m/s	1 fps = 0.3408 m/s
Volume rate of flow	cu ft/minute	liter/second	liter/s	1 ft^3/min = 0.4791 liter/s
	cu ft/second	meter^3/second	m^3/s	1 ft^3/sec = 0.02832 m^3/s
	cu in/second	milliliter/second	ml/s	1 in^3/sec = 16.39 ml/s
Temperature	°Fahrenheit	degree Celsius	°c	t °C = $^5/_9$ (t °F – 32)
	°Fahrenheit	degree Celsius	°c	1 °F = 0.5556 °C
Heat	British thermal unit (Btu)	joule	J	1 Btu = 1055 J
		kilojoule	kJ	1 Btu = 1.055 kJ
flow	Btu/hour	watt	W	1 Btu/hr = 0.2931 w
conductance	Btu·in/sf·hr·degF	watt/meter^2 ·degC	w/m^2 °C	1 Btu/ft^2 ·hr ·°F = 5.678 w/m^2· °C
resistance	ft^2·h·degF/Btu	meter^2 ·degK/W	m^2 °C/W	1 ft^2·h·°F/Btu = 0.176 m^2·°C/W
refrigeration	ton	watt	W	1 ton = 3519 W
Power	horsepower	watt	W	1 hp = 745.7 W
		kilowatt	kW	1 hp = 0.7457 kW
Light	candela	candela	cd	Basic SI unit of luminous intensity
lux	lumen	lumen	lm	1 lm = cd steradian
illuminance	footcandle	lux	lx	1 FC = 10.76 lx
	lumen/sf	lux	lx	1 lm/ft^2 = 10.76 lux
luminance	footlambert	candela/meter^2	cd/m^2	1 fL = 3.426 cd/m^2

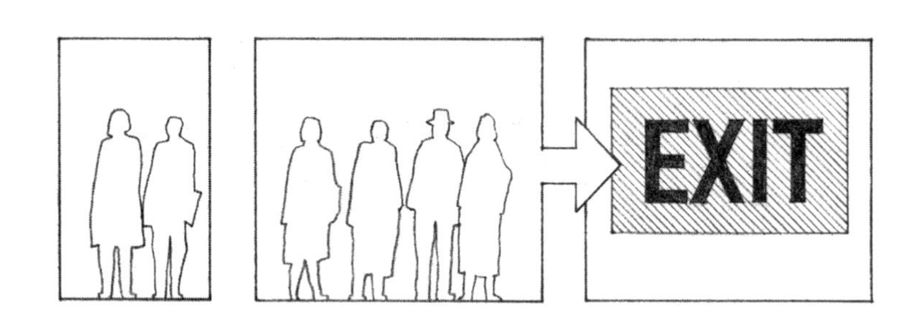

- Occupant load is the total number of persons that may occupy a building or portion thereof at any one time, determined by dividing the floor area assigned to a particular use by the square feet per occupant permitted in that use. Building codes use occupant load to establish the required number and width of exits for a building.

Building codes specify:

- The fire-resistance ratings of materials and construction required for a building, depending on its location, use and occupancy, and size (height and area per floor); see 2.06–2.07.
- The fire alarm, sprinkler, and other protection systems required for certain uses and occupancies; see 11.25.
- The required means of egress for the occupants of a building in case of a fire. A means of egress must provide safe and adequate access from any point in a building to protected exits leading to a place of refuge. There are three components to an egress system: exit access, exits, and exit discharge.

These requirements are intended to control the spread of fire and to allow sufficient time for the occupants of a burning building to exit safely before the structure weakens to the extent that it becomes dangerous. Consult the building code for specific requirements.

Exit Access

The path or passageway leading to an exit should be as direct as possible, be unobstructed by projections such as open doors, and be well lit.

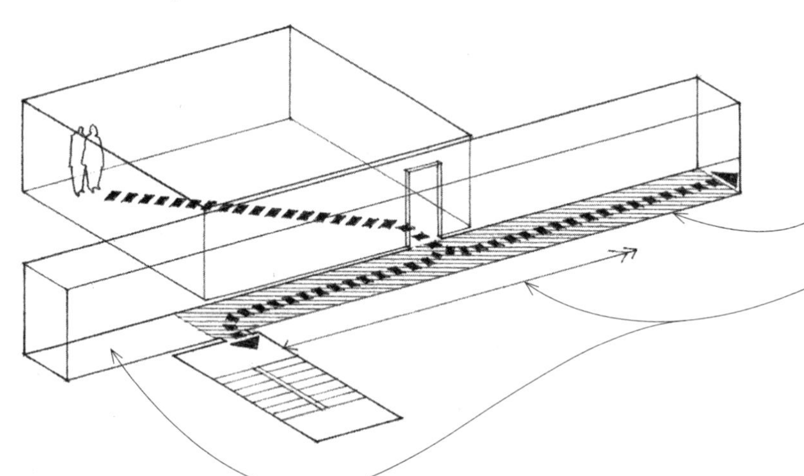

- Building codes specify the maximum distance of travel to an exit according to a building's use, occupancy, and degree of fire hazard.
- Building codes also specify the minimum distance between exits when two or more are required and limit the length of dead-end corridors. For most occupancies, a minimum of two exits is required to provide a margin of safety in case one exit is blocked.
- Exit paths for safe egress from a building should by illuminated by emergency lighting in the event of a power failure.
- Exits should be clearly identified by illuminated signs.

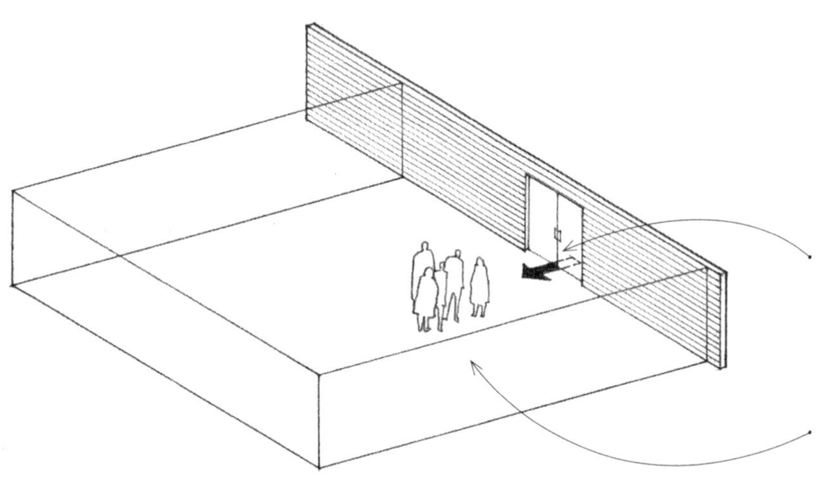

- A horizontal exit is a passage through or around a wall constructed as required for an occupancy separation, protected by an automatic-closing fire door, and leading to an area of refuge in the same building or on approximately the same level in an adjacent building.
- An area of refuge affords safety from fire or smoke coming from the area from which escape is made.

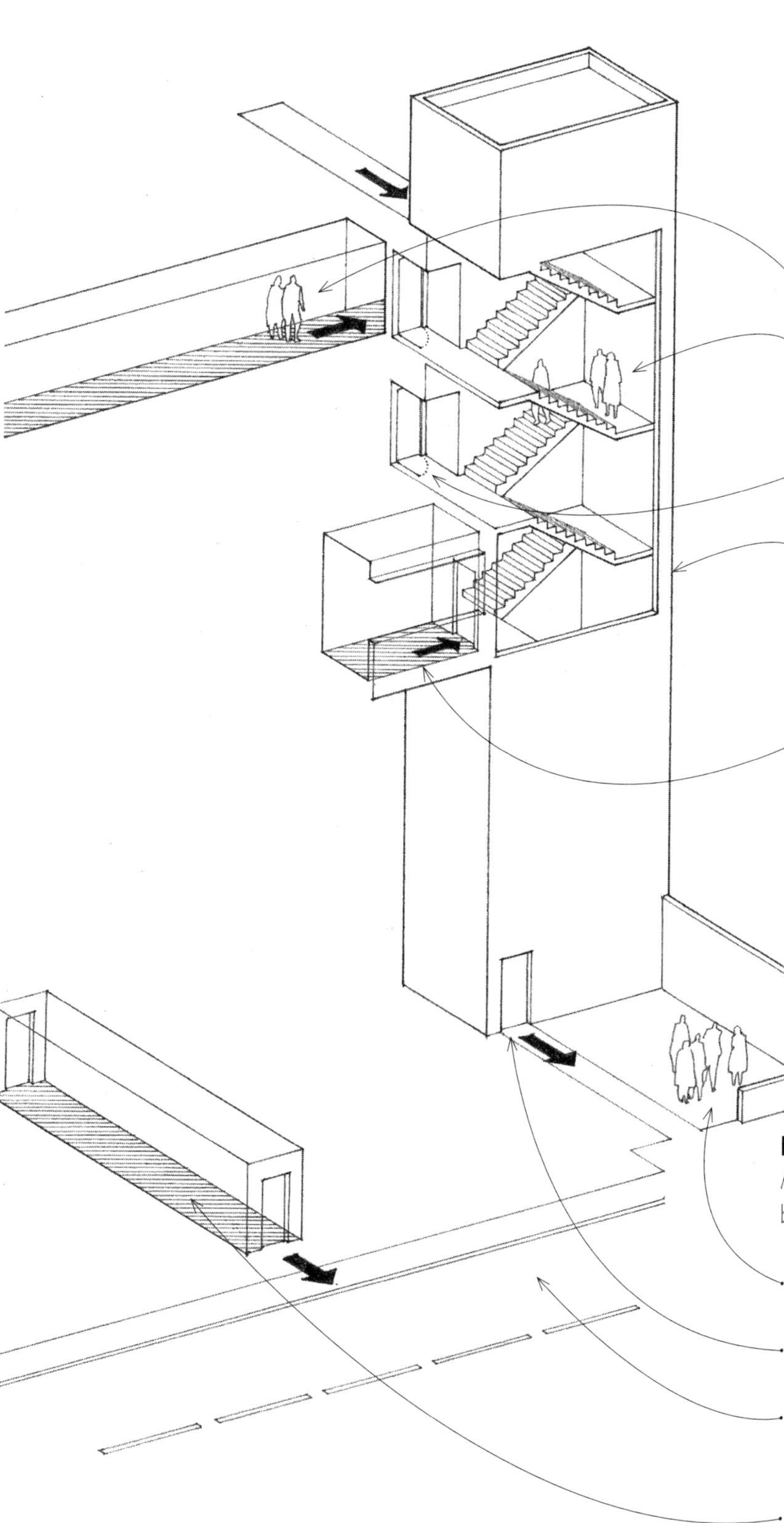

Exits

An exit must provide an enclosed and protected means of evacuation for the occupants of a building in the event of fire, leading from an exit access to an exit discharge. From a ground floor room or corridor, it may simply be a door opening directly to the outside. From a room or space above or below grade, a required exit usually consists of an exit stairway.

- Exit corridors must be enclosed by walls of fire-resistive construction in order to serve as required exits.
- Exit stairways lead to an exit passageway, an exit court, or public way, enclosed by fire-resistive construction with self-closing fire doors that swing in the direction of exit travel. See 9.04–9.05 for stairway dimensions and requirements.
- Exit doors provide access to a means of egress, swinging in the direction of exit travel, and usually equipped with a panic bar.
- Smokeproof enclosure is the enclosing of an exit stairway by walls of fire-resistive construction, accessible by a vestibule or by an open exterior balcony, and ventilated by natural or mechanical means to limit the penetration of smoke and heat. Building codes usually require one or more of the exit stairways for a high-rise building be protected by a smokeproof enclosure.
- An exterior exit balcony is a landing or porch projecting from the wall of a building and serving as a required means of egress.

Exit Discharge

All exits must discharge to a safe place of refuge outside of the building, such as an exit court or public way at ground level.

- Exit court is a yard or court providing egress to a public way for one or more required exits.
- Exterior exit is an exit door opening directly to an exit court or public way.
- Public way is a street, alley, or similar parcel of land open to the sky and deeded, dedicated, or otherwise permanently appropriated for the free passage and use of the general public.
- Exit passageway is a means of egress connecting a required exit or exit court with a public way, having no openings other than required exits and enclosed by fire-resistive construction as required for the walls, floors, and ceiling of the building served.

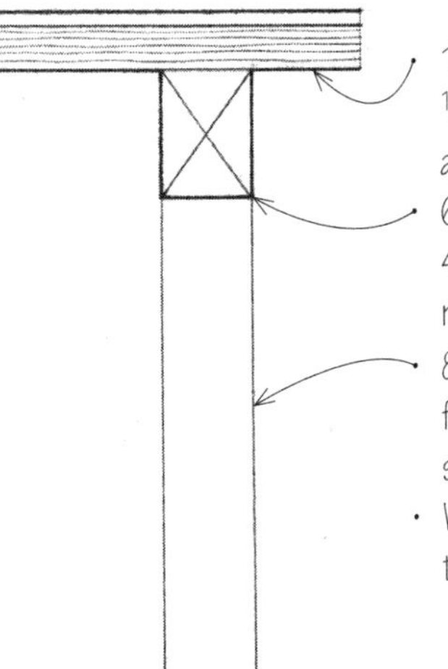

- 1" (25) tongue & groove flooring or 1/2" (13) plywood over wood planks at least 3" (75) thick
- 6x10 minimum for floor beams; 4x6 minimum for roof beams and truss members
- 8x8 minimum for columns supporting floor loads; 6x6 minimum for columns supporting only roof loads
- Wood may be chemically treated to reduce its flammability.

Heavy Timber (Type IV) Construction

- See 2.06.

- Reinforced concrete
- Thickness of the concrete cover and size of the steel member determine the fire rating.

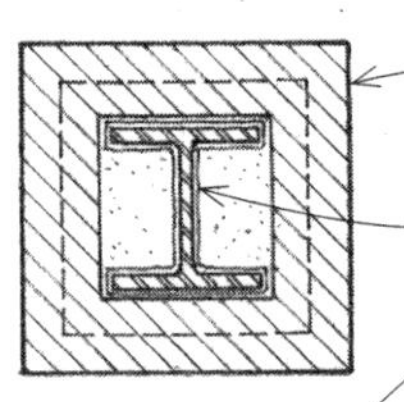

- Clay or shale brick with brick and mortar fill
- Building paper to break bond

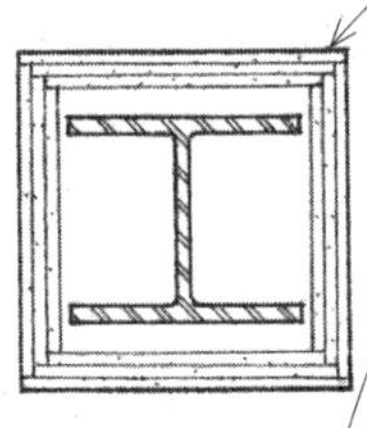

- Multiples layers of gypsum board, or perlite or vermiculite gypsum plaster on metal or gypsum lath

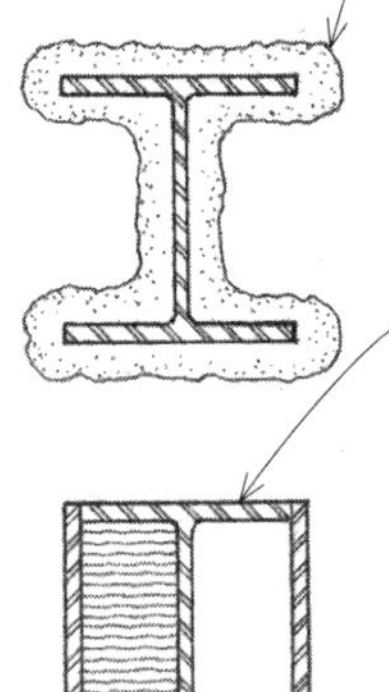

- Spray-on fireproofing is a mixture of gypsum plaster, mineral fibers with an inorganic binder, or magnesium oxychloride cement, applied by air pressure with a spray gun to provide a thermal barrier to the heat of a fire.
- Liquid-filled columns are hollow structural-steel columns filled with water to increase their fire resistance. If exposed to flame, the water absorbs heat, rises by convection to remove the heat, and is replaced with cooler water from a storage tank or a city water main.

Structural Steel

- Because structural steel can be weakened by the high temperatures of a fire, it requires protection to qualify for certain types of construction.

Fire-rated materials, assemblies, and construction have a fire-resistance rating required by their uses. This fire-resistance rating is determined by subjecting a full-size specimen to temperatures according to a standard time-temperature curve and establishing the length of time in hours the material or assembly can be expected to withstand exposure to fire without collapsing, developing any openings that permit the passage of flame or hot gases, or exceeding a specified temperature on the side away from the fire. Fire-resistant construction therefore involves both reducing the flammability of a material and controlling the spread of fire.

Materials used to provide fire protection must be inflammable and be able to withstand very high temperatures without disintegrating. They should also be low conductors of heat to insulate the protected materials from the heat generated by a fire. Such materials include concrete, often with lightweight aggregate, gypsum or vermiculite plaster, gypsum wallboard, and a variety of mineral fiber products.

On this and the facing page is a sampling of fire-resistance ratings for various construction assemblies. For more detailed specifications, consult the Underwriters' Laboratories, Inc. Materials List, or the governing building code. See also 2.06 for a table of the fire-resistance rating requirements for major building components.

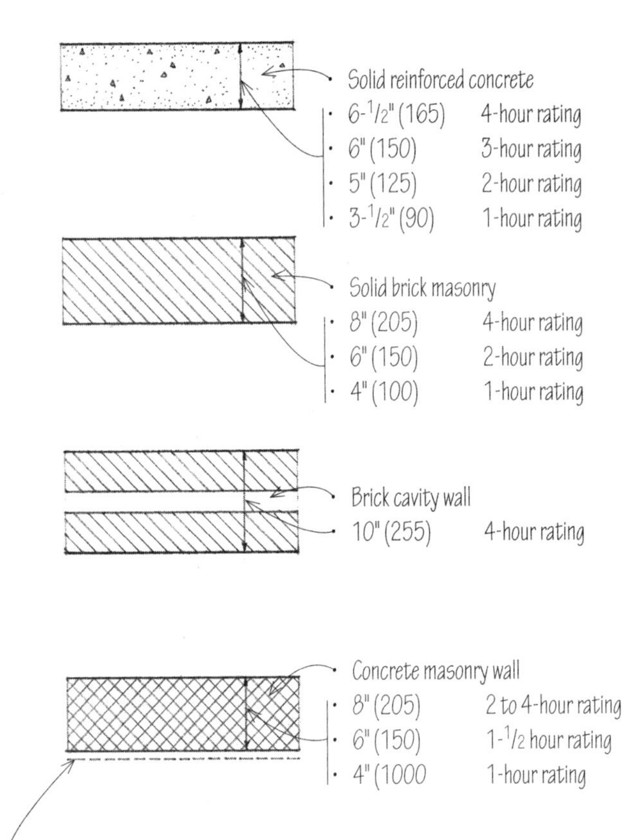

Concrete and Masonry Walls

- Ratings of all masonry walls may be increased with a coating of portland cement or gypsum plaster.

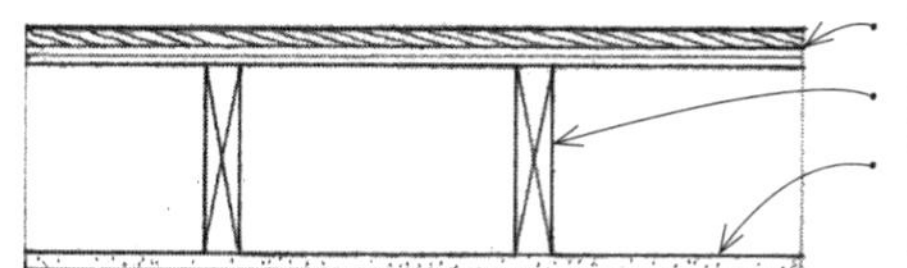

- Double wood floor
- Wood joists @ 16" (405) o.c.
- 1/2" (13) Type X gypsum board or 5/8" (16) gypsum plaster on metal lath

1-Hour Rating

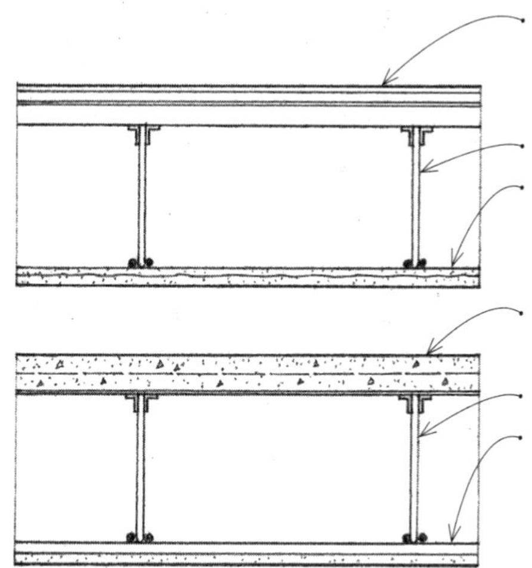

- Fire-resistant roofing over wood fiber insulation board and 1-1/2" (38) steel roof deck
- Steel joists
- 3/4" (19) gypsum plaster on metal lath

- 2" (51) reinforced concrete slab on steel form units
- Steel joists
- 5/8" (16) Type X gypsum board or perlite plaster on 3/8" (10) perforated gypsum lath attached to 3/4" (19) cold-rolled channels

2-Hour Rating

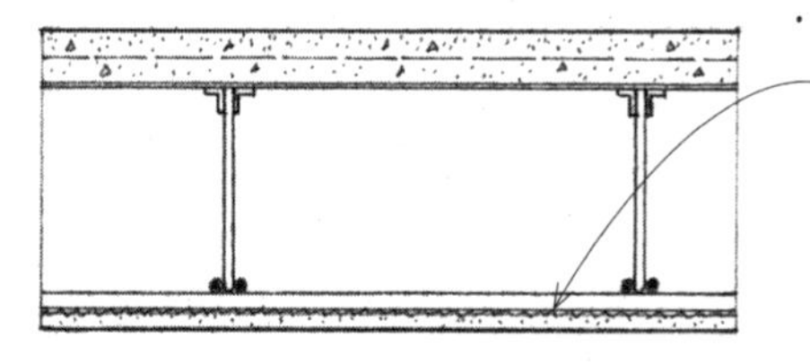

- Similar to above but with 2-1/2" (64) slab and 3/4" (19) vermiculite gypsum plaster on metal lath

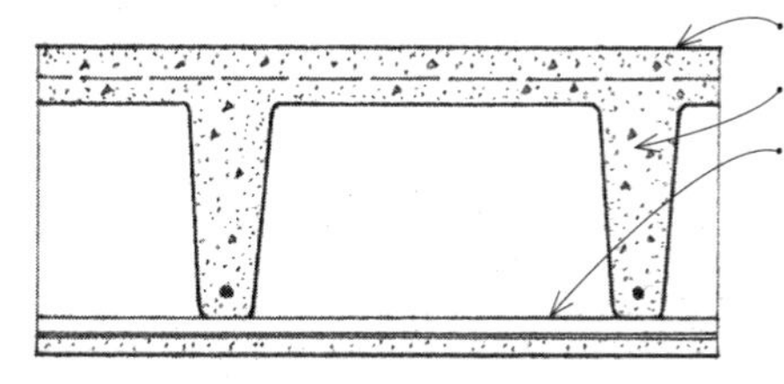

- 3" (75) slab
- Reinforced concrete joists
- 1" (25) vermiculite gypsum plaster on metal lath attached to 3/4" (19) cold-rolled channels @ 12" (305) o.c.

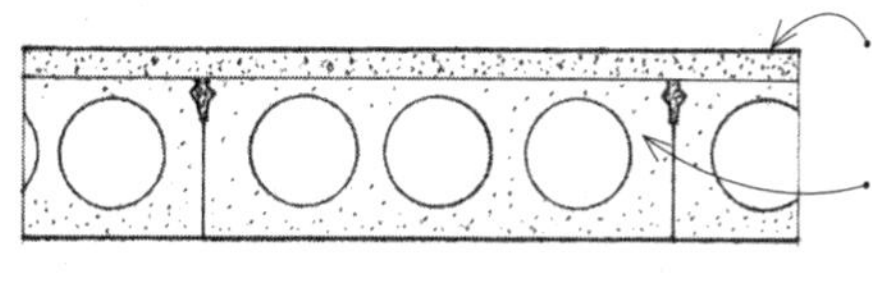

- 1-1/2" (38) sand-gravel concrete topping
- 8" (205) precast concrete slabs w/ all joints grouted

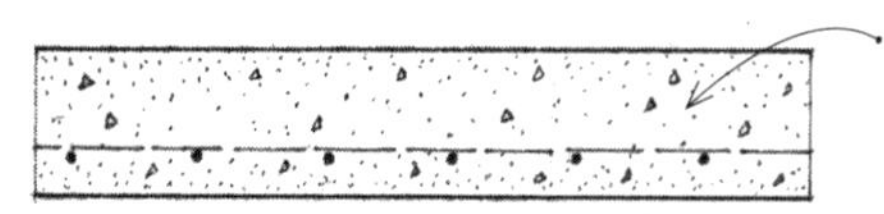

- 6-1/2" (165) regular or 5" (125) expanded shale concrete slab

4-Hour Rating

Floors and Roofs

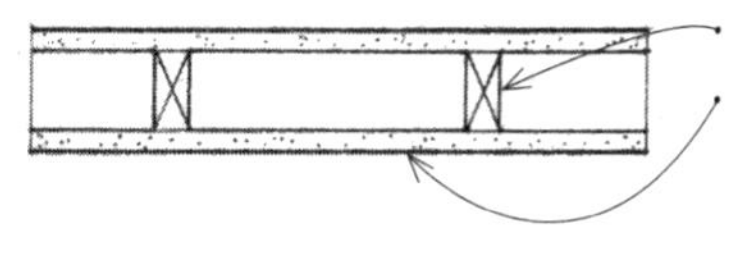

- 2x4 studs a 16" (405) o.c.
- 5/8" (16) gypsum plaster on metal lath, or two layers 1/2" (13) regular gypsum board or 5/8" (16) Type X gypsum board on each side

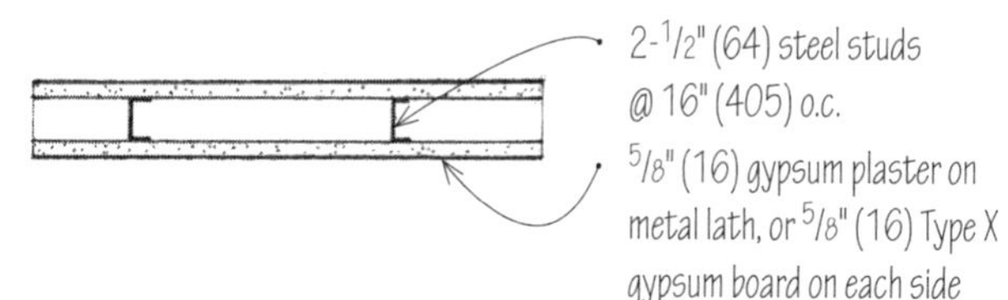

- 2-1/2" (64) steel studs @ 16" (405) o.c.
- 5/8" (16) gypsum plaster on metal lath, or 5/8" (16) Type X gypsum board on each side

- 2" (51) solid gypsum plaster partition with 3/4" (19) channels and 3/8" (10) gypsum lath

1-Hour Rating

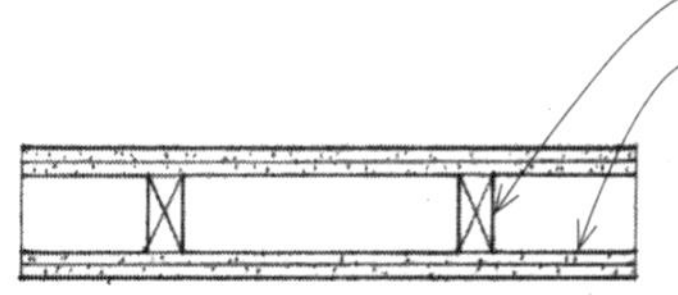

- 2x4 studs @ 16" (405) o.c.
- 7/8" (22) neat wood-fibered gypsum plaster on metal lath, or two layers 5/8" (16) Type X gypsum board on each side

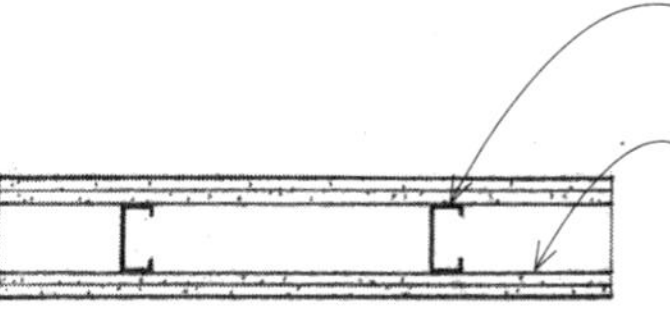

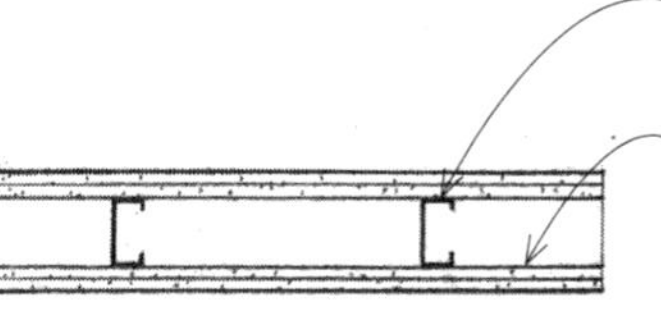

- Steel studs @ 16" or 24" (405 or 610) o.c.
- 3/4" (19) perlite gypsum plaster on 3/8" (10) perforated gypsum lath, or two layers 1/2" (13) Type X gypsum board on each side

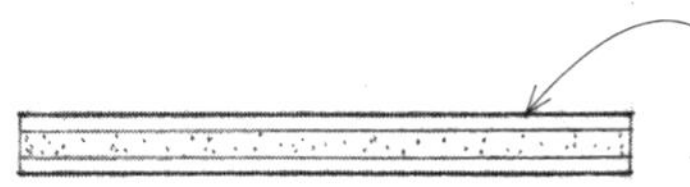

- 2" (51) solid gypsum plaster partition or 1/2" (13) Type X gypsum board on each side of 1" (25) gypsum coreboard

2-Hour Rating

Walls and Partitions

Acoustics is the branch of physics that deals with the production, control, transmission, reception, and effects of sound. Sound may be defined as the sensation stimulated in the organs of hearing by mechanical radiant energy transmitted as longitudinal pressure waves through the air or other medium.

- Sound waves are longitudinal pressure waves in air or an elastic medium producing an audible sensation.
- Sound travels through air at approximately 1087' (0.3 km) per second at sea level, through water at approximately 4500' (1.4 km) per second, through wood at approximately 11,700' (3.6 km) per second, and through steel at approximately 18,000' (5.5 km) per second.

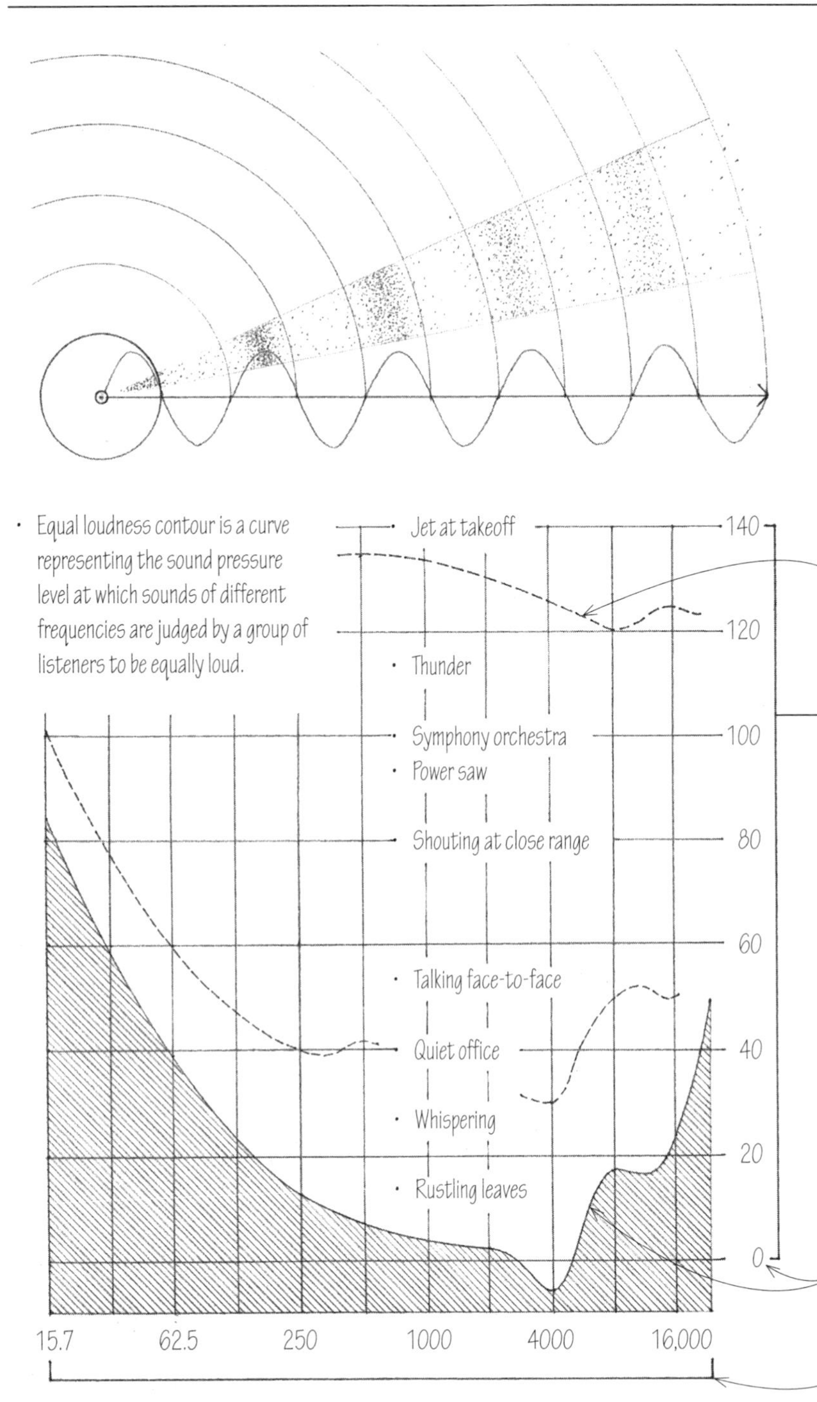

- Equal loudness contour is a curve representing the sound pressure level at which sounds of different frequencies are judged by a group of listeners to be equally loud.

- The threshold of pain is the level of sound intensity high enough to produce the sensation of pain in the human ear, usually around 130 dB.

- Decibel (dB) is a unit for expressing the relative pressure or intensity of sounds on a uniform scale from 0 for the least perceptible sound to about 130 for the average threshold of pain. Decibel measurement is based on a logarithmic scale since increments of sound pressure or intensity are perceived as equal when the ratio between successive changes in intensity remain constant. The decibel levels of two sound sources, therefore, cannot be added mathematically: e.g., 60 dB + 60 dB = 63 dB, not 120 dB.

- The threshold of hearing is the minimum sound pressure capable of stimulating an auditory sensation, usually 20 micropascals or zero dB.
- The audio frequency is a range of frequencies from 15 Hz to 20,000 Hz audible to the normal human ear. Hertz (Hz) is the SI unit of frequency, equal to one cycle per second.

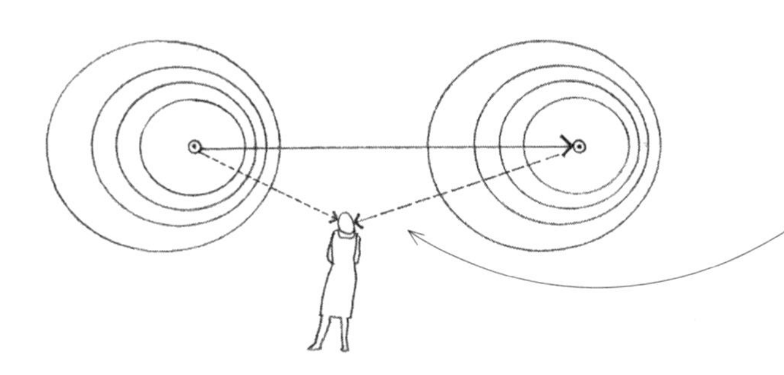

- Doppler effect is an apparent shift in frequency occurring when an acoustic source and listener are in motion relative to each other, the frequency increasing when the source and listener approach each other and decreasing when they move apart.

Acoustical design is the planning, shaping, finishing, and furnishing of an enclosed space to establish the acoustical environment necessary for distinct hearing of speech or musical sounds.

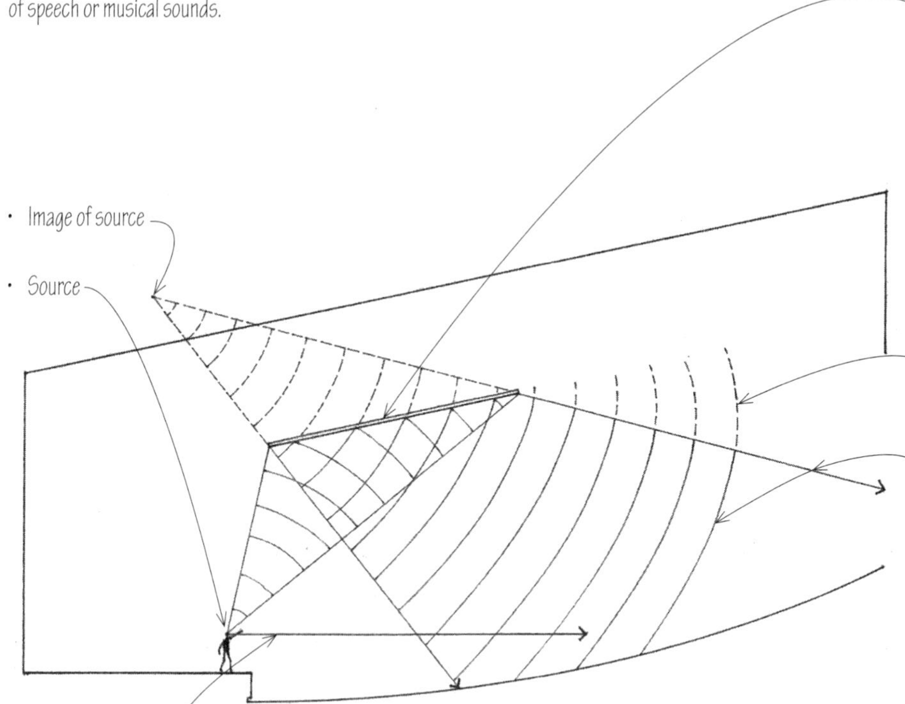

- Reflecting surfaces are nonabsorptive surfaces from which incident sound is reflected, used to redirect sound in a space. To be effective, a reflecting surface should have a least dimension equal to or greater than the wavelength of the lowest frequency of the sound being reflected.

- Diffracted sound is airborne sound waves bent by diffraction around an obstacle in their path.
- Reflected sound is the return of unabsorbed airborne sound after striking a surface, at an angle equal to the angle of incidence.

Airborne sound travels directly from a source to the listener. In a room, the human ear always hears direct sound before it hears reflected sound. As direct sound loses intensity, the importance of reflected sound increases.

- Attenuation is a decrease in energy or pressure per unit area of a sound wave, occurring as the distance from the source increases as a result of absorption, scattering, or spreading in three dimensions.

- Reverberation is the persistence of a sound within an enclosed space, caused by multiple reflection of the sound after its source has stopped. Reverberation time is the time in seconds required for a sound made in an enclosed space to diminish by 60 dB.
- Resonance is the intensification and prolongation of sound produced by sympathetic vibration, the vibration induced in one body by the vibrations of exactly the same period in a neighboring body.

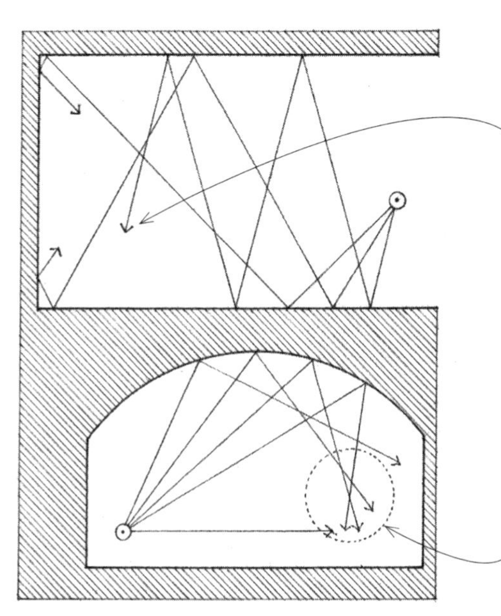

- Echoes are the repetitions of a sound produced by the reflection of sound waves from an obstructing surface, loud enough and received late enough to be perceived as distinct from the source; echoes may occur when parallel surfaces are more than 60' (18 m) apart.
- Flutter is a rapid succession of echoes caused by the reflection of sound waves back and forth between two parallel surfaces, with sufficient time between each reflection to cause the listener to be aware of separate, discrete signals.

- Focusing is the convergence of sound waves reflected from a concave surface.

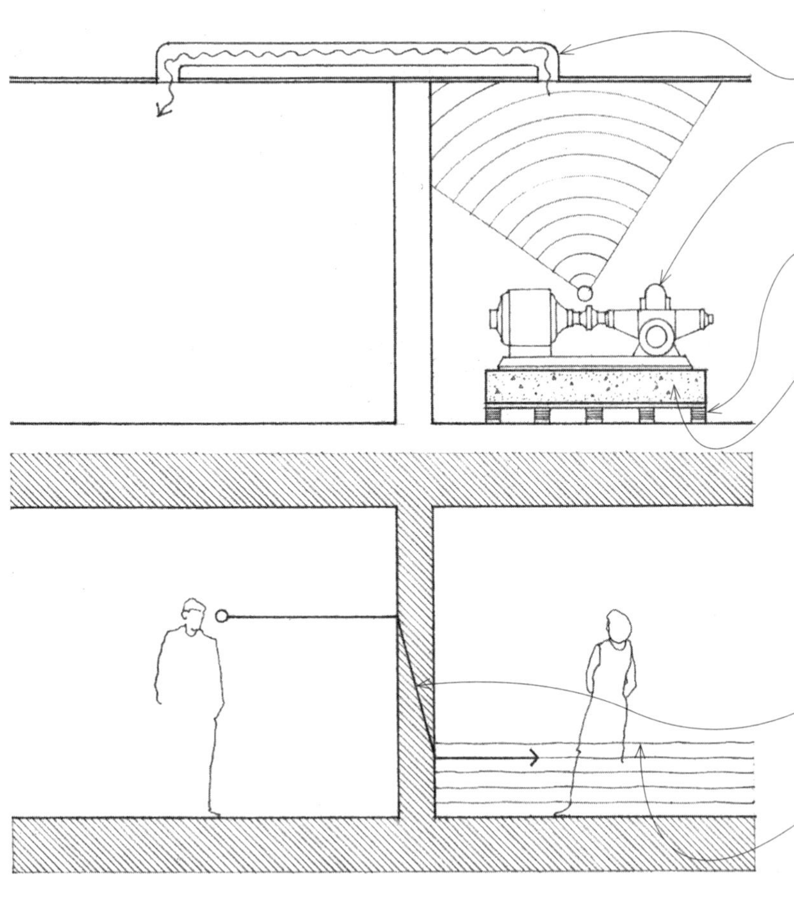

Noise is any sound that is unwanted, annoying, or discordant, or that interferes with one's hearing of something. Whenever possible, undesirable noises should be controlled at their source.

- Block flanking paths that transmit sound through plenum spaces and along such interconnecting structures as ductwork or piping.
- Select mechanical equipment with low sone ratings. Sone is a subjective unit of loudness equal to that of a 1000 Hz reference sound having an intensity of 40 dB.
- Use resilient mountings and flexible bellows to isolate equipment vibrations from the building structure and supply systems to reduce the transmission of vibration and noise to the supporting structure.
- Inertia block is a heavy concrete base for vibrating mechanical equipment, used in conjunction with vibration isolators to increase the mass of the equipment and decrease the potential for vibratory movement.

Noise Reduction

The required reduction in noise level from one space to another depends on the level of the sound source and the level of the sound's intrusion that may be acceptable to the listener. The perceived or apparent sound level in a space is dependent on:

- The transmission loss through the wall, floor, and ceiling construction;
- The absorptive qualities of the receiving space;
- The level of masking or background sound, which increases the threshold of audibility for other sounds in its presence.
- Background noise or ambient sound is the sound normally present in an environment, usually a composite of sounds from both exterior and interior sources, none of which are distinctly identifiable by the listener.

- White noise is an unvarying, unobtrusive sound having the same intensity for all frequencies of a given band, used to mask or obliterate unwanted sound.

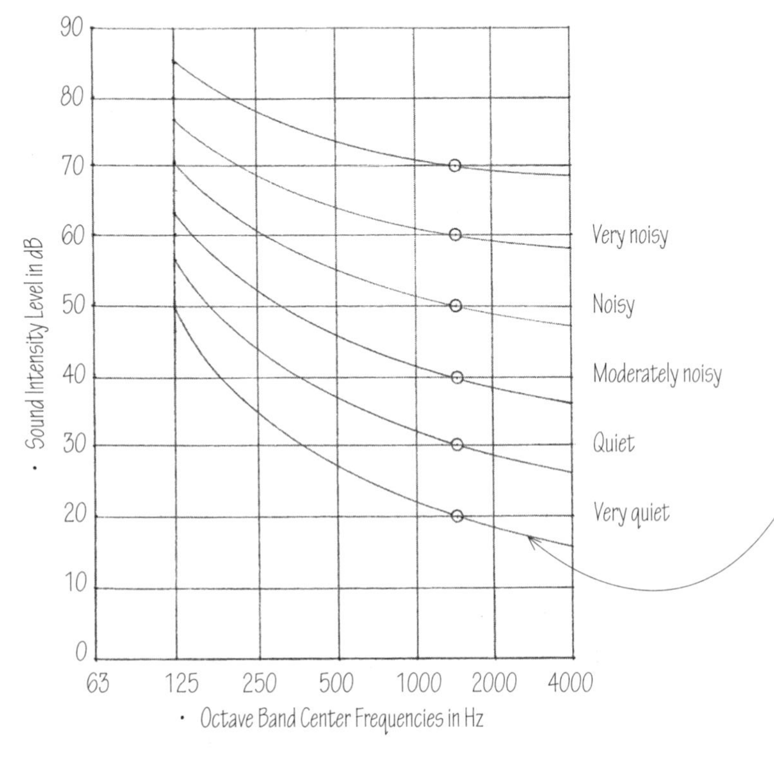

- Noise criteria curve is one of a series of curves representing the sound pressure level across the frequency spectrum for background noise that should not be exceeded in various environments. Higher noise levels are permitted at the lower frequencies since the human ear is less sensitive to sounds in this frequency region.

Transmission Loss

- Transmission loss (TL) is a measure of the performance of a building material or construction assembly in preventing the transmission of airborne sound, equal to the reduction in sound intensity as it passes through the material or assembly when tested at all 1/3 octave band center frequencies from 125 to 4000 Hz: expressed in decibels.
- Average TL is a single-number rating of the performance of a building material or construction assembly in preventing the transmission of airborne sound, equal to the average of its TL values at nine test frequencies.

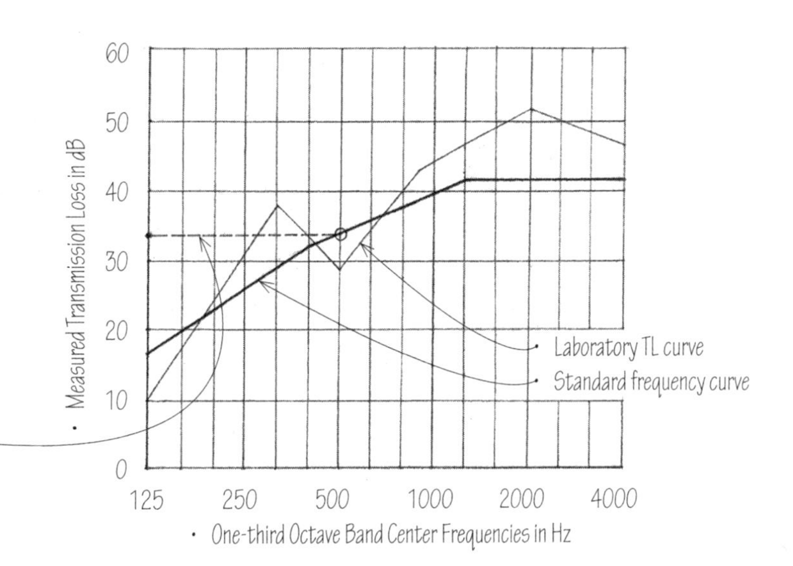

- Sound transmission class (STC) is a single-number rating of the performance of a building material or construction assembly in preventing the transmission of airborne sound, derived by comparing the laboratory TL test curve for the material or assembly to a standard frequency curve. The higher the STC rating, the greater the sound-isolating value of the material or construction. An open doorway has an STC rating of 10; normal construction has STC ratings from 30 to 60; special construction is required for STC ratings above 60.

Three factors enhance the TL rating of a construction assembly: separation into layers, mass, and absorptive capacity.

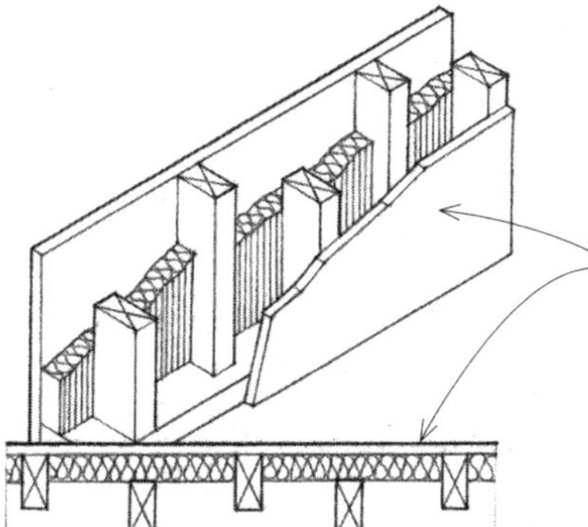

- Staggered-stud partitions for reducing sound transmission between rooms are framed with two separate rows of studs arranged in zigzag fashion and supporting opposite faces of the partition, sometimes with a fiberglass blanket between.
- Resilient mounting is a system of flexible supports or attachments, such as resilient channels and clips, that permits room surfaces to vibrate normally without transmitting the vibratory motions and associated noise to the supporting structure.
- Air spaces increase transmission loss.
- Seal pipe penetrations and other openings and cracks in walls and floors to maintain the continuity of sound isolation.

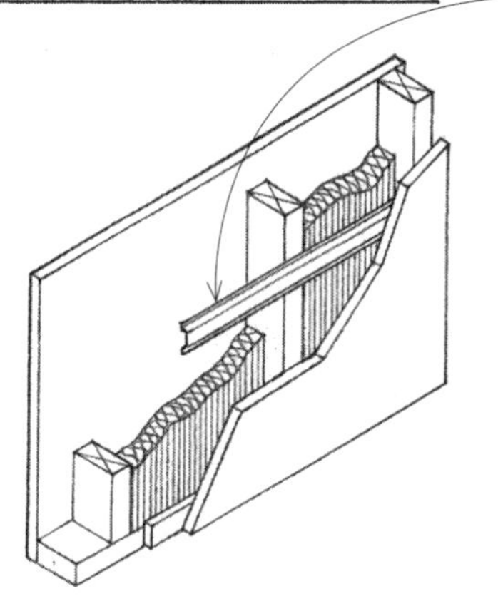

- Acoustic mass resists the transmission of sound by the inertia and elasticity of the transmitting medium. In general, the heavier and more dense a body, the greater its resistance to sound transmission.
- Absorption coefficient is a measure of the efficiency of a material in absorbing sound at a specified frequency, equal to the fractional part of the incident sound energy at that frequency absorbed by the material.

Impact Noise

Impact noise is structure-borne sound generated by physical impact, as by footsteps or the moving of furniture.

- Impact insulation class (IIC) is a single-number rating of the performance of a floor-ceiling construction in preventing the transmission of impact noise. The higher the IIC rating, the more effective is the construction in isolating impact noise. The IIC rating replaces the previously used Impact Noise Rating (INR) and is approximately equal to the INR rating + 51 dB for a given construction.

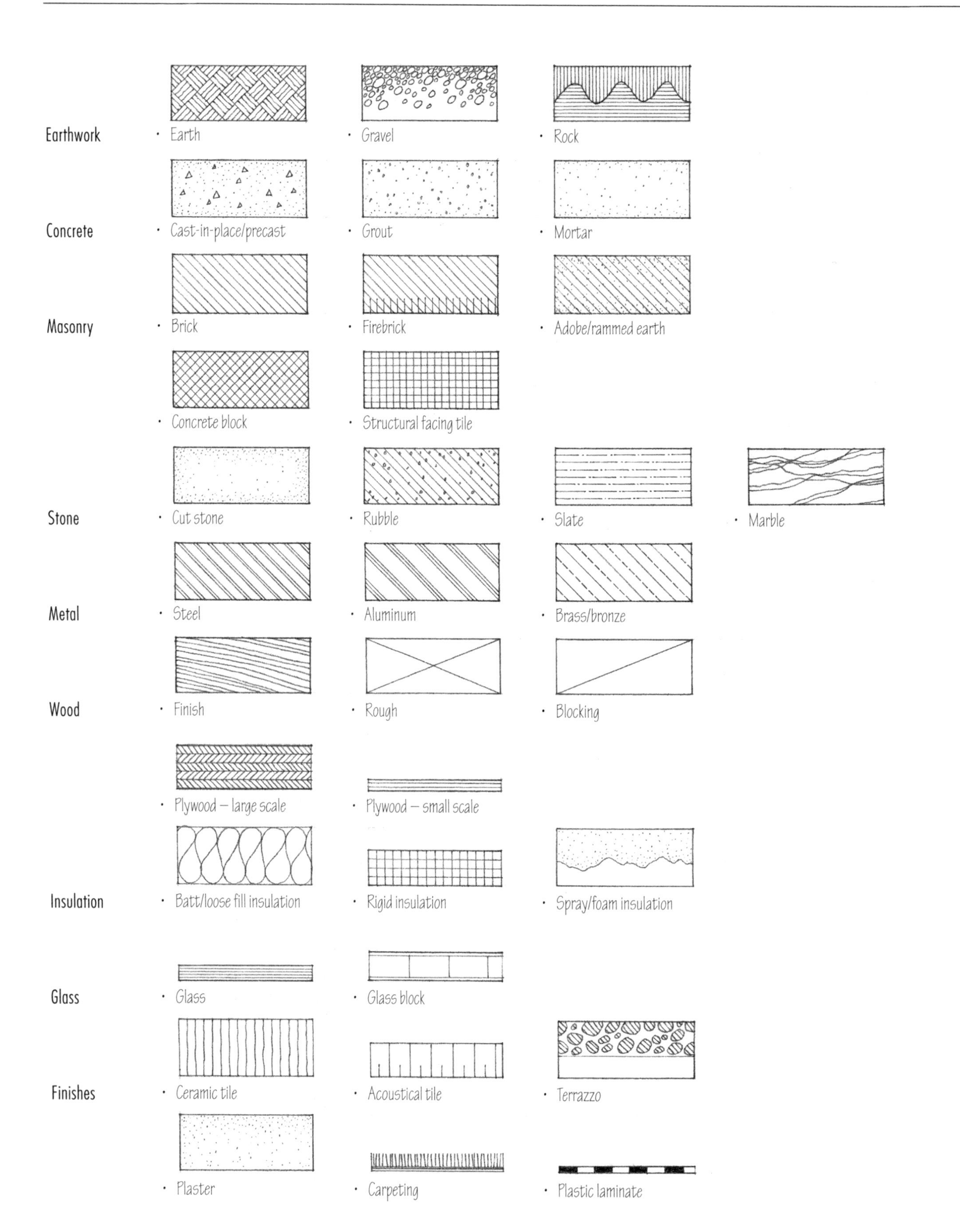
Earthwork
Earth
Gravel
Rock
Concrete
Cast-in-place/precast
Grout
Mortar
Masonry
Brick
Firebrick
Adobe/rammed earth
Concrete block
Structural facing tile
Stone
Cut stone
Rubble
Slate
Marble
Metal
Steel
Aluminum
Brass/bronze
Wood
Finish
Rough
Blocking
Plywood – large scale
Plywood – small scale
Insulation
Batt/loose fill insulation
Rigid insulation
Spray/foam insulation
Glass
Glass
Glass block
Finishes
Ceramic tile
Acoustical tile
Terrazzo
Plaster
Carpeting
Plastic laminate

The Construction Specifications Institute (CSI) created *MasterFormat*® to standardize information about construction requirements, products, and activities, and facilitate communication among architects, contractors, specifiers, and suppliers. *MasterFormat*® is the most widely adopted specification-writing standard for commercial design and construction projects in North America.

Along with her sister organization, Construction Specifications Canada (CSC), CSI issued a new edition of *MasterFormat*® in 2004 that adopted a six-digit numbering scheme to provide more flexibility and room for expansion than the five-digit numbers of the 1995 edition could provide. The 2004 edition also increased from 16 to 50 divisions to reflect innovations in and growing complexity of the construction industry, such as Building Information Modeling (BIM), life-cycle costing, and issues of pollution, remediation, and maintenance. Updates have been published in 2010 and 2012.

MasterFormat® organizes a master list of section numbers and subject titles into two major groups: the Procurement and Contracting Requirements Group (Division 00) and the Specifications Group, which is further subdivided into five subgroups:

- *General Requirements Subgroup: Division 01*
- *Facility Construction Subgroup: Divisions 02 through 19*
- *Facility Services Subgroup: Divisions 20 through 29*
- *Site and Infrastructure Subgroup: Divisions 30 through 39*
- *Process Equipment Subgroup: Divisions 40 through 49*

There are a total of 50 level-one titles or divisions, some of which are held in reserve for future use. Each division consists of sections defined by a number and title and arranged in levels depending on their breadth and depth of coverage.

SPECIFICATIONS GROUP
Facility Construction Subgroup
DIVISION 04 – MASONRY

- The first pair of digits represents the division or level one.
- The second pair of digits represents level two.
- The third pair of digits represents level three.

04 21 13.13: Brick Veneer Masonry

- When detail merits an additional level of classification, an additional pair of digits is attached to the end, preceded by a dot.

PROCUREMENT AND CONTRACTING REQUIREMENTS GROUP
DIVISION 00 – PROCUREMENT AND CONTRACTING REQUIREMENTS
00 10 00 Solicitation
00 20 00 Instructions for Procurement
00 30 00 Available Information
00 40 00 Procurement Forms and Supplements
00 50 00 Contracting Forms and Supplements
00 60 00 Project Forms
00 70 00 Conditions of the Contract
00 80 00 Unassigned
00 90 00 Revisions, Clarifications, and Modifications

SPECIFICATIONS GROUP
General Requirements Subgroup
DIVISION 01 – GENERAL REQUIREMENTS
01 00 00 General Requirements
01 10 00 Summary
01 20 00 Price and Payment Procedures
01 30 00 Administrative Requirements
01 40 00 Quality Requirements
01 50 00 Temporary Facilities and Controls
01 60 00 Product Requirements
01 70 00 Execution and Closeout Requirements
01 80 00 Performance Requirements
01 90 00 Life-Cycle Activities

Facility Construction Subgroup
DIVISION 02 – EXISTING CONDITIONS
02 00 00 Existing Conditions
02 10 00 Unassigned
02 20 00 Assessment
02 30 00 Subsurface Investigation
02 40 00 Demolition and Structure Moving
02 50 00 Site Remediation
02 60 00 Contaminated Site Material Removal
02 70 00 Water Remediation
02 80 00 Facility Remediation
02 90 00 Unassigned

DIVISION 03 – CONCRETE
03 00 00 Concrete
03 10 00 Concrete Forming and Accessories
03 20 00 Concrete Reinforcing
03 30 00 Cast-in-Place Concrete
03 40 00 Precast Concrete
03 50 00 Cast Decks and Underlayment
03 60 00 Grouting
03 70 00 Mass Concrete
03 80 00 Concrete Cutting and Boring
03 90 00 Unassigned

DIVISION 04 – MASONRY
04 00 00 Masonry
04 10 00 Unassigned
04 20 00 Unit Masonry
04 30 00 Unassigned
04 40 00 Stone Assemblies
04 50 00 Refractory Masonry
04 60 00 Corrosion-Resistant Masonry
04 70 00 Manufactured Masonry
04 80 00 Unassigned
04 90 00 Unassigned

DIVISION 05 – METALS
05 00 00 Metals
05 10 00 Structural Metal Framing
05 20 00 Metal Joists
05 30 00 Metal Decking
05 40 00 Cold-Formed Metal Framing
05 50 00 Metal Fabrications
05 60 00 Unassigned
05 70 00 Decorative Metal
05 80 00 Unassigned
05 90 00 Unassigned

DIVISION 06 – WOOD, PLASTICS, AND COMPOSITES
06 00 00 Wood, Plastics, and Composites
06 10 00 Rough Carpentry
06 20 00 Finish Carpentry
06 30 00 Unassigned
06 40 00 Architectural Woodwork
06 50 00 Structural Plastics
06 60 00 Plastic Fabrications
06 70 00 Structural Composites
06 80 00 Composite Fabrications
06 90 00 Unassigned

DIVISION 07 – THERMAL AND MOISTURE PROTECTION
07 00 00 Thermal and Moisture Protection
07 10 00 Dampproofing and Waterproofing
07 20 00 Thermal Protection
07 25 00 Weather Barriers
07 30 00 Steep Slope Roofing
07 40 00 Roofing and Siding Panels
07 50 00 Membrane Roofing
07 60 00 Flashing and Sheet Metal
07 70 00 Roof and Wall Specialties and Accessories
07 80 00 Fire and Smoke Protection
07 90 00 Joint Protection

DIVISION 08 – OPENINGS
08 00 00 Openings
08 10 00 Doors and Frames
08 20 00 Unassigned
08 30 00 Specialty Doors and Frames
08 40 00 Entrances, Storefronts, and Curtain Walls
08 50 00 Windows
08 60 00 Roof Windows and Skylights
08 70 00 Hardware
08 80 00 Glazing
08 90 00 Louvers and Vents

DIVISION 09 – FINISHES
09 00 00 Finishes
09 10 00 Unassigned
09 20 00 Plaster and Gypsum Board
09 30 00 Tiling
09 40 00 Unassigned
09 50 00 Ceilings
09 60 00 Flooring
09 70 00 Wall Finishes
09 80 00 Acoustic Treatment
09 90 00 Painting and Coating

DIVISION 10 – SPECIALTIES
10 00 00 Specialties
10 10 00 Information Specialties
10 20 00 Interior Specialties
10 30 00 Fireplaces and Stoves
10 40 00 Safety Specialties
10 50 00 Storage Specialties
10 60 00 Unassigned
10 70 00 Exterior Specialties
10 80 00 Other Specialties
10 90 00 Unassigned

DIVISION 11 – EQUIPMENT
11 00 00 Equipment
11 10 00 Vehicle and Pedestrian Equipment
11 15 00 Security, Detention and Banking Equipment
11 20 00 Commercial Equipment
11 30 00 Residential Equipment
11 40 00 Foodservice Equipment
11 50 00 Educational and Scientific Equipment
11 60 00 Entertainment Equipment
11 65 00 Athletic and Recreational Equipment
11 70 00 Healthcare Equipment
11 80 00 Collection and Disposal Equipment
11 90 00 Other Equipment

DIVISION 12 – FURNISHINGS
12 00 00 Furnishings
12 10 00 Art
12 20 00 Window Treatments
12 30 00 Casework
12 40 00 Furnishings and Accessories
12 50 00 Furniture
12 60 00 Multiple Seating
12 70 00 Unassigned
12 80 00 Unassigned
12 90 00 Other Furnishings

DIVISION 13 – SPECIAL CONSTRUCTION
13 00 00 Special Construction
13 10 00 Special Facility Components
13 20 00 Special Purpose Rooms
13 30 00 Special Structures
13 40 00 Integrated Construction
13 50 00 Special Instrumentation
13 60 00 Unassigned
13 70 00 Unassigned
13 80 00 Unassigned
13 90 00 Unassigned

DIVISION 14 – CONVEYING EQUIPMENT
14 00 00 Conveying Equipment
14 10 00 Dumbwaiters
14 20 00 Elevators
14 30 00 Escalators and Moving Walks
14 40 00 Lifts
14 50 00 Unassigned
14 60 00 Unassigned
14 70 00 Turntables
14 80 00 Scaffolding
14 90 00 Other Conveying Equipment

DIVISIONS 15–19 RESERVED

Facility Services Subgroup
DIVISION 20 – RESERVED
DIVISION 21 – FIRE SUPPRESSION
21 00 00 Fire Suppression
21 10 00 Water-Based Fire-Suppression Systems
21 20 00 Fire-Extinguishing Systems
21 30 00 Fire Pumps
21 40 00 Fire-Suppression Water Storage
21 50 00 Unassigned
21 60 00 Unassigned
21 70 00 Unassigned
21 80 00 Unassigned
21 90 00 Unassigned

DIVISION 22 – PLUMBING

22 00 00 Plumbing
22 10 00 Plumbing Piping
22 20 00 Unassigned
22 30 00 Plumbing Equipment
22 40 00 Plumbing Fixtures
22 50 00 Pool and Fountain Plumbing Systems
22 60 00 Gas and Vacuum Systems for Laboratory and Healthcare Facilities
22 70 00–22 90 00 Unassigned

DIVISION 23 – HEATING, VENTILATING, AND AIR-CONDITIONING (HVAC)

23 00 00 Heating, Ventilating, and Air-Conditioning (HVAC)
23 10 00 Facility Fuel Systems
23 20 00 HVAC Piping and Pumps
23 30 00 HVAC Air Distribution
23 40 00 HVAC Air Cleaning Devices
23 50 00 Central Heating Equipment
23 60 00 Central Cooling Equipment
23 70 00 Central HVAC Equipment
23 80 00 Decentralized HVAC Equipment
23 90 00 Unassigned

DIVISION 24 – RESERVED

DIVISION 25 – INTEGRATED AUTOMATION

25 00 00 Integrated Automation
25 10 00 Integrated Automation Network Equipment
25 20 00 Unassigned
25 30 00 Integrated Automation Instrumentation and Terminal Devices
25 40 00 Unassigned
25 50 00 Integrated Automation Facility Controls
25 60 00 Unassigned
25 70 00 Unassigned
25 80 00 Unassigned
25 90 00 Integrated Automation Control Sequences

DIVISION 26 – ELECTRICAL

26 00 00 Electrical
26 10 00 Medium-Voltage Electrical Distribution
26 20 00 Low-Voltage Electrical Distribution
26 30 00 Facility Electrical Power Generating and Storing Equipment
26 40 00 Electrical and Cathodic Protection
26 50 00 Lighting
26 60 00 Unassigned
26 70 00 Unassigned
26 80 00 Unassigned
26 90 00 Unassigned

DIVISION 27 – COMMUNICATIONS

27 00 00 Communications
27 10 00 Structured Cabling
27 20 00 Data Communications
27 30 00 Voice Communications
27 40 00 Audio-Video Communications
27 50 00 Distributed Communications and Monitoring Systems
27 60 00–27 90 00 Unassigned

DIVISION 28 – ELECTRONIC SAFETY AND SECURITY

28 00 00 Electronic Safety and Security
28 10 00 Electronic Access Control and Intrusion Detection
28 20 00 Electronic Surveillance
28 30 00 Electronic Detection and Alarm
28 40 00 Electronic Monitoring and Control
28 50 00 Unassigned
28 60 00 Unassigned
28 70 00 Unassigned
28 80 00 Unassigned
28 90 00 Unassigned

DIVISION 29 – RESERVED

Site and Infrastructure Subgroup

DIVISION 30 – RESERVED

DIVISION 31 – EARTHWORK

31 00 00 Earthwork
31 10 00 Site Clearing
31 20 00 Earth Moving
31 30 00 Earthwork Methods
31 40 00 Shoring and Underpinning
31 50 00 Excavation Support and Protection
31 60 00 Special Foundations and Load-Bearing Elements
31 70 00 Tunneling and Mining
31 80 00 Unassigned
31 90 00 Unassigned

DIVISION 32 – EXTERIOR IMPROVEMENTS

32 00 00 Exterior Improvements
32 10 00 Bases, Ballasts, and Paving
32 20 00 Unassigned
32 30 00 Site Improvements
32 40 00 Unassigned
32 50 00 Unassigned
32 60 00 Unassigned
32 70 00 Wetlands
32 80 00 Irrigation
32 90 00 Planting

DIVISION 33 – UTILITIES

33 00 00 Utilities
33 10 00 Water Utilities
33 20 00 Wells
33 30 00 Sanitary Sewerage Utilities
33 40 00 Storm Drainage Utilities
33 50 00 Fuel Distribution Utilities
33 60 00 Hydronic and Steam Energy Utilities
33 70 00 Electrical Utilities
33 80 00 Communications Utilities
33 90 00 Unassigned

DIVISION 34 – TRANSPORTATION

34 00 00 Transportation
34 10 00 Guideways/Railways
34 20 00 Traction Power
34 30 00 Unassigned
34 40 00 Transportation Signaling and Control Equipment
34 50 00 Transportation Fare Collection Equipment
34 60 00 Unassigned
34 70 00 Transportation Construction and Equipment
34 80 00 Bridges
34 90 00 Unassigned

DIVISION 35 – WATERWAY AND MARINE CONSTRUCTION

35 00 00 Waterway and Marine Construction
35 10 00 Waterway and Marine Signaling and Control Equipment
35 20 00 Waterway and Marine Construction and Equipment
35 30 00 Coastal Construction
35 40 00 Waterway Construction and Equipment
35 50 00 Marine Construction and Equipment
35 60 00 Unassigned
35 70 00 Dam Construction and Equipment
35 80 00 Unassigned
35 90 00 Unassigned

DIVISIONS 36–39 RESERVED

Process Equipment Subgroup

DIVISION 40 – PROCESS INTEGRATION

40 00 00 Process Integration
40 10 00 Gas and Vapor Process Piping
40 20 00 Liquids Process Piping
40 30 00 Solid and Mixed Materials Piping and Chutes
40 40 00 Process Piping and Equipment Protection
40 50 00 Unassigned
40 60 00 Unassigned
40 70 00 Unassigned
40 80 00 Commissioning of Process Systems
40 90 00 Instrumentation and Control for Process Systems

DIVISION 41 – MATERIAL PROCESSING AND HANDLING EQUIPMENT

41 00 00 Material Processing and Handling Equipment
41 10 00 Bulk Material Processing Equipment
41 20 00 Piece Material Handling Equipment
41 30 00 Manufacturing Equipment
41 40 00 Container Processing and Packaging
41 50 00 Material Storage
41 60 00 Mobile Plant Equipment
41 70 00 Unassigned
41 80 00 Unassigned
41 90 00 Unassigned

DIVISION 42 – PROCESS HEATING, COOLING, AND DRYING EQUIPMENT

42 00 00 Process Heating, Cooling, and Drying Equipment
42 10 00 Process Heating Equipment
42 20 00 Process Cooling Equipment
42 30 00 Process Drying Equipment
42 40 00 Unassigned
42 50 00 Unassigned
42 60 00 Unassigned
42 70 00 Unassigned
42 80 00 Unassigned
42 90 00 Unassigned

DIVISION 43 – PROCESS GAS AND LIQUID HANDLING, PURIFICATION, AND STORAGE EQUIPMENT

43 00 00 Process Gas and Liquid Handling, Purification, and Storage Equipment
43 10 00 Gas Handling Equipment
43 20 00 Liquid Handling Equipment
43 30 00 Gas and Liquid Purification Equipment
43 40 00 Gas and Liquid Storage
43 50 00 Unassigned
43 60 00 Unassigned
43 70 00 Unassigned
43 80 00 Unassigned
43 90 00 Unassigned

DIVISION 44 – POLLUTION AND WASTE CONTROL EQUIPMENT

44 00 00 Pollution and Waste Control Equipment
44 10 00 Air Pollution Control
44 20 00 Noise Pollution Control
44 40 00 Water Treatment Equipment
44 50 00 Solid Waste Control
44 60 00 Waste Thermal Processing Equipment
44 70 00 Unassigned
44 80 00 Unassigned
44 90 00 Unassigned

DIVISION 45 – INDUSTRY-SPECIFIC MANUFACTURING EQUIPMENT

45 00 00 Industry-Specific Manufacturing Equipment
45 60 00 Unassigned
45 70 00 Unassigned
45 80 00 Unassigned
45 90 00 Unassigned

DIVISION 46 – WATER AND WASTEWATER EQUIPMENT

46 00 00 Water and Wastewater Equipment
46 10 00 Unassigned
46 20 00 Water and Wastewater Preliminary Treatment Equipment
46 30 00 Water and Wastewater Chemical Feed Equipment
46 40 00 Water and Wastewater Clarification and Mixing Equipment
46 50 00 Water and Wastewater Secondary Treatment Equipment
46 60 00 Water and Wastewater Advanced Treatment Equipment
46 70 00 Water and Wastewater Residuals Handling and Treatment
46 80 00 Unassigned
46 90 00 Unassigned

DIVISION 47 – RESERVED

DIVISION 48 – ELECTRICAL POWER GENERATION

48 00 00 Electrical Power Generation
48 10 00 Electrical Power Generation Equipment
48 20 00 Unassigned
48 30 00 Unassigned
48 40 00 Unassigned
48 50 00 Unassigned
48 60 00 Unassigned
48 70 00 Electrical Power Generation Testing
48 80 00 Unassigned
48 90 00 Unassigned

DIVISION 49 – RESERVED

UNIFORMAT II (ASTM STANDARD E1557-09) provides a consistent reference for the description, economic analysis, and management of buildings during all phases of their life cycle, including planning, programming, design, construction, operations, and disposal. The format is based on the classification of elements, which are defined as "major components, common to most buildings, that perform a given function, regardless of the design specification, construction method, or materials used." Examples of functional building elements are foundations, superstructure, stairs, and plumbing. UNIFORMAT II therefore differs from and complements the MasterFormat© classification system, which is based on products and building materials for detailed quantity takeoffs of materials and tasks associated with the construction, operation, and maintenance of buildings.

The UNIFORMAT II organizational structure assumes that design information at the schematic phase can be communicated more effectively by functional building elements rather than by building material or product, and that an elemental classification system would be more easily understood by clients and others who do not have a technical background. A comprehensive and consistent classification of functional elements also enables the necessary cost information to be evaluated in the early stages of the design process, thus assuring faster and more accurate economic analysis of alternative design decisions early in the beginning stages of the project.

UNIFORMAT II classifies building elements into three hierarchical levels using an alphanumeric designation. There are seven Level 1 groups:

- Group A: Substructure, including Foundations & Basement Construction
- Group B: Shell, including Superstructure, Exterior Enclosure & Roofing
- Group C: Interiors, including Interior Construction, Stairs & Interior Finishes
- Group D: Conveying, Plumbing, HVAC, Fire Protection & Electrical Systems
- Group E: Equipment & Furnishings
- Group F: Special Construction & Demolition
- Group G: Building Sitework

Each Major Group Element is broken down into Level 2 Group Elements (B10, B20. . .) and Level 3 Individual Elements (B1010, B1020, B2010, B2020. . .). A Level 4 is proposed to break the individual elements into yet smaller Sub-elements B1011, B1012, B1013. . .).

ASTM UNIFORMAT II Classification for Building Elements (E1557-09)

Level 1 Major Group Elements	Level 2 Group Elements	Level 3 Individual Elements
A. SUBSTRUCTURE	A10 Foundations	A1010 Standard Foundations
		A1020 Special Foundations
		A1030 Slab on Grade
	A20 Basement Construction	A2010 Basement Excavation
		A2020 Basement Walls
B. SHELL	B10 Superstructure	B1010 Floor Construction
		B1020 Roof Construction
	B20 Exterior Enclosure	B2010 Exterior Walls
		B2020 Exterior Windows
		B2030 Exterior Doors
	B30 Roofing	B3010 Roof Coverings
		B3020 Roof Openings
C. INTERIORS	C10 Interior Construction	C1010 Partitions
		C1020 Interior Doors
		C1030 Specialties
	C20 Stairs	C2010 Stair Construction
		C2020 Stair Finishes
	C30 Interior Finishes	C3010 Wall Finishes
		C3020 Floor Finishes
		C3030 Ceiling Finishes
		C3040 Interior Coatings and Special Finishes

ASTM UNIFORMAT II Classification for Building Elements (E1557-09)

Level 1 Major Group Elements	Level 2 Group Elements	Level 3 Individual Elements
D. SERVICES	D10 Conveying	D1010 Elevators & Lifts
		D1020 Escalators & Moving Walks
		D1090 Other Conveying Systems
	D20 Plumbing	D2010 Plumbing Fixtures
		D2020 Domestic Water Distribution
		D2030 Sanitary Waste
		D2040 Rain Water Drainage
		D2090 Other Plumbing Systems
	D30 HVAC	D3010 Energy Supply
		D3020 Heat Generating Systems
		D3030 Cooling Generating Systems
		D3040 Distribution Systems
		D3050 Terminal & Package Units
		D3060 Controls & Instrumentation
		D3070 Systems Testing & Balancing
		D3090 Other HVAC Systems & Equipment
	D40 Fire Protection	D4010 Fire Alarm and Detection Systems
		D4020 Fire Suppression Water Supply & Equipment
		D4030 Standpipe Systems
		D4040 Sprinklers
		D4050 Fire Protection Specialties
		D4090 Other Fire Protection Systems
	D50 Electrical	D5010 Electrical Service & Distribution
		D5020 Lighting and Branch Wiring
		D5030 Communications & Security
		D5090 Other Electrical Systems
E. EQUIPMENT & FURNISHINGS	E10 Equipment	E1010 Commercial Equipment
		E1020 Institutional Equipment
		E1030 Vehicular Equipment
		E1040 Government Furnished Equipment
		E1090 Other Equipment
	E20 Furnishings	E2010 Fixed Furnishings
		E2020 Movable Furnishings
F. SPECIAL CONSTRUCTION & DEMOLITION	F10 Special Construction	F1010 Special Structures
		F1020 Integrated Construction
		F1030 Special Construction Systems
		F1040 Special Facilities
		F1050 Special Controls & Instrumentation
	F20 Selective Building Demolition	F2010 Building Elements Demolition
		F2020 Hazardous Components Abatement

ASTM UNIFORMAT II Classification for Building-Related Sitework (E1557-09)

Level 1 Major Group Elements	Level 2 Group Elements	Level 3 Individual Elements
G. BUILDING SITEWORK	G10 Site Preparation	G1010 Site Clearing
		G1020 Site Demolition & Relocations
		G1030 Site Earthwork
		G1040 Hazardous Waste Remediation
	G20 Site Improvements	G2010 Roadways
		G2020 Parking Lots
		G2030 Pedestrian Paving
		G2040 Site Development
		G2050 Landscaping
	G30 Site Civil/Mechanical Utilities	G3010 Water Supply
		G3020 Sanitary Sewer
		G3030 Storm Sewer
		G3040 Heating Distribution
		G3050 Cooling Distribution
		G3060 Fuel Distribution
		G3090 Other Site Mechanical Utilities
	G40 Site Electrical Utilities	G4010 Electrical Distribution
		G4020 Site Lighting
		G4030 Site Communications & Security
		G4090 Other Site Electrical Utilities
	G90 Other Site Construction	G9010 Service and Pedestrian Tunnels
		G9090 Other Site Systems & Equipment
Z. GENERAL CONDITIONS	Z10 General Requirements	
	Z20 Bidding Requirements	
	Z90 Project Cost Estimate	

LEED® 2009
For New Construction & Major Renovations
Version 2.2

Sustainable Sites (26 Possible Points)

SS Prerequisite 1 Construction Activity Pollution Prevention Required
SS Credit 1 Site Selection 1
SS Credit 2 Development Density & Community Connectivity 5
SS Credit 3 Brownfield Redevelopment 1
SS Credit 4.1 Alternative Transportation – Public Transportation Access 6
SS Credit 4.2 Alternative Transportation – Bicycle Storage & Changing Rooms 1
SS Credit 4.3 Alternative Transportation – Low Emitting & Fuel Efficient Vehicles 3
SS Credit 4.4 Alternative Transportation – Parking Capacity 2
SS Credit 5.1 Site Development – Protect or Restore Habitat 1
SS Credit 5.2 Site Development – Maximize Open Space 1
SS Credit 6.1 Stormwater Design – Quantity Control 1
SS Credit 6.2 Stormwater Design – Quality Control 1
SS Credit 7.1 Heat Island Effect – Non-Roof 1
SS Credit 7.2 Heat Island Effect – Roof 1
SS Credit 8 Light Pollution Reduction 1

Water Efficiency (10 Possible Points)

WE Prerequisite 1 Water Use Reduction—20% Reduction
WE Credit 1.2 Water Efficient Landscaping 2–4
WE Credit 2 Innovative Wastewater Technologies 2
WE Credit 3.1 Water Use Reduction 2–4

Energy & Atmosphere (35 Possible Points)

EA Prerequisite 1 Fundamental Commissioning of the Building Energy Systems Required
EA Prerequisite 2 Minimum Energy Performance Required
EA Prerequisite 3 Fundamental Refrigerant Management Required
EA Credit 1 Optimize Energy Performance 1–19
EA Credit 2 On-Site Renewable Energy 1–7
EA Credit 3 Enhanced Commissioning 2
EA Credit 4 Enhanced Refrigerant Management 2
EA Credit 5 Measurement & Verification 3
EA Credit 6 Green Power 2

Materials & Resources (14 Possible Points)

MR Prerequisite 1 Storage & Collection of Recyclables Required
MR Credit 1.1 Building Reuse – Maintain Existing Walls, Floors and Roof 1–3
MR Credit 1.2 Building Reuse – Maintain Existing Interior Nonstructural Elements 1
MR Credit 2 Construction Waste Management 1–2
MR Credit 3 Materials Reuse 1–2
MR Credit 4 Recycled Content 1–2
MR Credit 5 Regional Materials 1–2
MR Credit 6 Rapidly Renewable Materials 1
MR Credit 7 Certified Wood 1

Indoor Environmental Quality (15 Possible Points)

EQ Prerequisite 1 Minimum Indoor Air Quality (IAQ) Performance Required
EQ Prerequisite 2 Environmental Tobacco Smoke (ETS) Control Required
EQ Credit 1 Outdoor Air Delivery Monitoring 1
EQ Credit 2 Increased Ventilation 1
EQ Credit 3.1 Construction IAQ Management Plan, During Construction 1
EQ Credit 3.2 Construction IAQ Management Plan, Before Occupancy 1
EQ Credit 4.1 Low-Emitting Materials – Adhesives & Sealants 1
EQ Credit 4.2 Low-Emitting Materials – Paints & Coatings 1
EQ Credit 4.3 Low-Emitting Materials – Flooring Systems 1
EQ Credit 4.4 Low-Emitting Materials – Composite Wood & Agrifiber Products 1
EQ Credit 5 Indoor Chemical & Pollutant Source Control 1
EQ Credit 6.1 Controllability of Systems – Lighting 1
EQ Credit 6.2 Controllability of Systems – Thermal Comfort 1
EQ Credit 7.1 Thermal Comfort – Design 1
EQ Credit 7.2 Thermal Comfort – Verification 1
EQ Credit 8.1 Daylight & Views – Daylight 1
EQ Credit 8.2 Daylight & Views – Views 1

Innovation & Design Process (6 Possible Points)

ID Credit 1 Innovation in Design 1–5
ID Credit 2 LEED Accredited Professional 1

Regional Priority (4 Possible Points)

RP Credit 1 Regional Priority 1–4

100 base points +
6 possible Innovation in Design and 4 Regional Priority points

To receive LEED certification, a building project must meet certain prerequisites and performance benchmarks or credits within each category. Projects are awarded Certified, Silver, Gold, or Platinum certification depending on the number of credits they achieve.

- Certified 40–49 points
- Silver 50–59 points
- Gold 60–79 points
- Platinum 80 points and above

Professional & Trade Associations

American Institute of Architects
www.aia.org

American Institute of Building Design
www.aibd.org

American Society of Civil Engineers
www.asce.org

American Society of Interior Designers
www.asid.org

American Society of Landscape Architects
www.asla.org

Architecture 2030
www.architecture2030.org

Associated General Contractors of America
www.agc.org

Building Research Establishment
www.bre.co.uk

Canadian Construction Association
www.cca-acc.com

Construction Management Association of America
www.cmaanet.org

Construction Specifications Canada
www.csc-dcc.ca

Construction Specifications Institute
www.csinet.org

Environmental Protection Agency
www.epa.gov

Green Building Institute
http://greenbuildingnetwork.groupsite.com/

Home Innovation Research Labs
www.homeinnovation.com

Insurance Services Office
www.iso.com

McGraw-Hill Construction
www.construction.com

National Council of Architectural Registration Boards
www.ncarb.org

National Institute of Building Sciences
www.nibs.org

National Society of Professional Engineers
www.nspe.org

Partnership for Advancing Technology in Housing
www.pathnet.org

Royal Architectural Institute of Canada
www.raic.org

Society of American Registered Architects
www.sara-national.org

Structural Engineers Association of California
www.seaoc.org

Superintendent of Documents
U.S. Government Printing Office
www.gpoaccess.gov

Urban Land Institute
www.uli.org

U.S. Department of Energy
Energy Efficiency and Renewable Energy
www.eere.energy.gov

U.S. Department of Housing and Urban Development
www.portal.hud.gov

U.S. Department of Justice
2010 ADA Standards for Accessible Design
http://www.ada.gov

U.S. Department of Labor
Occupational Safety and Health Administration
www.osha.gov

U.S. Green Building Council
new.usgbc.org

CSI Division 03 • Concrete

American Concrete Institute
www.concrete.org

American Society for Concrete Contractors
www.ascconline.org

Architectural Precast Association
www.archprecast.org

Concrete Reinforcing Steel Institute
www.crsi.org

National Precast Concrete Association
http://precast.org

Portland Cement Association
www.cement.org

Post-Tensioning Institute
www.post-tensioning.org

Precast/Prestressed Concrete Institute
www.pci.org

Wire Reinforcement Institute
www.wirereinforcementinstitute.org

CSI Division 04 • Masonry

Brick Industry Association
www.brickinfo.org

Expanded Shale, Clay and Slate Institute
www.escsi.org

Indiana Limestone Institute of America
www.iliai.com

International Masonry Institute
www.imiweb.org

Marble Institute of America
www.marble-institute.com

Masonry Institute of America
www.masonryinstitute.org

National Concrete Masonry Association
www.ncma.org

CSI Division 05 • Metals

Aluminum Association
www.aluminum.org

American Institute of Steel Construction
www.aisc.org

American Iron and Steel Institute
www.steel.org

American Welding Society
www.aws.org

American Zinc Association
www.zinc.org

Cold-Formed Steel Engineers Institute
www.cfsei.org

Copper Development Association
www.copper.org

National Association of Architectural Metals Manufacturers
www.naamm.org

Specialty Steel Institute of North America
www.ssina.com

Steel Deck Institute
www.sdi.org

Steel Joist Institute
http://steeljoist.org

CSI Division 06 • Wood, Plastics and Composites

American Forest & Paper Association
www.afandpa.org

American Institute of Timber Construction
www.aitc-glulam.org

American Plywood Association
www.apawood.org

American Wood Council
www.awc.org

American Wood-Preservers Association
www.awpa.com

Architectural Woodwork Institute
www.awinet.org

Canadian Wood Council
www.cwc.ca

Ceiba Foundation for Tropical Conservation
www.ceiba.org

Composite Panel Association
http://compositepanel.org

Forest Products Laboratory
USDA Forest Service
www.fpl.fs.fed.us

Forest Stewardship Council
www.us.fsc.org

National Hardwood Lumber Association
www.nhla.org

Northeastern Lumber Manufacturers Association
www.nelma.org

Society of the Plastics Industry
www.plasticsindustry.org

Southern Forest Products Association
www.sfpa.org

Structural Building Components Association
www.sbcindustry.com

Western Red Cedar Lumber Association
www.wrcla.org

Western Wood Products Association
www2.wwpa.org

CSI Division 07 • Thermal and Moisture Protection

Adhesive and Sealant Council
www.ascouncil.org

Asphalt Roofing Manufacturers Association
www.asphaltroofing.org

Cellulose Insulation Manufacturers Association
www.cellulose.org

EIFS Industry Members Association
www.eima.com

National Roofing Contractors Association
www.nrca.net

North American Insulation Manufacturers Association
www.naima.org

Perlite Institute
www.perlite.org

Polyisocyanurate Insulation Manufacturers Association
www.polyiso.org

Roof Consultants Institute
www.rci-online.org

Structural Insulated Panel Association
www.sips.org

Stucco Manufacturers Association
www.stuccomfgassoc.com

Vermiculite Association
www.vermiculite.org

CSI Division 08 • Openings

American Architectural Manufacturers Association
www.aamanet.org

American Hardware Manufacturers Association
www.ahma.org

Builders Hardware Manufacturers Association
www.buildershardware.com

Door and Hardware Institute
www.dhi.org

Glass Association of North America
www.glasswebsite.com

National Fenestration Rating Council
www.nfrc.org

National Glass Association
www.glass.org

Steel Door Institute
www.steeldoor.org

Steel Window Institute
www.steelwindows.com

Window and Door Manufacturers Association
www.wdma.com

CSI Division 09 • Finishes

American Coatings Association
www.paint.org

Association of the Wall and Ceiling Industries International
www.awci.org

Carpet and Rug Institute
www.carpet-rug.com

Ceilings and Interior Systems Construction Association
www.cisca.org

Ceramic Tile Distributors Association
www.ctdahome.org

Gypsum Association
www.gypsum.org

Hardwood Manufacturers Association
www.hardwood.org

Hardwood Plywood and Veneer Association
www.hpva.org

Maple Flooring Manufacturers Association
www.maplefloor.org

National Council of Acoustical Consultants
www.ncac.com

National Terrazzo and Mosaic Association
www.ntma.com

National Wood Flooring Association
www.woodfloors.org

Painting and Decorating Contractors of America
www.pdca.org

Porcelain Enamel Institute
www.porcelainenamel.com

Resilient Floor Covering Institute
www.rfci.com

Terrazzo Tile and Marble Association of Canada
www.ttmac.com

Tile Council of North America
www.tcnatile.com

Vinyl Institute
www.vinylinfo.org

Wallcoverings Association
www.wallcoverings.org

CSI Division 10 • Specialties

Kitchen Cabinet Manufacturers Association
www.kcma.org

National Kitchen and Bath Association
www.nkba.org

CSI Division 11 • Equipment

American Society of Safety Engineers
www.asse.org

Association of Home Appliance Manufacturers
www.aham.org

Commercial Food Equipment Service Association
www.cfesa.com

National Solid Wastes Management Association
www.nswma.org

Solid Waste Association of North America
www.swana.org

CSI Division 12 • Furnishings

American Society of Furniture Designers
www.asfd.com

Business and Institutional Furniture Manufacturers Association
www.bifma.org

Home Furnishings International Association
www.hfia.com

Industrial Fabrics Association International
www.ifai.com

International Furnishings and Design Association
www.ifda.com

International Interior Design Association
www.iida.org

Specialty Steel Industry of North America
www.ssina.com

CSI Division 13 • Special Construction

American Fire Sprinkler Association
www.firesprinkler.org

Fire Suppression Systems Association
www.fssa.net

Metal Building Manufacturers Association
www.mbma.com

Modular Building Institute
www.modular.org

National Fire Protection Association
www.nfpa.org

Steel Construction Institute
www.steel-sci.org

CSI Division 14 • Conveying Systems

Conveyor Equipment Manufacturers Association
www.cemanet.org

Material Handling Institute
www.mhia.org

National Elevator Industry
www.neii.org

National Association of Elevator Safety Authorities
www.naesai.org

National Association of Elevator Contractors
www.naec.org

CSI Division 23 • Heating, Ventilating, and Air-Conditioning (HVAC)

American Gas Association
www.aga.org

American Society of Heating, Refrigeration, and Air-Conditioning Engineers
www.ashrae.org

American Society of Mechanical Engineers
www.asme.org

Home Ventilating Institute
www.hvi.org

CSI Division 26 • Electrical

Illuminating Engineering Society of North America
www.iesna.org

International Association of Lighting Designers
www.iald.org

National Electrical Manufacturers Association
www.nema.org

CSI Division 32 • Exterior Improvements

American Concrete Pavement Association
www.pavement.com

American Concrete Pipe Association
www.concrete-pipe.org

American Nursery and Landscape Association
www.anla.org

American Society of Sanitary Engineering
www.asse-plumbing.org

Asphalt Institute
www.asphaltinstitute.org

Asphalt Recycling & Reclaiming Association
www.arra.org

Construction Materials Recycling Association
www.cdrecycling.org

Deep Foundations Institute
www.dfi.org

International Association of Foundation Drilling
www.adsc-iafd.com

Plumbing and Drainage Institute
www.pdionline.org

Sponsoring Organizations for Model Codes and Standards

American National Standards Institute
www.ansi.org

American Society for Testing and Materials
www.astm.org

American Society of Safety Engineers
www.asse.org

International Association of Plumbing and Mechanical Officials
www.iapmo.org

International Code Council
www.iccsafe.org

International Organization for Standardization
www.iso.org

National Institute of Standards and Technology
www.nist.gov

National Research Council of Canada
www.nrc-cnrc.gc.ca

Underwriters' Laboratories
www.ul.com

Allen, Edward, Joseph Iano, and Patrick Rand. *Architectural Detailing*, 2nd Edition. John Wiley & Sons, 2006.
Allen, Edward. *Fundamentals of Building Construction*, 5th Edition. John Wiley & Sons, 2008.
Allen, Edward, and Joseph Iano. *The Architect's Studio Companion*, 5th Edition. John Wiley & Sons, 2011.
Ambrose, James, and Dmitri Vergun. *Simplified Building Design for Wind and Earthquake Forces*, 3rd Edition. John Wiley & Sons, 1997.
Ambrose, James, and Jeffrey E. Ollswang. *Simplified Design for Building Sound Control*. John Wiley & Sons, 1995.
Ambrose, James, and Patrick Tripeny. *Simplified Design of Concrete Structures*, 8th Edition. John Wiley & Sons, 2007.
Ambrose, James. *Simplified Design of Masonry Structures*. John Wiley & Sons, 1997.
Ambrose, James, and Patrick Tripeny. *Simplified Design of Steel Structures*, 8th Edition. John Wiley & Sons, 2007.
Ambrose, James, and Harry Parker. *Simplified Design of Wood Structures*, 6th Edition. John Wiley & Sons, 2009.
Ambrose, James, and Patrick Tripeny. *Simplified Engineering for Architects and Builders*, 11th Edition. John Wiley & Sons, 2010.
American Concrete Institute. *Building Code Requirements for Structural Concrete*. ACI, 2008.
American Concrete Institute. *Manual of Concrete Practice*. ACI, 2012.
American Institute of Architects. *The Environmental Resource Guide*. John Wiley & Sons, 1996. With supplements (1997, 1998). Also on CD-Rom, 1999.
American Institute of Timber Construction. *Timber Construction Manual*, 6th Edition. John Wiley & Sons, 2012.
American Society of Heating, Refrigeration, and Air-conditioning Engineers. *ASHRAE GreenGuide*, 3rd Edition. ASHRAE, 2010.
American Society of Heating, Refrigeration, and Air-conditioning Engineers. *ASHRAE Handbook – HVAC Applications*. ASHRAE, 2011.
Ballast, David Kent. *Handbook of Construction Tolerances*, 2nd Edition. John Wiley & Sons, 2007.
Barrie, Donald S., and Boyd C. Paulson. *Professional Construction Management*, 3rd Edition. McGraw-Hill, 2001.
Bockrath, Joseph T. *Contracts and the Legal Environment for Engineers and Architects*, 7th Edition. McGraw-Hill, 2010.
Butler, Robert Brown. *Standard Handbook of Architectural Engineering*. McGraw-Hill, 1999.
Joseph A. Wilkes, and William J. Cavanaugh, editor. *Architectural Acoustics: Principles and Practice*, 2nd Edition. John Wiley & Sons, 2009.
Ching, Francis D.K. *Architectural Graphics*, 5th Edition. John Wiley & Sons, 2009.
Ching, Francis D.K. *Architecture: Form, Space and Order*, 3rd Edition. John Wiley & Sons, 2007.
Ching, Francis D.K. *A Visual Dictionary of Architecture*, 2nd Edition. John Wiley & Sons, 2011.
Ching, Francis D.K., and Steven R. Winkel. *Building Codes Illustrated: A Guide to Understanding the 2012 International Building Code*, 4th Edition. John Wiley & Sons, 2012.
Ching, Francis D.K., Barry Onouye, and Doug Zuberbuhler. *Building Structures Illustrated: Patterns, Systems, and Design*. John Wiley & Sons, 2009.
Cote, Ron. *NFPA 101: Life Safety Code Handbook 2009*. National Fire Protection Association, 2009.
Crosbie, Michael J., and Donald Watson. *Timer-Saver Standards for Architectural Design*, 8th Edition. McGraw-Hill, 2004.
DeChiara, Joseph, Julius Panero, and Martin Zelnik. *Time-Saver Standards for Interiors and Space-Planning*, 2nd Edition. McGraw-Hill, 2001.
Dilaura, David, Kevin Houser, Richard Mistrick, and Gary Steffy, Eds. *IESNA Lighting Handbook*, 9th Edition. Illuminating Engineering Society of North America, 2011.
Dykstra, Alison. *Construction Project Management: A Complete Introduction*. Kirshner Publishing, 2011.
Evan Terry Associates. *Pocket Guide to the ADA: Americans with Disabilities Act Accessibility Guidelines for Buildings and Facilities*, 3rd Edition. John Wiley & Sons, 2006.
Hacker, John, and Julie A. Gorges. *Residential Steel Design and Construction: Energy Efficiency, Cost Savings, Code Compliance*. McGraw-Hill, 1997.
Hewlett, Peter C., editor. *Lea's Chemistry of Cement and Concrete*, 4th Edition. John Wiley & Sons, 2004.
Hornbostel, Caleb. *Building Design/Materials and Methods*. Kaplan Publishing, 2007.
Hurd, M.K. *Formwork for Concrete*, 7th Edition. American Concrete Institute, 2005.
International Code Council. *2012 IBC Structural/Seismic Design Manual, Volume 1*. International Code Council, 2006.
International Code Council. *2012 International Building Code*. International Code Council, 2011.
International Code Council. *2012 International Energy Conservation Code*. International Code Council, 2011.
International Code Council. *2012 International Fire Code*. International Code Council, 2011.
International Code Council. *2012 International Mechanical Code*. International Code Council, 2011.
International Code Council. *2012 International Plumbing Code*. International Code Council, 2011.
International Code Council. *2012 International Residential Code for One- and Two- Family Dwellings*. International Code Council, 2011.

International Code Council. *Recommended Lateral Force Requirements and Commentary*, 7th Edition. International Code Council, 2002.

Liebing, Ralph W. *Architectural Working Drawings*, 4th Edition. John Wiley & Sons, 1999.

Martin, Leslie D., and Christopher J. Perry. *PCI Design Handbook*, 6th Edition. Prestressed Concrete Institute, 2004.

Masonry Society. *Masonry Designers Guide*, 6th edition. Masonry Society, 2010.

Masonry Institute of America. *Masonry Design Manual*, 4th Edition. Masonry Institute of America, 2006.

Masonry Institute of America. *Reinforced Masonry Engineering Handbook*, 6th Edition. Masonry Institute of America, 2009.

Meisel, Art. *LEED Materials: A Resource Guide to Green Building*. Princeton Architectural Press, 2010.

Miller, Rex. *Electrician's Pocket Manual*. McGraw-Hill, 2005.

Moore, Fuller. *Environmental Control Systems: Heating, Cooling, Lighting*. McGraw-Hill, 1992.

National Roofing Contractors Association. *NRCA Roofing and Waterproofing Manual*, 5th Edition. NRCA, 2006.

O'Brien, James, and Fredric L. Plotnick. *CPM in Construction Management*, 7th Edition. McGraw-Hill, 2009.

Onouye, Barry, and Kevin Kane. *Statics and Strength of Materials for Architecture and Building Construction*, 4th Edition. Prentice Hall, 2011.

Onouye, Barry. *Statics and Strength of Materials: Foundations for Structural Design*. Prentice Hall, 2004.

Patterson, James. *Simplified Design for Building Fire Safety*. John Wiley & Sons, 1993.

Puerifoy, Robert L., Clifford J. Shexnayder, and Aviad Shapira. *Construction Planning, Equipment, and Methods*, 8th Edition. McGraw-Hill, 2010.

Ramsey, Charles George, Harold Sleeper, and AIA Staff. *Architectural Graphic Standards*, 11th Edition. John Wiley & Sons, 2007. Also on CD-Rom, 2007.

Reynolds, Donald E., and R.S. Means Staff. *Residential & Light Commercial Construction Standards*, 3rd Edition. R.S. Means, 2008.

Richter, H.P., and F. P. Hartwell. *Wiring Simplified: Based on the 2011 National Electrical Code*, 43rd Edition. Park Publishing, 2011.

Salter, Charles. *Acoustics: Architecture, Engineering, the Environment*. William Stout Publishers, 1998.

Schodek, Daniel L., and Martin Bechtold. *Structures*, 6th Edition. Prentice Hall, 2007.

Scott, James G. *Architectural Building Codes: A Graphic Reference*. John Wiley & Sons, 1997.

Simmons, H. Leslie. *Olin's Construction: Principles, Materials, and Methods*, 9th Edition. John Wiley & Sons, 2011.

Stein, Benjamin. *Building Technology: Mechanical and Electrical Systems*, 2nd Edition. John Wiley & Sons, 1997.

Stein, Benjamin, Walter T. Grondzik, Alison G. Kwok, and John Reynolds. *Mechanical and Electrical Equipment for Buildings*, 10th Edition. John Wiley & Sons, 2005.

Underwriters' Laboratory. *UL Fire Resistance Directory*, 3 vols. UL, 2007.

Wakita, Osamu A., and Richard M. Linde. *Professional Practice of Architectural Working Drawings*, 4th Edition. John Wiley & Sons, 2011.

Western Woods Products Association. *Western Woods Use Book*, 4th Edition. WWPA, 2005.

G

H

Q

R

S

T

U

V

W